NATURAL ZEOLITES

Occurrence, Properties, Use

NATURAL ZEOLITES

Occurrence, Properties, Use

"A selection of papers presented at Zeolite, 76, an International Conference on the Occurrence, Properties, and Utilization of Natural Zeolites, Tucson, Arizona, June 1976"

Edited by

L. B. SAND

Worcester Polytechnic Institute
Worcester, Massachusetts 01609, U.S.A.

and

F. A. MUMPTON

State University College
Brockport, New York 14420, U.S.A.

PERGAMON PRESS

OXFORD · NEW YORK · TORONTO · SYDNEY · PARIS · FRANKFURT

U.K.	Pergamon Press Ltd., Headington Hill Hall, Oxford OX3 0BW, England
U.S.A.	Pergamon Press Inc., Maxwell House, Fairview Park, Elmsford, New York 10523, U.S.A.
CANADA	Pergamon of Canada Ltd., 75 The East Mall, Toronto, Ontario, Canada
AUSTRALIA	Pergamon Press (Aust.) Pty. Ltd., 19a Boundary Street, Rushcutters Bay, N.S.W. 2011, Australia
FRANCE	Pergamon Press SARL, 24 rue des Ecoles, 75240 Paris, Cedex 05, France
FEDERAL REPUBLIC OF GERMANY	Pergamon Press GmbH, 6242 Kronberg-Taunus, Pferdstrasse 1, Federal Republic of Germany

First edition 1978

British Library Cataloguing in Publication Data

Zeolite '76 (Conference), Tucson
Natural zeolites.
1. Zeolites – Congresses
I. Title II. Sand, L B III. Mumpton, F A
549'.68 QE391.Z5 77-30439

ISBN 0-08-021922-5

Printed in Great Britain by William Clowes & Sons Limited London, Beccles and Colchester

To Trudy and Ardeth,
the better halves

CONTENTS

Physical and Chemical Properties

Applications

PREFACE

This volume consists of papers selected from those presented at ZEOLITE '76, an International Conference on the Occurrence, Properties, and Utilization of Natural Zeolites held at the Doubletree Inn in Tucson, Arizona, June 6 to 14, 1976. The conference was the first of its kind and was sponsored by the State University College, Brockport, New York, with local assistance by the University of Arizona. It was organized by Professor F. A. Mumpton who, with other members of the organizing committee, recognized the need for bringing all persons interested in natural zeolites together at a time when significant new uses were being found for them. T. H. Eyde, R. B. Laudon, L. B. Sand, and R. A. Sheppard formed the remainder of the Organizing Committee which received enthusiastic support from all areas of academe, government, and industry.

The conference was attended by nearly 200 scientists and engineers from more than two dozen countries and consisted of seven technical sessions and two special symposia on Deep-sea Zeolites and Pollution-abatement Applications. A one-day field trip was made by all participants to the Bowie zeolite deposit 120 miles east of Tucson where the mining operations of several companies were examined. A three-day optional field trip was held following the formal meeting to other zeolite deposits in northern Arizona and southern California. The conference received the official endorsement of the Mineralogical Society of America, the Clay Minerals Society, and the Society of Economic Geologists. Financial support is gratefully acknowledged from many industrial organizations and the U.S. Geological Survey, the U.S. Bureau of Mines, the National Science Foundation, and the American Chemical Society – Petroleum Research Fund.

The objective of this volume is to serve as a useful, basic reference in the field of natural zeolites. Credit and appreciation go to the authors of the papers contained herein and to the many workers who served as referees. Special thanks are due to the members of the conference's Program and Publications Committee – D. W. Breck, Miriam Kastner, R. A. Sheppard, and D. E. W. Vaughan.

INTRODUCTION

NATURAL ZEOLITES: A NEW INDUSTRIAL MINERAL COMMODITY

FREDERICK A. MUMPTON

Department of the Earth Sciences, State University College, Brockport, New York 14420

Abstract – In less than 20 years time, the status of the zeolite group of minerals changed from that of museum curiosity to an important industrial mineral commodity. This remarkable transformation is due in large part to the belated recognition in the late 1950s of the widespread occurrence of zeolite minerals as major constituents of Cenozoic sedimentary rocks of volcanic origin and to the research efforts of several industrial organizations during this period leading to the development of commercial applications for synthetic molecular sieves. The discovery that zeolite minerals formed on a large scale by reactions of volcanic tuffs and tuffaceous sedimentary rocks in marine and lacustrine environments was in itself a milestone in the geological sciences; however, the realization that such materials were also capable of being utilized in numerous areas of industrial and agricultural technology provided the impetus for the exploration and development programs that have since taken place on natural zeolites in many countries.

Although the commercial use of natural zeolites is still in a state of infancy, more than 300,000 tons of zeolitic tuff is currently mined each year in the United States, Japan, Italy, Hungary, Yugoslavia, Bulgaria, Germany, Korea, and Mexico. Natural zeolites are used for filler in the paper industry, in pozzolanic cements and concrete, as lightweight aggregate, in fertilizer and soil conditioners, as ion exchangers in wastewater treatment, as dietary supplements in animal husbandry, in the separation of oxygen and nitrogen from air, as reforming petroleum catalysts, and as acid-resistant adsorbents in gas drying and purification. In this era of environmental concern and of energy and resource conservation, the attractive physical and chemical properties of natural zeolites will be utilized worldwide even more in the years to come in the solutions to these problems.

INTRODUCTION

During the last 20 years, a success story has unfolded that chronicles the efforts of scientists and engineers in many parts of the world to demonstrate the academic and technological importance of a previously neglected group of natural materials. As the title of this volume indicates, the subject of this persistent effort is the group of minerals known as zeolites. Zeolites have been recognized for more than 200 years, but it was not until the middle of this century that the scientific community became generally aware of their attractive chemical and physical properties and of their geological significance in the paragenesis of tuffaceous sedimentary rocks. Since 1950 hundreds of articles dealing with zeolites have appeared in the technical literature, and it is currently difficult to pick up any of the leading chemical or geological journals without finding at least one paper devoted to these intriguing materials. Most of the effort has centered around ***synthetic*** zeolites (molecular sieves), an area that has grown into multimillion dollar businesses in several countries. However, in recent years an increasing amount of effort has been directed towards the ***natural*** zeolites – their occurrence in sedimentary rocks of volcanic origin, their valuable physical and chemical properties, and their present and potential applications in diverse areas of industrial and agricultural technology. Since their discovery in large mineable deposits in the western United States and in a number of other countries, interest in natural zeolites has grown steadily; within two decades the status of this group of minerals has changed from that of a museum curiosity to one of a full-fledged industrial mineral commodity.

Historical

The story began, of course, in 1756 when Freiherr Axel Fredrick Cronstedt, a Swedish mineralogist, collected a suite of well-formed crystals from the Svappavari Copper Mine, Lappmark, Sweden. He named these new minerals "zeolites" from the Greek ζειν and λιθοσ, meaning "boiling stones", in allusion to their peculiar frothing characteristics when heated before the blowpipe. Zeolites soon became recognized by geologists and mineralogists the world over as minor, but ubiquitous constituents in the vugs and cavities of almost every basalt and traprock formation. Here they form beautiful crystals that are greatly admired by

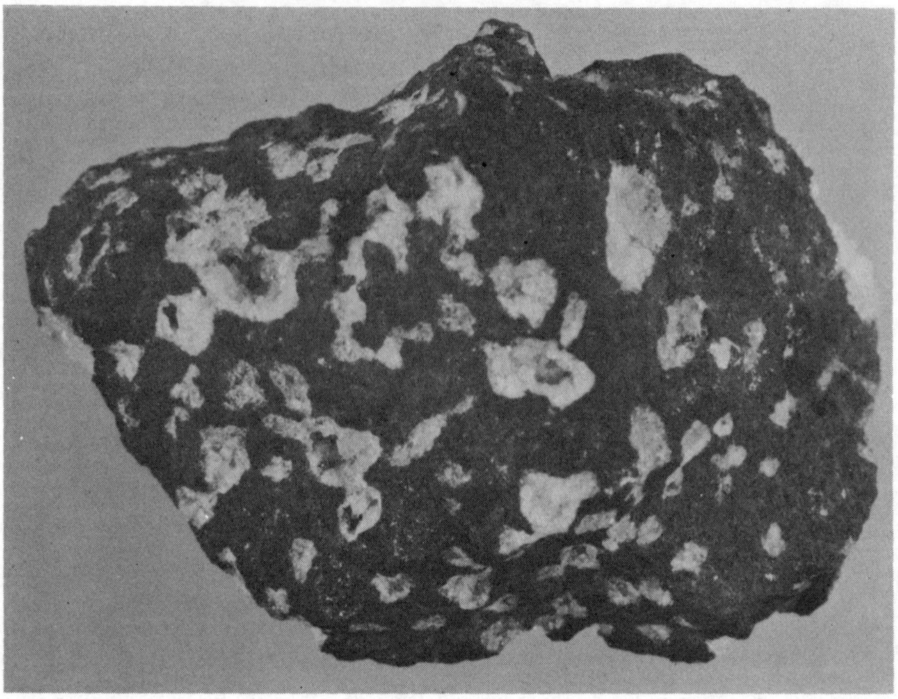

Fig. 1. Typical occurrence of zeolites in a basalt vug. Chabazite from New Mexico.

mineral collectors and are part of the museum collections of every country (Fig. 1). Here they remained, however, for 200 years gathering dust both literally and in a scientific sense. Over the years more than thirty-five individual species of zeolites have been recognized, to be sure, but at the same time relatively few investigations on them of geological importance were carried out during this period.

Chemists, however, were not so reticent in studying zeolites as evidenced by the pioneering work of Weigel and Steinhoff, who noted in 1925 that dehydrated zeolite crystals would adsorb small organic molecules but reject larger ones. This phenomenon was described in 1932 by McBain as "molecular sieving". In the two decades that followed, dozens of papers on the dehydration, adsorption, and ion-exchange properties of zeolite minerals appeared in the chemical literature, mostly from the laboratories of Professors R. M. Barrer in London and J. Sameshima in Japan. Much of this work focused on the zeolites chabazite and mordenite which appeared to have the most interesting adsorption properties of any of the zeolites known at the time. The rarity of these zeolites, or of *any* zeolite for that matter, precluded the development of large-scale industrial processes based upon the natural materials. Zeolites were sufficiently abundant in basalt cavities for experimental purposes, but there appeared to be no way in which they could be profitably extracted from these bodies in amounts large enough to support a commercial operation.

The non-availability of natural zeolites prompted the chemists to turn to synthesis as a means of obtaining a steady supply of these intriguing materials, and between 1944 and 1960 considerable effort went into the low-temperature hydrothermal synthesis of crystalline zeolites. In 1949, based upon the earlier innovative experiments of Barrer, R. M. Milton of Union Carbide Corporation's Linde Division began experiments in zeolite synthesis in Tonawanda, New York. Although his initial experiments were aimed at making chabazite because of its potential in the separation of oxygen and nitrogen from air (Barrer, 1938), early attempts to synthesize this mineral were unsuccessful. What was produced, however, was an entirely new type of zeolite which appeared to have adsorption and molecular sieve properties even better than those of chabazite. This zeolite was designated Linde type A zeolite and is presently the most important synthetic molecular sieve sorbent (Milton, 1959, 1968).

The 1950s found Union Carbide Corporation and other industrial laboratories deeply involved in developing processes and markets for their fledgling molecular sieve business. It was during this period that the geological profession began to hear reports and rumors from various parts of the globe that zeolites could be found in rocks other than typical basalts, and that in some places zeolite minerals made up almost 100% of the entire rock. Hints of such non-basalt, non-igneous

Fig. 2. Analcime-rich tuffs in a saline-lake deposit near Wikieup, Arizona. This is probably the locality described by Ross (1928). See also Sheppard and Gude (1973).

occurrences had found their way into the geological literature before this, but in general these reports were ignored by the profession. For example, in 1914 Albert Johannsen described what he thought were extremely fine-grained zeolites making up a substantial portion of Eocene tuff beds in the Uintah Basin of Colorado, Wyoming, and Utah. In 1928 W. H. Bradley and C. S. Ross recognized zeolites in saline-lake beds of Wyoming and Arizona (Fig. 2), respectively; and in 1933 two U.S. Geological Survey geologists, M. N. Bramlette and E. Posjnak, listed several occurrences of clinoptilolite in vitric tuffs in the western part of the country. Kerr in 1931 also noted zeolite minerals in California bentonites, and in 1936 Fenner described extensive zones of heulandite in Yellowstone drill cores. The deep-sea zeolites described by Murray and Renard in 1891 in the report of the H.M.S. *Challenger* predate all of these reports, but this work also was not extensively known until recently. Probably the earliest record of zeolites in sedimentary rocks was published in 1876 by Loew, who described chabazite from a tuff bed near Bowie, Arizona, site of current zeolite mining operations (Fig. 3); however, this article was only unearthed in 1975 by T. L. Finnell and R. A. Sheppard.

Although 20 years ago most of the geological world was still quite unaware of it, zeolites were being found in rocks other than igneous basalts, mostly in bedded tuffs in lacustrine and marine environments. Undoubtedly influenced by Coombs' (1954) discovery of several kinds of zeolites in low-grade metamorphic rocks of New Zealand, Sudo's work (1950) on the Green Tuff zeolites in Japan, the several Russian articles on "mordenite-like" zeolites in sedimentary rocks of the Soviet Union, and Sersale's (1958) early work on the zeolitic "tufo giallo napolitano" in Italy (Fig. 4), a few more far-sighted geologists in the United States began to examine tuffaceous sediments and altered volcanic ashes by X-ray diffraction techniques. The results were astonishing! What at first glance appeared to be a nondescript, fine-grained, altered glass turned out to contain upwards to 95% of a single zeolite phase (see Ames *et al.*, 1958). Several million-ton size deposits of the "rare" zeolite erionite were discovered in 1957 (Deffeyes, 1959), and the zeolite clinoptilolite suddenly became one of the most common authigenic silicates to be found in sedimentary rocks (Mumpton, 1960).

Thus, in the later 1950s the stage was set. The chemists were busy synthesizing zeolites and finding uses for them, and the geologists were beginning to realize that many zeolites could be found in mineable quantities in nature. Once each group learned what the other was doing, the zeolite story picked up its pace. The geologists were stimulated because, not only were their discoveries of significance in deciphering the origin of

Fig. 3. Surface mining operations of Union Carbide Corporation north of Bowie, Arizona. The 15-cm thick chabazite-rich tuff is carefully scraped and swept before mining to remove adhering clays and other impurities. This is probably the same sequence of tuff beds mentioned by Loew (1875).

Fig. 4. Seaside exposure of the "Tufo giallo napolitano" near Pozzouli, Italy. Note the local use of this zeolitic tuff as dimension stone in older buildings.

tuffaceous sedimentary rocks, but the zeolites themselves were actually "good for something". At the same time the chemists could now begin thinking in terms of inexpensive materials to complement the more costly synthetic products, thereby opening up entirely new areas of industrial applications. There is little question but what the side-by-side efforts of the chemists and the geologists have had a synergistic effect on the overall development of natural zeolite technology. Neither would have been able to achieve this success on its own.

Current status

Since their "rediscovery" in the 1950s, more than 1000 occurrences of zeolite minerals have been reported from sedimentary rocks of volcanic origin in more than forty countries. The flat-lying nature and high purity of some deposits have aroused strong commercial interest; and today efforts are being made to utilize natural zeolites in many of the same ways as their synthetic counterparts, as well as to develop new applications which take advantage of the low mining costs of the near-surface, natural deposits (Figs. 5–7). The commercial use of natural zeolites is still in its infancy; however, a total of more than 300,000 tons of zeolitic tuff is mined each year in the United States, Japan, Italy, Hungary, Bulgaria, Yugoslavia, Mexico, Korea, and Germany. Natural zeolites are used as filler in the paper industry, in pozzolanic cements and concrete, as lightweight aggregate, in fertilizer and soil conditioners, as ion-exchangers in water purification and pollution-abatement processes, as dietary supplements in animal husbandry, in the separation of oxygen and nitrogen from air, as reforming petroleum catalysts, and as acid-resistant adsorbents in gas drying and purification. In this area of environmental concern and of energy and resource conservation, the attractive physical and chemical properties of natural zeolites show great promise of being utilized even more in the future in the solution to many of these problems.

As the table of contents of this volume suggests, natural zeolites are of considerable interest today in several seemingly unrelated areas of technology and are the subject of numerous geological, mineralogical, chemical, and agricultural research projects. Many of the geological aspects of natural zeolites are treated in detail in other sections of this volume and will therefore be touched upon only briefly in this review. The utilization of natural zeolites, on the contrary, is a rapidly expanding area, and new uses appear on the scene almost on a monthly basis. Much of the following discussion will therefore be devoted to the industrial and agricultural applications that have been or are being developed for natural zeolite materials.

Fig. 5. Clinoptilolite mine of Zeaklite Chemical & Mining Company, Ltd., Itaya, Yamagata Prefecture, Japan.

Fig. 6. Mečkovac clinoptilolite deposit north of Vranje, Yugoslavia. A lithic tuff is mined by Nemetali Company as a pozzolanic material.

Fig. 7. Clinoptilolite ore stockpiled by Double Eagle Petroleum & Mining Company from tuff beds near Buckhorn, New Mexico.

GEOLOGICAL SIGNIFICANCE

Not only are zeolites widespread in the vugs and cavities of basaltic igneous rocks, they are currently recognized as one of the most common groups of authigenic silicates in sedimentary rocks of volcanic origin. They occur in rocks deposited in diverse geological environments and age, and have been classified by Sheppard (1973), Mumpton (1973a), and Munson and Sheppard (1974) into several types, including:

(1) deposits formed from volcanic materials in "closed", saline-lake systems;
(2) deposits formed in "open", freshwater-lake or groundwater systems;
(3) deposits formed in marine environments;
(4) deposits formed by low-grade, burial metamorphism;

(5) deposits formed by hydrothermal or hot-spring activity;
(6) deposits formed from volcanic materials in alkaline soils;
(7) deposits formed without direct evidence of volcanic precursors.

The most common zeolites in sedimentary rocks are analcime, chabazite, clinoptilolite, erionite, heulandite, laumontite, mordenite, phillipsite, and wairakite, with clinoptilolite ranking "number one" in abundance. Except for heulandite, laumontite, and wairakite, "sedimentary" zeolites are generally alkalic and are commonly more silicic than their igneous counterparts. Commercial interest is mainly in deposits of the first four types. Zeolitic tuffs in saline-lake deposits are generally a few centimeters to a few meters thick, but they commonly contain nearly monomineralic zones of the larger pore zeolites erionite and chabazite, which are relatively uncommon in other types of deposits. "Open-system" deposits and marine tuffs deposited close to their source are generally characterized by clinoptilolite and/or mordenite and may be several hundred meters thick. Small to moderate amounts of montmorillonite and cristobalite/tridymite may also be present, along with varying amounts of unreacted pyroclastic materials.

Most zeolites in sedimentary rocks formed from volcanic ash or other pyroclastic material by reaction of the amorphous, aluminosilicate glass with pervading pore waters (Fig. 8). Others originated by the alteration of pre-existing feldspars, feldspathoids, biogenic silica, or poorly crystalline clay minerals. Although the exact mechanisms of formation are still under investigation, zeolites in sedimentary rocks probably formed by means of dissolution–reprecipitation reactions, with or without an intermediate gel stage (Mumpton, 1973b). The factors controlling whether a zeolite or a clay mineral will form or which of several zeolite species will form from a given starting material are also only poorly understood, although temperature, pressure, reaction time, and the activities of dissolved species such as H^+, silica, alumina, and alkali and alkaline earth cations seem to be paramount in importance. Early formed species also tend to react further with pore fluids of different composition, thereby yielding even more complex assemblages.

Saline-lake deposits

Sheppard and Gude (1968, 1969a, 1973) correlated zeolite mineralogy with lake-water chemistry in saline-lake deposits of southern California and northwestern Arizona. They found that tuffs deposited in such hydrologically "closed" basins are characterized by

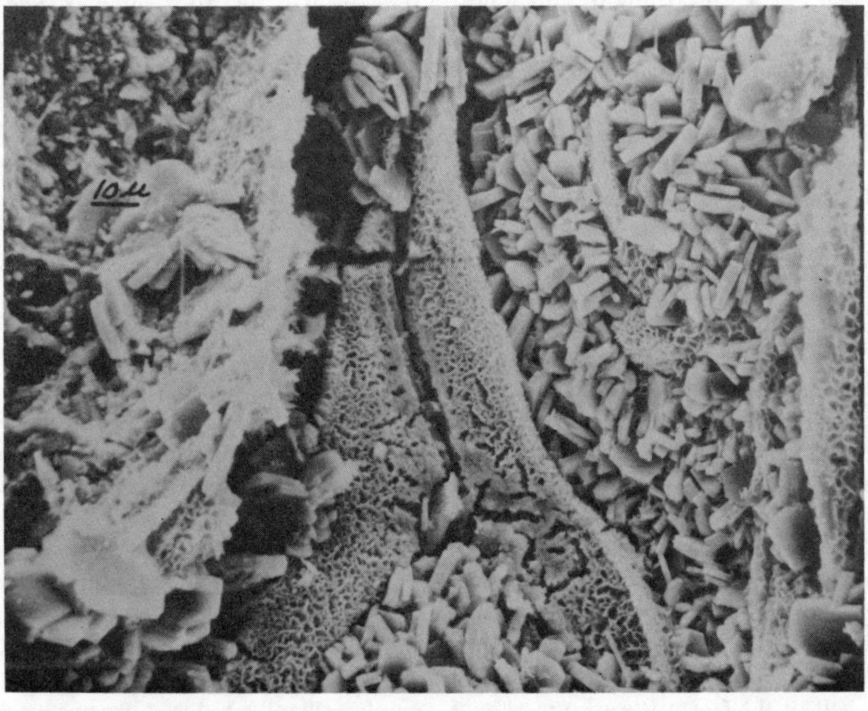

Fig. 8. Scanning electron micrograph showing clinoptilolite laths and montmorillonite-coated volcanic glass in tuff from Narrows, Oregon.

Fig. 9. Zeolitic tuffs in a lacustrine sequence near Rome, Oregon.

a lateral zonation: fresh ash near the margins of the basin, followed by a zone of almost complete zeolitization, a zone of analcime and finally, a central core of authigenic potassium feldspar. Similar facies relationships have been found in several other saline-lake deposits of the United States (Fig. 9), most notably the Green River Formation of Wyoming, Utah, and Colorado (Surdam and Parker, 1972). In East Africa zeolitic tuffs in the vicinity of Lake Magadi, Kenya (Hay, 1968; Surdam and Eugster, 1976), Lake Natron, Tanzania (Hay, 1966), and Olduvai Gorge, Tanzania (Hay, 1970) display this same lateral zonation of zeolite minerals. The dependence of zeolite mineralogy on alkalinity was recently demonstrated by Höller *et al.* (1974) who found that volcanic ash transformed into chabazite, phillipsite, and analcime successively with increasing concentrations of KOH and NaOH. Although the temperatures and concentrations in these experiments were considerably greater than natural conditions, the results suggest that solution geochemistry is a viable means of investigating the genesis of zeolites in saline, alkaline-lake environments.

Other species have also been found; but erionite, chabazite, and phillipsite seem to characterize the saline, alkaline-lake deposits and as such are important indicators of depositional environments and paleoclimates. Wolfbauer (1973) used the zeolite content of tuffs in the lower half of the Green River Formation to distinguish areas of fresh to brackish water from those of saline or alkaline water. The former seems to be characterized by mordenite/clinoptilolite assemblages; the latter by analcime and potassium feldspar. Carl and Van Diver (1975) postulated that Grenville alaskite bodies in northern New York State may originally have been ash-flow tuffs and that their apparent Na_2O-enrichment is due to a complex diagenetic history not unlike the transition of zeolitic tuff to analcimic tuff to potassium feldspar in saline-lake deposits.

The presence of erionite, chabazite, or phillipsite might also be a useful clue to finding buried deposits of saline minerals, fluorite (Sheppard and Gude, 1969b), or other commodities of commercial value in suspected "closed" basins.

Open-system deposits

Clinoptilolite and mordenite are the principal zeolites of so-called "open-system" deposits (Fig. 10), as well as of the thick, marine tuffs of Japan and Eastern Europe. Hay (1963) related the sharpness of the transition between an overlying, fresh glass/montmorillonite facies of the non-marine John Day Formation of central Oregon and a lower clinoptilolite facies to a significant increase in pH and ionic strength of descending pore waters as they reacted with volcanic glass. Where the water remained relatively fresh, the glass altered to montmorillonite, but as it became more and more alkalic from dissolved glass, clinoptilolite formed. A basal zone contains significant amounts of analcime which seems to

Fig. 10. Clinoptilolite-rich tuff about 25 m thick from an "open-system" deposit in the Fish Creek Mountains, Nevada.

have formed by dissolution of the earlier-formed clinoptilolite and/or montmorillonite. A similar vertical zonation of zeolites due to solution and hydrolysis of vitric ash by descending meteoric waters was described by Walton (1975) for the Oligocene Vieja Group in Texas. The Campanian Tuff and the Neopolitan Yellow Tuff near Naples, Italy, are both thought to have formed by the alteration of volcanic ash by descending groundwaters (Sersale, 1958). The latter consists of several hundred meters of chabazite/phillipsite overlain by a similar thickness of glassy tuff, known locally as pozzolana.

Burial metamorphic deposits

Burial metamorphic deposits, often described as burial diagenetic deposits, are widespread in the Green Tuff region of Japan, and commonly reach thicknesses of several thousand meters. In several areas Iijima and Utada (1972) and Iijima (1974) correlated vertical zones of (1) fresh ash, (2) alkali clinoptilolite, (3) clinoptilolite/mordenite, (4) analcime, and (5) albite with depth of burial and concomitant increases in temperature and pressure. The zones are consistent although the transition from one to another is not sharp. The role of pore-water composition in such deposits is not well understood and is still under investigation. As Höller *et al.* (1974) pointed out, the formation of zeolites from parent glass can be accelerated by changing *either* the temperature *or* the alkalinity of the solution. The upper zones of such sequences commonly contain large thicknesses of high-grade clinoptilolite and mordenite, and currently ten such deposits are mined in Japan.

Deep-sea deposits

Although the zeolite phillipsite was reported by Murray and Renard in 1891 from bottom sediments collected during the first *Challenger* expedition, one of the surprises of the recent JOIDES deep-sea drilling project has been the abundance of phillipsite and clinoptilolite in cores from dozens of locations in the Atlantic, Pacific, and Indian Oceans (Fig. 11). Numerous data have been collected on the distribution of zeolites in the deep-sea sediments, but this area is still one of the least understood insofar as zeolite genesis is concerned. Neither zeolite appears to be in equilibrium with sea water, according to Rex and Fan (1974), but are probably transient phases ultimately altering to quartz and feldspar. Kolla and Biscaye (1973) indicated from their studies of cores from the Indian Ocean that phillipsite is more abundant in Miocene to recent sediments, but sediments of lower Tertiary and Cretaceous age are more apt to contain clinoptilolite. The available data are insufficient; however, to decide whether this distribution reflects variations in sediment composition, sea-water composition, or a stability relationship with time. In any case, our understanding of zeolite formation in deep-sea environments is just now in

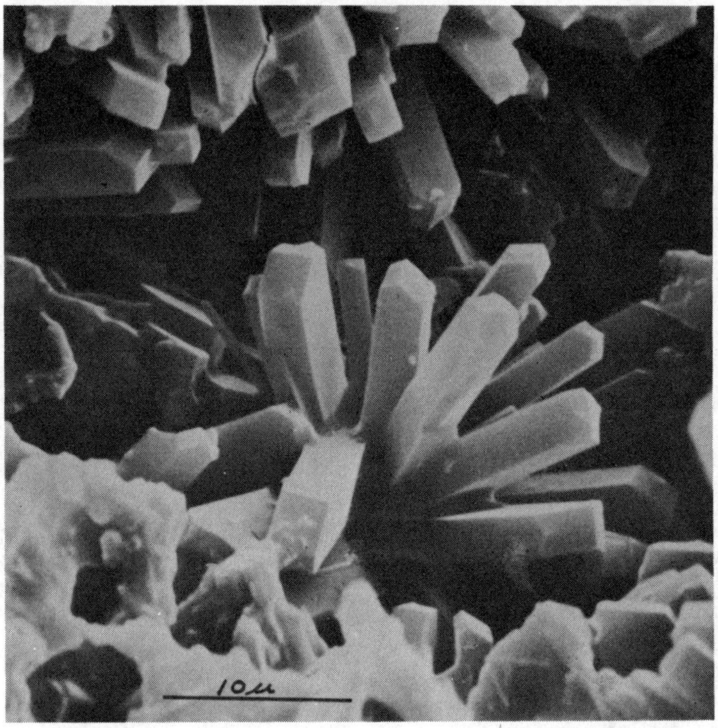

Fig. 11. Phillipsite rods from deep-sea sediments collected on Leg 42A of the JOIDES Deep Sea Drilling Project in the Tyrrhenian Sea. Scanning electron micrograph by R. E. Garrison.

the formative stage. If we are eventually able to decipher these distribution and genesis problems, we should have a valuable tool to understand the role of vulcanism in Cenozoic times, as well as to aid in understanding the genesis of zeolites in marine-laid tuffs, now found on dry land.

POLLUTION-CONTROL SIGNIFICANCE

In recent years zeolite minerals have found increasing application in the field of pollution abatement, and they are fast becoming standard components in the design and construction of such facilities. Both the ion-exchange and adsorption properties of zeolites can be utilized, however, most applications that have been developed are based on the ability of certain zeolites to exchange large cations selectively from aqueous solutions.

Radioactive-waste disposal

In 1959 L. L. Ames of the Hanford Laboratory of Battelle-Northwest, Richland, Washington, demonstrated the specificity of clinoptilolite for the removal of radioactive cesium and strontium from low-level waste streams of nuclear installations (Ames, 1959, 1960). The ions can be extracted with high efficiency from effluents and either stored indefinitely on the zeolite or removed by chemical means for subsequent purification and recovery. Using clinoptilolite crushed and screened to 20 X 50 mesh from the Hector, California deposit, solutions containing the radioactive cations were passed through columns packed with the zeolite until breakthrough occurred. The "saturated" columns then were removed from the system and buried as solid waste. The process was actively investigated at Hanford and at several other nuclear-test stations in the United States in the 1960s (Knoll, 1963; Hawkins and Short, 1965), but supply problems and engineering reluctance to use the somewhat variable natural material that was available at that time hindered its complete acceptance by the industry. Millions of gallons of low-level Cs^{137} wastes, however, have been processed through zeolite ion-exchangers at Hanford since that time, and a similar process was developed to recover this species from high-level effluents using a chabazite-rich ore from Bowie, Arizona (Nelson and Mercer, 1963; Mercer *et al.*, 1970a). At the National Reactor Testing Station, Arco, Idaho, steel drums filled with granular clinoptilolite also were used as ion-exchange columns for Sr^{90} and Cs^{137} having half-lives of 25 and 33 years, respectively. Once the capacities of the drums were reached, they were removed, buried and replaced with new drums containing fresh clinoptilolite (Wilding and Rhodes, 1963, 1965).

Similar processes have been investigated in several other countries, including Canada, Great Britain, France, Bulgaria, Hungary, Mexico, Japan, Germany and the Soviet Union. Canadian workers devised a scheme for the removal and fixation of long-lived fission products using zeolite ion-exchange (Mathers and Watson, 1962). More than 700 curies of Cs^{137} and Sr^{90} were extracted from 8200 liters of solution by 14 X 65 mesh clinoptilolite and placed in burial sites at Chalk River, Ontario. In Hungary, clinoptilolite from the Tokaj region has been used to encapsulate Sr^{90} for solid-waste disposal (Kakasy *et al.*, 1973) and for the removal of both Sr^{90} and Cs^{137} from low-level effluents (Adam *et al.*, 1971). A mixture of vermiculite and clinoptilolite was found by Daiev *et al.* (1970) to take up more than 98% of the Cs^{137}, Ti^{204}, As^{110}, Sr^{90}, and Ca^{45} in waste solutions, using zeolite ore from the large deposit at Kurdžali in south-eastern Bulgaria. The use of clinoptilolite- or mordenite-rich tuff combined with a small amount of a non-ionic polymer coagulant was found to be particularly effective for Cs^{137} removal in Japan (Kato, 1974).

Clinoptilolite from the Georgian S.S.R. has been shown by Nikashina *et al.* (1974) to sorb strontium selectively from radioactive solutions. Vdovina *et al.* (1976) found clinoptilolite able to extract 94% of the radioactive Cs present in a waste stream containing 10 g Cs/liter. According to a report of the International Atomic Energy Agency in Vienna (IAWA, 1972a), an altered volcanic tuff from the Eifel Region of Germany, marketed for more than 30 years as "Filtrolit", is useful in the removal of Cs^{137} from waste streams. Tuffs rich in chabazite and phillipsite are known from this region, and it is likely that the tuff in question contains a large proportion of these minerals. The zeolite-rich Neopolitan Yellow Tuff from Naples, Italy, has also been used routinely to remove radionuclides from contaminated effluents at Casaccia, Italy (IAEA, 1972b). The abundance of chabazite and phillipsite in this tuff gives rise to its overall ion-exchange capacity for Cs of 2.1 meq/g and for Sr of 0.7 meq/g. The loaded tuff can be stored directly or incorporated into concrete for long-term burial.

Studies are currently in progress at the Los Alamos Laboratory of the University of California on the addition of powdered clinoptilolite along with flocculating agents to primary effluents to extract Cs^{137}. The solid material is later removed by anthracite filtration and placed in burial storage (Rohrer, 1976). The escalation of nuclear-power plant construction in this and other countries in the next two decades will result in the production of large quantities of radioactive wastes. Natural zeolites capable of extracting species such as Sr^{90}, Cs^{137}, Co^{60}, Ca^{45}, and Cr^{51} selectively in the presence of high concentrations of competing ions and which are capable of retaining their ion-exchange properties in high-flux environments may well play major roles in the safe development of nuclear power. Natural zeolites are not only considerably less expensive than organic ion-exchange resins, they are much more resistant to nuclear degradation. As silicates, they also react rapidly in cement and glass systems, entraining the radioactive species in the final concrete products.

Sewage-effluent treatment

As a spin-off of their work in radioactive waste disposal, Ames and Mercer showed that clinoptilolite is also highly selective for ammonium ions and suggested that it could be useful in the extraction of ammoniacal nitrogen from sewage and agricultural effluents (Ames, 1967; Mercer, 1969; Mercer *et al.*, 1970b). Not only is NH_4^+ toxic to fish and other forms of aquatic life, it

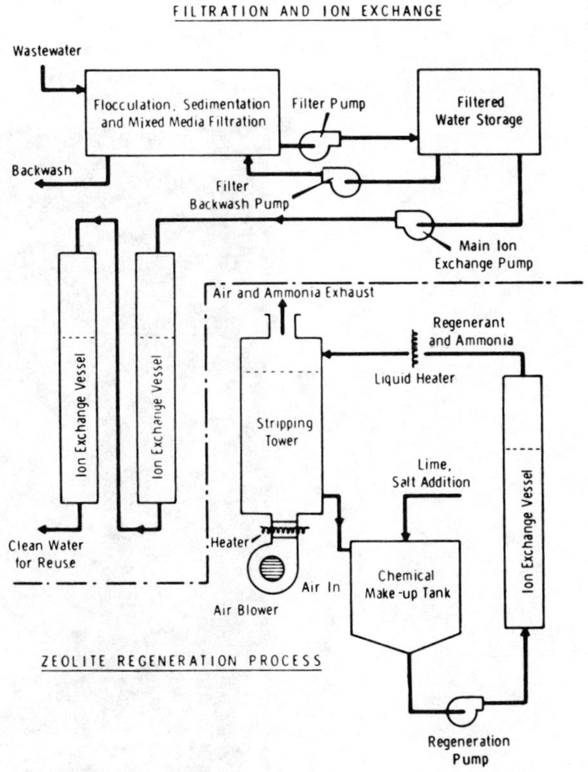

Fig. 12. Flow chart of wastewater treatment process using clinoptilolite ion-exchange processes. (Fig. 7 in Mercer *et al.*, 1970b)

also contributes greatly to the rapid growth of algae and leads to eutrophication of lakes and streams. Increasingly stringent regulations on the amount of nitrogen that is permissible in municipal and industrial wastewater effluents have been promulgated in recent years by local and national environmental protection agencies. In general, a limit of about one part per million has been imposed, creating an urgent need to develop processes which will reduce the nitrogen content of streams to this level. Based upon test data from a mobile ion-exchange unit at Lake Tahoe in 1970 (Battelle Northwest, 1971) and subsequent engineering studies by Koon and Kaufman (1971), several large-scale sewage-treatment plants were designed for communities in several parts of the United States. Plans call for clinoptilolite ion-exchange processes that would remove up to 99% of the contained ammonium ions from tertiary sewage effluents (Fig. 12).

A 0.6-Mgd (million gallons per day) experimental unit went on stream at Rosemont, Minnesota, in the middle of 1975 and utilizes a total of 90 tons of 20 X 50 mesh clinoptilolite in six 300-ft^3 columns (Fig. 13). In Virginia, 54-Mgd and 10-Mgd plants are in the construction stage for the Alexandria and Reston areas and will require 2000 and 400 tons of zeolite, respectively, The cost of the crushed and screened clinoptilolite is reported to be between $200 and $300 per ton, f.o.b. minesite. A 6-Mgd facility was scheduled to begin operations in mid-1976 in the North Lake Tahoe Sewage District in California and will use several hundred tons of clinoptilolite (Eyde, 1976). Two other units are contemplated for Waukegan, Illinois (Wilson, 1975), and Garabaldi, Oregon (Kapranos, 1976). The latter community overlooks Tillamook Bay, an important oyster production area in the Pacific Northwest, where special precautions are warranted to prevent contamination of the beds. A single physical-chemical unit for the removal of nutrients from combined sewer overflows has been designed by Murphy *et al.* (1977) and will employ a clinoptilolite ion-exchange process to extract NH_3-N.

If the performance of these first plants is satisfactory, the annual demand for clinoptilolite may exceed several hundred thousand tons in the United States alone within the next 20 years. During this period the country will also experience a sharp rise in the demand for water, giving added incentive for municipalities to conserve and reuse their existing water supplies. Studies in this area are in progress by CH_2 M-Hill, Inc. (1975) for the Denver Board of Water Commissioners to develop a means to

Fig. 13. Clinoptilolite-filled ion-exchange columns in the Rosemont, Minnesota, Experimental Wastewater Treatment Plant.

purify sewage effluent to potable standards. Preliminary plans have been proposed for a 1-Mgd prototype facility that will include a clinoptilolite ion-exchange step for removing the bulk of the ammoniacal nitrogen. Instead of releasing ammonia to the atmosphere during the regeneration steps, sulfuric acid will be added to produce ammonium sulfate to be sold locally as a fertilizer. An added benefit of clinoptilolite ion-exchange is that this zeolite appears to be able to extract trace amounts of heavy metals present in the wastewater which normally are detrimental to anticipated end uses (Chelischev *et al.*, 1974; Hashimoto, 1974; Yoshida *et al.*, 1976). The addition of zeolites to activated sludge also appears to aid oxidation and settling, according to Liles and Schwartz (1976).

In other countries the use of natural zeolites in wastewater treatment is only in the planning stages. The one exception is Japan where no less than two dozen scientific articles and patents on the use of clinoptilolite and other zeolites for ammonium removal have been published since 1973. Some schemes call for powdered zeolite to be added to the effluent and then filtered or sedimented out (e.g. Sato and Fukagawa, 1976), while others employ ion-exchange columns filled with sized zeolite. According to Torii (1974), a 0.1-Mgd unit to remove nitrogen from the effluent of a soap and detergent factory has been in operation since 1971. Also, the 21,000 gallons of wastewater produced each day from a tourist observatory at Toba, Mie Prefecture, is now purified by means of zeolite ion-exchange, using vessels packed with 2.5 tons of crushed clinoptilolite. Many more facilities of this type are planned in Japan for hotels, small factories, fish farms, and small communities lacking large-scale sewage-treatment systems.

Agricultural-wastewater treatment

In addition to the pollution from municipal and industrial wastewaters, streams and rivers in many parts of the United States are becoming more and more contaminated with nitrogen from irrigation run-off, seepage from animal feedlots, and waste streams from food processing plants. The use of natural zeolites to combat such pollution is a relatively untouched field, although some progress has been made, again mainly in Japan. Here farmers have sprinkled crushed zeolite on farmyard wastes for years (Minato, 1968), although the zeolitic nature of the adsorbent tuffs was only recently recognized. In the United States more than 700 million tons of animal wastes are produced each day, and its disposal is becoming a major problem. The problem is especially acute in stockyards and dairy farms located close to large population centers where both air and water pollution must be rigidly controlled. Clinoptilolite appears to be doubly useful in the treatment of such materials, as it would not only remove most of the ammoniacal nitrogen from the liquid portion of the wastes and thereby decrease the ammonia-laden aerosols that travel many miles downwind of the feedlots and confining pens, but it would also retain much of this nitrogen in the solid form thereby enhancing the fertilizer value of the manure. Such applications would most likely not require coarsely sized zeolite; powdered materials should be quite satisfactory and be readily available at prices of $50.00 to $75.00 per ton.

Stack-gas cleanup

Another area of pollution control involves the use of natural zeolites in the removal of SO_2 and other pollutants from stack gases of oil- and coal-burning power plants. About 25 million tons of sulfur are dumped into the atmosphere each year by power stations in the United States, an amount that could double or triple in the next few years as the country turns to high-sulfur coals to alleviate its energy problems. Many federal and local air-pollution regulations now limit the amount of SO_2 in stack-gas emissions to about 100 ppm; however, few processes now in use can meet these specifications. The relatively low concentrations of SO_2 in such emissions (about 3000 ppm) elude efficient scrubbing operations, but zeolite adsorption might be an economical means of concentrating this gas, as well as NO_x, CO_2, and various hydrocarbons for subsequent removal (Blodgett, 1972). Certain natural mordenites and clinoptilolites are capable of selectively adsorbing up to 200 mg of SO_2 per gram of zeolite under static conditions and up to 40 mg/g under dynamic conditions, even in the presence of copious amounts of CO_2 (Ishikawa *et al.*, 1971; Terui *et al.*, 1974; Anurov *et al.*, 1974; Smola *et al.*, 1975). They are especially suited to the low pH and high temperature conditions of exhaust-gas systems (Roux *et al.*, 1973). Linde Division of Union Carbide Corporation recently introduced a proprietary process for SO_2-cleanup of tailgases in sulfuric acid plants based upon adsorption by an undisclosed molecular sieve (Miller, 1973).

Oil-spill cleanup

A novel use of zeolites in pollution control is as a sorbent in oil-spill cleanup. Miki *et al.* (1974) pelletized a mixture of activated zeolite, expanded perlite, sodium carbonate, tartaric acid and a binder consisting of 20% methylsiloxane solution. The product had a bulk density of 0.5 g/cm^3 and an oil-sorption capacity of 0.97 g/gram. The lightweight material was able to float on water for more than 200 hours and sorb oil from the surface.

Oxygen production

Air and water pollution generally involve the presence of objectionable compounds and/or particles; however, they may also be caused by the *absence* of desirable ones, such as oxygen. Oxygen deficiencies in lakes and streams result in the rapid extinction of fish and plant life; in the atmosphere of a closed room its depletion is uncomfortable at best to human beings and hazardous at worst. Zeolite adsorption processes can be utilized to produce inexpensive, oxygen-enriched streams of varying degrees of purity. Based on earlier work of Barrer (1938), Dominé and Häy (1968) showed that nitrogen gas is selectively adsorbed from air by several zeolite materials yielding products containing up to 95% oxygen. A pressure-swing-adsorption process was developed in Japan (Tamura, 1970) and has been in operation since 1968 at Toyohashi City, a few miles north of Osaka, producing up to 500 m^3 of 90% O_2 per hour for use in secondary smelting operations (Tamura, 1971) (Fig. 14). The plant consists of three towers, each filled with 13 tons of acid-washed mordenite from the Itado Mine, Minase, Akita Prefecture, and operates at room temperature on a 9-minute cycle of adsorption–desorption-standby. It is reported to be competitive with air distillation in situations where enormous liquefaction facilities are not warranted. Portable and file-cabinet-size units have also been marketed in Japan to provide oxygen-enriched air for windowless restaurants where ventilation is poor. Larger units are in use to furnish oxygen to fish-breeding ponds and during the transportation of live fish (*vide infra*).

The potential markets for low-cost oxygen generators based on the selective adsorption properties of zeolite minerals are numerous and include river and pond aeration, pollution control in the paper and pulp industry by reoxygenation of downstream waters, and the generation of oxygen for secondary sewage treatment. The last application would find natural zeolites in direct competition with the more expensive synthetics currently used in several commercial systems, including Union Carbide's LINDOX process. This latter system is capable of producing up to 30 tons of 95% oxygen per day as the aeration gas in activated sludge systems (Union Carbide Corporation, 1975). The process has been successful for sewage-treatment plants operating at less than about 10-Mgd capacity.

Mordenite appears to be the best natural zeolite for oxygen generation, competing with the synthetic Ca–A zeolite, but certain clinoptilolites and chabazites also appear to be useful (Tamura, 1971; Torii *et al.*, 1971, 1973; Tsitsishvili *et al.*, 1972; Haralampiev *et al.*, 1975).

Fig. 14. Oxygen generation facility of Koyo Kaihatsu, Ltd., Toyocity, Japan. Pressure-swing adsorption of nitrogen from air using natural mordenite is employed.

Hagiwara (1974) found that the selectivity of mordenite for nitrogen is enhanced by transformation of the zeolite to a mixed Na–H form.

ENERGY-CONSERVATION SIGNIFICANCE

Coal gasification

To cope with the world's growing energy requirements, whether it be from fossil fuels, nuclear, solar, or some other heretofore untapped source, modern technology will be forced to develop new processes or upgrade old ones to do a more efficient job. New or modified technology invariably calls for new or modified materials with properties to match. Natural zeolites are a group of "new" materials that will undoubtedly be thoroughly evaluated in the development of new energy sources or in the conservation of old ones. In addition to their potential in the removal of SO_2 from stackgas emissions (*vide supra*), thereby allowing high-sulfur coals to be used in the production of electric power, the ability of certain zeolites to adsorb nitrogen selectively from air and produce oxygen-rich products might find zeolites at the forefront of such fields as in-place coal gasification. More than 25% of the coal reserves of the United States are in deep-lying, unmineable seams and will require underground gasification if they are to be utilized. Pumping air into the formations accelerates combustion but it also produces large quantities of nitrogenous oxides and hydrocarbons which are dangerous and difficult to handle. The use of oxygen-enriched air is preferred, but the cost of pure oxygen by conventional liquefaction may be prohibitive. On-site zeolite adsorption units, however, might be the answer in that they would be capable of producing any desired concentration of oxygen at a lower price than liquefaction. In this manner combustion could be optimized while minimizing equipment-corrosion problems that attend pure oxygen systems.

Natural gas purification

Although the use of zeolites in coal gasification may be some years in the future, natural zeolites *have* been employed successfully since 1968 to remove carbon dioxide from contaminated and sour natural gas. Utilizing natural zeolites from its Bowie, Arizona, deposit, NRG Corporation developed a pressure-swing-adsorption process (p.s.a.) to extract up to 25% CO_2, H_2S, and H_2O from well-gas in the Los Angeles area (see Wearout, 1971). With predicted shortages and rising prices of natural gas, attempts to exploit heretofore unrecoverable sources become increasingly important. In 1975 this company opened a methane-recovery and purification plant to treat gas produced by decaying organic matter in the Palos Verde sanitary landfill in southern California (Fig. 15). Raw gas containing about 50% CH_4 and 40% CO_2 is routed through two pre-treatment vessels to remove moisture,

Fig. 15. Palos Verde sanitary landfill, Los Angeles, California.

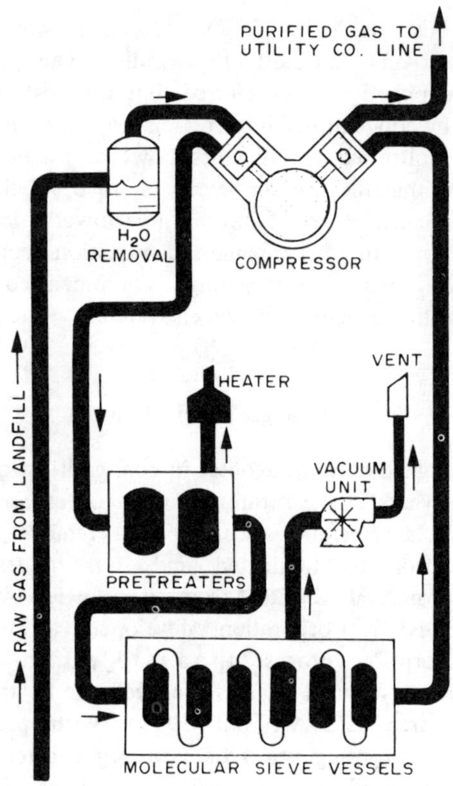

Fig. 16. Methane-purification scheme of NR6-Nufuels, Inc. at the Palos Verde landfill near Los Angeles, California.

hydrogen sulfide, mercaptans, etc. The dry gas is then fed into three parallel adsorption columns packed with pellets of the chabazite/erionite ore, and CO_2 is removed by p.s.a. techniques (Fig. 16). Approximately 1 million cubic feet of methane meeting pipeline specifications is produced each day and delivered to utility company pipelines (NRG, 1975).

That landfills generate methane gas has been known for years, but the CO_2 content is usually too high for efficient use of the gas. Zeolite adsorption processes to remove CO_2 and thereby raise the BTU content of the gas represent major breakthroughs in this area. They should also open up many additional low-grade methane sources for exploitation, such as the methane-rich gas produced in municipal sewage-treatment processes, in solid-waste incineration schemes, and during the digestion of feedlot manures. Not only has the low BTU content of such gases limited their use, but H_2S and other acidic components present severe corrosion problems for valves and pumps. Zeolite adsorption rids the gas of these contaminants also.

Goeppner and Hasselmann (1974) estimated that as much as 2.5 ft^3 of 70% methane gas is produced by the digestion of 100 gallons of municipal sewage. Commonly the impure product is used in-house as a source of heat for digesters or to operate pumps, but these uses tend to be inefficient and result in frequent equipment breakdown. If impurity CO_2 and H_2S could be removed from the gas, a 1000 BTU product could be prepared. Similarly, the methane-rich gas formed by the digestion of animal wastes is also a potential source of energy, *if* the BTU content can be increased by CO_2 removal. Goeppner and Hasselmann (1974) also suggest that a billion cubic feet of 700 BTU gas could be produced by treating the 500 million pounds of manure produced each day by the 12 million cattle now penned in feedlots throughout the United States. In this same area, Jewell (1974) indicates that if impurities could be removed, the methane produced by the digestion of the organic wastes on a typical New York State dairy farm of sixty cows would be equivalent to the farm's entire fossil-fuel requirements. The production of methane gas from municipal solid wastes has been examined by Kispert *et al.* (1974), and it appears as if pure CH_4 could be produced at a cost of about \$2.09 per thousand cubic feet by bacterial digestion, much like the anaerobic treatment of sewage sludge. A critical step in the proposed process is the scrubbing of the 50–50 methane–carbon dioxide product to remove CO_2. Zeolite-adsorption would seem to be competitive with standard monoethanolamine adsorption, and costs might actually be less.

Solar energy use

Zeolites might also make contributions in the solar energy field. Detailed schemes for utilizing energy from the sun's rays are commonly stymied by the lack of efficient heat exchangers. During the last few years considerable success has been achieved at the University of Texas and the Massachusetts Institute of Technology in using natural chabazite and clinoptilolite to adsorb and release heat from solar radiation for both air-conditioning and water heating. The dehydration of the zeolite by day and its rehydration at night results in the exchange of several hundred BTUs per pound of zeolite, sufficient to cool small buildings (Tchernev, 1975). Tchernev estimates that 1 ton of zeolite in solar panels spread over 200 square feet of roof surface will produce 1 ton of air-conditioning. The extremely non-linear adsorption isotherms of crystalline zeolites, in contrast with other sorbent materials, makes possible cooling efficiencies greater than 50%. If continued experimentation supports these initial results, solar energy uses may create a market for natural zeolites of several hundred thousand tons each year.

Petroleum production

In addition to providing valuable information to exploration programs on the paleo-environments of suspected oil and gas fields, natural zeolites have been shown to be potentially useful in the purification and production side of the business. Generally, natural zeolites cannot compete with certain synthetic molecular sieves in many adsorption and catalytic applications because of their inherently smaller pore sizes and adsorption capacities. Iron impurities also act as a poison in many reactions. However, several natural zeolites have been useful in specific applications. Mordenite, chabazite and clinoptilolite are generally capable of withstanding the rigors of continuous cycling in acid environments and have been used successfully to remove water and carbon dioxide from gaseous hydrocarbons (*vide supra* and Tsitsishvili *et al.*, 1976). Union Carbide Corporation (1962) claims its chabazite-rich AW-500 product from Bowie, Arizona, to be effective in reformer-hydrogen drying, chlorine drying, chlorinated and fluorinated hydrocarbon purification, and in the removal of HCl from by-product hydrogen streams. Ohtani *et al.* (1972a, 1972b) showed that a modified mordenite could be used as a conversion catalyst in hydrocarbon disproportionation.

The selective forming process recently developed by Mobil R & D Corporation (Chen, 1971) makes use of catalysts derived from its erionite/clinoptilolite ore from Jersey Valley, Nevada. Neel *et al.* (1973) reported that erionite could be used as a carrier in the catalytic reforming of hydrocarbons. A Bulgarian clinoptilolite was shown by Nikolova and Ivanov (1975) to have potential in olefin hydrogenation, and Scherzer, Vaughan, and Albers (1975) recently patented a hydrocarbon cracking catalyst consisting of a mixture of H–Y zeolite and H-ferrierite, the latter from Lovelock, Nevada. Since 1970 several articles on the potential applications of clinoptilolite and mordenite from Tokaj, Hungary, as catalysts in cracking, isomerization, hydrogenation, alkylation and polymerization have been published (Papp *et al.*, 1971; Kallo *et al.*, 1971; Detrekoy *et al.*, 1974; Beyer, 1975). All make use of the hydrogen form of the zeolite.

AGRICULTURAL SIGNIFICANCE[1]

Fertilizer uses

Although little has appeared in the scientific literature, there is growing interest in the application of zeolite minerals in the field of agriculture. Farmers in Japan have used natural zeolites for years to control the moisture content and malodor of animal manure and have added them to their fields in an attempt to neutralize low pH soils. Based upon their high-ion-exchange capacity and water retentivity, they have been used extensively as soil conditioners, and small tonnages have been exported to Taiwan for this purpose (Minato, 1968). Also, in Japan, the ion-exchange selectivity of clinoptilolite has been exploited in the preparation of chemical fertilizers which tend to improve the nitrogen retention of soils by promoting a slower release of ammonium ions. The ion-trapping capabilities of this zeolite have also been demonstrated by Nishita and Haug (1972) who showed that the addition of clinoptilolite to soils contaminated with Sr^{90} resulted in a marked decrease in the strontium uptake by plants. Fujii (1974) also reported that heavy metals such as Cd, Pb, Zn and Cu can be ion-exchanged from soils by natural zeolites added along with fertilizers, thereby preventing their movement from the soil to the food chain. Similarly, clinoptilolite and other zeolites showing high selectivity for heavy metal ions might find application as scavengers for undesirable species in soils contaminated by the addition of sludge from sewage-treatment plants or from industrial wastes. Median values of 31 ppm Cd, 1230 ppm Cu, 830 ppm Pb and 2780 ppm Zn have been reported for sludges from typical treatment plants (Cohen, 1977). Zeolite additives may be the key to the safe disposal and use of the ever-increasing quantities of sewage sludge that are produced each year.

Herbicide and pesticide carriers

Similar to many of their synthetic counterparts, the high ion-exchange and adsorption capacities of natural zeolites make them effective carriers of herbicides and pesticides. Yoshinaga *et al.* (1973) found clinoptilolite to be an excellent carrier of benzylphosphorothioate used to control stem blasting in rice. Using a natural zeolite as a base, Hayashizaki and Tsuneji (1973) developed a lime/nitrogen acaricide with good results.

Animal nutrition

Since 1965 experiments have been in progress in Japan on the use of clinoptilolite and mordenite as dietary supplements in animal husbandry. Up to 10% zeolite has been added to the normal rations of pigs, chickens and ruminants, resulting in significant increases in feed conversion values and in the general health of the animals. Onagi (1965) found that Leghorn chickens on a diet containing 10% clinoptilolite gained as much weight

[1] See also Mumpton, F. A. and Fishman, P. H. (1977) The application of natural zeolites in animal science and aquaculture: *J. Anim. Sci.* **45**, 1188-1203.

as did birds on a normal diet, with no adverse effects on the vitality of the chickens or on their egg production. In addition, the droppings of test birds contained up to 27% less moisture, making the manure considerably easier to handle. Arscott (1975) found that broiler chickens fed a diet containing 5% clinoptilolite gained slightly less weight over a 2-month period than birds on a normal diet, but that the average feed-conversion values (weight gain divided by nutritional feed intake) were significantly higher. Of equal or greater import was the fact that none of the forty-eight test birds died during the experiment, whereas three birds on the control diet and two on a control + antibiotic diet succumbed.

Kondo and Wagai (1968) observed similar results with swine. They found that pigs on diets containing 5% clinoptilolite gained up to 25% more weight than those on normal diets. No toxic or other adverse effects were noted. On the contrary, the presence of zeolite in the rations appeared to contribute measurably to the well-being of the animals. According to Torii (1974), the death rate and incidence of sickness among swine fed a diet containing 6% clinoptilolite was markedly lower than for control animals in experiments conducted on a large swine-raising facility in Japan. Scours and other intestinal diseases were greatly reduced. Apparently the vitalizing effect of a zeolite diet can also be transferred from mother to offspring. Torii (1974) reported that the growth rate of newborn pigs whose mothers were fed 600 g of clinoptilolite each day of the 35-day weaning period was 65–85% greater than those of control animals. In addition, the young pigs in the test group suffered almost no attacks of diarrhea, as opposed to those in control groups who were severely afflicted with scours, inhibiting their normal growth.

Similar studies were conducted at Oregon State University (England, 1975) on young pigs using rations containing 5% clinoptilolite from the Hector, California, deposit. Although lesser increases in growth rate were observed, the incidence of scours was significantly reduced for animals receiving the zeolite diet. Currently heavy doses of prophylactic antibiotics are used to control intestinal diseases which, unchecked, can result in high mortality rates in young swine after they are weaned. If federal regulations prohibit such usage, other means must be found, and natural zeolites might be the answer. The exact functions of the zeolite in both dietary and antibiotic phenomena are not well understood and await serious physiological and biochemical examination. The ammonium selectivity of clinoptilolite suggests that it might act as a nitrogen reservoir in the digestive system of the animal, allowing a slower release and more efficient use of ammonium ions produced by the breakdown of ingested rations in the development of animal protein. Zeolite particles might also stimulate the linings of the stomach and intestinal tract causing the animal to produce more antibodies, and thus inhibiting diseases such as scours.

In an attempt to reduce the toxic effects of high NH_4^+ concentrations in ruminal fluids when non-protein nitrogen compounds (NPN), such as urea and diuret, are added to the diets of cattle, sheep, and goats, White and Ohlrogge (1974) introduced both natural and synthetic zeolites into the rumen of test animals. Ammonium ions formed by the enzyme decomposition of NPN were immediately ion exchanged into the zeolite structure and held there for several hours until released by the regenerative action of sodium ions entering the rumen in saliva. Thus, the gradual release of ammonium ions allowed rumen micro-organisms to synthesize cellular protein continuously for easy assimilation into the animal's digestive system. The zeolite could be introduced as pellets or beads in a foraminous bolus or as a finely divided powder mixed with the normal rations. Both natural chabazite and clinoptilolite were found to be effective, although synthetic zeolite F worked best of all. Considering the possible price of about $1.00 per pound for zeolite F, if and when it becomes commercially available, it would seem that certain NH_4^+-selective *natural* zeolites, such as clinoptilolite, would be especially useful in this application. Prices of a few pennys a pound for the natural zeolites suggest that they could be fed to the animal as part of its normal rations.

In this same area Kondo *et al.* (1969) found that zeolites added to the feed improved the growth rate of calves by stimulating their appetite and improving digestion. Watanabe *et al.* (1971) raised six bullocks for 329 days on a diet containing 2% clinoptilolite. Little difference in weight gain or feed conversion was noted between these control animals, but the test steers showed slightly larger body dimensions and dressed out to give higher-quality meat. The differences were reflected in a 20% larger profit compared to animals on a normal diet. Diarrhea and other intestinal diseases were also less noticeable in the test animals. It is unfortunate that a much higher level of zeolite was not used in these experiments; earlier experiments in the United States added as much as 40% clay to the rations of domestic animals without adverse effects (Ousterhout, 1970).

Animal-waste treatment

The rising demand for animal protein in worldwide diets is attended by several major problems, not the least of which is what to do with the mountains of animal wastes produced each day on farms and in feedlots of every country. Livestock production in the United States alone creates more than one billion tons of solid waste and nearly 400 million tons of liquid waste each year (Laporte, 1975, p. 365). Accumulations of such materials pose serious health problems to man and animal alike and are significant sources of pollution for rivers and streams in the area. Large amounts of undigested protein remaining in the excrement represents a valuable resource that for the most part is now wasted, due to our growing dependence on chemical fertilizers. Natural zeolites have potential application in several areas of manure treatment, including (1) reducing the malodor of the excrement, (2) controlling the moisture content for ease of handling and (3) purifying methane gas produced by the anaerobic digestion of such materials (*vide supra*).

The semi-fluid droppings in large poultry houses commonly emit a stench that is discomforting to farm workers and to the chickens themselves. Lower egg production is a common result, as many birds suffer from respiratory diseases caused by the noxious fumes of ammonia and hydrogen sulfide. Torii (1974) reported that about 100 tons of clinoptilolite is used each year in the chicken houses of Japan. The zeolite is either mixed with the droppings or packed in boxes which are suspended from the ceilings. In either case, ammonia is removed, and egg production is increased. Similar results were obtained by Onagi (1968) and Kondo and Wagai (1968) when zeolites were added to the rations of chickens and pigs, respectively. About 25 tons of clinoptilolite are spread on the floors of a Sapporo pig farm each month to adsorb urine and other liquid wastes. The buildings are said to be dry, clean and considerably less odoriferous. An innovative use of natural zeolites in the treatment of animal excrement was developed by Komakine (1974) and involves mixing chicken droppings with zeolite and ferrous sulfate. The ferrous sulfate inhibits zymosis and decomposition of the droppings, while the zeolite stabilizes the hygroscopic nature of this compound and captures ammonium ions produced by the manure. The mixture is dried and can be used as an odorless fertilizer, or it can be used as an odorless feedstuff, rich in protein, for fowl, fish, and domestic animals. It has been used successfully in Japan to substitute for as much as 20% of the normal rations of swine, poultry, and edible carp.

AQUACULTURAL SIGNIFICANCE

Aquaculture may not be the number one food industry in the United States nor in very many other nations for that matter, but as the world's protein requirements continue to rise, more and more fish products may find their way to the dinner table or to the feeding trough in future years. Although progressive farmers in several countries are beginning to learn that catfish or trout can bring in more dollars per acre than wheat or rice, fish raising is a tricky business, and the chemical and biological environment of aquacultural systems must be maintained within close limits at all times. Natural zeolites have been found to be valuable materials in controlling these environments, and as more fish hatcheries turn to self-contained, water-reuse systems, a sizeable market for natural zeolites may develop.

Using processes similar to those employed in sewage-treatment plants for the extraction of ammoniacal nitrogen, Konikoff (1974) and Johnson and Sieburth (1974) showed clinoptilolite to be effective in the removal of ammonium ions from recirculating fish-culture systems. Ammonium is one of the most significant toxic metabolites in aquaculture and is extremely harmful to fish in concentrations exceeding a few parts per million. In oxygen-poor environments such concentrations can lead to damage of gill tissue, various gill diseases, and a reduction in the growth rate (Larmoyeaux and Piper, 1973). Unpublished results of tests conducted in 1973 at a working hatchery near Newport, Oregon, indicated that 97–99% of the ammonium produced in a recirculating system was removed by clinoptilolite ion-exchange columns from waters containing 0.34–1.43 mg NH_3–N per liter (Kapranos, 1976). Peters and Bose (1975) substantiated these results and found that trout remained perfectly healthy during a 4-week trial when zeolite ion-exchange was used to remove ammonium from recirculating tank-water. They concluded that selective ion-exchange using clinoptilolite may be a viable alternative to biological oxidation processes which are highly susceptible to minor changes in the temperature and chemistry of such systems. At the present time Becker Industries of Newport, Oregon, in conjunction with the U.S. Army Corps of Engineers, is developing a complete purification unit for hatchery-water reuse. The system will incorporate a zeolite ion-exchange unit for ammoniacal nitrogen removal and is designed to handle concentrations of 20–30 ppm NH_3–N at flow rates of 10–15-Mgd, typical of the more than 200 fish hatcheries now operating in the Pacific Northwest (Kapranos, 1976).

Jungle Laboratories of Comfort, Texas, is in the process of developing ammonia-removal systems for hatchery waters using clinoptilolite ion-exchange and for fish haulage operations where brain damage due to excess ammonium ions commonly results in sterility, stunted growth, and high mortality rates (Nichols, 1976). Throw-away cartridges and filters containing granular clinoptilolite are also being designed to regulate the ammoniacal nitrogen content of home aquaria and tanks used to transport tropical fish from the time they are caught until they reach the hobbyist. The U.S. Fish and Wildlife Service is investigating similar processes for ammonia removal from recirculating waters in tank trucks used to transport channel catfish from Texas to the Colorado River in Arizona (McCraren, 1976). Currently about 3500 pounds of 8–10-inch fish are carried in 12,000 gallons of water, or about 2.2 lb/ft^3. If the ammonium ions produced by decomposing fish excrement and waste food can be extracted the number of fish hauled in trucks of this size can be doubled.

Slone *et al.* (1975) were able to achieve biomass concentrations of more than 7 lb/ft^3 for catfish (*Ictaluras punctatus*) after 7 months in a vertical raceway system using recycled water and a zeolite ion-exchange process to remove ammoniacal nitrogen following biofiltration. NH_4-N concentrations of 5 mg/l were maintained easily with clinoptilolite from the Hector, California deposit.

As discussed above, zeolite adsorption units using natural mordenite are currently marketed in Japan and have been used to provide oxygen for aeration in fish culture and in the transportation of live fish (Minato, 1974). Small generators capable of producing 15 liters of 50% O_2 per hour are manufactured by the Koyo Development Company, Ltd. Carp and goldfish raised in oxygen-aerated water are said to be livelier and to have greater appetites (Koyo Kaihatsu, 1974). Little work has been carried out on the use of natural zeolites in fish food, but the protein-rich mixture of chicken droppings, ferrous sulfate and zeolite discussed above (Komakine, 1974) has been fed to carp in Japan with no adverse effects. Much as the excrement of pigs, chickens, and cattle is rendered less odoriferous by the addition of clinoptilolite to the rations of the animals, it would seem that natural zeolites incorporated into normal fish food would reduce the ammonia buildup in closed tanks or in large holding facilities where water is recirculated. Such possibilities bear investigating.

In the area of fish processing, Araki and Honda (1974) found that zeolitic tuff is able to deodorize protein processing odors. Gas containing large proportions of ammonia, volatile amines, mercaptans, and acetic acid were purified when passed through a column of pelletized zeolite at 250–400°C.

SIGNIFICANCE IN MINING AND METALLURGY

Exploration aids

Recent studies in Japan indicate that zeolite assemblages in altered tuffs can not only delineate the conditions of formation of certain ore deposits, but also serve as exploration tools, especially in areas of thick overburden. Yoshida and Utada (1969) and Utada *et al.* (1974) noted that analcime-rich aureoles surrounding Kuroko-type mineralization in Neogene sediments of the Green Tuff region are thickest in the vicinity of major ore deposits. Supported by the experimental results of Aoki (1974) on the conversion of clinoptilolite to analcime in sodium carbonate solutions, they suggest that hydrothermal alteration was superimposed on zeolitic alteration that took place earlier during burial diagenesis of the marine-tuff host rocks. The thickness of analcimic alteration in such rocks, therefore, may be a clue to the location of deep-seated ore bodies.

Also in Japan, Katayama *et al.* (1974) attributed the concentration of uranium to the presence of a heulandite-clinoptilolite zeolite in tuffaceous sandstones of Miocene age near Tono, Gigu Prefecture. Oxidized uranium in groundwater is presumed to have been adsorbed on the zeolite, which in some zones contains as much as 0.9% U. Although considerable experimentation is still required, these studies suggest that natural zeolites may be used to extract and concentrate uranium and other ions from low-level processing solutions, such as those encountered during the in-place or heap-leaching of uranium and copper ores and tailings.

Hydrometallurgical uses

Similarly, the ion-exchange properties of many natural zeolites lend themselves to the concentration of heavy metals from wastewater effluents of mining and metallurgical operations. Torii (1974) reported that sodium-exchanged clinoptilolite and mordenite removed almost all of the Cd^{++} from 10-ppm solutions of that ion. Unpublished results of F. F. Aplan (1975) also point out the effectiveness of certain zeolites for extracting heavy metal ions from aqueous media. He found that erionite and chabazite are able to remove more than 90% of the Pb^{++} and Cu^{++} from 2.5×10^{-3} molar solutions. Chelishchev *et al.* (1974) successfully extracted large, heavy-metal cations from complex solutions using clinoptilolite from the Georgian S.S.S.R. They reported a selectivity of Pb > Ag > Cd, Zn, Cu > Na. Fujimori and Moriya (1973) noted that clinoptilolite removed heavy metal ions from industrial wastewater in the sequence Pb >> Cd > Cu >> Zn. Although the overall specificities of individual zeolites for heavy metals have not yet been established, the preliminary data suggest that natural zeolites can contribute heavily towards eliminating a large part of the pollution from industrial wastewaters and also provide a means of recovering heretofore lost values from many hydrometallurgical effluents.

MISCELLANEOUS APPLICATIONS

Paper products

Approximately 3000 tons of high-brightness zeolite from the Itaya Mine, Yamagata Prefecture, Japan, is sold each month by the Zeeklite Chemical Company under the trade names SGW and HiZ as filler in the paper industry (Minato, 1975). Clinoptilolite ore is finely ground and classified by wet or dry cycloning into a 10-micrometer-size product having an abrasion index of less than 3% and a brightness of about 80 (Takasaka, 1975). According to Kobayashi (1970), kraft papers filled with clinoptilolite are bulkier, more opaque, easier to cut and less susceptible to ink blotting than those filled with clay. Hayakawa and Kobayashi (1973) patented a lightweight paper made from a mixture of 17% pulp, 25% chemically ground pulp, 28% bleached kraft, 1% sizing, 1% aluminium sulfate, and 28% zeolite powder. The product has a density of 0.68 g/cm^3 compared with 0.73 g/cm^3 for conventional paper. A few hundred tons of clinoptilolite from the Tokaj district of eastern Hungary (Nemecz and Varju, 1962) are also mined each month for paper-filling applications, although Kobor and Hegedus (1968) found Hungarian natural zeolites to be unsuited for the manufacture of wood-free papers.

Kato (1976) combined clinoptilolite and mordenite with organic dyes, such as C.I. Basic Red 34, to give heat-, light- and acid-resistant coloring composites useful in coating copying papers and in coloring plastics. Breck (1975) found that the addition of from 5 to 30% natural chabazite or synthetic A or X zeolites to conductive paper used in electrostatic reproduction resulted in a product having resistivities of 10^7–10^8 ohm centimeters. The strong hydration properties of the zeolites allowed these values to be maintained over a range of relative humidities from 5 to 90%.

Construction uses

Much as perlite and other volcanic glasses can be expanded into low-density pellets for use as lightweight aggregate in cements and concretes, natural zeolites also can be frothed into stable products by calcining at elevated temperatures. Stojanović (1972) reported densities as low as 0.8 g/cm^3 and porosities of up to 65% for expanded clinoptilolite from several Serbian deposits after firing at 1200–1400°C. Similar materials were prepared by Ishimaru and Ozata (1975) from Japanese zeolites at 1250°C. Kasai *et al.* (1973) prepared foaming agents by calcining clinoptilolite at about 550°C and cooling in air. The product was mixed with equal parts water and dolomite plaster, molded and hardened for 2 hours in an autoclave to obtain a product having a bulk density of 0.75 g/cm^3 and a compressive strength of about 47 kg/cm^2. In a search for perlite substitutes, Bush (1974) found that high-grade clinoptilolite from the Barstow Formation, California, expanded 4 to 6 times when heated for 5 minutes at from 1150–1250°C. Synthetic CaA zeolite loaded with 1-butene or butadiene was used by Ulisch (1975) to prepare a porous foam having a density of 0.95 g/cm^3; natural erionite or chabazite should work equally as well. The temperatures required for expanding zeolitic tuffs is significantly higher than those needed for perlite or other expandable materials (1200°C vs. 760°C); however, the foamed zeolite products are considerably stronger and more resistant to abrasion.

Also in the construction area, Torii (1974) reported that about 100 tons of clinoptilolite are consumed each month as filler in wheaten paste used to adhere plywood prior to hot-pressing. He also listed 200 tons per month of mordenite as being used to fabricate lightweight bricks (0.7–1.0 g/cm^3) of high physical strength and high acid- and alkali-resistance. By firing a mixture of 100 parts glass, 70 parts clinoptilolite, 3 parts carbon, and 3 parts H_3PO_4 at 800°C, Tamura (1974) was able to produce a high-strength porous glass having a density of only 0.22 g/cm^3.

Medical applications

In the medical field, Kato *et al.* (1969, 1970) found that clinoptilolite is useful as a polishing agent in fluoride-containing toothpaste. It is no more abrasive than the commonly used $CaHPO_4$ and allows more of the fluoride ion to remain in the anionic form. Andersson *et al.* (1975) were able to separate ammoniacal nitrogen from hemodialysis liquids in recycle-dialysis systems using a natural phillipsite. The zeolite was found to be superior in selectivity and ion-exchange capacity to other exchangers, including zirconium phosphate, Dowex 8 X 50, clinoptilolite, and Union Carbide Corporation's AW-500 chabazite/erionite product.

ACKNOWLEDGMENTS

Literature reviews of rapidly expanding subjects, such as the geological significance and the industrial utilization of natural zeolites, can never be up to date; however, with the generous help of many individuals in academy, industry, and government, this author has tried to make this review as current as possible. The author's gratitude is extended to L. L. Ames, G. H. Arscott, F. F. Aplan, D. C. England, W. Kapranos, M. Koizumi, R. B. Laudon, R. T. Mandeville, H. Minato, E. Nemecz, L. B. Sand, R. A. Sheppard, D. Tchernev, and D. E. W. Vaughan for numerous discussions and for providing unpublished information. Special thanks are due to Mr. Kazuo Torii of the National Industrial Research Institute, Tohoku, Sendai, Japan, for his many courtesies in providing data on the use of natural zeolites in Japan and for his translations of several articles.

REFERENCES

Adam, L., Kakasy, Gy. and Pallos, I. (1971) Removal of radioactive pollutants from water using mineral substances of natural domestic origin: *Hung. Mining Res. Inst.* **14**, 209-212.

Ames, L. L. (1959) Zeolitic extraction of cesium from aqueous solutions: Unclass. Rept. HY-62607, US Atomic Energy Comm., 23 pp.

Ames, L. L. (1960) The cation sieve properties of clinoptilolite: *Amer. Mineral.* **45**, 689-700.

Ames, L. L. (1967) Zeolitic removal of ammonium ions from agricultural wastewaters: *Proc. 13th Pacific Northwest Indust. Waste Conf., Washington State Univ.*, pp. 135-152.

Ames, L. L., Jr., Sand, L. B. and Goldich, S. S. (1958) A contribution on the Hector, California bentonite deposit: *Econ. Geol.* **53**, 22-37.

Andersson, Sten; Grenthe, Ingmar; Jonsson, Ernst and Naucler, Lars (1975) Separation of poisons from the dialysis liquid of a recycle dialysis system: Ger. Offen. 2,512,212, Sept. 25, 1975, 30 pp.

Anurov, S. A., Keltsev, N. V., Smola, V. I. and Torocheshnikov, N. S. (1974) Adsorption of SO_2 by natural zeolites: *Zh. Fiz. Khim.* **48**, 2124-2125.

Aoki, Morihiro (1974) Synthesis of analcime from clinoptilolite tuff in $NaCO_3$ solution. Experimental consideration related to the formation of analcime zone surrounding Kuroko deposits: *Ganseki Kobutso Kosho Gakkaishi* **69**, 171-180.

Aplan, F. F. (1975) Div. Mineral Processing, Penna. State Univ., University Park, Penna. (personal communication).

Araki, Kunio and Honda, Shigekazu (1974) Deodorizing process: Japan Kokai 74,034,898, Sept. 18, 1974, 2 pp.

Arscott, G. H. (1975) Dept. Poultry Science, Oregon State University, Corvallis, Oregon (personal communication).

Barrer, R. M. (1938) The sorption of polar and non-polar gases on zeolites: *Proc. Roy. Soc. Lond.* **167A**, 392-419.

Battelle-Northwest (1971) Wastewater ammonia removal by ion exchange: Rept. Project No. 17010 ECZ for US Environ. Protect. Agency, 62 pp.

Beyer, H. (1975) Cracking reactions. II. Kinetics of the catalytic cracking of propane, butane, and isobutane on H-clinoptilolite: *Acta Chim. Acad. Sci., Hung.* **84**, 25-43.

Blodgett, G. A. (1972) SO_2 adsorption on ion-exchanged mordenites: M.S. Thesis, Worcester Polytechnic Institute, 96 pp.

Bradley, W. H. (1928) Zeolite beds in the Green River Formation: *Science* **67**, 73-74.

Bramlette, M. N. and Posniak, E. (1933) Zeolitic alteration of pyroclastics: *Amer. Mineral.* **18**, 167-171.

Breck, D. W. (1975) Electrostatic printing process; U.S. Patent 3,884,687, May 20, 1975, 5 pp.

Bush, A. L. (1974) National self-sufficiency in lightweight aggregate resources: Minutes – 25th ann. meet. Perlite Inst., Colorado Springs, Colo., April 18-23, 1974.

Carl. J. D. and Van Diver, B. B. (1975) Precambrian Grenville alaskite bodies as ash-flow tuffs, northwest Adirondacks, New York: *Bull. Geol. Soc. Amer.* **86**, 1691-1707.

Chelishchev, N. F., Martynova, N. S., Fakina, L. K. and Berenshtein, B. G. (1974) Ion exchange of heavy metals on clinoptilolite: *Dokl. Akad. Nauk SSSR* **217**, 1140-1141.

Chen, N. Y. (1971) Shaped selective transition metal zeolite hydrocracking catalysts: U.S. Patent 3,630,966, Dec. 28, 1971.

CH_2M-Hill (1975) *Conceptual Design Report – Potable Water Reuse Plant, Successive Use Program:* for Board of Water Commissioners, Denver, Colo., Vol. I, Vol. II.

Cohen, J. M. (1977) Trace metal removal by wastewater treatment: Tech. Transfer, U.S. Env. Prot. Agen. January, 1977, pp. 2-7.

Coombs, D. S. (1954) The nature and alteration of some Triassic sediments from Southland, New Zealand: *Trans. Royal Soc. New Zealand* **82**, 65-109.

Cronstedt, A. F. (1756) Om en obekant barg art, som kallas Zeolites. *Akad. Handl. Stockholm* **17**, 120-123.

Daiev, Ch., Delchev, G., Zhelyazkov, V., Gradev, G. and Simov, S. (1970) Immobilization of radioactive wastes attached to natural sorbents in bitumen molds. Int. Atomic Energy Agency, Vienna, *Symp. Management of Low- and Intermediate-level Radioactive Wastes,* pp. 739-746.

Deffeyes, K. S. (1959) Erionite from Cenozoic tuffaceous sediments, central Nevada: *Amer. Mineral.* **44**, 501-509.

Detrekoy, E. J., Jacobs, P. A., Kallo, D. and Uytterhoeven, J. B. (1974) Catalytic activity of hydroxyl groups in clinoptilolite: *J. Catalysis* **32**, 442-451.

Dominé, D. and Häy, L. (1968) Process for separating mixtures of gases by isothermal adsorption: possibilities and application: In: *Molecular Sieves*, Soc. Chem. Industry, pp. 204-216.

England, D. C. (1975) Effect of zeolite on incidence and severity of scouring and level of performance of pigs during suckling and early postweaning: *Rept. 17th Swine Day*, Spec. Rep. 447, Ag. Ex. Stat., Ore. State Univ., pp. 30-33.

Eyde, T. H. (1976) Zeolites: *Mining Engineering*, Mar. 1976, pp. 51-53.

Fenner, C. N. (1936) Bore-hole investigations in Yellowstone Park: *J. Geol.* **44**, 225-315.

Fugii, Shigeharu (1974) Heavy metal adsorption by pulverized zeolites: Japan. Kokai 74,079,849, Aug. 1, 1974, 2 pp.

Fujimori, Keiichi and Moriya, Yoshio (1973) Removal and treatment of heavy metals in industrial wastewater. I. Neutralizing method and solidification by zeolite: *Asahi Garasukogyo Gijutsu Shoreikai Kenkyu, Hohoku* **23**, 243-246.

Goeppner, J. and Hasselmann, D. E. (1974) Digestion by-product may give answer to energy problem: *Water & Wastes Engineer* **11**, No. 4, 30-35.

Hagiwara, Zenzi and Yamamoto, Tatsuo (1974) Selective enrichment of oxygen and nitrogen: Japan. Kokai 74,054,289, May 27, 1974, 9 pp.

Haralampiev, G. A. *et al.* (1975) Possibilities of oxygen enrichment of air for metallurgy using Bulgarian clinoptilolite: Annual report HIC-HCTI-Sofia, pp. 60-75.

Hashimoto, Akio (1974) Adsorption of heavy metals and application to watersupply systems. II. Removal of Ca, Cd, and N as ammonia: *Mizu shori Gijutau* **15**, 1085-1091.

Hashimoto, Sadao and Miki, Kanji (1975) Water-purifying materials: Japan. Kokai 75,098,044, May 1, 1975, 5 pp.

Hawkins, D. B. and Short, H. L. (1965) Equations for the sorption of cesium and strontium on soil and clinoptilolite: U.S. Atom. Energy Comm. IDO-12046, 33 pp.

Hay, R. L. (1963) Stratigraphy and zeolite diagenesis of the John Day Formation of Oregon: Univ. Calif. Publ. in *Geol. Sci.* **42**, 199-262.

Hay, R. L. (1966) Zeolites and zeolite reactions in sedimentary rocks: *Geol. Soc. Amer.* Spec. Paper **85**, 130 pp.

Hay, R. L. (1968) Chert and its sodium–silicate precursors in sodium–carbonate lakes of East Africa: *Contr. Mineral.*

Petrol. **17**, 255-274.

Hay, R. L. (1970) Silicate reactions in three lithofacies of a semi-arid basin, Olduvai Gorge, Tanzania: *Mineral. Soc. Amer.* Spec. Paper **3**, 237-255.

Hayakawa, Junshiro and Kobayashi, Takeo (1973) Fillers for paper: Japan. Kokai 73,099,402, Dec. 15, 1973, 3 pp.

Hayashizaki, T. and Tsuneji, N. (1973) Acaricidal composition containing lime–nitrogen: Japan. Kokai 73,031,888, Oct. 1973, 4 pp.

Höller, H., Wirsching, U. and Fakhuri, M. (1974) Experiments zur Zeolithbildung durch hydrothermale Umwandlung: Zur Enstehung von Chabasit, Phillipsit, und Analcim aus den glasigen Bestandteilen der Bims-Staubtuffe des Laacher Vulkangebietes: *Contr. Mineral. & Petrol.* **46**, 49-60.

IAEA (1972a) Use of local minerals in the treatment of radioactive wastes: *Int. Atomic Energy Agency, Vienna, Tech. Rept.* **136**, 97-B.

IAEA (1972b) Use of local minerals in the treatment of radioactive wastes: *Int. Atomic Energy Agency, Vienna, Tech. Rept.* **136**, 68, 98.

Iijima, A. (1974) Genetical connection of sedimentary zeolites in between burial diagenesis and alkali–saline lake deposits: *Seminar on the Occurrence, Origin, and Utilization of Sedimentary Zeolites in the Circum–Pacific Region*; U.S.–Japan Cooperative Science Program, Menlo Park, Cal., July 1974 (unpublished abstract).

Iijima, A. and Utada, M. (1972) A critical review on the occurrence of zeolites in sedimentary rocks in Japan. *Jap. J. Geol. Geog.* **42**, 61-84.

Ishikawa, H., Tanaka, H., Uchiyama, K., Kwon, S., Asayama, T. and Morimoto, T. (1972) Adsorption and desorption of SO_2 gas on natural zeolites: *Waseda Diag. R.K. Kark. Hok.* **51**, 46-53.

Ishimaru, Hirohisa and Ozata, Katsuya (1975) Ceramic foam: Japan. Kokai 75,028,510, Mar. 24, 1975, 5 pp.

Jewell, W. J. (1974) Methane . . . the energy-sufficient farm: State Univ. New York – *The News* **3**, No. 2, 4.

Johannsen, A. (1914) Petrographic analysis of the Bridger, Washikie and other Eocene Formations of the Rocky Mountains: *Bull. Amer. Mus. Nat. Hist.* **33**, 209-222.

Johnson, P. W. and Sieburth, J. M. (1974) Ammonia removal by selective ion exchange, a backup system for microbiological filters in closed-system aquaculture: *Aquacul.* **4**, 61-68.

Kakasy, I., Pallos, I., and Adam, L. (1973) Étude des roches clinoptilolitiques de la Hongrie au point de vue de la fixation des déchets radioactifs liquides: *Publ. Hung. Mining Res. Inst.*, No. 16, pp. 169-175.

Kallo, D. *et al.* (1971) Catalytic isomerization of butene: *Kem. Kozlem.* **36**, 239 245.

Kapranos, W. (1976) Becker Industries, Portland, Oregon (personal communication).

Kasai, J., Urano, T., Kuji, T., Takeda, Y. and Machinaga, O. (1973) Inorganic foaming agents by calcination of natural silica-high zeolites: Japan. Kokai 73,066,123, Sept. 11, 1973, 5 pp.

Katayama, N., Kubo, K. and Hirono, S. (1974) Genesis of uranium deposits of the Tono Mine, Japan: *Proc. Symp. Formation of Uranium Ore Deposits, Athens, Greece, May 6-10, 1974,* pp. 437-452.

Kato, Chuzo (1976) Coloring composites: U.S. Patent 3,950180, April 13, 1976, 5 pp.

Kato, Kazuo; Nagata, Maohiro; Sakai, Yoshihiro; Shiba, Mitsue and Onishi, Masao (1969) Zeolite as polishing agent for dentrifice. I. (Chemical) properties of zeolite: *Iwate Daigaku Nagakubu Hokoku* **3**, 22-35.

Kato, Kazuo; Nagata, Naohiro; Sakai, Yoshihiro; Shaba, Mitsue; Kojima, Masahumi and Onishi, Masao (1970) Zeolites as polishing agent for dentrifice. II. Abrasive action and active fluoride ion in zeolite-containing fluoride: *Iyokizikai Kenkyushu Hokoku* **4**, 115-128.

Kato, Kiyoshi (1974) Removal of Cs^{137} from aqueous solutions by zeolite: *Hoken Butsuri* **9**, 11-16.

Kerr, P. E. (1931) Bentonite from Ventura, California: *Econ. Geol.* **26**, 153-168.

Kispert, R. G., Anderson, L. C., Walker, D. H., Sadek, S. E. and Wise, D. L. (1974) Fuel gas production from solid waste: NSF/RANN/SE/C Report 827/PR/74/2, 176 pp.

Knoll, K. C. (1963) Removal of cesium from aqueous solutions by adsorption: U.S. Patent 3,097,920.

Kobayashi, Y. (1970) Natural zeolite-fillers for paper: Japan. Kokai 70,041,044.

Kobor, Lidia and Hegedus, Istuan (1968) Hungarian zeolite in the paper industry: *Papiripar* **12**, 44-50.

Kolla, V. and Biscaye, P. E. (1973) Deep-sea zeolites: Variations in space and time in sediments of the Indian Ocean: *Mar. Geol.* **15**, 11-17.

Komakine, Chukei (1974) Feedstuff for fowl, fish, and domestic animals: U.S. Patent 3,836,676, Sept. 17, 1974, 3 pp.

Kondo, Kanae; Fujishiro, Seiji; Suzuki, Fumio; Taga, Teiki; Morinaga, Hideo; Wagai, Bunsaku and Kondo, Tonosuke (1969) Effect of zeolites on calf growth: *Chikusan No Kenikyu* **23**, 987-988.

Kondo, N. and Wagai, B. (1968) Experimental use of clinoptilolite-tuff as dietary supplements for pigs: *Yotonkai*, May, pp. 1-4.

Konikoff, M. A. (1973) Comparison of clinoptilolite and biofilters for nitrogen removal in recirculating fish culture systems: Ph.D. Diss., Southern Illinois Univ., Univ. Microfilms No. 74-6222, 98 pp.

Koon, J. H. and Kaufman, W. J. (1971) *Optimization of Ammonia Removal by Ion Exchange Using Clinoptilolite*: Water Pollution Control Research Series 17080, DAR 09/71, 189 pp.

Koyo Kaihatsu Co., Ltd. (1974) Adsorption Oxygen Generator: Company brochure, 4 pp.

Lamoyeux, J. D. and Piper, R. G. (1973) Effects of waste reuse on rainbow trout in hatcheries: *Prog. Fish Cult.* **35**, 2-8.

Laporte, L. F. (1975) *Encounter with the Earth*: Canfield Press, San Francisco, 538 pp.

Liles, A. W. and Schwartz, R. D. (1976) Method of treating wastewater: U.S. Patent 3,968,036, July 6, 1976, 10 pp.

Loew, O. (1875) Report upon mineralogical, agricultural, and chemical conditions observed in portions at Colorado, New Mexico, and Arizona in 1873: Pt. 6 of *U.S. Geog. and Geol. Explor. and Surv. West of 100th Meridian* **3**, 569-661.

Mathers, W. G. and Watson, L. C. (1962) A waste disposal experiment using mineral exchange on clinoptilolite: Atom. Energy of Canada, Ltd., AECL-1521.

McBain, J. W. (1932) *The Sorption of Gases and Vapors by Solids*: Rutledge & Sons, London.

McCraren, J. (1976) U.S. Fish & Wildlife Service, San Marcos, Texas (personal communication).

Mercer, B. W. (1969) Clinoptilolite in water-pollution control: *Ore Bin* **31**, 209-213.

Mercer, B. W., Ames, L. L. and Smith, P. W. (1970a) Cesium purification by zeolite ion exchange: *Nucl. Appl. & Tech.* **8**, 62-69.

Mercer, B. W., Ames, L. L., Touhill, C. J., Van Slyke, W. J. and Dean R. B. (1970b) Ammonia removal from secondary effluents by selective ion exchange. *J. Wat. Pollut. Control Fed.* **42**, R95-R107.

Miki, Kenji; Oyama, Ryoji and Kitagawa, Hidenari (1974) Oilspill adsorbent containing zeolite and perlite: Japan. Kokai 74,007,184, April 12, 1972, 6 pp.

Miller, W. C. (1973) Adsorption cuts SO_2, NO_x, Hg: *Chem. Engineer*, Aug. 6, 1973, pp. 62-63.

Milton, R. M. (1959) Molecular sieve adsorbents: U.S. Patent 2,882,243.

Milton, R. M. (1968) Commercial development of molecular sieve technology. In *Molecular Sieves*; Soc. Chem. Ind., London, pp. 199-203.

Minato, Hideo (1968) Characteristics and uses of natural zeolites: *Koatsugasu* **5**, 536-547.

Minato, Hideo (1974) Properties of sedimentary zeolites for

industrial application: *Seminar on the Occurrence, Origin, and Utilization of Sedimentary Zeolites in the Circum–Pacific Region*; U.S.–Japan Cooperative Science Program, Menlo Park, Cal., July 1974 (unpublished abstract).

Minato, Hideo (1975) Zeolite, its natural resources and utilization: *Ceramics (Japan)* **10**, 941-957.

Mumpton, F. A. (1960) Clinoptilolite redefined: *Amer. Mineral.* **45**, 351-369.

Mumpton, F. A. (1973a) Worldwide deposits and utilization of natural zeolites: *Industrial Minerals*, No. 73, pp. 30-45.

Mumpton, F. A. (1973b) Scanning electron microscopy and the origin of sedimentary zeolites: *Molecular Sieves: Proc. 3rd Int. Molecular Sieve Conf.*, pp. 159-161.

Munson, R. A. and Sheppard, R. A. (1974) Natural zeolites: their properties, occurrence, and uses: *Minerals Science Engineering* **6**, 19-34.

Murphy, C. B., Hrycyk, O. and Gleason, W. T. (1977) Single P/C unit removal of nutrients from combined sewer overflows: *J. Wat. Pollut. Control Fed.* **49** (in press).

Murray, J. and Renard, A. F. (1891) *Deep-sea Deposits*: Vol. 5, *Report on the Scientific Results of the Voyage of "HMS Challenger" during the years 1873-76*: Eyre & Spottiswoode, London, 525 pp.

Neel, E. E. A. *et al.* (1973) Catalytic manufacture of gasoline for automobile engines: Ger. Offen. 2,259,794, June 14, 1973, 25 pp.

Nelson, J. L. and Mercer, B. W. (1963) Ion exchange separation of cesium from alkaline waste supernatant solutions: U.S. Atomic Energy Comm. Doc. HY-76449.

Nemecz, Erno and Varju, Gyula (1962) Sodium bentonization, clinoptilolitization, and adularization in the rhyolitic tuffs of the Szerencs Piedmont area: *Acta Geol.* **6**, 389-426.

Nichols, R. (1976) Jungle Laboratories, Comfort, Texas (personal communication).

Nikashina, V. A., Zaborskaya, Ye. Yu., Mahalov, Ye. M. and Rubinshteen, R. N. (1974) Selective sorption of strontiom by natural clinoptilolite from aqueous solutions. *Radiokhim*, **16**, 753-756.

Nikolova, R. and Ivanov, D. (1975) Possible use of natural clinoptilolite as a catalyst carrier: *Khim. Ind. (Sofia)* **47**, 175-177.

Nishita, H. and Haug, R. M. (1972) Influences of clinoptilolite on Sr^{90} and Cs^{137} uptakes by plants: *Soil Sci.* **114**, 149-157.

NRG (1975) NuFuel, a new source of energy: brochure of NRG NuFuel Company, 6 pp.

Ohtani, Seiya *et al.* (1972a) Conversion catalyst for hydrocarbons: Japan. Kokai 72, 046,667, Nov. 1972, 5 pp.

Ohtani, Seiya *et al.* (1972b) Conversion catalyst for hydrocarbons: Japan. Kokai 72, 046, 668, Nov. 1972, 6 pp.

Onogi, T. (1966) 1. Treating experiments of chicken droppings with zeolitic tuff powder. 2. Experimental use of zeolite-tuffs as dietary supplements for chickens: Rept. Yamagata Stock Raising Inst., pp. 7-18.

Ousterhout, L. E. (1970) Nutritional effects of clays in feed: *Feedstuffs* **42**, 34-36.

Papp, J., Kallo, D. and Schay, G. (1971) Hydromethylation of toluene on clinoptilolite: *J. Catalysis* **23**, 168-182.

Peters, M. D. and Bose, R. J. (1975) Clinoptilolite – a physico-chemical approach to ammonia removal in hatchery and aquaculture water reuse systems: *Fish. and Marine Serv. Tech. Rept.* **535**, 12 pp.

Rex, R. and Fan, P. F. (1974) Zeolites in deep-sea sediments: *Seminar on the Occurrence, Origin, and Utilization of Sedimentary Zeolites in the Circum–Pacific Region*; U.S.–Japan Cooperative Science Program, Menlo Park, Cal., July 1974 (unpublished abstract).

Rohrer, D. M. (1976) Los Alamos Scientific Lab., Los Alamos, New Mexico (personal communication).

Ross, C. S. (1928) Sedimentary analcite: *Amer. Mineral.* **26**, 627-629.

Roux, A., Huang, A. A., Ma, Y. H. and Zwiebel, I. (1973) Sulfur dioxide adsorption on mordenites: *Amer. Inst. Chem. Engng Symposium Ser.* **134**, Gas Purification by Adsorption, pp. 46-53.

Sato, Mikio and Fukagawa, Kazuyuki (1976) Treatment for ammoniacal nitrogen-containing water: Japan. Kokai 76,068,967, June 15, 1976, 4 pp.

Scherzer, J., Vaughan, D. E. W. and Albers, E. W. (1975) Hydrocarbon cracking catalysts with promoter mixtures: U.S. Patent 3,894,940, April 15, 1975, 4 pp.

Sersale, R. (1958) Genesi e constituzione del tufo giallo napolitano: *Rend. Accad. Sci. Fische Mat. (Napoli)* **25**, 181-207.

Sheppard, R. A. (1973) Zeolites in sedimentary rocks: Brobst, D. A. and Pratt, E. P., eds., *United States Mineral Resources*, U.S. Geol. Survey Prof. Paper **820**, pp. 689-695.

Sheppard, R. A. and Gude, A. J., 3rd (1968) Distribution and genesis of authigenic silicate minerals in tuffs of Pleistocene Lake Tecopa, Inyo County, California: *U.S. Geol. Survey Prof. Paper* **597**, 39 pp.

Sheppard, R. A. and Gude, A. J., 3rd (1969a) Diagenesis of tuffs in the Barstow Formation, Mud Hills, San Bernardino County, California: *U.S. Geol. Survey Prof. Paper* **634**, 35 pp.

Sheppard, R. A. and Gude, A. J., 3rd (1969b) Authigenic fluorite in Pliocene lacustrine rocks near Rome, Malheur County, Oregon: *U.S. Geol. Prof. Pap.* **650-D**, D69-74.

Sheppard, R. A. and Gude, A. J., 3rd (1973) Zeolites and associated authigenic silicate minerals in tuffaceous rocks of the Big Sandy Formation, Mohave County, Arizona: *U.S. Geol. Survey Prof. Paper* **830**, 36 pp.

Slone, W. J., Turner, P. R. and Jester, D. B. (1975) A new closed vertical raceway fish culture system containing clinoptilolite as an ammonia stripper: *Proc. 54th Conf. West. Ass. Game, Fish, Conserv. Commissioners, July 1974, Albuquerque, N.M.*, pp. 381-394.

Smola, V. I., Anusov, S. A., Zihkovskii, V. A., Ket'tsev, N. V. and Torocheshnikov, N. S. (1975) Use of natural zeolites for removing SO_2 from industrial waste gases: *Prom. i San. Ochistka Gazov. Nauch.-tekhn. Sb.* **5**, 12-14.

Stojanović, D. (1972) Zeolite-containing volcanic tuffs and sedimentary rocks in Serbia: *Proc. Serb. Geol. Soc. for 1968-70*, pp. 9-20.

Sudo, T. (1950) Mineralogical studies of the zeolite-bearing pumice tuffs near Yokotemachi, Akita Prefecture: *J. Geol. Soc. Japan* **56**, 13-16.

Surdam, R. C. and Eugster, H. P. (1976) mineral reactions in the sedimentary deposits of the Lake Magadi region, Kenya: *Geol. Soc. Amer. Bull.* **87**, 1739-1752.

Surdam, R. C. and Parker, R. B. (1972) Authigenic alumino-silicate minerals in the tuffaceous rocks of the Green River Formation, Wyoming: *Geol. Soc. Amer. Bull.* **83**, 689-700.

Takasaka, Ayao (1975) Properties and use of the Itaya zeolite from Japan: *Eunsai* **20**, 127-134, 142.

Tamura, Kyoichi (1974) High strength porous glass: Japan. Kokai 74,098,817, Sept. 18, 1974, 3 pp.

Tamura, Takaaki (1970) Oxygen concentration process: U.S. Patent 3,533, 221, Oct. 13, 1970, 10 pp.

Tamura, T. (1971) Gas adsorption properties and industrial applications of Japanese tuff: *Seminar on Occurrence and Mineralogy of Sedimentary Zeolites in the Circum–Pacific Region; Nikko, Japan*; U.S.–Japan Cooperative Science Program (unpublished abstract).

Tchernev, D. I. (1975) Lincoln Lab., Mass. Inst. Tech., Lexington, Mass. (personal communication).

Terui, Akio *et al.* (1974) Adsorbing sulfur compounds, dust, soot from flue gas on zeolitic kieselguhr: Japan. Kokai 74,036,580, April 14, 1974, 2 pp.

Torii, Kazuo (1974) Utilization of sedimentary zeolites in Japan: *Seminar on the Occurrence, Origin, and Utilization of Sedimentary Zeolites in the Circum–Pacific Region*; U.S.–Japan Cooperative Science Program, Menlo Park, Cal., July 1974 (unpublished abstract).

Torii, Kazuo; Hotta, Masami; Onodera, Yoshio and Asaka,

Makoto (1973) Adsorption of nitrogen-oxygen on zeolite tuff: *J. Chem. Soc. Japan* No. 2, pp. 225-232.
Torii, Kazuo; Yoshio, Onodera; Makoto, Asaka and Hiroshi, Ihara (1971) Character of gas adsorption and separation in mordenite tuff: *Kogyo Kagaku Zasshi* **74**, 2018-2024.
Tsitsishvili, G. V., Urotadze, S. L., Lukin, V. D. and Bagirov, R. M. (1976) Drying and treatment of natural gas by clinoptilolite in an experimental pilot plant: *Soobshch. Akad. Nauk Gruz. SSR* **81**, 369-371.
Tsitsishvili, G. V., Andronikashvili, T. G., Sabelashvili, Sh. D. and Koridzy, Z. I. (1972) Gas chromatographic separation of Ar–O–N on clinosorb: *Zak. Rad.* **38**, 57-58.
Ulisch, Guenter (1975) Inorganic porous materials: Ger. Offen. 2,334, 224, Jan. 23, 1975, 11 pp.
Union Carbide Corporation (1962) Linde molecular sieve type AW-500: Linde Molecular Sieves Bulletin F-1617, Aug. 15, 1962, 4 pp.
Union Carbide Corporation (1975) Modular UNOX Systems: Sales brochure, Environmental Systems Department F-3577A, 8 pp.
Utada, M., Minato, H., Ishikawa, T. and Yoshizaki, Y. (1974) The alteration zones surrounding Kuroko-type deposits in Nishi–Aizu District, Fukushima Prefecture, with emphasis on the analcime zone as an indicator in exploration for ore deposits: *Mining Geol.*, Spec. Issue No. 6, pp. 291-302.
Vdovina, E. D., Radyuk, R. I. and Sultanov, A. S. (1976) Use of Uzbekistan natural zeolites for the purification of low-level wastewaters. I. Sorption of radioactive cesium: *Radiokhimiya* **18**, 422-423.
Walton, A. W. (1975) Zeolitic diagenesis in Oligocene volcanic sediments, Trans-Pecos, Texas: *Geol. Soc. Amer. Bull.* **86**, 615-624.
Watanabe, S., Kanaka, Y. and Kuroda, A. (1971) Report on the experimental use of zeolite-tuff as dietary supplements for cattle: Rept. Okayama Prefecture Fed. Agric. Coop. Ass., April 1971, 18 pp.
Wearout, J. D. (1971) Purification of gases by adsorption. U.S. Patent 3,594,983, Jan. 1, 1971, 27 pp.
Wiegel, O. and Steinhof, E. (1925) Adsorption of organic liquid vapors by chabazite: *Z. Kristallogr.* **61**, 125-154.
White, J. L. and Ohlroggi, A. J. (1974) Ion exchange materials to increase consumption of non-protein nitrogen in ruminants: Canadian Patent 939186, Jan. 2, 1974, 30 pp.
Wilding, M. W. and Rhodes, D. W. (1963) Removal of radioisotopes from solution by earth materials from eastern Idaho: U.S. Atomic Energy Comm. Doc. IDO-14624.
Wilding, M. W. and Rhodes, D. W. (1965) Decontamination of radioactive effluent with clinoptilolite. U.S. Atomic Energy Comm. Doc. IDO-14657.
Wilson, T. E. (1975) Greeley and Hanson Engineers, Chicago, Ill. (personal communication).
Wolfbauer, C. A. (1973) Criteria for recognizing paleo-environments in a playalake complex: *The Green River Formation of Wyoming*: Guidebook, 25th Field Conf., Wyoming Geol. Ass., pp. 87-91.
Yoshida, Hideo; Kurata, Akira and Sanga, Seiji (1976) Removal of heavy metal ions from wastewater using zeolites: *Mizu Shori Gijutsu* **17**, 219-226.
Yoshida, K. and Utada, M. (1969) A study on alteration of Miocene Green Tuffs in the Kuroko-type mineralization area in Odate Basin, Akita Prefecture: *Mining Geol.* **18**, 332-342.
Yoshinaga, E. *et al.* (1973) Organophosphate-containing agricultural and horticultural granule formation: U.S. Patent 3,708,573.

CRYSTAL STRUCTURE AND MINERALOGY

MINERALOGY AND CRYSTAL CHEMISTRY OF ZEOLITES

GLAUCO GOTTARDI

Istituto di Mineralogia, Università di Modena, Modena, Italy

Abstract – A zeolite ("boiling stone" in ancient Greek) is an alumino-silicate, whose framework structure of $(Si,Al)O_4$ tetrahedra contains pores filled with water molecules and exchangeable cations; a zeolite "boils" at about 200°C, giving off water which is readily readsorbed at room temperature.

Zeolites are classified by crystal structure into groups having the same units but which are linked to form different frameworks. Each group may be further subdivided into subgroups of minerals having the same framework, but different chemical compositions and/or different symmetries. Although there is as yet no general agreement on nomenclature for minerals with the same framework, detailed and accurate single crystal refinements of natural zeolites having the same framework and different compositions are known so far in five subgroups: (1) natrolite, mesolite, and scolecite; (2) thomsonite and gonnardite; (3) stellerite, stilbite, and barrerite; (4) heulandite and clinoptilolite; and (5) phillipsite and harmotome.

The general chemical formula for zeolites is $(Na,K)_x(Mg,Ca,Sr,Ba)_y[Al_{x+2y}Si_{n-(x+2y)}O_{2n}].mH_2O$. Framework atoms are enclosed by the square brackets. One important chemical parameter of every zeolite is the fraction of SiO_4 tetrahedra, R = Si:(Si + Al). R ranges from 0.5 to 0.87 and remains constant during most cation exchanges, whereas most or all non-framework cations are exchangeable. The number of water molecules indicates the pore volume and usually $n/2 < m < n$. In other words, the oxygens of the waters normally do not exceed half of the oxygens of the framework. Some zeolites have identical chemical compositions, and additional analytical techniques, such as X-ray powder diffraction, are required for identification.

All accurate structure refinements of zeolites are performed on natural crystals. Mazzite, the natural counterpart of synthetic molecular sieve Omega, is a case in point. Barrer and Villiger (1969) proposed a model for this structure based on studies of the microcrystalline synthetic phase before the mineral was discovered. Galli (1974) was able to correct the previous model with data obtained from a single crystal of the mineral. The frameworks of all zeolites are known, so that most current crystal chemical studies in the field determine how a given framework is modified by its own composition or by the contents of its pores.

INTRODUCTION

The term zeolite was created by Cronstedt (1756) from the Greek ζειν = "to boil" and λιθοσ = "stone" for minerals, which expel water when heated and hence seem to boil. They are very common and may be found as fine crystals of hydrothermal genesis in geodes and in fissures of volcanic rocks, or as microcrystalline masses of sedimentary origin. A zeolite is a framework alumino-silicate whose structure contains channels filled with water and exchangeable cations. The term has also been applied to similar synthetic compounds, which are also called molecular sieves, the term zeolite refers mainly to the natural minerals. As usual, however, when trying to arrive at a definition for the classification of minerals, the first essential is to define one's terms, in the above "definition", for example, one must clarify the meaning of "exchangeable", for all the framework silicates may be cation-exchanged, given the right condition $[P(H_2O),T]$. In zeolites exchange should be possible at "low temperature", namely below 100°C or, at the outside, 150°C. A precise limit, however, cannot be set.

As is well known, the structure of a framework silicate is composed of $(Si,Al)O_4$ tetrahedra which share all their oxygen vertices with nearby tetrahedra. The chemical formula of this framework is therefore:

$$[(Si,Al)(O_{0.5})_4]_n = [(Si,Al)O_2]_n$$

with the number of negative charges equal to the number of Alions. The negative charge of the framework is balanced by alkalies and alkaline earths located outside the tetrahedra, in the channels. The sundry types of zeolite framework (nearly 23 in all) have been discussed and classified in many general papers, for instance by Meier (1968) and by Meier and Olson (1971), and their description need not be repeated here. One particular point, however, calls for attention, namely the presence in different structures of the same group of tetrahedra, or secondary building unit, which may be assembled in different ways to produce different types of framework. As shown in Figs. 1–3, the main

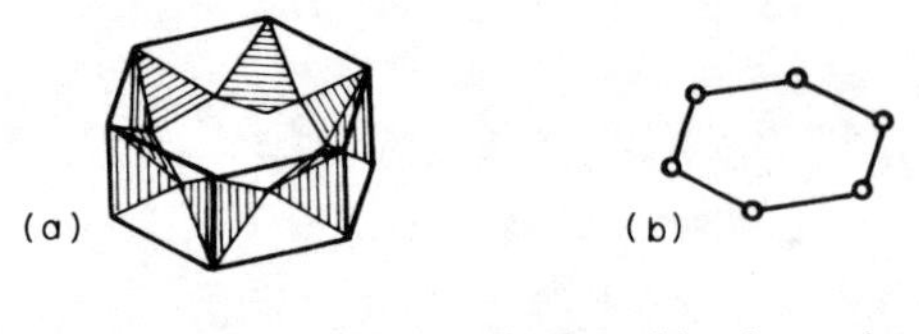

Fig. 1. (a) a single ring of six tetrahedra, (b) schematic drawing of the same ring showing the positions of Si-atoms only.

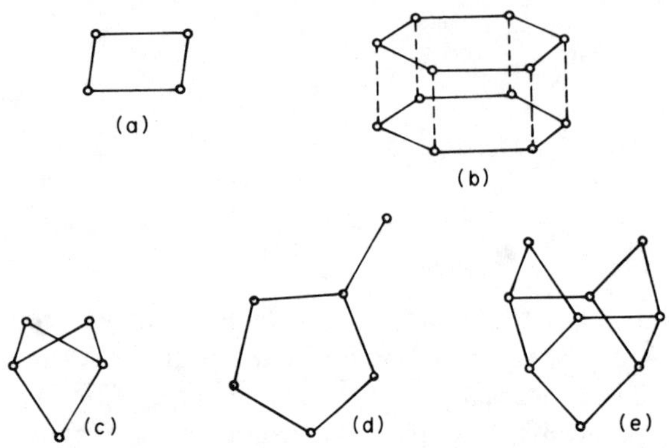

Fig. 2. The secondary building units (SBU) of the zeolite frameworks. (a) The single four-ring (S4R), (b) the six-ring, single or double (S6R and D6R), (c) the natrolite-unit (4–1), (d) the mordenite-unit (5–1), (e) the stilbite-unit (4–4–1).

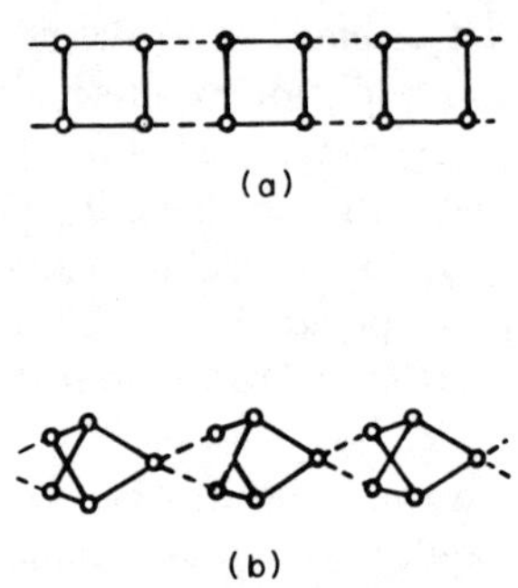

Fig. 3. Examples of interconnection of SBU. (a) The phillipsite ladder, as a connection of S4R, (b) the natrolite chain, as a connection of 4–1.

secondary building units (SBU) are:

- the "four-ring" (S4R),
- the "six-ring", single or double (S6R and D6R),
- the "fibrous-zeolite-unit" (4-1),
- the "mordenite-unit" (5-1),
- the "stilbite-unit" (4-4-1).

In accordance with these structural peculiarities, zeolites may be classified into groups with common SBU; within each group genera may be formed that consist of zeolites with the same framework. On this basis, I have listed all natural zeolites in Table 1. One of the questions which may be raised when reading this table is why phillipsite and harmotome are considered different zeolite species. The fact is, they have the same framework and the same symmetry, but in harmotome the most abundant cation is Ba, while in phillipsite it is Ca or Na or K. This poses the question of whether we should have four names, or maybe even only one, and of whether Ba is more important than Ca or Na or K. I could answer these questions from my particular point of view, but no general agreement has been reached so far on the matter and neither has it, therefore, on the classification of zeolites. In some cases, however, for example for stilbite, barrerite and stellerite, the same framework is associated not only with different cation contents, but also with different symmetries, and this could be a criterion for having different names.

The general chemical formula of a zeolite is:

$$M_xD_y\,[\mathrm{Al}_{x+2y}\mathrm{Si}_{n-(x+2y)}\mathrm{O}_{2n}]\,.m\mathrm{H_2O}$$

where M = Na, K, or other monovalent cations, and
D = Mg, Ca, Sr, Ba, and other divalent cations.

Al may be substituted by Fe^{+3}, but only to a small extent, by Be in some rare zeolite-like minerals, and by Ga in synthetic compounds. Si may be substituted by P in some rare zeolite-like minerals and by Ge in synthetic compounds. M and D are generally alkalies and alkaline earths in natural zeolites, but many other ions may be inserted instead by ion-exchange processes or during synthesis. An important chemical parameter in a zeolite is its ratio R = Si:(Si + Al + Fe), which represents the percentage of the tetrahedra occupied by Si (in zeolites $0.50 < R < 0.87$). The number m of water molecules gives an idea of the volume of the channels in relation to the total volume; this number does not normally exceed half the number of framework oxygens, and $n/2 < m < n$.

The dehydration of zeolites to activate them for use is usually complete at 250°C and is reversible: dehydrated zeolites are the strongest desiccants and other molecules other than water may fill the channels.

Apart from particular substitutions, the chemistry of zeolites is best represented in two dimensions with a triangular diagram

$$[\mathrm{Si_4O_8}] - D[\mathrm{Al_2Si_2O_8}] - M_2\,[\mathrm{Al_2Si_2O_8}] \text{ (see Fig. 4)}$$

The area covered by points representing zeolite analyses nearly coincides with the whole triangle, only two small parts near the upper and the right vertices remaining free. A full account of the complexity of zeolite chemistry may be found in Gottardi (1972); here only a few points will be touched on. Each zeolite group occupies an area in the triangular diagram, and this area may be further subdivided into smaller areas pertaining

TABLE 1. Zeolites

Group (SBU)	Framework type	Space group	Species name	Typical unit cell content	Most abundant cation
S4R	Phillipsite	$P2_1/m$	Phillipsite	$(Ca_{0.5},Na,K)_6[Al_6Si_{10}O_{32}]\cdot 12H_2O$	Ca or Na or K
	Phillipsite	$P2_1/m$	Harmotome	$Ba_2[Al_4Si_{12}O_{32}]\cdot 12H_2O$	Ba
	Gismondine	$P2_1/c$	Gismondine	$Ca_4[Al_8Si_8O_{32}]\cdot 16H_2O$	Ca
	Gismondine	$I4_1/amd$	Garronite	$NaCa_{2.5}[Al_6Si_{10}O_{32}]\cdot 14H_2O$	Ca
	Leucite	$Ia3d$	Analcime	$Na_{16}[Al_{16}Si_{32}O_{96}]\cdot 16H_2O$	Na
	Leucite	$I2/a$	Wairakite	$Ca_8[Al_{16}Si_{32}O_{96}]\cdot 16H_2O$	Ca
	Paulingite	$Im3m$	Paulingite	$(K_2,Na_2,Ca,Ba)_{76}[Al_{152}Si_{525}O_{1354}]\cdot 700H_2O$	K
	Laumontite	Am	Laumontite	$Ca_4[Al_8Si_{16}O_{49}]\cdot 16H_2O$	Ca
	Yugawaralite	Pc	Yugawaralite	$Ca_2[Al_4Si_{12}O_{32}]\cdot 8H_2O$	Ca
S6R and D6R	Chabazite	$R\bar{3}m$	Chabazite	$Ca_2[Al_4Si_8O_{24}]\cdot 13H_2O$	Ca or Na
	Gmelinite	$P6_3/mmc$	Gmelinite	$Na_8[Al_8Si_{16}O_{48}]\cdot 24H_2O$	Na or Ca
	Faujasite	$Fd3m$	Faujasite	$Na_{12}Ca_{12}Mg_{11}[Al_{59}Si_{133}O_{384}]\cdot 235H_2O$	Na
	Erionite	$P6_3/mmc$	Erionite	$(K_2,Ca,Mg,Na_2)_{4.5}[Al_9Si_{27}O_{72}]\cdot 27H_2O$	Ca or Na or K
	Offretite	$P\bar{6}m2$	Offretite	$(K_2,Mg,Ca,Na_2)_{2.5}[Al_5Si_{13}O_{36}]\cdot 15H_2O$	Ca or Mg
	Levyne	$R\bar{3}m$	Levyne	$Ca_3[Al_6Si_{12}O_{36}]\cdot 18H_2O$	Ca
	Mazzite	$P6_3/mmc$	Mazzite	$K_2Mg_2Ca_{1.5}[Al_9Si_{27}O_{72}]\cdot 28H_2O$	Mg
4-1	Natrolite	$Fdd2$	Natrolite	$Na_{16}[Al_{16}Si_{24}O_{80}]\cdot 16H_2O$	Na
	Natrolite	$I4_1/amd$	Tetr. Natrolite	idem	Na
	Natrolite	$Fdd2$	Mesolite	$Na_{16}Ca_{16}[Al_{48}Si_{72}O_{240}]\cdot 64H_2O$	Na or Ca
	Natrolite	Cc	Scolecite	$Ca_8[Al_{16}Si_{24}O_{80}]\cdot 24H_2O$	Ca
	Thomsonite	$Pnna$	Thomsonite	$Na_4Ca_8[Al_{20}Si_{20}O_{80}]\cdot 24H_2O$	Ca
	Thomsonite	$Pbmn$	Gonnardite	$Na_4Ca_2[Al_8Si_{12}O_{40}]\cdot 14H_2O$	Na
	Edingtonite	$P2_12_12$	Edingtonite	$Ba_2[Al_4Si_6O_{20}]\cdot 8H_2O$	Ba
5-1	Mordenite	$Cmcm$	Mordenite	$Na_8[Al_8Si_{40}O_{96}]\cdot 24H_2O$	Ca or Na
	Dachiardite	$C2/m$	Dachiardite	$Na_5[Al_5Si_{19}O_{48}]\cdot 12H_2O$	Ca or Na
	Ferrierite	$Immm$	Ferrierite	$Na_{1.5}Mg_2[Al_{5.5}Si_{30.5}O_{72}]\cdot 18H_2O$	Mg
	Epistilbite	$C2/m$	Epistilbite	$Ca_3[Al_6Si_{18}O_{48}]\cdot 16H_2O$	Ca
4-4-1	Stellerite	$Fmmm$	Stellerite	$Ca_4[Al_8Si_{28}O_{72}]\cdot 28H_2O$	Ca
	Stellerite	$C2/m$	Stilbite	$Na_2Ca_4[Al_{10}Si_{26}O_{72}]\cdot 34H_2O$	Ca or Na
	Stellerite	$Amma$	Barrerite	$Na_8[Al_8Si_{28}O_{72}]\cdot 26H_2O$	Na
	Brewsterite	$P2_1/m$	Brewsterite	$Sr_2[Al_4Si_{12}O_{32}]\cdot 10H_2O$	Sr
	Heulandite	$C2/m$	Heulandite	$Ca_4[Al_8Si_{28}O_{72}]\cdot 24H_2O$	Ca
	Heulandite	$C2/m$	Clinoptilolite	$Na_6[Al_6Si_{30}O_{72}]\cdot 24H_2O$	Na

See Fig. 2 for the code of the secondary building units (SBU).

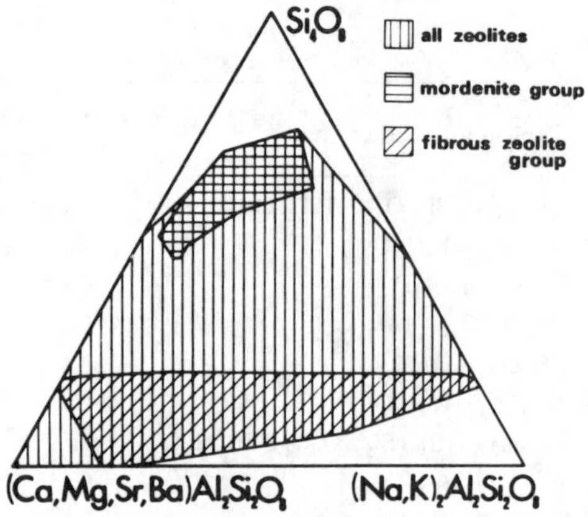

Fig. 4. The triangular diagram $[Si_4O_8]$–$D[Al_2Si_2O_8]$–$M_2[Al_2Si_2O_8]$. The area covered by all zeolites is shown, as well as the areas of two zeolite groups, the mordenite one and the fibrous zeolite one.

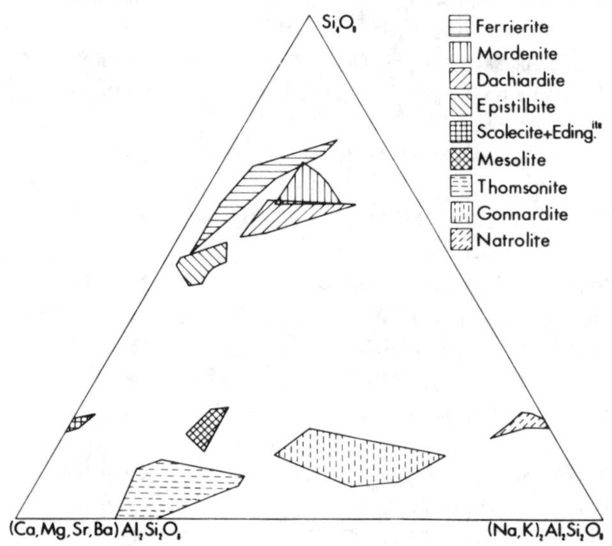

Fig. 5. Subdivision between the species of the area of the mordenite group and of the fibrous zeolite group.

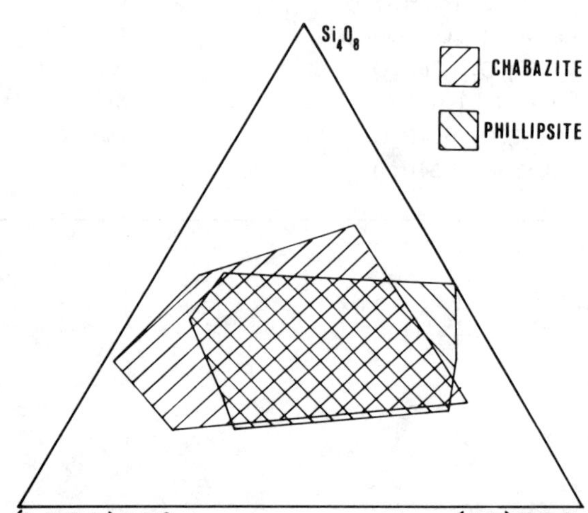

Fig. 6. Superposition of the fields of phillipsite (S4R group) and of chabazite (D6R group).

to the species of the group. Areas pertaining to the different species of the same group do not overlap to any great extent (see Fig. 5) with one exception, namely the six-ring group where overlapping is more marked. On the other hand, there are large areas of the triangle pertaining to the areas of different species of different groups (see Fig. 6). Overlapping would be reduced by the use of diagrams in more than two dimensions, for then a complete separation of the components would be possible. Overlapping is more probable in the central zone of the triangle; it is virtually impossible to guess to which zeolite a point in this zone is supposed to refer, while points outside this central zone are likely to be correctly attributed.

MINERALOGY OF ZEOLITES

The mineralogy of zeolites is only a part of the larger branch of science dealing with all zeolites, natural and synthetic. The importance of the study of natural zeolites, in comparison with the analogous synthetic compounds, may be viewed in two ways: first, considering the number of the natural zeolites as compared to the number of the synthetic varieties, and second, considering those studies in which better results can be obtained using the natural zeolites.

Table 1 contains twenty-three types of framework, each corresponding to one genus of zeolites. The synthetic phases corresponding to seventeen of these frameworks are described in the literature; but no one has synthesized compounds with the frameworks of dachiardite, epistilbite, stellerite, brewsterite, laumontite, or paulingite. On the other hand, the number of framework-types known only from their synthetic phases certainly exceeds six. It is worth noting that the first reported synthesis of a zeolite dates back to 1862 and was done by St. Claire Deville (see Breck, 1974): it therefore postdates the discovery of zeolites by a hundred years, while systematic research into synthesis started only in the 1940s. It may be confidently foreseen that all missing frameworks will be obtained in the next decade and that a simultaneous increase in the number of the exclusively synthetic framework types will take place. What happened for the first time in 1974, when a framework type, already known as synthetic, was found in nature, will occur more frequently; the new mineral mazzite (Galli *et al.*, 1974) almost certainly has the same framework as the molecular sieve Omega, patented in 1967 by Union Carbide Corporation. Other new zeolite species introduced into the nomenclature in the last 5 years or so are cowlesite (Wise and Tschernich, 1975), barrerite (Passaglia and Pongiluppi, 1974) and tetragonal natrolite

(Krog Andersen *et al.*, 1969). The framework of cowlesite is unknown, whereas both barrerite and tetragonal natrolite belong to known genera, as is given in Table 1.

To conclude the first part of this review, the history of zeolites is characterized by an initial prevalence in the number of the natural phases, followed by a steep increase in the number of synthetic molecular sieves, which today equals, but which in the future will far outstrip, the number of natural phases.

Let us compare the results obtainable from natural zeolites with those obtainable from molecular sieves with the same framework. Here, the difference is striking and helpful to the mineralogist, who can get results that are impossible when working on synthetic materials. All studies based on single crystals, which, apart from rare exceptions, occur in nature, belong to this category. Accurate crystal structure determinations or refinements have recently been published for offretite (Gard and Tait, 1972), levyne (Merlino *et al.*, 1975), erionite (Kawahara and Curien, 1969; Gard and Tait, 1973), yugawaralite (Kerr and Williams, 1969; Leimet and Slaughter, 1969), edingtonite (Galli, 1976) and for the others mentioned later in connection with the crystal chemistry of zeolites. The manner in which the structure of mazzite was determined is worthy of comment, for it demonstrates clearly the superiority of the data afforded by natural materials, yielding as they do wonderful single crystals. The crystal structure of the molecular sieve Omega, proposed by Barrer and Villiger (1969), was based on a synthetic microcrystalline material. Galli (1974), with access to a single crystal of mazzite, was able to correct and to improve on the previous model. He was at an advantage, for single crystal data yield more information than powder data.

One problem which can be tackled by the correct choice of natural single crystals is the Order-Disorder (O-D) distribution of Si and Al in the tetrahedra of a framework. Fully ordered (Si,Al) distributions are known so far for the following zeolites: [in square brackets the average ratio R = Si:(Si + Al + Fe)] gismondine [1:2 = 0.50], thomsonite [1:2 = 0.50], natrolite, mesolite, and scolecite [3:5 = 0.60], edingtonite [3:5 = 0.60], laumontite [4:6 = 0.66], yugawaralite [6:8 = 0.75]. For two of these minerals we probably also know the corresponding disordered phases: tetragonal natrolite (natrolite) and garronite (gismondine). In both cases the existence of the disordered phases is related to an "excessive" R-value: thus in ordered gismondine R = 0.50, but in disordered garronite R = 0.60, in ordered natrolite R = 0.60, but in disordered tetragonal natrolite R = 0.62. Generally speaking, order has never been detected with certainty in other natural zeolites, although some authors claim to have found partial enrichments of Al in certain tetrahedra.

For the sake of simplicity, the O-D relations in zeolites may be divided into three classes:

1. "Basic zeolites" with $0.50 < R < 0.625$. This class includes the ordered structures natrolite, mesolite, scolecite, edingtonite, thomsonite, and gismondine, and the disordered structures garronite, gonnardite, and many phillipsites.
2. "Intermediate zeolites" with $0.625 < R < 0.75$. This class includes three ordered structures, laumontite, analcime and yugawaralite, all the others phillipsite, harmotome, chabazite, gmelinite, mazzite, offretite, brewsterite, stilbite, faujasite, and paulingite) being disordered according to the available refinements.

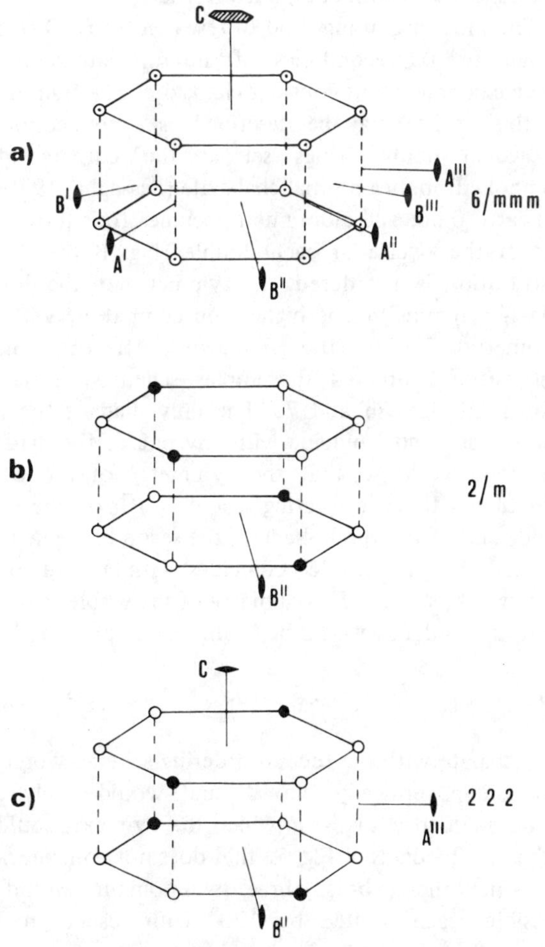

Fig. 7. Highest possible symmetry of the hexagonal doublering with 8Si (circles) and 4A1 (dots); the position of the oxygens is not shown; only the symmetry axes are shown in the drawing, but full symbols are given on the right side. (a) With disordered (Si,Al)-distribution, (b) with an ordered (Si,Al)-distribution, (c) with another ordered (Si,Al)-distribution. All other ordered (Si,Al)-distribution have lower symmetries, if the Al-O-Al avoidance rule is obeyed.

3. "Acid zeolites" with $R > 0.75$. This class, which contains disordered structures only, includes stellerite, barrerite, erionite, heulandite, mordenite, dachiardite, ferrierite, and epistilbite.

Ordered structures are most likely to occur in the first class; only gonnardite seems to have an exclusively disordered structure. In order to explain this exception, Gottardi and Alberti (1974) emphasized the possibility of an order with anti-phase domains, a hypothesis which needs experimental support.

In the second class there would appear to be little likelihood of ordered structures; long-range ordering probably does exist but it is camouflaged by fine pseudomeroedric or meroedric twinning.

In the third class the low number of Al-atoms may favor only short-range order, probably with a different kind of ordering in different domains: this scheme is almost identical with complete disorder.

The most promising field of research for O–D studies is therefore the second class of "intermediate zeolites". As an example, let us consider chabazite. The framework of this zeolite may be described as a connection of hexagonal double rings set at the corners of a hombohedron. In a normal chabazite (Passaglia, 1970), R is nearly 0.66, and aluminium occupies four tetrahedra out of the twelve of each double ring. If the (Si,Al) distribution is disordered, the symmetry of the double ring is 6/mmm; that is higher and compatible with the symmetry $\bar{3}$m of the framework. If the (Si,Al) distribution is ordered, the simplest schemes are the two shown in Figs. 7b and 7c. The only diad of the first scheme does not coincide with any one of the diads of the framework, so that the symmetry of the whole structure with such ordering would be P$\bar{1}$; on the other hand, one of the three diads of the second scheme (that marked A‴ in Fig. 7c) coincides with a diad of the framework, so that the symmetry of the whole structure with such ordering would be C2(monoclinic) with

$$a_{mon} = a_{hex}\sqrt{3} \qquad b_{mon} = a_{hex} \qquad c_{mon} = 2a_{rhom}$$

A chabazite with the second ordering scheme would be monoclinic, optically biaxial, and would easily give pseudomeroedric twins, in which the twin-axis could be either of the diads of Fig. 7a that does not coincide with A‴. In other words, three twin-domains would be possible, each rotated by 120° with respect to the others. Partial experimental evidence of the existence of these twin domains is afforded by the pictures in Fig. 8, taken with crossed nicols on thin sections of three optically biaxial chabazites. The cleavage traces which cross nearby individuals without deviation indicate the presence of pseudomeroedric twins. The problem will be resolved when it is possible to make a structure refinement of a twin-free fragment taken from a chabazite crystal with well-developed twins. In my opinion, the study of an apparently untwinned chabazite would not be rewarding, because of the high probability of submicroscopic twinning.

As an example of this kind of research, it is useful to recall the results of the structure refinement of leucite (Mazzi, *et al.*, 1976) and of analcime (Mazzi and Galli, in preparation). At room temperature both crystals are tetragonal (pseudo-cubic) exhibiting cubic macrocrystals which are but an intergrowth due to a complex twinning. The deviation from the cubic system is definite for leucite (not a zeolite) and can also be detected in powder patterns. Analcime (a zeolite) is metrically very near to cubicity, so that its X-ray diffraction powder pattern shows only some line broadening and in one case a line splitting. In thin section both minerals are clearly anisotropic and twinned (for analcime see Fig. 9). Refinement was made possible by careful selection of an untwinned fragment. In leucite the lowering of symmetry from the cubic to the tetragonal system is not due to any (Si,Al) ordering, but to a squeezing of a cage centered by potassium, which otherwise would be too large for such a cation. In analcime the lowering of symmetry is due for (Si,Al) ordering.

This kind of research is possible only when working with single crystals which are readily available in nature. However, single crystals of zeolites also have been grown in the laboratory; indeed, Ciric (1967) was the first to obtain single crystals of zeolite A (no natural equivalent), P (genus gismondine), and X (genus faujasite). Charnell (1971) grew single crystals of zeolite A (with $R = 50\%$) and of zeolite X (with $R = 54\%$); the first was refined by Grämlich and Meier (1971), the second by Olson (1970), and both were found to have an order given by an alternation of Si and Al in the tetrahedra. Synthetic single crystals with higher percentages of Si in the tetrahedra will probably either be disordered or have their order camouflaged by submicroscopic twinning. This discouraging forecast is based on what is known for feldspars, another class of framework silicates; in their case, it took years of research on synthesis to obtain microcrystalline samples of the ordered phase that were recognizable as such by comparison with their natural counterparts. Growth of ordered single crystals of feldspars was successful only for anorthite (with $R = 50\%$) but never for feldspars with $R > 50\%$.

CRYSTAL CHEMISTRY OF ZEOLITES

Hitherto I have put forward hypotheses which may

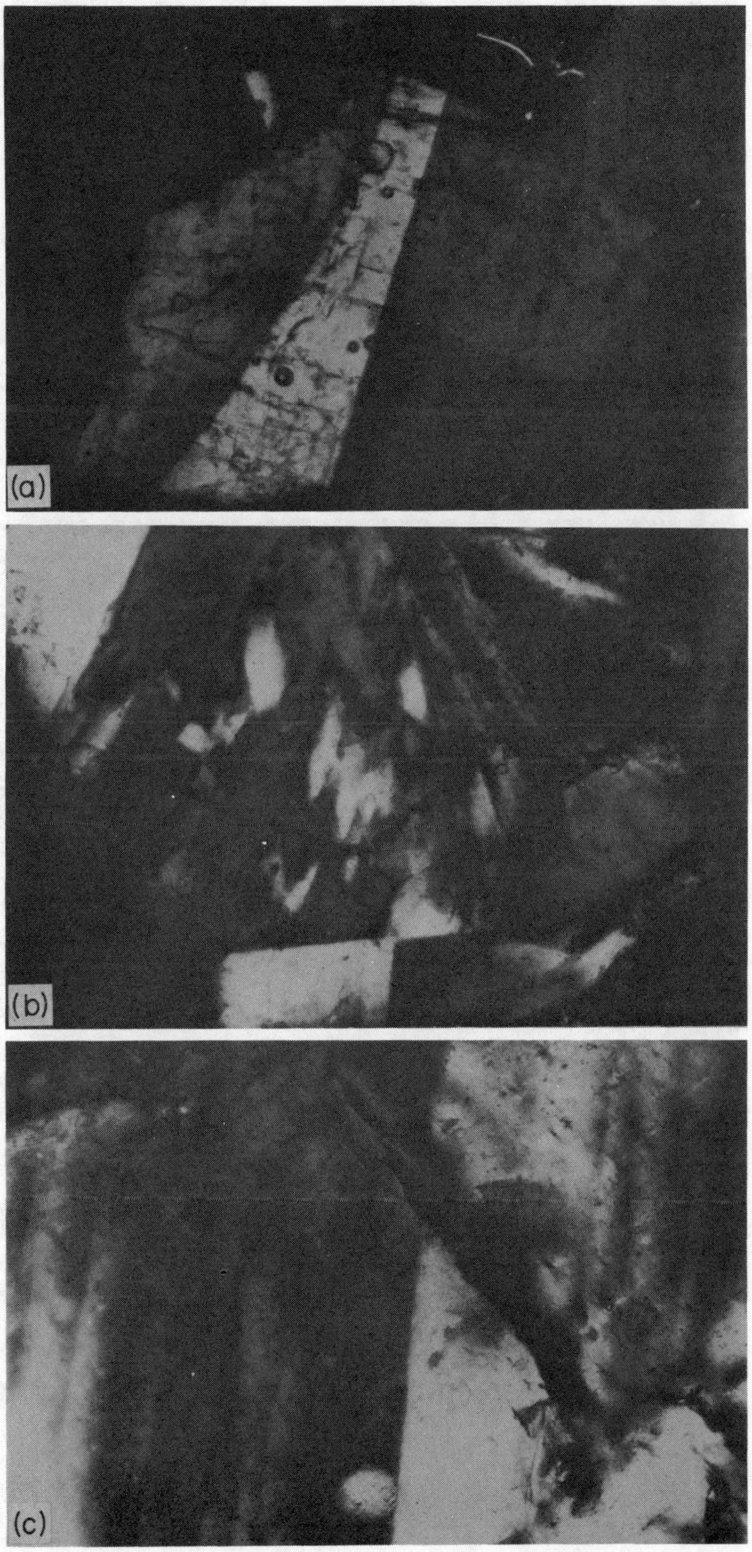

Fig. 8. Photomicrographs of chabazite with crossed nicols, to show the existence of pseudomeroedric twins. (a) Wasson's Bluff, Nova Scotia (75×); (b) Flinders, Tasmania (75×); (c) Repcice, Bohemia (200×).

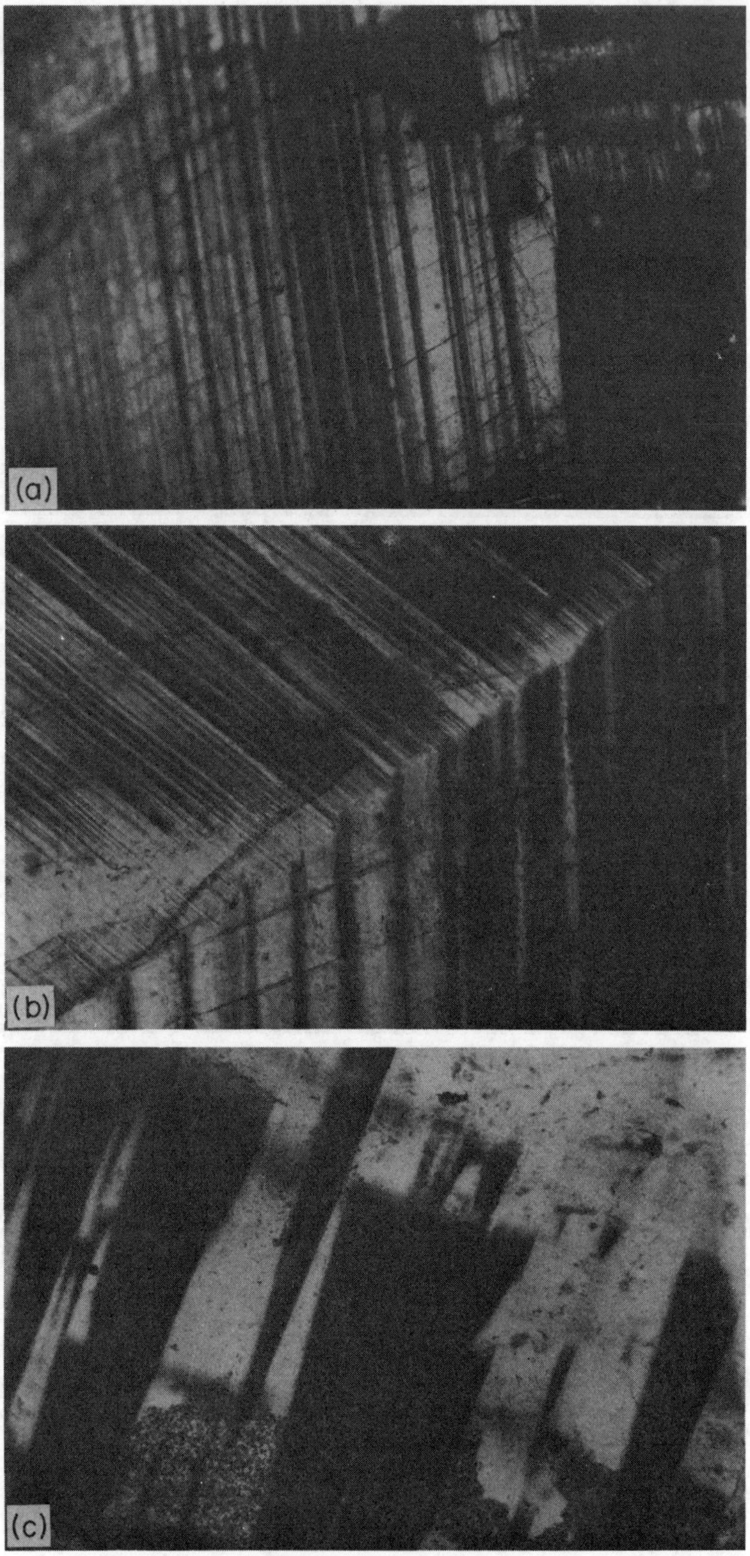

Fig. 9. Photomicrographs of analcime with crossed nicols, to show the existence of pseudomeroedric twins. (a) Valle di Fassa, Italy (200×); (b) Valle di Fassa, Italy (200×); (c) Aussig, Bohemia (200×).

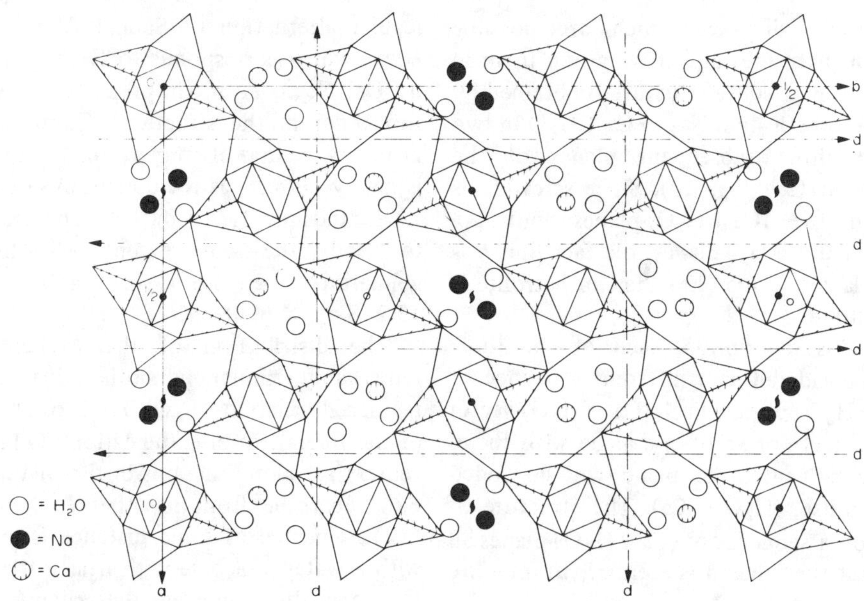

Fig. 10. The structure of mesolite viewed along *c*. The *c*-channels can be grouped in (010)-planes; in each plane all channels are occupied either by Na_2 or by $Ca+H_2O$.

be used as a basis for future structural work. In what follows, the results of the detailed refinements of different minerals with the same framework will be summarized. In my opinion, this kind of research is best described under the heading "crystal chemistry" rather than "crystal structure", since it deals with the influence of the changing chemistry on one and the same structural type. When speaking of zeolites, it involves detecting the influence on the structure of the ratio $R = \text{Si}:(\text{Si} + \text{Al})$ in the framework and of the cations and water molecules contained in the channels. In only five cases are detailed refinements available for different natural zeolites with the same framework (in brackets, the space group and another of the refinement):

- I: natrolite (*Fdd*2 : Meier, 1960), mesolite (*Fdd*2 : Adiwidjaja, 1972), scolecite (*F*1*d*1 or *Cc* : Adiwidjaja, 1972);
- II: gonnardite (*Pbmn* : Amirov *et al.*, 1972), thomsonite (*Pcnn* : Dal Negro, Rossi, Sabelli, and Tazzoli, in preparation);
- III: stellerite (*Fmmm* : Galli and Alberti, 1975), stilbite (*C*2/*m* : Slaughter, 1970; Galli, 1971), barrerite (*Amma* : Galli and Alberti, 1976);
- IV: phillipsite, harmotome } ($P2_1/m$: Rinaldi *et al.*, 1974),
- V: heulandite (*Cm* : Merkle and Slaughter, 1968; *C*2/*m* : Alberti, 1972; Bartl, 1973), clinoptilolite (*C*2/*m* : Alberti, 1975).

I. Tetragonal Natrolite, Natrolite, Mesolite, and Scolecite

The symmetry of the disordered framework of the natrolite type is $I\bar{4}2d$ (Alberti and Gottardi, 1975); this is also the most probable symmetry of tetragonal natrolite according to Krog-Andersen *et al.* (1969). Thus, although no refinement has so far been published for this mineral, there is sufficient evidence to state that tetragonal natrolite is the disordered equivalent of natrolite. Pabst (1971) reached the same conclusion when describing a natrolite intergrowth akin to twinning, which was interpreted as an ordering in macrodomains of a previous disordered crystal with natrolite framework, namely of a tetragonal natrolite crystal.

If a natrolite framework has a natrolite type order, its symmetry is *Fdd*2, which was in fact found by Meier (1960) in his refinement of the natrolite structure. Mesolite (Adiwidjaja, 1972) belongs to the same space group but has a *b* three times larger than the corresponding *b* of natrolite. Scolecite (Adiwidjaja, 1972) has symmetry *F*1*d*1, which is usually written *Cc*, and a unit cell nearly as large as natrolite. All three minerals have the same type of ordering in the structure. The relations between them have already been given by Meier (1968) and the new refinements by Adiwidjaja confirm his description, which can be summarized as follows:

1. In space group *Fdd*2, there are diads and screw-diads along *c*; in natrolite Na-atoms are located at a short distance from these screw-diads, thus forming a zig-zag pattern.
2. In scolecite, each Na-atom on one side of the screw-diad is substituted with a Ca-atom, while the Na-atoms on the other side are substituted with H_2O-molecules, so that the screw-diads disappear and the symmetry becomes *F*1*d*1 (or *Cc*).

3. If we consider all axes which are possible screw-diads in natrolite, and if we group them in (010)-planes, the mesolite structure is obtained by operating the substitution $Na_2 \rightarrow Ca + H_2O$ in two out of the three subsequent planes of the (010)-succession (see Fig. 10). Since screw-diads disappear in the ($Ca + H_2O$)-planes, but are maintained in the (Na_2)-planes, the fact that *b* is three times larger in mesolite than in natrolite is easily understood.

To generalize this description, each of the three minerals can be considered as a different sequence of (Na_2)– and ($Ca + H_2O$)-planes. Each ($Ca \rightarrow H_2O$)-vector may determine an acute or an obtuse angle with the *a*-axis; if the angle is acute, the plane will be coded ($Ca + H_2O$), if obtuse, ($H_2O + Ca$). The structure of scolecite is given by a sequence of ($Ca + H_2O$)-planes all iso-oriented, so that the mineral is acentric. In mesolite the sequence of planes is

$$(Na_2)\,(Ca + H_2O)\,(Ca + H_2O)\,(Na_2)\,(H_2O + Ca)$$
$$(H_2O + Ca)\,(Na_2)\,(Ca + H_2O)\ldots$$

The alternation of the direction of the ($Ca + H_2O$)-planes allows the centricity and orthorhombicity to be maintained. There is a remarkable constancy in the average T-O distance: if T = Si, the value is very near to 1.62 Å, while if T = Al the value is very near to 1.74 Å, the standard deviations for all three refinements being less than 0.01 Å.

Meier (1968) also emphasized the stretching of the chains, which is maximal in natrolite and minimal in scolecite, and the periods of the chains, which coincide with *c*, being 6.60 Å in natrolite, 6.547 Å in mesolite, and 6.531 Å in scolecite. This shortening is a consequence of the substitution of (Na_2) with ($Ca + H_2O$), which deforms the chains slightly.

II. Gonnardite and Thomsonite

The two minerals have the following formulae:

$Na_4Ca_2\,[Al_8Si_{12}O_{40}]\cdot 12H_2O$ Z = 1 Gonnardite

$Na_2Ca_4\,[Al_{10}Si_{10}O_{40}]\cdot 12H_2O$ Z = 2 Thomsonite

According to Alberti and Gottardi (1975), the highest symmetry for a mineral with a thomsonite framework and a disordered (Si,Al)-distribution is Pman, which corresponds to Pmna (standard setting) and also to Pbmn, given for gonnardite by Amirov *et al.* (1972) who in fact found a (Si,Al) disorder. The highest symmetry for a mineral with the same framework, but with a regular alternation of Si and Al in the tetrahedra, is *Pcnn*, which corresponds to *Pnna*, given for thomsonite by Dal Negro *et al.* (in preparation) who in fact found this order in the structure. Thomsonite has a *c* value nearly twice that of gonnardite. The order in thomsonite is nearly perfect, as the dimensions of the tetrahedra are 1.62 ± 0.01 or 1.74 ± 0.01 Å. The chains are stretched to nearly the same extent in both minerals; *c* in gonnardite is 6.63 Å, and in thomsonite 13.218 = 2 × 6.609 Å.

The distribution of the exchangeable cations is remarkably similar in both minerals, there being a regular alternation of two kinds of channel along *c* (axis of the fibers). In one, the cation C(1) is on the inversion centers set along the axis in the middle of the channel, 6.6 Å apart one from the other. In the other, each cation C(2) is located at some distance from the central axis, with a water molecule with nearly the same *z/c*, but on the opposite side of the channel. These couples C(2)-H_2O being repeated at 3.3 Å intervals. In gonnardite C(1) accounts for all Ca_2, and C(2) for all Na_4. In thomsonite C(1) accounts for half of the total Ca_4, and C(2) for the remaining Ca_2 plus all Na_2. It is necessary to point out that in thomsonite C(1) is not exactly on the inversion centers, but in two sites, a fraction of an angstrom apart, each with an occupancy of 50%. This detail does not change the previous picture.

III. Stellerite, Stilbite, Barrerite

The symmetry of the framework in this group (Fmmm) is also present in the whole structure of one mineral, stellerite, whose empirical chemical formula is

$$Ca_4\,[Al_8Si_{28}O_{72}]\cdot nH_2O.$$

The commonest of the three minerals is stilbite:

$$Na_2Ca_4\,[Al_{10}Si_{26}O_{72}]\cdot nH_2O$$

with space group *C*2/*m*, or *F*1 2/*m* 1, if the pseudo-orthorhombicity has to be emphasized. Very rare is barrerite, nearly

$$Na_8\,[Al_8Si_{28}O_{72}]\cdot nH_2O$$

with space group $A\,2_1/m\,2/m\,2/a$, hence orthorhombic but with less symmetry than stellerite. The total number of cations grows from four to six to eight in the sequence given. The refinements show no clear-cut (Si,Al)-ordering. There are five cation sites, which are coded here, and in Galli and Alberti (1976) as 1, 1P, 2, 2P, and 3. In stellerite sites 1 and 1P (the two are

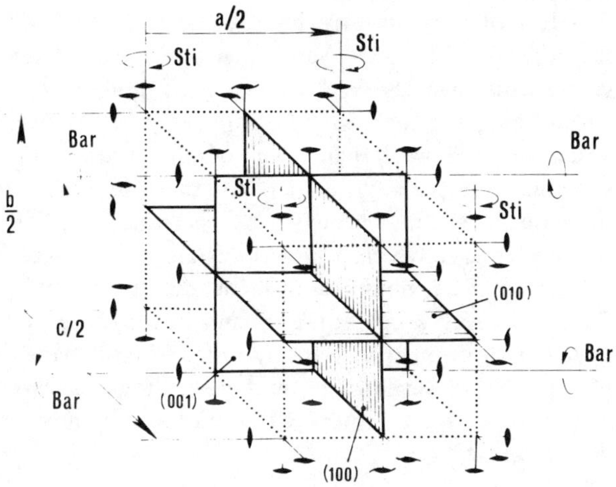

Fig. 11. Symmetry elements in the space group of stellerite, *Fmmm*. In stilbite the framework is forced to rotate slightly around the diads coded "Sti"; the sense of rotation is always the same; as a consequence all (100)- and (001)- mirrors disappear. In barrerite the framework is forced to rotate slightly around the screw-diads coded "Bar"; the sense of rotation is alternated; as a consequence the (001)-mirrors disappear.

symmetrically equivalent in stellerite and stilbite but not in barrerite) are nearly completely filled by Ca_4, while the other sites are empty; in stilbite sites 1 and 1P are again filled by Ca_4, and Na_2 occupies in part site 2P, while sites 2 and 3 are empty; in barrerite all five sites are occupied by Na_8, which is, however, insufficient to fill them.

Galli and Alberti (1975, 1976) explain the relationship between the symmetry of the framework and that of the overall structure on the basis of electrostatic repulsion between exchangeable cations:

(a) In stellerite the number of cations is low, so that they are far away from one another (minimal distance ~6 Å).

(b) In stilbite the insertion of sodium in 2P between two calciums pulls them apart a little, causing a slight rotation around diads coinciding with *b*. The sense of rotation is always the same, and this destroys all diads coinciding with *a* and *c*. The symmetry becomes $F1\ 2/m\ 1$, that is, monoclinic.

(c) In barrerite sodium is inserted not only in site 2P, but also in site 2, and the repulsive forces generated by the two sites are opposite, hence they do not cause any rotation around the *b*-diads. The repulsive forces are weaker because sites 1 and 1P in barrerite are occupied by monovalent sodium instead of divalent calcium, as in stilbite. The sodium inserted in site 3 causes a rotation around the screw-diads parallel to *a*, the sense of rotation being alternated along a row of screw-diads so that orthorhombicity is maintained, although some elements of symmetry, such as the diads parallel to *a*, are lost (see Fig. 11).

It would be very interesting to know whether a sodium-exchanged stellerite has the same structure as barrerite, and vice versa. The question cannot be couched in such simple terms for stilbite, because of its higher Al-content. It would also be interesting to know where the cations are in the sodium-rich stilbite described by Harada and Tomita (1967).

IV. Phillipsite and Harmotome

Rinaldi *et al.* (1974) give accurate data on the structures of these two zeolites with the same framework. Symmetry, lattice constants, and atomic parameters both within and outside the framework are similar in both minerals, which have the following empirical chemical formulae:

phillipsite	$K_2Ca_{1.7}[Al_6Si_{10}O_{32}]\cdot nH_2O$
harmotome	$Ba_2Ca_{0.5}[Al_5Si_{11}O_{32}]\cdot nH_2O$

The difference in the framework is minimal; barium occupies the same position as potassium; calcium sites are the same in both minerals, but the occupancies decrease from ~40% (phillipsite) to 15% (harmotome). The close resemblance of these two structures explains why Galli and Loschi Ghittoni (1972) found a regression equation for phillipsite that is also valid for harmotome (the regression equation gives an approximate chemical composition on the basis of unit cell data).

One point that emerges from this study is worth detailed description. Phillipsite and harmotome are both monoclinic but pseudo-orthorhombic (with the unit cell data of Rinaldi *et al.* (1974), one obtains beta = 89.52° for phillipsite and beta = 90.19° for harmotome). In an earlier structure determination, Steinfink (1962) described phillipsite as truly orthorhombic. As usual in such cases, pseudomeroedric twins are common, and they had to work hard to find an untwinned fragment. This is the typical situation described in the second part of this paper, and it seemed likely that refinement would reveal an (Si,Al)-ordering; but no order at all was found in phillipsite or harmotome. The symmetry of the phillipsite framework is $B2/m\,2_1/m\,2/b$, and the symmetry of the structure is $P2_1/m$, or $B1\,2_1/m\,1$ if the pseudo-orthorhombicity is emphasized. Thus one is bound to ask what causes this lowering of symmetry and the disappearance of the two diads. Certainly not an ordering, as said before, and if the framework itself does not contain something to explain the lowering of symmetry, a solution to the problem must be sought in

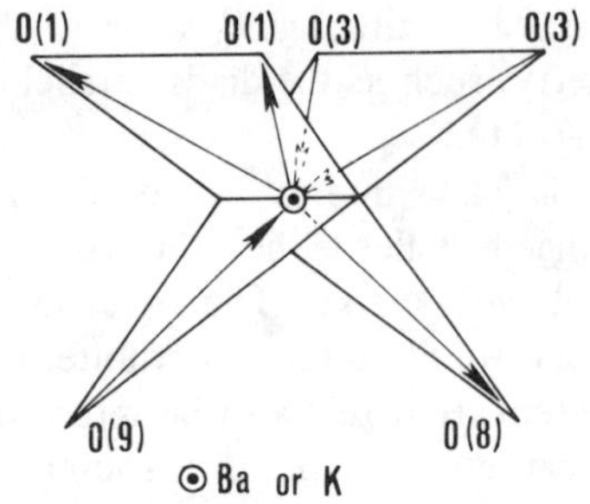

Fig. 12. Deformation of the framework in phillipsite and harmotome. Distances from the cation to O(3) and to O(1) should be equal in the orthorhombic system; nevertheless with this symmetry the site would be too large for the cation involved, and hence the two oxygens O(3) are attracted (see arrows) at a shorter distance, while the other two O(1) are averted at a larger distance. The same is true for O(8) and O(9).

the position of the cations:

The Ca-position is definitely monoclinic, but calcium atoms cannot cause the deviation from orthogonality, because in harmotomes the obliquity is always definite, although their Ca-content is commonly very low.

The Ba–K-position shows a nearly orthorhombic geometry, but there is a definite deviation from the higher symmetry in the framework oxygens which are coordinated by the large Ba (or K) atoms, as shown in Fig. 12. Each large cation coordinates six framework oxygens, two O(1) and two O(3), one O(8), and one O(9); in the orthogonal space group the first two are equivalent to the second two, as are O(8) and O(9). But in the ideal orthogonal framework the hole would be too large even for such cations as Ba or K, so that they attract three of these six oxygens [two O(3) and one O(9)] until they are at a bond distance (~3 Å), while the other three [two O(1) and one O(8)] are averted at a greater distance. So, in the case of phillipsite and harmotome, the fact that the symmetry of the overall structure is lower than that of the framework is explained in a way similar to that described for tetragonal leucite, namely, by a squeezing of a framework cage which otherwise would be too large for its cation. The non-orthorhombic distribution of the Ca-sites is only a consequence, and not a cause of the lowering of symmetry. For any order to be detected in phillipsite and harmotome, refinement should be carried out either in space groups with even lower symmetries, or in larger cells.

V. Heulandite and Clinoptilolite

Three refinements are available for the structure of heulandite and two for clinoptilolite. The results are remarkably similar, as is seen from Table 2. Significantly an Al-enrichment in T(2) is shown by every refinement, so that nearly 4Al and 4Si should occupy this eight-fold position. In clinoptilolite these 4Al practically exhaust all Al-atoms, and it is little wonder that no other T–O distance is great enough to afford evidence of other Al-enrichments. The other T–O distances found in samples 1, 2, and 3 (heulandites) give evidence of a dispersion of the remaining Al in the remaining sites.

Data on cation sites are collected in Table 3. There is good agreement between the parameters given by the five refinements for sites C(1) and C(2), although there are some differences in the occupancies. The main one is observed in sites 2 and 2B, which are equivalent in space

TABLE 2. Average T–O distances in zeolites with the heulandite framework

Refinement number	Space group	R	T(1)–O	T(2)–O	T(3)–O	T(4)–O	T(5)–O
1	*Cm*	0.75	1.65[4] 1.64[4]	1.66[4] 1.66[4]	1.62[4] 1.62[4]	1.66[4] 1.63[4]	1.60[4]
2	*C*2/*m*	0.75	1.63[8]	1.66[8]	1.64[8]	1.62[8]	1.64[4]
3	*I*2/*m*	0.75	1.63[8]	1.66[8]	1.64[8]	1.63[8]	1.65[4]
4	*C*2/*m*	0.83	1.63[8]	1.65[8]	1.61[8]	1.61[8]	1.62[4]
5	*C*2/*m*	0.81	1.61[8]	1.67[8]	1.62[8]	1.62[8]	1.62[4]

1. Heulandite, Giebelsbach, by Merkle and Slaughter (1968).
2. Heulandite, Faröer, by Alberti (1972).
3. Heulandite, Iceland, by Bartl (1973) by neutron diffraction.
4. Clinoptilolite, Agoura, by Alberti (1975).
5. Clinoptilolite, Alpe di Siusi, by Alberti (1975).

T-site code and origin as in Alberti (1972)
Multiplicity in square brackets. *R* = Si:(Si+Al).

TABLE 3. Cation sites in zeolites with the heulandite framework

Refinement number	C(1) [and C(1B)]				C(2) [and C(2B)]				C(3)			
	x	y	z	Occ	x	y	z	Occ	x	y	z	Occ
1	0.15 0.35	0 0.50	0.67 0.34	38 34	0.04 0.46	0.50 0	0.21 0.79	0 48				
2	0.15	0	0.66	70	0.04	0.50	0.20	46				
3												
4	0.14	0	0.67	34	0.04	0.50	0.21	34	0	0	0	22
5	0.15	0	0.66	54	0.04	0.50	0.22	44	0	0	0	10

Occ = occupancy %.

group *C2/m* but non-equivalent in space group *Cm*, used by Merkle and Slaughter (1968) and which gave 0% and 48%. Alberti (1972) also tried a refinement in *Cm* and obtained 19% and 77% for the same occupancies; he states that the "most probable" space group is indeed *Cm*, but, in view of the strong pseudo-centricity of the structure, every least square refinement in the acentric group gives unreliable data. Apart from this possible splitting of C(2), the range of occupancies for C(1) and for C(2) (as a whole) in the three samples of heulandite is small, or at least reasonable for a zeolite. In clinoptilolites, the exchangeable cations are mainly monovalent and hence their number is larger. C(1) and C(2) would in fact be sufficient to locate all cations in clinoptilolite as well, because the occupancies are usually well under the maximum value allowed; however, this possibility is not exploited, and clinoptilolites "prefer" to create a cation site C(3) with low occupancy (10–20%), while the two other sites maintain occupancy levels as low as in heulandite.

No relevant structural differences were revealed between heulandite and clinoptilolite, but it should be pointed out that, apart from its chemical composition, one of the characteristic features of clinoptilolite is its microcrystallinity. A refinement of clinoptilolite was made possible by the finding of two macrocrystalline clinoptilolites, which therefore are not typical clinoptilolites. On the other hand, it is also true to say that a typical microcrystalline clinoptilolite will never be refined.

CONCLUSIONS

The description given for the five groups exhausts the review of the crystal chemistry of zeolites if we consider only those natural zeolites which have been recently the object of structural refinements by single crystals. Other structural data are available for many more groups. Smith (1964), Barrer *et al.* (1969) and Passaglia (1970) described how the chemistry influences the lattice constants and cation sites in chabazites. A correlation between chemistry and lattice constants was ascertained by Sheppard and Gude (1969) and by Rinaldi (1976) for erionite and offretite, by Galli and Loschi Ghittoni (1972) for phillipsite, by Alietti (1972) and Boles (1972) for heulandite and clinoptilolite, by Galli and Rinaldi (1974) for epistilbite, by Passaglia (1975) for mordenite, and by Rinaldi, Smith, and Jung (1975) for faujasite. Some expected features of these correlations demonstrate the usefulness of these studies, which are a necessary premise for more detailed research.

REFERENCES

Adiwidjaja, G. (1972) *Strukturbeziehungen in der Natrolithgruppe und das Entwässerungsverhalten des Skolezits.* Dissertation Univ. Hamburg.

Alberti, A. (1972) On the crystal structure of the zeolite heulandite: *Tsch. Min. Petr. Mitt.* **18**, 129-146.

Alberti, A. (1975) The crystal structure of two clinoptilolites: *Tsch. Min. Petr. Mitt.* **22**, 25-37.

Alberti, A. and Gottardi, G. (1975) Possible structures in fibrous zeolites: *Neues Jb. Miner., Mh.* **1975**, 396-411.

Alietti, A. (1972) Polymorphism and crystal chemistry of heulandite and clinoptilolites. *Amer. Mineral.* **57**, 1448-1458.

Amirov, S. T., Asratkulu, M. A., Mamedov, Kh. S. and Belov, N. V. (1972) Crystal structure of the zeolite $Na_2Ca[Al_2Si_3O_{10}]_2 \cdot 6H_2O$ (Gonnardite): *Sov. Phys. Dokl.* **17**, 316-317.

Barrer, R. M., Davies, J. A. and Rees, L. V. C. (1969) Thermodynamics and thermochemistry of cation exchange in chabazite: *J. Inorg. Nucl. Chem.* **31**, 219-232.

Barrer, R. M. and Villiger, H. (1969) Probable structure of zeolite Omega: *Chem. Comm.* **1969**, 659-660.

Bartl, H. (1973) Neutronenbeugungsuntersuchung des Zeolithes Heulandit: *Z. Kristallogr.* **137**, 440-441.

Boles, J. R. (1972) Composition, optical properties, cell dimensions, and thermal stability of some heulandite group zeolites: *Amer. Mineral.* **57**, 1463-1493.

Breck, D. W. (1974) *Zeolite Molecular Sieves*, John Wiley & Sons, New York.

Charnell, J. F. (1971) Gel growth of large crystals of sodium A and sodium X zeolites: *J. Cryst. Growth* **8**, 291-294.

Ciric, J. (1967) Synthetic zeolites: growth of large single crystals: *Science* **155**, 689.

Cronstedt, A. F. (1756) Rön och beskrifning om en obekant barg art, som kallas zeolites: *Kongl. Vet. Acad. Handl.* **17**, 120-123.

Galli, E. (1971) Refinement of the crystal structure of stilbite: *Acta Cryst.* **B27**, 833-841.

Galli, E. (1974) Mazzite, a zeolite: *Cryst. Struct. Comm.* **3**, 339-344.

Galli, E. (1976) Crystal structure refinement of edingtonite: *Acta Cryst.* **B32.**

Galli, E. and Alberti, A. (1975) The crystal structure of stellerite: *Bull. Soc. Fr. Minér. Cristallogr.* **98**, 11-18.

Galli, E. and Alberti, A. (1975) The crystal structure of barrerite: *Bull. Soc. Fr. Minér. Cristallogr.* **98**, 331-340.

Galli, E. and Loschi Ghittoni, A. G. (1972) The crystal chemistry of phillipsites: *Amer. Mineral.* **57**, 1125-1145.

Galli, E., Passaglia, E., Pongiluppi, D. and Rinaldi, R. (1974) Mazzite, a new mineral, the natural counterpart of the synthetic zeolite Omega: *Contr. Miner. Petrol.* **45**, 99-105.

Galli, E. and Rinaldi, R. (1974) The crystal chemistry of epistilbites: *Amer. Mineral.* **59**, 1055-1061.

Gard, J. A. and Tait, J. M. (1972) The crystal structure of the zeolite offretite: *Acta Cryst.* **B28**, 825-834.

Gard, J. A. and Tait, J. M. (1973) Refinement of the crystal structure of erionite: *Molecular Sieves*, Leuven Univ. Press, pp. 94-99.

Gottardi, G. (1972) Fortschritte auf dem Gebiet der Kristallographie und Kristallchemie der Zeolithe: *Fortschr. Miner.* **49**, 13-30.

Gottardi, G. and Alberti, A. (1974) Domain structure in garronite: a hypothesis: *Miner. Mag.* **39**, 898-899.

Grämlich, V. and Meier, W. M. (1971) The crystal structure of hydrated NaA: a detailed refinement of a pseudosymmetric zeolite structure: *Z. Kristallogr.* **133**, 134-149.

Harada, K. and Tomita, K. (1967) A sodian stilbite from Onigajo, Mie prefecture, Japan, with some experimental studies concerning the conversion of stilbite to wairakite at low water vapor pressures: *Amer. Mineral.* **52**, 1438-1450.

Kawahara, A. and Curien, H. (1969) La structure cristalline de l'erionite: *Bull. Soc. Fr. Miner. Cristallogr.* **92**, 250-256.

Kerr, I. S. and Williams, D. J. (1969) The crystal structure of yugawaralite: *Acta Cryst.* **B25**, 1183-1190.

Krogh Andersen, E., Danø, M. and Petersen, O. V. (1969) A tetragonal natrolite: *Medd. Grønland* **181**, No. 10, 1-19.

Leimer, H. W. and Slaughter, M. (1969) The determination and refinement of the crystal structure of yugawaralite: *Z. Kristallogr.* **130**, 88-111.

Mazzi, F., Galli, E. and Gottardi, G. (1976) The crystal structure of tetragonal leucite: *Amer. Mineral.* **61**, 108-115.

Meier, W. M. (1960) The crystal structure of natrolite: *Z. Kristallogr.* **113**, 430-444.

Meier, W. M. (1968) Zeolite structure: *Molecular Sieves*, Soc. of Chem. Industry, London, pp. 10-27.

Meier, W. M. and Olson, D. H. (1971) Zeolite frameworks: *Adv. Chem. Series* **101**, *Molecular Sieve Zeolites*, pp. 155-170.

Merkle, A. B. and Slaughter, M. (1968) Determination and refinement of the structure of heulandite: *Amer. Mineral.* **53**, 1120-1138.

Merlino, S., Galli, E. and Alberti, A. (1975) The crystal structure of levyne: *Tsch. Miner. Petrol. Mitt.* **22**, 117-129.

Olson, D. H. (1970) A reinvestigation of the crystal structure of the zeolite hydrated NaX: *J. Phys. Chem.* **74**, 2758-2764.

Pabst, A. (1971) Natrolite from the Green River Formation, Colorado, showing an intergrowth akin to twinning: *Amer. Mineral.* **56**, 560-569.

Passaglia, E. (1970) The crystal chemistry of chabazites: *Amer. Mineral.* **55**, 1278-1301.

Passaglia, E. (1975) The crystal chemistry of mordenites: *Contr. Miner. Petrol.* **50**, 65-77.

Passaglia, E. and Pongiluppi, D. (1975) Barrerite, a new natural zeolite: *Miner. Mag.* **40**, 208.

Rinaldi, R. (1976) Crystal chemistry and structural epitaxy of offretite-erionite from Sasbach, Kaiserstuhl: *N. Jb. Miner., Mh* **1976**, 145-156.

Rinaldi, R., Pluth, J. J. and Smith, J. V. (1974) Zeolites of the phillipsite family. Refinement of the crystal structures of phillipsite and harmotome: *Acta Cryst.* **B30**, 2426-2433.

Rinaldi, R., Smith, J. V. and Jung, G. (1975) Chemistry and paragenesis of faujasite, phillipsite and offretite from Sasbach, Kaiserstuhl, Germany: *Neues Jb. Miner., Mh* **1975**, 433-443.

Sheppard, R. A. and Gude, A. J. (1969) Chemical composition and physical properties of the related zeolites offretite and erionite: *Amer. Mineral.* **54**, 875-886.

Slaughter, M. (1970) Crystal structure of stilbite: *Amer. Mineral.* **55**, 387-397.

Smith, J. V. (1964) Crystal structures with the chabazite framework. Part V. Interpretation of physico-chemical properties in terms of the crystal structure: *J. Chem. Soc.* **1964**, 3759-3770.

Steinfink, H. (1962) The crystal structure of the zeolite phillipsite: *Acta Cryst.* **15**, 644-651.

Union Carbide Corporation (1967) Dutch Pat. 6,710,729, August 3.

Wise, W. S. and Tschernich, R. W. (1975) Cowlesite, a new Ca-zeolite: *Amer. Mineral.* **60**, 951-956.

LATTICE-CONSTANT VARIATIONS IN CATION-EXCHANGED CHABAZITES

ELIO PASSAGLIA

Istituto di Mineralogia e Petrologia, Università di Modena, 41100 Modena, Italy

Abstract – Li-, Na-, K-, Rb-, Cs-, Ca-, Sr- and Ba-exchanged forms were obtained from three natural chabazites of different Si/(Si + Al) ratios. The unit-cell dimensions measured on both the hydrated and partially dehydrated (at 250°C) exchanged forms suggest that site I at the center of the double 6-ring is occupied in the hydrated phases by Sr and by Sr and K in the samples with Si/(Si + Al) ratios of 0.667 and 0.59, respectively. Li, Na and Ca occupy the same position upon dehydration. This behavior is explained on the basis of assumptions that need confirmation by single crystal structure refinements of cation-exchanged forms.

INTRODUCTION

The crystal structure of both hydrated and dehydrated chabazite, a rhombohedral zeolite, has been determined by Dent and Smith (1958), Nowacki *et al.* (1958), Smith (1962), and Smith *et al.* (1963, 1964). Chemically chabazite shows a large variability of both the Si/Al ratio and the prevalent exchangeable cations (Passaglia, 1970). Its ion-exchange properties have been extensively studied by Barrer (1958, 1974), Barrer *et al.* (1969), and others. These properties are influenced by the framework charge density and therefore by the Si/Al ratio (Barrer and Klinowski, 1972). According to Barrer and Sammon (1955) in the Na-, Ag-, Ca- and Ba-exchanged forms, the unit-cell dimensions are very similar. Smith (1964) explains the physical chemistry of ion exchange in chabazites in terms of crystal structure.

The aim of this work is to investigate the correlations between lattice-constant variations and chemistry in the cation exchanged forms. For this purpose three samples of natural chabazites of different Si/Al ratios were exchanged with the following cations: Li, Na, K, Rb, Cs, Ca, Sr, and Ba. Attempts to prepare a Mg form were unsuccessful. The unit-cell dimensions were measured both on the hydrated forms and on the partially dehydrated ones at 250°C.

The following symbols are used: R for $Si/(Si + Al + Fe^{+3})$ ratio; cardinal number for the sample number.

EXPERIMENTAL

Starting materials

Sample No. 1. Sedimentary chabazite from John Day, Grant County, Oregon (Sheppard and Gude, 1970). Its framework is silica-rich or "acid" (R = 0.80); Ca prevails among the exchangeable cations. Attempts to obtain a pure fraction required for the exchanges were carried out by sample disaggregation in an ultrasonic bath and repeated centrifuging in a heavy-liquid mixture of bromoform and acetone. Unfortunately the obtained material still contained small impurities of glass and montmorillonite. For this reason, the exchangeable-cation contents of the exchanged forms could not be determined reliably and are not reported here.

Sample No. 2. Chabazite from Keller, Nidda, Germany. Clear, white, transparent rhombohedra. Its silica content (Table 1) is intermediate (R = 0.667); Ca prevails among the exchangeable cations, but its Sr content is also appreciable.

Sample No. 3. Chabazite from Osa, Roma, Italy. Milky-white crystals twinned to form aggregates with spherical shape, diameter 1–5 mm. Its framework is silica-poor or "basic" (R = 0.59) (Table 1); Ca and Sr prevail among the exchangeable cations, but the K content is also high.

Ion exchange

The samples were exchanged as follows: about 100 mg of powdered material were placed in a Teflon tube containing 100 cm^3 of a 1 N-solution of the

TABLE 1. Exchangeable-cation contents in wt.% of natural and exchanged forms of the chabazites from Keller, Germany, and Osa, Italy*

	Natural	Exchanged forms							
		Li	Na	K	Rb	Cs	Ca	Sr	Ba
					Sample 2 – Keller, Germany				
Na_2O	0.44	0.40	10.63	0.14	0.10	0.30	0.20	0.26	0.03
K_2O	0.69	0.38	0.22	16.14	0.21	0.26	0.08	0.06	0.04
MgO	0.06	tr.	tr.	tr.	tr.	tr.	tr.	tr.	tr.
CaO	8.47	0.47	0.46	0.42	0.54	0.91	9.97	0.21	0.18
SrO	2.99	0.14	0.41	0.36	0.09	0.13	0.15	18.10	tr.
BaO	0.01	tr.	tr.	tr.	tr.	tr.	tr.	tr.	27.15
					Sample 3 – Osa, Italy				
Na_2O	0.43	0.24	13.07	0.04	0.02	0.57	0.04	0.06	tr.
K_2O	4.59	0.11	0.04	18.63	0.18	0.34	0.05	0.04	0.02
MgO	0.01	tr.	tr.	0.24	0.04	0.17	tr.	0.05	0.05
CaO	6.43	0.18	0.19	0.26	0.95	1.30	12.58	0.44	0.35
SrO	5.86	0.12	0.10	0.20	0.14	0.17	0.25	20.58	tr.
BaO	0.45	0.10	0.14	0.30	0.32	0.30	0.12	tr.	31.40

*The complete chemical analyses of natural samples include for sample 2: SiO_2 49.50, Al_2O_3 19.34, Fe_2O_3 0.04, $H_2O_{tot.}$ 22.31, Sum 99.25; and for sample 3: SiO_2 38.36, Al_2O_3 22.63, Fe_2O_3 0.03, $H_2O_{tot.}$ 21.11, Sum 99.90.

Quantitative data for Li_2O, Rb_2O, and Cs_2O are not available; qualitative tests showed their presence only in the corresponding exchanged forms.

exchanging cation salt. The tubes were slowly oscillated in an oven at ~110°C. After 6 days the 1 N-solution was removed and the treatment repeated until a chemical determination of the exchangeable cations on the powder carefully washed with a Gooch filter showed the exchange to be nearly complete. The 1 N-solutions were prepared with Merck "suprapur" chlorides and deionized water. Exchangeable cations contents were determined by an atomic absorption spectrophotometer (Perkin-Elmer 303) according to the method described by Loschi Ghittoni and Passaglia (1974). Powder tracings of the hydrated exchanged forms were recorded on a Philips diffractometer using $Pb(NO_3)_2$ as internal standard. Powder tracings of the dehydrated exchanged forms were recorded on an X-ray diffractometer with a high temperature attachment. For technical reasons the samples were kept in the open air so that during the tests the temperature oscillations were about 5–10°C. Quartz was used as internal standard, taking due account of its coefficients of linear expansion (Frondel, 1962). Cell dimensions were derived by a least-squares computer program using the reflections from 5 to 1.70 Å; spacings both too far from the standard lines and/or not unequivocally referable to a set of indices were discarded.

CHEMICAL RESULTS

The thermodynamic affinity sequence, based on the standard free energies of exchange in chabazites (Barrer *et al.*, 1969), states that all cations, except Cs, exchange almost completely with Na. In this study this rule has proved true for sample 2 and probably for sample 1, but not for sample 3.

Sample No. 1

The exchange procedure was the same as that applied to No. 2. Chemical analyses of the exchanged forms were unreliable due to contamination of the starting material. Chemical tests on the exchanging cations, however, showed remarkable increases in the guest species and corresponding decreases of the original cations contents.

Sample No. 2

The natural sample was treated with 1 N Li-, Na-, K-, Rb-, Cs-, Ca-, Sr- and Ba-solutions; only the exchanges with Na, K and Ba completely occurred. All the other forms were prepared from the Na-form. Some forms were also obtained as follows: Sr and Rb from K; Ba and K from Na. Since the Na-form was reobtained from all the other forms, the resulting exchanges were reversible. The exchangeable-cation contents are reported in Table 1. Due to experimental difficulties, a precise determination of Li, Rb and Cs in the corresponding forms was impossible; however, the detected high contents of these cations and the corresponding low contents of the other cations argue for the indicated exchanges.

Sample No. 3

The natural sample was treated with Na 1 N-solution. In spite of several and lengthened treatments, the exchange was incomplete, and the obtained form contains 4.50% SrO. The same material was therefore treated with Ca in N-solution, and after several and lengthened treatments, complete exchange resulted. The Na-form was therefore obtained from the Ca-form. Li, K, Rb, Cs, Sr and Ba forms were prepared from the Na-form. Attempts to reobtain the Na-form from all the others were successful, except for Sr: in this case the final form still contains 4.30% SrO. The Ca-form was prepared also from the K-form. The exchangeable-cation contents are reported in Table 1. For the Li, Rb and Cs forms, the comments made for sample 2 are also applicable.

CRYSTALLOGRAPHIC RESULTS

The hexagonal and rhombohedral lattice constants and the calculated hexagonal volumes of the hydrated exchanged forms are reported in Table 2. It is clear that c varies more than a, when considering the hexagonal cell, and α varies more than a, when considering the rhombohedral cell. From here on, only the hexagonal cell is considered, keeping in mind that analogous conclusions can be drawn by considering the rhombohedral cell. Since the study of twenty-eight polycationic natural chabazites showed that c depends mainly on R (Passaglia, 1970), in the interpretation of the results the influence of R on the lattice constants was distinguished from that of the exchangeable cations. As the volume variations at constant R (that is in the same sample) are remarkably less than those at variable R (Fig. 1), it is clear that the volume is more influenced by R than by the exchangeable cations. This is confirmed also by the correlation line between volume and R, calculated on twenty-eight polycationic chabazites, which shows almost the same slope as the lines pertinent to the

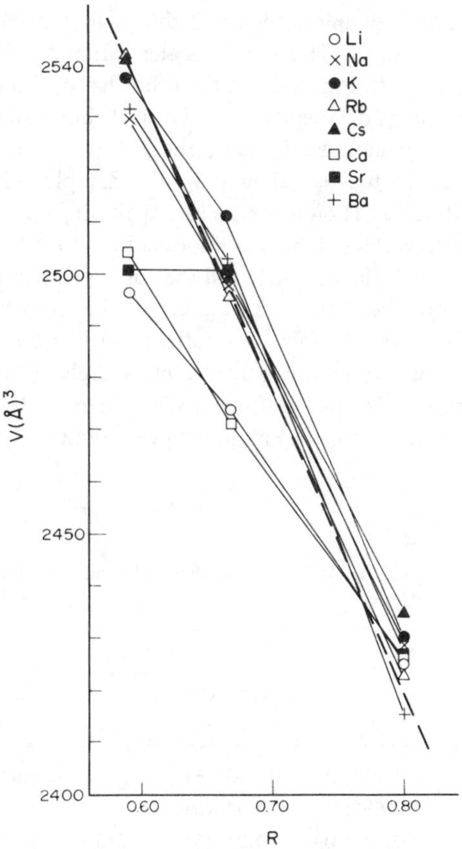

Fig. 1. Volumes of cation-exchanged chabazites against Si/(Si+Al) ratios (R).

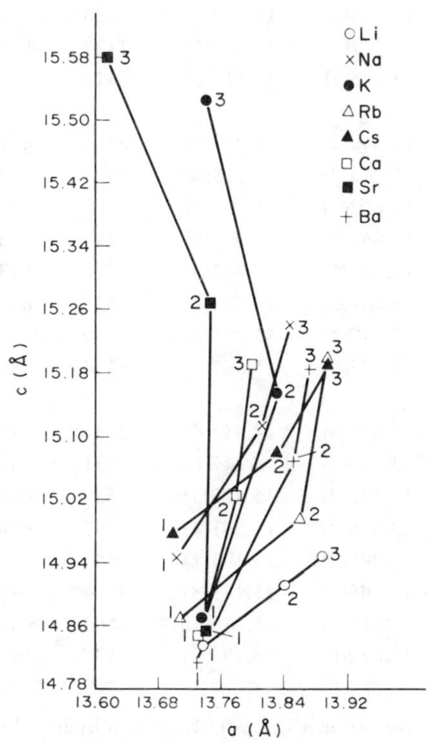

Fig. 2. Unit-cell parameters of cation-exchanged chabazites. Digits correspond to samples numbers.

monocationic samples. Figure 2 shows that c increases with decreasing R. This confirms the earlier findings by Passaglia (1970). Figure 3, on the other hand, shows that the variation of c depends not only on R, but also on the exchangeable cations; in fact, the total c variation is almost equal to that shown by the sample with the lowest R value. Therefore, the earlier findings (Passaglia, 1970) were affected by the impossibility to distinguish between the influence of R and that of the exchangeable cations on the lattice constants of the polycationic natural samples. In fact, the correlation line between c and R drawn from the polycationic samples (Passaglia, 1970) shows a slope intermediate between those of the lines pertinent to the monocationic samples of this study.

Table 2 also reports the lattice constants of the dehydrated exchanged forms. The choice of 250°C as the dehydration temperature allows the observation in all forms, except the Na-form of sample 2, of their greatest lattice-constant variations; while, at a higher temperature, some forms are destroyed. As for the Na-form of sample 2, the greatest lattice-constant variations occur at 350°C. The X-ray diffraction powder tracings recorded at intervals of 50°C until the destruction of each form allow the following conclusions. (1) At constant R, the following is a series of decreasing thermal stability: Cs, Rb, K, Na, Ca, Li, Sr, Ba; exchangeable cation being the same, thermal stability increases with R. (2) All Ba forms are already destroyed at 200°C; this means that in these forms the

TABLE 2. Hexagonal and rhombohedral cells of hydrated and dehydrated exchanged chabazites (standard errors in brackets)*

Sample	Hydrated					Dehydrated at 250°C				
	Hexagonal cell			Rhombohedral cell		Hexagonal cell			Rhombohedral cell	
	a(Å)	c(Å)	V(Å)³	a(Å)	α°	a(Å)	c(Å)	V(Å)³	a(Å)	α°
No.1 Li	13.739(2)	14.839(6)	2426	9.348(2)	94.59	13.65(3)	14.94(3)	2410	9.32(2)	94.13(1)
Na	13.698(3)	14.943(6)	2428	9.346(2)	94.24	13.65(2)	14.88(1)	2403	9.31(1)	94.27(1)
K	13.737(2)	14.871(5)	2430	9.353(2)	94.51	13.67(2)	14.80(1)	2395	9.31(1)	94.50(1)
Rb	13.714(2)	14.877(5)	2423	9.342(2)	94.44	13.70(2)	14.86(2)	2416	9.33(1)	94.43(1)
Cs	13.700(5)	14.980(6)	2435	9.354(4)	94.16	13.65(2)	14.90(1)	2406	9.32(1)	94.22(1)
Ca	13.735(3)	14.850(6)	2426	9.348(2)	94.56	13.60(8)	15.00(4)	2401	9.31(5)	93.85(10)
Sr	13.737(2)	14.855(4)	2427	9.350(2)	94.55	13.59(3)	14.89(2)	2383	9.28(2)	94.08(1)
Ba	13.725(2)	14.812(5)	2416	9.336(2)	94.62	–	–	–	–	–
No.2 Li	13.840(2)	14.916(4)	2474	9.411(2)	94.66	13 56(2)	15.30(1)	2436	9.34(1)	93.28(1)
Na†	13.813(2)	15.115(2)	2498	9.433(1)	94.14	13.35(2)	15.75(2)	2431	9.33(1)	91.40(1)
K	13.832(3)	15.156(3)	2511	9.450(1)	94.09	13.45(1)	15.66(1)	2454	9.36(1)	91.90(1)
Rb	13.862(3)	15.006(5)	2497	9.438(2)	94.51	13.69(3)	15.12(2)	2456	9.37(2)	93.80(1)
Cs	13.834(2)	15.082(3)	2500	9.437(1)	94.27	13.68(2)	15.18(2)	2460	9.38(1)	93.63(1)
Ca	13.781(3)	15.029(4)	2472	9.402(2)	94.25	13.61(2)	15.29(1)	2452	9.36(1)	93.20(1)
Sr	13.747(2)	15.271(3)	2499	9.429(1)	93.60	13.63(3)	15.24(1)	2452	9.37(1)	93.37(1)
Ba	13.852(2)	15.067(2)	2504	9.444(1)	94.34	–	–	–	–	–
No.3 Li	13.886(5)	14.952(6)	2497	9.440(4)	94.70	13.63(2)	15.28(1)	2457	9.37(1)	93.20(1)
Na	13.844(4)	15.247(6)	2531	9.472(3)	93.91	13.44(9)	15.64(5)	2447	9.35(5)	91.75(10)
K	13.739(2)	15.526(5)	2538	9.471(2)	92.99	13.45(4)	15.75(3)	2467	9.37(2)	91.68(2)
Rb	13.896(3)	15.201(3)	2542	9.489(1)	94.16	13.64(1)	15.40(1)	2482	9.40(1)	93.03(1)
Cs	13.896(2)	15.193(2)	2541	9.487(1)	94.16	13.65(1)	15.33(2)	2472	9.39(1)	93.22(1)
Ca	13.798(5)	15.187(8)	2504	9.439(4)	93.93	13.41(1)	15.74(1)	2452	9.35(1)	91.60(1)
Sr	13.616(7)	15.576(10)	2501	9.421(5)	92.55	13.55(3)	15.50(2)	2465	9.37(2)	92.54(1)
Ba	13.872(6)	15.192(9)	2532	9.476(4)	94.11	–	–	–	–	–

* Standard errors on α's are ~0.01 for most hydrated samples.

† For this Na-exchanged form, the unit cell constants of the dehydrated phase were obtained at 350°C. The same obtained at 250°C are: $a_{hex.}$ = 13.64(2)Å; c = 15.40(1)Å.

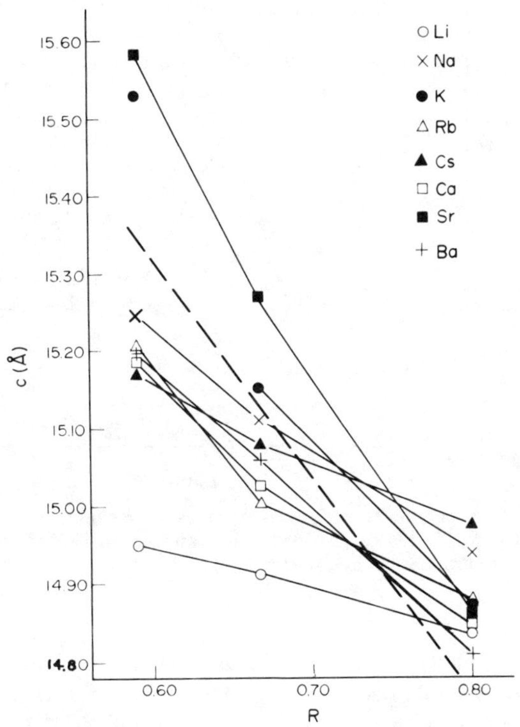

Fig. 3. c parameters of cation-exchanged chabazites against Si/(Si+Al) ratios (R).

framework breakdown is triggered by dehydration. This thermal behavior is similar to that found for heulandite-clinoptilolite (Alietti *et al.*, 1974) and for stilbite-stellerite-barrerite (Passaglia and Pongiluppi, unpublished data).

By comparing the lattice-constant values (Table 2), the volumes of the dehydrated phases appear remarkably smaller than those of the corresponding hydrated phases. The a parameter decreases also with dehydration, whereas the c parameter (except in some forms of sample 1 and in the Sr forms of samples 2 and 3) increases, often considerably, with dehydration. The behavior of the c parameter upon dehydration can be explained in terms of the positions of the exchangeable cations within the structures of hydrated and dehydrated chabazite.

INTERPRETATION OF RESULTS

The following structural description of chabazite is from Smith (1962) and Smith *et al.* (1963, 1964). In the rhombohedral unit cell, the possible exchangeable-cation positions are: (a) site I in the small cage A which can be entered through a six-membered ring whose "free" aperture has a radius of about 1.3 Å; (b) site II in the large ellipsoidal cage B which can be entered through six eight-membered rings of 3–4 Å in diameter. The same unit cell, moreover, contains twelve tetrahedral cavities which can be occupied by Si or Al. The framework charge density and therefore the exchangeable cation number per unit cell depends on the R value. For the samples studied here, exchangeable-cation numbers needed to satisfy electrical neutrality are: 2.38 monovalent or 1.19 divalent cations in sample 1; about four monovalent or two divalent cations in sample 2; and about five monovalent or 2.5 divalent cations in sample 3. The exchangeable cations tend towards those positions where they neutralize the framework negative field; these positions correspond approximately to the framework points with the highest density of tetrahedral cavities. Although these points change with the Si/Al order of the sample, in chabazite one of them surely corresponds to the center of the A cage. Hence this point represents a strong attractive pole, and, consequently, if the exchangeable cation radius is less than 1.3 Å, site I should be always occupied. However, from the present knowledge on the exchangeable-cation positions, their distribution is as follows: in hydrated Ca-chabazite, site I is not occupied and the two Ca atoms occupy site II near the hexagonal rings separated from the oxygen framework by water molecules; in dehydrated Ca-chabazite, site I is occupied by Ca for about two-thirds of the unit cell, whereas the remaining Ca occupies site II near the centers of the six-membered rings, or it is located on the internal surface of cage B.

The vacancy and occupancy of site I in the hydrated and dehydrated phases, respectively, have also been found in Ca-faujasite (Bennett and Smith, 1968a, b) and in synthetic Ca and La X (Pluth and Smith, 1972, 1973; Bennett *et al.*, 1969).

According to Smith (1962) six oxygens from the pseudohexagonal prism are in nearly octahedral coordination with Ca in site I. This actually enhances the ditrigonality of cage A with a remarkable decrease of α and a very small one of $a_{\mathrm{rh.}}$, and corresponds to an increase of c and a decrease of a hexagonal parameters. In brief, site I occupancy is revealed by a relatively high value of c, and this site will certainly be occupied, upon dehydration, by cations of radius less than 1.3 Å. In the light of these assumptions we can therefore interpret the results of the present work. Through Figs. 4, 5 and 6 we can visualize for each exchanged form the c variations upon dehydration. Empty circles pertain to hydrated forms, solid circles to dehydrated ones.

From Fig. 4 (sample 1), site I is not occupied in any of the hydrated forms. Dehydration makes Li, Ca, and, perhaps, Sr, occupy site I. The relatively high value of c for the Cs hydrated form cannot be ascribed to site I occupancy but was caused by simple volumetric increase as shown by the considerable c decrease upon

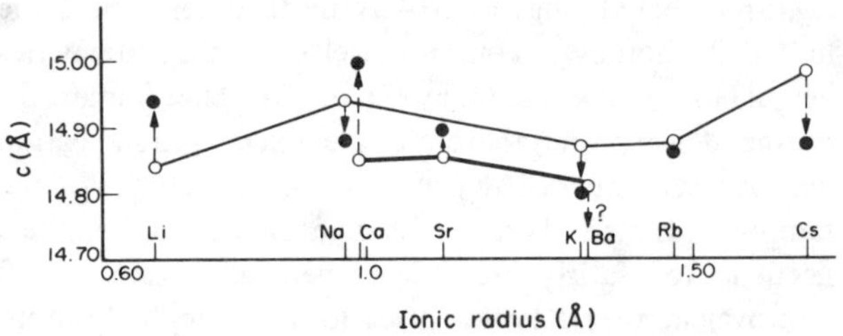

Fig. 4. *c* parameter variations upon dehydration of the cation-exchanged forms of John Day chabazite (sample 1). Empty circles = hydrated phases; solid circles = dehydrated phases.

dehydration.

From Fig. 5 (sample 2), the high value of *c* for the hydrated Sr form, which does not change upon dehydration, already reveals site I occupancy by Sr in the hydrated phase. Dehydration makes Li, Na, K, and Ca occupy site I. The small *c* increases upon dehydration for the Rb and Cs forms are due, most likely, to a site I occupancy by small amounts of Ca present in these forms as impurity.

From Fig. 6 (sample 3), the high *c* values for the hydrated Sr and K forms already reveal site I occupancy by these cations in the hydrated phases. The noticeable increase of *c* upon dehydration for the K-form reveals, moreover, that site I occupancy in the hydrated phase is only partial. Dehydration makes Li, Na, and Ca occupy site I. The considerations on the Rb- and Cs- forms of sample 2 are valid also for this sample.

The *c* parameter value depends on (1) site I occupancy per unit cell and (2) amount of deformation (ditrigonality) of cage A. Factor (1) clearly depends on both availability of exchangeable cations and framework charge distribution. Since the rhombohedral unit cell contains one cage A and a normal chabazite (R ~ 0.70) contains two divalent cations, the cation availability is higher than required for full occupancy of site I. Therefore the only partial occupancy (2/3 per unit cell) of site I observed in dehydrated Ca-chabazite (Smith, 1962) is probably due to electrostatic equilibrium reasons. It is therefore possible to ascribe the larger *c* values of the Na and K forms as compared to those of Ca and Sr (sample 2, Fig. 5) to total occupancy of site I in the former. Such total occupancy of site I also occurs for divalent cations when R has lower values. This corresponds to an increase, and probably different distribution, of the negative charges in the framework. The above explains also the near uniformity of *c* in the Na, K, Sr and Ca dehydrated forms of sample 3 (Fig. 6). Since the cation-oxygen distance in site I is of the same order of magnitude of a bond distance (Smith, 1962; Smith *et al.*, 1967; Eulenberger *et al.*, 1967; Olson and Sherry, 1968; Bennett and Smith, 1968a; Olson, 1968, 1970; Bennett *et al.*, 1969; Mortier and Bosmans, 1971; Pluth and Smith, 1972) and since the rigidity of the framework decreases with increasing Al content, the amount of deformation of cage A (factor 2) depends on both the exchangeable cation radius and the R value of the sample. The small differences in *c* observed in the same sample (at constant R) and with the same degree of occupancy of site I are therefore a consequence of the different ionic radii of the various cations. As to the "anomalous" behavior of the Li-forms whose *c* values are always less than expected from consideration of its ionic radius, it is possible that

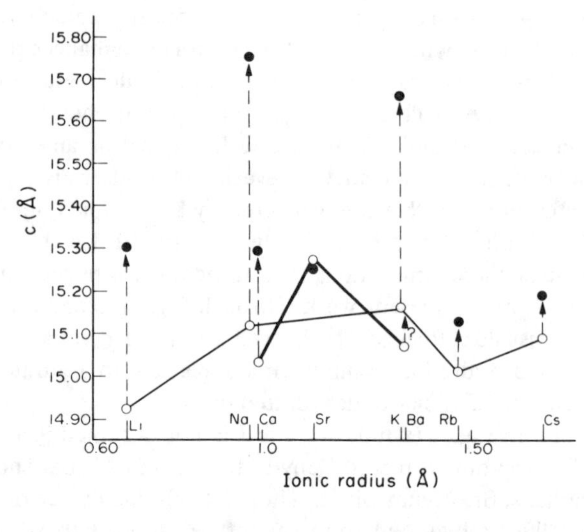

Fig. 5. *c* parameter variations upon dehydration of the cation-exchanged forms of Keller chabazite (sample 2). Empty circles = hydrated phases; solid circles = dehydrated phases.

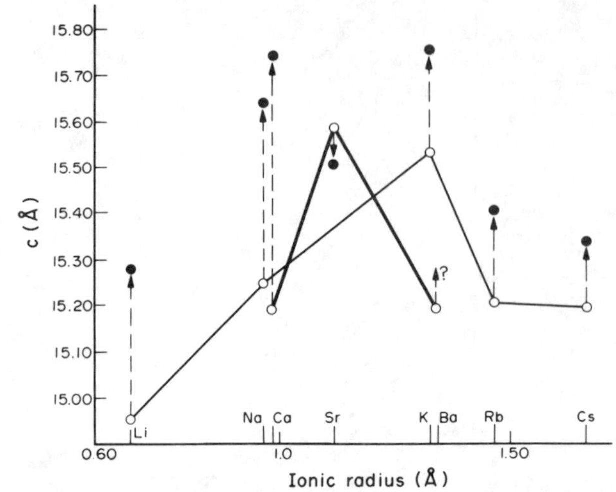

Fig. 6. *c* parameter variations upon dehydration of the cation-exchanged forms of Osa chabazite (sample 3). Empty circles = hydrated phases; solid circles = dehydrated phases.

because of its rather short bond distance (2.16 Å), Li occupies a different position in the A cage than all other cations, thus producing a different deformation of the cage. Finally, the smaller increase of *c* upon dehydration of the K-form of sample 3, as compared to that for the same form of sample 2, already confirms a partial K occupancy of site I in the hydrated phase of the former.

CONCLUSIONS

The results of this work suggest that in chabazite, site I inside the pseudohexagonal prism is occupied already in the hydrated phases by Sr and K. Li, Na and Ca enter site I upon dehydration; Rb and Cs are not allowed in it because of their ionic radii; and Ba by framework destruction upon dehydration.

Sites I and II are the only possible exchangeable-cations sites in chabazite. In site II the cation is coordinated mainly to water molecules, whereas in site I it is coordinated only to the oxygens of the double 6-ring. Since the radius of the "free" aperture to enter site I is about 1.3 Å, only when "bare" can cations smaller than 1.3 Å in radius occupy site I. The present work seems to indicate that water-molecule stripping occurs for Li, Na and Ca only through dehydration, whereas it occurs spontaneously for Sr and K. This behavior can be schematized by the influence on the cation of two antagonist forces: the first (F_1) pulls the cation toward site I, which is a negative pole of attraction on account of the high tetrahedral cavity density; the other (F_2), due to the hydration energy, inhibits water stripping from the cation and hence its migration to site I. On the other hand, an increase of F_2 is attained by the increase of the ionic potential (charge/radius) of the cation, whereas F_1 increases also with an increase of the charge density of the framework, i.e. for decreasing R. For every cation there will be a value of R below which $F_1 > F_2$, with consequent occupancy of site I already in the hydrated phases of the cations with radius less than 1.3 Å. The three values of R for the samples studied here are evidently too high in order that $F_1 > F_2$ for Li, Na and Ca, but sufficiently low for K and Sr. In particular, a value of R = 0.667 is sufficient for Sr (sample 2), whereas for K it is necessary a valve of R = 0.59 (sample 3). Moreover, for K the value of F_2 may be less than expected due to lack of space in site II for a full K-water coordination. Some analogies with faujasite-type zeolites are worth noting: site I is always occupied in the dehydrated phases (Smith *et al.*, 1967; Eulenberger *et al.*, 1967; Olson, 1968; Bennett and Smith, 1968a, b; Bennett *et al.*, 1969; Pluth and Smith, 1972), whereas in the hydrated phases it is occupied by Na, K, and Sr and only in the more "basic" X zeolite (Mortier and Bosmans, 1971; Olson, 1970; Olson and Sherry, 1968). Moreover, K occupancy of site I in the latter case increases with Al content (Mortier and Bosmans, 1971).

In conclusion, when $F_2 > F_1$, the exchangeable cation which occupies site I only owing to dehydration ($F_2 = 0$), must regain its original position in site II following rehydration as a result of attraction by water molecules. This results from the almost identical powder diffractometer tracings taken at room temperature before and after dehydration.

In the light of the results of this study, it is also possible to explain: (a) the incomplete exchange of Sr by Na in sample 3, (b) the anomalous increase of Ca, as compared with those of the starting Na forms, in the Cs form of sample 2 and in the Rb and Cs forms of sample 3. Case (a) indicates that the exchange energy of Na is insufficient to remove Sr from site I in the sample with lowest R(0.59). The same energy is in fact sufficient for the same purpose in sample 2 (R = 0.667). Case (b) indicates that, for the large cations Rb and Cs and with decreasing values of R, the volume of cage B becomes insufficient to host the number necessary to neutralize the framework anionic charge. During the exchange experiments of these forms, the Ca impurities, although present only in trace amounts in Rb and Cs chlorides and in deionized water, are readily captured because of their greater ability to neutralize the negative framework charge by having a smaller radius and higher positive charge.

ACKNOWLEDGMENTS

The original manuscript was improved through helpful discussion with G. Gottardi and R. Rinaldi whom I wish to thank also for critical reading of the final manuscript. Thanks are also due to P. Gallitelli for kind permission to use the high-temperature X-ray diffractometer and R. A. Sheppard for providing the John Day sample.

This work was made possible through financial support of the CNR, Rome.

REFERENCES

Alietti, A., Gottardi, G. and Poppi, L. (1974) The heat behaviour of the cation exchanged zeolites with heulandite structure: *Tschermaks Min. Petr. Mitt.* **21**, 291-298.

Barrer, R. M. (1958) Crystalline ion-exchangers: *Proc. Chem. Soc.*, p. 99.

Barrer, R. M. (1974) Isomorphous replacement by ion exchange: some equilibrium aspects: *Bull. Soc. Fr. Miner. Crystallogr.* **97**, 89-100.

Barrer, R. M., Davies, J. A. and Rees, L. V. C. (1969) Thermodynamics and thermochemistry of cation exchange in chabazite: *J. Inorg. Nucl. Chem.* **31**, 219-232.

Barrer, R. M. and Klinowski, J. (1972) Influence of framework charge density on ion-exchange properties of zeolites: *J. Chem. Soc., Faraday Trans. I* **68**, 1956-1963.

Barrer, R. M. and Sammon, D. C. (1955) Exchange equilibria in crystals of chabazite: *J. Chem. Soc.*, pp. 2838-2849.

Bennett, J. M. and Smith, J. V. (1968a) Positions of cations and molecules in zeolites with the faujasite-type framework. I. Dehydrated Ca-exchanged faujasite: *Mater. Res. Bull.* **3**, 633-642.

Bennett, J. M. and Smith, J. V. (1968b) Positions of cations and molecules in zeolites with the faujasite-type framework. III. Hydrated Ca-exchanged faujasite: *Mater. Res. Bull.* **3**, 933-940.

Bennett, J. M., Smith, J. V. and Angell, C. L. (1969) Positions of cations and molecules in zeolites with the faujasite-type framework. V. La-exchanged type X zeolite at 25, 425 and 735°C: *Mater. Res. Bull.* **4**, 77-86.

Dent, L. S. and Smith, J. V. (1958) Crystal structure of chabazite, a molecular sieve: *Nature* **181**, 1794-1796.

Eulenberger, G. R., Shoemaker, D. P. and Keil, J. K. (1967) The crystal structures of hydrated and dehydrated synthetic zeolites with faujasite aluminosilicate frameworks. I. The dehydrated sodium, potassium, and silver forms: *J. Phys. Chem.* **71**, 1812-1819.

Frondel, C. (1962) The system of mineralogy (Dana's): Vol. III. *Silica Minerals*, Wiley, New York.

Loschi Ghittoni, A. G. and Passaglia, E. (1974) Microanalysis of zeolites: *Studia Geologica* 7, 161-165.

Mortier, W. J. and Bosmans, H. J. (1971) Location of univalent cations in synthetic zeolites of the Y and X type with varying silicon to aluminium ratio. I. Hydrated potassium exchanged forms: *J. Phys. Chem.* **75**, 3327-3333.

Nowacki, W., Koyama, H. and Mladeck, M. H. (1958) Über die kristall structure des zeolithes chabasit: *Experientia* **14**, 396.

Olson, D. H. (1968) Crystal structure of the zeolite nickel faujasite: *J. Phys. Chem.* 72, 4366-4373.

Olson, D. H. (1970) A reinvestigation of the crystal structure of the zeolite hydrated NaX: *J. Phys. Chem.* **74**, 2758-2764.

Olson, D. H. and Sherry, H. S. (1968) An X-ray study of strontium-sodium exchange in Linde X. An example of a two-phase zeolite system: *J. Phys. Chem.* **72**, 4095-4104.

Passaglia, E. (1970) The crystal chemistry of chabazites: *Amer. Mineral.* **55**, 1278-1301.

Pluth, J. J. and Smith, J. V. (1972) Positions of cations and molecules in zeolites with the faujasite-type framework. VII. Dehydrated Ca-exchanged X: *Mater. Res. Bull.* **7**, 1311-1322.

Pluth, J. J. and Smith, J. V. (1973) Positions of cations and molecules in zeolites with the faujasite-type framework. VIII. Hydrated Ca-exchanged X: *Mater. Res. Bull.* **8**, 459-468.

Sheppard, R. A. and Gude, A. J., 3rd (1970) Calcic siliceous chabazite from the John Day Formation, Grant County, Oregon: *U.S. Geol. Survey Prof. Paper* **700-D**, 176-180.

Smith, J. V. (1962) Crystal structures with a chabazite framework. I. Dehydrated Ca-chabazite: *Acta Crystallogr.* **15**, 835-845.

Smith, J. V. (1964) Crystal structures with a chabazite framework. Part V. Interpretation of physicochemical properties in terms of the crystal structure: *J. Chem. Soc.*, pp. 3759-3770.

Smith, J. V., Rinaldi, F. and Dent Glasser, L. S. (1963) Crystal structures with a chabazite framework. II. Hydrated Ca-chabazite at room temperature: *Acta Crystallogr.* **16**, 45-53.

Smith, J. V., Knowles, C. R. and Rinaldi, F. (1964) Crystal structure with a chabazite framework. III. Hydrated Ca-chabazite at +20° and −150°C: *Acta Crystallogr.* **17**, 374-384.

Smith, J. V., Bennet, J. M. and Flanigen, E. M. (1967) Dehydrated lanthanum-exchanged type Y zeolite: *Nature* **215**, 241-244.

POSITIONS OF CATIONS AND MOLECULES IN ZEOLITES WITH THE MORDENITE-TYPE FRAMEWORK. IV. DEHYDRATED AND REHYDRATED K-EXCHANGED "PTILOLITE"

W. M. MORTIER,* J. J. PLUTH and J. V. SMITH

Department of the Geophysical Sciences, University of Chicago, Chicago, Illinois, 60637, U.S.A.

Abstract – The structures of hydrated and dehydrated K-exchanged ptilolite $K_8Al_8Si_{40}O_{96}$ were determined at room temperature by least-squares refinement of 3D X-ray diffraction data. Cell dimensions are: *a* 18.167(±0.009), *b* 20.611(±0.015), *c* 7.59(±0.006); and *a* 18.030(±0.007), *b* 20.406(±0.010), *c* 7.457(±0.005) Å, respectively. The hydrated structure has 3.0K in site IV, 1.9K in site I, and water (and perhaps some intermixed K) in five other sites obeying *Cmcm*. The dehydrated structure has lowered symmetry, *Pbcn*, probably because 3.3K in site II distort the framework by bonding to 0(9). Site IV contains 3K and site VI 0.9K.

Comparison with hydrated-Na, dehydrated- and hydrated-Ca, and dehydrated-H "ptilolites" shows only small changes in the dimensions of the cylindrical channels. Site occupancies vary widely, probably in response to minimization of electrostatic potential. Al atoms probably occur in all types of tetrahedral sites, and the dispersed distribution of exchangeable cations may be related. The framework geometry depends partly on molecular–orbital overlap of framework atoms and partly on bonding effects from the exchangeable cations.

INTRODUCTION

Zeolites of the mordenite group have large channels, are stable to high temperature, and are resistant to acids (Breck, 1974). Mordenite is one of the eight natural zeolites abundant enough for commercial exploitation, and synthetic varieties have been used for two decades. All structure determinations have been made on the only known specimen coarse enough for detailed single-crystal X-ray analysis – a ptilolite from Challis Valley, Idaho, in the Linde mineral collection. The aluminosilicate framework (Meier, 1961) obeys space group *Cmcm* and contains columns of linked 5-rings joined to give cylindrical channels bounded by near-circular 12-rings with "free diameter" near 7 Å (stereo-plot in Meier and Olson, 1971). Zig-zag interconnected passages bounded by near-circular 8-rings and flattened 10-rings cross-link the cylindrical channels. The columns of 5-rings can be cross-linked in four ordered ways (Sherman and Bennett, 1973) which preserve the cylindrical channels but change the interconnections. Although most mordenites have the *Cmcm* stacking type, electron and X-ray diffraction studies showed that the *Immm*, *Cmmm*, and *Imcm* types contribute to some specimens (Sherman and Bennett, 1973). The structure analysis of dehydrated hydrogen-ptilolite revealed weak electron-density peaks consistent with 2% stacking other than the Cmcm type (Mortier *et al.*, 1975a); probably all crystals from the Challis Valley specimens have similar stacking changes. Although some mordenites (*large port* type) have sorption properties consistent with 12-rings, others (*small port* type) have sorption consistent only with 8-rings (Sand, 1968). So far no structural explanation has proved convincing.

After reporting new structure analyses for dehydrated and rehydrated K-exchanged ptilolite, we compare them with other structure analyses (hydrated Na-exchanged ptilolite, Meier, 1961; refinement by Gramlich, 1971; dehydrated Ca-exchanged ptilolite, Mortier *et al.*, 1975b; dehydrated H-ptilolite, Mortier *et al.*, 1975a; rehydrated Ca-exchanged ptilolite, Mortier *et al.*, 1976). For brevity the types are designated dK, hK, hNa, dCa, dH, and hCa, respectively.

EXPERIMENTAL

Single crystals of natural ptilolite were ion-exchanged for 11 days in 1 M KCl solution at 90°C, yielding the composition $K_{8.0}Al_{8.0}Si_{40.0}O_{96}\cdot nH_2O$ from electron

* Now at Centrum voor Oppervlaktescheikunde en Colloidale Scheikunde, K.U.L., B-3030 Heverlee, Belgium.

microprobe analysis (solid-state detector, Reed-Ware program, feldspar standards).

A needle 0.08 X 0.08 X 0.5 mm was dehydrated in a silica capillary at 10^{-3} Pa for 2 days at 100°C and 26 hours at 510°C before sealing. During structure refinement it became obvious that the capillary had leaked and the crystal was really hK. A similar crystal was successfully converted to dK under the same conditions.

Systematic absences for hK indicated the space group *Cmcm*, but those for dK indicated *Pbcn*, a sub-group of *Cmcm*. Cell dimensions obtained at room temperature from fifteen Friedel pairs with MoK_α are: hK = *a* 18.167(9), *b* 20.611(15), *c* 7.529(6) Å; dK = *a* 18.030(7), *b* 20.406(10), *c* 7.457(5).

Data were collected on a Picker FACS-1 diffractometer at room temperature using $\theta-2\theta$ scans (2°/min) to $\sin\theta/\lambda = 0.65$. For hK four equivalent sets totalling 6910 diffractions and for dK two equivalent sets of 6879 diffractions were averaged, and all intensities with $F \geq 3\sigma_F$ (1533 for hK and 2407 for dK) were used in the refinement. No absorption correction was applied to either data set ($\mu = 10.4\ cm^{-1}$).

Extra-framework electron density was located and the structures were refined by alternating least-squares and Fourier methods. For dK in *Pbcn* the framework parameters for hK had to be adjusted by the DLS program (Villiger, 1969) before the least-squares refinement would converge. The stacking faults were treated as described by Mortier *et al.* (1975b), but for dK only the populations of T(3) and T(4) were refined with the population of T(3)′ and T(4)′ set to (1–p). Cations and water molecules were assigned positions on the basis of observed electron density, coordination, and chemical composition. Final values of R, wR, and S for hK and dK are 0.061, 0.052, 4.1 and 0.076, 0.046, 2.3, respectively (see Mortier *et al.*, 1975a, for definitions and details of structure refinements).

Tables 1, 2, and 3 list population and positional parameters, anisotropic thermal parameters, and inter-

TABLE 1. Atomic position and population parameters

(a) *Hydrated K-ptilolite (hK)*

Type		Population	x/a	y/b	z/c	B
T(1)	(16h)	16	0.19850(7)	0.42742(6)	0.54128(15)	–
T(2)	(16h)	16	0.19656(7)	0.19071(6)	0.54487(15)	–
T(3)	(8g)	7.61(5)	0.08755(10)	0.38181(8)	¼	–
T(4)	(8g)	7.76(6)	0.08671(11)	0.22693(8)	¼	–
O(1)	(16h)	16	0.1240(2)	0.4148(2)	0.4295(4)	–
O(2)	(16h)	16	0.1225(2)	0.1945(2)	0.4271(4)	–
O(3)	(16h)	16	0.2631(2)	0.3776(2)	0.4890(4)	–
O(4)	(8g)	7.69	0.1001(3)	0.3052(3)	¼	–
O(5)	(8g)	8	0.1690(3)	0.1946(3)	¾	–
O(6)	(8g)	8	0.1774(3)	0.4210(2)	¾	–
O(7)	(8e)	8	0.2297(2)	½	½	–
O(8)	(8d)	8	¼	¼	½	–
O(9)	(4c)	3.85	0	0.4011(3)	¼	–
O(10)	(4c)	3.85	0	0.2064(4)	¼	–
K(I)	(4b)	1.92(3)	0	½	0	–
W(II)*	(8f)	4.5(1)	0	0.4321(6)	0.8076(14)	–
W(III)*	(8g)	4.7(2)	0.0361(15)	0.3062(12)	¾	23(2)
K(IV)	(8f)	3.02(5)	0	0.1825(5)	0.788(3)	–
W(V)*	(8f)	8.4(2)	0	0.0884(7)	0.986(2)	–
S(VI)*	(8g)	5.8(2)	0.1238(12)	0.0258(10)	¾	20(1)
S(VIII)*	(8e)	6.5(3)	0.1057(13)	0	0	25(2)
T(3)′	(8g)	0.20(3)	0.0876	0.3818	¾	1.7
T(4)′	(8g)	0.35(4)	0.0867	0.2269	¾	1.7
O(4)′	(8g)	0.28	0.1001	0.3052	¾	3.2
0(9)′	(4c)	0.14	0	0.4011	¾	3.2
O(10)′	(4c)	0.14	0	0.2064	¾	3.2

TABLE 1 (*cont.*)

(b) *Dehydrated K-ptilolite (dK)*

T(1)	(8d)	8	0.18981(9)	0.42347(7)	0.57996(19)	–
T(1A)†	(8d)	8	0.20164(9)	0.43084(7)	0.99512(20)	–
T(2)	(8d)	8	0.19200(9)	0.18419(7)	0.56823(19)	–
T(2A)†	(8d)	8	0.20321(8)	0.19549(7)	0.98235(17)	–
T(3)	(8d)	7.72(4)	0.08488(8)	0.38184(7)	0.28388(20)	–
T(4)	(8d)	7.81(4)	0.08472(8)	0.22444(7)	0.27668(20)	–
O(1)	(8d)	8	0.1102(2)	0.4137(2)	0.4775(5)	–
O(1A)†	(8d)	8	0.1346(3)	0.4122(3)	0.1262(6)	–
O(2)	(8d)	8	0.1136(2)	0.1921(2)	0.4646(5)	–
O(2A)†	(8d)	8	0.1322(2)	0.1970(2)	0.1105(5)	–
O(3)	(8d)	8	0.2475(2)	0.3680(2)	0.5233(6)	–
O(3A)†	(8d)	8	0.2738(2)	0.3865(2)	0.0290(6)	–
O(4)	(8d)	7.76	0.0896(2)	0.3034(2)	0.2885(6)	–
O(5)	(8d)	8	0.1707(2)	0.1913(2)	0.7795(5)	–
O(6)	(8d)	8	0.1702(2)	0.4212(2)	0.7930(5)	–
O(7)	(8d)	8	0.2748(2)	0.4938(2)	0.5265(5)	–
O(8)	(8d)	8	0.2488(2)	0.2393(2)	0.5031(6)	–
O(9)	(4c)	3.88	0	0.4078(3)	¼	–
O(10)	(4c)	3.88	0	0.1999(3)	¼	–
K(II)	(8d)	3.34(3)	0.0125(4)	0.4517(2)	0.8090(9)	–
K(IV)	(8d)	3.04(3)	0.0128(4)	0.1823(3)	0.8007(9)	–
K(VI)	(8d)	0.91(3)	0.1606(7)	0.0238(10)	0.7906(20)	–
T(3)′	(8d)	0.24	0.0849	0.3818	0.7839	1.3
T(4)′	(8d)	0.24	0.0847	0.2244	0.7767	1.3
O(4)′	(8d)	0.24	0.0896	0.3034	0.7885	3.8
O(9)′	(4c)	0.12	0	0.4078	¾	3.2
O(10)′	(4c)	0.12	0	0.1999	¾	3.2

*O^- scattering factor used for these sites. W means H_2O, S means unspecified. Primed atoms result from stacking change.

†Atom pairs such as T(1) and T(1A) are equivalent in *Cmcm* and are related by mirrors at $z = ¼, ¾$.

TABLE 2. Anisotropic thermal parameters

(a) *Hydrated K-ptilolite (hK)*

Atom	β_{11}	β_{22}	β_{33}	β_{12}	β_{13}	β_{23}
T(1)	112(4)	66(3)	60(2)	5(3)	–28(8)	–10(7)
T(2)	120(4)	80(3)	62(2)	13(3)	–18(8)	8(7)
T(3)	58(6)	59(4)	39(3)	–6(4)	0	0
T(4)	77(6)	62(5)	55(4)	4(4)	0	0
O(1)	174(12)	228(11)	97(7)	–11(10)	–99(23)	–104(22)
O(2)	193(12)	206(10)	112(7)	–8(11)	–108(23)	117(25)
O(3)	245(14)	126(9)	161(9)	62(9)	82(28)	–42(22)
O(4)	212(18)	80(11)	206(12)	3(14)	0	0
O(5)	313(20)	230(16)	78(9)	70(17)	0	0
O(6)	243(19)	198(15)	64(8)	–33(14)	0	0
O(7)	153(16)	91(11)	130(11)	0	0	38(26)

TABLE 2 (*cont.*)

O(8)	222(20)	173(14)	181(13)	−80(14)	−157(41)	190(33)
O(9)	74(21)	131(18)	91(13)	0	0	0
O(10)	72(20)	149(21)	137(15)	0	0	0
K(I)	139(16)	223(17)	195(14)	0	0	459(37)
W(II)	364(78)	436(78)	241(85)	0	0	502(169)
K(IV)	276(20)	888(48)	481(70)	0	0	−1014(159)
W(V)	1493(111)	933(79)	1590(90)	0	0	−2127(224)

(b) *Dehydrated K-ptilolite (dK)*

T(1)	144(5)	83(4)	68(3)	11(4)	−84(10)	−47(8)
T(1A)	160(5)	98(3)	63(3)	2(4)	13(10)	27(8)
T(2)	177(5)	90(4)	68(3)	21(4)	−53(10)	−21(8)
T(2A)	137(4)	91(3)	62(3)	23(4)	21(9)	14(9)
T(3)	97(5)	78(4)	66(3)	−12(3)	−42(10)	2(8)
T(4)	77(5)	77(3)	72(3)	5(3)	−27(10)	1(8)
O(1)	214(15)	250(13)	120(9)	−27(11)	−102(29)	−123(29)
O(1A)	474(24)	458(21)	136(10)	−206(18)	382(40)	−99(38)
O(2)	266(15)	217(13)	139(9)	−24(12)	−166(29)	104(31)
O(2A)	289(17)	312(15)	152(9)	−22(15)	360(32)	−146(35)
O(3)	316(20)	190(12)	215(11)	94(13)	9(39)	−189(32)
O(3A)	335(19)	119(11)	241(12)	71(11)	−284(38)	2(30)
O(4)	198(14)	102(9)	343(12)	−7(11)	−349(36)	−26(34)
O(5)	305(16)	257(13)	88(7)	16(13)	−60(29)	−75(29)
O(6)	302(17)	186(11)	98(8)	−11(11)	−104(31)	12(25)
O(7)	171(13)	133(10)	164(9)	23(10)	41(31)	23(25)
O(8)	271(17)	133(11)	198(11)	−34(11)	34(38)	66(29)
O(9)	166(21)	117(15)	311(19)	0	−423(51)	0
O(10)	110(17)	117(14)	247(15)	0	−103(41)	0
K(II)	359(32)	384(14)	593(34)	17(16)	533(73)	876(51)
K(IV)	286(29)	613(23)	339(27)	−46(19)	338(51)	−504(58)
K(VI)	142(5)	862(91)	206(39)	−19(49)	−131(104)	233(147)

All numbers should be divided by 10^5 except for β_{33} which should be divided by 10^4. Anisotropic temperature factor defined as

$$\exp - \sum_{i=1}^{3} \sum_{j=1}^{3} \beta_{ij} h_i h_j.$$

TABLE 3. Interatomic distances (Å) and angles (°)

(a) *Hydrated K-ptilolite (hK)*

Framework

T(1)–O(1)	1.615(4)	O(1)–T(1)–O(3)	112.4(2)	T(2)–O(2)	1.614(4)	O(2)–T(2)–O(3)	109.6(2)
–O(3)	1.608(4)	–O(6)	107.1(2)	–O(3)	1.608(4)	–O(5)	105.2(2)
–O(6)	1.623(2)	–O(7)	109.9(2)	–O(5)	1.626(2)	–O(8)	110.8(2)
–O(7)	1.630(2)	O(3)–T(1)–O(6)	110.9(2)	–O(8)	1.597(2)	O(3)–T(2)–O(5)	109.5(2)
Mean	1.619	–O(7)	106.5(2)	Mean	1.611	–O(8)	111.1(2)
		O(6)–T(1)–O(7)	109.9(2)			O(5)–T(2)–O(8)	110.5(2)

TABLE 3 (*cont.*)

T(3)–2 O(1)	1.651(3)	O(1)–T(3)–O(1)	109.9(3)	T(4)–2 O(2)	1.627(3)	O(2)–T(4)–O(2)	110.1(3)
– O(4)	1.595(6) 2	–O(4)	110.5(2)	– O(4)	1.632(6) 2	–O(4)	110.3(2)
– O(9)	1.639(3) 2	–O(9)	106.8(2)	– O(10)	1.631(3) 2	–O(10)	106.2(2)
Mean	1.634	O(4)–T(3)–O(9)	112.3(3)	Mean	1.628	O(4)–T(4)–O(10)	113.6(3)

T(1)–O(1)–T(3)	146.0(2)	T(1)–O(6)–T(1)	151.1(4)
T(2)–O(2)–T(4)	143.4(2)	T(1)–O(7)–T(1)	139.2(3)
T(1)–O(3)–T(2)	158.5(3)	T(2)–O(8)–T(2)	180
T(3)–O(4)–T(4)	163.1(4)	T(3)–O(9)–T(3)	151.9(5)
T(2)–O(5)–T(2)	143.6(4)	T(4)–O(10)–T(4)	149.9(5)

Cation-Water

K(I)–2 W(II)	2.015(13)*	W(II)–W(II)	0.87(2)*	W(V)–K(IV)	2.45(2)
–2 W(II)	2.706(11)	–K(I)	2.706(11)	–S(VIII)	2.65(2)
–2 O(9)	2.774(5)	–W(III)	2.711(3)	–K(IV)	2.83(2)
–4 O(1)	2.906(4)	–2 O(1)	3.020(8)	–O(10)	3.142(13)
		–2 O(6)	3.260(6)	–W(VI)	3.14(2)
K(IV)–K(IV)	0.58(4)*	– O(9)	3.391(11)	–2 O(2)	3.187(10)
–W(V)	2.45(2)	– O(9)	3.466(15)		
–W(III)	2.64(3)			S(VI)–2 S(VIII)	1.983(8)*
–W(V)	2.83(2)	W(III)–W(III)	1.31(5)*	–2 W(V)	3.14(2)
–2 O(5)	3.093(6)	–K(IV)	2.64(3)	–2 O(7)	3.30(2)
–2 O(2)	3.099(15)	–W(II)	2.71(3)	–2 O(3)	3.38 (2)
		–O(5)	3.33(3)		
		–O(6)	3.49(3)	S(VIII)–2 W(V)	2.65(2)
				– O(7)	2.99(2)
				–2 O(3)	3.47(2)

(b) *Dehydrated K-ptilolite (dK)*

Framework

T(1)–O(1)	1.638(4)	O(1)–T(1)–O(3)	111.2(2)	T(1A)–O(1A)	1.600(5)	O(1A)–T(1A)–O(3A)	112.5(3)
–O(3)	1.594(4)	–O(6)	105.1(2)	–O(3A)	1.604(4)	–O(6)	105.9(2)
–O(6)	1.628(4)	–O(7)	109.6(2)	–O(6)	1.623(4)	–O(7)	109.8(2)
–O(7)	1.622(4)	O(3)–T(1)–O(6)	112.4(2)	–O(7)	1.612(4)	O(3A)–T(1A)–O(6)	111.2(2)
Mean	1.621	–O(7)	107.7(2)	Mean	1.610	–O(7)	107.6(2)
		O(6)–T(1)–O(7)	110.7(2)			O(6)–T(1A)–O(7)	109.8(2)
T(2)–O(2)	1.620(4)	O(2)–T(2)–O(3A)	109.9(2)	T(2A)–O(2A)	1.599(4)	O(2A)–T(2A)–O(3)	110.2(2)
–O(3A)	1.596(4)	–O(5)	104.3(2)	–O(3)	1.601(4)	–O(5)	105.6(2)
–O(5)	1.628(4)	–O(8)	110.2(2)	–O(5)	1.624(4)	–O(8)	111.1(2)
–O(8)	1.597(4)	O(3A)–T(2)–O(5)	110.4(2)	–O(8)	1.594(4)	O(3)–T(2A)–O(5)	109.6(2)
Mean	1.610	– O(8)	109.4(2)	Mean	1.605	–O(8)	110.9(2)
		O(5)–T(2)–O(8)	112.5(2)			O(5)–T(2A)–O(8)	109.3(2)
T(3)–O(1)	1.648(4)	O(1)–T(3)–O(1A)	109.6(2)	T(4)–O(2)	1.634(4)	O(2)–T(4)–O(2A)	110.6(2)
–O(1A)	1.604(4)	–O(4)	111.1(2)	–O(2A)	1.607(4)	–O(4)	109.8(2)
–O(4)	1.604(4)	–O(9)	105.4(2)	–O(4)	1.616(4)	–O(10)	106.3(2)
–O(9)	1.639(2)	O(1A)–T(3)–O(4)	111.9(3)	–O(10)	1.620(2)	O(2A)–T(4)–O(4)	111.2(2)
Mean	1.624	– O(9)	106.5(2)	Mean	1.619	–O(10)	107.4(2)
		O(4)–T(3)–O(9)	112.0(2)			O(4)–T(4)–O(10)	111.5(2)

TABLE 3 (*cont.*)

T(1)–O(1)–T(3)	134.3(3)	T(2)–O(5)–T(2A)	145.2(3)
T(1A)–O(1A)–T(3)	164.2(4)	T(1)–O(6)–T(1A)	145.8(3)
T(2)–O(2)–T(4)	136.7(3)	T(1)–O(7)–T(1A)	140.9(3)
T(2A)–O(2A)–T(4)	153.5(3)	T(2)–O(8)–T(2A)	164.8(3)
T(1)–O(3)–T(2A)	170.9(3)	T(3)–O(9)–T(3)	142.3(4)
T(2)–O(3A)–T(1A)	148.5(3)	T(4)–O(10)–T(4)	144.0(4)
T(3)–O(4)–T(4)	172.5(3)		

Cation

K(II)–K(II)	0.989(12)*	K(IV)–K(IV)	0.885(11)*	K(VI)–O(7)	2.728(14)
–O(1)	2.834(9)	–O(5)	2.858(8)	–O(7)	2.919(16)
–O(9)	2.908(7)	–O(2)	2.880(9)	–O(3A)	2.925(17)
–O(6)	2.912(9)	–O(2)	3.102(9)	–O(3)	3.260(18)
–O(1)	3.133(10)	–O(2A)	3.172(7)	–O(5)	3.423(21)
–O(1A)	3.330(8)	–O(5)	3.367(7)		
–O(9)	3.416(7)	–O(10)	3.378(7)		
–O(6)	3.437(8)				

*These abnormal distances result only for simultaneous occupancy of adjacent sites.

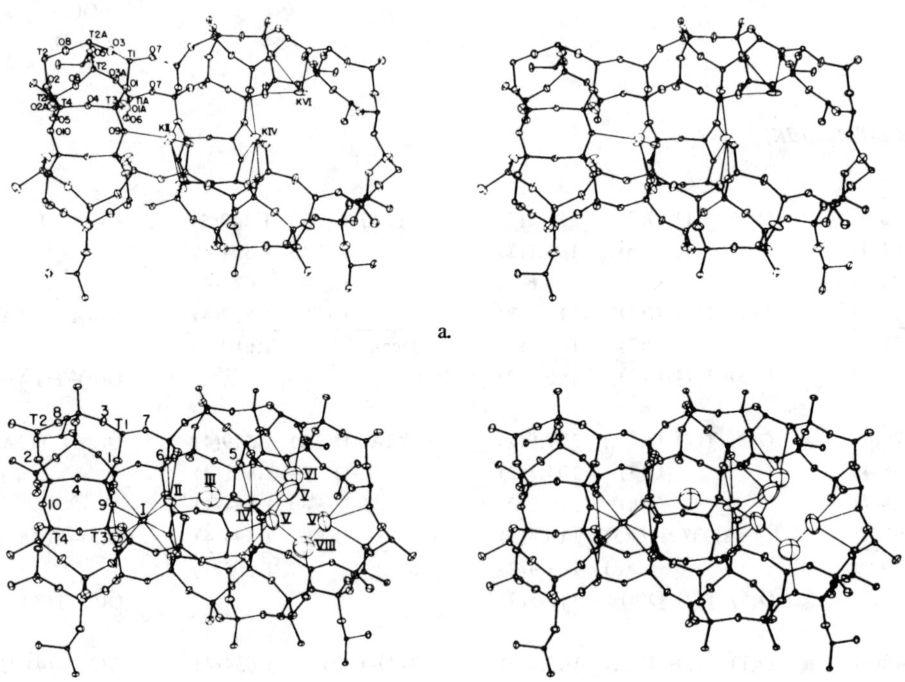

Fig. 1. Stereoplots of dehydrated, potassium-exchanged ptilolite (a) and hydrated potassium-exchanged ptilolite (b) with ellipsoids at 30% probability level. Viewed down *c*-axis with *b* to right. (a) Only K–O distances less than 3.4 A are shown.

atomic distances and angles. Observed and calculated structure amplitudes were deposited as NAPS-02652 with Microfiche Publications.*

DISCUSSION OF dK AND hK FRAMEWORK

Figures 1(a,b) and 2(a,b) are stereoplots of the framework and coordination of cations and water molecules. The remarkable feature of dK is the puckering of the framework associated with reduction of symmetry from *Cmcm* to *Pbcn*. Such reduction involves loss of the following symmetry elements: *m* perpendicular to the *a*-axis, $n \perp b$, $m \perp c$, 2 parallel to *a*, 2_1 to *b*, 2_1 to *c*, and one set of centers of symmetry. The 16-fold general site (h) in *Cmcm* splits up into two 8-fold general sites (d) in *Pbcn* while each 8-fold special site (d or g) goes into an 8-fold general site by reduction of point symmetry. During dehydration the T–O–T framework angles change by up to 18° with a mean change of 8°. Most of the change results from a zig-zag in the framework developed as atoms move to left and right of the positions constrained by *Cmcm* symmetry. For the pseudo-symmetric average framework of dK, the mean angular changes from hK are 5°.

In dK, 3.3K occupy 8-fold site II, 3.0K occupy 8-fold site IV, and 0.9K occupy 8-fold site VI. No cations were assigned to site II for other exchanged forms of mordenite, but about four water molecules were placed there in the hydrated varieties hNa, hCa, and hK. In hNa and hCa each center of a set of four 8-rings lying perpendicular to *b* is occupied by a water molecule, but in hK each position is split into two positions 1.0 Å apart. Only one out of each pair can be occupied, and the 3.3K account for 82% of full occupancy. Displacement of 0.5 Å away from the center of the ring allows the K ion to bond to 0(9) at 2.91 Å while retaining bonding to six oxygens from the ring at 2.83–3.44 Å. (Note that all distances in disordered structures are subject to an unknown error because of automatic averaging.) Without the displacement, the K ion would have a strongly waisted coordination, whereas with the displacement the bonding becomes more isotropic. Because the latter is favorable electrostatically, it provides the simplest explanation for the reduction of symmetry. The framework cannot have an inherent tendency to lower symmetry because dK and the hydrated varieties all obey the higher symmetry.

The 3.0K(IV) occupy an 8-fold site 0.45 Å from the center of a 4-fold set of 8-rings perpendicular to *b*. Again, spatial restriction limits the maximum occupancy to 4. In this site, the only possible coordination is to oxygen atoms from the 8-ring, six of which lie at 2.86–3.38 Å. The 0.9K(VI) lie in an 8-fold site displaced over 1 Å from the center of a 6-ring sub-perpendicular to *a*. The bonding is strongly one-sided to four oxygens at 2.73–3.26 Å and perhaps to a fifth oxygen at 3.42 Å.

The total population of 7.29 ± 0.09 is considerably less than the value of 8.0 from electron microprobe analysis, but it must be emphasized that small amounts of K in other sites may be undetected and that refinement of disordered structures is rather uncertain. Furthermore, the electron microprobe analysis may be biased.

The large channel system is almost free from cations but the interconnecting channels are almost completely blocked by K(II) and K(IV). The cations in sites IV and especially VI provide opportunity for attachment to sorbed molecules. Because sites II and IV could accommodate all the K ions, the only simple explanation of the occupancy of site VI is to allow bonding to undersaturated oxygens from Al-centered tetrahedra. Indeed the same explanation would be valid for dCa.

Turning now to hK, the interpretation is very uncertain. Some electron-density peaks may result from combined occupancy of K and H_2O because the K ion and the H_2O molecule should have similar distances to framework oxygens. Although sites II, III and IV are formally 8-fold in Table 1, they actually each occur as four pairs of sites too close for simultaneous occupancy (Table 3). Furthermore, sites VI and VIII cannot be occupied simultaneously unless major displacements raise their separation from 1.98 to ~2.7 Å. The four sites I occur 3.7 Å apart and there should be no electrostatic restriction on simultaneous occupancy though alternate occupancy may be preferred to minimize cation–cation repulsion. The displacement ellipsoids for all extra-framework sites are very large, especially for sites III, V, VI and VIII. Bearing all these factors in mind, one can attempt to place K and H_2O into the sites on the assumption that only one species goes into a site. The placement utilizes the greater scattering power of K over H_2O (approximately 2.5) and the likelihood of a stronger coordination around K than around H_2O.

Site IV must contain K because it is effectively 4-fold and the observed electron density corresponds to either 3.0K or ~$7H_2O$. This assignment is confirmed by the strong coordination to species in site V at 2.45 Å, site III

*** See NAPS document No. 02652 for 9 pages of supplementary material. Order from ASIS/NAPS c/o Microfiche Publications, 440 Park Avenue South, New York, New York 10016. Remit in advance for each NAPS accession number. Make checks payable to Microfiche Publications. Photocopies are $5.00. Microfiche are $3.00. Outside of the United States and Canada, postage is $2.00 for a photocopy or $1.00 for a fiche.**

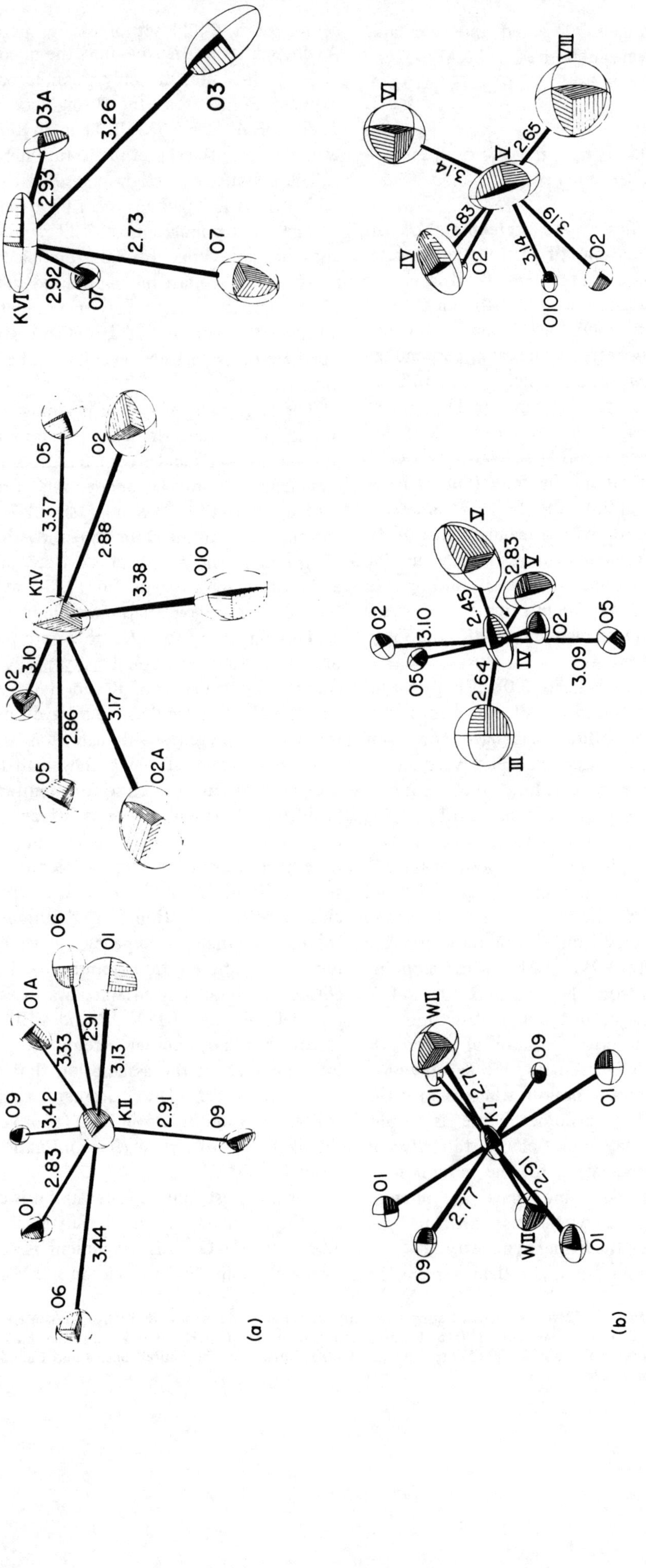

Fig. 2. Coordination of extra-framework species of dehydrated (a) and hydrated (b) potassium exchanged ptilolite with ellipsoids at 50% probability level. (b) Only the three best-defined sites are shown.

at 2.64, site V′ at 2.83, and to four framework oxygens at 3.09 and 3.10 Å. Because K occupies site IV, the sites III and V should be occupied by H_2O which is consistent with the poorer coordination at these sites.

The electron density at site I corresponds to either 1.9K or ~5H_2O and that at site II to either 4.5H_2O or ~1.8K. Both sites are effectively 4-fold, and assignment of only one species leads either to deficiency of K or excess of H_2O. The distances are no help in assignment of species, but the coordination at site I is more suitable for K than H_2O. In the tables, K is assigned to I and H_2O to II, but a mixed occupancy may be better. Sites VI and VIII lie in the large channels and are poorly defined. Perhaps both K and H_2O occupy them.

The inconclusive assignment is typical of hydrated zeolites with large univalent cations, as has been demonstrated by many structure analyses. Diffusion should be very easy, and the X-ray data must provide merely a time and space average of many structural complexions.

COMPARISON OF MORDENITE STRUCTURES

Table 4 compares the site occupancies in the six modifications of ptilolite. Partial occupancy of several sites is a common feature, even though full occupancy of fewer sites is mathematically possible. Almost certainly, the partial occupancy can be attributed to minimization of electrostatic potential, probably in response to distribution of Al over all the tetrahedral sites. Detailed analysis of each individual structure shows that the cation distribution and bond lengths are reasonable (e.g., in dCa), but prediction of cation occupancy from first principles is not possible. Indeed the structure of dK came as a surprise to the authors even though it makes sense in retrospect. Further analyses of other exchanged forms are needed: for example, will dSr have the same structure as dCa, and will dRb resemble dK? How can the structure accommodate the very small ion Mg or the highly charged REE ions? The following general comments are highly tentative but point the way to detailed understanding of mordenite modifications.

Activation energies for diffusion (Breck, 1974) are 44, 36, and 30 kJ/mole for Ca, Na, and K, respectively, in mordenite. These correlate with the ionic potential and are consistent with the proposed site occupancies in Table 4, where Ca and Na maintain hydration complexes in the large channel, but K probably does not.

Sites I and II are close, as are III and IV. The formal charge distribution for dCa and dK is fairly similar when these pairs are considered:

	I	II	I + II	III	IV	III + IV	VI
Ca	3.3		3.3	1.2	0.9	2.1	1.3
K		3.3	3.3		3.0	3.0	0.9

Whether this results from maximization of cation separation, complying with Al substitution, or both, is unknown. Earlier the occupancy of site II by K rather than site I as in dCa was explained by the bonding to 0(9). Another possible effect is increase of cation separation as the 1.7Ca in I are replaced by the 3.3K in

TABLE 4. Comparison of site occupancies for ptilolite (mordenite)

	Type	hNa†	hCa	hK†	dCa	dK	dH
I	4b	4Na	1.70Ca	1.92K	1.66Ca		0.65Na
II	4c	4W	3.95W				
	8f*			4.5W			
	8d*					3.34K	
III	8g*	4W		4.7W	0.62Ca		
	4c		4.7W				
IV	8f*			3.02K	0.45Ca		
	4c	4 ?	3.4W				
	8d*					3.04K	
V	16h		10.3W				
	8f	8W		8.4W			
VI	8g	8W	8.7W	5.8W,K	0.63Ca		
	8d					0.91K	
VII	4c	1Na	1.68Ca				
VIII	8c			6.5W,K			

*These sites are effectively 4-fold because of close approach if fully occupied.

†Cation and water assignments are uncertain and mixed occupancy may be important.

II. Because sites II and IV are highly occupied in dK it is not surprising that III is not occupied.

Table 5 compares the dimensions straddling the large rings. Changes are minor except for the 8-rings in dK which become strongly distorted, though their areas are quite similar to those in other varieties. The dH variety has the most open-channel system. Table 6 compares the T-O-T and O-T-O angles. Major zig-zagging results from the reduced symmetry in dK, but the mean angles are not greatly dissimilar from those in other varieties. Because the charge distribution in dH and the hydrated varieties should be fairly uniform, it is not surprising that all these varieties have similar framework angles. The largest deviations in dCa from this "relaxed" configuration can be explained by bonding to the highly polarizing Ca cation; e.g., the Ca(I) cation is strongly bonded to 0(9) causing a strong reduction in the T(3)-0(9)-T(3) angle and a corresponding increase in the adjacent T(3)-0(4)-T(3) angle.

A strong correlation exists between T-O-T angles and observed T-O distances in all mordenite varieties. The deviation of an individual T-O distance from the mean of four distances for each tetrahedron (denoted Δ(T-O)) is linearly related to secant (T-O-T) in accordance with prediction from semi-empirical molecular-orbital calculations for feldspar frameworks by Gibbs *et al.* (1974). Least-squares fitting to the line $\sec(\text{T-O-T}) = m\Delta(\text{T-O}) + b$ gave the following:

	dH	dK	hK
m	−6.67	−7.78	−4.63
b	−1.149	−1.188	−1.159
correlation coefficient	0.94	0.83	0.68

The slope for dK is steeper than for other mordenites, but not as steep as for feldspars.

Calculations identical to those performed by Mortier *et al.* (1975a, 1976) using average T-O-T angles (T(1) 148.7°, T(2) 156.4, T(3) 151.8, T(4) 150.0 for hK) (T(1) 148.0, T(1A) 149.9, T(2) 148.8, T(2A) 158.6, T(3) 153.3, T(4) 151.7 for dK) yielded adjusted T-O distances of 1.612, 1.621, 1.634, 1.624 Å for hK and 1.615, 1.608, 1.606, 1.612, 1.626, 1.619 for dK. Using these T-O distances and mean T-O distances of 1.622 for hK and 1.614 for dK, the predicted fractional Al contents at the T sites are 0.10, 0.16, 0.25, 0.18 and 0.17, 0.13, 0.11, 0.15, 0.25, 0.20, respectively. These Al distributions are close to values previously calculated, and indicate partial Si,Al ordering.

ACKNOWLEDGMENTS

We thank the following: NSF-GH33636Al(MRL), ACS-Petroleum Research Fund 6276AC2 and the Union Carbide Corporation for financial aid. W. J. M. thanks Belgisch National Fonds voor Wetenschappelijk Onderzoek for a research grant as Aangesteld Navorser.

REFERENCES

Breck, D. W. (1974) *Zeolite Molecular Sieves:* Wiley, New York.

Gibbs, G. V., Louisnathan, S. J., Ribbe, P. H. and Phillips, M. W. (1974) Semi-empirical molecular orbital calculations for the atoms of the tetrahedral framework in anorthite, low albite, maximum microcline and reedmergnerite. In: *The Feldspars*, Mackenzie, W. S. and Zussman, J., eds., Manchester University Press, Manchester.

Gramlich, V. (1971) Untersuchung und Verfeinerung pseudo-symmetrischer Structuren: Diss. No. 4633, ETH, Zürich.

TABLE 5. Dimensions across channels (Å)

	dH	dCa	dK	hCa	hNa	hK
			12-ring			
O(7)–O(7)	9.95	9.83	9.92	9.80	9.69	9.82
O(10)–O(10)	9.17	8.93	8.97	9.19	9.26	9.30
			Large-channel 8-ring			
O(5)–O(5)	6.26	6.18	6.17	6.13	6.14	6.14
O(10)–O(10)	7.53	7.46	7.46	7.51	7.53	7.53
O(2)–O(2)	6.66	6.60	5.91,7.52*	6.61	6.59	6.59
			Site II 8-ring			
O(6)–O(6)	6.43	6.28	6.17	6.41	6.41	6.45
O(1)–O(1)	6.64	6.52	5.68,7.42*	6.58	6.58	6.60
O(9)–O(9)	7.53	7.46	7.46	7.51	7.53	7.53

*Two values because of lower symmetry.

Meier, W. M. (1961) The crystal structure of mordenite (ptilolite): *Z. Kristallogr.* **115**, 439-450.

Meier, W. M. and Olson, D. H. (1971) Zeolite frameworks: *Adv. Chem. Ser.* **101**, 155-170.

Mortier, W. J., Pluth, J. J. and Smith, J. V. (1975a) Positions of cations and molecules in zeolites with the mordenite-type framework. II. Dehydrated hydrogen-ptilolite: *Mater. Res. Bull.* **10**, 1319-1326.

Mortier, W. J., Pluth, J. J. and Smith, J. V. (1975b) Positions of cations and molecules in zeolites with the mordenite-type framework. I. Dehydrated Ca-exchanged ptilolite: *Mater. Res. Bull.* **10**, 1037-1046.

Mortier, W. J., Pluth, J. J. and Smith, J. V. (1976) Positions of cations and molecules in zeolites with the mordenite-type framework. III. Rehydrated Ca-exchanged ptilolite: *Mater. Res. Bull.* **11**, 15-22.

Sand, L. B. (1968) Synthesis of large-port and small port mordenites: *Molecular Sieves*, Soc. Chem. Ind., London, pp. 71-77.

Sherman, J. D. and Bennett, J. M. (1973) Framework structures related to the zeolite mordenite: *Adv. Chem. Ser.* **121**, 52-65.

Villiger, H. (1969) DLS program and manual: ETH Zürich.

TABLE 6. Comparison of T–O–T and O–T–O angles

	dH	dCa	dK		hCa	hNa	hK
T(1)–O(1)–T(3)	148.7(2)	146.5(3)	134.3(3)	164.2(4)	146.0(3)	144.9(4)	146.0(2)
T(2)–O(2)–T(4)	146.0(2)	147.5(3)	136.7(3)	153.5(3)	144.4(3)	143.6(4)	143.4(2)
T(2)–O(3)–T(1)	158.9(2)	160.2(3)	170.9(3)	148.5(3)	158.9(3)	158.7(3)	158.5(3)
T(3)–O(4)–T(4)	168.0(2)	174.9(3)	172.5(3)		167.2(4)	166.3(5)	163.1(4)
T(2)–O(5)–T(2)	146.0(2)	145.8(4)	145.3(3)		143.7(4)	144.2(6)	143.6(4)
T(1)–O(6)–T(1)	151.4(3)	148.4(4)	145.8(3)		150.5(4)	149.9(5)	151.1(4)
T(1)–O(7)–T(1)	141.2(2)	140.8(3)	140.9(3)		138.8(3)	136.5(4)	139.2(3)
T(2)–O(8)–T(2)	180	180	164.8(3)		180	180	180
T(3)–O(9)–T(3)	146.9(4)	136.4(5)	142.3(4)		146.2(5)	146.5(7)	151.9(5)
T(4)–O(10)–T(4)	148.3(4)	143.4(5)	144.0(4)		148.3(5)	148.0(8)	149.9(5)
O(1)–T(1)–O(3)	111.7(2)	112.2(2)	111.2(2)	112.5(3)	112.1(2)	113.3(2)	112.4(2)
O(1) O(6)	108.4(2)	106.2(2)	105.1(2)	105.9(2)	107.2(2)	106.5(3)	107.1(2)
O(1) O(7)	108.7(1)	108.3(2)	109.6(2)	109.8(2)	109.2(2)	109.6(3)	109.9(2)
O(3) O(6)	111.2(2)	111.3(3)	112.4(2)	111.2(2)	111.5(3)	111.6(4)	110.9(2)
O(3) O(7)	106.8(2)	107.5(2)	107.7(2)	107.6(2)	106.2(2)	105.4(3)	106.5(2)
O(6) O(7)	110.0(1)	111.3(2)	110.7(2)	109.8(2)	110.7(2)	110.6(3)	109.9(2)
O(2)–T(2)–O(3)	109.7(1)	110.8(2)	109.9(2)	110.2(2)	109.7(2)	109.2(3)	109.6(2)
O(2) O(5)	106.9(1)	106.8(2)	104.3(2)	105.6(2)	105.9(2)*	105.9(4)	105.2(2)
O(2) O(8)	110.0(1)	109.7(2)	110.2(2)	111.1(2)	110.7(2)	110.5(2)	110.8(2)
O(3) O(5)	109.3(2)	108.9(3)	110.4(2)	109.6(2)	109.2(3)	109.4(4)	109.5(2)
O(3) O(8)	110.3(1)	110.1(2)	109.4(2)	110.9(2)	110.5(2)	111.0(2)	111.1(2)
O(5) O(8)	110.4(2)	110.4(3)	112.5(2)	109.3(2)	110.7(2)	110.7(4)	110.5(2)
O(1)–T(3)–O(1)	109.4(2)	109.9(3)	109.6(2)		109.9(3)	110.2(4)	109.9(3)
O(1) O(4)	110.3(1)	112.7(2)	111.1(2)	111.9(3)	111.3(2)	111.2(2)	110.5(2)
O(1) O(9)	107.3(1)	103.4(2)	105.4(2)	106.5(3)	105.5(2)	105.2(3)	106.8(2)
O(4) O(9)	112.4(2)	113.8(2)	112.0(2)		113.1(3)	113.4(4)	112.3(3)
O(2)–T(4)–O(2)	108.7(2)	108.8(3)	110.6(2)		108.7(3)	109.0(4)	110.1(3)
O(2) O(4)	110.2(1)	110.7(2)	109.8(2)	111.2(2)	111.1(2)	111.6(3)	110.3(2)
O(2) O(10)	107.9(1)	107.7(2)	106.3(2)	107.4(2)	106.6(2)	105.6(3)	106.2(2)
O(4) O(10)	112.0(2)	110.9(3)	111.5(2)		112.4(3)	113.0(5)	113.6(3)

*Misprinted as 109.9 in Mortier *et al.* (1976).

HYDRATED POTASSIUM PTILOLITE FOBS AND FCALC (X10) PAGE 1

L= 0

K	H	FOBS	FCALC
0	2	1061	1105
0	4	251	275
0	6	45	22
0	8	547	458
0	10	2051	1956
0	12	142	146
0	14	140	114
0	16	601	594
0	18	713	727
0	20	511	503
0	22	98	124
1	1	479	531
1	3	539	521
1	5	479	398
1	7	302	305
1	9	254	312
1	11	401	403
1	13	133	152
1	15	301	297
1	17	183	172
1	19	45	19
1	21	118	116
2	0	507	500
2	2	58	68
2	4	330	355
2	6	380	425
2	8	150	106
2	10	166	181
2	12	366	320
2	14	165	174
2	16	67	24
2	18	279	255
2	20	73	79
3	1	711	725
3	3	953	907
3	5	1931	1822
3	7	155	141
3	9	536	572
3	11	141	129
3	13	199	220
3	15	565	551
3	17	174	172
3	21	210	200
3	23	172	186
4	0	207	190
4	2	147	113
4	4	109	108
4	6	175	140
4	8	118	84
4	10	56	90
4	12	87	117
4	18	110	115
5	1	1875	1871
5	3	1985	1958
5	5	284	246
5	9	283	243
5	11	997	1027
5	13	859	798
5	15	919	927
5	17	455	473
5	19	134	137
5	21	174	189
6	0	851	769
6	2	152	173
6	6	171	128
6	8	216	177
6	10	218	216
6	12	79	27
6	18	336	349
6	22	55	38
7	1	253	190
7	3	651	645
7	5	444	396
7	7	57	46
7	11	280	268
7	13	576	560
7	15	229	218
7	17	108	83
7	19	92	82
7	21	59	60
8	0	1362	1218
8	2	295	369
8	4	708	651
8	6	1644	1676
8	8	1028	940
8	10	1144	1156
8	12	1130	1174
8	16	645	646
8	18	241	225
8	20	394	378
8	22	431	413
9	1	127	37
9	3	351	320
9	5	186	164
9	7	321	315
9	9	44	47
9	13	185	181
9	15	82	70
9	17	127	140
10	0	1843	1867
10	2	98	53
10	4	305	267
10	6	381	335
10	8	127	103
10	10	728	729
10	12	185	169
10	14	85	101
10	16	248	234
10	20	87	74
11	1	368	277
11	3	164	146
11	5	432	394
11	7	95	55
11	9	292	299
11	11	59	20
11	13	438	410
11	15	228	229
11	17	67	89
11	19	67	63
11	21	66	27
12	0	153	180
12	2	353	314
12	4	296	288
12	6	582	557
12	10	75	64
12	12	383	381
12	16	357	325
12	18	57	54
13	1	285	286
13	3	1320	1274
13	5	1751	1768
13	7	146	168
13	9	362	349
13	11	230	243
13	13	617	814
13	15	733	740
13	17	294	283
14	0	345	365
14	2	190	193
14	4	134	120
14	6	230	252
14	8	224	211
14	10	297	311
14	12	235	212
14	14	73	124
14	16	71	82
14	18	58	21
14	20	82	83
15	1	191	200
15	3	43	9
15	5	84	67
15	7	409	419
15	9	94	92
15	11	273	277
15	13	108	111
15	15	51	56
15	17	300	280
16	0	1278	1250
16	2	835	865
16	4	378	375
16	6	54	13
16	8	320	296
16	10	473	477
16	12	203	175
16	14	61	69
16	16	222	202
16	18	304	298
17	3	114	118
17	5	91	73
17	7	63	60
17	11	113	120
17	13	106	112
17	17	82	78
18	0	509	494
18	2	574	587
18	4	64	51
18	6	538	522

HYDRATED POTASSIUM PTILOLITE FOBS AND FCALC (X10) PAGE 2

K	H	FOBS	FCALC
18	8	204	211
18	10	254	240
18	12	567	565
18	14	155	194
18	16	444	438
19	1	86	62
19	3	65	73
19	9	77	50
19	11	72	31
19	13	154	157
19	15	78	57
20	0	338	383
20	2	116	127
20	4	134	111
20	6	223	237
20	8	366	374
20	10	255	260
20	12	72	86
21	1	347	341
21	3	666	659
21	5	382	393
21	7	490	507
21	11	459	446
21	13	509	492
22	0	116	136
22	2	188	187
22	4	89	94
22	8	140	127
22	10	101	112
22	12	110	97
23	3	143	142
23	5	158	153
23	7	73	78
23	11	88	71
24	0	184	147
24	4	136	133
24	6	117	119
24	8	354	333
24	10	309	293
25	1	251	218
25	3	107	109
25	5	143	124
25	7	182	206
26	0	837	790
26	2	290	281
26	4	81	80

L= 1

K	H	FOBS	FCALC
1	1	572	518
1	3	133	149
1	5	1590	1610
1	7	244	240
1	9	488	493
1	11	212	202
1	13	311	307
1	15	213	211
1	17	260	255
1	19	148	141
1	21	52	65
1	23	307	297
2	0	575	603
2	2	259	261
2	4	269	296
2	6	273	297
2	8	207	201
2	10	227	230
2	12	257	267
2	14	61	55
2	16	130	123
2	18	62	70
2	20	59	52
2	22	102	122
3	1	180	214
3	3	198	215
3	5	116	123
3	7	171	181
3	9	40	10
3	11	203	196
3	13	42	57
3	15	39	8
3	17	65	48
3	21	43	26
4	0	432	427
4	2	614	640
4	4	389	428
4	6	668	716
4	8	601	600
4	10	294	304
4	14	137	140
4	16	122	109
4	18	42	6
4	20	197	193
4	22	128	122
5	1	38	17
5	3	236	252
5	5	159	149
5	7	51	54
5	9	45	42
5	13	113	120
5	19	49	18
5	21	40	8
5	23	72	44
6	0	591	564
6	2	593	587
6	4	54	25
6	6	479	478
6	8	131	130
6	10	264	264
6	12	377	373
6	14	180	185
6	16	168	168
6	18	228	229
6	20	45	46
6	22	230	225
7	1	638	632
7	5	31	44
7	7	186	186
7	9	68	63
7	11	460	474
7	13	251	258
7	15	50	63
7	17	251	247
7	19	95	87
7	21	144	143
8	0	251	250
8	8	177	197
8	10	118	128
8	12	152	150
8	14	38	24
8	16	48	54
8	18	121	112
8	22	70	48
9	1	85	73
9	3	304	307
9	5	622	660
9	7	440	424
9	9	342	339
9	11	398	406
9	17	223	219
9	19	76	74
9	21	93	95
10	0	38	74
10	2	209	195
10	6	54	89
10	8	167	168
10	18	54	44
11	1	459	455
11	5	213	201
11	7	169	170
11	9	225	225
11	11	358	351
11	13	117	117
11	15	41	6
11	17	142	140
11	21	108	114
12	0	972	1008
12	2	55	51
12	4	123	112
12	6	136	138
12	8	284	311
12	10	504	530
12	12	344	349
12	14	64	53
12	16	134	133
12	18	260	255
12	20	144	129
13	1	147	127
13	3	39	43
13	5	138	145
13	7	99	92
13	9	43	24
13	11	156	138
13	13	106	96
13	15	64	63
13	17	64	72
14	0	618	637
14	2	62	40
14	4	233	233

HYDRATED POTASSIUM PTILOLITE FOBS AND FCALC (X10) PAGE 3

K	H	FOBS	FCALC
14	6	346	360
14	8	398	424
14	10	369	366
14	12	373	379
14	14	54	56
14	16	265	262
14	18	268	279
15	1	38	70
15	3	45	50
15	5	134	155
15	7	108	119
15	11	82	115
15	15	91	92
15	17	47	77
15	19	46	27
16	0	77	85
16	4	165	157
16	6	170	174
16	10	45	72
16	12	40	6
16	14	45	51
16	16	94	86
17	1	227	206
17	3	166	185
17	5	291	319
17	7	208	218
17	9	69	58
17	11	273	279
17	15	116	123
17	17	126	122
18	0	78	64
18	2	50	62
18	6	65	94
18	8	60	38
18	10	47	34
18	12	49	59
18	16	57	53
19	1	222	199
19	3	37	53
19	5	234	243
19	7	286	312
19	9	70	75
19	11	261	275
19	13	186	185
19	15	113	111
20	2	113	121
20	4	123	112
20	6	243	233
20	12	100	158
21	1	43	79
21	5	152	155
21	7	104	102
21	11	40	52
21	13	65	71
22	0	120	125
22	2	125	135
22	4	85	80
22	6	182	172
22	8	115	132
22	10	107	100
22	12	54	47
23	1	42	15
23	7	45	27
24	0	152	136
24	2	94	104
24	4	150	131
24	6	208	209
24	10	74	80
25	1	109	94
25	3	90	93
25	5	305	310
26	0	176	158
26	2	49	21
26	4	102	98

L= 2

K	H	FOBS	FCALC
0	0	1130	1156
0	2	2602	2523
0	4	1116	1109
0	6	82	41
0	8	139	103
0	10	84	48
0	12	1497	1469
0	14	113	99
0	16	834	821
0	18	248	280
0	20	329	324
0	22	564	537
1	1	170	225
1	3	374	371
1	5	397	384
1	7	76	42
1	9	141	164
1	11	235	211
1	13	255	262
1	15	375	371
1	19	119	138
1	23	147	128
2	0	867	874
2	2	661	631
2	4	46	34
2	6	178	151
2	8	65	71
2	10	441	411
2	12	352	351
2	14	234	261
2	16	263	252
2	18	68	63
2	22	174	193
3	1	632	568
3	3	829	794
3	5	343	350
3	7	768	759
3	9	607	609
3	13	322	343
3	15	247	215
3	17	149	128
3	19	96	91
3	21	238	228
4	0	30	22
4	2	193	193
4	4	181	176
4	6	69	62
4	8	67	67
4	10	74	62
4	12	221	238
4	14	51	56
4	18	80	68
4	22	85	101
5	1	378	369
5	3	1182	1164
5	5	117	42
5	7	323	312
5	9	41	33
5	11	283	287
5	13	1421	1407
5	15	684	701
5	17	222	230
5	19	102	105
5	21	116	94
6	0	58	25
6	2	321	309
6	4	37	33
6	6	91	73
6	8	113	135
6	10	182	161
6	12	414	425
6	14	64	99
6	16	143	145
6	20	104	117
6	22	274	277
7	1	222	222
7	3	348	352
7	5	236	234
7	7	529	532
7	9	89	50
7	11	265	261
7	13	52	40
7	15	101	94
7	17	377	384
7	19	122	125
7	21	96	101
8	0	308	355
8	2	114	121
8	4	842	833
8	6	138	122
8	8	1172	1158
8	10	1089	1105
8	12	699	699
8	14	175	179
8	16	69	53
8	18	746	730
8	20	291	273
8	22	132	124
9	1	44	23
9	3	176	189
9	5	129	103
9	7	175	212

HYDRATED POTASSIUM PTILOLITE FOBS AND FCALC (X10) PAGE 4

K	H	FOBS	FCALC
9	9	82	41
9	13	149	141
9	15	59	68
9	17	113	134
9	19	48	22
10	0	257	247
10	2	364	366
10	6	134	111
10	10	206	179
10	12	515	517
10	14	69	85
10	16	182	170
10	18	56	67
11	1	296	273
11	3	181	163
11	5	97	87
11	7	275	280
11	9	74	94
11	11	430	425
11	13	197	210
11	15	127	119
11	17	256	250
11	19	75	55
12	0	749	738
12	2	106	119
12	4	131	104
12	6	193	185
12	8	59	77
12	10	428	444
12	12	91	118
12	14	113	117
12	16	42	57
12	18	205	189
12	20	136	126
13	1	920	902
13	3	825	840
13	5	191	183
13	7	1092	1078
13	9	677	683
13	11	480	456
13	13	224	218
13	15	319	319
13	17	465	459
13	19	209	187
14	0	325	300
14	2	55	51
14	4	50	60
14	6	66	61
14	8	79	92
14	10	283	306
14	12	153	154
14	14	141	162
14	18	58	44
15	1	174	175
15	3	110	103
15	5	609	609
15	7	139	141
15	9	43	14
15	11	83	38
15	15	187	192
15	17	112	117
16	0	826	825
16	2	699	698
16	4	517	556
16	6	84	71
16	8	219	203
16	10	342	330
16	12	209	185
16	16	313	300
16	18	247	231
17	3	65	34
17	5	139	138
17	15	58	42
18	0	625	623
18	2	141	142
18	4	186	184
18	6	213	195
18	8	337	334
18	10	457	438
18	12	84	86
18	14	49	43
18	16	98	75
19	1	49	12
19	5	149	150
20	0	86	106
20	2	251	259
20	4	198	218
20	6	296	315
20	8	62	51
20	12	298	284
20	14	55	30
21	1	71	34
21	3	406	390
21	5	568	557
21	7	185	178
21	11	160	146
21	13	585	566
22	0	104	116
22	2	150	153
22	8	140	134
22	10	93	77
22	12	83	82
23	1	167	165
23	3	50	55
23	5	88	70
23	7	63	55
23	9	143	138
23	11	70	59
24	2	122	92
24	4	42	34
24	6	258	248
24	8	293	274
25	1	123	131
25	5	295	294
26	0	186	159
26	2	369	358

L= 3

K	H	FOBS	FCALC
1	1	316	275
1	3	44	12
1	5	292	310
1	7	1125	1167
1	9	221	193
1	11	780	789
1	13	38	36
1	15	171	171
1	17	603	614
1	19	48	64
1	21	390	399
2	0	206	215
2	2	235	250
2	4	162	161
2	6	229	218
2	8	57	46
2	10	257	279
2	12	182	195
2	14	75	101
2	16	198	198
2	18	69	64
2	20	98	100
2	22	63	57
3	1	117	146
3	3	139	118
3	5	434	428
3	7	184	196
3	9	38	18
3	11	55	47
3	13	43	31
3	15	128	133
3	17	115	122
3	19	85	95
3	21	78	66
4	0	63	85
4	2	464	447
4	4	901	918
4	6	41	45
4	8	1033	1064
4	10	725	767
4	12	126	119
4	14	264	261
4	16	52	34
4	18	583	564
4	20	472	474
5	5	180	196
5	7	213	220
5	9	42	46
5	11	97	91
5	15	127	139
5	17	149	143
5	19	44	45
5	21	72	46
6	0	133	132
6	2	230	216
6	4	135	145
6	6	317	295
6	8	521	515
6	10	419	435
6	12	361	372

HYDRATED POTASSIUM PTILOLITE FOBS AND FCALC (X10) PAGE 5

K	H	FOBS	FCALC
6	14	152	168
6	16	215	208
6	18	344	354
6	20	98	107
7	1	390	382
7	3	287	294
7	5	548	565
7	7	66	82
7	9	74	76
7	11	146	138
7	13	500	501
7	15	265	275
7	17	86	71
7	19	94	89
8	0	267	256
8	2	168	167
8	4	208	232
8	6	66	105
8	8	127	128
8	10	37	22
8	12	124	113
8	20	112	114
9	1	272	279
9	3	103	118
9	5	515	532
9	7	624	673
9	11	231	266
9	13	386	391
9	15	397	397
9	17	312	305
9	19	136	161
9	21	249	243
10	0	88	82
10	2	73	83
10	4	187	183
10	6	71	68
10	8	71	95
10	10	131	122
10	12	39	38
10	14	96	90
10	18	67	71
10	20	109	120
11	1	293	288
11	3	219	218
11	5	245	252
11	7	73	73
11	9	120	98
11	11	297	290
11	13	264	264
11	15	139	141
11	17	53	64
11	19	48	50
12	0	125	100
12	2	743	748
12	4	510	541
12	6	434	464
12	8	495	503
12	10	159	171
12	12	595	606
12	16	290	309

K	H	FOBS	FCALC
12	18	175	181
13	1	189	163
13	3	155	156
13	5	119	129
13	7	132	135
13	9	92	81
13	11	89	60
13	17	97	114
14	0	423	424
14	2	390	405
14	4	703	727
14	6	477	487
14	8	239	244
14	12	297	295
14	14	69	61
14	16	265	260
15	1	114	146
15	3	209	213
15	5	40	77
15	7	149	149
15	9	37	29
15	11	73	98
15	13	64	82
15	15	43	35
15	17	52	47
16	0	136	159
16	2	115	113
16	6	53	36
16	8	141	132
16	10	142	138
16	12	57	84
16	16	48	13
17	1	68	52
17	3	351	356
17	5	619	628
17	7	390	409
17	9	47	56
17	11	115	127
17	13	418	407
17	15	318	316
18	0	171	165
18	2	46	55
18	8	64	69
18	10	142	159
18	12	94	80
18	14	94	73
19	1	324	334
19	3	125	130
19	5	333	330
19	7	178	170
19	9	77	85
19	11	175	170
19	13	140	130
20	0	406	417
20	2	151	152
20	4	40	36
20	8	310	297
20	10	330	323
20	12	48	48
21	5	81	104

K	H	FOBS	FCALC
21	7	173	153
21	9	54	41
21	11	77	84
22	0	169	169
22	2	217	221
22	4	99	109
22	6	44	23
22	8	382	369
22	10	256	257
23	5	45	22
23	7	75	61
23	9	50	34
24	0	304	306
24	2	56	25
24	4	151	156
24	6	166	159
25	1	199	208
25	3	285	292

L= 4

K	H	FOBS	FCALC
0	0	2550	2483
0	2	564	525
0	4	481	443
0	6	354	375
0	8	325	315
0	10	839	860
0	12	619	635
0	16	93	63
0	18	395	382
0	20	279	294
1	1	346	351
1	3	129	109
1	5	159	189
1	7	251	229
1	9	131	145
1	11	341	342
1	13	54	24
1	15	80	66
1	17	121	123
1	21	106	100
2	0	328	304
2	2	151	151
2	4	238	212
2	6	324	292
2	8	82	73
2	10	118	102
2	12	299	295
2	14	78	81
2	16	98	99
2	18	182	183
2	20	64	50
3	1	131	98
3	3	235	232
3	5	819	813
3	7	194	172
3	9	97	101
3	11	56	32
3	13	44	49

HYDRATED POTASSIUM PTILOLITE FOBS AND FCALC (X10) PAGE 6

K	H	FOBS	FCALC
3	15	119	132
3	17	85	98
3	21	80	96
4	2	93	95
4	8	151	148
4	10	48	44
4	16	48	49
5	1	535	552
5	3	369	341
5	5	269	241
5	7	354	360
5	9	189	175
5	11	785	795
5	13	105	88
5	15	230	238
5	17	410	405
5	19	51	24
5	21	123	135
6	0	384	355
6	2	57	70
6	4	35	44
6	6	182	182
6	8	200	218
6	10	205	178
6	12	166	151
6	14	37	25
6	16	59	40
6	18	216	230
6	20	52	47
7	1	346	335
7	3	286	273
7	5	326	331
7	9	121	118
7	11	243	247
7	13	366	377
7	15	97	86
7	17	87	98
7	19	93	73
8	0	143	136
8	2	227	225
8	4	91	70
8	6	869	872
8	8	609	622
8	10	332	342
8	12	474	494
8	14	48	43
8	16	213	207
8	20	190	195
9	3	180	185
9	5	62	45
9	7	154	146
9	9	40	49
9	13	91	101
9	17	50	62
9	19	46	32
10	0	604	610
10	2	66	56
10	4	143	136
10	6	165	167
10	8	71	56

K	H	FOBS	FCALC
10	10	357	370
10	12	156	157
10	14	94	97
10	18	92	67
11	1	256	266
11	3	48	64
11	5	152	126
11	7	107	99
11	9	184	220
11	11	94	101
11	13	237	244
11	15	98	104
11	19	70	47
12	0	101	113
12	2	200	219
12	4	286	272
12	6	362	354
12	8	47	45
12	10	47	40
12	12	188	213
12	14	39	1
12	16	189	184
12	18	87	91
13	1	185	196
13	3	363	356
13	5	1135	1130
13	9	43	15
13	11	212	212
13	13	257	270
13	15	249	252
13	17	262	261
14	0	162	167
14	2	96	92
14	4	95	91
14	6	134	159
14	8	112	108
14	10	147	152
14	12	92	91
14	16	59	48
15	1	237	216
15	3	85	109
15	5	150	175
15	7	228	252
15	9	120	112
15	11	176	150
15	13	57	39
15	17	175	178
16	0	557	557
16	2	395	406
16	4	212	224
16	6	64	81
16	8	82	76
16	10	164	152
16	12	53	3
16	14	54	40
16	16	54	40
17	1	100	109
17	9	68	71
17	11	57	67
17	13	75	68

K	H	FOBS	FCALC
18	0	86	63
18	2	285	289
18	4	103	132
18	6	349	350
18	8	153	140
18	12	218	226
18	14	155	142
19	1	80	69
19	3	48	51
19	5	58	30
19	9	69	67
19	13	124	124
20	0	249	269
20	4	142	151
20	6	215	232
20	8	257	228
20	10	159	162
20	12	95	94
21	1	166	147
21	3	247	254
21	5	68	73
21	7	269	268
21	9	95	97
21	11	291	272
22	0	77	68
22	2	115	103
22	6	43	13
22	8	65	45
23	5	139	145
24	0	49	21
24	2	51	42
24	4	129	130

L= 5

K	H	FOBS	FCALC
1	1	340	338
1	3	598	615
1	5	1306	1350
1	7	303	302
1	9	277	289
1	11	41	17
1	13	536	541
1	15	381	383
1	19	74	78
2	0	329	314
2	2	145	137
2	4	132	158
2	6	212	240
2	8	138	129
2	10	56	58
2	12	203	207
2	14	66	74
2	18	49	67
3	1	95	65
3	3	238	265
3	5	296	293
3	7	299	317
3	11	147	146
3	13	133	143

HYDRATED POTASSIUM PTILOLITE FOBS AND FCALC (X10) PAGE 7

K	H	FOBS	FCALC
3	15	103	106
3	17	142	129
4	0	728	752
4	2	984	1025
4	4	445	421
4	6	618	646
4	8	1020	1056
4	10	631	642
4	12	413	398
4	14	62	54
4	16	227	233
4	18	379	365
4	20	423	412
5	3	199	208
5	5	305	311
5	7	139	162
5	11	65	86
5	13	151	154
5	15	130	124
5	17	88	79
5	19	59	50
6	0	353	347
6	2	354	358
6	4	207	223
6	6	541	516
6	8	231	239
6	10	107	93
6	12	338	326
6	14	95	102
6	16	161	165
6	18	109	102
7	1	425	436
7	3	321	324
7	5	139	150
7	7	383	379
7	11	430	446
7	13	53	43
7	15	183	190
7	17	310	296
7	19	103	85
8	0	187	189
8	2	141	133
8	6	71	58
8	8	277	273
8	10	147	148
8	12	51	42
8	16	55	44
8	18	161	155
9	1	195	200
9	3	675	711
9	5	879	912
9	7	745	749
9	9	178	186
9	11	503	507
9	13	347	359
9	15	343	336
9	17	427	417
10	0	153	162
10	2	266	274
10	4	112	128

K	H	FOBS	FCALC
10	6	45	51
10	8	203	221
10	10	108	116
10	12	83	92
10	18	102	101
11	1	290	298
11	3	164	180
11	5	100	106
11	7	234	249
11	9	142	137
11	11	224	230
11	15	66	58
11	17	182	181
12	0	1110	1108
12	2	430	451
12	8	621	648
12	10	688	701
12	14	77	53
13	1	82	69
13	3	95	101
13	5	126	141
13	7	42	13
13	11	104	108
13	13	128	118
13	15	88	86
14	0	707	718
14	2	208	212
14	6	131	134
14	8	491	498
14	10	442	456
14	12	225	237
14	16	109	108
15	3	63	79
15	5	112	119
15	7	51	35
15	11	83	97
15	15	64	69
16	0	214	219
16	2	117	118
16	4	120	115
16	6	163	156
16	8	108	111
16	10	176	175
16	12	53	68
16	14	48	51
17	1	304	315
17	3	496	500
17	5	648	667
17	7	463	464
17	11	394	387
17	13	252	256
18	0	172	165
18	2	113	124
18	4	68	73
18	6	115	127
18	8	104	105
18	10	147	141
18	12	127	123
19	1	130	137
19	5	190	183

K	H	FOBS	FCALC
19	7	237	247
19	9	91	89
19	11	238	242
20	0	221	232
20	2	252	251
20	4	122	120
20	6	309	293
20	8	168	166
20	10	220	212
21	1	53	69
21	3	58	27
21	5	133	129
22	0	287	293
22	2	298	302
22	4	122	119
22	6	242	245

L= 6

K	H	FOBS	FCALC
0	0	895	837
0	2	336	375
0	4	860	855
0	6	104	103
0	8	188	170
0	10	455	469
0	12	184	204
0	14	46	2
0	16	75	84
0	18	231	209
1	1	179	176
1	3	120	124
1	5	198	215
1	7	56	65
1	9	40	11
1	11	113	107
1	17	47	48
2	0	441	440
2	2	118	132
2	4	110	101
2	6	135	116
2	10	243	244
2	12	140	141
2	14	78	97
2	16	99	94
3	1	53	6
3	3	180	169
3	7	138	160
3	9	131	147
3	13	188	185
3	15	88	109
4	0	165	169
4	2	57	45
4	6	40	17
4	8	80	70
4	10	54	62
4	14	41	21
4	16	70	30
5	1	661	654
5	3	320	293

HYDRATED POTASSIUM PTILOLITE FOBS AND FCALC (X10) PAGE 8

K	H	FOBS	FCALC
5	5	96	107
5	7	100	73
5	9	239	258
5	11	235	230
5	13	170	196
5	15	51	2
6	0	329	317
6	2	149	146
6	4	105	100
6	6	89	72
6	10	152	161
6	12	153	156
6	14	71	84
7	1	190	198
7	5	216	213
7	7	223	223
7	9	43	27
7	11	201	210
7	13	137	145
7	15	44	47
7	17	219	205
8	0	328	323
8	2	387	374
8	4	147	167
8	6	250	235
8	8	248	284
8	10	156	171
8	12	195	165
8	14	142	127
8	16	297	280
9	3	50	30
9	11	52	32
9	13	42	30
10	0	382	388
10	2	85	31
10	4	208	200
10	8	50	75
10	10	99	107
10	14	61	54
10	16	95	100
11	1	121	123
11	5	203	204
11	9	55	20
11	11	145	148
12	0	231	230
12	2	156	147
12	6	166	181
12	10	156	180
12	12	199	206
12	14	54	71
13	1	182	203
13	3	179	174
13	5	270	261
13	7	375	394
13	9	161	171
13	11	213	225
13	13	359	356
14	0	58	44
14	2	67	55
14	8	45	19

K	H	FOBS	FCALC
14	14	81	69
15	5	252	256
15	7	198	196
15	9	58	66
15	11	102	105
15	13	70	49
16	0	50	75
16	2	42	64
16	4	204	233
16	6	49	46
16	8	124	140
16	10	58	66
16	12	54	59
17	7	92	78
17	11	48	46
18	0	82	54
18	2	162	142
18	4	58	46
18	6	284	274
18	8	120	124
19	3	51	5
19	5	81	80
19	7	59	48
20	0	193	173
20	2	113	87
20	4	90	92
20	6	185	187
21	1	175	159

L= 7

K	H	FOBS	FCALC
1	1	495	461
1	3	307	314
1	5	186	166
1	7	525	541
1	9	256	253
1	11	385	397
1	13	107	98
1	15	270	256
2	0	173	179
2	2	84	101
2	4	73	78
2	6	90	99
2	8	61	61
2	10	169	174
2	14	87	88
2	16	113	115
3	1	63	61
3	3	104	100
3	5	245	232
3	7	112	111
3	9	114	93
3	13	98	76
3	15	154	151
4	0	743	734
4	2	379	368
4	4	196	203
4	6	122	122
4	8	494	453

K	H	FOBS	FCALC
4	10	639	637
4	12	282	287
4	14	197	191
4	16	189	181
5	1	85	78
5	3	83	87
5	5	146	146
5	7	152	129
5	9	80	68
5	11	69	50
5	13	56	58
5	15	139	136
6	0	325	330
6	2	123	115
6	4	114	139
6	6	195	200
6	8	195	204
6	10	283	292
6	12	155	158
6	14	93	90
7	1	122	110
7	3	187	190
7	5	408	419
7	11	67	75
7	13	348	343
7	15	248	241
8	0	37	13
8	2	114	115
8	4	88	78
8	6	99	102
8	12	137	130
8	14	50	9
9	1	99	78
9	3	353	360
9	5	571	565
9	7	288	297
9	9	166	143
9	11	64	75
9	13	384	397
10	0	140	136
10	2	140	129
10	4	90	78
10	6	42	20
10	8	86	72
10	10	92	87
10	12	92	99
11	1	127	133
11	3	126	121
11	5	237	236
11	11	129	139
11	13	190	186
12	0	233	210
12	2	516	517
12	4	271	249
12	6	321	318
12	8	273	274
12	10	244	214
12	12	473	479
13	1	74	67
13	3	93	78

HYDRATED POTASSIUM PTILOLITE FOBS AND FCALC (X10) PAGE 9

K	H	FOBS	FCALC
13	5	69	33
13	7	142	135
14	0	41	24
14	2	288	285
14	4	299	296
14	6	296	302
14	8	52	43
15	1	67	77
15	3	107	110
15	5	44	34
15	7	101	105
16	0	178	165
16	2	101	92
16	4	48	13
16	6	51	27
16	8	47	80
17	1	55	80
17	3	353	339
17	5	464	451
18	0	152	149
18	2	72	54

L= 8

K	H	FOBS	FCALC
0	2	354	309
0	4	52	17
0	6	402	408
0	8	125	105
0	10	153	97
0	12	666	656
1	1	72	87
1	3	84	65
1	5	70	73
1	11	136	138
1	13	135	132
2	0	55	76
2	2	117	110
2	4	111	107
2	6	148	162
2	12	144	152
3	1	119	107
3	3	199	171
3	7	139	129
3	9	78	49
3	11	114	121
3	13	251	220
4	0	57	31
4	4	47	24
4	6	52	37
4	8	41	47
4	12	51	52
5	3	278	267
5	5	365	334
5	9	200	182
5	11	208	232
6	0	63	79
6	2	75	58
6	4	59	46
6	6	111	112
6	8	101	100
6	10	62	51
6	12	190	175
7	1	205	224
7	5	85	104
7	7	155	135
7	9	83	97
7	11	176	163
8	0	677	631
8	2	196	172
8	6	177	193
8	8	149	106
8	10	405	343
9	5	61	55
9	9	62	53
10	0	84	35
10	2	130	142
10	4	68	76
10	10	55	14
11	1	155	144
11	3	72	43
11	5	81	87
11	7	149	143
11	9	67	75
12	0	196	192
12	4	117	121
12	6	132	143
12	8	79	54
13	1	348	323
13	3	246	232
13	5	91	145
13	7	327	306
14	2	63	25
14	4	45	24
15	1	141	131
15	3	86	54

L= 9

K	H	FOBS	FCALC
1	1	66	69
1	3	315	295
1	5	389	380
2	0	61	76
2	2	103	89
2	4	50	45
2	6	120	135
2	8	61	37
3	1	81	62
3	3	150	152
3	5	106	105
3	7	103	87
4	0	406	386
4	2	355	350
4	6	234	235
4	8	213	192
5	3	105	105
5	5	106	108
5	7	64	48
6	0	161	162
6	2	135	136
6	4	54	29
6	6	229	237
7	1	211	215
7	3	103	101
7	5	58	48
8	0	150	145
8	4	59	60
9	1	211	201
9	3	366	366
10	0	121	132
10	2	135	122

DEHYDRATED POTASSIUM PTILOLITE FOBS AND FCALC (X10) PAGE 1

L= 0

K	H	FOBS	FCALC
0	2	3290	3299
0	4	380	351
0	8	306	356
0	10	3231	3173
0	12	802	816
0	14	374	309
0	16	386	399
0	18	1335	1332
0	20	435	439
1	1	2220	2176
1	3	1026	1035
1	5	546	505
1	7	582	588
1	9	616	574
1	11	795	793
1	15	244	218
1	17	215	202
1	19	231	214
1	21	245	216
2	0	1951	1940
2	2	445	417
2	4	813	775
2	6	1248	1180
2	10	424	426
2	12	651	649
2	14	151	41
2	18	611	610
3	1	2093	2194
3	3	2010	1955
3	5	3207	3076
3	7	215	244
3	9	1132	1129
3	11	271	204
3	13	769	728
3	15	818	812
3	21	168	104
3	23	185	225
4	0	888	895
4	2	286	263
4	4	411	394
4	6	227	206
4	8	178	173
4	10	258	234
4	14	147	189
4	16	176	229
4	18	150	151
5	1	4214	4192
5	3	3115	3009
5	5	1174	1119
5	7	179	39
5	9	272	277
5	11	2058	2060
5	13	878	871
5	15	1154	1202
5	17	643	641
5	19	240	240
5	21	366	325
6	0	1549	1510
6	2	288	280
6	4	496	466
6	6	322	312
6	8	269	272
6	10	413	382
6	12	391	394
6	14	176	92
6	18	387	433
7	1	168	75
7	3	668	658
7	5	1126	1117
7	7	553	588
7	9	412	388
7	11	254	190
7	13	1043	1042
7	15	241	254
7	17	383	364
7	19	235	197
7	21	119	123
8	0	3000	2949
8	2	361	369
8	4	943	823
8	6	3250	3192
8	8	1470	1420
8	10	2179	2190
8	12	1426	1441
8	14	191	107
8	16	946	962
8	18	214	224
8	20	650	657
8	22	420	402
9	1	156	164
9	3	135	152
9	5	531	531
9	7	382	343
9	9	193	196
9	11	102	86
9	13	126	94
9	15	239	229
9	17	217	212
9	21	125	15
10	0	3112	3017
10	2	450	484
10	4	921	885
10	6	140	143
10	8	384	384
10	10	1185	1152
10	12	186	221
10	14	366	365
10	20	121	39
11	1	175	123
11	5	1245	1204
11	7	311	315
11	9	115	29
11	13	607	607
11	15	413	427
11	17	132	44
11	19	132	190
12	0	95	70
12	2	1101	1073
12	4	228	156
12	6	1176	1190
12	8	297	342
12	12	646	657
12	14	539	519
12	16	610	580
12	18	251	235
13	1	364	362
13	3	2536	2435
13	5	3065	3044
13	7	218	189
13	9	497	497
13	11	340	327
13	13	1501	1509
13	15	797	819
13	17	473	484
13	19	427	380
14	0	411	417
14	2	553	527
14	4	129	51
14	8	413	435
14	10	367	340
14	12	292	249
14	16	115	6
15	1	267	249
15	3	406	369
15	5	134	106
15	7	965	985
15	9	128	76
15	11	429	435
15	15	147	202
15	17	261	307
15	19	356	345
16	0	1829	1625
16	2	1357	1339
16	4	465	520
16	6	173	133
16	8	664	636
16	10	769	757
16	12	420	409
16	16	320	242
16	18	376	366
17	1	133	11
17	3	224	228
17	5	134	153
17	7	301	288
17	9	178	188
17	11	136	115
17	13	224	251
17	17	312	353
18	0	760	726
18	2	885	867
18	6	1018	1019
18	8	194	163
18	10	352	347
18	12	623	631
18	14	362	374
18	16	563	611
19	1	111	46

DEHYDRATED POTASSIUM PTILCLITE FCBS AND FCALC (X10) PAGE 2

K	H	FOBS	FCALC
19	3	136	156
19	5	114	17
19	9	120	56
19	13	131	80
20	0	787	754
20	2	236	202
20	4	414	343
20	6	214	259
20	8	711	762
20	10	362	360
20	12	388	424
20	14	228	220
21	1	519	497
21	3	1056	983
21	5	602	642
21	7	667	682
21	11	566	593
21	13	612	591
22	2	259	243
22	8	206	181
22	10	189	168
22	12	284	246
23	5	411	410
23	7	232	216
23	9	240	226
23	11	288	254
24	0	326	293
24	2	225	180
24	8	330	310
25	1	555	496
25	3	324	282
25	5	298	291
25	7	367	402
26	0	1150	1072
26	2	333	329
26	4	153	127

L= 1

K	H	FOBS	FCALC
1	1	1417	1393
1	2	514	544
1	3	573	608
1	4	491	503
1	5	2905	2888
1	6	188	201
1	7	677	700
1	8	663	641
1	9	707	716
1	10	469	476
1	11	397	385
1	13	967	944
1	14	219	231
1	15	94	53
1	16	495	522
1	17	651	667
1	18	373	406
1	19	316	308
1	20	414	416
1	21	140	123
1	23	556	533
2	0	1125	1126
2	2	562	565
2	3	310	332
2	4	392	379
2	5	232	234
2	6	626	618
2	7	400	381
2	10	382	379
2	11	233	222
2	12	420	414
2	13	335	336
2	14	202	236
2	15	146	121
2	16	298	298
2	17	105	107
2	18	222	213
2	19	136	74
3	1	566	594
3	2	211	244
3	3	324	357
3	4	224	217
3	5	308	319
3	6	160	147
3	7	274	247
3	8	411	403
3	9	87	128
3	10	84	61
3	11	451	447
3	12	92	94
3	14	185	13
3	15	101	92
3	16	261	264
3	18	163	171
3	19	125	149
3	20	168	123
3	21	117	78
3	23	116	35
4	0	1207	1234
4	1	305	345
4	2	773	756
4	3	615	660
4	4	730	752
4	5	211	255
4	6	1383	1394
4	7	388	389
4	8	1060	1064
4	9	308	322
4	10	493	500
4	11	314	318
4	12	127	99
4	13	851	857
4	14	542	536
4	15	871	897
4	16	234	251
4	17	122	111
4	21	302	321
4	22	160	107
4	23	424	386
5	1	151	200
5	2	320	357
5	3	78	108
5	4	456	472
5	5	409	413
5	6	425	453
5	7	163	186
5	8	231	220
5	9	107	131
5	10	160	209
5	11	91	131
5	13	108	47
5	14	198	138
5	16	231	232
5	17	145	75
5	18	100	134
5	20	110	86
6	0	1276	1245
6	2	1163	1143
6	3	246	246
6	4	243	220
6	5	362	369
6	6	723	735
6	7	404	405
6	8	283	290
6	10	469	493
6	11	185	180
6	12	712	708
6	13	575	574
6	14	520	527
6	15	295	303
6	16	272	227
6	17	208	228
6	18	476	456
6	19	112	120
6	20	145	170
6	22	340	337
7	1	1298	1332
7	2	239	240
7	3	133	123
7	4	75	74
7	5	363	381
7	6	409	456
7	7	195	165
7	8	359	364
7	9	80	57
7	10	338	351
7	11	872	860
7	12	205	117
7	13	454	463
7	14	209	177
7	16	293	263
7	17	380	356
7	18	214	169
7	19	250	244
7	20	141	144
7	22	136	92
8	0	323	309
8	1	154	139
8	2	258	285
8	3	518	525

DEHYDRATED POTASSIUM PTILCLITE FCBS AND FCALC (X10) PAGE 3

K	H	FOBS	FCALC
8	4	340	359
8	5	521	516
8	6	95	31
8	7	499	499
8	8	631	644
8	9	603	590
8	10	171	150
8	11	510	516
8	12	287	269
8	14	97	104
8	15	252	243
8	17	131	156
8	18	249	243
8	21	99	56
8	22	133	99
9	1	108	99
9	2	112	136
9	3	905	915
9	4	450	485
9	5	856	875
9	6	464	507
9	7	659	660
9	8	194	226
9	9	690	699
9	10	548	562
9	11	593	606
9	12	722	738
9	13	279	309
9	16	440	470
9	17	130	101
9	18	385	392
9	19	202	239
9	20	250	241
10	0	208	185
10	1	328	335
10	2	444	441
10	3	209	240
10	4	169	181
10	5	124	80
10	6	139	145
10	7	156	107
10	8	307	297
10	9	223	202
10	10	175	160
10	11	180	162
10	13	258	277
10	14	105	181
10	16	103	16
10	17	131	34
10	21	112	102
11	1	1058	1087
11	3	106	101
11	4	256	246
11	5	495	481
11	6	78	76
11	7	403	392
11	8	423	421
11	9	115	101
11	10	178	181
11	11	779	773
11	12	191	228
11	13	373	345
11	14	246	283
11	16	117	14
11	17	288	297
11	19	280	296
11	20	99	60
11	21	185	180
12	0	1865	1881
12	1	276	314
12	2	153	140
12	3	139	131
12	4	243	234
12	5	296	326
12	6	268	249
12	7	138	126
12	8	707	711
12	9	182	162
12	10	713	720
12	12	830	808
12	13	312	326
12	14	130	106
12	15	458	468
12	16	203	191
12	18	502	511
12	19	115	159
12	20	110	49
13	1	359	332
13	2	486	509
13	3	224	245
13	4	734	761
13	5	288	298
13	6	599	605
13	7	446	451
13	8	439	435
13	10	161	148
13	11	384	368
13	12	269	285
13	13	143	121
13	14	404	408
13	16	161	141
13	17	187	198
13	18	139	89
13	19	209	181
13	20	102	46
14	0	978	995
14	1	138	182
14	2	175	169
14	3	141	157
14	4	634	627
14	5	269	271
14	6	845	845
14	7	216	177
14	8	956	980
14	9	115	110
14	10	471	490
14	11	258	262
14	12	624	626
14	13	130	105
14	15	257	267
14	16	399	396
14	18	521	523
14	19	205	221
15	2	135	74
15	3	284	264
15	5	323	337
15	7	237	255
15	9	190	195
15	11	116	39
15	12	161	75
15	14	131	11
15	15	120	135
16	0	311	332
16	1	1009	1055
16	3	617	642
16	5	244	241
16	6	277	290
16	7	98	89
16	8	250	233
16	9	224	221
16	11	176	174
16	15	200	192
16	16	223	208
16	17	218	220
17	3	515	531
17	4	129	106
17	5	242	271
17	6	175	198
17	7	419	419
17	8	203	233
17	9	327	280
17	10	197	185
17	11	467	447
17	12	235	212
17	14	126	95
17	16	375	351
17	17	168	133
18	0	207	168
18	1	463	516
18	2	195	158
18	3	228	310
18	7	167	141
18	9	326	320
18	10	108	113
18	11	348	335
18	12	142	119
18	13	202	208
18	15	164	78
18	17	224	93
19	1	370	378
19	2	171	163
19	4	129	146
19	5	563	550
19	6	240	242
19	7	582	593
19	8	96	117
19	9	129	106
19	11	388	408
19	12	105	88
19	13	313	321

DEHYDRATED POTASSIUM PTILOLITE FOBS AND FCALC (X10) PAGE 4

K	H	FOBS	FCALC
19	16	135	141
20	1	168	141
20	3	208	166
20	4	153	140
20	5	82	20
20	6	312	307
20	9	132	146
20	11	240	203
20	12	154	155
20	13	170	155
20	15	214	85
21	2	442	430
21	3	150	179
21	4	512	507
21	5	394	375
21	6	395	377
21	7	182	156
21	8	210	170
21	9	117	107
21	10	259	217
21	12	460	417
21	13	252	228
21	14	424	397
22	0	263	260
22	6	160	159
22	8	308	297
22	10	118	102
23	4	142	152
23	5	139	51
23	6	139	127
23	8	123	115
23	10	156	106
24	0	308	258
24	1	133	99
24	2	257	241
24	3	172	145
24	4	217	231
24	5	232	223
24	6	599	605
24	7	366	374
24	9	451	429
25	1	183	144
25	5	377	380
25	7	152	137
26	0	208	199
26	1	539	517
26	2	184	184
26	3	211	191

L= 2

K	H	FOBS	FCALC
0	0	1453	1480
0	1	1100	1096
0	2	5260	5142
0	3	1513	1489
0	4	1716	1687
0	5	762	748
0	6	797	744
0	7	362	385
0	8	93	89
0	9	201	181
0	10	416	412
0	11	324	363
0	12	2874	2868
0	13	567	605
0	14	540	561
0	15	907	903
0	16	929	931
0	17	409	422
0	19	225	193
0	20	663	665
0	22	767	798
1	1	414	389
1	2	146	149
1	3	844	791
1	4	624	632
1	5	618	595
1	6	96	69
1	7	318	333
1	8	134	123
1	9	203	173
1	10	107	66
1	11	355	349
1	12	126	60
1	13	541	566
1	14	327	324
1	15	164	142
1	19	118	145
1	22	145	128
2	0	750	709
2	1	279	297
2	2	750	684
2	3	165	158
2	4	386	394
2	5	328	324
2	6	621	603
2	7	229	227
2	8	471	439
2	9	450	443
2	10	602	582
2	12	513	499
2	14	326	307
2	15	164	151
2	16	432	443
2	17	89	5
2	18	105	146
2	20	234	224
2	22	358	384
3	1	513	475
3	2	574	601
3	3	1924	1869
3	4	452	474
3	5	277	265
3	6	178	84
3	7	1822	1798
3	8	396	419
3	9	730	725
3	10	288	284
3	11	103	86
3	13	636	635
3	14	278	264
3	15	366	322
3	17	480	483
3	20	322	325
3	21	144	86
3	22	142	44
4	0	171	173
4	1	387	423
4	2	64	38
4	3	220	236
4	4	250	214
4	5	88	54
4	7	192	186
4	8	71	47
4	10	118	139
4	13	125	46
4	14	129	158
4	17	193	156
4	20	120	62
4	22	147	37
5	1	565	554
5	2	402	422
5	3	1541	1474
5	4	687	759
5	5	138	229
5	6	88	52
5	7	630	611
5	8	417	420
5	9	373	392
5	10	85	29
5	11	229	219
5	12	295	340
5	13	2118	2118
5	14	754	763
5	15	843	864
5	16	115	137
5	17	194	191
5	18	202	206
5	19	379	393
5	20	263	271
5	22	258	301
6	1	77	100
6	2	460	437
6	4	179	204
6	5	227	207
6	6	149	187
6	7	238	241
6	8	200	186
6	9	179	164
6	10	98	35
6	11	203	209
6	12	406	377
6	14	299	312
6	15	117	133
6	16	239	271
6	17	191	141
6	20	157	255
6	22	181	255
7	1	548	578

DEHYDRATED POTASSIUM PTILOLITE FOBS AND FCALC (X10) PAGE 5

K	H	FOBS	FCALC
7	2	316	356
7	3	173	166
7	4	166	146
7	5	471	476
7	6	236	242
7	7	690	690
7	8	287	273
7	9	121	97
7	10	411	402
7	11	858	877
7	12	353	329
7	13	139	14
7	14	146	144
7	15	202	225
7	16	111	196
7	17	465	452
7	18	98	64
7	19	306	324
7	20	239	261
7	21	143	205
7	22	127	36
8	0	602	676
8	1	364	387
8	2	300	327
8	3	181	183
8	4	1392	1356
8	5	476	477
8	6	122	65
8	7	354	371
8	8	1907	1895
8	9	720	765
8	10	1840	1813
8	11	148	164
8	12	1040	1050
8	14	219	273
8	15	297	262
8	17	472	466
8	18	913	920
8	19	313	343
8	20	369	385
8	21	228	191
9	2	96	34
9	3	345	323
9	4	191	170
9	5	385	376
9	6	155	124
9	7	361	358
9	8	113	93
9	12	88	25
9	13	183	193
9	14	104	20
9	15	205	216
9	16	103	52
9	17	236	192
9	19	142	135
10	0	176	177
10	1	224	253
10	2	739	728
10	3	206	214
10	5	229	241
10	6	164	128
10	7	273	272
10	8	180	161
10	9	294	297
10	10	459	451
10	11	196	198
10	12	1066	1091
10	13	122	155
10	17	125	157
10	18	140	125
11	1	323	345
11	2	88	69
11	3	506	473
11	4	76	9
11	5	136	151
11	7	686	658
11	8	273	245
11	9	129	81
11	10	330	341
11	11	441	452
11	13	99	92
11	15	243	262
11	17	411	421
11	18	105	72
11	19	110	133
11	20	187	133
12	0	1168	1131
12	1	453	598
12	2	286	276
12	3	151	123
12	4	349	351
12	5	72	53
12	6	554	559
12	7	267	293
12	8	383	387
12	9	661	658
12	10	680	696
12	11	152	201
12	12	455	446
12	13	154	143
12	17	297	310
12	18	397	395
12	19	167	190
13	1	1373	1392
13	2	791	832
13	3	1317	1347
13	4	202	177
13	5	132	123
13	6	390	420
13	7	2138	2115
13	8	409	419
13	9	957	930
13	10	307	331
13	11	759	738
13	12	361	331
13	13	299	249
13	14	112	42
13	15	437	441
13	16	313	341
13	17	892	961
13	18	205	213
13	19	309	288
14	0	408	353
14	2	194	242
14	4	272	229
14	6	178	124
14	8	182	179
14	9	98	54
14	10	286	324
14	11	110	94
14	12	244	274
14	14	109	166
14	17	102	96
15	1	309	308
15	4	392	404
15	5	959	977
15	6	311	339
15	7	218	210
15	11	239	225
15	12	151	177
15	13	150	102
15	14	203	216
15	15	155	167
15	17	323	335
16	0	1186	1157
16	1	256	263
16	2	1080	1089
16	3	248	275
16	4	556	563
16	5	95	2
16	7	222	178
16	8	484	453
16	9	133	155
16	10	563	541
16	12	439	441
16	15	199	166
16	16	243	228
16	17	143	134
16	18	319	320
17	4	291	273
17	5	287	259
17	6	287	273
17	10	186	198
17	11	213	224
17	12	207	181
17	14	133	120
17	15	135	124
17	16	177	107
18	0	1020	1006
18	1	265	273
18	3	191	142
18	4	281	272
18	6	517	526
18	7	112	111
18	8	471	467
18	9	265	273
18	10	693	659
18	13	102	102
18	15	107	26
19	11	98	33

DEHYDRATED POTASSIUM PTILOLITE FOBS AND FCALC (X10) PAGE 6

K	H	FOBS	FCALC
19	12	101	64
20	0	169	205
20	1	252	237
20	2	255	229
20	3	315	344
20	4	373	377
20	5	163	127
20	6	660	669
20	7	360	385
20	9	194	203
20	11	273	247
20	12	252	224
20	14	306	234
21	1	114	130
21	3	464	472
21	4	265	276
21	5	786	767
21	6	136	184
21	7	260	276
21	9	169	154
21	11	192	153
21	12	259	236
21	13	645	603
22	0	261	261
22	1	166	136
22	2	161	178
22	4	181	171
22	8	172	160
22	9	136	145
22	10	259	229
22	12	156	180
23	1	202	220
23	3	174	138
23	5	269	280
23	7	400	379
23	9	150	190
24	0	222	176
24	2	258	212
24	4	147	89
24	6	227	228
24	7	116	45
24	8	174	193
25	1	343	320
25	4	490	459
25	5	404	408
25	6	261	255

L= 3

K	H	FOBS	FCALC
1	1	693	675
1	2	173	164
1	3	104	139
1	4	285	253
1	5	378	408
1	6	789	816
1	7	1816	1841
1	8	1141	1117
1	9	178	148
1	10	694	681
1	11	1237	1272
1	15	275	301
1	16	739	724
1	17	670	675
1	18	451	469
1	19	431	415
1	20	199	215
1	21	442	467
1	22	218	221
2	0	550	542
2	1	207	176
2	2	343	333
2	3	205	183
2	4	256	260
2	5	251	229
2	6	335	301
2	7	141	124
2	8	106	55
2	9	599	597
2	10	191	174
2	11	122	103
2	12	425	432
2	13	88	18
2	15	267	284
2	16	98	19
2	17	159	104
2	18	172	198
2	21	178	163
2	22	170	159
3	1	160	144
3	2	550	526
3	3	228	199
3	4	212	132
3	5	645	628
3	6	257	278
3	7	182	178
3	9	114	60
3	10	196	204
3	12	181	104
3	13	290	276
3	14	343	342
3	16	157	174
3	17	176	167
3	19	204	178
3	21	167	155
4	0	285	321
4	1	89	7
4	2	682	732
4	3	1263	1303
4	4	949	943
4	5	661	653
4	6	178	206
4	7	363	372
4	8	1580	1628
4	9	1188	1207
4	10	1072	1080
4	11	368	402
4	12	119	61
4	13	787	797
4	15	342	296
4	16	369	363
4	17	114	62
4	18	741	748
4	19	477	499
4	20	267	291
4	21	139	173
4	22	152	129
5	1	79	10
5	4	419	440
5	6	515	507
5	7	167	180
5	8	383	366
5	10	113	27
5	11	168	151
5	12	135	75
5	14	268	278
5	16	200	215
5	19	117	115
5	20	99	32
5	21	100	59
6	0	566	564
6	1	151	150
6	2	422	399
6	3	203	200
6	4	303	301
6	5	471	469
6	6	617	614
6	7	144	91
6	8	609	601
6	9	724	704
6	10	695	728
6	11	163	134
6	12	597	610
6	14	341	332
6	15	588	588
6	16	174	136
6	18	496	522
6	20	120	98
6	21	236	219
7	1	610	623
7	2	210	206
7	3	404	381
7	4	677	682
7	5	766	789
7	6	298	304
7	8	276	288
7	9	119	136
7	10	328	303
7	11	197	222
7	12	216	166
7	13	699	725
7	14	448	458
7	15	285	260
7	18	252	243
7	19	175	144
7	20	225	171
7	21	119	86
8	0	171	163
8	2	167	162
8	3	570	576

DEHYDRATED POTASSIUM PTILOLITE FOBS AND FCALC (X10) PAGE 7

K	H	FOBS	FCALC
8	4	171	167
8	5	495	491
8	6	604	587
8	7	154	125
8	8	290	288
8	9	228	215
8	10	199	234
8	11	134	86
8	12	240	228
8	14	234	201
8	16	401	391
8	17	108	42
8	19	114	22
9	1	151	142
9	2	402	422
9	3	285	282
9	4	629	630
9	5	1363	1382
9	6	1145	1189
9	7	698	734
9	8	583	618
9	9	110	68
9	10	177	159
9	11	599	637
9	12	101	63
9	13	656	615
9	14	309	320
9	15	538	554
9	16	395	411
9	18	279	304
10	1	221	236
10	2	121	84
10	3	302	310
10	4	89	43
10	5	166	167
10	7	269	292
10	8	130	122
10	9	229	249
10	10	153	161
10	11	322	326
10	12	138	104
10	13	138	125
10	16	112	62
10	17	184	190
10	18	154	86
11	1	426	449
11	2	192	174
11	3	438	443
11	4	563	570
11	5	476	484
11	6	102	152
11	7	416	408
11	8	180	157
11	9	371	357
11	10	382	405
11	11	228	230
11	13	594	594
11	14	260	267
11	15	169	154
11	16	207	166
11	17	298	300
11	20	162	137
12	0	243	226
12	1	187	222
12	2	1129	1137
12	3	990	1010
12	4	636	668
12	6	741	743
12	7	373	347
12	8	561	544
12	9	151	113
12	10	153	149
12	11	527	533
12	12	772	763
12	13	558	561
12	14	179	238
12	16	521	535
12	18	132	85
12	19	156	168
13	1	155	143
13	2	640	644
13	4	312	321
13	5	589	581
13	6	337	359
13	7	165	152
13	8	482	462
13	12	126	140
13	13	232	247
13	15	126	128
13	16	255	267
13	17	235	214
13	18	147	146
14	0	343	368
14	1	229	214
14	2	528	529
14	3	919	957
14	4	937	966
14	5	230	212
14	6	890	895
14	7	353	347
14	8	170	173
14	9	306	285
14	10	111	151
14	11	177	177
14	12	523	508
14	13	279	245
14	15	232	230
14	16	472	446
14	17	106	69
14	18	199	180
15	3	235	247
15	5	127	90
15	6	101	138
15	7	250	238
15	8	161	149
15	11	126	4
15	12	92	51
15	13	100	88
15	17	164	128
16	0	615	634
16	1	461	466
16	2	466	453
16	3	377	376
16	5	184	211
16	6	186	150
16	7	152	155
16	8	341	317
16	9	106	37
16	10	260	252
16	11	103	56
16	12	112	13
16	14	248	210
16	16	187	228
17	3	450	470
17	4	466	458
17	5	941	915
17	6	561	570
17	7	273	305
17	8	188	220
17	10	134	166
17	11	228	203
17	12	172	177
17	13	576	559
17	14	381	356
17	15	311	348
17	16	344	355
18	0	145	115
18	1	498	505
18	3	109	126
18	5	80	85
18	6	139	140
18	7	183	147
18	9	301	302
18	11	297	308
18	12	107	67
19	1	391	438
19	2	375	382
19	3	240	270
19	4	332	326
19	5	505	502
19	6	198	186
19	7	393	399
19	8	114	128
19	10	230	216
19	11	327	294
19	12	116	127
19	13	227	207
19	14	160	154
20	0	261	273
20	1	257	292
20	4	117	111
20	8	363	357
20	9	331	314
20	10	278	250
20	11	197	231
20	13	177	128
21	1	204	263
21	2	104	135
21	3	278	250
21	4	292	291

DEHYDRATED POTASSIUM PTILOLITE FOBS AND FCALC (X10) PAGE 8

K	H	FOBS	FCALC
21	5	162	108
21	6	170	187
21	7	228	219
21	9	198	135
21	10	101	17
21	11	402	388
21	12	186	172
22	0	313	295
22	2	232	232
22	3	133	142
22	4	149	192
22	5	86	49
22	6	224	208
22	7	175	164
22	8	333	326
22	9	223	226
22	10	332	308
23	6	148	156
23	8	148	140
24	0	505	516
24	1	243	263
24	4	309	308
24	5	253	240
25	1	174	231
25	2	237	232

L= 4

K	H	FOBS	FCALC
0	0	3247	3341
0	1	2081	2079
0	2	917	872
0	3	131	101
0	4	861	842
0	5	302	297
0	6	1165	1174
0	7	629	614
0	8	398	374
0	9	1164	1204
0	10	1432	1439
0	11	286	299
0	12	497	546
0	13	870	868
0	14	421	409
0	15	205	251
0	16	453	424
0	17	625	643
0	18	241	251
0	19	330	313
0	20	410	381
0	21	480	479
1	1	336	333
1	2	497	491
1	3	134	69
1	4	115	103
1	6	334	315
1	7	500	490
1	8	262	287
1	9	409	394
1	10	329	347
1	11	197	180
1	12	197	199
1	13	106	86
1	16	109	151
1	17	205	185
1	20	194	216
2	0	767	771
2	1	88	23
2	2	188	182
2	3	570	592
2	4	104	60
2	5	126	134
2	6	448	422
2	7	204	199
2	8	176	101
2	9	293	293
2	10	282	278
2	11	152	180
2	12	458	499
2	13	229	248
2	15	183	153
2	16	99	33
2	17	170	184
2	18	267	301
2	19	135	154
2	21	159	167
3	1	229	248
3	2	205	197
3	3	482	445
3	4	764	771
3	5	952	952
3	6	606	619
3	7	559	550
3	8	304	290
3	9	422	426
3	11	125	35
3	12	287	294
3	14	136	116
3	15	248	250
3	17	105	96
3	18	287	298
3	19	313	270
3	20	182	87
4	0	139	151
4	1	375	364
4	2	247	242
4	3	205	184
4	4	93	43
4	5	108	108
4	6	78	44
4	7	118	98
4	8	105	65
4	9	270	265
4	10	128	93
4	11	96	47
4	12	101	16
4	15	107	105
4	17	142	170
4	19	139	136
4	21	106	45
5	1	994	1013
5	2	844	890
5	4	361	368
5	5	233	218
5	6	409	418
5	7	718	714
5	8	227	146
5	9	376	366
5	10	1081	1090
5	11	985	999
5	12	599	598
5	13	105	60
5	15	247	281
5	16	322	324
5	17	494	502
5	18	245	219
5	19	293	266
5	20	511	479
6	0	514	520
6	1	128	159
6	2	118	146
6	3	230	231
6	6	438	431
6	7	130	22
6	8	114	29
6	9	396	402
6	10	323	292
6	11	133	143
6	12	190	220
6	16	159	79
6	17	138	101
6	18	204	217
6	19	133	144
7	1	427	435
7	3	132	40
7	4	773	790
7	5	312	294
7	6	323	337
7	7	441	438
7	8	401	423
7	9	88	23
7	11	344	329
7	12	336	319
7	13	345	366
7	14	436	434
7	15	138	123
7	16	136	128
7	17	159	147
7	19	333	309
7	20	108	88
8	0	119	110
8	1	218	226
8	2	166	133
8	3	330	378
8	4	182	166
8	5	687	701
8	6	972	985
8	7	1220	1263
8	8	504	504
8	9	189	217

DEHYDRATED POTASSIUM PTILOLITE FOBS AND FCALC (X10) PAGE 9

K	H	FOBS	FCALC
8	10	778	819
8	11	305	313
8	12	720	711
8	13	183	167
8	14	221	227
8	15	526	519
8	17	349	328
8	18	246	213
8	20	422	438
9	1	197	197
9	3	288	269
9	4	122	85
9	5	364	364
9	6	186	144
9	7	329	308
9	8	131	127
9	9	133	93
9	11	126	112
9	12	130	50
9	14	110	53
9	15	235	226
9	16	110	62
9	17	284	248
9	19	155	60
10	0	742	736
10	1	367	375
10	2	111	71
10	3	84	57
10	4	168	107
10	5	93	6
10	6	130	81
10	7	101	108
10	8	210	203
10	9	386	411
10	10	632	652
10	11	193	185
10	13	159	113
10	16	230	230
10	18	115	161
11	1	232	215
11	2	126	125
11	3	235	235
11	4	242	210
11	5	530	510
11	6	160	134
11	7	161	143
11	8	329	329
11	10	178	141
11	11	144	110
11	12	217	216
11	13	247	288
11	14	106	125
11	15	220	230
11	19	123	133
12	0	527	547
12	1	298	278
12	2	118	81
12	3	730	758
12	4	207	184
12	5	397	403
12	6	219	244
12	7	731	753
12	8	302	301
12	9	190	132
12	10	323	319
12	11	305	319
12	12	263	286
12	13	156	190
12	15	450	439
12	18	300	279
13	1	164	163
13	3	389	378
13	4	1300	1328
13	5	1291	1320
13	6	525	567
13	8	560	556
13	9	455	430
13	10	101	47
13	11	418	417
13	12	157	210
13	13	365	389
13	14	476	505
13	15	243	260
13	16	351	368
13	17	161	111
14	0	362	363
14	2	194	165
14	4	140	142
14	7	104	39
14	8	188	179
14	10	214	256
14	12	134	159
14	14	164	176
14	17	132	31
15	1	162	185
15	2	485	507
15	4	100	87
15	5	226	244
15	6	242	259
15	7	290	277
15	8	341	356
15	10	138	169
15	12	186	174
15	13	208	206
15	14	116	88
15	15	109	80
15	16	180	169
16	0	695	737
16	1	368	425
16	2	564	591
16	3	121	158
16	4	373	391
16	5	112	22
16	6	104	49
16	7	182	210
16	8	146	145
16	9	174	191
16	10	276	277
16	12	250	243
16	13	161	157
16	14	145	102
17	1	190	223
17	3	255	233
17	4	545	540
17	5	352	350
17	6	351	334
17	7	325	302
17	9	126	76
17	11	142	163
17	12	216	199
17	13	129	73
17	14	287	239
17	15	275	220
18	2	271	275
18	3	389	417
18	4	239	248
18	5	356	341
18	6	345	351
18	7	425	444
18	8	117	103
18	10	103	26
18	12	336	321
18	14	133	121
19	10	112	51
19	11	148	67
20	0	126	104
20	1	533	556
20	2	213	201
20	3	119	87
20	4	224	265
20	6	421	426
20	7	235	199
20	9	605	588
21	2	239	288
21	3	236	261
21	4	126	163
21	5	183	190
21	7	351	315
21	10	383	358
22	0	221	245
22	1	146	144
22	2	303	267
22	4	165	143
22	6	141	124
22	7	154	161
22	8	227	204
23	4	98	109
23	5	185	208
23	6	111	9
24	0	267	246

L= 5

K	H	FOBS	FCALC
1	1	266	281
1	2	222	155
1	3	915	944
1	4	1807	1805
1	5	1410	1452
1	6	1312	1280

DEHYDRATED POTASSIUM PTILOLITE FOBS AND FCALC (X10) PAGE 10

1	7	162	170
1	8	508	497
1	9	275	293
1	11	151	156
1	12	526	537
1	13	527	555
1	14	775	768
1	15	436	435
1	16	151	81
1	20	287	314
2	0	292	317
2	2	149	84
2	3	555	551
2	5	273	281
2	6	142	64
2	7	536	522
2	8	264	236
2	9	116	32
2	10	318	302
2	11	227	213
2	13	215	218
2	15	237	254
2	17	122	22
2	19	133	73
3	1	157	144
3	2	380	386
3	3	250	241
3	4	617	638
3	6	478	499
3	7	131	126
3	8	159	71
3	10	338	350
3	11	146	60
3	12	344	350
3	13	94	36
3	14	217	210
3	15	120	5
3	16	120	176
3	17	109	10
3	19	105	159
4	0	756	763
4	1	1356	1387
4	2	1138	1158
4	3	1312	1323
4	4	204	208
4	5	181	178
4	6	874	881
4	7	1638	1636
4	8	900	949
4	9	1071	1107
4	10	400	417
4	11	831	835
4	12	408	452
4	13	172	147
4	15	113	153
4	16	365	353
4	17	389	394
4	18	292	348
4	19	276	274
5	1	115	73

5	2	315	308
5	4	558	569
5	5	153	87
5	6	486	474
5	7	136	90
5	8	125	104
5	9	107	28
5	11	111	42
5	12	195	184
5	13	117	101
5	14	138	189
5	15	102	56
5	16	118	156
6	0	138	93
6	1	261	243
6	2	600	597
6	3	615	793
6	4	333	305
6	5	297	277
6	6	653	669
6	7	714	679
6	8	197	198
6	9	229	192
6	11	341	334
6	12	435	417
6	13	334	311
6	14	116	105
6	15	219	215
6	16	171	197
6	17	200	242
6	19	186	194
7	1	354	361
7	2	795	798
7	3	427	470
7	4	134	174
7	5	267	273
7	6	529	512
7	7	536	537
7	8	537	520
7	10	324	283
7	11	492	500
7	12	265	257
7	15	239	231
7	16	457	437
7	17	198	248
7	18	107	112
7	19	167	105
8	0	228	263
8	1	560	555
8	2	95	69
8	3	209	193
8	4	268	234
8	5	119	21
8	7	274	271
8	8	318	309
8	9	305	290
8	10	179	152
8	11	139	105
8	12	177	142
8	13	179	206

8	17	206	198
8	18	255	249
9	1	278	301
9	2	709	737
9	3	843	825
9	4	1273	1309
9	5	841	880
9	6	1406	1395
9	7	493	501
9	8	353	348
9	9	271	256
9	10	444	458
9	11	358	380
9	12	551	552
9	13	324	344
9	14	281	306
9	15	420	418
9	16	480	432
9	17	306	330
9	18	105	62
10	0	158	122
10	1	455	446
10	2	99	49
10	3	268	300
10	5	116	84
10	6	163	144
10	7	375	374
10	8	99	77
10	9	324	351
10	11	170	135
10	15	127	56
11	1	304	287
11	2	771	786
11	4	116	57
11	5	283	249
11	6	252	248
11	8	435	420
11	9	111	10
11	10	359	344
11	11	371	385
11	12	234	234
11	16	287	267
12	0	1083	1141
12	1	1291	1317
12	2	452	444
12	3	160	138
12	4	91	81
12	5	184	166
12	6	99	45
12	7	470	452
12	8	620	636
12	9	1089	1066
12	10	561	619
12	11	337	327
12	12	138	112
12	13	150	104
12	15	275	234
12	17	404	354
13	1	192	142
13	2	92	94

DEHYDRATED POTASSIUM PTILOLITE FOBS AND FCALC (X10) PAGE 11

13	3	197	190
13	4	308	307
13	6	197	203
13	7	192	187
13	8	137	71
13	10	343	332
13	11	155	143
13	14	233	236
14	0	760	753
14	1	731	725
14	2	311	325
14	5	264	272
14	7	179	153
14	8	503	540
14	9	778	780
14	10	325	340
14	11	151	125
14	12	105	48
14	13	269	280
14	15	238	222
15	2	102	69
15	3	113	133
15	5	205	208
15	10	195	202
15	11	130	116
15	15	137	113
16	0	302	307
16	1	185	208
16	2	255	235
16	4	153	156
16	5	316	324
16	6	108	95
16	7	412	427
16	8	136	103
16	9	290	328
16	10	113	73
16	12	105	4
17	1	341	339
17	2	636	669
17	3	448	470
17	4	750	760
17	5	466	513
17	6	725	735
17	7	412	444
17	8	314	317
17	9	117	40
17	10	410	369
17	11	283	344
17	12	405	404
17	13	164	173
18	1	224	266
18	3	299	308
18	4	160	97
18	7	127	173
18	8	131	92
18	9	117	142
18	10	120	4
18	11	212	222
19	1	201	164
19	2	226	211

19	3	122	103
19	4	338	332
19	5	220	237
19	7	296	291
19	8	270	305
19	10	259	251
19	11	156	157
20	0	220	224
20	1	246	248
20	2	162	159
20	3	256	271
20	5	87	107
20	6	278	274
20	7	284	273
20	8	179	186
20	9	200	174
21	3	116	120
21	4	246	227
21	5	194	200
22	0	275	260
22	1	315	351
22	2	248	270
22	3	287	234

L= 6

K	H	FOBS	FCALC
0	0	256	124
0	1	916	906
0	2	481	514
0	3	1320	1394
0	4	626	623
0	5	286	307
0	6	363	309
0	7	293	304
0	8	437	421
0	9	660	666
0	10	234	240
0	12	638	624
0	13	294	314
0	14	330	327
0	15	513	554
0	16	392	336
0	18	286	221
1	2	418	414
1	3	163	115
1	4	169	184
1	5	144	72
1	6	116	144
1	7	107	73
1	8	154	158
1	9	172	158
1	10	250	250
1	11	193	195
1	12	121	103
1	14	103	107
1	15	108	33
1	17	149	60
1	18	106	20
2	0	362	364

2	1	379	365
2	2	142	132
2	3	305	287
2	4	205	226
2	5	250	246
2	6	94	58
2	7	302	256
2	9	289	286
2	10	214	212
2	11	107	42
2	15	316	346
2	16	113	108
3	1	152	71
3	2	139	118
3	3	125	92
3	4	114	116
3	5	92	96
3	6	153	146
3	7	203	226
3	8	159	188
3	9	312	301
3	10	175	159
3	11	298	266
3	12	117	59
3	14	131	148
3	17	248	229
4	0	126	104
4	2	254	214
4	3	248	243
4	4	205	190
4	5	118	77
4	6	147	100
4	8	117	87
4	9	116	77
4	11	101	6
4	14	150	70
4	16	128	126
4	17	103	34
5	1	321	372
5	2	1175	1200
5	3	185	125
5	5	104	39
5	6	102	92
5	8	500	481
5	9	260	207
5	10	112	96
5	11	510	505
5	13	115	107
5	14	332	375
5	16	226	226
5	17	247	245
6	0	123	47
6	1	379	381
6	2	255	270
6	3	262	243
6	4	103	33
6	6	126	39
6	7	97	116
6	8	130	73
6	9	298	280

DEHYDRATED POTASSIUM PTILOLITE FOBS AND FCALC (X10) PAGE 12

K	H	FOBS	FCALC
6	11	140	101
6	12	232	206
6	13	137	135
6	16	145	24
6	17	177	60
7	1	260	253
7	2	323	310
7	3	121	90
7	4	387	395
7	5	353	367
7	6	104	42
7	7	121	76
7	8	388	396
7	9	177	117
7	10	299	297
7	11	108	65
7	13	211	205
7	14	137	169
7	16	384	364
7	17	266	239
8	0	101	93
8	1	405	399
8	2	242	220
8	3	488	461
8	4	104	92
8	5	199	213
8	6	326	297
8	7	97	62
8	9	468	507
8	10	238	244
8	11	267	238
8	12	281	256
8	13	206	175
8	14	436	408
8	15	183	199
8	16	485	444
9	3	255	246
9	4	172	143
9	5	441	400
9	6	152	148
9	7	180	187
9	9	119	65
9	12	117	64
9	13	230	226
9	14	105	50
9	15	115	184
10	0	217	220
10	1	411	421
10	3	338	339
10	4	95	33
10	5	112	144
10	6	170	162
10	7	124	116
10	9	136	145
10	12	230	211
10	14	111	127
10	16	176	214
11	1	167	130
11	2	213	232
11	3	116	107

K	H	FOBS	FCALC
11	7	169	138
11	8	108	116
11	10	263	291
11	13	114	79
11	14	120	46
11	15	106	52
12	0	135	100
12	1	253	242
12	2	304	309
12	3	299	295
12	4	214	159
12	5	228	256
12	6	355	362
12	7	110	130
12	8	311	316
12	9	486	487
12	10	194	140
12	12	181	158
12	14	136	114
12	15	301	349
13	1	112	112
13	2	141	173
13	3	113	8
13	4	589	595
13	5	304	286
13	6	113	115
13	7	414	460
13	8	530	567
13	9	406	355
13	10	370	397
13	11	197	164
13	12	174	72
13	13	257	249
13	14	427	456
14	0	173	116
14	1	150	111
14	2	247	259
14	7	148	124
14	8	177	154
14	9	165	118
14	10	158	146
14	12	201	234
15	3	131	86
15	4	388	388
15	5	123	125
15	6	148	103
15	7	170	119
15	8	108	63
15	10	242	256
15	11	162	188
16	0	194	195
16	1	106	26
16	2	152	172
16	3	184	147
16	4	197	191
16	6	217	209
16	10	102	68
16	11	139	85
17	1	133	69
17	2	149	150

K	H	FOBS	FCALC
17	3	391	366
17	5	479	434
17	6	213	173
17	7	263	291
17	8	192	178
18	1	130	124
18	2	112	124
18	3	330	356
18	4	139	92
18	5	289	283
18	6	409	353
18	7	219	176
18	8	129	49
18	9	207	231
19	4	115	141
19	5	120	2
19	6	101	58
19	7	135	47
20	0	427	417
20	1	106	46
20	2	108	149
20	3	380	352
20	5	133	145

L= 7

K	H	FOBS	FCALC
1	1	249	243
1	2	932	927
1	3	160	178
1	4	300	270
1	5	267	246
1	6	634	630
1	7	262	252
1	8	799	801
1	9	166	201
1	10	451	444
1	12	349	361
1	14	209	192
1	15	195	214
1	16	702	672
2	0	267	246
2	1	220	242
2	2	154	132
2	3	187	203
2	5	127	53
2	6	123	78
2	7	96	9
2	8	216	202
2	9	322	312
2	10	239	206
2	11	99	11
2	12	117	121
2	14	133	52
2	15	211	177
3	1	159	126
3	2	183	206
3	3	110	127
3	4	381	385
3	5	90	14

DEHYDRATED POTASSIUM PTILOLITE FOBS AND FCALC (X10) PAGE 13

K	H	FOBS	FCALC
3	6	319	304
3	7	171	159
3	10	130	50
3	11	216	189
3	12	189	180
3	13	130	126
3	14	287	291
3	15	141	79
4	0	357	373
4	1	1160	1152
4	2	113	124
4	3	463	453
4	4	161	165
4	7	574	570
4	8	159	127
4	9	1111	1083
4	10	161	210
4	11	631	662
4	12	124	174
4	13	401	390
4	14	333	333
5	2	175	158
5	4	221	239
5	5	160	128
5	6	285	277
5	8	205	184
5	12	111	105
5	14	150	179
5	15	119	4
6	0	312	354
6	1	329	333
6	2	140	108
6	3	464.	449
6	5	245	211
6	7	131	163
6	8	103	139
6	9	652	611
6	10	144	186
6	11	105	61
6	14	162	137
6	15	459	403
7	2	99	84
7	3	153	136
7	4	721	702
7	5	333	332
7	6	426	397
7	7	105	80
7	10	276	272
7	12	236	209
7	13	191	171
7	14	560	532
8	0	336	318
8	1	209	174
8	2	120	108
8	3	351	310
8	4	112	1
8	5	121	104
8	6	147	133
8	7	243	219
8	10	164	154

K	H	FOBS	FCALC
8	11	125	143
8	12	114	23
8	13	131	97
8	14	132	6
9	2	340	378
9	3	106	112
9	4	985	971
9	5	261	295
9	6	778	761
9	8	306	306
9	9	116	86
9	12	298	300
9	14	688	668
10	0	130	67
10	1	165	187
10	2	158	125
10	3	126	164
10	4	157	118
10	5	96	28
10	6	139	130
10	7	125	113
10	9	196	191
10	13	141	130
11	1	102	114
11	2	115	81
11	3	190	171
11	4	588	571
11	5	161	175
11	6	254	253
11	7	198	155
11	8	180	167
11	9	152	121
11	10	203	159
11	12	210	173
12	0	176	140
12	1	558	553
12	2	272	262
12	3	871	848
12	4	235	206
12	5	161	150
12	6	151	133
12	7	546	532
12	8	107	97
12	9	261	257
12	10	117	69
12	11	634	632
12	12	229	243
13	1	98	68
13	2	222	192
13	8	313	288
13	9	106	27
13	10	143	39
14	2	168	188
14	3	683	651
14	4	155	182
14	5	103	139
14	7	361	340
14	9	147	170
15	2	174	156
15	3	147	111

K	H	FOBS	FCALC
15	5	239	192
15	8	168	108
16	1	140	191
16	2	124	29
16	3	115	74
16	5	163	137
16	6	174	145
16	7	242	245
17	1	113	60
17	2	222	230
17	3	103	119
17	4	713	704
17	5	214	234
17	6	518	517
18	0	116	22
18	1	218	214
18	2	101	59

L= 8

K	H	FOBS	FCALC
0	0	161	218
0	1	171	97
0	2	165	116
0	3	594	615
0	4	101	86
0	5	342	379
0	6	259	246
0	7	451	483
0	8	234	256
0	10	150	64
0	11	814	784
0	12	351	210
0	13	449	436
1	3	132	122
1	4	220	238
1	5	162	115
1	6	159	61
1	7	111	62
1	8	127	147
1	9	154	157
1	11	110	67
1	12	234	157
1	13	110	95
2	3	309	325
2	9	172	207
2	10	140	106
2	11	187	194
2	12	174	148
2	13	291	270
3	2	259	268
3	4	149	124
3	5	110	81
3	6	155	116
3	7	133	96
3	8	211	198
3	9	119	113
3	12	214	216
4	0	181	172
4	8	124	105

DEHYDRATED POTASSIUM PTILOLITE FOBS AND FCALC (X10) PAGE 14

K	H	FOBS	FCALC
4	12	134	103
5	4	709	696
5	5	150	101
5	6	332	326
5	9	271	235
5	10	482	468
5	11	209	164
5	12	294	229
6	0	308	317
6	2	143	124
6	3	260	257
6	5	112	118
6	7	122	151
6	10	194	179
6	11	116	123
6	12	162	90
7	2	390	387
7	4	167	194
7	5	115	4
7	7	123	34
7	8	277	305
7	9	162	84
7	10	357	325
8	0	125	26
8	1	718	704
8	5	185	198
8	6	108	93
8	7	122	105
8	8	105	45
8	9	602	562
8	11	311	263
9	1	136	158
9	3	214	184
9	5	211	178
9	7	263	254
9	9	154	64
10	1	167	128
10	3	106	145
10	5	89	71
10	8	128	139
10	9	109	28
11	1	126	142
11	2	173	166
11	3	111	64
11	6	126	83
11	8	238	212
12	0	146	144
12	1	205	233
12	2	166	128
12	3	256	281
12	5	112	87
12	6	179	160
12	7	110	71
13	2	650	642
13	6	204	170
14	0	310	286
14	1	156	87
14	2	172	130

L= 9

K	H	FOBS	FCALC
1	1	119	90
1	3	200	113
1	4	801	770
1	6	342	320
1	7	223	200
1	8	209	182
2	0	111	143
2	1	136	111
2	2	206	186
2	3	176	178
2	4	134	8
2	6	232	177
2	8	127	79
3	1	168	92
3	2	100	96
3	3	314	284
3	4	116	120
3	5	277	226
3	6	160	145
4	0	344	296
4	1	642	612
4	2	146	45
4	3	470	437
4	4	208	169
4	5	249	229
4	6	268	214
4	7	485	437
5	1	168	91
5	4	160	155
5	6	111	130
6	2	122	79
6	3	378	358
6	5	258	259
6	6	127	62
7	2	518	478
7	4	107	84
7	5	130	79
7	6	186	176
8	0	373	317
8	1	139	139
8	2	246	223
9	1	228	178
9	2	472	466
9	3	272	186

CHARACTERIZATION OF THE OFFRETITE—LEVYNITE INTERGROWTH FROM BEECH CREEK, OREGON, BY ADSORPTION AND ELECTRON DIFFRACTION

J. M. BENNETT and R. W. GROSE

Union Carbide Corporation, Tarrytown, New York 10591

Abstract – The X-ray powder diffraction pattern of the zeolite intergrowth from Beech Creek, Grant County, Oregon, contains diffraction lines which cannot be indexed using the unit cell of either levynite or offretite. The two zeolites differ greatly in morphology, and selected-area electron diffraction patterns show that the symmetry of the fibrous crystals is hexagonal with $a = 13.25–13.35$ Å and $c = 15.10–15.20$ Å and systematic absences for $hh.l$ of $l = 2n+1$. These data are compatible either with the erionite structure or with an ordered silicon–aluminium distribution in the offretite structure.

The oxygen adsorption capacity of another sample indicates that the material is greater than 90% zeolite. This sample slowly adsorbs n-butane but does not adsorb neopentane. The offretite crystal structure with its "12-membered ring" adsorption channel permits the adsorption of molecules with a kinetic diameter greater than 6 Å, such as neopentane. The adsorption data are incompatible with the offretite crystal structure; thus the fibrous zeolite is erionite and not offretite.

INTRODUCTION

Offretite is a rare zeolite that was first found near Montbrison, France (Gonnard, 1890). The limited amount of sample available from this locality has prevented the characterization of the physical adsorption properties. An occurrence near Beech Creek, Grant County, Oregon, has been described (Sheppard *et al.*, 1974) and sufficient material from this occurrence was supplied by Dr. R. A. Sheppard for adsorption, X-ray diffraction, and electron diffraction studies.

Several occurrences of levynite have been described where levynite crystals are found as thin lamallae coated on both sides by a compact fibrous material. The fibers grow epitaxially on the (0001) plane of levynite and are elongated in the c direction. The fibers, although from different localities, have been described as offretite (Sheppard *et al.*, 1974) and erionite (Passaglia *et al.*, 1974). To distinguish between the two zeolite species requires careful use of several characterization techniques.

MATERIALS

Beech Creek, Oregon, crystal intergrowth

Three samples from Beech Creek, Oregon, were obtained from Dr. R. A. Sheppard. Two samples were used for the X-ray diffraction and electron diffraction examination. One of these samples was that studied by Sheppard *et al.* (1974). One gram of crystal was hand-picked from vesicles in the zeolite-bearing basalt of the third sample. Optically, the sample consists of platy and fibrous crystals, but no attempt was made to separate the zeolite species. The crystal composite was lightly ground to pass through a 325 mesh screen for powder X-ray diffraction and adsorption studies.

Erionite

Erionite mineral from a Rome, Oregon sedimentary occurrence was ground to pass through a 325 mesh screen for powder X-ray diffraction and adsorption studies. Oxygen adsorption studies indicate that the sample is 90% erionite. Standard wet chemical analyses gave the following weight-percent composition: 2.4 Na_2O, 4.3 K_2O, 1.8 CaO, 0.7 MgO, 0.6 Fe_2O_3, 14.0 Al_2O_3, 60.5 SiO_2, 16.3 H_2O; $Si/Al_2 = 7.35$.

TMA-offretite

Synthetic TMA-offretite was prepared by the methods of Whyte *et al.* (1971). Oxygen adsorption studies indicate that the synthetic zeolite is greater than 90% offretite. Standard wet chemical analyses gave the following weight-percent composition: 1.8 Na_2O, 8.4 K_2O, 10.1 $[(CH_3)_4N]_2O$, 11.1 Al_2O_3, 56.2

TABLE 1. Powder X-ray diffraction data of erionite, offretite, and levynite

							Beech Creek, Oregon				
TMA-offretite			Erionite Rome, Oregon			Erionite*	Zeolite intergrowth		Levynite†		
hkl	*d*, Å	*I*	*hkl*	*d*, Å	*I*	*d* cal, Å	*d*, Å	*I*	*hkl*	*d*, Å	*I*
100	11.33	100	100	11.12	100	11.52	11.65	100	100	11.57	‡
							10.66	11			
							10.46	21	101	10.36	11
			101	8.93	7	9.18	9.16	5			
							8.46	8			
							8.18	53	102	8.15	20
							7.62	20	003	7.63	‡
001	7.50	21	002	7.40	12	7.60	7.39	10			
110	6.58	39	110	6.52	55	6.65	6.73	26	110	6.69	24
101	6.22	12	102	6.22	7	6.34	6.28	1	103	6.37	‡
							5.81	2	200	5.78	‡
200	5.70	16	200	5.66	15	5.76	5.64	1	201	5.61	1
			201	5.29	15	5.38	5.40	4			
							5.20	11	202	5.17	11
							5.05	6	113	5.03	1
			103	4.55	10	4.64	4.54	8			
201	4.53	32	202	4.52	12	4.59					
210	4.30	46	210	4.29	57	4.35	4.37	32	211	4.30	4
							4.24	24	105	4.25	15
			211	4.12	21	4.19	4.19	8			
							4.11	50	212	4.08	100
300	3.79	36	300	3.79	32	3.84	3.88	16	300	3.85	18
									006	3.81	4
									301	3.80	4
211	3.73	83	212	3.72	55	3.78	3.77	16	213	3.79	7
									302	3.66	1
102	3.57	58	104	3.56	27	3.61	3.60	4	205	3.59	5
							3.51	12			
							3.49	15	214	3.47	7
301	3.39	2	302	3.38	3	3.43	3.46	7	303	3.44	2
220	3.29	17	220	3.29	24	3.32	3.36	11	220	3.34	16
112	3.27	6	213/	3.26	13	3.30	3.27	4	116	3.31	2
			114								
			221	3.20	4	3.25	3.23	5	304	3.20	2
310	3.16	12	310	3.15	11	3.19	3.16	16	215	3.16	17
202	3.14	22	204	3.13	15	3.17	3.14	26			
			311	3.10	4	3.13	3.10	15	312	3.09	7
			303	3.05	2	3.06					
221	3.07	2	222	3.02	2	3.05					
311	2.92	6	312	2.91	7	2.94					
400	2.84	56	400	2.85	42	2.88	2.88	23	216	2.89	15
									401	2.89	9
212	2.83	67	214	2.83	39	2.86			207	2.84	10
			401	2.80	32	2.83	2.81	64	402	2.80	60

*Calculated from electron diffraction unit cell of Beech Creek crystals, a=13.3 Å, b=15.2 Å.

†R. A. Sheppard *et al.* (1974).

‡Calculated *d* values not observed by Sheppard *et al.* (1974).

TABLE 2. Adsorption characterization

Adsorbate	O_2		n-Butane				Neopentane	
Kinetic diameter, Å*	3.46		4.3				6.2	
Pressure, torr	750		750				750	
Temperature, °C	−183		23				23	
	g/g	cm^3/g	g/g 8 hr	cm^3/g 8 hr	g/g 21 hr	cm^3/g 21 hr	g/g	cm^3/g
Zeolite intergrowth, Beech Creek, Oregon	.216	.189	.023	.040	.040	.069	.003	.005
Erionite, Rome, Oregon (90% purity)	.181	.159	.051	.073	.088	.126	.005	.008
TMA-offretite‡	.230	.202	.122	.211			.069	.112

*Kinetic diameters from Breck (1974, p. 636).
‡Adsorption data from Breck (1974, p. 623).

SiO_2; Si/Al_2 = 8.6. Powder X-ray diffraction data are given in Table 1 for the as-synthesized zeolite. Adsorption data were obtained after calcination at 550°C in air and are listed in Table 2.

EXPERIMENTAL TECHNIQUES

Powder X-ray diffractograms were obtained using copper Kα radiation with a graphite monochromator and a scanning speed of one degree two-theta per minute. A small sample from each of two specimens was dispersed in isopropyl alcohol and a drop of the suspension allowed to dry on a carbon- and gold-coated standard electron microscope grid. The samples were examined using an RCA EMU4A electron microscope fitted with a ±30° double-tilt stage.

For adsorption studies, one-half gram of the zeolite powder was hand-pressed into a one-half inch diameter pellet and placed in a conventional gravimetric apparatus of the McBain–Bakr type. The zeolite samples were outgassed at 350°C and 10^{-5} torr for 16 hours before the initial adsorbate was introduced and again before introduction of other adsorbates.

RESULTS AND DISCUSSION

The X-ray powder diffraction patterns of the Beech Creek zeolite mixture contained diffraction lines which could be indexed using the unit cells of offretite and levynite (Table 1, Fig. 1B) and are consistent with the pattern expected from a mixture of these two zeolite species. In addition, all the patterns examined contained certain diffraction lines which cannot be indexed using the unit cell of either offretite or levynite.

The X-ray powder diffraction patterns of offretite (Fig. 1A) and erionite (Fig. 1C) are similar. The erionite pattern, however, contains diffraction lines with odd ℓ values, such as the 101, 201, and the 211, which are not present in an offretite pattern due to differences in unit cell and space group (Gard and Tait, 1971). A comparison of the Beech Creek zeolite pattern with the pattern of erionite indicated that the lines which cannot be indexed using the unit cell of offretite or levynite can be indexed using the unit cell of erionite (Table 1). The intensities of these reflections suggested that an ordered erionite was present in the crystal mixture. The odd ℓ reflections of a disordered erionite would be very weak or undetectable (Gard and Tait, 1971, 1972). Two shoulders on peaks at 10.66 and 8.46 Å and peaks at 7.62 and 3.51 Å are unidentified.

The X-ray powder diffraction pattern of the fibrous crystals separated from the Beech Creek intergrowth reported by Sheppard *et al.* (1974) did not contain reflections forbidden by the offretite structure. The discrepancy in X-ray data cannot be readily resolved. To confirm the presence of erionite in the crystal intergrowth, electron diffraction and adsorption data were obtained.

The electron optical examination of different specimens showed that, as expected, the samples contained both platy and fibrous crystals (Fig. 2). Electron diffraction studies showed that the symmetry of both materials is hexagonal. The unit-cell dimensions of the platy crystals were determined as a = 13.3–13.4 Å and c = 23 Å, which, in conjunction with the X-ray data, is sufficient to identify the material as being levynite. Figure 3 shows a typical electron diffraction pattern from a fibrous crystal with [010] parallel to the electron beam. Tilting this and similar crystals around c^* gave

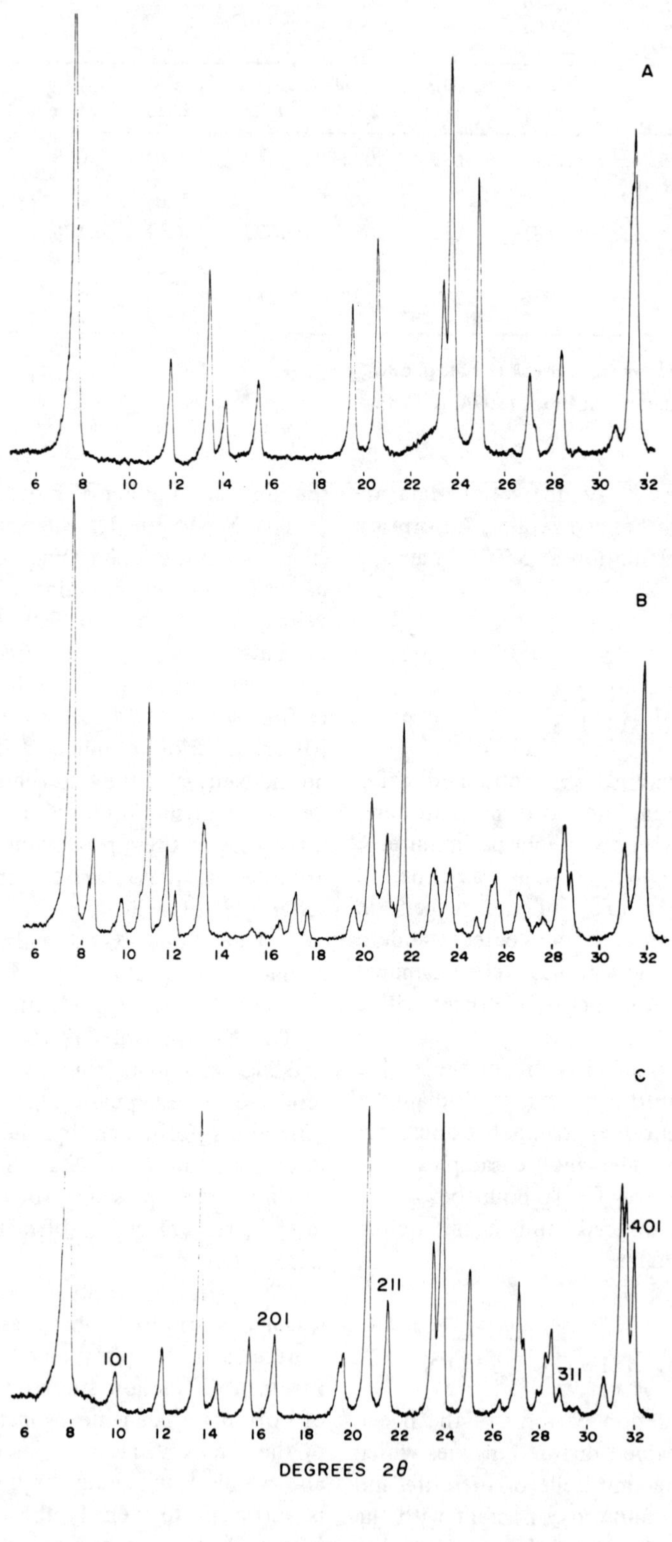

Fig. 1. X-ray powder diffraction patterns using CuKα radiation. (A) Synthetic TMA-offretite. (B) Beech Creek, Oregon, zeolite composite. (C) Erionite, Rome, Oregon.

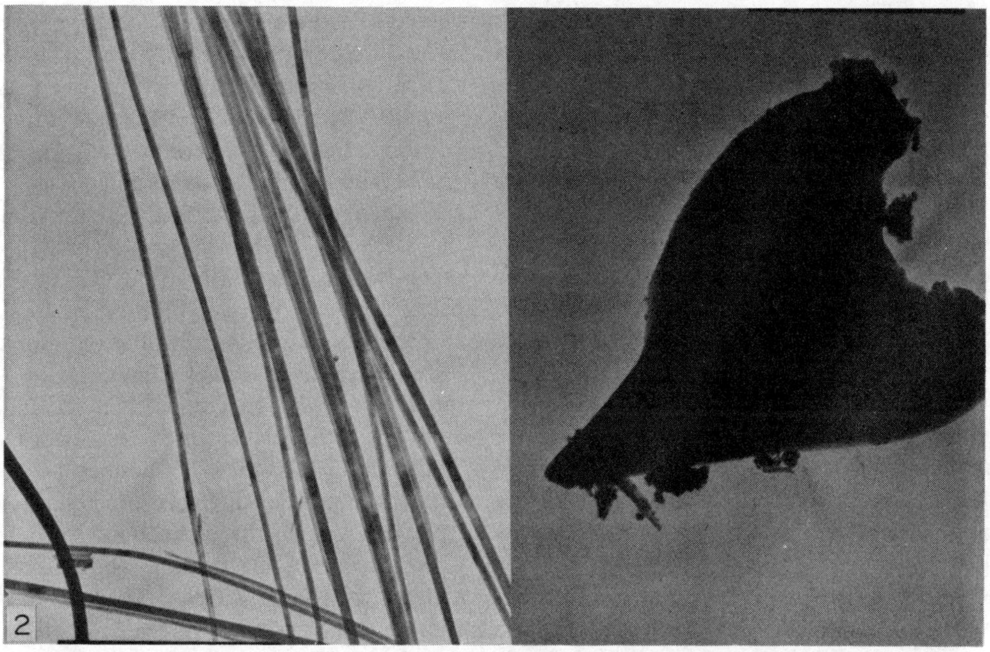

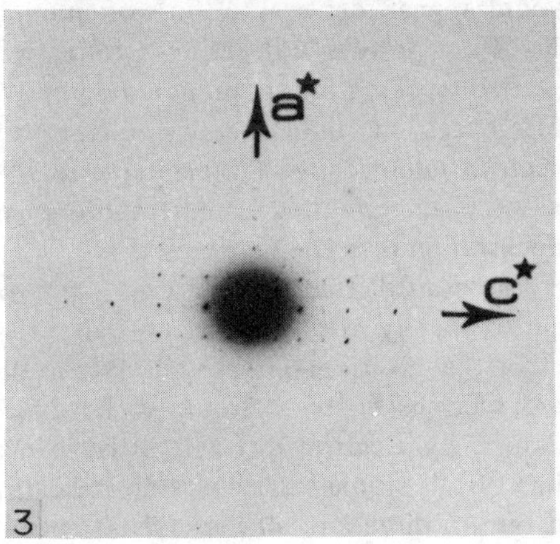

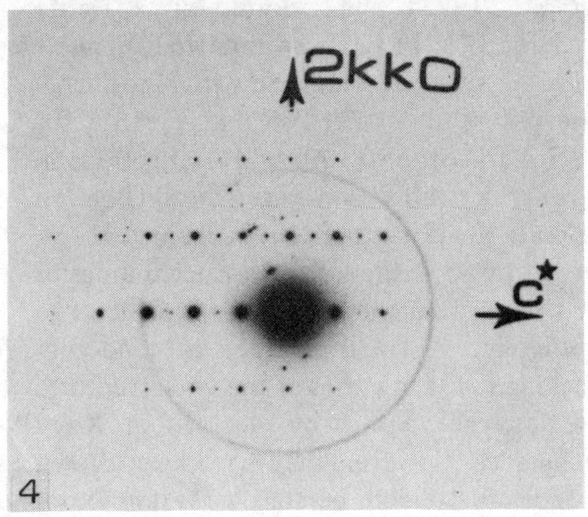

Fig. 2. Typical morphology of the fibrous and platy crystals of erionite and levynite, respectively, from Beech Creek, Oregon. Scale 1 mm. is equivalent to 400 Å.

Fig. 3. Electron diffraction pattern of a fibrous crystal of erionite, Beech Creek, Oregon, with [010] parallel to the electron beam.

Fig. 4. Electron diffraction pattern of offretite intergrown with erionite, Beech Creek, Oregon, with [120] parallel to the electron beam from the only crystal showing streaking in the c^* direction.

unit-cell dimensions of $a = 13.25-13.35$ Å and $c = 15.10-15.20$ Å with systematic absences for $hh.\ell$ of $\ell = 2n+1$. These unit-cell dimensions and symmetry are more consistent with the erionite structure than with the offretite structure and tend to confirm the X-ray diffraction observations.

The offretite crystal structure, having a free channel diameter of 6.3 Å, is capable of adsorbing large hydrocarbon molecules such as neopentane (Gard and Tait, 1972). Erionite, with a smaller pore diameter, will adsorb molecules up to the size of normal paraffins, about 4.3 Å, while levynite has an effective pore size of about 3.6 Å which excludes normal paraffins (Barrer and Peterson, 1964; Barrer and Kerr, 1959).

The oxygen adsorption capacity of the Beech Creek zeolite mixture indicates that the sample is greater than 90% zeolite (Table 2). Erionite, levynite, and synthetic TMA-offretite have a similar saturation capacity for oxygen of about 0.20–0.24 g/g of activated zeolite (Breck, 1974). The slow normal-butane adsorption by the Beech Creek zeolites to a capacity of 0.04 g/g after 21 hours is attributed to the erionite fraction of the intergrowth because levynite does not adsorb normal paraffins. A comparison of the Beech Creek zeolite normal-butane capacity with that of the Rome, Oregon, erionite indicates that the Beech Creek crystal intergrowth is about 50% erionite. The synthetic TMA-offretite with its larger adsorption pores rapidly adsorbs normal-butane to saturation capacity. The Beech Creek crystals and Rome, Oregon, erionite do not adsorb neopentane, which is consistent with the smaller pore size of the erionite structure. The TMA-offretite rapidly adsorbs neopentane to a saturation capacity of 0.069 g/g. The rejection of the neopentane molecule by the Beech Creek crystal mixture established the absence of a large-pore adsorbent such as offretite.

A random distribution of stacking faults in the erionite structure will have little effect on the adsorption properties. In contrast, a stacking fault in the offretite structure will block the large channel. Two or more such faults along a channel will reduce the adsorption capacity, giving the appearance of a small-pore zeolite by adsorption studies. This faulting can be observed directly by transmission electron microscopy and indirectly by streaking of the electron diffraction pattern along c^* (Bennett, 1966).

Erionite and offretite have very similar structures, and stacking faults both of offretite in erionite (Kokotailo *et al.*, 1972) and erionite in offretite (Bennett and Gard, 1967) have been reported. Bennett and Gard (1967) showed that the aluminosilicate frameworks are differently stacked so that $c \sim 15$ Å for erionite, but 7.5 Å for offretite. All odd ℓ reflections in the X-ray powder patterns and electron diffraction patterns of erionite are therefore absent from those of offretite. The sensitivity of the diffraction techniques to the number of stacking faults in a structure is difficult to determine absolutely. A small number of randomly distributed stacking faults in either of the two structures would not be observable either by electron or X-ray diffraction techniques. However, they have been shown to occur in individual small crystals of erionite by transmission electron microscopy (Kokotailo *et al.*, 1972). As the number of stacking faults increases, the electron diffraction patterns are affected. The observable effect is dependent both on the number and thickness of the stacking faults. Where there are isolated stacking faults, the electron diffraction patterns are streaked in the c^* direction for both offretite and erionite. As the thickness of the stacking fault or intergrowth increases to about 60 Å, substantial broadening of the streak at the position of the odd ℓ reflections occurs. Further thickening of the intergrowth reduces the broadening of these odd ℓ reflections until they are indistinguishable in shape from the even ℓ reflections. Some streaking may still be observed and is caused by the thin fault region which arises when offretite changes to erionite and vice versa. This would appear identical to a small number of isolated offretite faults in an erionite structure.

X-ray diffraction is insensitive to thin regions of intergrowths of offretite in erionite and vice versa; as the regions of intergrowth increase in thickness, the effect observed in the electron diffraction pattern is paralleled in the X-ray powder pattern. The X-ray powder diffraction pattern from offretite crystals containing erionite-type stacking faults would show broad peaks corresponding to reflections forbidden by the space group. The breadth of these reflections is dependent on the coherency of the faulted region. As the thickness of the stacking faults increases, there is a corresponding reduction in line width of the broad peaks, until the pattern appears like a typical erionite pattern but with an altered intensity distribution. In erionite with offretite-type stacking faults, the "forbidden" reflections with odd ℓ values broaden and then sharpen again as the thickness of the intergrowth increases.

Electron diffraction techniques are more sensitive than X-ray powder diffraction techniques when used to study stacking faults in zeolites. Both techniques are insensitive when the stacking faults result from a small number of intergrowths of large blocks of offretite in erionite and vice versa. The electron diffraction patterns would appear identical to those of an unfaulted erionite; the X-ray powder diffraction patterns would be similar to that expected for an unfaulted erionite but with some differences in the intensity distribution. Dark-field electron microscopy is a technique that will distinguish between an unfaulted erionite and one containing large blocks of an offretite intergrowth.

Electron diffraction patterns were obtained from approximately fifty fibrous crystals oriented to bring either the [010] or [120] zone axes to the beam. These two sections enable offretite to be distinguished from erionite by electron diffraction data. In all but one of the crystals studied there was no indication of streaking in the c^* direction; all the crystals examined appear to be erionite and not offretite. A few crystals gave electron diffraction patterns in which the odd reflections were weak, indicating that the erionite was faulted with thick offretite intergrowths. Without dark-field microscopy it is difficult to distinguish erionite from a faulted erionite; the weakening of the odd ℓ reflections is an indication, but great care must be exercised in the use of the intensity of electron diffraction spots because of secondary scattering. However, there is strong indication from several crystals that the erionite is in fact faulted. This would be difficult to observe using both X-ray powder diffraction techniques and adsorption properties.

One crystal gave a streaked electron diffraction pattern indicating disorder in the c direction, which would indicate an intergrowth of offretite in erionite (Fig. 4). One other crystal gave an electron diffraction pattern that was streaked both in the c^* and a^* direction. The reason for the streaking in the a^* direction is not understood.

CONCLUSIONS

Interpretation of the adsorption data and X-ray powder diffraction data identifies the fibrous material as erionite and not offretite. The electron diffraction data indicate that the erionite contains intergrowths of offretite and that the material is mainly erionite containing offretite faults and not offretite containing erionite faulting.

This result is contrary to that obtained by Sheppard *et al.* (1974) who identified the fibrous material as offretite. It is difficult to see how they could have failed to identify erionite by X-ray techniques. It is possible that the difference between the X-ray results is due to the different techniques used for sample preparation. Sheppard *et al.* (1974) used a collodion-membrane mount, while glass-slide and aluminium sample-holder techniques were used in this study. A comparison of the observed X-ray diffraction data for levynite by Sheppard *et al.* (1974) and in this study (Table 1) shows differences in the intensity of many different reflections, and not all reflections permitted for levynite were observed in both studies. The reflections used to distinguish offretite from erionite by X-ray diffraction techniques are weak and not easily observed. One other possibility is that even though the samples came from the same locality, both offretite and erionite can grow topotaxial on levynite, i.e., the sample studied by Sheppard *et al.* (1974) was from a levynite–offretite intergrowth, while that used in this study was from a levynite–erionite intergrowth. However, this is contradicted by some of the electron diffraction data which were obtained in this study, which clearly showed erionite in the same sample used by Sheppard *et al.* (1974). Passaglia *et al.* (1974) have identified a levynite with a similar morphology to that found at Beech Creek and which also contains a topotaxial growth of a fibrous material identified as erionite by single crystal rotation photographs. In their experience, when the levynite crystals are lamellar, there is a strong possibility that the levynite might be intergrown with erionite.

ACKNOWLEDGMENTS

We thank R. A. Sheppard (U.S. Geological Survey, Denver, Colorado) for providing the Beech Creek zeolite specimens. The assistance of the Central Scientific Laboratory of Union Carbide Corporation in determining X-ray analyses is gratefully acknowledged. We also thank Miss E. M. Flanigen and Dr. D. W. Breck for helpful discussions during the preparation of this manuscript.

REFERENCES

Barrer, R. M. and Kerr, I. S. (1959) Intracrystalline channels in levynite and some related zeolites: *Trans. Faraday Soc.* **55**, 1915-1923.

Barrer, R. M. and Peterson, D. L. (1964) Kinetics of n-paraffin sorption in the natural zeolite erionite: *J. Phys. Chem.* **68**, 3437-3429.

Bennett, J. M. (1966) Electron microscope studies on some zeolites and other phases: PhD Thesis, University of Aberdeen.

Bennett, J. M. and Gard, J. A. (1967) Non-identity of the zeolites erionite and offretite: *Nature* **214**, 1005.

Breck, D. W. (1974) *Zeolite Molecular Sieves*: Wiley-Interscience, N.Y., pp. 143, 159, 166.

Gard, J. A. and Tait, J. M. (1971) Structural studies on erionite and offretite: *Adv. Chem. Ser.* **101**, 230-236.

Gard, J. A. and Tait, J. M. (1972) The crystal structure of the zeolite offretite: *Acta Crystallogr.* **B28**, 825-834.

Gonnard, M. J. (1890) Sur l'offrétite espèce minérale nouvelle: *C. R. Acad. Sci. (Paris)* **111**, 1002-1003.

Kokotailo, G. T., Saworuk, S. and Lawton, S. L. (1972) Direct observation of stacking faults in the zeolite erionite: *Amer. Mineral.* 57, 429-444.

Passaglia, E., Galli, E. and Rinaldi, R. (1974) Levynes and erionites from Sardinia, Italy: *Contr. Mineral. & Petrol.* **43**, 253-259.

Sheppard, R. A., Gude, A. J. III, Desborough, G. A. and White, J. S., Jr. (1974) Levynite–offretite intergrowths from basalt near Beech Creek, Grant County, Oregon: *Amer. Mineral.* **59**, 837-842.

Whyte, T. E., Jr., Wu, E. L., Kerr, G. T. and Venuto, P. B. (1971) Synthetic offretite: *J. Catal.* **20**, 88-96.

CRYSTAL STRUCTURES OF HEAT-COLLAPSED PHASES OF BARRERITE

A. ALBERTI and G. VEZZALINI

Istituto di Mineralogia e Petrologia, Università di Modena, 41100 Modena, Italy

Abstract – Barrerite is orthorhombic, with space group *Amma* and unit-cell dimensions of a = 13.643, b = 18.200, c = 17.842 Å. On heating at 200–380°C it transforms in a few hours to barrerite B, with space group *Amma* and unit-cell dimensions of a = 13.61, b = 17.12, c = 16.03 Å. After a few days barrerite B rehydrates to barrerite C, with space group *Amma* and unit-cell dimensions of a = 13.61, b = 17.61, c = 16.24 Å. On heating at 400–450°C barrerite transforms in a few hours to a D phase which is also orthorhombic and has unit-cell dimensions of a = 12.90, b = 16.87, c = 16.18 Å. The heat-collapsed phases B and C are characterized by major changes in the framework. A statistical breaking of an oxygen bridge gives rise to two half-occupied, face-sharing tetrahedra. A rotation of the fundamental building unit, which is responsible for the contraction of the unit cell, results in a remarkably different geometry of the channels parallel to a whose nearly circular section in the natural phase A becomes markedly elliptical in B and C.

INTRODUCTION

It is known that thermal dehydration produces changes in the framework of zeolites. In the case of platy zeolites, dehydration is a complex phenomenon involving, in most cases, several polymorphic, stepwise transformations taking place at different temperatures and which correspond to different percent reductions of unit-cell volume. Despite the close structural similarities between platy zeolites, these polymorphic transformations occur at different temperatures and by different methods in each species. The complexity of such phenomena and the abundance of platy zeolites in nature makes the detailed study of their thermal behavior particularly interesting.

The aim of the present work is to determine the structural modifications that occur in barrerite as a result of thermal treatment. One similar study is the structure determination of the B-phase of heulandite (Alberti, 1973), and future studies are planned for stilbite and stellerite.

When barrerite is held for a few hours at any temperature in the range 200–380°C and then cooled to room temperature, it collapses to form a new phase which we call barrerite B. This new phase is affected by a volume decrease of about 16% and has, most probably, the same space group (Amma) as the natural phase. Barrerite B is metastable and at room temperature within a short time (some hours for the powder, about 2 weeks for a single crystal) inverts to a new phase, barrerite C, which is stable for a long time and which has a 4% larger unit-cell volume with respect to barrerite B. Its space group is also *Amma*. If barrerite is heated above 400°C it transforms into a phase D, which is also stable at room temperature and which has a 6% smaller unit-cell volume with respect to barrerite B. The polymorphic changes are therefore easily detected by the large changes of unit cell. The framework topology of barrerite is identical to that of stilbite and stellerite. Despite the rarity of the mineral we deemed that structure determinations of these phases would be very useful. One other reason for studying barrerite was the complexity of the heat behavior with the formation of three different phases compared to the formation of two that have been so far detected for stilbite and stellerite. A preliminary Weissenberg study of phase D yielded poor-quality photographs which discouraged the collection of intensity data on the crystal. Further appropriate work on phase D is planned.

EXPERIMENTAL

Barrerite from Capo Pula (Passaglia and Pongiluppi, 1974, 1975) was used in the present work. Its crystal structure was refined by Galli and Alberti (1976). The unit cell contains $(Ca_{1.68}Mg_{0.34}Na_{10.9}K_{2.12})_{tot\,=\,15.04}(Al_{16.38}Fe_{0.02})_{tot\,=\,16.40}Si_{55.44}O_{144} \times 51.56\,H_2O$.

Intensity data for phase B on a single crystal of 0.2 × 0.1 × 0.2 mm were collected immediately following the heating of the crystal at 250°C. The intensity data for phase C were collected on another single crystal

(0.2 × 0.1 × 0.2 mm) about 5 months after heating at 250°C. A Philips PW 1100 diffractometer (MoKα radiation) of the Istituto di Mineralogia di Perugia was used to collect intensity data for both crystals and for the precise determination of the unit-cell parameters. About 100 independent reflections of the type $k+\ell = 2n+1$ were found to be extinct for the phase B, thus confirming an A-centered cell. A total of 1020 diffraction intensities for phase B and 1051 for phase C were examined. The reflections of the type hk0 with $h = 2n+1$ were systematically absent in both the crystals, thus confirming the space group as *Amma* for the heat-collapsed phases also; 978 and 1004 unique reflections were obtained for phases B and C, respectively (approximately 5% in the MoKα limiting sphere). Of these, 796 and 804 reflections with $I > 2\sigma(I)$ were considered observed for barrerite B and C, respectively.

The unit-cell data, obtained with the single-crystal diffractometer, are for phase B, a = 13.609(4), b = 17.119(6), c = 16.029(6); and for phase C, a = 13.611(4), b = 17.607(6), c = 16.236(6). Parameters a, b, and c of phase B are thus 0.3%, 5.9%, and 10.2% shorter than the corresponding parameters of phase A, natural barrerite (Passaglia and Pongiluppi, 1974), while the cell volume is 15.7% smaller. Parameters a, b, and c of phase C are 0.0%, 2.8%, and 1.3% greater than the corresponding parameters of phase B, while the cell volume is 4.2% larger.

A thermal gravimetric analysis executed on macrocrystals immediately after heating for 16 hours at 250°C indicates that the unit cell contains a residual of 13.8 water molecules. A TG made on powder after transformation to phase C indicates a water content of 33.0 molecules versus 51.6 water molecules in phase A.

The cell parameters of phase D were measured from oscillating and equatorial Weissenberg photographs on a crystal heated at 400°C for 16 hours and gave the following unit-cell dimensions: a = 12.90, b = 16.87, c = 16.18, with a variation of −5.4%, −7.3%, −9.3% in parameters and of −20.5% in volume with respect to phase A; these data represent variations of −5.2%, −1.5%, +0.9%, and −5.7%, respectively, in comparison with phase B.

STRUCTURE DETERMINATIONS AND REFINEMENTS

The structure determinations and refinements were carried out on a CDC CYBER 76 computer. The atomic scattering curves for neutral atoms used in the calculation of the structure factors were provided by Cromer and Mann (1968), while the atomic scattering curve for H_2O was provided by Hajdu (1972). For the exchangeable cations an average curve was used obtained from the chemical analysis with 72.5% Na, 14.1% K, 11.2% Ca, and 2.2% Mg; for the T atoms an average curve was obtained with 77.2% Si and 22.8% Al. No absorption correction was made because of the MoKα radiation used and the small crystal sizes. Because of the negligible values of $\Delta f'$ and $\Delta f''$ for the atoms involved (Cromer, 1965), no correction for anomalous scattering was made.

An attempt to refine the data for phase B using the coordinates of phase A (Galli and Alberti, 1976) by the least-squares method resulted in improbable T−O distances and too high a value (0.50) of the reliability index R. A three-dimensional Fo-synthesis revealed a structure whose maxima generally differed from the atomic positions indicated by the least-squares method. The T−O distances were possible but the R value was too high. Attempts to refine the coordinates obtained by the DLS program (Meier and Villiger, 1969) also resulted in improbable distances and very high R values. A determination of the structure by a direct methods program (LSAM, Germain *et al.*, 1971) was therefore attempted. One solution indicated a possible framework, but the refinement of these coordinates resulted in a minimum R value of 0.33. A difference Fourier synthesis, however, indicated that T(1P) and T(4) are disordered between two positions symmetrically situated with respect to the face O(4P), O(7P), O(1P) for T(1P) [T(1PD) is the corresponding site] and O(3), O(8), O(10) for T(4) [T(4D) is the corresponding site]. The framework is therefore alternatively bonded and unbonded at O(3P) and in this latter case a vertex of the tetrahedra T(1P) and T(4) is considered to be hydroxyl group. If we consider that the water molecule W(1) of barrerite A can be the fourth vertex of these tetraheda, then according to the occupancy of this site in phase A, the two related sites in phases C and B have an occupancy close to 50%. A refinement carried out according to this hypothesis reduced the R value to 0.15 and 0.17 for phases B and C, respectively. At this point the refinements were carried out using a combination of three-dimensional Fourier synthesis (to locate cations and water molecules) and least-squares techniques using weights according to counting statistics, individual isotropic temperature factors, and variable occupancy factors of exchangeable cations and water molecules. The reliability indices for the observed reflections were R = 8.1% for phase B and R = 12.2% for phase C. Finally, individual anisotropic temperature factors were used. The reliability indices for the observed reflections resulted in values of R = 0.053 and wR = 0.064 for phase B and R = 0.075 and wR = 0.092 for phase C. A final refinement confirmed that the occupancy factors

TABLE 1. Atomic coordinates and occupancy factors for barrerite B and C*

	Barrerite B				Barrerite C			
Atom	x/a	y/b	z/c	Occupancy (%)	x/a	y/b	z/c	Occupancy (%)
T(1)	.1375(2)	.3601(1)	.1375(1)	100	.1362(2)	.3532(2)	.1375(2)	100
T(1P)	.1514(4)	.2777(3)	.3899(3)	50	.1484(5)	.2806(4)	.3830(4)	50
T(3)	.0409(2)	.4101(1)	.3046(1)	100	.0422(3)	.4109(2)	.3016(2)	100
T(4)	.1523(4)	.1955(3)	.2153(4)	50	.1483(5)	.1946(4)	.2161(5)	50
T(5)	0	.3237(2)	0	100	0	.3160(3)	0	100
T(1PD)	.1361(3)	.3139(3)	.4590(3)	50	.1385(5)	.3122(4)	.4550(5)	50
T(4D)	.1439(3)	.1725(3)	.1495(4)	50	.1457(5)	.1730(4)	.1497(6)	50
O(1)	.0597(4)	.3833(3)	.0624(4)	100	.0653(7)	.3723(5)	.0594(7)	100
O(1P)	.0770(5)	.2289(4)	.4503(4)	100	.0729(8)	.2350(6)	.4474(6)	100
O(3)	.1306(4)	.2666(3)	.1540(3)	100	.1269(6)	.2627(5)	.1617(6)	100
O(3P)	.1795(11)	.2313(8)	.3054(7)	50	.1761(15)	.2253(12)	.3056(11)	50
O(4)	.1120(5)	.4096(3)	.2219(4)	100	.1058(9)	.4054(6)	.2179(7)	100
O(4P)	.0971(6)	.3686(5)	.3829(5)	100	.1020(9)	.3688(6)	.3794(9)	100
O(7)	¼	.3877(5)	.1128(5)	100	¼	.3769(7)	.1172(7)	100
O(7P)	¼	.2957(7)	.4438(6)	100	¼	.2927(9)	.4386(8)	100
O(8)	.0637(5)	.1322(4)	.2148(4)	100	.0662(7)	.1276(5)	.2106(6)	100
O(9)	.0218(8)	½	.3329(6)	100	.0283(9)	½	.3255(7)	100
O(10)	¼	.1458(5)	.1804(5)	100	¼	.1484(7)	.1780(8)	100
OH(1)	.1237(8)	.1422(6)	.0527(6)	50	.1291(12)	.1504(10)	.0529(8)	50
C(1)	.063(2)	0	.005(1)	24(1)	.073(3)	0	.003(2)	27(2)
C(2)	0	.022(1)	0	27(1)	0	.023(3)	0	22(2)
C(3)	.025(1)	0	.480(1)	32(1)	.021(3)	0	.473(3)	19(2)
C(4)	¼	½	.225(1)	76(1)	¼	½	.223(1)	48(2)
C(5)	.159(3)	.150(3)	.341(3)	8(1)	.186(3)	.116(2)	.375(2)	21(1)
W(1)	¼	½	.389(4)	21(2)	¼	½	.376(2)	98(4)
W(2)	.178(5)	0	.141(4)	14(2)	.167(4)	0	.136(3)	23(2)
W(3)	.133(4)	.043(3)	.022(3)	21(2)	.109(4)	.085(4)	.034(4)	21(2)
W(4)	.132(5)	0	.363(4)	17(2)	.089(4)	0	.376(5)	35(6)
W(5)					.065(4)	0	.419(5)	39(5)
W(6)					¼	0	.380(2)	70(3)
W(7)					.218(3)	0	.254(2)	32(2)
W(8)					.192(4)	0	.049(3)	37(3)
W(9)					¼	.100(3)	.438(3)	41(3)
W(10)					¼	½	.019(2)	66(4)

*T sites host 77.2% Si and 22.8% Al; C sites host 72.5% Na, 14.1% K, 11.2% Ca, and 2.2% Mg.

of T(1P), T(4), T(1PD), T(4D), O(3P), and OH(1) were close to 50%. The C(3), C(4), and C(5) positions were assigned as cations because of the distances between these sites and framework oxygens. The C(1) and C(2) sites were assigned to cations, because of the too short C(1)-C(2), C(1)-C(1), and C(2)-C(2) distances, which restrict the sum of the occupancies of these two sites to less than 50%. This sum is approximately 50% if these sites are occupied by cations, but goes up to about 100% if these sites are occupied by water molecules because of their lower scattering power. Obviously the sites present only in phase C can be assigned only to water molecules.

Agreement between this model and the data obtained by chemical and thermal gravimetric analysis is summarized in the following table:

		Phase B	Phase C
		atoms per unit cell	
Exchangeable cations	chem. analysis	15.04	15.04
	struct. refinement	11.00	10.66
Water molecules	T.G. analysis	13.8	33.0
	struct. refinement	6.7	29.3

TABLE 2. Thermal parameters ($\beta_{ij} \times 10^4$) and root-mean-square thermal vibrations (Å) along the ellipsoid axes for barrerite B and C*

Barrerite B

Atom	β_{11} or B	β_{22}	β_{33}	β_{12}	β_{13}	β_{23}	R.m.s.		
T(1)	17(2)	21(1)	20(1)	−1(1)	−1(1)	−2(1)	.12(1)	.16(1)	.18(1)
T(1P)	27(3)	25(2)	20(3)	−4(2)	−8(2)	6(2)	.13(1)	.16(1)	.22(1)
T(3)	25(2)	16(1)	20(1)	−2(1)	3(1)	3(1)	.14(1)	.16(1)	.17(1)
T(4)	21(3)	21(2)	23(3)	−1(2)	−2(2)	0(2)	.14(1)	.17(1)	.18(1)
T(5)	19(2)	16(1)	33(2)	0	−8(2)	0	.12(1)	.15(1)	.21(1)
T(1PD)	17(3)	21(2)	12(2)	−2(2)	0(2)	0(2)	.12(1)	.13(1)	.18(1)
T(4D)	18(3)	23(2)	16(3)	−5(2)	−2(2)	0(2)	.12(1)	.14(1)	.19(1)
O(1)	38(4)	20(2)	38(3)	−1(3)	−20(3)	−3(2)	.14(1)	.17(1)	.25(1)
O(1P)	53(5)	44(3)	48(4)	−21(3)	−5(4)	12(3)	.18(1)	.23(1)	.31(1)
O(3)	34(4)	17(3)	30(3)	2(2)	−3(3)	2(2)	.16(1)	.18(1)	.20(1)
O(3P)	89(11)	44(7)	23(6)	14(8)	−19(7)	−12(5)	.14(3)	.24(2)	.32(2)
O(4)	41(4)	29(3)	33(3)	−7(3)	17(3)	−11(2)	.14(1)	.19(1)	.26(1)
O(4P)	80(6)	51(4)	61(4)	−5(4)	−19(4)	24(3)	.20(1)	.27(1)	.35(1)
O(7)	15(5)	31(4)	23(4)	0	0	4(3)	.12(2)	.17(2)	.22(1)
O(7P)	39(7)	67(6)	38(5)	0	0	−6(5)	.18(2)	.22(1)	.32(1)
O(8)	34(4)	44(3)	45(4)	−17(3)	5(3)	9(3)	.14(2)	.24(1)	.28(1)
O(9)	93(9)	21(4)	39(5)	0	37(5)	0	.16(2)	.18(2)	.33(1)
O(10)	35(6)	25(4)	41(4)	0	0	2(3)	.19(2)	.19(2)	.23(1)
OH(1)	41(8)	28(5)	10(5)	8(5)	0(5)	5(4)	.11(3)	.18(2)	.23(2)
C(1)	4.0(4)								
C(2)	3.4(4)								
C(3)	2.2(3)								
C(4)	2.1(1)								
C(5)	5.2(10)								
W(1)	5.5(17)								
W(2)	4.5(17)								
W(3)	9.9(17)								
W(4)	6.8(16)								

Barrerite C

Atom	β_{11} or B	β_{22}	β_{33}	β_{12}	β_{13}	β_{23}	R.m.s.		
T(1)	19(2)	16(2)	35(2)	0(2)	6(2)	−4(1)	.13(1)	.16(1)	.22(1)
T(1P)	19(4)	19(3)	13(3)	−4(3)	−2(3)	0(3)	.12(2)	.14(2)	.18(1)
T(3)	21(2)	13(2)	35(2)	−1(2)	2(2)	1(1)	.14(1)	.15(1)	.22(1)
T(4)	25(5)	20(3)	14(4)	−3(3)	−4(4)	−11(3)	.08(3)	.16(2)	.21(1)
T(5)	18(3)	13(2)	46(3)	0	−2(3)	0	.13(1)	.14(1)	.25(1)
T(1PD)	12(5)	14(3)	20(4)	−1(3)	3(3)	3(3)	.10(2)	.14(2)	.17(2)
T(4D)	17(5)	12(3)	36(5)	−3(3)	0(4)	5(3)	.11(2)	.14(2)	.22(2)
O(1)	43(7)	22(4)	76(6)	−9(4)	−27(5)	10(4)	.16(2)	.19(2)	.34(1)
O(1P)	47(7)	45(5)	55(5)	−21(5)	4(5)	−7(4)	.17(2)	.26(1)	.31(1)
O(3)	45(6)	14(4)	48(5)	8(4)	3(5)	3(4)	.13(2)	.21(1)	.26(1)
O(3P)	71(16)	46(10)	31(9)	1(11)	−4(10)	−3(8)	.20(3)	.26(3)	.27(3)
O(4)	98(10)	30(4)	82(7)	−9(5)	68(7)	−16(5)	.15(2)	.21(2)	.43(2)
O(4P)	82(9)	32(5)	115(9)	−10(6)	−35(7)	27(5)	.19(2)	.25(2)	.42(2)
O(7)	25(8)	27(5)	21(5)	0	0	1(4)	.15(2)	.16(2)	.21(2)
O(7P)	51(11)	49(7)	30(7)	0	0	−10(6)	.19(2)	.22(2)	.29(2)
O(8)	43(7)	28(4)	56(6)	−8(4)	23(5)	10(4)	.13(2)	.23(2)	.30(1)
O(9)	59(10)	16(5)	24(5)	0	12(6)	0	.16(2)	.16(2)	.25(2)

TABLE 2. *(cont.)*

O(10)	28(8)	21(5)	37(6)	0	0	1(5)	.16(2)	.18(2)	.22(2)
OH(1)	47(12)	24(7)	0(7)	−2(8)	2(7)	−1(5)	.01(4)	.20(3)	.22(3)
C(1)	8.0(10)								
C(2)	9.5(15)								
C(3)	5.0(9)								
C(4)	4.6(5)								
C(5)	8.9(11)								
W(1)	10.8(9)								
W(2)	3.2(12)								
W(3)	7.5(17)								
W(4)	4.7(15)								
W(5)	4.6(9)								
W(6)	4.6(6)								
W(7)	4.5(12)								
W(8)	10.4(18)								
W(9)	7.9(12)								
W(10)	8.6(11)								

*The form of the anisotropic temperature factors is:

$\exp[-(h^2\beta_{11} + k^2\beta_{22} + l^2\beta_{33} + 2hk\beta_{12} + 2hl\beta_{13} + 2kl\beta_{23})]$.

TABLE 3. Interatomic distances (Å) within the frameworks of barrerite B and C

Barrerite B							
T(1) tetrahedron		T(1P) tetrahedron		T(3) tetrahedron		T(4) tetrahedron	
T(1)–O(1)	1.65(1) Å	T(1P)–O(1P)	1.63(1) Å	T(3)–O(4)	1.64(1) Å	T(4)–O(3)	1.59(1) Å
T(1)–O(3)	1.62(1)	T(1P)–O(3P)	1.61(1)	T(3)–O(4P)	1.63(1)	T(4)–O(3P)	1.61(1)
T(1)–O(4)	1.63(1)	T(1P)–O(4P)	1.73(1)	T(3)–O(8)	1.63(1)	T(4)–O(8)	1.62(1)
T(1)–O(7)	1.65(1)	T(1P)–O(7P)	1.62(1)	T(3)–O(9)	1.63(1)	T(4)–O(10)	1.67(1)
Average	1.64	Average	1.65	Average	1.63	Average	1.62
O(1)–O(3)	2.66(1)	O(1P)–O(3P)	2.71(2)	O(4)–O(4P)	2.68(1)	O(3)–O(3P)	2.59(1)
O(1)–O(4)	2.69(1)	O(1P)–O(4P)	2.64(1)	O(4)–O(8)	2.70(1)	O(3)–O(8)	2.66(1)
O(1)–O(7)	2.71(1)	O(1P)–O(7P)	2.62(1)	O(4)–O(9)	2.66(1)	O(3)–O(10)	2.66(1)
O(3)–O(4)	2.69(1)	O(3P)–O(4P)	2.88(2)	O(4P)–O(8)	2.69(1)	O(3P)–O(8)	2.73(2)
O(3)–O(7)	2.72(1)	O(3P)–O(7P)	2.66(2)	O(4P)–O(9)	2.60(1)	O(3P)–O(10)	2.66(2)
O(4)–O(7)	2.59(1)	O(4P)–O(7P)	2.61(1)	O(8)–O(9)	2.66(1)	O(8)–O(10)	2.60(1)

T(5) tetrahedron		T(1PD) tetrahedron		T(4D) tetrahedron	
T(5)–O(1)	1.65(1) [×2] Å	T(1PD)–O(1P)	1.67(1) Å	T(4D)–O(3)	1.62(1) Å
T(5)–O(1P)	1.59(1) [×2]	T(1PD)–OH(1)	1.69(1)	T(4D)–OH(1)	1.66(1)
Average	1.62	T(1PD)–O(4P)	1.63(1)	T(4D)–O(8)	1.66(1)
		T(1PD)–O(7P)	1.60(1)	T(4D)–O(10)	1.59(1)
O(1)–O(1)	2.58(1)	Average	1.65	Average	1.63
O(1P)–O(1P)	2.63(1)				

TABLE 3. (*cont.*)

O(1)–O(1P)	2.68(1) [×2]	O(1P)–OH(1)	2.82(1)	O(3)–OH(1)	2.68(1)
O(1)–O(1P)	2.64(1) [×2]	O(1P)–O(4P)	2.64(1)	O(3)–O(8)	2.66(1)
		O(1P)–O(7P)	2.62(1)	O(3)–O(10)	2.66(1)
		OH(1)–O(4P)	2.76(1)	OH(1)–O(8)	2.73(1)
		OH(1)–O(7P)	2.67(1)	OH(1)–O(10)	2.67(1)
		O(4P)–O(7P)	2.61(1)	O(8)–O(10)	2.60(1)

Barrerite C

T(1) tetrahedron		T(1P) tetrahedron		T(3) tetrahedron		T(4) tetrahedron	
T(1)–O(1)	1.63(1) Å	T(1P)–O(1P)	1.67(1) Å	T(3)–O(4)	1.62(1) Å	T(4)–O(3)	1.52(1) Å
T(1)–O(3)	1.65(1)	T(1P)–O(3P)	1.63(2)	T(3)–O(4P)	1.68(1)	T(4)–O(3P)	1.60(2)
T(1)–O(4)	1.65(1)	T(1P)–O(4P)	1.68(1)	T(3)–O(8)	1.64(1)	T(4)–O(8)	1.63(1)
T(1)–O(7)	1.64(1)	T(1P)–O(7P)	1.67(1)	T(3)–O(9)	1.63(1)	T(4)–O(10)	1.72(1)
Average	1.64	Average	1.66	Average	1.64	Average	1.62
O(1)–O(3)	2.68(1)	O(1P)–O(3P)	2.70(2)	O(4)–O(4P)	2.71(2)	O(3)–O(3P)	2.52(2)
O(1)–O(4)	2.69(2)	O(1P)–O(4P)	2.63(1)	O(4)–O(8)	2.68(1)	O(3)–O(8)	2.64(1)
O(1)–O(7)	2.69(1)	O(1P)–O(7P)	2.62(1)	O(4)–O(9)	2.64(1)	O(3)–O(10)	2.63(1)
O(3)–O(4)	2.69(1)	O(3P)–O(4P)	2.97(2)	O(4P)–O(8)	2.72(2)	O(3P)–O(8)	2.76(2)
O(3)–O(7)	2.71(1)	O(3P)–O(7P)	2.66(2)	O(4P)–O(9)	2.67(1)	O(3P)–O(10)	2.68(2)
O(4)–O(7)	2.60(1)	O(4P)–O(7P)	2.60(2)	O(8)–O(9)	2.66(1)	O(8)–O(10)	2.58(1)

T(5) tetrahedron		T(1PD) tetrahedron		T(4D) tetrahedron	
T(5)–O(1)	1.64(1) [×2] Å	T(1PD)–O(1P)	1.63(1) Å	T(4D)–O(3)	1.61(1) Å
T(5)–O(1P)	1.59(1) [×2] Å	T(1PD)–OH(1)	1.73(2)	T(4D)–OH(1)	1.64(2)
Average	1.615	T(1PD)–O(4P)	1.65(1)	T(4D)–O(8)	1.67(1)
		T(1PD)–O(7P)	1.58(1)	T(4D)–O(10)	1.55(1)
O(1)–O(1)	2.62(2)	Average	1.65	Average	1.62
O(1P)–O(1P)	2.61(2)				
O(1)–O(1P)	2.67(1) [×2]	O(1P)–OH(1)	2.76(2)	O(3)–OH(1)	2.66(2)
O(1)–O(1P)	2.62(1) [×2]	O(1P)–O(4P)	2.63(1)	O(3)–O(8)	2.64(1)
		O(1P)–O(7P)	2.62(1)	O(3)–O(10)	2.63(1)
		OH(1)–O(4P)	2.86(2)	OH(1)–O(8)	2.73(2)
		OH(1)–O(7P)	2.67(2)	OH(1)–O(10)	2.61(2)
		O(4P)–O(7P)	2.60(2)	O(8)–O(10)	2.58(1)

The positional and thermal parameters are given in Tables 1 and 2, respectively; interatomic distances are reported in Tables 3 and 4. Observed and calculated structure factors are available from the Istituto di Mineralogia e Petrologia di Modena. X-ray powder diffraction data for barrerite B, C, and D are reported in Tables 5, 6, and 7, respectively.

DISCUSSION OF THE STRUCTURES

Tetrahedral framework

The fundamental unit of the framework in natural barrerite, which is the same in all platy zeolites, can be described as being formed of rings of four and five

TABLE 4. Cations, oxygens, and water molecules: distances less than 3.30 Å, for barrerite B and C

Barrerite B

C(1) polyhedron		C(2) polyhedron		C(3) polyhedron		C(4) polyhedron	
C(1)–C(1)	1.73(5) Å	C(2)–C(1)	0.94(2) [×2] Å	C(3)–C(3)	0.94(2) Å	C(4)–O(4)	2.43(1) [×4] Å
C(1)–C(2)	0.94(2) [×2]	C(2)–C(2)	0.74(3)	C(3)–O(1)	2.45(1) [×2]	C(4)–O(7)	2.63(1) [×2]
C(1)–O(4P)	3.02(2) [×2]	C(2)–O(4P)	2.97(1) [×2]	C(3)–O(1)	2.40(1) [×2]	C(4)–W(1)	2.60(7)
C(1)–O(9)	2.85(2)	C(2)–O(9)	2.72(1) [×2]	C(3)–W(4)	2.40(8)		
C(1)–O(9)	2.81(2)	C(2)–OH(1)	2.79(2) [×2]			W(3) polyhedron	
C(1)–OH(1)	2.68(2) [×2]	C(2)–W(3)	1.84(6) [×2]	W(2) polyhedron		W(3)–C(1)	1.20(5)
C(1)–W(1)	3.16(5)	C(2)–W(3)	2.11(6) [×2]	W(2)–C(1)	2.68(8)	W(3)–C(1)	2.76(7)
C(1)–W(2)	2.68(8)			W(2)–O(8)	2.98(5) [×2]	W(3)–C(2)	1.84(6)
C(1)–W(3)	1.20(5) [×2]	W(1) polyhedron		W(2)–O(9)	2.73(8)	W(3)–C(2)	2.11(6)
C(1)–W(3)	2.76(7) [×2]	W(1)–C(1)	3.16(5) [×2]	W(2)–O(10)	2.76(3) [×2]	W(3)–O(4P)	2.72(5)
		W(1)–C(4)	2.60(7)	W(2)–OH(1)	2.91(4) [×2]	W(3)–O(9)	3.21(5)
C(5) polyhedron		W(1)–O(4P)	3.06(1) [×4]	W(2)–W(2)	2.00(2)	W(3)–OH(1)	1.78(5)
C(5)–C(5)	2.45(9)	W(1)–O(9)	3.23(2) [×2]	W(2)–W(3)	2.17(8) [×2]	W(3)–OH(1)	3.21(5)
C(5)–O(1P)	2.49(4)	W(1)–W(3)	2.79(7)			W(3)–W(1)	2.79(7)
C(5)–O(3P)	1.51(5)			W(4) polyhedron		W(3)–W(2)	2.17(8)
C(5)–O(3P)	2.64(5)			W(4)–C(3)	2.40(8)	W(3)–W(3)	3.26(11)
C(5)–O(7P)	3.23(4)			W(4)–C(5)	2.63(5)	W(3)–W(3)	1.46(9)
C(5)–O(8)	2.42(4)			W(4)–W(4)	3.18(15)		
C(5)–O(10)	2.83(4)						
C(5)–W(4)	2.63(5)						

Barrerite C

C(1) polyhedron		C(2) polyhedron		C(3) polyhedron		C(4) polyhedron	
C(1)–C(1)	2.00(9) Å	C(2)–C(1)	1.07(4) [×2] Å	C(3)–C(3)	1.11(7) Å	C(4)–O(4)	2.58(1) [×4] Å
C(1)–C(2)	1.07(4) [×2]	C(2)–C(2)	0.78(11)	C(3)–O(1)	2.72(2) [×2]	C(4)–O(7)	2.76(2) [×2]
C(1)–O(4P)	3.08(3) [×2]	C(2)–O(4P)	3.07(4) [×2]	C(3)–O(1)	2.60(2) [×2]	C(4)–W(1)	2.49(4)
C(1)–O(9)	2.94(4)	C(2)–O(9)	2.88(1) [×2]	C(3)–W(4)	1.88(7)		
C(1)–O(9)	3.12(4)	C(2)–OH(1)	2.98(4) [×2]	C(3)–W(4)	2.99(8)	W(3) polyhedron	
C(1)–OH(1)	2.87(2) [×2]	C(2)–W(2)	3.18(5) [×2]	C(3)–W(5)	2.25(6)	W(3)–C(1)	1.63(7)
C(1)–W(1)	3.16(5)	C(2)–W(3)	1.92(7) [×2]	C(3)–W(5)	1.14(4)	W(3)–C(1)	2.95(8)
C(1)–W(2)	2.50(7)	C(2)–W(3)	2.45(8) [×2]	C(3)–W(10)	3.18(4)	W(3)–C(2)	1.92(7)
C(1)–W(3)	1.63(7) [×2]	C(2)–W(8)	2.77(6) [×2]			W(3)–C(2)	2.45(8)
C(1)–W(3)	2.95(8) [×2]			W(2) polyhedron		W(3)–O(4P)	2.63(6)
C(1)–W(8)	1.79(7)	W(1) polyhedron		W(2)–C(1)	2.50(7)	W(3)–O(7P)	3.28(7)
C(1)–W(8)	3.26(8)	W(1)–C(1)	3.16(5) [×2]	W(2)–C(2)	3.18(5) [×2]	W(3)–O(8)	3.04(7)
		W(1)–C(4)	2.49(4)	W(2)–O(8)	2.91(4) [×2]	W(3)–O(10)	3.23(7)
C(5) polyhedron		W(1)–O(4P)	3.07(1) [×4]	W(2)–O(9)	2.74(6)	W(3)–OH(1)	1.24(7)
C(5)–C(5)	1.73(9)	W(1)–O(9)	3.13(2) [×2]	W(2)–O(10)	2.93(3) [×2]	W(3)–W(2)	2.35(8)
C(5)–O(1P)	2.86(5)	W(1)–W(8)	2.91(7) [×2]	W(2)–OH(1)	3.01(3) [×2]	W(3)–W(3)	3.17(13)
C(5)–O(3P)	2.24(5)			W(2)–W(2)	2.26(11)	W(3)–W(3)	2.95(14)
C(5)–O(3P)	2.91(5)	W(5) polyhedron		W(2)–W(3)	2.35(8) [×2]	W(3)–W(8)	1.88(7)
C(5)–O(8)	3.15(5)	W(5)–C(3)	1.14(4)	W(2)–W(7)	2.05(7)	W(3)–W(8)	3.07(9)
C(5)–W(4)	2.42(6)	W(5)–C(5)	2.67(5) [×2]	W(2)–W(7)	2.48(7)		
C(5)–W(5)	2.67(5)	W(5)–O(1)	3.27(3)	W(2)–W(8)	1.45(7)	W(7) polyhedron	
C(5)–W(6)	2.22(5)	W(5)–O(1)	2.93(3)	W(2)–W(8)	2.36(8)	W(7)–C(5)	3.12(6) [×2]
C(5)–W(7)	2.87(6)	W(5)–W(4)	0.74(6)			W(7)–C(5)	2.87(6) [×2]
C(5)–W(7)	3.12(6)	W(5)–W(6)	2.53(4)	W(6) polyhedron		W(7)–O(8)	3.14(3) [×2]

TABLE 4. (*cont.*)

C(5)–W(9)	1.35(5)
C(5)–W(10)	3.20(5)
W(4) polyhedron	
W(4)–C(3)	1.88(7)
W(4)–C(3)	2.99(8)
W(4)–C(5)	2.42(6) [×2]
W(4)–W(5)	0.74(6)
W(4)–W(6)	2.16(6)
W(4)–W(7)	2.57(8)
W(4)–W(7)	3.21(8)
W(4)–W(9)	2.98(6) [×2]
W(4)–W(10)	3.21(7)
W(5)–W(7)	3.29(6)
W(5)–W(9)	3.06(4) [×2]
W(5)–W(10)	3.00(4)
W(8) polyhedron	
W(8)–C(1)	3.26(8)
W(8)–C(1)	1.79(7)
W(8)–C(2)	2.77(6) [×2]
W(8)–OH(1)	2.78(3) [×2]
W(8)–W(1)	2.91(7)
W(8)–W(2)	2.36(8)
W(8)–W(2)	1.45(7)
W(8)–W(3)	1.88(7) [×2]
W(8)–W(3)	3.07(9) [×2]
W(8)–W(8)	1.55(12)
W(6)–C(5)	2.22(5) [×4]
W(6)–W(4)	2.16(6) [×2]
W(6)–W(5)	2.53(4) [×2]
W(6)–W(7)	2.09(5) [×2]
W(6)–W(9)	1.99(5) [×2]
W(6)–W(10)	2.24(5)
W(9) polyhedron	
W(9)–C(5)	1.35(5) [×2]
W(9)–O(1)	3.24(3) [×2]
W(9)–O(3P)	3.23(5) [×2]
W(9)–O(7)	2.95(5)
W(9)–W(4)	2.98(6) [×2]
W(9)–W(5)	3.06(4) [×2]
W(9)–W(6)	1.99(5)
W(9)–W(10)	2.20(5)
W(7)–O(10)	2.92(2) [×2]
W(7)–W(2)	2.05(7)
W(7)–W(2)	2.48(7)
W(7)–W(4)	2.57(8)
W(7)–W(4)	3.21(8)
W(7)–W(5)	3.29(6)
W(7)–W(6)	2.09(5)
W(7)–W(7)	0.85(9)
W(10) polyhedron	
W(10)–C(3)	3.18(4) [×2]
W(10)–C(5)	3.20(5) [×4]
W(10)–O(7)	2.70(3) [×2]
W(10)–W(4)	3.21(7) [×2]
W(10)–W(5)	3.00(4) [×2]
W(10)–W(6)	2.24(5)
W(10)–W(9)	2.20(5) [×2]

TABLE 5. X-ray powder diffraction data for barrerite B*

h k l	I/I_o	d_{obs}(Å)	d_{calc}(Å)	*h k l*	I/I_o	d_{obs}(Å)	d_{calc}(Å)
	20	8.77		4 2 0 0 1 5	10	3.173	3.162 3.151
0 2 0	100	8.52	8.56		10	3.074	
1 0 2	1	6.95	6.91	2 5 1 3 0 4	3	3.003	3.004 3.003
2 0 0	2	6.80	6.80	4 2 2 0 4 4	5	2.935	2.941 2.925
0 2 2	1	5.86	5.85	3 4 2 0 5 3	7	2.903	2.902 2.883
2 0 2	2	5.22	5.19	1 4 4 2 1 5 0 6 0	4	2.854	2.860 2.859 2.853
	15	4.40		1 3 5	4	2.745	2.740
0 4 0	30	4.27	4.28		3	2.705	
	6	4.14		3 5 1 0 6 2 2 4 4	4	2.694	2.694 2.688 2.687
0 0 4	2	3.99	4.01	5 1 1 1 6 2 2 6 0	2	2.635	2.651 2.637 2.631
0 3 3	6	3.88	3.90	4 3 3	4	2.561	2.564
1 0 4	15	3.84	3.84	1 2 6 2 6 2	2	2.505	2.507 2.500
1 3 3	15	3.73	3.75		5	2.201	
0 2 4 2 4 0 1 4 2	4	3.64	3.63 3.62 3.64	0 8 0	3	2.141	2.140
3 2 2	4	3.583	3.585				
1 2 4	5	3.515	3.507				
2 0 4	1	3.449	3.453				
4 0 0	4	3.403	3.402				
0 5 1	4	3.344	3.348				
4 1 1 1 5 1	2	3.257	3.267 3.251				
2 2 4	2	3.211	3.202				

*Ni-filtered CuKα_1 radiation (λ = 1.5405 Å), angular speed ½° per minute, $Pb(NO_3)_2$ internal standard. The powder was heated for 16 hours at 250°C and then cooled to room temperature. The recording of the diffraction pattern was started 30 minutes after heating and continued for 90 minutes. About 20% of barrerite already was

cont. p. 93

TABLE 6. X-ray powder diffraction data for barrerite C*

h k l	I/I_o	d_{obs}(Å)	d_{calc}(Å)	*h k l*	I/I_o	d_{obs}(Å)	d_{calc}(Å)
0 2 0*	100	8.76	8.80	3 2 4*	2	2.862	2.860
0 0 2*	1	8.10	8.11	4 1 3 }*	2	2.845	2.842
1 0 2*	3	6.98	6.97	0 3 5 }			2.840
2 0 0*	2	6.79	6.80	0 6 2*	4	2.758	2.759
0 2 2*	5	5.97	5.97	3 5 1 }*	4	2.740	2.741
1 2 2*	2	5.463	5.463	2 4 4 }			2.732
2 2 0*	1	5.381	5.382	2 5 3 }			2.707
0 1 3*	5	5.172	5.170	0 0 6 }			2.705
1 1 3*	1	4.836	4.833	1 6 2 }	5	2.705	2.704
2 2 2*	7	4.482	4.485	2 6 0 }			2.694
0 4 0*	70	4.393	4.401	4 4 0 }			2.691
2 3 1*	7	4.283	4.285	2 5 3 }			2.707
3 1 1*	8	4.239	4.239	0 0 6 }			2.705
2 1 3*	10	4.117	4.116	1 6 2 }	5	2.701	2.704
0 0 4*	3	4.057	4.057	2 6 0 }			2.694
0 3 3*	2	3.978	3.977	4 4 0 }			2.691
1 0 4*	15	3.890	3.888	2 3 5 }*			2.621
0 4 2*	15	3.872	3.868	3 1 5 }	2	2.618	2.610
1 3 3*	25	3.816	3.817	4 0 4 }			2.606
1 4 2*	28	3.720	3.721	2 3 5 }			2.621
1 2 4*	10	3.557	3.556	3 1 5 }	2	2.607	2.610
2 0 4*	7	3.487	3.484	4 0 4 }*			2.606
0 5 1 }	7	3.437	3.440	0 2 6 }			2.585
2 3 3 }			3.433	4 3 3 }	1	2.581	2.585
4 0 0*	5	3.401	3.401	5 0 2 }*			2.580
2 4 2*	3	3.361	3.363	2 6 2 }*	3	2.557	2.557
1 5 1*	7	3.333	3.335	4 4 2 }			2.554
0 1 5*	6	3.192	3.192	3 4 4*	2	2.494	2.492
4 2 0*	8	3.174	3.173	0 7 1*	2	2.488	2.485
2 5 1*	10	3.070	3.070	0 5 5 }*	2	2.385	2.386
3 0 4*	1	3.029	3.024	0 6 4 }			2.377
3 3 3 }*	2	2.988	2.990	0 8 0*	15	2.203	2.200
0 4 4 }			2.983	6 2 0 }	15	2.198	2.196
0 6 0 *	15	2.936	2.934	5 2 4 }			2.189
2 1 5 }*			2.889				
4 3 1 }	4	2.891	2.895				
1 5 3 }			2.884				

*Ni-filtered $CuK\alpha_1$ radiation, angular speed ¼° per minute, $Pb(NO_3)_2$ internal standard. The powder was heated for 16 hours at 280°C and then cooled to room temperature. The diffraction pattern was recorded ~1 year after the heating experiment. The unit-cell parameters were refined from the powder data according to the method of Alberti (1976) which takes into account also the structure factors of the reflections for the indexing of the observed peaks. The unit-cell parameters are a = 13.605(7) Å, b = 17.602(7) Å, c = 16.227(8) Å in good agreement with the values obtained with the single-crystal diffractometer. Only the reflections marked with a * have been used for the refinement.

inverted to phase C at the start and about 40% at the end of the recording. The intensities were not corrected for the variation of the phase B/phase C ratio which occurred during the recording. The unit-cell data used for the calculation of the d_{calc} are reported in the text. The table reports only intensities and d_{obs} values for the peaks ascribed to phase C. The 2.705 Å and 2.201 Å peaks can be indexed also with lattice planes of the B phase. Nevertheless, the very low values of the Fo's of these planes and the presence of strong lines in phase C suggest that these peaks are more probably due to phase C.

TABLE 7. X-ray powder diffraction data for barrerite D*

h k l	I/I_o	d_{obs}(Å)	d_{calc}(Å)	*h k l*	I/I_o	d_{obs}(Å)	d_{calc}(Å)
0 2 0	100	8.43	8.43	1 4 2	15	3.594	3.592
1 0 2	7	6.88	6.85	2 4 0 }	2	3.528	{ 3.530
0 2 2	6	5.82	5.84	1 2 4 }			{ 3.510
2 1 1	2	5.67	5.65	3 2 2	5	3.454	3.462
1 2 2 }	8	5.33	{ 5.32	2 4 2 }	2	3.230	{ 3.235
0 3 1 }			{ 5.31	4 0 0 }			{ 3.225
2 2 0	12	5.10	5.12	1 5 1 }			{ 3.200
1 1 3	2	4.75	4.77	0 1 5 }	8	3.192	{ 3.178
2 2 2	7	4.34	4.33	2 2 4 }			{ 3.175
0 4 0	70	4.21	4.22	3 0 4 }	2	2.938	{ 2.946
2 3 1	10	4.10	4.10	2 5 1 }			{ 2.940
0 0 4 }			{ 4.05	0 6 0 }			{ 2.812
3 1 1 }	10	4.05	{ 4.04	0 3 5 }	7	2.801	{ 2.805
2 1 3 }			{ 4.02	1 5 3 }			{ 2.793
1 0 4	4	3.84	3.86	2 4 4 }	5	2.661	{ 2.660
0 4 2 }	10	3.75	{ 3.74	0 6 2 }			{ 2.656
1 3 3 }			{ 3.73				

*Ni-filtered CuKα_1 radiation, angular speed ½° per minute, quartz external standard. The powder was heated for 16 hours at 450°C then cooled to room temperature. The diffraction pattern was recorded ~1 month after the heating experiment. The unit-cell data used for the calculation of the d_{calc} are reported in the text. The reflections used are those possible in the space group *Amma*. However, it is possible that the correct space group is different.

tetrahedra (Meier, 1968) (see Fig. 1). These units form chains parallel to [001] by sharing their T(5) atoms (Fig. 2). The chains are linked laterally through oxygen bridges to form sheets parallel to (010). Two oxygen bridges per unit provide the links between layers.

In the collapsed phases two kinds of building units have to be considered as a result of the breaking of the O(3P) oxygen bridge. One kind has the same topology as the original unit, the other is formed by two five-membered rings and two very irregular seven-membered rings (see Fig. 1d). A similar statistical occupancy was observed by Townes *et al.* (1967) in a ferrimagnetic barium ferrite for a cation in two half-occupied face-sharing FeO_4 tetrahedra.

In barrerite A the fundamental unit is nearly parallel to *c* (see Fig. 2) and its length, or the distance between two successive atoms T(5) in the direction [001], is 8.93 Å. In barrerite B, if the unit remains parallel to *c* this distance reduces to 8.02 Å. To compensate, at least in part, for this contraction, the unit rotates about 18° around its center of gravity, and as a result its length increases to 8.40 Å. In this way the chain of these single units develops in the *c* direction with a pronounced zig-zag (see Figs. 2b, 2c). Furthermore, as a consequence of this crushing, the unit widens, and its T(1)–T(1P) distance increases to 4.61 Å. At the same time the unit rotates about 5° around its T(5)–T(5) axis. In phase C, as a result of the rehydration and the increase in the unit cell, the building unit tends to regain its original disposition of the natural phase A. The length increases to 8.44 Å, and there is a decrease both in the rotation around the center of gravity to 16° and the amount of rotation around the T(5)–T(5) axis.

The channels in the structure of barrerite A are formed by ten-membered rings parallel to [100] and by eight-membered rings parallel to [001]. The channels parallel to *a* are of two different types (see Fig. 3); the first type at y/b = 0 has ellipsoidal access, with axes approximately 5.0 Å (parallel to *c*) and 4.3 Å; the second, at y/b = ½, is roughly circular, with a minimum free diameter of about 4.6 Å. The channels parallel to *c* also are of two different types (see Fig. 4). In both types the minimum diameter is about 2.7 Å. The rotation of the fundamental unit causes marked shifts in the oxygens bridging between different chains, namely, O(9), O(1), O(1P). In particular, oxygens O(1) and O(1P), which are bonded to T(5), are pulled out of position and move away from the origin when T(5) shifts. As a consequence of the increased distance between the two T(3) atoms related by the binary axis,

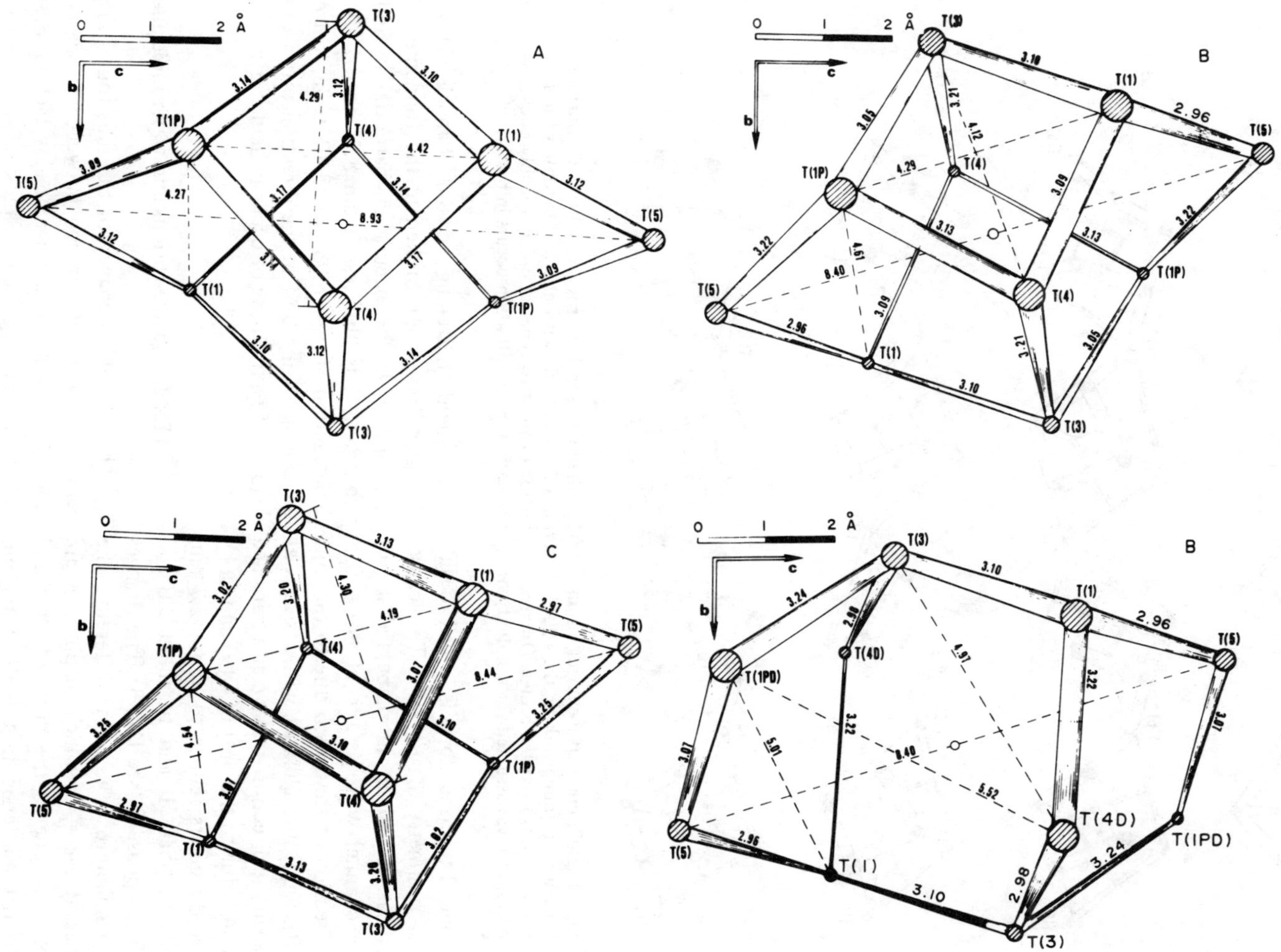

Fig. 1. Top left, framework building unit for barrerite A; top right, for barrerite B; bottom left, for barrerite C. Bottom right, second type of building unit for barrerite B with broken O(3P) oxygen link. The same unit is present also in barrerite C.

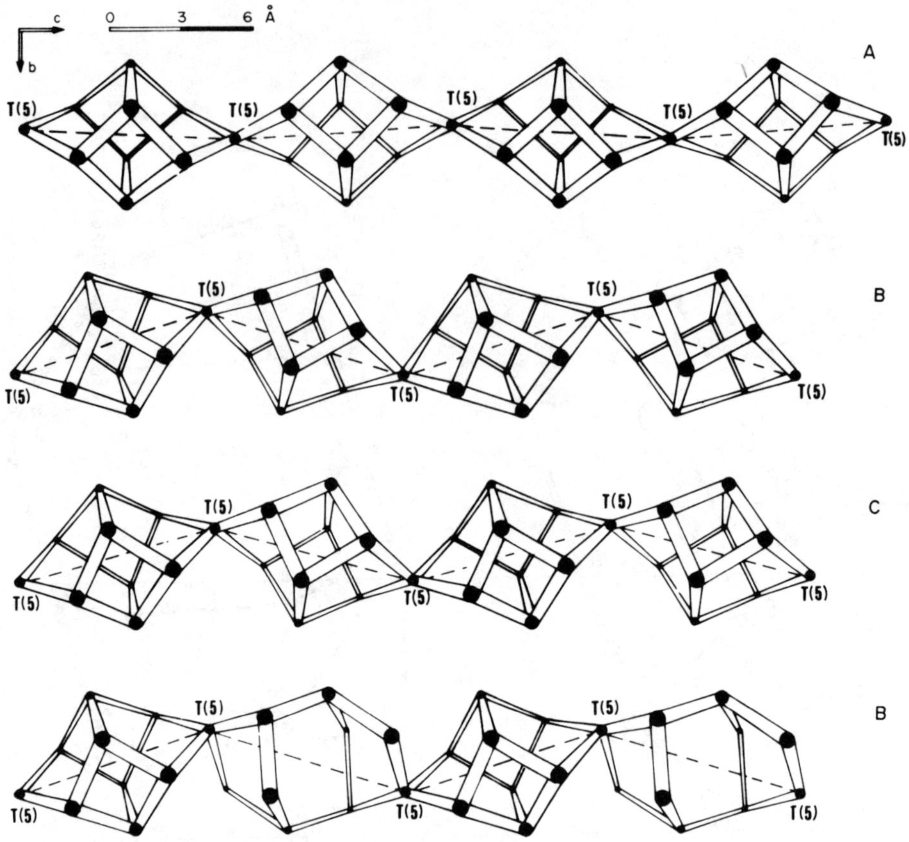

Fig. 2. Building unit behavior in the [001] direction for barrerite A, B, and C. Bottom, behavior of framework in the [001] direction of barrerite B, with alternating units with and without O(3P) oxygen bridge. Ordered distribution of the two types of units probably does not occur, but was assumed for simplicity of representation. Same situation occurs in barrerite C.

the oxygen O(9) bonded to these atoms is forced to slide along the plane of symmetry at $y/b = ½$. This causes considerable modification both in the shape and size of the channels. The channel with its axis at $y/b = 0$ is elongated in the direction of the axis b due to the shift of O(1P) and O(9), while the channel with its axis at $y/b = ½$ is elongated in the direction of axis c and flattened in the other direction, due to the shift of O(9) and O(1). The minimum free diameters reduce to about 2.7 and 1.8 Å, respectively, in phase B and to about 3.0 and 2.2 Å, respectively, in phase C (see Fig. 3). The changes in the channels parallel to c are very different. The channels with axes at $z/c = 0$, which in phase A are elongated parallel to b, due to the shifts of O(7) and O(9), become strongly flattened in this direction and elongated in the direction of a. The channel at $z/c = ½$, on the contrary, is only slightly modified due to the shift of O(10). The minimum free diameters reduce to about 1.5 and 2.5 Å in phase B and 1.7 and 2.5 Å in phase C.

In heulandite (Alberti, 1973) the situation is similar; in this case heating also causes a rotation of the fundamental units around their center of gravity and as a consequent flattening of the channels. However, no oxygen bridges are broken in the fundamental unit in heulandite B, as occurs in barrerite B and C, the contraction is also less (from 8.77 to 8.57 Å versus a contraction from 8.93 to 8.40 Å in barrerite B). Widening of the unit in the direction orthogonal to the T(5)–T(5) axis does not occur in heulandite.

Exchangeable cations and water molecules

The partial occupancy of the sites found in phases B and C prevents distinction between cations and water molecules from the site occupancy factors. Besides, the complete rearrangement of the exchangeable cations and water molecules in the polymorphic transformations of barrerite prevents useful information being drawn from the comparison of the sites occupied in natural and collapsed phases. The short distances C(3)–O(1) = 2.40 Å, C(4)–O(4) = 2.43 Å, and C(5)–O(1P) = 2.49 Å in phase B suggest occupancy of these sites by cations rather than water molecules. The distances C(1)–C(1) = 1.73 Å, C(1)–C(2) = 0.94 Å, and C(2)–C(2) = 0.74 Å

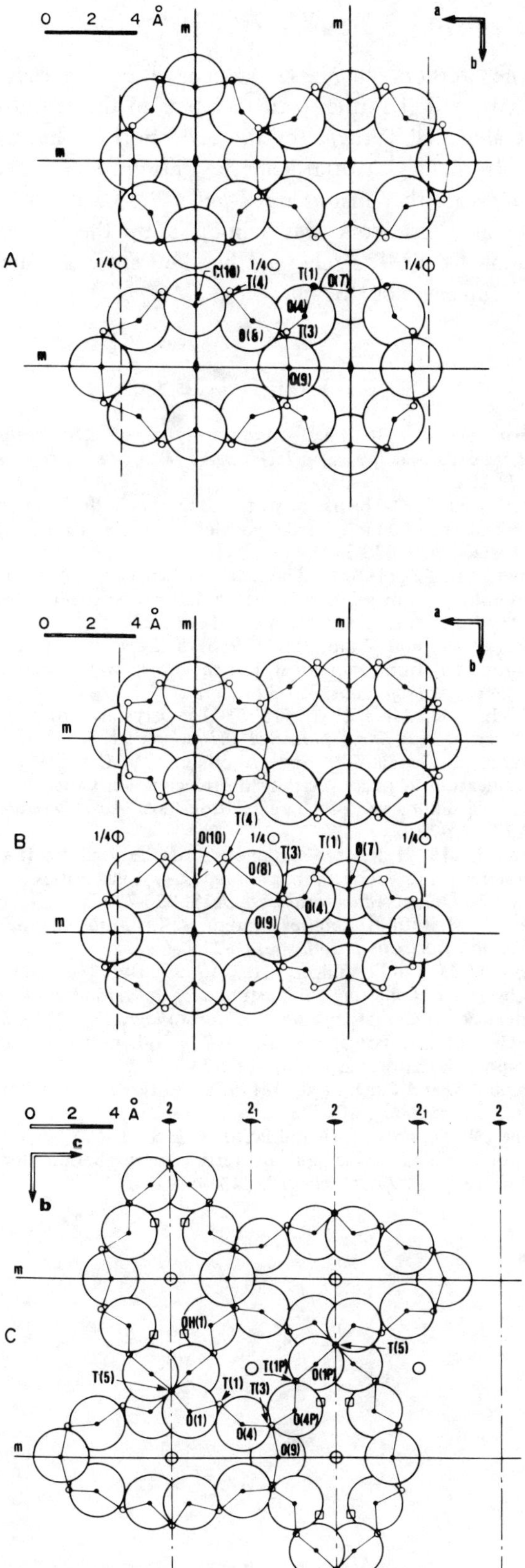

Fig. 3. Top to bottom projection on the (100) plane of the ten-membered rings (channels) of barrerite A, B, and C.

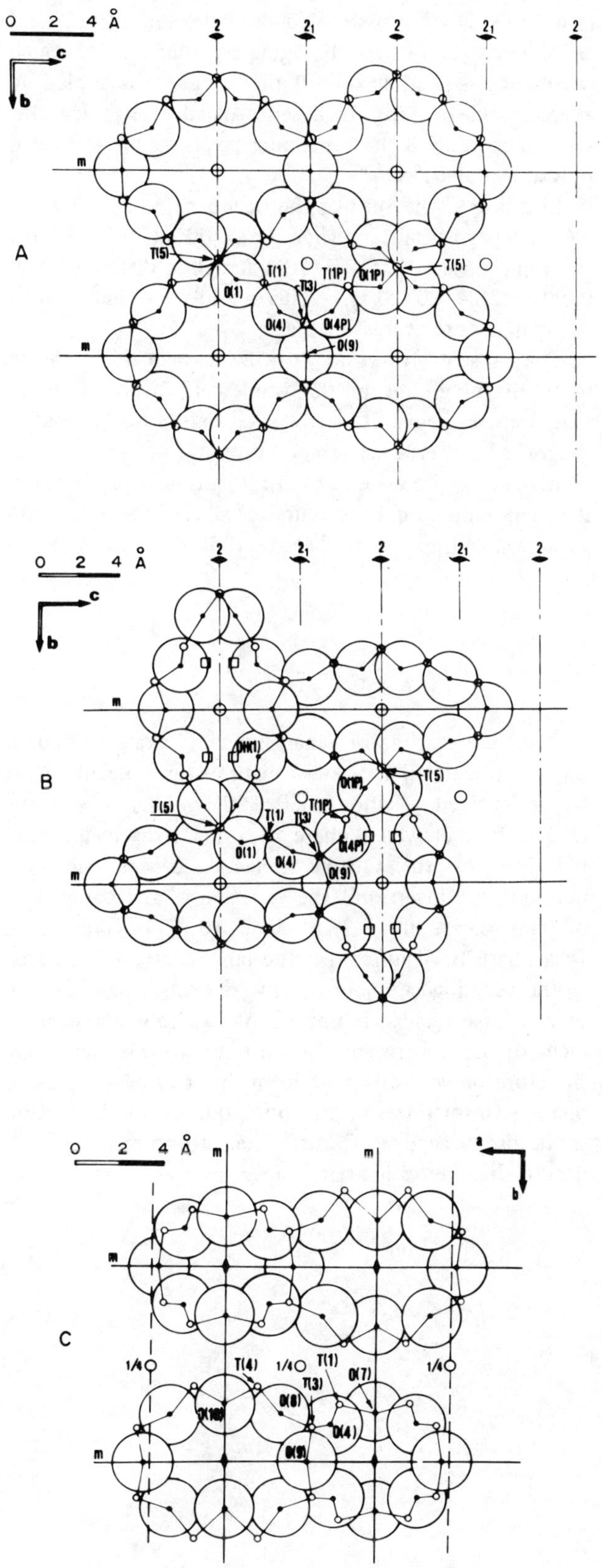

Fig. 4. Top to bottom projection on the (001) plane of the eight-membered rings (channels) of barrerite A, B, and C.

(see Table 4) are too short to allow the sum of C(1) and C(2) occupancies to be greater than 50%. Such occupancy is possible only if these sites are occupied by exchangeable cations. In absence of adverse indications we believe that all the remaining positions are occupied by water molecules.

In phase C the sum of the occupancies of W(6) and W(10) is significantly higher than 100% (70% ± 3% for W(6) and 66% ± 4% for W(10)) and their distance is too short (2.24 ± 0.05 Å). This implies that these occupancies are overestimated.

The present structural work has shown the presence of hydroxyl groups in the framework of barrerite at temperatures above 200°C. This accounts for the loss of weight of barrerite at temperatures up to 450–500°C. Similarly, the presence of hydroxyl groups is probable in the high-temperature structures of all zeolites that retain a fraction of their "water" at about 500°C.

BARRERITE D

Thermal gravimetric analysis on crystals heated at 250°C indicates that the last eight water molecules start to be evolved at about 380–390°C. This is also the temperature at which phase D forms. This means that the loss of the OH(1) molecule causes the new polymorphic inversion. The variation of the parameters of phase D are of −5.2%, −1.5%, +0.9% in comparison with phase B. Significant is the large decrease in *a*. Due to the very high number of oxygen bridges parallel to *a*, such a large change is improbable without a rearrangement of the framework. The structure of this phase may therefore be very different from that of the A, B, and C phases. Unfortunately, the poor quality of diffraction spots discouraged us from further structure determination of this interesting structure.

ACKNOWLEDGMENTS

The authors thank G. Cocco, P. F. Zanazzi, L. Fanfani, and the other staff members of the Istituto di Mineralogia di Perugia for the collection of the X-ray intensity data; G. Gottardi and R. Rinaldi for the critical reading of the manuscript; and W. Lugli for the drawings. This work was supported by the Consiglio Nazionale delle Ricerche and by the Centro di Calcolo della Università di Modena.

REFERENCES

Alberti, A. (1973) The structure type of heulandite B (heat-collapsed phase): *Tschermaks Min. Petr. Mitt.* **19**, 173-184.

Alberti, A. (1976) The use of structure factors in the refinement of unit-cell parameters from powder diffraction data: *J. appl. Crystallogr.* 9, 373-374.

Cromer, D. T. (1965) Anomalous dispersion corrections computed from self-consistent field relativistic Dirac–Slater wave functions: *Acta Crystallogr.* **18**, 17-23.

Cromer, D. T. and Mann, J. B. (1968) X-ray scattering factors computed from numerical Hartree–Fock wave functions: *Acta Crystallogr.* **A24**, 321-324.

Galli, E. and Alberti, A. (1976) The crystal structure of barrerite: *Bull. Soc. fr. Minéral. Cristallogr.* **99**, 331-340.

Germain, G., Main, P. and Woolfson, M. M. (1971) The application of phase relationships to complex structures. III. The optimum use of phase relationships: *Acta Crystallogr.* **A27**, 368-376.

Hajdu, F. (1972) Revised parameters of the analytic fits for coherent and incoherent scattered X-ray intensities of the first 36 atoms: *Acta Crystallogr.* **A28**, 250-252.

Meier, W. M. (1968) Zeolites structures. In: *Molecular Sieves*, London, *Soc. Chem. Ind.*, pp. 10-27.

Meier, W. M. and Villiger, H. (1969) Die Methode der Abstandsverfeinerung zur Bestimmung der Atomkoordinaten idealisierter Gerüststrukturen: *Z. Kristallogr.* **129**, 411-423.

Passaglia, E. and Pongiluppi, D. (1974) Sodian stellerite from Capo Pula, Sardegna: *Lithos* 7, 69-73.

Passaglia, E. and Pongiluppi, D. (1975) Barrerite, a new natural zeolite: *Min. Mag.* **40**, 208.

Townes, W. D., Fang, J. H. and Perrotta, A. J. (1967) The crystal structure and refinement of ferrimagnetic barium ferrite, $BaFe_{12}O_{19}$: *Z. Kristallogr.* **125**, 437-449.

CONSTITUENT SHEETS IN THE ZEOLITE FRAMEWORKS OF THE MORDENITE GROUP

W. M. MEIER

Institut für Kristallographie und Petrographie der ETH,
8092 Zurich, Switzerland

Abstract – A common feature of the frameworks of mordenite and related zeolites is the existence of 6-ring sheets of tetrahedra (6^3 nets of T-atoms) in these structures. In mordenite, dachiardite, and epistilbite these 6-ring sheets are linked to each other through single 4-membered rings; in ferrierite through single 6-membered rings; 5-rings are thereby generated. The framework structures of bikitaite and of synthetic zeolite Li-ABW, like that of some related aluminosilicates, such as petalite, are also composed of 6-ring sheets. None of the other known zeolite structures contain any 6-ring sheets. In all framework structures of the mordenite group of zeolites, the 6-ring sheets appear highly corrugated and are generally parallel to a dominant crystal face. In these 3-connected sheets of tetrahedra an equal number of apical oxygens are pointing up and down, and, depending on the pattern of the tetrahedral apices pointing one way, four types of 6-ring sheets can be observed, while others can be postulated. At least in mordenite, ferrierite, and petalite, the 6-ring sheets appear to contain silicon predominantly, if not exclusively.

Various hypothetical structures related to mordenite have been proposed but not observed despite specific investigations. It is noteworthy that none of these hypothetical frameworks contain 6-ring sheets.

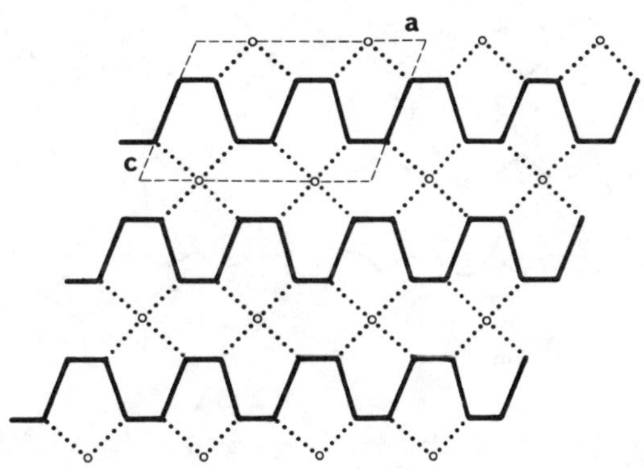

Fig. 1. Skeletal framework structure of petalite, $LiAlSi_4O_{10}$ (Z=2), in projection. Oxygen atoms are not shown. The puckered 6-ring sheets (marked by heavy lines) that are parallel (001) and contain only Si atoms are interconnected through single tetrahedra comprising Al atoms (indicated by circles).

INTRODUCTION

Zeolite structures have been classified on the basis of secondary building units. Structure types assigned to the mordenite group can be built from 5-1 units consisting of a single 5-ring and an attached tetrahedron. Thus defined, the mordenite group, or group 6 according to Breck (1974), comprises the frameworks of mordenite, dachiardite, epistilbite, ferrierite, and bikitaite (Meier, 1968). It does not include the framework type of Li-ABW (Kerr, 1974), which clearly resembles that of bikitaite. The latter also shows a remarkable likeness to the structure of petalite which, like the Li-ABW structure, cannot be based on 5-1 units. The tetrahedral framework structure of petalite determined by Zemann-Hedlick and Zemann (1955) and refined by Liebau (1961) can be considered to consist of puckered 6-ring silica sheets which are linked to each other through single alumina tetrahedra, as indicated schematically in Fig. 1. The silica sheets in the petalite structure are consistent with the presence of perfect (001) cleavage and the lamellar habit of the crystals. The aluminosilicate framework of petalite has several features in common with most structure types of the mordenite group, such as the relatively large number of 5-rings. Close examination of the mordenite-group structures shows that they all contain 6-ring sheets which in most instances are highly puckered. An investigation of this common structural feature appeared to be of considerable mineralogical interest.

ZEOLITE FRAMEWORKS CONTAINING 6-RING SHEETS

In 6-ring sheets of tetrahedra, the T-atoms form a 3-connected net involving 6-rings only (6^3 net). In the

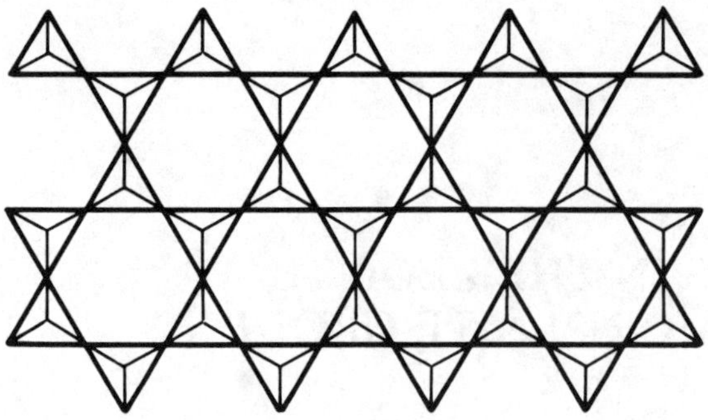

Fig. 2. Classic tetrahedral sheet in the mica structures.

classic 6-ring sheet of tetrahedra, which is a basic component of mica layers, the apices of the tetrahedra are all located on the same side of the essentially planar sheet (Fig. 2). Three-dimensional frameworks cannot be produced from such sheets, of course, unless some of the tetrahedra point the other way. The ideal sheet-like unit for building three-dimensional frameworks contains equal numbers of tetrahedra pointing in either direction. Such 6-ring sheets occur in the zeolite frameworks shown in skeletal form in Fig. 3. As indicated in Fig. 3, the 6-ring sheets in these frameworks appear highly puckered, and for this reason they are not readily recognizable in customary wire models.

The 6-ring sheets in the zeolite frameworks of Fig. 3 are of four types. Sheets of type I (Fig. 4) with tetrahedral apices alternatively pointing up and down occur in a number of derivative structures of tridymite and cristobalite. Examples are the aluminosilicates nepheline and carnegieite in which these sheets are directly interconnected. In more corrugated form, type I sheets are found in the structure of bikitaite, $LiAlSi_2O_6 \cdot H_2O$ (Z=2), determined by Appleman (1960) and Kocman *et al.* (1974). Puckered sheets of type I, parallel to (001) in this framework structure, are linked to each other through single tetrahedral chains oriented parallel to [010]. Alternatively, the bikitaite framework could be considered to consist of type III sheets (Fig. 6) parallel to (100) with the same kind of single chains in between. The former assignment seems preferable, however, in view of the morphology of the crystals and the perfect $\{001\}$ cleavage.

Sheets of type II, shown in Fig. 5, are found in synthetic zeolite Li-ABW, $LiAlSiO_4 \cdot H_2O$ (Z=4). Its framework structure, determined by Kerr (1974), can be set up by directly interconnecting sheets of type II. It has maximum symmetry Imam and appears to be one of the least complex zeolite frameworks. It is of interest

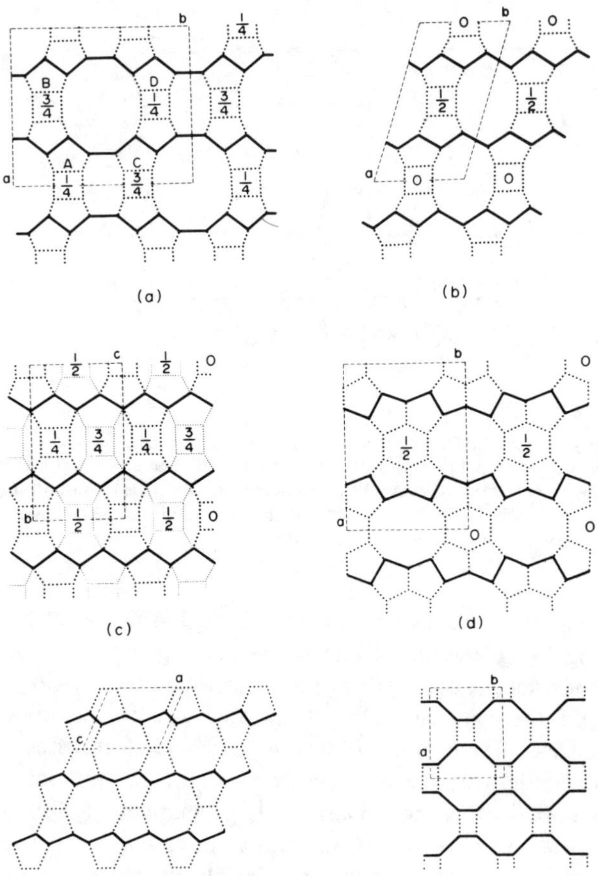

Fig. 3. Skeletal framework structures of mordenite (a), dachiardite (b), epistilbite (c), ferrierite (d), bikitaite (e), and synthetic zeolite Li-ABW (f). The puckered 6-ring sheets, marked by heavy lines, are directly interconnected in Li-ABW (f). In mordenite (a), dachiardite (b), and epistilbite (c), the 6-ring sheets are linked through single 4-rings; in ferrierite (d), through single 6-rings (dotted). Heights of the connecting rings are indicated. In bikitaite (e) the 6-ring sheets are linked through single chains of tetrahedra parallel to [001]. The sheets are of type I in (e), II in (f), III in (b, c, d), and IV in (a).

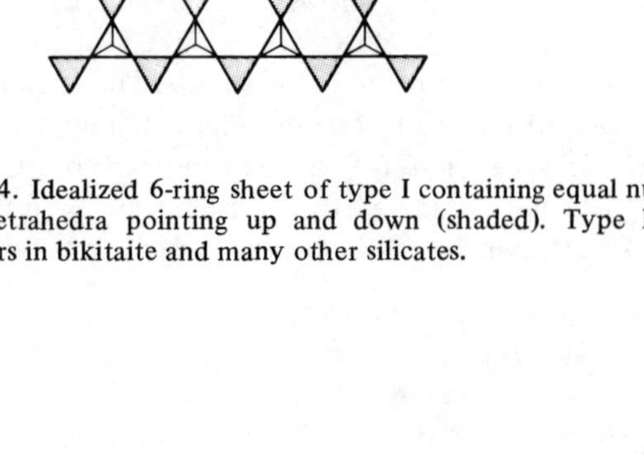

Fig. 4. Idealized 6-ring sheet of type I containing equal numbers of tetrahedra pointing up and down (shaded). Type I sheet occurs in bikitaite and many other silicates.

that the synthetic aluminosilicates, $RbAlSiO_4$ and $CsAlSiO_4$, first described by Barrer and McCallum (1953), are isotypes of Li-ABW according to studies by Chung (1972).

Sheets of type III (Fig. 6) form part of the framework structures of dachiardite (Gottardi and Meier, 1963), epistilbite (Merlino, 1965; Perrotta, 1967; Slaughter and Kane, 1969), and ferrierite (Vaughan, 1966). The connecting elements are single 4-rings in dachiardite and epistilbite and single 6-rings in ferrierite as shown in Fig. 3. The more complex 6-ring sheet of type IV (Fig. 7), connected through single 4-rings as shown in Fig. 3(a), can be located in the structure of mordenite (Meier, 1961).

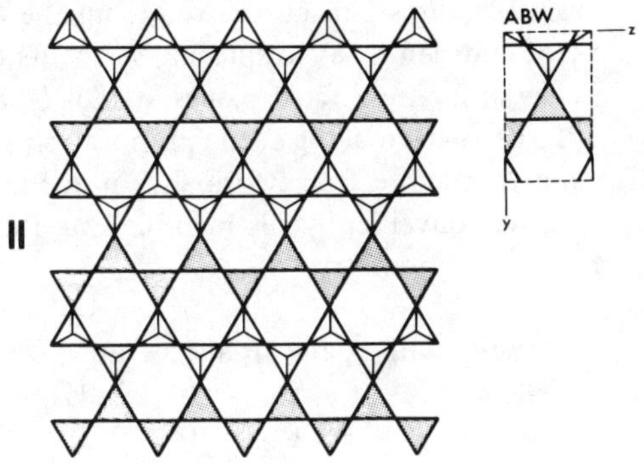

Fig. 5. Idealized 6-ring sheet of type II found in synthetic zeolite Li-ABW and in petalite, $LiAlSi_4O_{10}$.

Other types of 6-ring sheets made up of equal numbers of tetrahedra pointing each way can be readily postulated but have not been found to occur in any of the known zeolite structures. An example of such a 6-ring sheet is shown in Fig. 8. In highly corrugated form this type of sheet can be found in paracelsian, $BaAl_2Si_2O_8$. In this structure, determined by Smith (1953), the directly interconnected 6-ring sheets are parallel to (001). It is worth noting that the contracted framework of paracelsian could in principle be expanded to a possible zeolite structure (comparable in openness to phillipsite or gismondine).

A relatively large number of 5-rings are generated in all the frameworks described here in which 6-ring sheets

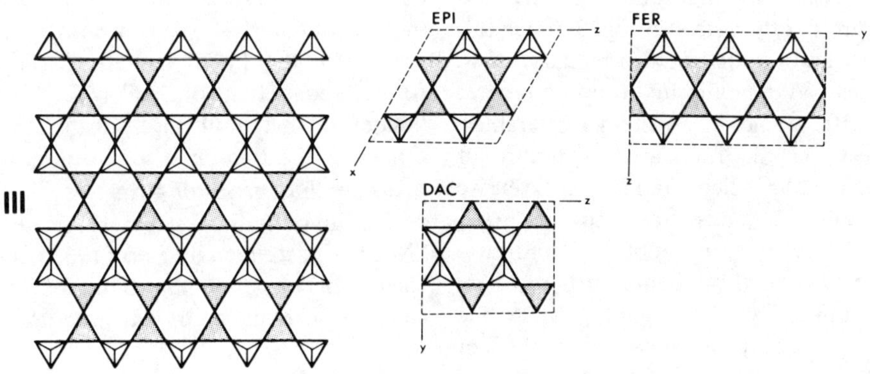

Fig. 6. Idealized 6-ring sheet of type III present in dachiardite, epistilbite, and ferrierite.

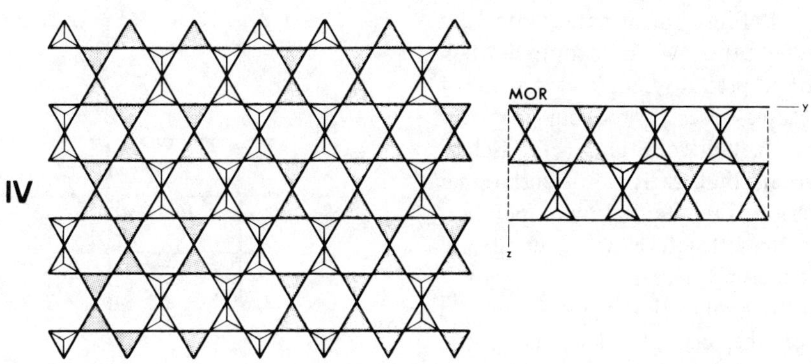

Fig. 7. Idealized 6-ring sheet of type IV constituting the major part of the framework structure of mordenite.

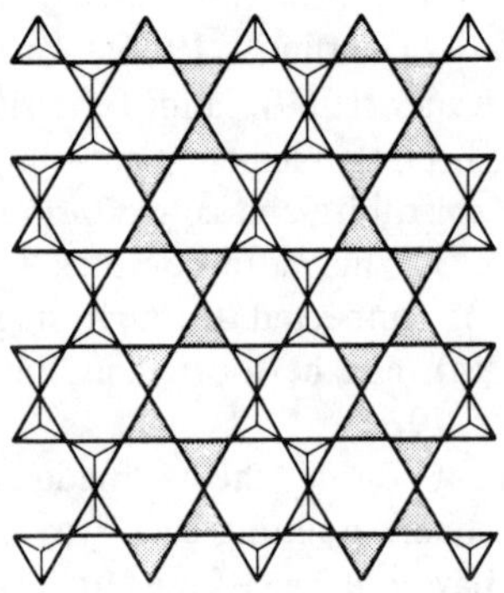

Fig. 8. Idealized 6-ring sheet existing in paracelsian, $BaAl_2Si_2O_8$, in which puckered sheets of this type are directly interconnected to a framework containing "double crankshaft" chains (readily recognizable in this illustration of a single sheet).

are interconnected through additional elements; 5-rings are absent in the frameworks consisting of stacked 6-ring sheets only. Thus, there are no 5-rings in the Li-ABW framework which can alternatively be derived by directly linking up 3-connected 4,8-ring sheets ($4{\cdot}8^2$ nets of T-atoms). These 4,8-ring sheets are oriented parallel (001) and are clearly visible in Fig. 3(f). On this basis the Li-ABW structure type can also be assigned to the set of structures of the phillipsite group of zeolites (comprising phillipsite, gismondine, and yugawaralite).

The 6-ring sheets in the framework structure of petalite contain essentially silicon only. An extensive structure analysis of hydrated sodium mordenite (V. Gramlich and W. M. Meier, to be published) indicates that the 6-ring sheets in this zeolite structure are probably also occupied by silicon only, while the connecting 4-rings presumably contain equal numbers of Si and Al atoms. Somewhat analogous results have been obtained in a recent structure refinement of hydrated ferrierite (R. Meier, unpublished results), with aluminum presumably occupying half the tetrahedral sites in the connecting 6-rings, while the 6-ring sheets appear to contain silicon only. Definite information on the Al-containing sites in zeolite frameworks is quite limited, however, since the marked pseudosymmetry of most of these framework structures poses formidable problems in investigations of the Si,Al-distributions (cf. Fischer, 1973). It must be noted that each non-translational pseudosymmetry operation is a potential twin operation giving rise to a "symmetrized structure" in which Si- and Al-sites are very likely superimposed.

The 6-ring sheets in these minerals can also be related to the morphology of the crystals. The sheets are generally parallel to dominant crystal faces. Furthermore, perfect cleavage has been noted for $\{100\}$ in bikitaite, $\{010\}$ in epistilbite, and $\{001\}$ in petalite.

HYPOTHETICAL STRUCTURES RELATED TO MORDENITE AND DACHIARDITE

Model structures related to mordenite have been proposed by Kerr (1963) and Sherman and Bennett (1973) but have not been found to occur. The three hypothetical structures can be derived from the *Cmcm* framework of mordenite by displacing segments of the framework by $c/2$ along fault planes $(010)_{0,\frac{1}{2}}$ and/or $(100)_{\frac{1}{4},\frac{3}{4}}$. In projection along c, the proposed structures look like the mordenite framework shown in Fig. 3(a), yet they can be conveniently distinguished on the basis of their maximum symmetry.

		A	B	C	D
(i)	*Immm* with	A at 0,	B at ½,	C at 0,	D at ½
(ii)	*Imcm*	0	0	½	½
(iii)	*Cmmm*	0	0	0	0

It is noteworthy that none of these three hypothetical structures contain any 6-ring sheets. By replacing the c-glide planes in the Cmcm framework of mordenite by mirror planes of the Immm structure (i) the 6-ring sheet is transformed to a more complicated sheet containing 4-, 6-, and 8-rings as demonstrated in Fig. 9. In the Imcm structure (ii) a peculiar sheet is formed consisting of ladder-like segments of 4-rings besides elongated 14-rings. Finally, by transforming the *Cmcm* to the *Cmmm* structure the original 6-ring sheet breaks up into bands of 4-rings.

The hypothetical structure related to dachiardite, "No. 6", postulated by Kerr (1963), resembles model (i) above. In place of the 6-ring sheet it contains the complex sheet of the Immm structure depicted in Fig. 9.

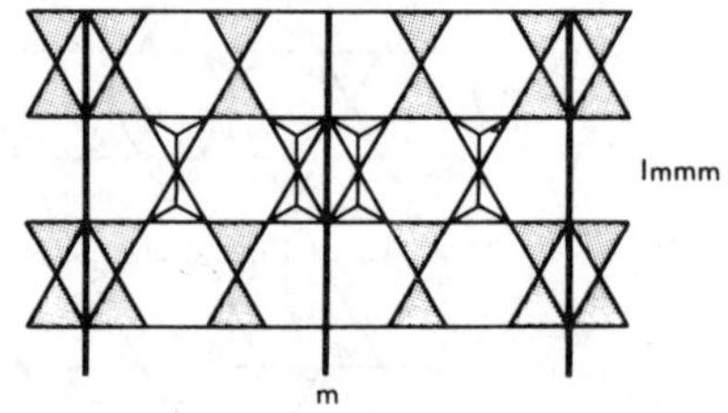

Fig. 9. The sheet of tetrahedra in the hypothetical *Immm* framework derived from the mordenite structure compared to the corresponding 6-ring sheet in the *Cmcm* framework of mordenite.

CONCLUSIONS

The tetrahedral frameworks containing 6-ring sheets constitute a set of structures comprising all the presently known members of the mordenite group, as well as Li-ABW. No 6-ring sheets can be found in any of the other known zeolite structures. These 6-ring sheets therefore appear to be a remarkable feature of the mordenite-group zeolites, and are a structural characteristic that the mordenite-group has in common with related non-zeolitic silicates, such as petalite. The fact that hypothetical frameworks resembling the mordenite structure but not containing 6-ring sheets have not been observed despite special investigations may also indicate that the 6-ring sheets in this set of structures are of some chemical significance.

ACKNOWLEDGMENTS

I am indebted to Dr. V. Gramlich and R. Meier for permission to quote unpublished results and to Dr. Ch. Baerlocher for helpful discussions. This work has been part of a research program supported by the Schweizerische Nationalfonds.

REFERENCES

Appleman, D. E. (1960) The crystal structure of bikitaite (abstr.): *Acta Crystallogr.* **13**, 1002.

Barrer, R. M. and McCallum, N. (1953) Hydrothermal chemistry of silicates. IV. Rubidium and cesium aluminosilicates: *J. Chem. Soc.* **1953**, 4029-4035.

Breck, D. W. (1974) *Zeolite Molecular Sieves*: Wiley-Interscience.

Chung, S. J. (1972) Tetraedergerüste bei Fluoberyllaten und Sulfaten: Ph.D. Thesis, Rheinisch-Westfälische Technische Hochschule Aachen.

Fischer, K. F. (1973) Zeolite structure refinement: *Adv. Chem. Ser.* **121**, 31-38.

Gottardi, G. and Meier, W. M. (1963) The crystal structure of dachiardite: *Z. Kristallogr.* **119**, 53-64.

Kerr, I. S. (1963) Possible structures related to mordenite: *Nature* **197**, 1194-1195.

Kerr, I. S. (1974) Crystal structure of a synthetic lithium zeolite: *Z. Kristallogr.* **139**, 186-195.

Kocman, V., Gait, R. I. and Rucklidge, J. (1974) The crystal structure of bikitaite: *Amer. Mineral.* **59**, 71-78.

Liebau, F. (1961) Untersuchungen an Schichtsilikaten des Formeltyps $A_m(Si_2O_5)_n$. III. Zur Kristallstruktur von Petalit: *Acta Crystallogr.* **14**, 399-406.

Meier, W. M. (1961) The crystal structure of mordenite (ptilolite): *Z. Kristallogr.* **115**, 439-450.

Meier, W. M. (1968) Zeolite structures: *Soc. Chem. Ind. Monograph on "Molecular Sieves"*, pp. 10-27.

Merlino, S. (1965) Struttura dell'epistilbite: *Mem. Soc. Tosc. Sci. nat.* (A) **72**, 480-483.

Perrotta, A. J. (1967) The crystal structure of epistilbite: *Mineral. Mag.* **36**, 480-490.

Sherman, J. D. and Bennett, J. M. (1973) Framework structures related to the zeolite mordenite: *Adv. Chem. Ser.* **121**, 52-65.

Slaughter, M. and Kane, W. T. (1969) The crystal structure of disordered epistilbite: *Z. Kristallogr.* **130**, 68-87.

Smith, J. V. (1953) The crystal structure of paracelsian, $BaAl_2Si_2O_8$: *Acta Crystallogr.* **6**, 613-620.

Vaughan, P. A. (1966) The crystal structure of the zeolite ferrierite: *Acta Crystallogr.* **21**, 983-990.

Zemann-Hedlick, A. and Zemann, J. (1955) Die Kristallstruktur von Petalit, $LiAlSi_4O_{10}$: *Acta Crystallogr.* **8**, 781-787.

DACHIARDITE-BEARING ZEOLITE ASSEMBLAGES IN THE PACIFIC NORTHWEST

WILLIAM S. WISE

Department of Geological Sciences, University of California, Santa Barbara, California 93106, U.S.A.

and

R. W. TSCHERNICH

532 Avenue A, Snohomish, Washington 98290, U.S.A.

Abstract – Dachiardite has been found in cavity-lining, zeolitic assemblages in tholeiitic basalts at Altoona, southwestern Washington, and at Cape Lookout and Agate Beach, Oregon. At Altoona dachiardite was the final phase to have formed, following siderite, an iron smectite, mordenite, ferrierite, and clinoptilolite. At Cape Lookout it occurs in the sequence iron smectite, erionite, phillipsite, mordenite, dachiardite, and clinoptilolite. In the Agate Beach dachiardite is associated only with clinoptilolite. The dachiardite at all localities occurs as small bundles of tiny blades, a habit similar to ferrierite, but crystal growths tend to be more complex.

Individual crystals, commonly 0.25 mm long and 0.05 mm wide, are blade-shaped with dominant (010) and (100) faces and (001), ($\bar{1}01$), and ($\bar{4}03$) terminations. The optic plane is perpendicular to (010) with $c \wedge X = 2°$. The refractive indices are $\alpha = 1.48$ $\beta = 1.481$, and $\gamma = 1.485$ with $2V_Z = 55°–60°$. The density is 2.16 g/cm³.

Electron microprobe analyses coupled with separate water determinations yield the following composition for the Altoona dachiardite: $Ca_{1.0}Na_{0.6}K_{0.3}[Al_{3.4}Si_{20.6}]O_{48} \cdot 14.2H_2O$. The Cape Lookout is similar but has a Si/Al ratio of 5. Some Altoona clusters are strongly zoned from alkali-rich cores to Ca-rich rims.

Analyses of associated mordenite and dachiardite from both localities indicate that compositional parameters such as the activity of silica, do not play a role in controlling the crystallization of one zeolite in preference to the other. The controlling factor may be one of nucleation.

The Si/Al ratio, probably related to the a_{SiO_2} of the crystallizing solutions, varies from 3.6 to 6.0 in the four described dachiardites. Because clinoptilolite or heulandite and mordenite occur with dachiardite at each locality, the a_{SiO_2} value necessary for dachiardite crystallization must have a specific value at each temperature. Therefore, lower values of Si/Al occur from crystallization at high temperatures.

INTRODUCTION

Dachiardite is a rare high-silica zeolite; its cation content is similar to that of mordenite, a commonly associated mineral. The type dachiardite is from hydrothermally altered zones in a pegmatite near San Piero di Campo, Elba, Italy, where it occurs with tourmaline, pollucite, stilbite, heulandite, and abundant mordenite. It forms small (2 mm) colorless crystals, appearing as eight-sided prisms, which have been shown by Berman (1925) to be twin-groups consisting of eight sectors. The dachiardite composition determined by Gottardi (1960) is:

$$Ca_{1.6}Mg_{0.1}Na_{1.1}K_{0.7}(Al_{5.2}Si_{18.8})O_{48} \cdot 12.7H_2O$$

and the Si/Al ratio is 3.62. On the basis of structural similarities Gottardi (1960) proposed that dachiardite and mordenite were dimorphous, despite minor differences in composition, especially in the Si/Al ratio. The crystal structure was solved by Gottardi and Meier (1963), who demonstrated the structural relationship with mordenite.

Recently dachiardite in clusters of fibers was found at Alpe di Siusi, Bolzano, Italy (Alberti, 1975), where it is associated with abundant mordenite and sparse clinoptilolite. The fibers are orange to colorless and up to 15 mm in length. The dachiardite, which occurs in vesicles of a porphyritic basalt, has the composition

$$Ca_{0.5}Na_{2.6}K_{0.7}(Al_{4.3}Fe_{0.1}Si_{19.6})O_{48} \cdot 13.4H_2O$$

with Si/Al = 4.48

Dachiardite has apparently been found in Japan but is as yet undescribed (Alberti, 1975).

During a detailed study of zeolite assemblages in basalt cavities in Oregon and Washington, dachiardite was found associated with mordenite and other high-silica zeolites at two localities: Altoona, southwestern Washington, and Cape Lookout, Oregon. The tiny crystals have a crystal form and optical properties distinctly different from those of the dachiardite from Elba. This paper will describe the three new occurrences and the crystal form, physical properties, and chemical compositions of the dachiardite.

DESCRIPTION OF THE NEW DACHIARDITE LOCALITIES

Altoona, Washington

Dachiardite occurs in cavities in a basalt breccia exposed along the Columbia River at and east of Altoona in western Wahkiahum County, Washington. The breccia, mapped as the Altoona Basalt by Wolfe and McKee (1968), is intercalated with the Miocene Astoria shale. It is "... chemically similar to the upper unit of Miocene volcanic rocks of the central Oregon Coast Ranges and to the late Yakima type of the Columbia River Group of the Columbia Plateau" (Wolfe and McKee, 1968, p. 4). The zeolite-bearing rock originally come from a small quarry one-half mile east of the settlement of Altoona, but most specimens were collected from blocks mined from the quarry but now used as riprap along the Columbia River in Altoona and in an area about 1 mile west of the town.

Zeolites form linings and radial groups of white or colorless blades in small vesicles and spaces between fragments in the brecciated basalt. The basalt consists of abundant glass with less than 20% plagioclase and clinopyroxene. Some fragments contain vesicles (up to 1 cm) commonly lined with siderite which is covered with greenish-black smectite.

Mordenite was the first zeolite to crystallize in most of the cavities on a 0.5 mm thick lining of exceedingly fine fibers. Clusters of ferrierite blades up to 2 mm long have grown on the mordenite, and clinoptilolite has overgrown both earlier minerals. The genetic relations of dachiardite to the other zeolites are not always clear, but we have found it having overgrown clinoptilolite in several specimens. At the Altoona locality clinoptilolite and mordenite are abundant with ferrierite less common. The dachiardite is relatively scarce.

Cape Lookout, Oregon

Zeolites, including dachiardite, occur in vesicular basalt from Cape Lookout along Netarts Bay north to Cape Meares, Tillamook County, Oregon. The best specimens of the complex assemblages were collected from a small quarry 4 miles north of Cape Lookout State Park. The basalt, mapped as Miocene Columbia River Basalt, is intercalated with Astoria Formation sandstones (Mangum, 1967). It apparently has a tholeiitic composition, similar to the basalt at the Altoona locality. The zeolites form linings or fibrous fillings of vesicular cavities in the dense, glassy basalt. The basalt consists of about 35% plagioclase, 30% clinopyroxene, 5% magnetite, and 30% glass. Zeolitic vesicles range in size up to 2 cm.

All cavities were first lined with an iron-rich smectite. The first zeolite to crystallize was erionite, which formed sprays or masses of fine needles. Phillipsite and clinoptilolite crystals and clusters of mordenite fibers have been found suspended on the erionite needles. Mordenite at first formed porcelaneous cavity linings and small balls impaled on the erionite; at a later stage tiny cottonball aggregates formed. Dachiardite grew on the porcelaneous cavity, lining mordenite. Because of its large size, clinoptilolite may give the false impression of having formed earlier. However, large (several millimeters) clusters of pale-green clinoptilolite were found to include all the other minerals. Filiform and cubic pyrite and calcite were the final products to crystallize in the cavities. At Cape Lookout erionite, mordenite, and clinoptilolite are abundant; phillipsite is scarce; and dachiardite is rare.

Agate Beach, Oregon

A few small clusters of dachiardite have been found in basalt vesicles in a quarry at Yaquina Head near Agate Beach, Lincoln County, Oregon. The basalt, petrographically and chemically similar to that at Altoona and Cape Lookout, is interbedded with yellow sandstones, correlative with the Miocene Astoria Formation.

Zeolitic cavities occur both in vesicles and breccias at the base of the flows. Most cavities are lined with iron-rich smectite. Fibrous erionite and clinoptilolite are fairly abundant (see Wise and Tschernich, 1976b) along with calcite and pyrite. Dachiardite has been found only with clinoptilolite and is very scarce.

CRYSTALLOGRAPHY

The common habits of dachiardite and the associated ferrierite and mordenite at Altoona are illustrated in Fig. 1. Figures 2 through 5 illustrate some of the crystal forms of the dachiardite. The crystal habit of dachiardite at the Altoona and Cape Lookout localities differs significantly from that at the type locality of San Piero

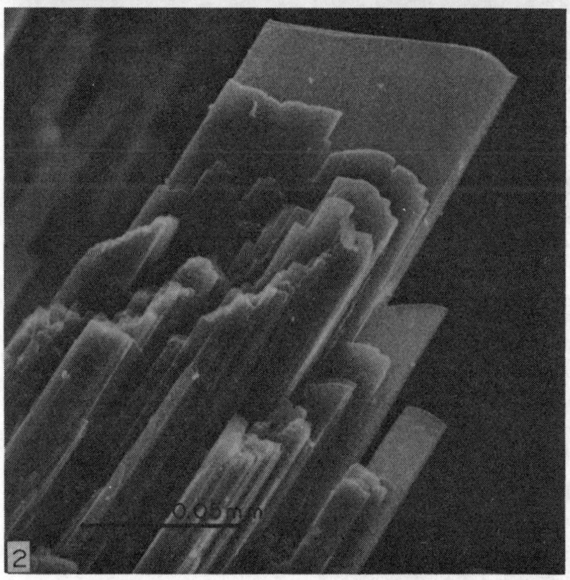

Fig. 1. Scanning electron micrograph of dachiardite (D), ferrierite (F), and mordenite (M) from Altoona, Washington.

Fig. 2. Scanning electron micrograph of dachiardite crystals stacked on the (010) faces. Terminations on the large crystal are ($\bar{4}03$) with a small (001).

TABLE 1. Comparison of the optical properties of dachiardite from Elba and the Pacific Northwest

Elba (Bonatti and Gottardi, 1960)	Altoona and Cape Lookout
$\alpha = 1.494$	$\alpha = 1.480$
$\beta = 1.496$	$\beta = 1.481$ all $\pm.002$
$\gamma = 1.499$	$\gamma \approx 1.485$
$2V_z = 73°$	$2V_z = 55°$ to $60°$
OP // 010	OP ⊥ 010
$X \parallel b$	$Z \parallel b$
$c \wedge Z = +38°$	$c \wedge X = +2°$

di Campo, Elba. The Elba dachiardite forms eight-sided, blocky, cyclic twins (Fig. 6A), while in the Pacific Northwest it forms bush-like groups of thin blades (Fig. 6B).

As illustrated in Figs. 2, 3, and 6C and D, dachiardite from both localities displays dominant (010) forms with smaller (100) faces, both elongated parallel to the *c*-axis, and is terminated by (001), ($\bar{1}01$), and ($\bar{4}03$) faces. A few crystals found at Altoona show a dominant (100) face with vertical striations due to offsets in the crystal faces (Figs. 6F and G). Evidence of other forms is seen by the presence of a tapering face (Fig. 4) and by irregularities in the terminations (Figs. 2 and 3).

No evidence of cyclic twinning has been found on the dachiardite. However, there is some suggestion that polysynthetic twinning may have developed on (100) twin planes. The major evidence is the appearance of numerous re-entrant angles on the terminations of some crystals (Fig. 6E) from repeated ($\bar{4}03$) faces. Because the blades show extinction angles of about 2°, optical evidence of twinning is vague.

The optical orientation of the Pacific Northwest dachiardite is also substantially different from that of the type locality (see Table 1 and Fig. 6H). The density, determined on several 10-mg clusters with a Berman Balance, averaged 2.16 g/cm^3. This value agrees well with those obtained by Berman (1925), 2.165, on the type material, and Alberti (1975), 2.17, on the Alpe di Siusi material, as well as with the calculated density 2.16.

The X-ray powder diffraction patterns of dachiardite from both localities are essentially identical with that given by Alberti (1975). The refined cell parameters are

$$a = 18.62 \pm 0.02 \text{ Å}; b = 7.50 \pm 0.01 \text{ Å};$$
$$c = 10.24 \pm 0.01 \text{ Å}; \beta = 108°16' \pm 5'.$$

CHEMICAL COMPOSITION

All of the dachiardite analysed in this study occur as clusters of small blades. Entire clusters with as many of the associated zeolites as possible were embedded in epoxy and ground until the center of the cluster and crystals were exposed. The mount was then polished and coated with 250 Å of carbon. Analyses were made on an ARL electron microprobe, using an accelerating voltage of 15 kilovolts for all elements with a sample current of 10 nanoamps. Standards used are andesine (Ca and Al), K-feldspar (K and Si), albite (Na), olivine (Mg), hematite (Fe), sanbornite (Ba), and celestite (Sr). Emission data were reduced with a modified version of EMPADR 7 (Rucklidge and Gasparrini, Department of Geology,

University of Toronto, written communication, 1969). Several difficulties arise in the microanalysis of zeolites. The precautions and techniques followed here are described in Wise and Tschernich (1976a, p. 63). Selected microprobe analyses and calculated cell contents of the dachiardite are presented in Table 2, and all analyses are plotted on Fig. 7. Cell contents of the associated minerals are presented in Table 3.

The remainder in the analyses is water; however, because the heat of the electron beam drives some of the water into the probe vacuum, this value is less than the amount in the structure at atmospheric conditions. A

Fig. 3. Scanning electron micrograph of a single dachiardite blade with (001) and ($\bar{1}$01) terminations. Note that the blade is tapered.

Fig. 4. Scanning electron micrograph of dachiardite crystals with some of the terminations shown in Figs. 6 C and D.

Fig. 5. Scanning electron micrograph of dachiardite illustrating the parallel growth of crystals, giving rise to the complex blades with striations parallel to the *c*-axis.

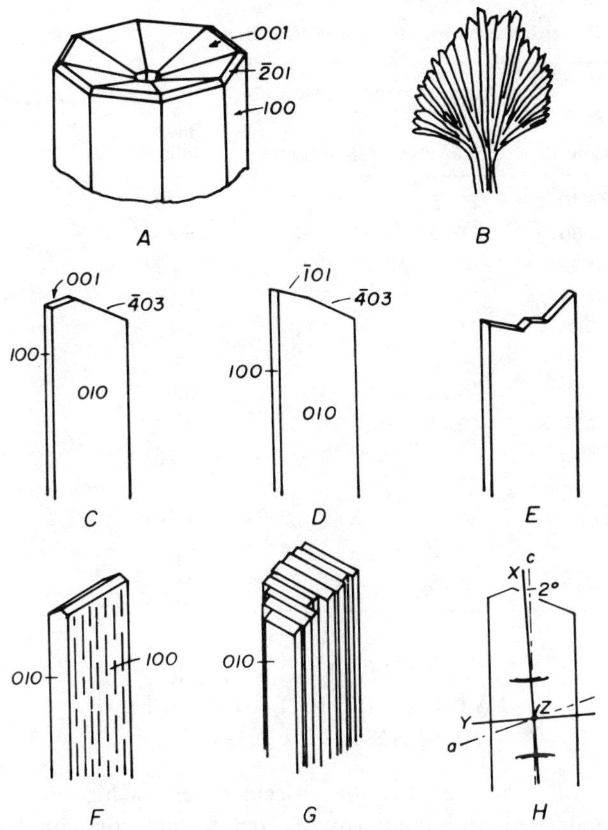

Fig. 6. Crystal forms, habits, and optical orientation of dachiardite. A. Eight sectored, cyclic twins from Elba (Berman, 1925). Height of growth is about 2 mm. B. Bush-like growths at Altoona, Washington, and Cape Lookout, Oregon, localities. Height of growth is about 2 mm. C and D. Dominant forms of dachiardite blades at both Pacific Northwest localities. E. Terminations suggesting polysynthetic twinning on (100). F and G. Crystals with dominant (100) faces with striations. H. Orientation of principal vibration directions in the Pacific Northwest dachiardites.

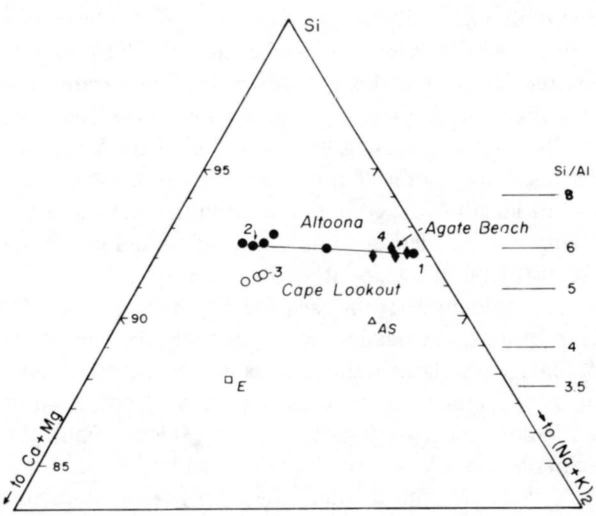

Fig. 7. Si–Ca+Mg–(Na+K)$_2$ plot of the molecular proportion compositions of dachiardite from Altoona (solid circles), Cape Lookout (open circles), and Agate Beach (diamonds). Numbered points refer to analyses in Table 2. Other points are: AS–Alpe di Siusi (Alberti, 1975) and E–Elba (Gottardi, 1960).

TABLE 2. Selected microprobe analyses of dachiardite from the Pacific Northwest. Column (1), base of a crystal cluster and column (2), tip of crystals from Altoona, Washington. Column (3) center of cluster, Cape Lookout, Oregon. Column (4) center of cluster, Agate Beach, Oregon

	(1)	(2)	(3)	(4)
SiO_2	72.57	72.26	71.33	75.09
Al_2O_3	10.12	10.18	12.15	10.98
Fe_2O_3	0.08	0.08	0.02	0.01
CaO	0.29	3.28	3.77	0.70
Na_2O	4.99	1.10	1.06	4.01
K_2O	1.66	0.94	2.55	2.69
Total*	89.71	87.84	90.88	93.48
Cell contents on the basis of 48 oxygens				
Si	20.57	20.67	20.03	20.48
Al	3.38	3.43	4.02	3.53
Fe	0.02	0.02	–	0.01
Ca	0.09	1.01	1.14	0.21
Na	2.74	0.61	0.58	2.12
K	0.60	0.34	0.91	0.93
Si/Al	6.05	5.99	4.98	5.80

*Remainder is H_2O; Mg, Ba, and Sr looked for, but not found.

better measure of the water content was obtained by heating the small samples to 600°C and measuring the weight loss with a microbalance. The value 14.5% H_2O, obtained on Altoona clusters in this manner, is equivalent to 14.2 H_2O molecules in the cell. This amount of water is more than that reported by Gottardi (1960) or Alberti (1975), but must represent the true value for the water content of our samples, because the calculated density is 2.16 g/cm^3 equal to the measured density. The dachiardite, heated to 600°C, retained the framework structure and rehydrated in the atmospheric humidity.

Of the Altoona samples prepared for analysis, one cluster was sectioned such that the compositional changes during growth could be determined. The earliest crystals at the base of the cluster are alkali-rich (col. 1, Table 2), but the composition becomes increasingly Ca-rich toward the ends of the blades (col. 2, Table 2). This sequence, plotted on Fig. 7, is also characterized by a constant Si/Al ratio. Analyses of other cluster, also plotted on Fig. 7, show high Ca-contents. The dachiardite from Cape Lookout shows much less variability. A typical analysis is given in col. 3, Table 2,

TABLE 3. Cell contents of minerals associated with dachiardite from the Pacific Northwest

	Altoona, Washington			Cape Lookout, Oregon			
	Mordenite	Ferrierite	Clinop-tilolite	Erionite	Phillipsite	Mordenite	Clinop-tilolite
Si	20.41	30.62	29.53	27.30	24.25	20.25	29.68
Al	3.65	5.43	6.43	8.66	7.62	3.83	6.86
Fe	0.03	–	–	0.03	–	–	0.02
Mg	0.06	0.96	0.42	0.59	–	0.03	0.04
Ca	0.85	0.44	2.17	2.60	1.11	1.00	2.10
Sr	0.03	–	0.06	–	–	–	–
Ba	0.01	0.06	0.07	–	0.27	–	0.07
Na	1.37	1.26	0.39	0.17	2.13	1.79	0.50
K	0.13	1.05	0.79	2.16	3.24	0.41	1.36
O	48	72	72	72	64	48	72
Si/Al	5.55	5.64	4.59	3.14	3.18	5.29	4.26

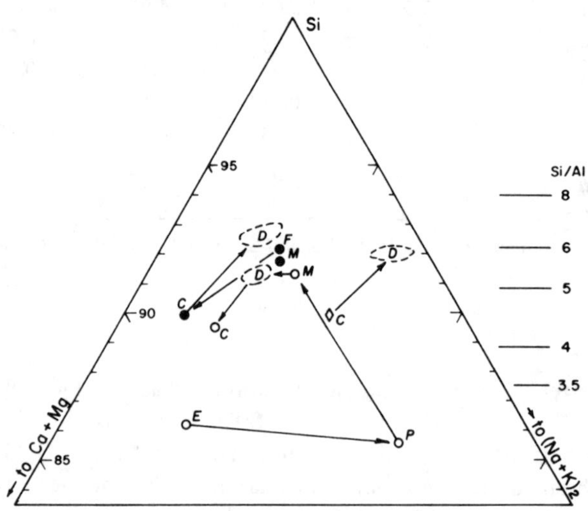

Fig. 8. Si–Ca+Mg–$(Na+K)_2$ plot of the molecular proportion compositions of all associated zeolites at Altoona (solid circles), Cape Lookout (open circles), Agate Beach (diamond). Arrows show the most commonly observed sequence of phase crystallization.

and three analyses are plotted on Fig. 7. These points represent Si/Al ratios that average 5.1, while the Altoona samples average near 6.

The new compositional data, when combined with those from the other two localities on Fig. 7, illustrate that both the Si/Al ratio of dachiardite and the exchangeable cation content can vary within wide limits. The Si content of dachiardite from different localities can vary at least from 18.8 to 20.7 atoms per unit cell. However, the analyses of the Pacific Northwest dachiardites show that the Si/Al is remarkably consistent at any one locality, even if the cation content is variable.

ORIGIN OF THE DACHIARDITE-BEARING ASSEMBLAGES

In the five described occurrences dachiardite is associated with abundant mordenite and common to scarce clinoptilolite or heulandite. The compositions of these and other associated phases at Altoona and Cape Lookout have been plotted on Fig. 8. Comparison of Tables 2 and 3 with Fig. 8 reveals several interesting relationships.

The zeolite phases at Cape Lookout all have lower Si/Al ratios than the corresponding phases at Altoona. Moreover, the fluids in the cavities at Cape Lookout apparently were K-rich, because the first phase was erionite, which Wise and Tschernich (1976b) suggest requires K to crystallize, followed by phillipsite. The crystallization of these two phases may have increased the silica activity causing the formation of the high-silica zeolites. Comparison of the K contents of these zeolites at both localities suggests that even in the late stages the activity of K in the Cape Lookout fluids could have been greater than in those at Altoona.

The mordenites from each locality have very similar compositions, especially with respect to the silica content, even though the associated dachiardites show larger differences. This results in the Cape Lookout mordenite having slightly more silica than the dachiardite, while at Altoona the dachiardite has more silica than the mordenite. Therefore, there does not appear to be any consistent compositional differences between mordenite and dachiardite. It appears doubtful, then, that a compositional parameter, such as the activity of silica, played a dominant role in controlling the crystallization of one structure or the other. Certainly the cation content of the solution appears to have little effect. Judging from the much smaller crystal size of mordenite, it may be that mordenite nucleates more easily in supersaturated fluids, while dachiardite crystallizes more slowly from solutions with less supersaturation.

It is interesting to note the silica contents of the mordenite–clinoptilolite–dachiardite association at Altoona is greater than at Cape Lookout. Clearly the phases at Cape Lookout crystallized in response to lower activities of silica than did those at Altoona. If a specific value of a_{SiO_2} in the fluids is required for each phase, it is clearly not the same at each locality. These values of the activity of silica will decrease with increasing temperature, which may explain the different Si/Al ratios of dachiardite at the five localities. The Altoona dachiardite formed at the lowest temperatures while the Elba material crystallized at the highest.

REFERENCES

Alberti, A. (1975) Sodium-rich dachiardite from Alpe di Siusi, Italy: *Contr. Mineral. Petrol.* **49**, 63-66.

Berman, H. (1925) Notes on dachiardite: *Amer. Mineral.* **10**, 421-428.

Bonatti, S. and Gottardi, G. (1960) Dati ottici e strutturali sulla dachiardite: *Periodico Mineral. (Rome)* **29**, 103-107.

Gottardi (1960) Sul dimorfismo mordenite–dachiardite: *Periodico Mineral. (Rome)* **29**, 183-191.

Gottardi, G. and Meier, W. M. (1963) The crystal structure of dachiardite: *Z. Kristallogr.* **119**, 53-64.

Mangum, D. (1967) Geology of the Cape Lookout State Park, near Tillamook, Oregon: *The Ore Bin* **29**, 85-109.

Rucklidge and Gasparrini (1969) written communication.

Wise, W. S. and Tschernich, R. W. (1976a) Chemical composition of ferrierite: *Amer. Mineral.* **61**, 60-66.

Wise, W. S. and Tschernich, R. W. (1976b) The chemical compositions and origin of zeolites – offretite, erionite, and levyne: *Amer. Mineral.* **61**, 853-863.

Wolfe, E. W. and McKee, E. H. (1968) Geology of the Grays River Quadrangle, Wahkiakum and Pacific Counties, Washington: Wash. Div. Mines and Geology, Geologic Map GM-4.

MORPHOLOGY OF ZEOLITES IN SEDIMENTARY ROCKS BY SCANNING ELECTRON MICROSCOPY

FREDERICK A. MUMPTON

Department of the Earth Sciences, State University College, Brockport, NY 14420, U.S.A.

and

W. CLAYTON ORMSBY

Office of Research, Federal Highway Administration, Washington, DC 20590, U.S.A.

Abstract – In the nearly 2000 occurrences of zeolites in sedimentary rocks of volcanic origin about fifteen zeolite minerals have been identified. The mode of occurrence of six of these, clinoptilolite, erionite, chabazite, phillipsite, analcime, and mordenite, is described, and their morphology is illustrated with scanning electron micrographs.

INTRODUCTION

Although zeolites have been known since 1756 as minor constituents in vugs and cavities of basalts and other mafic igneous rocks, they have just recently emerged from the dusty drawers of mineral collections to be recognized as one of the most abundant and significant silicates in sedimentary rocks. No longer mere mineralogical curiosities, zeolites are now mineral commodities in the true sense of the word. Nearly 2000 occurrences have been reported since the late 1950s in sedimentary rocks of volcanic origin in more than forty countries. Their intriguing adsorption, ion-exchange, catalytic and dehydration properties are the basis for dozens of commercial applications in many areas of industrial and agricultural technology, and more than 300,000 tons of zeolitic tuff are mined each year and used in the paper industry, in pozzolanic cements and concrete, as lightweight aggregate, in fertilizer and soil conditioners, as ion exchangers in pollution-abatement processes, as dietary supplements in animal husbandry, in the separation of oxygen and nitrogen from air and as acid-resistant adsorbents in gas drying and purification (Mumpton, 1975).

Most sedimentary occurrences of zeolite minerals are in rocks of Cenozoic age. They are thought to have formed chiefly by the alteration of volcanic glass in aqueous environments, shortly after deposition. Such deposits are generally flat-lying and range in thickness from less than a centimeter to more than 100 meters. Several distinct types of deposits are recognized including (1) deposits formed from pyroclastic materials in "closed" saline-lake basins, (2) deposits formed in "open", freshwater lake or groundwater systems, (3) deposits formed in true marine environments, (4) deposits formed by low-grade, burial metamorphism, (5) deposits formed by hydrothermal or hot-spring activity, (6) deposits formed from volcanic debris or clay minerals in alkaline soils, and (7) deposits formed without direct evidence of volcanic precursors (Mumpton, 1973a; Sheppard, 1973; Munson and Sheppard, 1974). About fifteen different zeolites have been identified in sedimentary rocks; however, analcime, chabazite, clinoptilolite, erionite, heulandite, laumontite, mordenite, and phillipsite are the most common, with clinoptilolite ranking "number one" in

Editors' note: To satisfy requests for a collection of scanning electron micrographs of typical zeolites as they occur in sedimentary rocks, this article by F. A. Mumpton and W. C. Ormsby published in *Clays and Clay Minerals,* **24**, 1-23 (1976), is being reprinted here.

Fig. 1. Scanning electron micrograph of clinoptilolite from a lacustrine tuff near Castle Creek, Idaho. Note the characteristic monoclinic symmetry of the blades and laths, some of which are similar to the coffin-shape of megascopic heulandite that occurs in vugs in basalts. (Sample 25-5-25.)

Fig. 2. Scanning electron micrograph of clinoptilolite from Castle Creek, Idaho. Thin scales of montmorillonite are also present, although from X-ray diffraction analysis the sample is nearly pure clinoptilolite. (Sample 25-5-25.)

abundance.

Many zeolitic sedimentary rocks are quite pure and beds containing up to 95% of a single species are common. Generally two or more zeolites are present as well-formed, micron-size crystals, along with varying amounts of fresh volcanic ash, cristobalite, montmorillonite, potassium feldspar, and other autigenic minerals. Minor amounts of pyrogenic and detrital quartz, feldspar, mica, etc., are also present but are commonly much coarser grained. The very small particle size of authigenic zeolites makes their examination by ordinary light microscopy extremely difficult. Normal transmission electron microscopy also yields less than satisfactory results, especially if information on the textural relationships of coexisting phases is desired. The ease of sample preparation and the great depth of field obtainable by *scanning electron microscopy* combine to make SEM an ideal technique for the examination of sedimentary zeolite assemblages. The form, habit, size, and spatial relationships of constituent minerals can easily be observed *without* disrupting the aggregate. SEM data are also useful from the point of view of understanding growth mechanisms which may have been responsible for the transformation of volcanic glass to

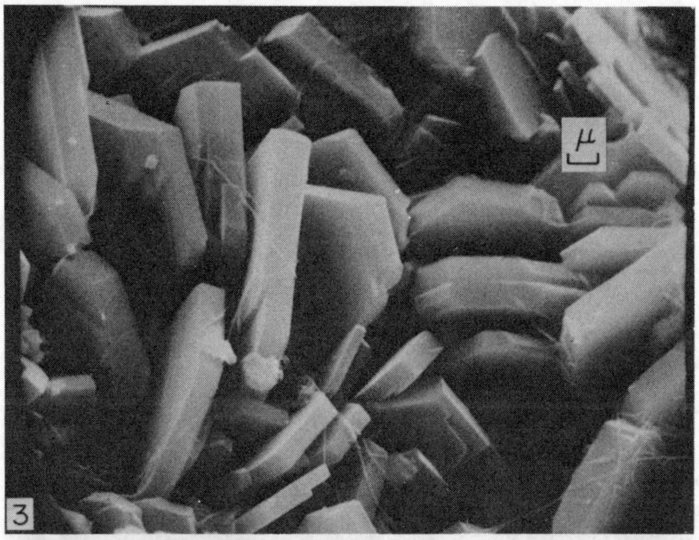

Fig. 3. Scanning electron micrograph of clinoptilolite from a lacustrine tuff near Hector, California, being mined by NL Industries. Filiform mordenite, detectable by X-ray diffraction analysis, crisscross among the clinoptilolite plates. (Sample 25-5-23.)

Fig. 4. Scanning electron micrograph of clinoptilolite from a lacustrine tuff near Toponica, Serbia, Yugoslavia. The crystals resemble laths rather than plates but still display monoclinic symmetry. Mordenite fibers are also present. (Sample 25-13-19.)

euhedral zeolite crystals (Mumpton, 1973b). The textures of such assemblages are also of considerable value, for example in the development of beneficiation processes for commercial deposits.

EXPERIMENTAL

Using a Cambridge Stereoscan model S-4 scanning electron microscope, the morphological characteristics of the most important sedimentary zeolites listed above were determined from as many genetically different types of deposit as possible. More than 100 specimens were examined in this investigation and nearly 500 photographs taken; however, only a few illustrating the common morphologies of these zeolites are shown. Samples were broken to expose fresh surfaces, and chips about 2–10 mm in size were cemented with silver paste to half-inch diameter aluminum discs and dried at about 55°C in a vacuum oven. The samples were then coated with a composite film of carbon (100 Å) and

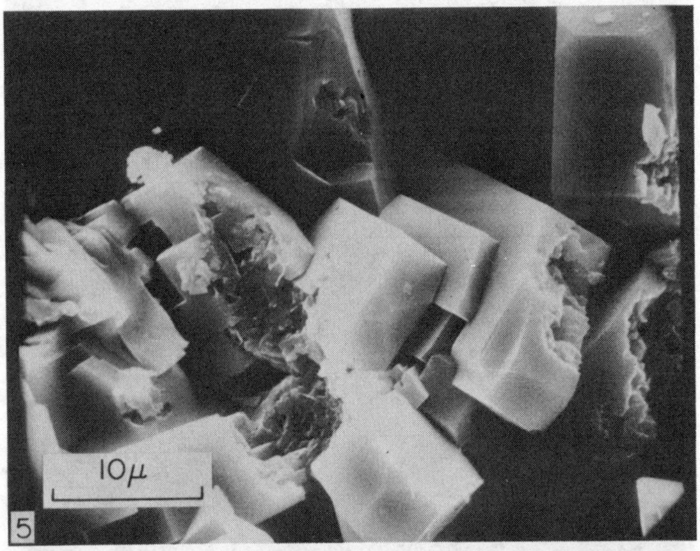

Fig. 5. Scanning electron micrograph of clinoptilolite from Hungry Valley, Nevada. The severely etched crystals are probably due to pore-water dissolution, after formation in a saline-lake environment. (Sample 42-41-20-1.)

Fig. 6. Scanning electron micrograph of a coffin-shaped crystal of clinoptilolite from saline-lake tuff near Durkee, Oregon. The fibers coating the crystal are probably mordenite. (Sample 25-53-69.)

gold–palladium alloy (200 Å) to insure electrical conductivity and thus prevent charging effects in the microscope. Images generated by the emission of secondary electrons were examined on the phosphorescent screen of the instrument and photographed with a polaroid camera. Selected areas also were examined by non-dispersive X-ray spectroscopy using a detector and analyser manufactured by EDAX Instrumental, Inc. to obtain qualitative elemental analyses of individual crystals.

DESCRIPTION OF ZEOLITES

Clinoptilolite

Clinoptilolite in sedimentary rocks generally occurs as euhedral plates and laths, several micron in length and 1–2 microns thick. Most crystals display characteristic monoclinic symmetry, and many are coffin-shaped, mimicking megascopic heulandite that occurs in vugs and cavities of basalt. Typical assemblages of coarse-

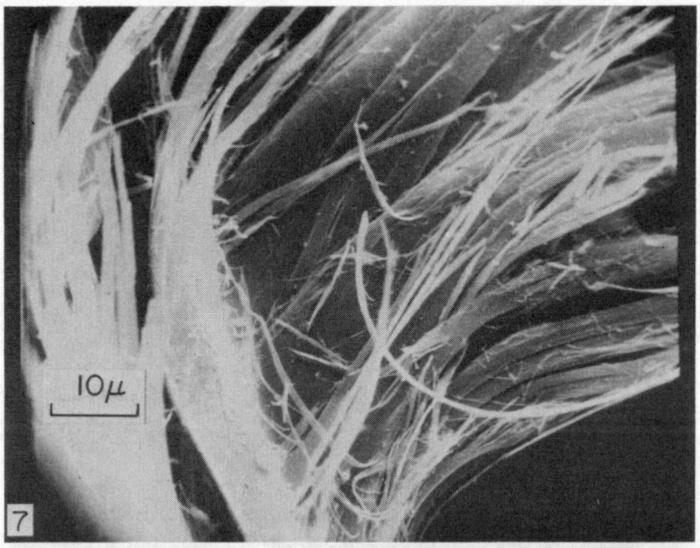

Fig. 7. Scanning electron micrograph of erionite from the type locality near Durkee, Oregon. The woolly fibers are several hundred microns long and split readily into fibrils less than 1μm in diameter. (Sample 25-43-161.)

Fig. 8. Scanning electron micrograph of erionite needles from a saline-lake tuff near Eastgate, Nevada. The needles are 10–20 μm in length and about 1 μm thick. (Sample 42-60-66.)

grained clinoptilolite are shown in Figs. 1 and 2, scanning electron micrographs of a lacustrine tuff from Castle Creek, Idaho, that was probably zeolitized under freshwater conditions. X-ray powder diffraction studies indicate that this sample is almost pure clinoptilolite; however, several thin scales of probable montmorillonite (see Borst and Keller, 1969) are in the fields of view. Similar clinoptilolite plates are shown in Fig. 3, a micrograph of a sample from the well-known locality at Hector, California (Ames, Sand, and Goldich, 1958). Here, thin threads of a second phase lie atop and criss-cross the clinoptilolite plates. As discussed below, this material is mordenite, a zeolite that is easily recognized in X-ray diffraction patterns of Hector ore. Similar fibrous mordenite has been observed in most of the clinoptilolite samples examined in this study.

Clinoptilolite from Taponica, Yugoslavia, is much narrower and resembles laths rather than plates or blades. As seen in Fig. 4, it occurs in clusters of closely spaced, well-oriented crystals. Filiform mordenite is also present. The susceptibility of many sedimentary zeolites to changes in pore-water composition is illustrated by the severely etched clinoptilolite from Hungry Valley, Nevada, shown in Fig. 5. The ragged appearance of the crystals is probably due to partial dissolution after the zeolite formed from parent volcanic glass. The hole or

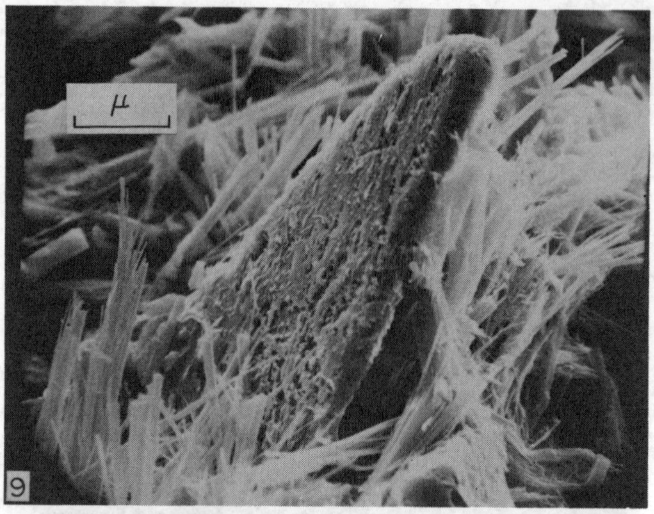

Fig. 9. Scanning electron micrograph of erionite and parent volcanic glass from a saline-lake tuff in Jersey Valley, Nevada, mined by Mobil Petroleum Corporation. The pitted nature of the glass shard is probably the result of arrested pore-water dissolution. (Sample 25-43-127.)

Fig. 10. Scanning electron micrograph of megascopic erionite from a basalt cavity. The hexagonal symmetry is readily apparent. The rods are several hundred μm in length. (Fig. 4, Waltinger and Zirkl, 1974.)

pit in the large, coffin-shaped crystal in Fig. 6 may also be due to a secondary dissolution process.

In general, the monoclinic habit of clinoptilolite is easily recognized in the scanning electron microscope; however, many specimens contain masses of anhedral particles which, from X-ray powder diffraction analyses, must be clinoptilolite; however, these particles cannot be identified on the basis of morphology alone.

Erionite

Until about 1958, erionite was a rare mineral, found as woolly masses in only one locality near Durkee, Oregon. A scanning electron micrograph of the type material (Fig. 7) shows it to consist of closely packed fibers, wherein even the thinnest fiber seems to consist of even thinner members. The many sedimentary erionites examined in this study contain not what could be termed fibers, but rather clusters of acicular needles, as illustrated in Figs. 8 and 9. Most of the needles in Fig. 8, a micrograph of an erionite-rich tuff from a saline-lake deposit near Eastgate, Nevada, are 10–20 μm in length and 1–3 μm thick. The splintery nature of erionite needles is readily seen in Fig. 9, a micrograph of a similar tuff from Jersey Valley, Nevada. Note the "whisk-

Fig. 11. Scanning electron micrograph of sedimentary erionite from a saline-lake tuff near Shoshone, California. Note the hexagonal symmetry of the rods and their similarity with the erionite shown in Fig. 10. (Sample 25-7-5.)

Fig. 12. Scanning electron micrograph of bean-shaped bundles of erionite from a lacustrine tuff near Hector, California. The bundles are about 20 μm long and about 10 μm thick. (Sample 25-43-45.)

broom" manner by which needles split apart at their ends. A large fragment of volcanic glass is also shown in this figure. The pitted nature of the glass shard probably is due to arrested pore-water dissolution, and the absence of crystal growth on the surface of the residual glass fragment suggests a simple dissolution–precipitation mechanism for the formation of the zeolites in this sample (Mumpton, 1973b).

Figure 10 is a reproduction of Fig. 4 in Waltinger and Zirkl (1974) of megascopic erionite from a basalt cavity. The needles are several hundred microns in length and are characterized by hexagonal symmetry. The same hexagonal form is displayed by rods of sedimentary erionite from Shoshone, California, shown in Fig. 11 and seems characteristic of this phase. A somewhat different morphology is shown by erionite from a tuff collected near Hector, California (Figs. 12 and 13). Here bean-shaped bundles of needles are common. Each bundle is about 20 μm in length and about 10 μm thick and consists of hundreds of individual needles, each slightly less than a micron thick. A similar habit for erionite is shown in Fig. 14, a micrograph of an erionite–chabazite assemblage from Bowie, Arizona. Note the sheaf of erionite needles embracing another, both within a small cavity lined with micron-size crystals of chabazite.

Fig. 13. Scanning electron micrograph of erionite from Hector, California, at higher magnification. The bundles consist of hundreds of individual needles, each about a half-micron thick. (Sample 25-43-45.)

Fig. 14. Scanning electron micrograph of erionite bundles from a saline-lake tuff north of Bowie, Arizona. The bundles are surrounded by intergrowths of micron-size chabazite crystals. (Sample 25-43-152-1.)

Chabazite

Sedimentary chabazite generally occurs as cube-like rhombohedra, as illustrated in Figs. 15–19. Intergrown "cubes" or "rhombs", varying in size from less than a micron to several microns, are not uncommon and are shown in Figs. 15 and 16, respectively. Isolated crystals up to 5 or 10 μm on a side also can be found (see Figs. 17 and 18). The cube-like habit is well developed in samples from both saline-lake and marine environments and, as pointed out in an earlier publication (Mumpton, 1973a), is similar to that displayed by chabazite in most of the world's traprock bodies; only the scale is different. Pore-water dissolution is again suggested by the mottled appearance of the chabazite rhombs shown in Fig. 19, a micrograph of a saline-lake tuff from Reese River Valley, Nevada.

Phillipsite

Although phillipsite is a common constituent of many sedimentary zeolite deposits, beds of the pure mineral are rare and have been found in only a few localities of the world. It occurs as stout prisms and stubby laths, 3–30 μm in length and 0.3–3 μm thick, as shown in Figs. 20–24. The pseudo-orthorhombic

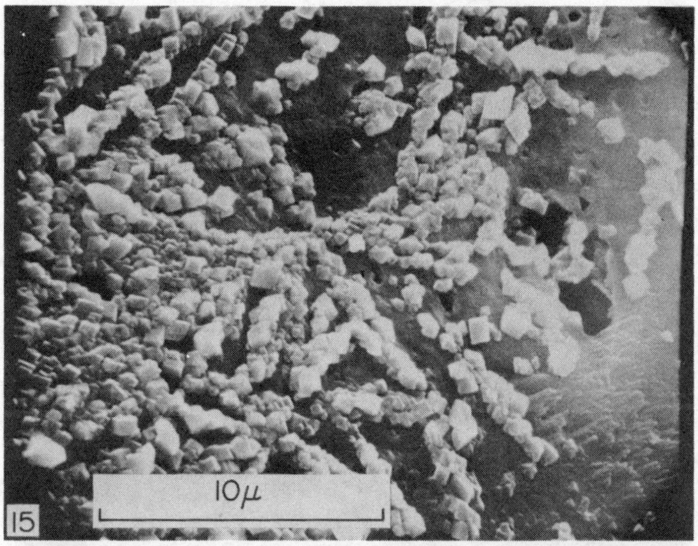

Fig. 15. Scanning electron micrograph of micron-size "cubes" or "rhombs" of chabazite from the Bowie, Arizona, zeolite deposit. The crystals commonly form polycrystalline crusts or strings of crystals as seen in the upper right-hand corner of the figure. (Sample 25-43-152-2.)

Fig. 16. Scanning electron micrograph of chabazite "cubes" or "rhombs" from a lacustrine tuff along Beaver Divide, Wyoming. The crystals are several microns in diameter and are commonly intergrown. (Sample 25-7-4.)

symmetry is generally evident, especially where the crystals are capped by two-sided "domes", as illustrated in Fig. 21. A common characteristic of many phillipsites is a "cracking" along cleavage surfaces parallel to the axis of elongation, breaking the crystal into smaller sections (see Fig. 22). Phillipsite from saline-lake deposits also seems mottled or etched, also probably due to partial dissolution in response to changing pore-water composition. Rosettes or bundles of radiating crystals, as illustrated in Figs. 20 and 24 respectively, are not uncommon habits for this phase.

Analcime

Analcime is easily recognized in the scanning electron microscope by its characteristic cubo-octahedral and trapezohedral habits, as shown in Figs. 25–30. Figures 25 and 26 are micrographs of a marine tuff from Ischia, Italy. Euhedral crystals of 10–30 μm in diameter are common. Crystals of similar size and intergrowths of crystals in a green, saline-lake tuff from Wikieup, Arizona (Sheppard and Gude, 1973) are somewhat less well developed and are coated with submicron-size flakes

Fig. 17. Scanning electron micrograph of well-formed chabazite crystals from a marine tuff near Kruft, Germany, currently mined as a pozzolan. (Sample 42-51-4.)

Fig. 18. Scanning electron micrograph of an isolated chabazite crystal in a saline-lake tuff near Durkee, Oregon. The surrounding matrix is probably montmorillonite. (Sample 25-53-128.)

of a probable ferruginous clay mineral and/or potassium feldspar (see Figs. 27 and 28). The cubic habit of sedimentary analcime is recognizable even in subhedral fragments or in crystals that are partially coated with montmorillonite, as illustrated in Figs. 29 and 30, micrographs of analcime-rich tuffs from lacustrine deposits near Barstow, California, and Tejupan, Mexico, respectively.

Mordenite

As mentioned above, mordenite is a constituent of almost every specimen of clinoptilolite-rich tuff examined in this study. It occurs as thin, curved fibers, a few tenths of a micron in diameter, coating grains and bridging gaps between larger grains of the platy zeolite. (See Figs. 3, 4, 5, 31, and 32.) The fibers are extremely delicate and length:width ratios of 100 or more are common. The filaments commonly intertwine into closely packed mesh structures (Fig. 33) and commonly form "rats' nests" of the type shown in Fig. 34, an electron micrograph of monomineralic mordenite from an ash-flow tuff near Beatty, Nevada. Coarser-grained mordenite is the major constituent of a marine tuff from Pismo Beach, California, and is shown in Fig. 35. The mordenite is needle-like and about 1 μm in diameter. A

Fig. 19. Scanning electron micrograph of chabazite crystals, somewhat mottled in appearance, from a saline-lake tuff in the Reese River Valley, Nevada.

Fig. 20. Scanning electron micrograph of rosettes of phillipsite from saline-lake tuff in Pine Valley, Nevada. The prisms and laths are about 3–30 μm long and 0.3 to 3 μm thick. (Sample 25-7-6.)

monomineralic tuff from a saline-lake deposit near Lovelock, Nevada, also contains relatively thick mordenite needles, arranged in radiate clusters (see Fig. 36).

A major problem plaguing the recognition of mordenite and erionite with the scanning electron microscope is the occurrence of both in the form of thin needles. If there is enough of either zeolite to be detected by X-ray powder diffraction, one can be reasonably confident of identification inasmuch as the two zeolites rarely occur together. If only trace amounts are present, the problem becomes difficult, and it may not be possible to distinguish these two zeolites, especially where the fibers are in the 1-μm-thick range. Thinner fibers have been observed only in mordenite samples, but thicker fibers are invariably erionite and may display characteristic hexagonal symmetry. By way of illustration, Fig. 37 is a scanning electron micrograph of a clinoptilolite-rich tuff from a large deposit in Dzegvi, Georgian S.S.R., described by Gvakhariya and others (1972). Neither erionite nor mordenite was reported in this deposit, nor can either zeolite be detected in X-ray diffraction patterns of the sample. Fibers from 0.5 to 0.8 μm in diameter, however, are plainly visible in the sample but cannot be identified unequivocally from the data available. If the fibers are

Fig. 21. Scanning electron micrograph of phillipsite prisms from Pine Valley, Nevada, displaying characteristic pseudo-orthorhombic symmetry. (Sample 25-7-6.)

Fig. 22. Scanning electron micrograph of phillipsite prisms from a marine tuff near Naples, Italy (*Tufo giallo napolitano*). Note the "cracking" of many crystals along cleavage surfaces parallel to the axis of elongation. Montmorillonite is also present in this sample. (Sample 25-37-106.)

erionite, this is the first occurrence of erionite to be recognized in sedimentary zeolite deposits of the U.S.S.R. Hopefully, further examination of zeolite samples from this deposit will resolve this question.

ACKNOWLEDGMENTS

The authors are grateful to the Research Foundation of the State University of New York for early support of this project (Grant No. 228600A-6) and to the Office of Research, Federal Highway Administration, for making its facilities available for this study. Thanks are also due to R. A. Sheppard, G. V. Gvakhariya, and E. Passaglia for providing samples, and to R. A. Sheppard for having reviewed the manuscript.

REFERENCES

Ames, L. L., Jr., Sand, L. B. and Goldich, S. S. (1958) A contribution to the Hector, California bentonite deposit: *Econ. Geol.* **53**, 22-37.

Borst, R. L. and Keller, W. D. (1969) Scanning electron micrographs of API reference clay minerals and other selected samples: *Proc. Int. Clay Conf., Tokyo*, 1969 Vol. I.

Gvakhariya, G. V., Skhirtladze, N. I. and Batiashvili, T. V. (1972) Clinoptilolite-bearing volcanic tuff of the north slope

Fig. 23. Scanning electron micrograph of phillipsite laths and prisms from a saline-lake tuff near Kirkland, Arizona. Montmorillonite is also abundant in this sample. (Sample 25-7-7.)

Fig. 24. Scanning electron micrograph of radiating rods of phillipsite in a marine tuff near Marino, Rome, Italy. (Sample 25-37-107.)

of the Trialet Range: *Doklady Akad. Nauk, SSSR* **205**, 179-180.

Mumpton, F. A. (1973a) Worldwide deposits and utilization of sedimentary zeolites: *Industrial Minerals* (London) No. 73, 30-45.

Mumpton, F. A. (1973b) Scanning electron microscopy and the origin of sedimentary zeolites: Molecular Sieves: *Proc. 3rd Int. Molecular Sieve Conf.*, pp. 159-161.

Mumpton, F. A. (1975) Commercial uses of natural zeolites: Section of 4th edition of *Industrial Minerals and Rocks* (edited by S. J. Lefond), pp. 1262-1274. Am. Inst. Min. Metall. Petrol. Engrs.

Munson, R. A. and Sheppard, R. A. (1974) Natural zeolites: their properties, occurrences, and uses: *Min. Sci. & Engng*, **6**, 19-34.

Sheppard, R. A. (1973) Zeolites in sedimentary rocks: *U.S. Geol. Surv. Prof. Pap.* **820**, 689-695.

Sheppard, R. A. and Gude, A. J., I II (1973) Zeolites and associated authigenic silicate minerals in tuffaceous rocks of the Big Sandy Formation, Mohave County, Arizona: *U.S. Geol. Surv. Prof. Pap.* **830**, 36 pp.

Waltinger, H. and Zirkl, E. J. (1974) Rasterelektronenmikroskopische Aufnahmen von Erionit aus Kollnitz, Lavanttal, Kärnten: *Carinthia II* **164/84,** 125-135.

Fig. 25. Scanning electron micrograph of euhedral analcime from a marine tuff from Ischia, Italy. Note the characteristic cubo-octahedral and trapezohedral symmetry. (Sample 42-51-53.)

Fig. 26. Scanning electron micrograph of an assemblage of analcime crystals from Ischia, Italy. Crystals range from 5 to 25 μm in size. (Sample 42-51-53.)

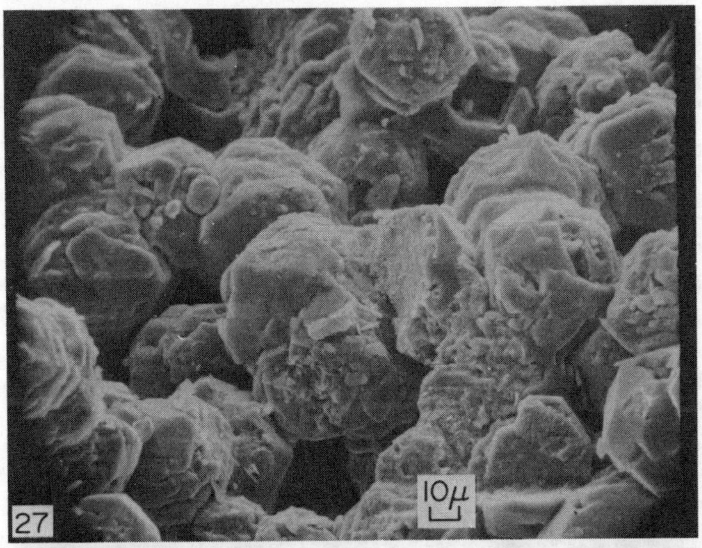

Fig. 27. Scanning electron micrograph of analcime crystals from a saline-lake tuff near Wikieup, Arizona. Subhedral crystals are coated with a probable ferruginous clay mineral. (Sample 25-53-6.)

Fig. 28. Scanning electron micrograph of a well-formed analcime crystal from Wikieup, Arizona, about 30 μm in diameter. (Sample 25-53-6.)

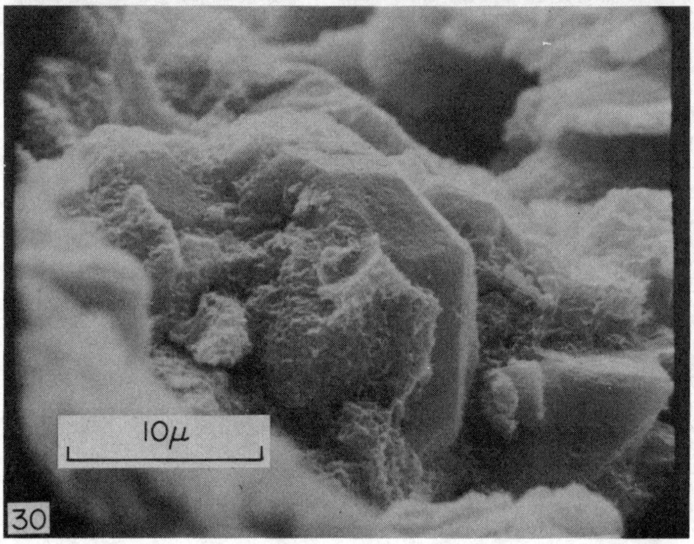

Fig. 29. Scanning electron micrograph of large analcime crystals in a matrix of micron-size crystals of potassium feldspar from a saline-lake tuff near Barstow, California. (Sample 25-43-136.)

Fig. 30. Scanning electron micrograph of a large analcime crystal coated with montmorillonite from a lacustrine tuff near Tejupan, Oaxaca, Mexico. (Sample 42-47-18.)

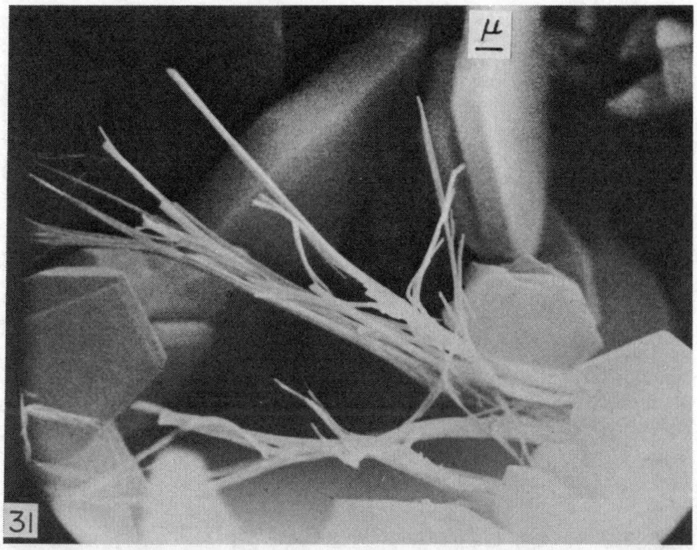

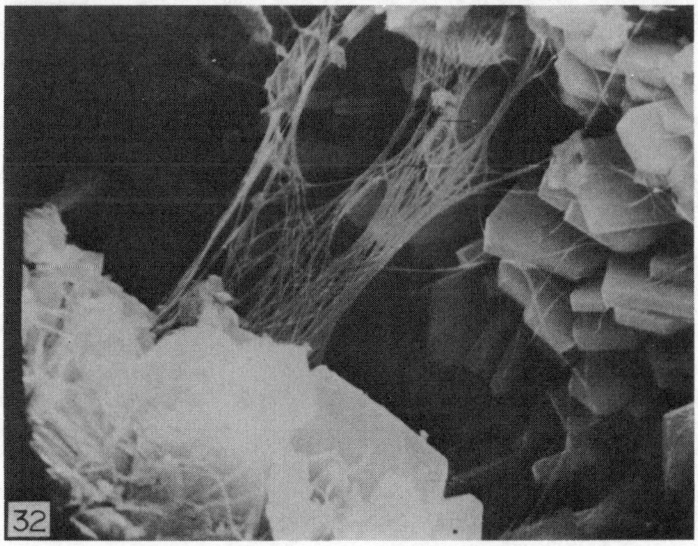

Fig. 31. Scanning electron micrograph of mordenite fibers and platy clinoptilolite from a lacustrine tuff near Castle Creek, Idaho. The fibers are from 0.1 to 0.2 μm in diameter and 10 to 20 μm in length. (Sample 25-5-25.)

Fig. 32. Scanning electron micrograph of filiform mordenite bridging the gaps between clinoptilolite crystals in spider-web fashion in a lacustrine tuff near Hector, California. Individual fibrils are a few tenths of a micron in diameter. (Sample 25-5-23.)

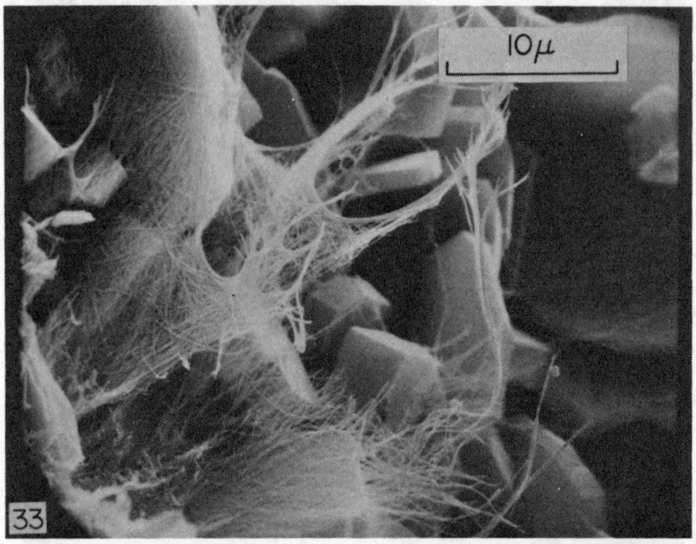

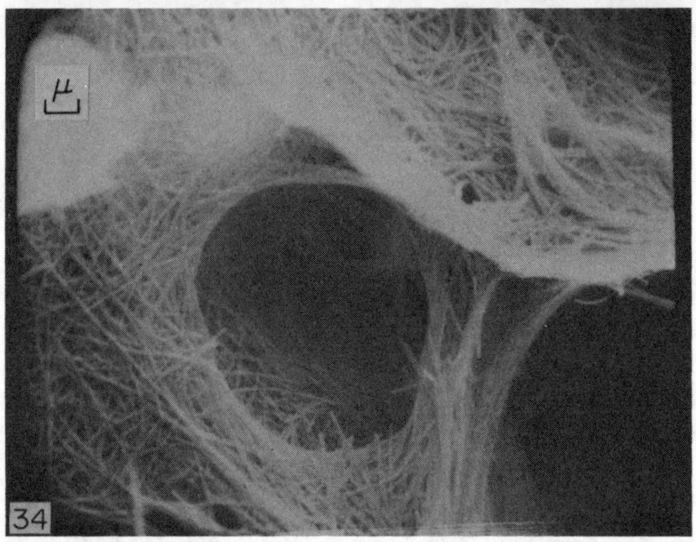

Fig. 33. Scanning electron micrograph of intertwined mordenite fibers with clinoptilolite from an ash-flow tuff near Monolith, California. This tuff has been mined by Monolith Portland Cement Company as a pozzolan. (Sample 25-5-5.)

Fig. 34. Scanning electron micrograph of interlaced mordenite fibers in a monomineralic ash-flow tuff near Beatty, Nevada. Such "rats' nests" of fibers are not uncommon in many mordenite-rich tuffs. (Sample 42-41-11.)

Fig. 35. Scanning electron micrograph of needle-shaped mordenite in a marine tuff near Pismo Beach, California. Individual needles are slightly less than one micron in diameter. (Sample 42-41-7.)

Fig. 36. Scanning electron micrograph of bundles of radiating mordenite needles from a lacustrine tuff near Lovelock, Nevada. (Sample 42-60-69.)

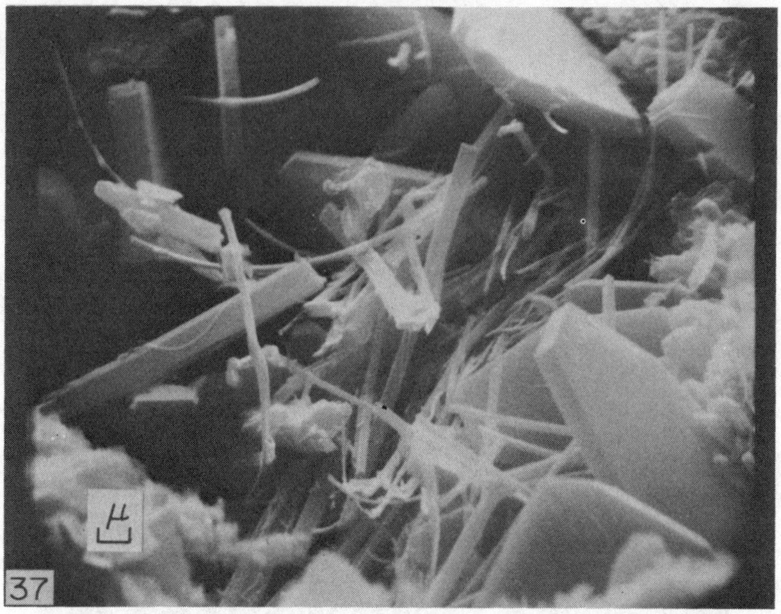

Fig. 37. Scanning electron micrograph of needles and fibers associated with clinoptilolite from a marine tuff near Dzegvi, Georgia, U.S.S.R. Although erionite or mordenite is suspected, neither can be identified from morphology alone, nor has either been detected in the sample by X-ray diffraction examination. (Sample 25-23-6.)

GEOLOGY AND OCCURRENCES

GEOLOGIC OCCURRENCE OF ZEOLITES

RICHARD L. HAY

Department of Geology and Geophysics, University of California
Berkeley, California 94720, U.S.A.

Abstract – Zeolites occur principally in unmetamorphosed sedimentary rocks and are particularly widespread in volcaniclastic strata. The most common naturally occurring zeolites are analcime, clinoptilolite, heulandite, laumontite, and phillipsite. The species and amount of zeolite vary primarily as a function of texture and composition of the host rock, composition of interstitial water, age, and temperature. Zeolites form during and after burial, generally by reaction of pore waters with solid aluminosilicate materials (e.g. volcanic glass, feldspar, biogenic silica, and clay minerals). Early formed zeolites (e.g. clinoptilolite and phillipsite) are commonly precursors to other zeolites (e.g. analcime, heulandite, laumontite). The pH is the most important chemical parameter in determining the rate of zeolitic reactions, and at a pH of 9.5, as in many desert lakes, substantial amounts of zeolites can form in a few thousand years.

Most zeolitic occurrences can be assigned to one of several types of geologic environments or hydrologic systems. These are (1) saline, alkaline lakes; (2) saline, alkaline soils and land surfaces; (3) sea-floor sediments; (4) percolating water in an open hydrologic system; (5) hydrothermal alteration; and (6) burial diagenesis. These types of occurrence result in characteristic types of zeolite zoning. Large, nearly monomineralic deposits of zeolites are found as altered tuffs of saline, alkaline lakes and of open hydrologic systems.

INTRODUCTION

Zeolites occur in rocks of many types, ages, and geologic settings. Most early work was on zeolites in igneous rocks, particularly the well-formed crystals in cavities of basalt flows. Later work, using powder X-ray diffractometry, showed that the greater abundance of zeolites is in sedimentary rocks, and many strata consist chiefly of zeolites. The history of investigation and discovery of zeolites and their origin have been reviewed elsewhere (Hay, 1966) and will be only briefly mentioned here. A major early discovery was of vast quantities of phillipsite in deep-sea sediments during explorations by the *H.M.S. Challenger* (Murray and Renard, 1891). Among the studies of rocks on land, the work of Coombs (1954) was notable in recognizing a vertical zoning of zeolites and associated minerals in a 10,000-m column of rocks. This led to greater emphasis on stratigraphic relations of zeolites and their interpretation in terms of burial depth, pressure, and temperature. A zeolite metamorphic facies, of lower grade than the greenschist facies, was established by some petrologists on the basis of these zeolite zones.

Studies of saline-lake deposits showed that variations in pH, salinity, and cation content in lake water can produce mineralogic effects comparable to major changes in pressure and temperature (Hay, 1966). Moreover, time proved to be an additional variable in determining the zeolite species and associated minerals in saline-lake deposits. For example, phillipsite and clinoptilolite are generally early zeolites to form from glass, and they may be replaced successively by analcime and K-feldspar (Hay, 1966; Sheppard and Gude, 1968, 1969a). Zeolite mineral assemblages are determined by several variables: composition of the host rock, composition of the pore water, pressure and temperature, and time (Hay, 1966).

This review paper describes and illustrates the main types of zeolite occurrences in sedimentary rocks. More comprehensive listings of occurrences are given in other review papers (Coombs *et al.*, 1959; Deffeyes, 1959; Hay, 1966; Iijima and Utada, 1966, 1972; Sheppard, 1971).

ZEOLITE MINERALOGY AND ORIGIN

Five zeolites occur most commonly in sedimentary rocks: analcime, clinoptilolite, heulandite, laumontite, and phillipsite. An additional group of zeolites ranks next in abundance: chabazite, erionite, mordenite, natrolite (with gonnardite), and wairakite. Salient compositional aspects of these zeolites are listed in Table 1, where the zeolites are listed in order of decreasing $Si/Al + Fe^{3+}$ ratio. It can be seen that the more siliceous zeolites are also the more hydrated ones. The

TABLE 1. Compositional features of common zeolites*

Name	Si/Al + Fe^{3+}	Dominant cations	H_2O molecules per Al atom
Clinoptilolite	4.0–5.1	K > Na	3.0–3.5
Mordenite	4.3–5.3	Na > K	3.2–3.5
Heulandite	2.9–4.0	Ca, Na	2.5–3.0
Erionite	2.9–3.7	Na, K	3.0–3.4
Chabazite	1.7–3.8	Ca, Na	2.7–4.1
Phillipsite	1.3–3.4	K, Na, Ca	1.7–3.3
Analcime	1.7–2.9	Na	1.0–1.3
Laumontite	2.0	Ca	2.0
Wairakite	2.0	Ca	1.0
Natrolite	1.5	Na	1.0

*Data from Hay (1966) modified by data from the following sources: clinoptilolite and heulandite (Boles, 1972), mordenite (Sheppard and Gude, 1969a), erionite (Sheppard and Gude, 1969b), chabazite (Gude and Sheppard, 1966), phillipsite (Sheppard and Gude, 1970), and analcime (Iijima and Harada, 1968).

$Si/Al+Fe^{3+}$ ratio varies widely for several of the alkali-rich zeolites such as phillipsite and chabazite. The more siliceous varieties of these zeolites contain chiefly Na^+ and K^+, whereas the least siliceous ones generally contain chiefly Ca^{2+}.

Zeolites are formed by reaction of pore water with solid materials. A common reactant is volcanic glass but other reactants are X-ray-amorphous and poorly crystalline clay, montmorillonite, plagioclase, nepheline, biogenic silica, and quartz. Both clay minerals and zeolites can form from the same materials, but whether a clay mineral or zeolite is formed depends on the physical environment and on the activities of dissolved species such as H^+, alkali- and alkaline-earth ions, H_4SiO_4, and $Al(OH)_4^-$. Phyllosilicates are generally favored by high ratios of H^+ to Na^+, K^+, and Ca^{2+} and by high activities of Mg^{2+}.

The species of zeolite will depend on the temperature, pressure, and various chemical parameters: activity ratio of Si to Al, activities of Ca^{2+}, Na^+, and K^+, and the activity (or partial pressure) of $H_2O(P_{H_2O})$. It should be noted that the chemical activity of water can be lowered and the less hydrous zeolites favored by increasing temperature, ionic strength, or partial pressure of carbon dioxide (P_{CO_2}). These matters are discussed more fully elsewhere (Hay, 1966; Miyashiro and Shido, 1970).

Early formed zeolites commonly react with pore fluid to form other zeolites in sedimentary rocks. This can result from a change in the physical or chemical environment, or simply from the availability of sufficient time for a less-stable zeolite to transform to a more stable one (Hay, 1966; Sheppard and Gude, 1968, 1969a). For example, phillipsite and clinoptilolite are commonly replaced by analcime in saline-lake deposits, as they are in low-temperature hydrothermal experiments (Boles, 1971). The clinoptilolite-mordenite assemblage is replaced by a heulandite-analcime assemblage at depth in thick sequences of marine or fresh-water deposits. At high salinities or temperatures analcimic assemblages are replaced by alkali feldspars and associated minerals.

TYPES OF ZEOLITE OCCURRENCE

Most zeolite occurrences can be assigned one of several types of geologic environments or hydrologic systems. They are (1) saline, alkaline lakes; (2) saline, alkaline soils and land surfaces; (3) marine sediments; (4) percolating water in an open hydrologic system; (5) hydrothermal alteration; and (6) burial diagenesis or metamorphism (see Hay, 1966; Iijima and Utada, 1972; Sheppard, 1973; Munson and Sheppard, 1974; Kossovskaya, 1975). These types of occurrences result in characteristic types of zeolite zoning, which are diagrammatically illustrated in Figs. 1 and 2. Zoning in saline-lake deposits is illustrated best by mineral

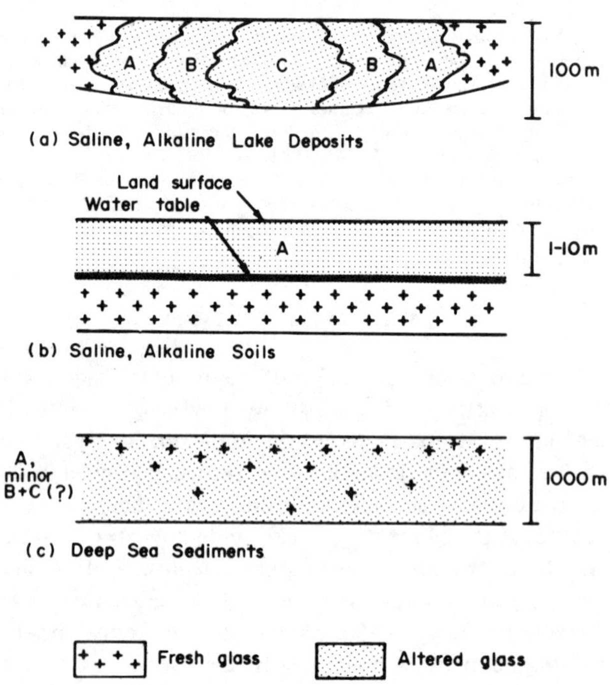

Fig. 1. Diagrams showing patterns of authigenic zeolites and feldspars in tuffs of saline, alkaline lakes; saline, alkaline soils, and deep-sea sediments. Zone A is characterized by non-analcimic, alkali-rich zeolites, zone B by analcime, and zone C by feldspars.

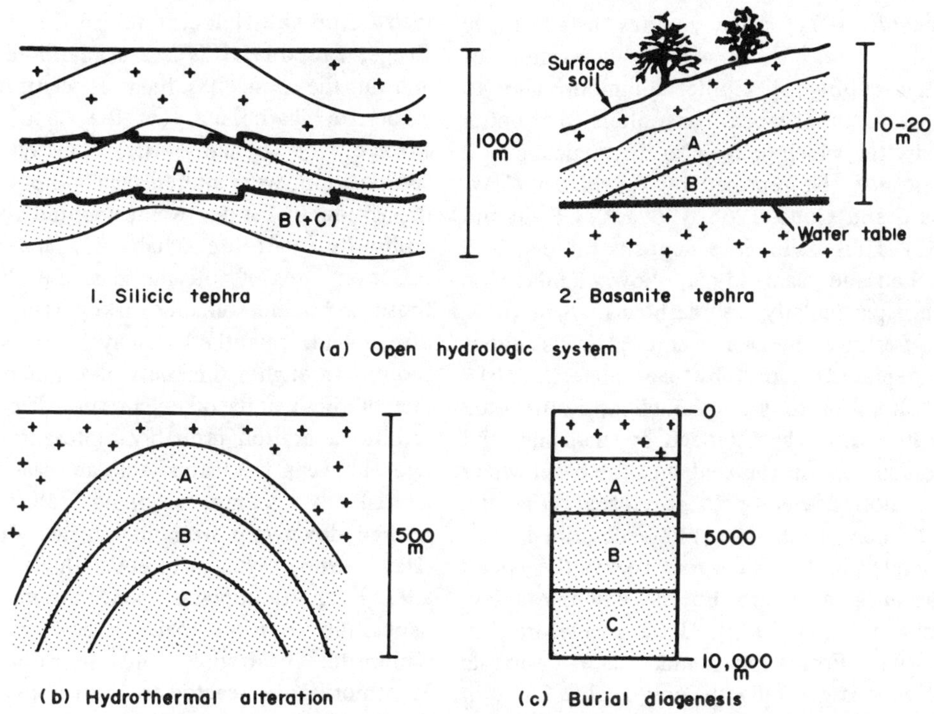

Fig. 2. Diagrams showing patterns of authigenic zeolites and feldspars in tuffs where the zonation is of (a) open-system type, (b) hydrothermal, and (c) a result of burial diagenesis. Zone A is characterized by non-analcimic alkali-rich zeolites, zone B by analcime or heulandite, and zone C by K-feldspar in (a_1) and by albite with or without laumontite in (b) and (c). Symbols are the same as in Fig. 1.

assemblages in altered vitric tuffs (see especially Sheppard and Gude, 1968, 1973). The saline, alkaline surface type of occurrence may involve only the topmost 30 cm of sediment, but it can extend downward to the water table at depths of 10 to 15 m or more. The marine-sediment type is based on those zeolites in deep-sea sediments, which commonly increase in abundance as a function of age and depth in tuffaceous deposits. The open-system type concerns accumulations of reactive material, generally vitric tephra, which show a vertical zonation of authigenic silicate minerals that is mainly attributable to the progressive change in meteoric water by reaction with solid materials along its flow path (Hay, 1963a, 1966; Walton, 1975). The hydrothermal alteration type refers to that resulting from localized, abnormally high geothermal gradients, as in areas of volcanic activity, and is commonly associated with hot springs. Burial diagenesis refers to the vertical zonation of authigenic minerals in thick sedimentary accumulations where the increase of temperature with depth seems the probable cause of the zoning (Coombs, 1954; Coombs *et al.*, 1959; Iijima and Utada, 1972). Obviously, these categories are not mutually exclusive, and burial diagenesis can modify the assemblages produced by other mechanisms.

Zeolites of saline, alkaline lakes

Zeolites are both common and widespread in deposits of saline, alkaline lakes, and vitric tuffs may alter to yield nearly pure deposits of zeolites and K-feldspar. Lakes of this type are in closed basins of arid and semiarid regions. Lake water is rich in dissolved sodium carbonate-bicarbonate and has a pH of about 9.5. Trona ($Na_2CO_3 \cdot NaHCO_3 \cdot 2H_2O$) is among the saline minerals that crystallize at maximum salinities. A wide variety of materials react to form zeolites in this alkaline environment, reflecting the solubility of Al and Si at high pH. Some common reactants are volcanic glass, biogenic silica, poorly crystalline clay, montmorillonite, plagioclase, nepheline, and quartz. Zeolites are among the most common authigenic silicates formed in this environment, and other authigenic minerals include sodium silicates, K-feldspar, and borosilicates. Reactions are relatively rapid, and vitric tuff can alter to zeolites in 1000 years (Hay, 1964, 1966).

The most common zeolites formed from silicic glass in this environment are phillipsite, clinoptilolite, and erionite; less common ones are chabazite and mordenite. Monomineralic beds of phillipsite, clinoptilolite, and erionite that formed from rhyolitic tuffs are widespread in Plio-Pleistocene lake deposits of the western United

States (Sheppard, 1971). The zeolites listed above characteristically alter to analcime in water of moderately high salinity; all zeolites commonly alter to form K-feldspar. In other words, glass alters to a zeolite such as phillipsite, which may alter to analcime and which in turn may be replaced by K-feldspar (Hay, 1966; Sheppard and Gude, 1968, 1969a; Goodwin and Surdam, 1967). Thus, saline-lake deposits more than a few hundred thousand years old may show a mineralogic zoning, which is particularly well defined in vitric tuffs; (a) an outer fresh-water zone where glass is either unaltered or replaced in part by clay minerals; (b) a saline zone of alteration to phillipsite, clinoptilolite, etc.; (c) a more saline zone characterized by analcime; and (d) a highly saline zone in the center of the basin where tuffs may be altered chiefly to K-feldspar (Fig. 1). Sheppard and Gude (1968, 1973) have described some excellent examples in late Cenozoic rocks; the Green River Formation is a superb Eocene example (Hay, 1966; Surdam and Parker, 1972). Albite is widespread in the Green River Formation, unlike late Cenozoic deposits. Analcime is generally the only zeolite found in pre-Eocene saline-lake deposits, and albite becomes increasingly abundant as a function of age, partly at the expense of analcime.

Claystones are mineralogically zoned in the same general pattern as associated tuffs (e.g. Hay, 1966, p. 49). However, analcime is a common zeolite, or indeed the only zeolite, in non tuffaceous claystones interbedded with zeolitic tuffs containing phillipsite or clinoptilolite. This reflects the fact that clay minerals commonly react to form analcime. The analcime in claystones readily reacts to form K-feldspar in the highly saline zone where the zeolites in tuffs are altered to K-feldspar (Hay, 1970, p. 242).

Zeolites of soils and land surfaces

Zeolites are formed readily from suitable materials at the land surface where the pH is high as a result of the concentration of sodium carbonate-bicarbonate by evapotranspiration. Saline, alkaline soils are found in semiarid regions. Analcime has formed in alkaline soil of the San Joaquin Valley of California (Baldar and Whittig, 1968) and of Ruzizi, Burundi (Frankart and Herbillon, 1970). Montmorillonite is the principal reactant in both these examples, and in the San Joaquin Valley illite and analcime seem to be reaction products. The content of analcime in the latter area is highest at the surface and decreases to its lower limit at a depth of 1.2 m.

A wide variety of zeolites has formed at the land surface in tuffaceous and non-tuffaceous sediments of Pleistocene and Holocene age in the vicinity of Olduvai Gorge, Tanzania. This is a semiarid region which now, and in the past, has high concentrations of sodium carbonate-bicarbonate at the land surface. Wind-worked, or eolian tuffs, mostly of nephelinite composition, were altered penecontemporaneously with deposition to form a well-cemented rock consisting of phillipsite, natrolite, chabazite, analcime, and illite. Reaction rates of volcanic glass are about the same as those in saline, alkaline lakes (Hay, 1963b, 1976). Stream-laid, reworked trachytic tuffs are altered to zeolites through a thickness of as much as 18 m beneath one of these units of eolian tuff. The lower surface of zeolitic alteration probably represents the water table. Beds between the water table and the eolian tuffs were altered by sodium-carbonate solutions periodically flushed downward from the surface during heavy rains (Hay, 1966). Non-tuffaceous fluvial claystones were altered on floodplains penecontemporaneously with deposition to form analcime and smaller amounts of phillipsite, chabazite, and natrolite (Hay 1970). Montmorillonite reacted to form illite and zeolites, as in the California example. The claystones richest in zeolites (15–40%) are bright reddish-brown and may represent a relatively modern, small-scale example of the late Mesozoic analcimolites. These terrestrial deposits are rich in analcime and are dominantly reddish-brown (e.g., High and Picard, 1965; Vanderstappen and Verbeek, 1964; Joulia *et al.*, 1959). The analcimolites are the world's thickest and most widespread non-marine sedimentary zeolite deposits.

Zeolites in marine sediments

Zeolites occur in a wide variety of marine sediments and sedimentary rocks and are relatively common in many strata. Some zeolites were formed at shallow depths and low temperatures, but others were formed in response to substantial increases in temperature. Here only the zeolites formed at low temperature will be considered, the higher-temperature zeolites are discussed elsewhere under burial diagenesis. Many noteworthy contributions were made subsequent to the report of Murray and Renard (1891) on the *Challenger* expedition (e.g., Norin, 1958; Bonatti, 1963; Biscaye, 1964; Hathaway and Sacks, 1965; Heath, 1969). However, most of our present information is based on studies of cores obtained by the *Glomar Challenger* in the Deep-Sea Drilling Program (DSDP), which lasted from 1968 to 1975. The drill holes number in the hundreds, and most penetrated to depths of 400 to 700 m. Upper Cretaceous and Cenozoic sediments are widely represented in the cores, which were studied mineralogically on shipboard and on shore. Pore-water samples squeezed from the cores have been chemically analysed, so that information is now available about the water involved in forming the zeolites in marine sediments. Twenty-seven volumes of results, supplemented by many papers, have been published on the results of the DSDP so far. These data provide our best evidence about the origin of zeolites in the marine sediments exposed on land, and the following discussion is confined to the zeolites in deep-sea sediments.

Phillipsite and clinoptilolite are the dominant zeolites in deep-sea sediments. Analcime is next in abundance, and erionite and mordenite occur rarely. Authigenic minerals commonly associated with zeolites are palygorskite, smectites, sepiolite, cristobalite, and quartz. Zeolites are particularly abundant in volcaniclastic sediments, especially vitric ash, and they may form as much as 80% of altered tephra deposits. As might be expected, phillipsite is commonly associated with low-silica, generally basaltic, tephra, and clinoptilolite with siliceous tephra. However, phillipsite can be the principal or the only zeolite in certain silicic tuffs (e.g., Leg 21, Hole 205; Burns *et al.*, 1973) and clinoptilolite may be the only zeolite in mafic ash layers (e.g., Gieskes and Kastner, 1975). Analcime is common only in sediments containing mafic tephra. Clinoptilolite is widespread and can be a major component of pelagic detrital clays with biogenic silica (e.g., von Rad and Rösch, 1972). Small amounts of clinoptilolite are in most other types of fine-grained marine sediments, for example, siliceous ooze, carbonate ooze, and green clay (e.g., Biscaye, 1964; Heath, 1969). No systematic zoning of zeolite species with depth or age is apparent in Cenozoic rocks, but clinoptilolite seems to be the dominant zeolite in Cretaceous rocks. Analcime occurs at any level, unlike burial-diagenetic sequences where it underlies the clinoptilolite zone.

Small amounts of zeolites occur at shallow depth in Quaternary sediments, indicating that zeolites can form in much less than one million years. Czynscinski (1973) inferred on studies of Indian Ocean cores that phillipsite crystals grow to their full size ($\sim$45 μm) in as little as 150,000 years. Although rapid by marine standards, this is of course far slower than in saline, alkaline lakes. Overall, the proportion of zeolites increases as a function of age, at least in tuffaceous sediments, reflecting the progressive alteration of volcanic glass. The relation between age and alteration of glass is highly variable. In Leg 21, Hole 205 (southwest Pacific), Plio-Pleistocene silicic ash is partly altered to zeolites, whereas Upper Miocene silicic ashes are unaltered (Burns *et al.*, 1973, pp. 57-102, 723-750). Cretaceous sediments of the western Pacific contain unaltered mafic vitric shards of silt size (Churkin and Packham, 1973).

Much if not most of the zeolites in sea-floor sediments were formed by reaction of glass with pore fluid, with or without the addition of biogenic silica. Authigenic smectites and palygorskite are invariably associated with zeolites in altered tephra and the amount of clay minerals commonly exceeds that of zeolites. Palagonite and probably montmorillonite can represent intermediate stages in the alteration of mafic glass ("sideromelane") to phillipsite (Bonatti, 1963). Zeolites are also formed by reaction of poorly crystalline or X-ray-amorphous clays. As evidence, authigenic clinoptilolite is in many non-tuffaceous, brown, detrital pelagic clays. Biogenic silica contributes to the formation of clinoptilolite, judging from the common association of radiolarians or authigenic cristobalite with clinoptilolite (e.g., von Rad and Rösch, 1972). Moreover, clinoptilolite casts of radiolarians and forams have been widely noted in non-tuffaceous sediments (e.g., von Rad and Rösch, 1972). The X-ray amorphous clays formed in weathering of volcanic rocks, particularly vitric ashes, are highly reactive in a marine environment, and Ristvet (1973) found phillipsite, nontronite, montmorillonite-illite, and sepiolite forming from clays of this sort in Kaneohe Bay, Oahu, Hawaii.

It seems likely that at least some analcime has formed from a pre-existing zeolite. McKelvey and Fleet (1974) describe fibrous analcime in a sequence of pyroclastic rocks that elsewhere contain fibrous mordenite. The fibrous analcime may be a replacement of mordenite or another zeolite, as noted in non-marine tuffs by Sheppard and Gude (1973) and Walton (1975).

Pore fluid of the DSDP cores generally varies only slightly in composition except for silica, which commonly ranges from about 4 to 65 ppm. The pH is generally between 7 and 8, and salinity is generally close to that of sea water. The dissolved silica is highest ($\sim$40-65 ppm) in siliceous oozes and lowest (4-10 ppm) in sediments lacking biogenic silica. Unaltered silicic ash of Miocene age in the southwest Pacific has pore water with about 30 ppm silica (Presley *et al.*, 1974). The content of dissolved silica bears a variable but significant relation to the nature of the zeolites. With some exceptions, pore waters of clinoptilolite-bearing sediments have SiO_2 contents of 20-40 ppm (e.g. Gieskes and Kastner, 1975). In cores with significantly lower SiO_2 contents in the pore waters, clinoptilolite crystals may be partly dissolved (e.g., Leg 14, Hole 137; Berger and von Rad, 1972; Waterman *et al.*, 1972). Pore waters of phillipsite-bearing sediments generally have less than 20 ppm dissolved silica. Particularly significant is Hole 199 of Leg 20, in the western Pacific, where SiO_2 ranges from 4 to 10.5 ppm in pore waters of phillipsite-bearing tuffaceous brown clay of Miocene and Pliocene age at 96 to 174 m depths (Waterman, 1973;

Matti *et al.*, 1973). The low silica content of the pore water probably explains why phillipsite was formed from these silicic ashes, which might otherwise be expected to have yielded clinoptilolite. It should be noted that phillipsite dissolves with long exposure to sea water, having about 1-3 ppm dissolved silica (Rex, 1967).

In conclusion, studies of deep-sea sediments show that phillipsite, clinoptilolite, and other zeolites form at low temperatures by reaction of volcanic glass and poorly crystalline aluminosilicate materials with sea water enriched in silica. Reaction rates are slow, and vitric ashes of late Miocene age are commonly unaltered or only partly altered. Zeolites in altered vitric ash layers are generally associated with a large amount of authigenic clay minerals, chiefly smectites and palygorskite. Little evidence exists for widespread alteration of non-analcimic zeolites to analcime over a period of about 100 million years.

Authigenic feldspar is probably present in some DSDP cores, but studies have thus far not documented satisfactorily the nature and extent of its occurrence. Thus, Rex *et al.* (1971) report an unusual abundance of K-feldspar in the 2-20 μm fraction of samples from Pacific DSDP cores analysed by X-ray diffraction (Leg 6, Holes 44, 47, 53, and 60). An authigenic origin was suggested although not proved for these occurrences. K-feldspar of sanidine structure and possible authigenic origin is mentioned and illustrated by von Rad and Rösch (1972, p. 737; Pl. 2, Fig. 6). Very small amounts of authigenic K-feldspar and albite (?) occur in cores from the Indian Ocean (Leg 26, Holes 250, 253, and 254) according to Davies *et al.* (1974). These supposedly authigenic feldspars are not described. If present in the DSDP cores, authigenic feldspars must be relatively rare, which contrasts with their relatively common occurrence in Paleozoic and Mesozoic marine limestones, shales, sandstones, and tuffs (see Hay, 1966; Kastner, 1971; Buyce and Friedman, 1975). Zeolites progressively decrease in abundance as a function of age in Mesozoic and Paleozoic marine rocks, probably reflecting in large measure their alteration to feldspars (Hay, 1966). Further study of the DSDP cores can be expected to document the early stages of this process.

Zeolites of the open-system type

Tephra sequences commonly show a more or less vertical zonation of zeolites and other authigenic silicate minerals that reflects the chemical change in meteoric water moving through the system. Solution and hydrolysis of vitric ash by subsurface water in the upper part of the system increases the pH and dissolved solids to the point where glass alters rapidly and zeolites are precipitated. Flow in an open hydrologic system is either downward or has a downward component where meteoric water enters the system and results in a vertical to gently inclined zonation of water composition and authigenic minerals (Hay, 1963a; Walton, 1975).

Thick, non-marine accumulations of silicic tephra may contain an upper zone 200 to 500 m thick with fresh glass, montmorillonite, and possibly some opal. The next lower zone, as much as 500 or more meters thick, is generally characterized by clinoptilolite and may contain beds with more than 90% of this zeolite. Clinoptilolite of the zeolitic zone replaces glass and fills pore space in the rock. The contact with the overlying zone is generally sharp and may be marked by cavernous zone 2-20 mm thick in which shards have been dissolved (Hay, 1963a). The underlying zone is characterized by analcime, with or without authigenic K-feldspar and quartz (Fig. 2). The Vieja Group (Oligocene) of Texas is an excellent example with both the analcime and clinoptilolite zones (Walton, 1975). The analcime seems to have been formed by reaction of pore water with both clinoptilolite and montmorillonite. Miocene tuffs of the Nevada Test Site are more complexly zoned than this and may have been formed by percolating water above the permanent water table (Hoover, 1968).

Nephelinite and basanite tuffs may exhibit a comparable zonation of zeolites in thicknesses on the order of 10 meters. This zonation at very shallow depths reflects the highly reactive nature of alkali-rich, silica-poor glass. For example, the basanite vitric tuffs of Koko Crater, Oahu, Hawaii, exhibit the following vertical zonation: (a) a zone 1.5-12 m thick with fresh glass, opal and montmorillonite; (b) a zone 1.5-10 m thick of palagonite tuffs with phillipsite and chabazite; and (c) a lowermost zone, 8 or more meters thick, in which analcime is the dominant zeolite (Hay and Iijima, 1968, p. 355).

In the open-system type, zeolitic alteration of glass, once started, may go rapidly to completion. This probably reflects the high pH ($\sim$9.5) caused by hydrolysis. In the John Day Formation of Oregon (Oligocene-Miocene), authigenic K-feldspar and celadonite very near in the zeolitic rocks give nearly the same K-Ar ages (22, 24 $\times 10^6$ years), suggesting that the alteration was relatively rapid (Hay, 1963a). Koko Crater is only about 35,000 years old (Gramlich *et al.*, 1971), proving that the glass altered rapidly, by geologic standards, to palagonite and zeolites. Another probably comparable example is the Neapolitan yellow ashes of the Phlegraen Fields in southern Italy, which were erupted 10,000 years ago. Where thickest, the ashes contain chabazite, and isotopic evidence suggests that zeolitic alteration lasted at least 4000-5000 years and is not now active (Capaldi *et al.*, 1971).

Hydrothermal zeolites

Zeolites are widespread in areas of hydrothermal alteration and may exhibit a well-defined zonation. Clinoptilolite or mordenite characterizes the shallowest and coolest zones, and progressively deeper zones commonly contain analcime or heulandite, laumontite, and wairakite. Well-known examples are in Yellowstone Park, U.S.A. (Fenner, 1936; Honda and Muffler, 1970); Wairakei, New Zealand (Steiner, 1953, 1955); and Onikobe, Japan (Seki *et al.*, 1969). The mineralogic zonation represents primarily progressive dehydration with increasing temperature.

Alteration zones surrounding Kuroko-type ore deposits in Japan exhibit a complex zonation resulting from a combination of submarine hydrothermal alteration and low-temperature diagenesis (Iijima, 1974; Utada *et al.*, 1974). In the simplest form, an analcime zone extends laterally from a phyllosilicate hydrothermal zone and is progressively bordered by mordenite and by clinoptilolite–mordenite zones. Analcimic rocks may alternate with mordenite- and clinoptilolite-bearing rocks. Ferrierite, a relatively rare siliceous zeolite, is locally found in the clinoptilolite–mordenite zone (Utada *et al.*, 1974).

Curiously, laumontite is readily formed by hydrothermal alteration, yet has not been synthesized, to this author's knowledge, in unseeded hydrothermal experiments from anything but dehydrated laumontite (Coombs *et al.*, 1959). Nature's ease in forming laumontite is illustrated by the laumontite precipitated in radiators fed by natural hot spring water in Pauzhetka City, Kamchatka (Lebedev and Gorokhova, 1968)!

Zeolites of burial diagenetic type

This type, also termed burial metamorphic, comprises the zeolites formed at significantly increased temperatures on a regional scale in thick accumulations, of mainly volcaniclastic sediments. A vertical zonation of zeolites and associated minerals characterizes thicknesses of 3–12 km. In Cenozoic sequences, a surface zone as much as 2 km thick may contain fresh glass, and successively lower zones may have assemblages of (1) clinoptilolite and mordenite, (2) analcime and heulandite, and (3) laumontite and albite. The laumontite–albite zone grades into a zone with prehnite and pumpellyite, which represents a transition into the greenschist facies of metamorphism. Locally, a wairakite-bearing zone may be intercalated between rocks with laumontite and those with prehnite and pumpellyite. In pre-Cenozoic strata (e.g., Coombs, 1954) the uppermost zones are generally eroded.

The mineralogic zones represent a decrease in hydration with depth, as expected from an increase in temperature. Moreover, zeolitic assemblages correlate rather well with coal rank (Kisch, 1966, 1969), pointing to depth of burial and geothermal gradient as the principal controlling factors.

Probably the most instructive examples of burial-diagenetic zoning are in late Cenozoic volcaniclastic sediments of the Green Tuff region of Japan (Iijima and Utada, 1966, 1972). Here the thickness and mineralogy of the zones can be correlated to some extent with the geothermal gradient (Utada, 1971). Zeolitic diagenesis is presently proceeding in three areas which have been studied, allowing an estimate of the temperatures at which the zones are forming. The best-known example is the Niigata oil field, where drill cores are available for a 5-km thickness of marine strata (Iijima and Utada, 1971). An uppermost zone, 0.8–1.9 km thick, contains fresh glass, and the underlying zone, 1.6–2.5 km thick, contains mordenite and clinoptilolite. Successively lower are an analcimic zone, about 1 km thick, and a zone with albite of which as much as 0.7 km was penetrated. Analcime was formed by reaction of mordenite and clinoptilolite, as shown by pseudomorphs, and albite was formed by reaction of quartz and analcime. Temperatures at the top of the mordenite and clinoptilolite zone are 41–49°C, and those at the analcime–albite transition are 120–124°C. This temperature is significantly lower than that of the 190°C inferred on the basis of laboratory experiments (Liou, 1971; Thompson, 1971) which is used as a geological thermometer by some geologists.

The upper zones of burial diagenesis can be mineralogically similar to those of the open hydrologic system, and the two types may be difficult to distinguish in thick sequences of silicic tephra. Indeed, the distinction between the two types is to some extent artificial because any increase in temperature on burial will speed up open-system reactions. There are, however, a few fundamental differences in the two types which may be recognizable. In the open-system type, a sharp contact may separate the fresh-glass zone from the clinoptilolite zone. In the John Day Formation, a zone of solution 2–20 mm thick with unfilled shard cavities separates the two zones (Hay, 1963a). In burial diagenesis, by contrast, the contact is gradational, and a thick transitional zone may be present (Utada, 1971). In the open-system type, zeolitic alteration is generally short-lived and penecontemporaneous with deposition; it may not correspond to the time of deepest burial. In the

John Day Formation, K–Ar dates on authigenic phases show that alteration was penecontemporaneous with deposition, about 23×10^6 years ago, whereas the deepest burial was during Miocene time about 15×10^6 years ago, when the Columbia River basalts were erupted. K–Ar dates on authigenic materials in tuffs of the Sutsut Group, British Columbia (Read and Eisbacher, 1974) suggest that the alteration to zeolites and albite in these rocks is more likely of the open-system type than of the burial diagensis type.

ACKNOWLEDGMENTS

I have benefited by discussions with many workers in the field of zeolites and am particularly indebted to R. A. Sheppard and A. Iijima. For aid with the zeolites of DSDP cores, I thank H. E. Cook. Debbie Aoki is thanked for typing the manuscript.

REFERENCES

Baldar, N. A. and Whittig, L. D. (1968) Occurrence and synthesis of soil zeolites: *Proc. Soil Sci. Amer.* **32**, 235-238.

Berger, W. H. and von Rad, U. (1972) Cretaceous and Cenozoic sediments from the Atlantic Ocean. In: Hayes, D. E., ed., *Initial Reports of the Deep Sea Drilling Project* **14**, U.S. Govt. Printing Office, Washington, pp. 787-954.

Biscaye, P. E. (1964) Distinction between kaolinite and chlorite in recent sediments by X-ray diffraction: *Amer. Mineral.* **49**, 1281-1289.

Boles, J. R. (1971) Synthesis of analcime from natural heulandite and clinoptilolite: *Amer. Mineral.* **56**, 1724-1734.

Boles, J. R. (1972) Composition, optical properties, cell dimensions, and thermal stability of some heulandite-group zeolites: *Amer. Mineral.* **57**, 1452-1493.

Bonatti, E. (1963) Zeolites in Pacific pelagic sediments: *Trans. New York Acad. Sci.*, Ser. 11, **25**, 938-948.

Burns, R. E. *et al.* (1973) *Initial Reports of the Deep Sea Drilling Project*, Vol. **21**: U.S. Govt. Printing Office, Washington, pp. 57-102.

Buyce, M. R. and Friedman, G. R. (1975) Significance of authigenic K-feldspar in Cambrian–Ordovician carbonate rocks of the Proto-Atlantic shelf in North America: *J. Sediment. Petrol.* **45**, 808-821.

Capaldi, G., Civetta, L. and Gasparini, P. (1971) Fractionation of the ^{238}U decay series in the zeolitization of volcanic ashes: *Geochim. Cosmochim. Acta* **35**, 1067-1072.

Churkin, M. and Packham, G. H. (1973) Volcanic rocks and volcanic constituents in sediments, Leg 21, Deep Sea Drilling Project. In: Burns, R. E., editor, *Initial Reports of the Deep-Sea Drilling Project*, Vol. **21**: U.S. Govt. Printing Office, Washington, pp. 481-493.

Coombs, D. S. (1954) The nature and alteration of some Triassic sediments from Southland, New Zealand: *Trans. Roy. Soc. New Zealand* **82**, 65-103.

Coombs, D. S., Ellis, A. S., Fyfe, W. S. and Taylor, A. M. (1959) The zeolite facies, with comments on the interpretation of hydrothermal syntheses: *Geochim. Cosmochim. Acta* **17**, 53-107.

Czyscinski, K. (1973) Authigenic phillipsite formation rates in the central Indian Ocean and the equatorial Indian Ocean: *Deep-Sea Res.* **20**, 555-559.

Davies, T. A. *et al.* (1974) *Initial Reports of the Deep Sea Drilling Project*, Vol. **26**: U.S. Govt. Printing Office, Washington, pp. 26-29, 161, 237.

Deffeyes, K. S. (1959) Zeolites in sedimentary rocks: *J. Sediment. Petrol.* **29**, 602-609.

Fenner, C. N. (1936) Bore-hole investigations in Yellowstone Park: *J. Geol.* **44**, 225-315.

Frankart, R. and Herbillon, A. J. (1970) Presence et genese d'analcime dans les sols sodique de la Basse Ruzizi (Burundi): *Bull. Groupe franc. Argiles* **22**, 79-89.

Gieskes, J. M. and Kastner, M. (1975) Evidence for extensive diagenesis, Madagascar Basin, Deep Sea Drilling Site 245: *Geochim. Cosmochim. Acta* **39**, 1385-1393.

Goodwin, J. H. and Surdam, R. C. (1967) Zeolitization of tuffaceous rocks of the Green River Formation, Wyoming: *Science* **157**, 307-308.

Gramlich, J. W., Lewis, V. A. and Naughton, J. J. (1971) Potassium argon dating of recent basalts of the Honolulu Volcanic Series: *Geol. Soc. Amer. Bull.* **82**, 1399-1404.

Gude, A. J., 3rd and Sheppard, R. A. (1966) Silica-rich chabazite from the Barstow Formation, San Bernardino County, southern California: *Amer. Mineral.* **51**, 909-915.

Hathaway, J. C. and Sachs, P. L. (1965) Sepiolite and clinoptilolite from the Mid-Atlantic Ridge: *Amer. Mineral.* **50**, 852-867.

Hay, R. L. (1963a) Stratigraphy and zeolitic diagenesis of the John Day Formation of Oregon: *Calif. Univ. Pubs. Geol. Sci.* **42**, 199-262.

Hay, R. L. (1963b) Zeolitic weathering in Olduvai Gorge, Tanganyika: *Geol. Soc. Amer. Bull.* **74**, 1281-1286.

Hay, R. L. (1964) Phillipsite of saline lakes and soils: *Amer. Mineral.* **49**, 1366-1387.

Hay, R. L. (1966) Zeolites and zeolitic reactions in sedimentary rocks: *Geol. Soc. Amer. Spec. Paper* **85**, 130.

Hay, R. L. (1970) Silicate reactions in three lithofacies of a semiarid basin, Olduvai Gorge, Tanzania: *Mineral. Soc. Amer. Spec. Papers* **3**, 237-255.

Hay, R. L. (1976) *Geology of the Olduvai Gorge*: Calif. Univ., Berkeley, p. 203.

Hay, R. L. and Iijima, A. (1968) Nature and origin of palagonite tuffs of the Honolulu Group on Oahu, Hawaii: *Geol. Soc. Amer. Mem.* **116**, 331-376.

Heath, G. R. (1969) Mineralogy of Cenozoic deep-sea sediments from the equatorial Pacific Ocean: *Geol. Soc. Amer. Bull.* **80**, 1997-2018.

High, L. R. and Picard, M. D. (1965) Sedimentary petrology and origin of analcime-rich Popo Agie Member, Chugwater Formation (Triassic), west-central Wyoming: *J. Sediment. Petrol.* **35**, 49-70.

Honda, S. and Muffler, L. J. P. (1970) Hydrothermal alteration in core from research drill hole Y-1, Upper Geyser Basin, Yellowstone National Park, Wyoming: *Amer. Mineral.* **55**, 1714-1737.

Hoover, D. L. (1968) Genesis of zeolites, Nevada test site: *Geol. Soc. Amer. Mem.* **110**, 275-284.

Iijima, A. (1974) Clay and zeolitic alteration zones surrounding Kuroko deposits in the Hokuroku District, northern Akita, as submarine hydrothermal-diagenetic alteration products. In: *Geology of Kuroko Deposits*, Ishihara, S., Kanehira, K., Sasaki, A., Sato, T. and Shimazaki, Y., eds.: Soc. Mining Geologists Japan, Tokyo, pp. 267-289.

Iijima, A. and Harada, T. (1968) Authigenic zeolites in zeolitic palagonite tuffs on Oahu, Hawaii: *Amer. Mineral.* **54**, 182-197.

Iijima, A. and Utada, M. (1966) Zeolites in sedimentary rocks, with reference to the depositional environments and zonal distribution: *Sedimentology* **7**, 327-357.

Iijima, A. and Utada, M. (1971) Present-day zeolitic diagenesis of the Neogene geosynclinal deposits in the Niigata oil field, Japan. In: *Advances in Chemistry Series* 101, *Molecular Sieve Zeolites – 1*: Amer. Chem. Soc., pp. 342-349.

Iijima, A. and Utada, M. (1972) A critical review on the occurrence of zeolites in sedimentary rocks in Japan: *Jap. J. Geol. Geog.* **42**, 61-84.

Joulia, F., Bonifas, M., Camez, T., Millot, G. and Weil, R. (1959) Decouverte d'un important niveau d'analcimolite greseuse dans le Continental intercalaire de l'ouest de l'Air (Sahara central): *Service de Geol. et de Prospection Miniere, Dakar*, p. 40.

Kastner, M. (1971) Authigenic feldspars in carbonate rocks: *Amer. Mineral.* **56**, 1403-1442.

Kisch, H. J. (1966) Zeolite facies and regional rank of bituminous coals: *Geol. Mag.* **103**, 414-422.

Kisch, H. J. (1969) Coal-rank and burial-metamorphic mineral facies. In: *Advances in Organic Geochemistry*, 1968: Pergamon, Oxford, pp. 407-425.

Kossovskaya, A. G. (1975) Genetic types of zeolites in stratified deposits: *Lith. Mineral. Res.* **10**, 162-178.

Lebedev, L. M. and Gorokhova, L. M. (1968) Recent mineralization in technical structures of Pauzhetka City (Kamchatka): *Dokl. Akad. Nauk SSSR* **182**, 1399-1401.

Liou, J. G. (1971) Analcime equilibria: *Lithos* **4**, 389-402.

Matti, J. C., Zemmels, I. and Cook, H. E. (1973) X-ray mineralogy of sediments from the far western Pacific, Leg 20, DSDP. In: Heezen, B. C. *et al.*, eds., *Initial Reports of the Deep Sea Drilling Project*, Vol. **20**: U.S. Govt. Printing Office, Washington, pp. 323-334.

McKelvey, B. C. and Fleet, A. J. (1974) Eocene basaltic pyroclastics at Site 253, Ninetyeast Ridge. In: Davies, T. A. *et al.*, eds., *Initial Reports of the Deep Sea Drilling Project*, Vol. **26**: U.S. Govt. Printing Office, Washington, pp. 553-565.

Miyashiro, A. and Shido, F. (1970) Progressive metamorphism as zeolite assemblages: *Lithos* **3**, 251-260.

Munson, R. A. and Sheppard, R. A. (1974) Natural zeolites: their properties, occurrences, and uses: *Minerals Sci. Engng* **6**, 19-34.

Murray, J. and Renard, A. F. (1891) *Deep-Sea Deposits*: Volume 5 of report on the scientific results of the voyage of H.M.S. *Challenger* during the years 1873-76: Eyre & Spottiswoode, London.

Norin, E. (1958) The sediments of the central Tyrrhenian Sea. In: *Report of Swedish Deep-sea Expedition, 1947-1948*, Vol. 8, *Sediment Cores from the Mediterranean Sea and the Red Sea*; Pettersson, H., ed.: Göteborgs Kungl. Vetenkaps. och Vitterhets-Samhalle.

Presley, B. J., Trefry, J., Armstrong, D. and Nuzzo, M. (1974) Interstitial water chemistry: Deep Sea Drilling Project, Legs 21 and 22. In von der Borch, C. C., *et al.*, eds., *Initial Reports of the Deep Sea Drilling Project*, Vol. **22**: U.S. Govt. Printing Office, Washington, pp. 861-864.

Read, P. B. and Eisbacher, G. H. (1974) Regional zeolite alteration of the Sutsut Group, north-central British Columbia: *Can. Mineral.* **12**, 527-541.

Rex, R. W. (1967) Authigenic silicates formed from basaltic glass by more than 60 million years' contact with sea water, Sylvania Guyot, Marshall Islands. In *Clays and Clay Minerals – Proceedings of 15th Conference, 1966*, Bailey, S. W., ed.: Pergamon, New York, pp. 195-203.

Rex, R. W., Eklund, W. A. and Jamieson, I. M. (1971) X-ray mineralogy studies, Leg 6. In: Fisher, A. G. *et al.*, eds., *Initial Reports of the Deep Sea Drilling Project*, Vol. **6**: U.S. Govt. Printing Office, Washington, pp. 753-810.

Ristvet, B. L. (1973) Recent authigenic nontronite and montmorillonite–illite from Kaneohe Bay, Oahu, Hawaii. In: *Geol. Soc. Am. Abstracts with Programs* **5**, 781-782 (abstract).

Seki, Y. and Oki, Y. (1969) Wairakite–analcime solid solutions from low-grade metamorphic rocks of the Tanzawa Mountains, central Japan: *Mineral J.* **6**, 36-45.

Seki, Y., Onuki, H., Okumura, K. and Takashima, I. (1969) Zeolite distribution in the Katayama geothermal area, Onikobe, Japan: *Jap. J. Geol. Geog.* **40**, 63-79.

Sheppard, R. A. (1971) Zeolites in sedimentary deposits of the United States – a review. In: *Advances in Chemistry Series* 101, *Molecular Sieve Zeolites – 1*: Amer. Chem. Soc., pp. 279-310.

Sheppard, R. A. (1973) Zeolites in sedimentary rocks: *U.S. Geol. Surv. Profess. Paper* 820, pp. 689-695.

Sheppard, R. A. and Gude, A. J. 3rd (1968) Distribution and genesis of authigenic silicate minerals in tuffs of Pleistocene Lake Tecopa, Inyo County, California: *U.S. Geol. Surv. Profess. Paper* **597**, p. 38.

Sheppard, R. A. and Gude, A. J. 3rd (1969a) Diagenesis of tuffs in the Barstow Formation, Mud Hills, San Bernardino County, California: *U.S. Geol. Surv. Profess. Paper* **634**, p. 35.

Sheppard, R. A. and Gude, A. J. 3rd (1969b) Chemical composition and physical properties of the related zeolites offretite and erionite: *Amer. Mineral.* **54**, 875-886.

Sheppard, R. A. and Gude, A. J. 3rd (1970) Chemical composition and physical properties of phillipsite from the Pacific and Indian Oceans: *Amer. Mineral.* **55**, 2053-2062.

Sheppard, R. A. and Gude, A. J. 3rd (1973) Zeolites and associated authigenic silicate minerals in tuffaceous rocks of the Big Sandy Formation, Mohave County, Arizona: *U.S. Geol. Surv. Profess. Paper* **830**, p. 36.

Steiner, A. (1953) Hydrothermal rock alteration at Wairakei, New Zealand: *Econ. Geol.* **48**, 1-13.

Steiner, A. (1955) Hydrothermal rock alteration: *Dep. Sci. Indust. Res. Bull.* **117**, 21-26.

Surdam, R. C. and Parker, R. D. (1972) Authigenic aluminosilicate minerals in the tuffaceous rocks of the Green River Formation, Wyoming: *Geol. Soc. Amer. Bull.* **83**, 689-700.

Thompson, A. B. (1971) Analcite–albite equilibria at low temperatures: *Am. J. Sci.* **271**, 79-92.

Utada, M. (1971) Zeolitic zoning in the Neogene pyroclastic rocks of Japan: *Sci. Papers Coll. Gen. Educ. Univ. Tokyo* **21**, 189-221.

Utada, M., Minato, H., Ishikawa, T. and Yoshizaki, Y. (1974) The alteration zones surrounding Kuroko-type deposits in the Nishi–Aizu District, Fukushima Prefecture, with emphasis on the analcime zone as an indicator in exploration for ore deposits: *Mining Geol.*, Spec. Issue, pp. 291-302.

Vanderstappen, R. and Verbeek, T. (1959) Presence d'analcime d'origine sedimentarie dans le Mesozoique du Bassin du congo: *Soc. Belge Geologie Paleont. Hydrol. Bull.* **68**, 417-421.

von Rad, U. and Rösch, H. (1972) Mineralogy and origin of clay minerals, silica, and authigenic silicates in Leg 14 sediments. In: Hayes, D. E. *et al.*, eds., *Initial Reports of Deep Sea Drilling Project*, Vol. **14**: U.S. Govt. Printing Office, Washington, pp. 727-751.

Walton, A. W. (1975) Zeolitic diagenesis in Oligocene volcanic sediments, Trans-Pecos Texas: *Geol. Soc. Amer. Bull.* **86**, 615-624.

Waterman, L. S. (1973) Interstitial water studies on small core samples, Leg 20. In: Heezen, B. C. and MacGregor, I. D., eds., *Initial Reports of the Deep Sea Drilling Project*, Vol. **20**: U.S. Govt. Printing Office, Washington, pp. 423-425.

Waterman, L. S., Sayles, L. S. and Manheim, F. T. (1972) Interstitial water studies on small core samples, Leg 14. In: Hayes, D. E. *et al.*, eds., *Initial Reports of the Deep Sea Drilling Project*, Vol. **14**: U.S. Govt. Printing Office, Washington, pp. 753-762.

ZEOLITES IN SALINE, ALKALINE-LAKE DEPOSITS

RONALD C. SURDAM

Department of Geology, University of Wyoming, Laramie, Wyoming 82071, U.S.A.

and

RICHARD A. SHEPPARD

U.S. Geological Survey, Box 25046 Denver Federal Center, Denver, Colorado 80225, U.S.A.

Abstract – Zeolites were identified in saline, alkaline-lake deposits in the 1920s; however, until the late 1950s, these occurrences were considered rare. Saline, alkaline-lake deposits are now recognized as an ideal environment for zeolites, especially analcime, clinoptilolite, phillipsite, erionite, chabazite, and mordenite. Most saline, alkaline lakes result from a special geological setting characterized by block or rift faulting. Tectonic events of this type on a continent commonly have two important regional consequences: volcanism and closed hydrographic basins. Volcanism yields an abundance of glassy pyroclastic material, and closed hydrographic basins, especially where evaporation exceeds precipitation, develop saline and alkaline fluids. The combination of highly reactive volcanic material and saline, alkaline solutions is ideal for the formation of zeolites. Clay minerals and aluminosilicate gels also react locally to form zeolites.

The distinctive feature of saline, alkaline-lake deposits is the lateral zonation of diagenetic mineral facies. Recent studies of the distribution of authigenic silicate minerals in tuffaceous rocks of Cenozoic closed-basin deposits show a consistent pattern. Generally, there is a lateral zonation in a basinward direction of fresh glass to alkalic, silicic zeolites to analcime to potassium feldspar. This distribution pattern is probably due to the chemical zonation that existed in the ground, lake, and interstitial water during deposition of the tuffaceous rocks. The important reactions during diagenesis that produce these patterns are:

1. glass → alkalic, silicic zeolites,
2. alkalic, silicic zeolites → analcime,
3. analcime → potassium feldspar,
4. alkalic, silicic zeolites → potassium feldspar.

Some significant chemical controls on these reactions are cation ratios, pH, Si:Al ratios, activity of silica, activity of water, and salinity.

Modern closed hydrographic basins, characterized by saline, alkaline lakes and tuffaceous sediments, provide data for understanding ancient zeolite deposits. Current studies of modern environments such as Lake Magadi, Kenya, and Teels Marsh, Nevada, are leading to a determination of those chemical factors that control the individual reactions during diagenesis.

INTRODUCTION

In recent years it has become apparent that zeolites are among the most common authigenic silicate minerals in sedimentary rocks (Hay, 1966; Sheppard, 1971, 1973, 1975; Munson and Sheppard, 1974). These authors indicate that zeolites occur in sedimentary rocks that are diverse in lithology, age, and depositional environment. Several classifications of zeolite deposits have been offered; and although differing in detail, all investigators agree that one prominent type is the closed hydrographic basin or saline, alkaline-lake deposit (Hay, 1966; Sheppard, 1973; Kossovskaya, 1975; and Mumpton, 1975). This type of deposit is commonly the result of the reaction of volcanic glass with connate water trapped during sedimentation in a saline, alkaline lake (Sheppard, 1973). This paper describes in detail the geological, hydrological, and chemical processes responsible for the formation of zeolites in saline, alkaline-lake deposits.

Although there are nearly forty species of zeolites that occur in nature, only six are common in saline, alkaline-lake deposits; they are analcime, clinoptilolite, phillipsite, erionite, chabazite, and mordenite (see Table 1). These six zeolites show considerable range in Si:Al ratios and in cation and water contents. Their distribution in saline, alkaline-lake deposits in the United States has been reviewed by Hay (1966), Sheppard (1971), Surdam (1972), and Papke (1972); the reader is referred to these reports for extensive lists of deposits. Table 2 is a partial list of zeolitic saline, alkaline-lake

TABLE 1. Formulas of zeolites found in saline, alkaline-lake deposits

Name	Dominant cations	Formula[1]
Analcime	Na	$NaAlSi_{2.0-2.8}O_{6.0-7.6}\cdot 1.0\text{–}1.3\ H_2O$
Chabazite	Na, Ca	$NaAlSi_{1.7-3.8}O_{5.4-9.6}\cdot 2.7\text{–}4.1\ H_2O$
Clinoptilolite	Na, K, Ca	$NaAlSi_{4.2-5.0}O_{10.4-12.0}\cdot 3.5\text{–}4.0\ H_2O$
Mordenite	Na, Ca, K	$NaAlSi_{4.5-5.0}O_{11.0-12.0}\cdot 3.2\text{–}3.5\ H_2O$
Erionite	Na, K	$NaAlSi_{3.0-3.7}O_{8.0-9.4}\cdot 3.0\text{–}3.4\ H_2O$
Phillipsite	Na, K	$NaAlSi_{1.3-3.4}O_{4.6-8.8}\cdot 1.7\text{–}3.3\ H_2O$

[1] Formulas are standardized in terms of a sodium end member that has one aluminum atom (after Sheppard and Gude, 1969).

TABLE 2. Zeolites in saline, alkaline-lake deposits exclusive of the United States (Zeolites: A, analcime; Ch, chabazite; Cl, clinoptilolite; E, erionite; M, mordenite, P, phillipsite)

Locality	Zeolites	Occurrence	References
Tucuman Province, Argentina	A	Claystone in the Rio Sali Formation of Tertiary age	Teruggi and Andreis (1963)
Canadon Hondo, Patagonia, Argentina	Cl	Tuff in the Sarmiento Formation of Eocene to Oligocene age	Mason and Sand (1960)
San Jorge Gulf Basin, Argentina	A	Tuff and tuffaceous sandstone in the Chubutian Group of Cretaceous age	Teruggi (1964)
Anfiteatro de Tico, Santa Cruz, Argentina	Cl	Tuff in the Matilde Formation of Jurassic age	Andreis and Rodriquez (1968)
Tupungato, Mendoza, Argentina	A	Tuff in the Dark Victor Formation of Triassic age	Baldwin (1944)
Mikaliccik, Turkey	A	Tuff and tuffaceous mudstone of Miocene and Pliocene age	Echle (1974, 1975)
Golpazari, Turkey	A	Tuff of Oligocene age	Ataman and Beseme (1972)
Tuva, U.S.S.R.	A	Tuff and tuffaceous sandstone of Carboniferous age	Bur'yanova (1960)
Air, Niger	A	Sandstone and argillite in the Continental Intercalaire of Cretaceous Age	Joulia and others (1959)
Agadez, Niger	A	Sandstone and siltstone in the Serie des Gres d'Agades of Triassic and Jurassic age	Pacquet (1968)
Kisangant, Zaire	A	Sandstone and shale of Jurassic and Cretaceous age	Vanderstappen and Verbeek (1959, 1964); Vernet (1961)
Sterkspruit, South Africa	A	Sandstone and shale in the Stormberg Red Beds of Permian to Triassic age	Fuller and Moir (1967)
Nata, Botswana	Cl	Claystone of Quaternary age	Smale (1968)
Lake Magadi, Kenya	A, Ch, Cl, E, M, P	Tuff and claystone of Pleistocene and Holocene age	Hay (1968); Surdam and Eugster (1976)
Olduvai Gorge, Tanzania	A, Ch, Cl, E, P	Tuff and claystone in Bed I and Bed II of Pleistocene age	Hay (1970)
Lake Natron, Tanzania	A, Ch, P	Tuff and clay of Holocene age	Hay (1966)
North end of Lake Natron, Tanzania	A, Cl, E, M, P	Tuff and claystone in the High Natron Beds of Pleistocene age	Hay (1968)
West side of Lake Natron, Tanzania	A, E, P	Tuff and claystone in the Peninj Beds of Pleistocene age	Hay (1966)
Nzega, Tanzania	A	Claystone in the Manyonga Lake Beds of Pleistocene age	McKie (1958)
Sydney Basin, New South Wales, Australia	A	Tuff in the Newcastle Coal Measures of Permian age	Loughnan (1966)
Taylor Valley, Antarctica	P	Ash of Quaternary age	Linkletter (1974)

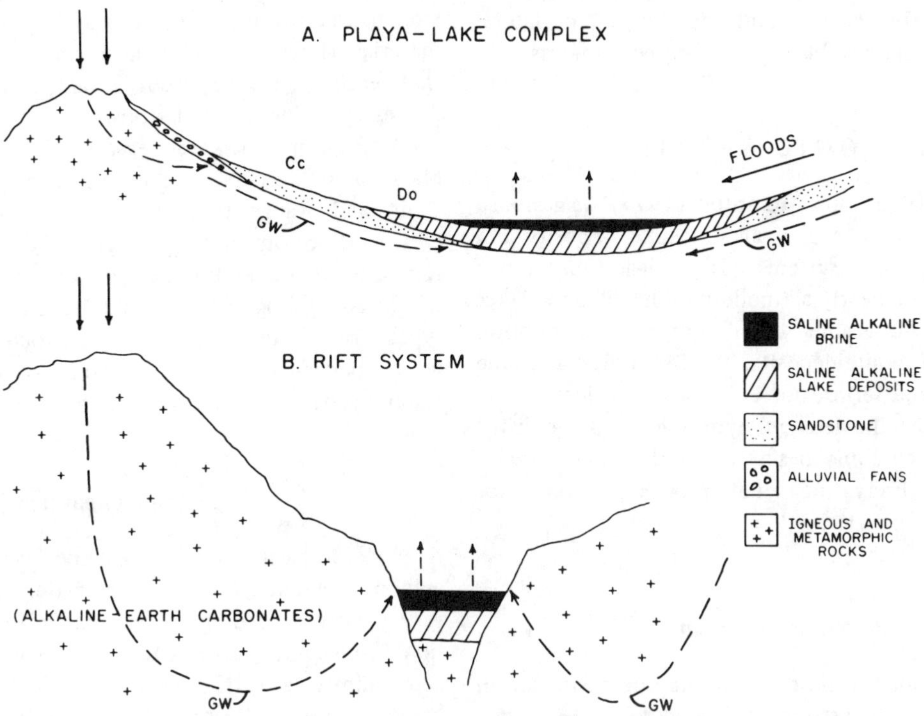

Fig. 1. Hydrology and brine evolution in a playa-lake complex (A) and a rift system (B). Solid arrows, rainwater; GW, ground-water circulation path; Cc, calcite precipitation; Do, dolomite formation; vertical dashed arrows indicate evaporation (modified from Eugster and Surdam, 1973).

deposits exclusive of the United States.

It is obvious from the large number of zeolitic rocks in saline, alkaline-lake deposits that this type of depositional environment is an excellent habitat for the formation of zeolites. The combination of highly reactive volcanic material and saline, alkaline solutions is apparently ideal for the formation of zeolites. In terms of understanding zeolite genesis, the closed hydrographic basin or saline, alkaline-lake deposits are particularly significant because there are several modern environments where zeolites are presently forming.

HISTORY OF INVESTIGATIONS

Zeolites were identified in saline, alkaline lake deposits in the 1920s; however, until the late 1950s these occurrences were considered rare. Of the more than 400 published reports that describe zeolites in sedimentary rocks throughout the world, more than 75% were published in the last 15 years. Much of this recent surge of interest in zeolites relates to saline, alkaline-lake deposits.

A partial list of papers which have brought us to our current level of understanding of saline, alkaline-lake deposits includes Ross (1928) and Bradley (1928) who made some of the initial identification of zeolites in saline, lacustrine deposits of the United States: C. S. Ross identified analcime in the Big Sandy Formation of Arizona, and W. H. Bradley identified analcime in the Green River Formation of Wyoming. In 1933 Bramlette and Posnjak documented the occurrence of zeolites as alteration products of vitric ash. Although this paper dealt with tuffaceous rocks deposited in a marine environment, it laid the groundwork for subsequent investigations of saline-lake deposits by K. S. Deffeyes, F. B. Van Houten, and R. L. Hay. Deffeyes (1959) stressed the abundance of zeolites in tuffaceous rocks and the genesis of these zeolites by the reaction of vitric material and saline, alkaline solutions. Van Houten (1964; 1965) documented these ideas with careful observation of zeolite occurrences in Cenozoic rocks of Wyoming and the Triassic rocks of New Jersey. Hay (1966) brought together all the existing observational, theoretical, and experimental data pertinent to natural zeolites. In a series of papers Sheppard and Gude (1968, 1969, 1973) documented the mineral distributions and authigenic mineral reactions in saline, alkaline-lake deposits in basins of the western United States and are responsible for our current understanding of the lateral zoning of mineral assemblages in such environments.

GEOLOGIC SETTING

Sheppard (1973) and Mumpton (1975) suggested a relationship between saline, alkaline-lake deposits and closed hydrographic systems. This suggestion is not surprising because nearly all modern saline, alkaline lakes are in closed hydrographic basins. For an interpretation of closed basin zeolite deposits, it is essential to examine at least in general terms, not only the mineralogy of the deposits, but also the geology, hydrology, and chemistry of closed hydrographic basins. With these data, many mineralogical observations that previously were not understood become clarified.

Modern closed basins

Nearly all modern closed hydrographic basins are in one of two tectonic settings: (1) block-faulted terrains, e.g., the Basin and Range provinces of the western United States, or (2) trough valleys associated with rifting, e.g., the East Rift Valley of Kenya. Figure 1 shows a schematic cross-section of the two end-member types of closed basins. Obviously, the examples shown in Fig. 1 are idealized, and we hasten to stress that it is possible to develop closed basins by combining elements from the two. One type, which we designate a "playa-lake complex", is a broad flat valley floor surrounded by high mountain ranges. The other, which we designate a "rift system", is a narrow, flat valley with steep walls. The important difference between these two is the geohydrology. In the playa-lake complex, a perennial playa-lake occupies the lowest portion of the valley floor. The ground-water flow is generally shallow, and much of it will discharge at the mud-flat or playa fringe adjacent to the lake (Fig. 1A). In contrast, the ground-water circulation in the rift system is deep, and discharge is by springs along faults at the edge of the lake (Fig. 1B).

An important feature of both types is that only a limited amount of clastic debris reaches the lake. In the playa-lake complex, the sharp break in slope at the foot of the mountains traps most of the clastic material in alluvial fans, and most streams flowing beyond the fans are intermittent. In the rift system most of the flow into the lake is by springs which do not carry clastic debris. During dilute or humid stages, or perhaps early in the evolution of a closed hydrographic basin, perennial streams or rivers may reach the valley floor, for example, Pyramid or Walker Lakes in Nevada. Such perennial streams may transport a relatively large supply of clastic material. However, in most arid regions the clastic debris that reaches the valley floor is transported by sheet-flow processes during times of storms.

Deep Springs Lake in eastern California and Teels Marsh in western Nevada are excellent examples of the Basin and Range type of closed hydrographic basin (Jones, 1966; Smith and Drever, 1976). Lake Magadi in Kenya is an outstanding example of the Rift Valley type of closed hydrographic basin (Eugster, 1970). In both types, significant systematic hydrochemical changes occur, particularly from the point of view of ground water discharge to the center of the basin.

Hydrographic features

A closed-basin lake is a dynamic feature because the area of the lake, the depth of water, and the salinity vary greatly according to seasonal inflow and evaporation (Langbein, 1961). Thus, fluctuations in lake level and salinity are the result of a delicate imbalance between precipitation and evaporation. All lakes in closed hydrographic basins are characterized by such fluctuations due to seasonal or short-term climatic variations (Fig. 2). There are also many examples of longer-term variations in closed basins. The level of Lake Rudolf, Kenya, has fallen approximately 46 m in the last 3000 years as a result of climatic factors (Fuchs, 1939). During the period 1904 to 1930 the average net decrease in lake level was 23–30 cm/yr (Fuchs, 1939). However, the imbalance between inflow and evaporation is a delicate one, and Fuchs (1939) estimated that an increase of 12.7 cm/yr in the rate of precipitation would be sufficient to cause a rise in the level of Lake Rudolf. The most significant control on the existence of closed lakes is *evaporation* (Langbein, 1961).

Lakes lacking outlets or lakes in closed hydrographic basins are exclusive features of the arid and semi-arid zones where annual evaporation exceeds rainfall (Langbein, 1961). Russell (1896, pp. 131-132) concluded: "The study of the present geography of the earth shows that in regions where the mean annual precipitation exceeds 20 or perhaps 25 inches, inclosed lakes do not occur, although the topographical conditions may be favorable." Langbein (1961, p. 4) interpreted this phenomenon as follows: "The greater the aridity, the greater is the rate of net evaporation. The greater the net evaporation, the smaller the area of a closed lake (larger the playa fringe). The smaller the lake area, the more numerous are the topographic opportunities for a closed lake." Thus, saline, alkaline lakes are dependent not only on a special geological setting but also on an arid or semi-arid climate.

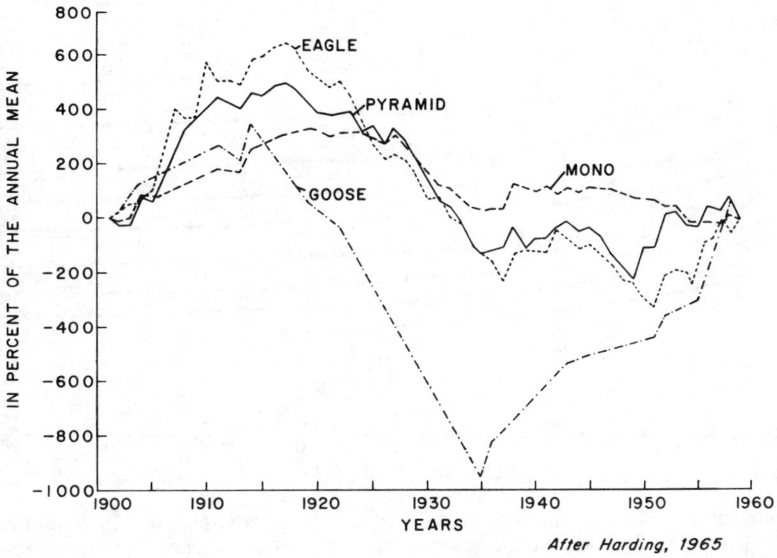

Fig. 2. Short-term fluctuations in runoff into closed basins in western United States (after Harding, 1965).

Brine evolution

Relative to zeolite genesis, a significant characteristic of a closed hydrographic basin is the brine evolution, an understanding of which simplifies the explanation of mineral patterns. Saline, *alkaline* brines, in contrast to merely saline brines, ideally develop in a closed hydrographic basin as follows: soft water forms by reaction of CO_2-charged groundwater with igneous and metamorphic rocks (Jones, 1966; Garrels and MacKenzie, 1967; Hardie and Eugster, 1970) and emerges at the foot of alluvial fans. The ground water flow from the alluvial fans to the center of the basin is shallow. Where the climate is arid, or where dry cycles are characterized by marked seasonal contrasts, evaporative concentration is intense; and calcite precipitates from the ground water in the capillary zone. Both pH and Mg:Ca ratio increase steadily as the ground water slowly moves toward the center of the basin while evaporation, evaporative pumping, and recycling of efflorescent crusts continues. After the Mg:Ca ratio through calcite precipitation has reached a value 12, protodolomite forms; and the salinity of the ground water continues to increase until the water reaches the lake at the center of the basin (Eugster and Surdam, 1973). The lake water is not only saline, but also alkaline.* If evaporation remains intense, a sodium carbonate mineral such as trona would eventually precipitate.

In contrast, a rift system does not exhibit alkaline-earth playa flats adjacent to the lake because of differences in hydrology. As already mentioned, in a rift system the ground water circulation is very deep; and the lake is fed by alkaline hot springs devoid of Ca and Mg, the precipitation of which must have occurred earlier during underground circulation (Eugster, 1970).

Most importantly, in a playa-lake complex the shallow ground water is characterized by systematic chemical changes; these changes become pronounced from the point of ground water discharge, across the mud flats, and to the lake at the center of the basin. Generally, the systematic change is from fresh water to saline, alkaline brines. The ground water in a rift system, however, is not characterized by shallow circulation; and when it emerges at the lake margin, it is in the form of concentrated saline, alkaline waters (Eugster, 1970). Thus, the shallow sediments in a playa-lake complex are subjected to ground waters that vary systematically from dilute water to concentrated brines; whereas, sediments in a rift system are subjected only to concentrated brines.

Weathering reactions. As Hardie and Eugster (1970) suggested, the compositional history of closed-basin waters can be separated into two phases: (1) acquisition of solutes by the dilute waters through weathering-type reactions with soil and bedrock (Jones, 1966; Garrels and MacKenzie, 1967) and (2) evaporative concentration which eventually causes the precipitation of very soluble minerals (Hardie and Eugster, 1970; Eugster, 1970). Jones (1966) suggested that the composition of saline, alkaline waters in closed basins is chiefly the result of

* In this context, alkalinity is defined (cf. Stumm and Morgan, 1970, p. 130) as:

$$\text{alkalinity} = \underset{\text{total}}{HCO_3^-} + 2\underset{\text{total}}{CO_3^=} + OH^- - H^+$$

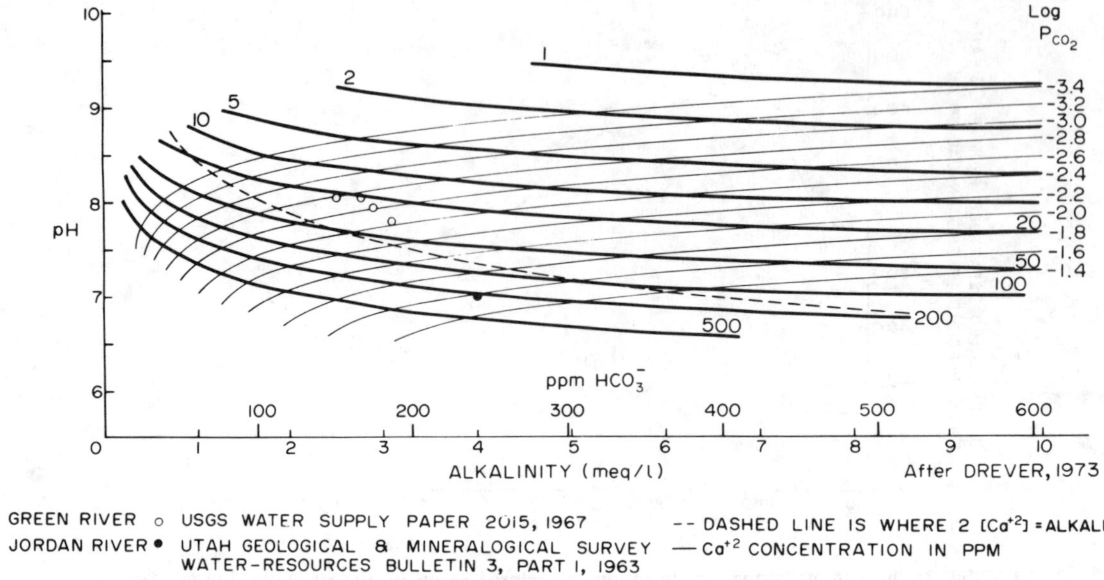

Fig. 3. Graphical display of one of the most significant chemical divides in brine evolution. The dashed line is where Ca^{++} exactly balances alkalinity. Water plotting above dashed line, upon further concentration, will precipitate calcite and evolve into an alkaline brine. Water plotting below the dashed line, upon precipitating calcite, will evolve into a saline but not an alkaline brine (after Drever, 1973).

weathering processes accompanied by simple evaporative concentration and selective mineral precipitation. For example, in the absence of older chemogenic deposits, water compositions are related to weathering reactions, or simply to the hydrolysis of silicates by CO_2-charged waters, as is shown schematically below:

$$\text{primary silicate} + H_2O + CO_2 = \text{clay} + Ca^{+2} + Na^+ + HCO_3^- + SiO_2.$$

The primary silicates react with CO_2-charged water, i.e. soil water; a clay mineral is formed that remains in the soil profile; and cations, HCO_3^-, and SiO_2, are released into solution. Garrels and MacKenzie (1967) have shown that if this type of water is isolated and concentrated by isothermal evaporation in equilibrium with the earth's atmosphere, it produces a highly alkaline $Na\text{-}HCO_3\text{-}CO_3$ water.

Chemical trends. To produce saline, alkaline brine by evaporation, the parent waters are severely restricted in terms of possible compositional variations. The parent waters must be rich in HCO_3^-, the $HCO_3^-/(Ca^{+2} + Mg^{+2})$ molar ratio must be distinctly greater than unity, and other anions such as Cl^- and SO_4^{-2} must be minor (for details see Hardie and Eugster, 1970). This is exactly the type of water derived as a result of the weathering of igneous or metamorphic rocks (Garrels and MacKenzie, 1967). However, the presence of abundant pyritic shales or older evaporites in the drainage basin precludes the development of alkaline brines (Jones, 1966).

During evaporation, the chemical evolution of the dilute waters can be understood in terms of "chemical divides". The most important chemical divide in the evolution of alkaline brines is the early precipitation of alkaline-earth carbonates. This divide, which mainly determines the major compositional trend a water follows, is determined at a very early stage (ionic strength <0.1) in the evaporation history, commonly by calcite precipitation (Hardie and Eugster, 1970). If alkalinity/$2\cdot mCa^{+2}>1$, the solution will be depleted in Ca^{+2} by calcite precipitation and will with further evaporation be enriched in CO_3^{-2}, resulting in a saline, *alkaline* brine, such as Lake Magadi. In contrast, if alkalinity/$2\cdot mCa^{+2}<1$, the solution will be depleted in CO_3^{-2} by calcite precipitation. Further evaporation will cause enrichment in Ca^{+2} and will result in a saline but *not* alkaline lake, such as the Great Salt Lake. Drever (1972) illustrated this point very well (Fig. 3; also Hardie and Eugster, 1970, Figs. 4 and 6). In Fig. 3, the open circles represent the composition of Green River water (alkalinity/$2\cdot mCa^{+2}<1$). During evaporation, calcite will be the first mineral to precipitate, and a CO_3^{-2}-rich water, and ultimately an alkaline brine, will result. The black circle in Fig. 3 represents the composition of the Jordan River which provides 20% of the inflow to the Great Salt Lake (alkalinity/ $2\cdot mCa^{+2}<1$). With evaporation, calcite again will be the first mineral to precipitate; however, a Ca^{+2}-rich water and ultimately a saline brine will result. In Fig. 3, evaporation of any dilute water that plots above the

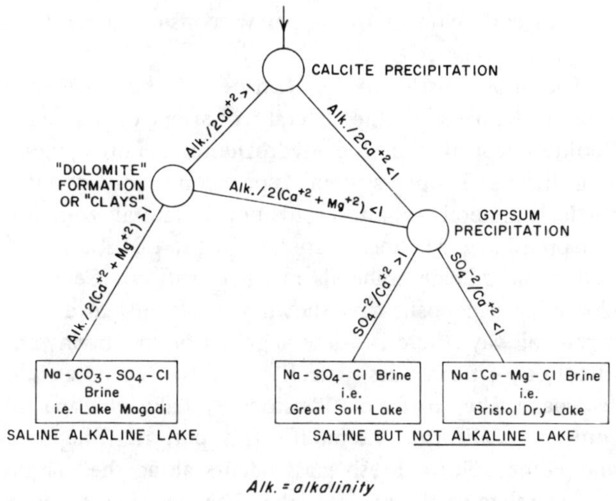

Fig. 4. Flow sheet for evaporative concentration of dilute waters (modified from Hardie and Eugster, 1970).

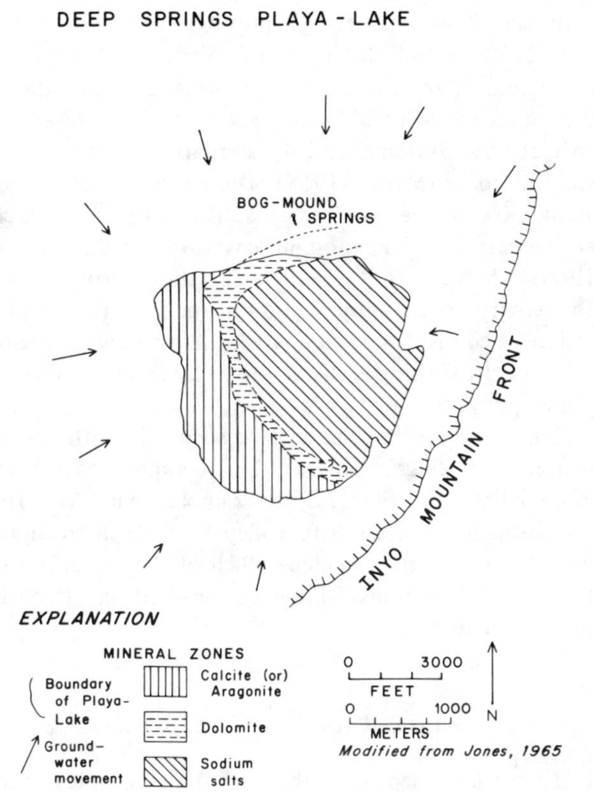

Fig. 5. Distribution of mineral zones at modern Deep Springs playa, California (Jones, 1965). Playa-lake deposits surrounded by marginal silts and sands.

dashed line (alkalinity = $2Ca^{+2}$) will result in a saline, *alkaline* brine, but evaporation of any water that plots below the dashed line will result in a saline but *not* alkaline brine.

Other important chemical divides are summarized in Fig. 4 (modified after Hardie and Eugster, 1970). The ratio of alkalinity/alkaline-earth ions determines the compositional trend early in the evaporation history of the water (ionic strength <0.1) and is controlled by the kinds of rocks exposed to weathering reactions in the hydrographic basin. For alkaline brines to develop, the hydrographic basin should be characterized by igneous or metamorphic rocks. If large quantities of pyritic shales or older evaporites are exposed in the hydrographic basin, the alkalinity/$2(Ca^{+2}Mg^{+2})$ ratio will be less than unity and will preclude the formation of alkaline brines. Thus a closed basin surrounded by igneous and metamorphic rocks, or by sediments other than pyritic shales and evaporites, is an ideal setting to develop saline, alkaline brines and lakes.

As already discussed, in a closed hydrographic basin precipitation of alkaline-earth carbonates profoundly affects the composition of the waters and hence the sequence and nature of the saline minerals formed in the center of the basin (Eugster and Hardie, 1975). The result is the formation of a mineral zonation with the most soluble minerals occupying the center of the basin and segregated from the less soluble phases (Hunt, 1960; Jones, 1965; Hardie, 1968). Jones (1965) and Hardie (1968), among others, have documented this type of mineral zonation at Deep Springs Valley and Saline Valley in eastern California (Fig. 5).

Efflorescent crusts. Another very important process in the formation of concentrated brines in closed basins is the recycling of efflorescent crusts (Hardie, 1968; Eugster and Hardie, 1975; Jones *et al.*, 1977). The preferential enrichment of NaCl in brines associated with playa lakes can readily be explained by recycling of efflorescent crusts. The brine enrichment in NaCl is the result of occasional rains that dissolve the most soluble material from the efflorescent crusts covering the playa surface, and subsequent flood water or ground water transports the solutes to the adjacent lacustrine environment. Hardie (1968) has shown that efflorescent crusts on modern playa surfaces are typically halite because the crusts are products of complete dehydration of brine drawn to the surface by capillary action. Recently, it has been recognized that re-solution of efflorescent crusts is a significant method of increasing the salinity of ground water and lacustrine water in closed basins (Jones *et al.*, 1976).

Volcanism

Volcanism is another significant geological consequence of block faulting or rifting. For example, both the Basin and Range province of the western United States and the East Rift Valley of Kenya are characterized by volcanic rocks. According to data of the Nevada Bureau of Mines, about 78% of Nevada is covered by Tertiary and Quaternary sediments and volcanic rocks (Maxey, 1968). The volcanic rocks cover about 30% of the state; whereas the sediments, which are in part of volcanic origin, cover 48%. Baker (1958; 1963) showed that nearly all the Tertiary and Quaternary stratigraphic section in the East Rift Valley of Kenya consists of volcanic rocks. A tectonic setting of block or rift faulting is characterized by an abundance of glassy, pyroclastic material.

The pyroclastic material associated with these tectonic settings ranges in composition from intermediate to silicic. For example, much of the volcanism in the East Rift Valley resulted in trachytic pyroclastic material, whereas much of the volcanism in the Basin and Range Province resulted in rhyolitic pyroclastic material.

MINERAL PATTERNS

Lacustrine deposits that contain zeolites also commonly show the following geological features: (1) fossil assemblages indicative of lacustrine environments, i.e. "fresh"-water diatoms, ostracodes, turtles, and fish; (2) laminated mudstone that interfingers laterally with fluviatile sandstone and conglomerate containing vertebrate fossils; (3) non-marine evaporites; (4) "chemical deltas" and tufa or spring deposits; and (5) strand-line deposits such as algal stromatolites, ostracode lag deposits, ooliths and pisoliths, flat-pebble conglomerates, and beach sands.

Several of these features strongly suggest that the lacustrine deposits characterized by zeolites were, at least in part, saline and alkaline. The most obvious indication is the occurrence of saline minerals. Some deposits contain bedded saline minerals, such as trona, nahcolite, and halite; whereas others contain disseminated crystal molds of gaylussite or pirrsonite. Dolomitic mudstone, which typifies many of the deposits, is a further indication of saline, alkaline conditions (Jones, 1965). The distribution of fossils also indicates abnormal salinity. Fresh-water diatoms and ostracodes are commonly found in sediments near inlets, but away from these areas there is an absence of fossils.

Indirect evidence of alkalinity is the abundant calcite in deposits at the lake margin. The calcite is best explained as precipitates formed where inflowing calcium-bearing "fresh" water mixed with alkaline water. This phenomenon can be observed at modern Mono Lake in California and Pyramid Lake in Nevada. Smith (1966) described similar chemical deltas of calcium carbonate in the upper Wisconsin sediments at the inlet of Searles Lake.

The most distinguishing feature of the saline-lake zeolite deposits is the lateral zonation of minerals. Zeolite deposits of the hydrothermal, burial metamorphic, and open-system types commonly show a vertical mineral zonation but not a lateral zonation (Sheppard, 1975). Recent studies on the distribution of authigenic silicate minerals in silicic tuffs of Cenozoic closed-basin deposits have shown a consistent pattern. In a general way, there is a lateral gradation in a basinward direction of fresh glass to zeolites and then to potassium feldspar. The tuffs of Pleistocene Lake Tecopa in southern California exemplify this pattern (Sheppard and Gude, 1968). Fresh glass occurs along the margin and at inlets of the ancient lake. The glass is succeeded inwardly by a zone of zeolites and then, in the central part of the lake basin, by potassium feldspar. The zeolites in the Lake Tecopa deposit are chiefly phillipsite, erionite, and clinoptilolite.

A variation in the distribution pattern of authigenic silicate minerals in tuffs of closed-basin deposits is illustrated by the Big Sandy Formation of northwestern Arizona (Sheppard and Gude, 1973). The Big Sandy Formation consists chiefly of lacustrine strata that were deposited in the valley of the Big Sandy River near Wikieup. A nonanalcimic zeolite facies occurs along the margin of the ancient lake deposit and is succeeded basinward by an analcime facies and then by a potassium-feldspar facies. Zeolites in the nonanalcimic zeolite facies are chiefly chabazite, clinoptilolite, erionite, and phillipsite. A similar distribution pattern has been recognized in the Miocene Barstow Formation of southern California (Sheppard and Gude, 1969) and in the Eocene Green River Formation of Wyoming (Surdam and Parker, 1972).

In addition to zeolites, potassium feldspar, clay minerals, and alkaline-earth carbonates, saline, alkaline-lake deposits locally contain opal or chalcedony, quartz, searlesite ($NaBSi_2O_6 \cdot H_2O$), fluorite (CaF_2), or dawsonite ($NaAl(CO_3)(OH)_2$) of authigenic origin. Saline minerals commonly associated with these deposits are trona ($Na_2CO_3 \cdot NaHCO_3 \cdot 2H_2O$), nahcolite ($NaHCO_3$), pirssonite ($Na_2CO_3 \cdot CaCO_3 \cdot 2H_2O$), gaylussite ($Na_2CO_3 \cdot CaCO_3 \cdot 5H_2O$), and halite ($NaCl$).

Pleistocene Lake Tecopa deposits

Three diagenetic mineral facies are recognized in the tuffs of the Lake Tecopa deposits (Fig. 6). Tuffs nearest the lake margin are characterized by fresh glass and are termed the "fresh-glass facies". Tuffs in the central part of the lake basin are characterized by potassium feldspar and (or) searlesite and are termed the "potassium

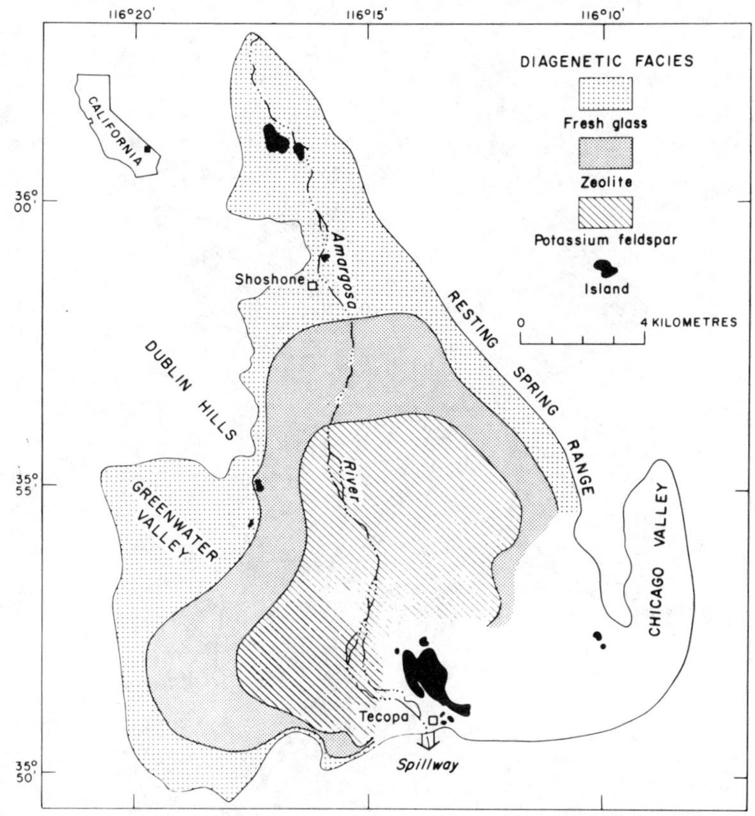

Fig. 6. Map of ancient Lake Tecopa near Shoshone, California, showing diagenetic facies for a composite of all the tuff beds in the deposit (after Sheppard and Gude, 1968).

feldspar facies". Those tuffs between the fresh-glass facies and the potassium-feldspar facies, and characterized by zeolite, are termed the "zeolite facies".

The boundaries between the facies are laterally gradational and difficult to recognize in the field. Tuffs of the fresh-glass facies generally can be distinguished from those of the zeolite facies in the field, but tuffs of the zeolite facies almost never can be distinguished from those of the potassium feldspar facies (Sheppard and Gude, 1968). X-ray diffractometer powder data of bulk samples, coupled with thin section study, are considered essential for positive identification and placement in the proper facies.

Figure 6 shows the distribution of the mineral facies in the Pleistocene Lake Tecopa deposits (Sheppard and Gude, 1968). It is particularly important to note that the pattern shown in Fig. 6 is in single tuff beds. The boundary between the fresh-glass facies and zeolite facies was determined by the first appearance of zeolite. The boundary between the zeolite facies and the potassium feldspar facies was placed at the first appearance of potassium feldspar or searlesite. The distribution of potassium feldspar and searlesite is similar throughout the basin; commonly the two minerals are associated. In a general way, the mineral facies boundary parallels the shape of the lake basin in plan (see Fig. 6). The zeolite facies is generally 0.8–1.6 km wide except near the Amargosa River inlet at the north end of the basin, where the facies is broadened to 2.0–2.8 km. This latter feature suggests that the chemical depositional environment affected, if not controlled, the distribution of the mineral facies.

Pliocene Big Sandy Formation

Three diagenetic mineral facies are recognized in the tuffs of the Pliocene Big Sandy Formation in Arizona (Fig. 7). Tuffs nearest the margin of the formation are characterized by zeolites other than analcime and are termed the "nonanalcimic zeolitic facies" (Sheppard and Gude, 1973). Tuffs in the central part of the lake basin are characterized by potassium feldspar and are termed the "potassium feldspar facies". Those tuffs between the

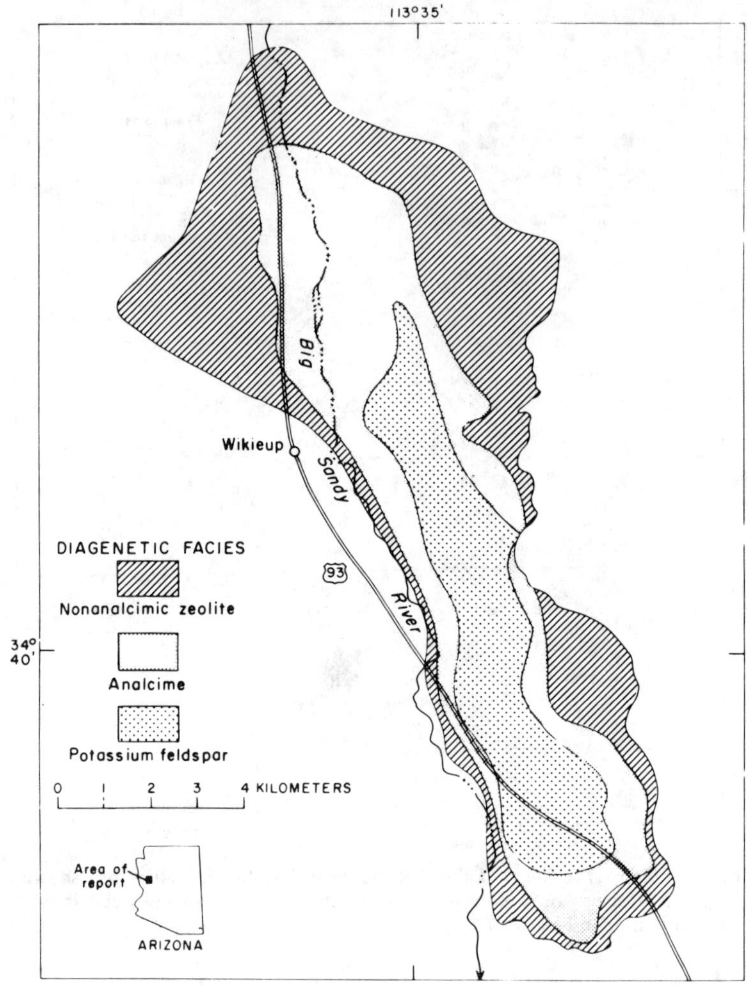

Fig. 7. Map of the Big Sandy Formation near Wikieup, Arizona, showing diagenetic facies for a composite of all the tuffaceous rocks (after Sheppard and Gude, 1973).

nonanalcimic zeolite facies and the potassium feldspar facies are characterized by analcime and are called the "analcime facies" (Sheppard and Gude, 1973). No relict fresh glass was recognized in any of the tuffaceous rocks of the Big Sandy Formation (Sheppard and Gude, 1973).

The boundaries between the facies are laterally gradational and difficult to recognize in the field. Figure 7 shows the diagenetic mineral facies for a composite of all the tuffaceous rocks in the Big Sandy Formation. This is the same pattern that is observed in individual tuff beds (Sheppard and Gude, 1973). The boundary between the nonanalcimic zeolite facies and the analcime facies was placed at the first appearance of analcime, and the boundary between the analcime facies and the potassium feldspar facies was placed at the first appearance of authigenic potassium feldspar (Sheppard and Gude, 1973).

The mineral facies in plan are elongated parallel to the depositional basin. The nonanalcimic zeolite facies and the analcime facies are broadest at the northern part of the basin which was the major inlet of the ancient lake. The nonanalcimic zeolite facies is also broad and extends basinward near the other inlets. In a general way, the nonanalcimic zeolite facies and the analcime facies narrow where the basin narrows (Fig. 7). About 5 km southeast of Wikieup, where the basin was narrowest, both the nonanalcimic zeolite and analcime facies are absent along the eastern margin (Fig. 7). Again, these features suggest that the chemical depositional environment affected, if not controlled, the distribution of the diagenetic mineral facies.

Miocene Barstow Formation and Eocene Green River Formation

Two other typical examples of lateral zonation of diagenetic mineral facies in zeolitic closed-basin deposits are the Miocene Barstow Formation in California (Sheppard and Gude, 1969) and the Eocene Green River Formation in Wyoming (Surdam and Parker, 1972). Both have a zone of analcime separating the other zeolites from the zone of potassium feldspar. Minor relict fresh glass occurs in the Barstow Formation, but none has been found in the Green River Formation. The zeolite facies of the Green River Formation grades laterally toward the basin margin into a clay facies (Surdam and Parker, 1972).

The sediments of Lake Tecopa, the Big Sandy Formation, the Barstow Formation, and the Green River Formation range in age from Pleistocene to Eocene and have several features in common: (1) they were subjected to *very shallow* burial and show only slight deformation; (2) exposures are good, and tuffs can be traced throughout most of the basins; (3) they all show an abundance and variety of authigenic silicate minerals in the tuffs; and (4) they show a pronounced lateral zonation of diagenetic minerals, summarized in Fig. 8.

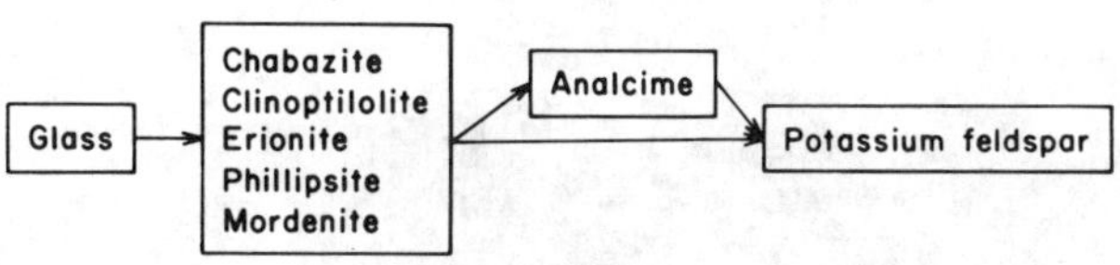

Fig. 8. Schematic reaction sequence that leads to lateral mineral zonation that characterizes many saline, alkaline-lake deposits (after Sheppard and Gude, 1973).

INTERPRETATION OF MINERAL PATTERNS

It is noteworthy that the boundaries of the above-mentioned mineral zones can be described by the following reactions:

1. glass → alkalic, silicic zeolites,
2. alkalic, silicic zeolites → analcime,
3. analcime → K-feldspar,
4. alkalic, silicic zeolites → K-feldspar.

A determination of the chemical parameters that significantly affect these reactions is important to an understanding of the genesis of the diagenetic mineral zones in saline, alkaline-lake deposits.

Formation of alkalic, silicic zeolites from silicic glass

Evidence supporting the reaction of glass to zeolites is unequivocal (Fig. 9). Not only is there a reciprocal relation between glass and zeolites in vitric tuffs containing zeolites, but commonly the zeolites are pseudomorphous after glass shards (Fig. 10). Deffeyes (1959) emphasized that zeolites form during diagenesis by solution of the shards and subsequent precipitation of zeolites from the solution, rather than by devitrification of the shards. Surdam and Eugster (1976) demonstrated that in the Lake Magadi region of Kenya erionite can form directly from trachytic glass by the addition of only H_2O. They also emphasized that the water in contact with the glass during the reaction was an alkaline brine. Hay (1966) showed that alkaline brines are ideal agents for the hydration and solution of volcanic glass. Mariner and Surdam (1970) suggested that an aluminosilicate gel may first precipitate from the solution, and then zeolites grow from the gel. A gel was recently recognized in Holocene tuffs at Teels Marsh, Nevada, associated with phillipsite and rhyolitic glass (Surdam and Mariner, 1971).

An individual tuff bed will commonly contain several zeolites that differ significantly in composition; this suggests that either initial compositional inhomogeneities existed, or more probable, that compositional parameters changed during the hydration and solution of the glass. Table 1 suggests that the important compositional parameters in zeolite genesis are cation ratios, Si/Al ratios, and H_2O activity. All of these parameters will be greatly affected by changes in salinity and/or alkalinity. For example, experimental work by Mariner and Surdam (1970) indicated that the Si/Al ratio of zeolites formed by the hydration and solution of glass is related to the pH of the solution. Hay (1966) showed that the rate of solution of silicic glass increases with increasing salinity and alkalinity. The activity of H_2O is related directly to salinity, ranging from 1.0 in distilled water to about 0.7 in the most concentrated modern saline, alkaline brines. The affect of alkalinity on the activity of the alkaline earths was previously demonstrated, i.e. the activity of the alkaline earths decreases rapidly with increasing alkalinity. Thus, the important chemical parameters during the glass-to-zeolite reaction are cation ratios, Si/Al ratios, and activity of H_2O, whereas the most significant parameters affecting the solution of glass are salinity and pH.

Reaction of alkalic, silicic zeolites to form analcime

Analcime is a very common constituent of saline, alkaline-lake deposits (Hay, 1966; Sheppard and Gude,

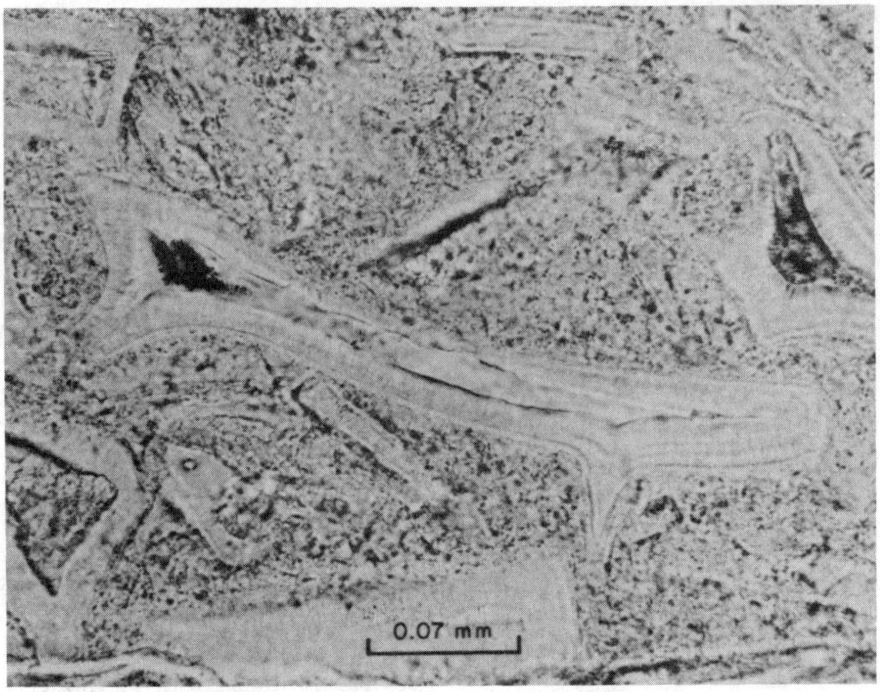

Fig. 9. Scanning electron micrograph showing glass fragment altered to acicular erionite in an ash bed from Jersey Valley, Nevada. Scanning electron micrograph from Mumpton, 1973.

Fig. 10. Photomicrograph of zeolitic tuff, showing pseudomorphs of acicular clinoptilolite after shards. The bands in the clinoptilolite are due to variations in the index of refraction along the length of the fibers. The large pseudomorph in center of the photograph has a hollow interior (after Sheppard and Gude, 1959).

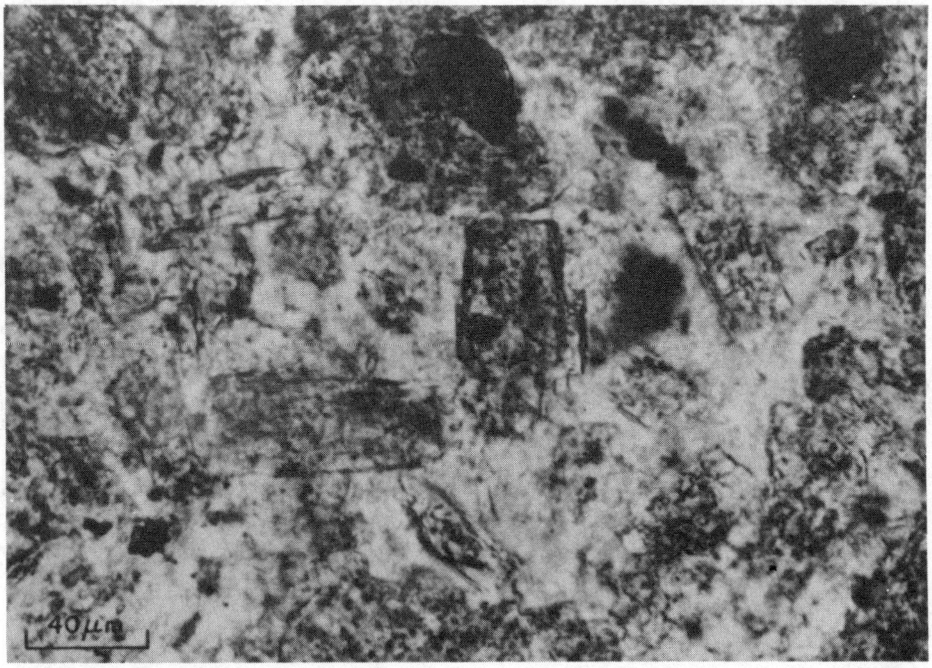

Fig. 11. Analcime pseudomorphs after prismatic clinoptilolite. Light areas are quartz (after Sheppard and Gude, 1973).

1969, 1973; Hay, 1970; Surdam and Parker, 1972; Surdam and Eugster, 1976). Data gathered from tuffaceous sediments strongly support the conclusion that analcime *does not* form directly from the alteration of glass because the association of analcime with glass has not been reported. Analcime is, however, commonly associated with a variety of alkalic, silicic zeolites (Fig. 11). Sheppard and Gude (1969, 1973) described petrographic evidence showing that clinoptilolite and phillipsite had altered to analcime, and Surdam and Eugster (1976) documented the reaction of erionite to analcime. Thus, analcime in tuffs of saline, alkaline-lake deposits formed from precursor zeolites and not directly from silicic glass (Hay, 1966).

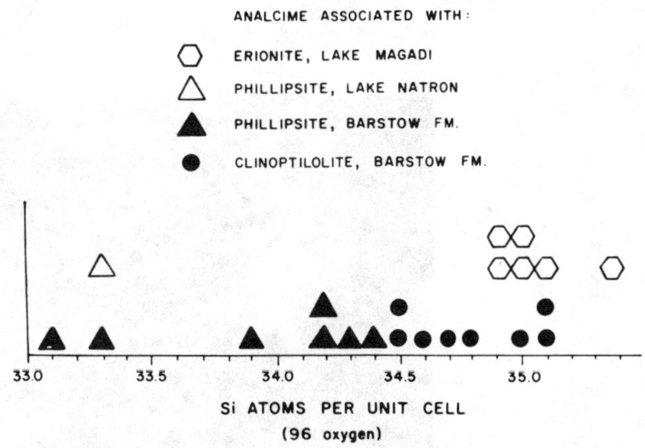

Fig. 12. Plot showing correlation between the silicon content of analcime and the precursor zeolite. Composition of analcime determined from X-ray diffractometer data by measurement of the displacement of the (639) peak of analcime. Data from Barstow Formation are after Sheppard and Gude (1969) and data from Lake Magadi are after Surdam and Eugster (1976).

Sheppard and Gude (1969) studying the diagenesis of tuffs in the Barstow Formation, and Surdam and Eugster (1976) studying the mineral reactions in tuffaceous rocks of the Lake Magadi region, suggested a correlation between the composition of the zeolite precursor and the Si/Al ratio of analcime. They observed (Fig. 12) that analcime associated with and derived from clinoptilolite and erionite has a relatively high number of silicon atoms per unit cell (34.5 to 35.3), whereas analcime associated with and derived from phillipsite (a silica-deficient zeolite) has a low number of silicon atoms per unit cell (33.1 to 34.4).

Boles (1971) studied the experimental reaction of clinoptilolite and heulandite to analcime and showed that the Si/Al ratio of the analcime product is chiefly a function of the Si/Al ratio of the zeolite reactant. From both field observations and laboratory experiments, there is a strong support for the reaction of zeolite to analcime; the analcime inherits its Si/Al ratio from the zeolite precursor.

The reaction of silicic, alkalic zeolites to analcime can

be illustrated by the following equations and equilibrium constants:

$$\begin{array}{ll}\text{clinoptilolite} & \text{analcime}\\ Na_2K_2CaAl_6Si_{30}O_{72}\cdot 24H_2O + 4Na^+ \rightarrow & 6NaAlSi_2O_6\cdot H_2O\end{array}$$
$$+ 18H_2O + 18SiO_2 + 2K^+ + Ca^{+2}$$

$$K_1 = \frac{(A_{H_2O})^{18}(A_{SiO_2})^{18}(A_{K^+})^2(A_{Ca^+})}{(A_{Na^+})^4} \quad (1)$$

$$\begin{array}{ll}\text{phillipsite} & \text{analcime}\\ Na_2K_2Al_4Si_{12}O_{32}\cdot 12H_2O + 2Na^+ \rightarrow & 4NaAlSl_2O_6\cdot H_2O +\end{array}$$
$$8H_2O + 4SiO_2 + 2K^+$$

$$K_2 = \frac{(A_{H_2O})^8(A_{SiO_2})^4(A_{K^+})^2}{(A_{Na^+})^2} \quad (2)$$

$$\begin{array}{ll}\text{erionite} & \text{analcime}\\ NaKAl_2Si_7O_{18}\cdot 6H_2O + Na^+ \rightarrow & 2NaAlSi_2O_6\cdot H_2O +\end{array}$$
$$4H_2O + 3SiO_2 + K^+$$

$$K_3 = \frac{(A_{H_2O})^4\,(A_{SiO_2})^3\,A_{K^+}}{A_{Na^+}} \quad (3)$$

The equations and equilibrium constants show that the activities of the cations, silica, and water will control the reactions. The major changes are a gain of sodium and losses of K^+, Ca^{+2}, SiO_2, and H_2O. Furthermore, experimental work by others, and theoretical considerations, indicate that the reaction of an alkalic, silicic zeolite to form analcime is favored by a high Na^+/H^+ ratio (Hess, 1966; Boles, 1971), a relatively low activity of SiO_2 (Coombs *et al.*, 1959), or a relatively low Si/Al ratio (Senderov, 1963), and a relatively low activity of H_2O in the pore fluid. Increased salinity and alkalinity favors the formation of analcime because an increase of these parameters decreases the activity of H_2O and the Si/Al ratio, generally affects the cation ratios, and specifically increases the Na^+/H^+ ratio. The above arguments are based on the assumption that equilibrium chemical factors alone are responsible for the formation of analcime; however, kinetic factors may be equally important.

Reaction of zeolites to form potassium feldspar

Formation of potassium feldspar from zeolite precursors in tuffs that were never deeply buried has been well documented in recent studies (Fig. 13).

Fig. 13. Scanning electron micrograph (from Mumpton and Ormsby, 1976) of large analcime crystals in a matrix of micrometer-sized crystals of potassium feldspar from a tuff near Barstow, California.

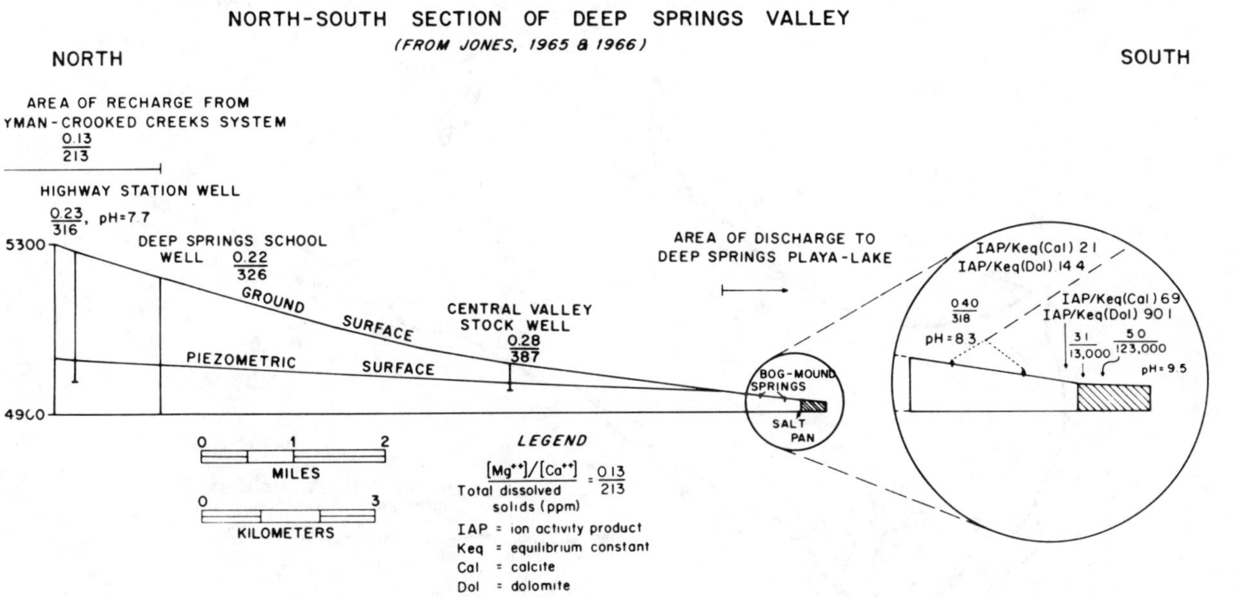

Fig. 14. Section showing hydrochemistry of a modern playa-lake complex, Deep Springs Valley, California. Note evolution of brines from north to south as they migrate from alluvial deposits to playa. Diagram constructed from Jones (1965, 1966).

Analcime is replaced by potassium feldspar in tuffs of the Eocene Green River Formation, Wyoming (Iijima and Hay, 1968; Surdam and Parker, 1972). Analcime and clinoptilolite are replaced by potassium feldspar in tuffs of the Miocene Barstow Formation, California (Sheppard and Gude, 1969). Potassium feldspar replaced analcime, clinoptilolite, and phillipsite in the Pliocene Big Sandy Formation, Arizona (Sheppard and Gude, 1973). Phillipsite is replaced by potassium feldspar in tuffs of Pleistocene Lake Tecopa, California (Sheppard and Gude, 1968). Potassium feldspar has *never* been reported to occur with glass; thereby precluding the possibility that it could form as a direct product of glass alteration. The formation of potassium feldspar in these examples has been correlated with a highly saline and alkaline depositional environment; the potassium feldspar is commonly associated with bedded saline minerals.

The reaction of zeolites to potassium feldspar can be illustrated by the following equations and equilibrium constants:

$$\underset{\text{analcime}}{NaAlSi_2O_6\cdot H_2O} + SiO_2 + K^+ \rightarrow \underset{\text{K-feldspar}}{KAlSi_3O_8} + H_2O + Na^+$$

$$K_4 = \frac{A_{H_2O}\,A_{Na^+}}{A_{SiO_2}\,A_{K^+}} \qquad (4)$$

$$\underset{\text{phillipsite}}{Na_2K_2Al_4Si_{12}O_{32}\cdot 12H_2O} + 2K^+ \rightarrow \underset{\text{K-feldspar}}{4KAlSi_3O_8} + 12H_2O + 2Na^+$$

$$K_5 = \frac{(A_{H_2O})^{12}(A_{Na^+})^2}{(A_{K^+})^2} \qquad (5)$$

$$\underset{\text{clinoptilolite}}{Na_2K_2CaAl_6Si_{30}O_{72}\cdot 24H_2O} + 4K^+ \rightarrow \underset{\text{K-feldspar}}{6KASi_3O_8} + 24H_2O + 12SiO_2 + 2Na^+ + Ca^{+2}$$

$$K_6 = \frac{(A_{H_2O})^{24}(A_{SiO_2})^{12}(A_{Na^+})^2\,A_{Ca^{+2}}}{(A_{K^+})^4} \qquad (6)$$

These equations and equilibrium constants show that the activities of the cations, silica, and water control the reactions. Specifically, the major changes resulting from the reactions are a gain of potassium and losses of sodium and water. Depending on the composition of the zeolite, SiO_2 for the reaction may be in excess or deficient in the zeolite precursor.

Saline and alkaline pore water trapped in the tuffs during deposition, or which evolved during sedimentation and burial, are significant factors responsible for the formation of potassium feldspar. Eugster (1970) studied the evolution of saline, alkaline brines and showed that the K^+ concentration is highest in the most saline and alkaline brines. Coupled with the high pH, the high potassium concentration would result

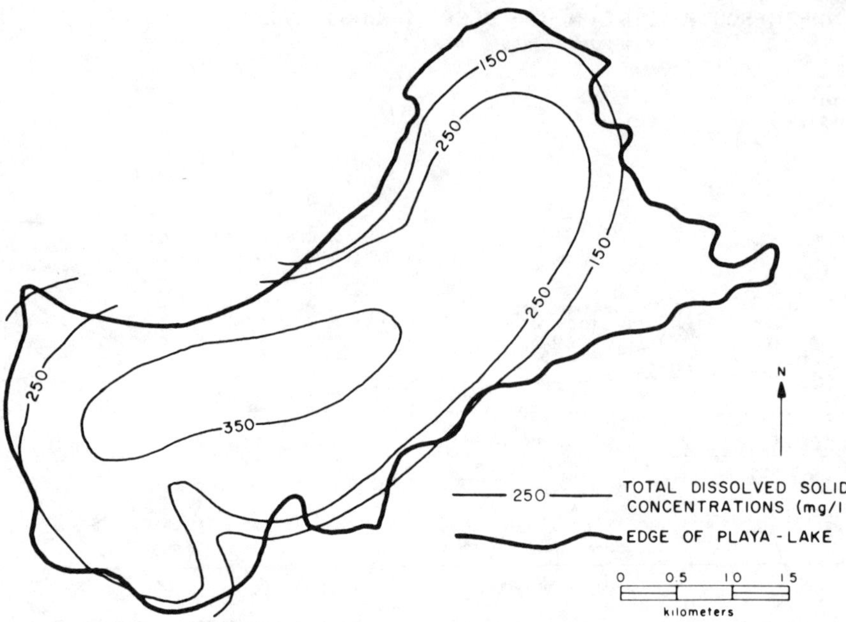

Fig. 15. Distribution of total dissolved solids in the interstitial brines of the playa-surface at Teels Marsh, Nevada. Numbers are concentrations (mg/l × 10^3) (after Smith, 1974).

in a relatively high K^+/H^+ ratio suitable for the crystallization of potassium feldspar. The high pH would also result in increased solubility of SiO_2 and would contribute to the stabilization of K-feldspar relative to analcime. A relatively high salinity would lower the activity of H_2O and favor the formation of anhydrous potassium feldspar from hydrous zeolites, including analcime.

Kinetic factors may be important in the zeolite to potassium feldspar reaction. Recently, a synthesis of potassium feldspar was accomplished from natural clinoptilolite in a saturated solution of KOH at 80°C after 44 hours (Sheppard and Gude, 1969). Although this experiment took place, it did not duplicate the chemical environment prevalent during the diagenesis of saline, alkaline-lake deposits; it does demonstrate the rapidity of the reaction of a zeolite to potassium feldspar in a favorable environment.

Hydrochemistry

The above discussion demonstrates that the chemistry of the pore water is a controlling factor in the mineral reactions that characterize saline, alkaline-lake deposits. Furthermore, a general trend toward increasing salinity and alkalinity favors the glass to zeolite to potassium feldspar reaction sequence.

The relation of brine evolution in closed hydrographic basins to the mineral patterns in saline, alkaline-lake deposits is direct. Brine evolution is gradual and results in systematic chemical gradients. Consequently, in the closed hydrology, the gradients are reflected in centripetal mineral patterns (Fig. 5). Figure 14 is a north-south cross-section through Deep Springs Valley, California, and summarizes the chemical evolution that takes place in a typical closed basin. It should also be noted that from the point of recharge to the saline mineral pan the Na^+ varies from about 12,000–100,000 ppm and that K^+ varies from about 5000–18,000 ppm (Table 3). The Ca^{+2} concentration, however, varies from about 40–3 ppm. Eugster (1970) showed that from the point of recharge to the interstitial brines in the saline mineral pan at Lake Magadi, the SiO_2 concentration varies from 25–1000 ppm or greater. In a closed basin, the concentration factor from dilute inflow in the recharge area to interstitial brines in the discharge area can be as high as 5000. Salinity gradients across the playa lake are also common (Fig. 15).

A probable solution to the problem of diagenetic silicate mineral patterns in tuffaceous rocks in closed basins is suggested by placing a tuff in the context of the hydrochemistry already discussed (Fig. 14). A significant fact is that the tuff will act as an aquifer because it is interbedded with relatively impermeable mudrock. As long as the tuff is above the piezometric surface and

TABLE 3. Average composition of dilute inflow water and concentrated brine at Deep Springs Valley,* California, and Lake Magadi,† Kenya (in ppm)

	Dilute inflow							
	pH	SiO_2	Na	K	HCO_3	CO_3	SO_4	Cl
Deep Springs	7.8	22	12	4	223	–	50	4
Lake Magadi	7.7	38	25	6	85	–	–	10
	Interstitial brines							
Deep Springs	9.6	301	99,650	17,550	9000	18,950	47,350	111,000
Lake Magadi	10.3	661	83,109	1167	8876	64,100	1925	49,772

Deep Springs interstitial brines average 300,000 total dissolved solids
Lake Magadi interstitial brines average 210,000 total dissolved solids

*From Jones (1965).
†From Eugster (1970).

interbedded with permeable alluvial deposits, it will undergo very little alteration. However, once the ash bed intersects the piezometric surface, it is susceptible to hydration and alteration. This alteration will be most pronounced where the ash bed enters the zone of capillary draw, evaporative pumping, and recycling of efflorescent crusts. At this position salinity and alkalinity of the pore water will increase significantly. This process of concentrating the pore water becomes even more pronounced across the playa from the point of ground-water discharge toward the lacustrine environment. The ash bed will be subjected to consistently high salinity and alkalinity in the saline lacustrine environment, because there the effects of evaporation are most pronounced.

Figure 16 is our interpretation of how the hydrochemistry and authigenic mineral patterns of saline, alkaline-lake deposits are related. In the alluvial zone above the water table, the ash will remain relatively unaltered. Near the point of ground-water discharge, particularly in the playa zone, pore water in the ash bed is subjected to capillary draw, and the alteration begins by hydration and solution of glass and eventually the formation of alkalic, silicic zeolites. Further concentration of the pore water due to evaporation and recycling of efflorescent crusts occurs near the interface of the playa and the saline, lacustrine environment, and the alkalic, silicic zeolites may react to analcime. In the saline, lacustrine setting, where evaporative concentration is most intense, the salinity and alkalinity

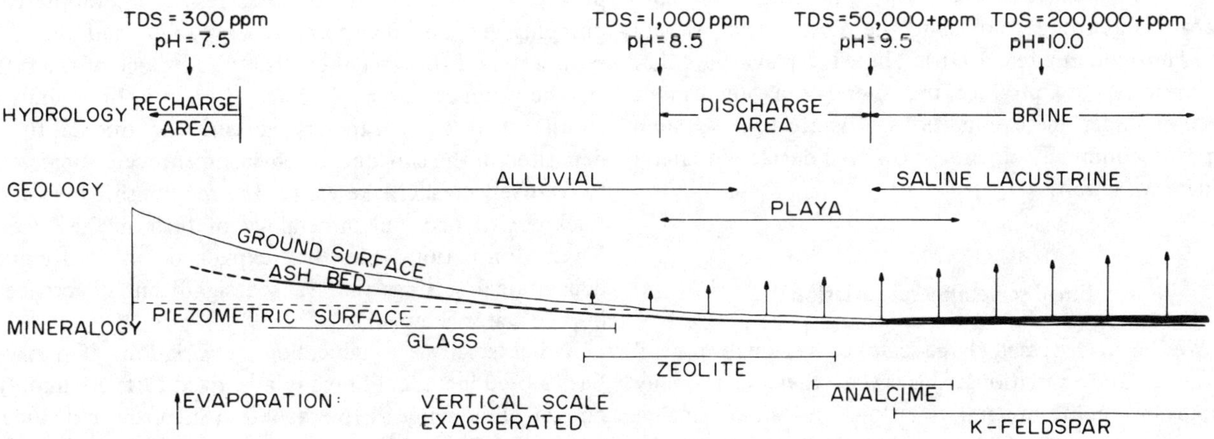

Fig. 16. Schematic diagram of the relationships among hydrology, geology, and authigenic mineral zones in a playa-lake complex.

are highest, and the alkalic, silicic zeolites or analcime may react to potassium feldspar. We do not imply that all the mineral reactions take place at the surface because many reactions take place only after burial. However, Fig. 16 demonstrates the development of centripetal salinity and alkalinity gradients in the pore water of a closed basin and also relates the subsequent authigenic silicate minerals to the hydrochemistry.

We emphasize that this setting is a dynamic one; and the position of lithologic, hydrologic, and mineralogic boundaries will vary with time, mainly as a function of climatic parameters. The dynamic nature of the hydrochemistry in a closed basin explains the fact that almost all the boundaries between authigenic mineral zones are gradational.

Hay (1966) has observed in the field, and Mariner (1971) has shown experimentally, that silicate mineral reaction rates increase with increasing salinity and alkalinity. These data are important to our interpretation of the mineral zones and patterns in saline, alkaline-lake deposits. They suggest that kinetic factors amplify the silicate mineral zones resulting from the centripetal salinity and alkalinity gradients. On the outer fringe of the playa-alluvial environment, where pore waters have not been greatly concentrated, ash beds suffer only incipient zeolitization. However, that same ash deposited in the saline, alkaline lake containing concentrated pore water undergoes a sequence of authigenic reactions culminating with the formation of the potassium feldspar.

Not all authigenic silicate minerals in saline, alkaline-lake deposits are characterized by pronounced centripetal patterns. The Pleistocene High Magadi Beds of Kenya, for example, have no lateral variations in the distribution of authigenic minerals. This is because the lake, in which these sediments were deposited, was in a narrow fault trough, similar to the present Lake Magadi (Surdam and Eugster, 1976). The wide playa flats that are necessary to produce the lateral zonation cannot develop under these conditions. Therefore, rift-system deposits commonly do not show well-developed lateral mineral variations.

Small-scale mineral variations

We have discussed large-scale or regional mineral patterns and variations; however, there are many important zeolite mineral variations on a much smaller scale. For example, Sheppard and Gude (1969) in the Barstow Formation showed a relationship between authigenic mineralogy and thickness of tuff beds, demonstrating that thin tuffs generally consist of analcime and/or potassium feldspar, whereas thick tuffs consist of zeolites other than analcime. Analcime and (or) potassium feldspar occur in 98% of the tuffs in the Barstow Formation that are 6 inches or less thick, whereas 94% of those zeolite tuffs that lack analcime or potassium feldspar are more than 6 inches thick.

A similar relationship between the authigenic mineralogy and thickness of tuffs was recognized by Hay (1966, p. 43) at Olduvai Gorge, Tanzania. Pleistocene tuffs less than 1 inch thick are chiefly potassium feldspar, but thicker tuffs contain phillipsite and potassium feldspar. The tuffs were originally similar in composition and were deposited in a saline, alkaline-lake environment.

Recently a similar relationship was recognized in the tuffs of the Green River Formation in Wyoming. In the Laney Member of the Green River Formation, the thin tuffs (<2 inches) consist of analcime, but the thicker tuffs (>4 inches) consist of clinoptilolite and/or mordenite. Both types of tuffs are interbedded with oil shale. Other lithologies associated with the oil shale, such as algal stromatolites, pisoliths, ooliths, flat-pebble conglomerates, and various lag deposits, suggest that the Laney Member was deposited in a very shallow playa-lake complex characterized by frequent and wide shifts in shoreline position. The sedimentary structures associated with each type of tuff are significantly different. The thin tuffs are characterized by mudcracks and are well sorted, but the thicker tuffs have a flat, sharp lower contact, are commonly graded, and have a gradational upper contact. The thin tuffs were deposited on a mudflat and may have been transported by aeolian processes, but the thick tuffs were deposited in a shallow, subaqueous environment. As a result, the two types of tuffs have been subjected to different hydrologic and hydrochemical conditions as summarized in Fig. 17. The thin tuffs, as a result of evaporative pumping, have been exposed to very saline and alkaline pore waters. Efflorescent crusts at the surface of the tuff are the product of complete dehydration of brine drawn to the surface by capillary action. The thicker tuffs deposited in the subaqueous environment were subjected to relatively fresher lake water. The relationship between thickness of bed and mineralogy of tuffs in the Green River Formation can be explained by different depositional settings that caused significant differences in pore water chemistry.

Other small-scale mineralogical variations that have been noted include: (1) many zeolitized tuffs commonly consist of as many as four zeolites (Sheppard and Gude, 1968, 1969, 1973), and (2) some thick, relatively monomineralic tuff beds have a thin rind of another zeolite at the top and bottom of the bed. These mineralogical variations can be explained by differences in the composition of the pore fluid, probably related to

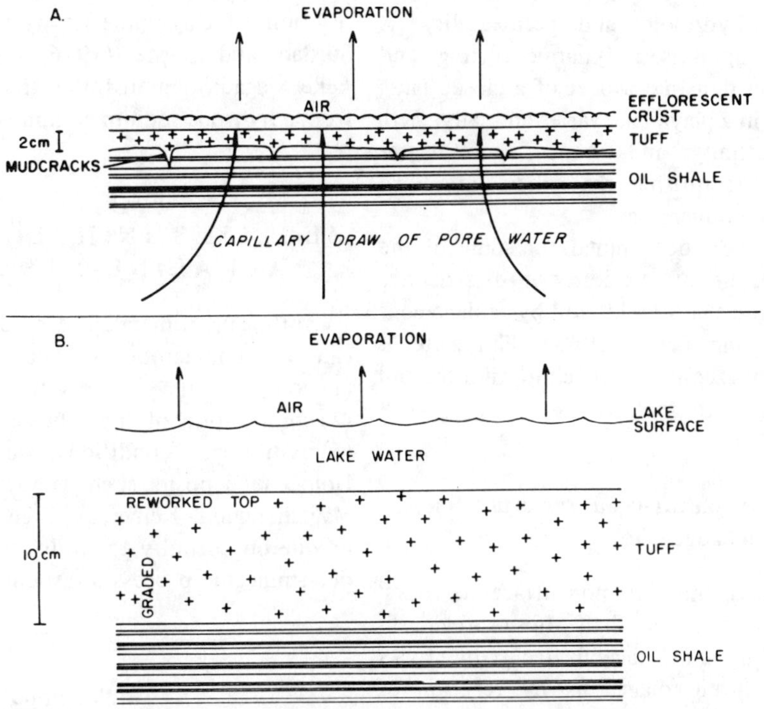

Fig. 17. Diagrammatic representation of two significantly different hydrologic settings in a playa-lake complex. The thin ash bed (top) has been deposited on the playa; and even though ground water in this setting is not as concentrated as the lake brine, the pore water can be greatly concentrated at or very near the surface by evaporative pumping. The thicker ash bed (bottom) has been deposited in the lake where the water is consistently more concentrated than the ground water in the playa. Evaporative pumping will not function until the lake water is evaporated to dryness. If *alteration* proceeds near the surface, the thin ash bed can develop a mineral assemblage indicative of higher salinities and alkalinities than those in the thicker ash bed.

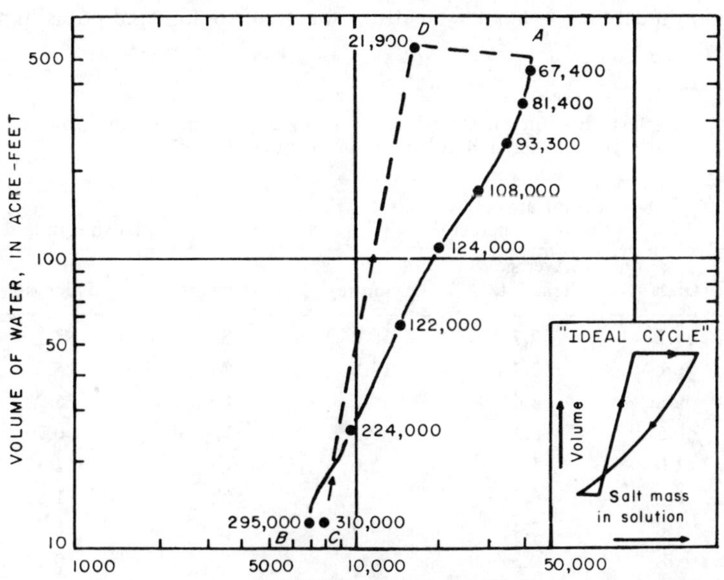

Fig. 18. Relation of lake volume to (total) dissolved solids at Deep Springs Lake, California for samples collected from October 1959 to January 1961. The numbers at the points represent total dissolved solids in ppm. Yearly sampling progression is A (January), B (June), C (October), D (December). The insert shows the "ideal cycle" for closed lake of Langbein (1961, p. 9). (Modified from Jones, 1965.)

subtle variations in hydrology and permeability. A closed hydrographic basin is a dynamic setting, and Fig. 18 illustrates one dynamic aspect of a closed lake, i.e. how the salt load in a playa lake varies annually. With this variation in salinity and, hence, pore water chemistry, it is not surprising that there are local variations in the zeolite mineralogy.

Age relationships were determined for some of the authigenic minerals in the tuffs by study of the sequence of filling of cavities and shard molds and by replacement relations (Sheppard and Gude, 1968). Phillipsite is consistently an early zeolite, and clinoptilolite and erionite are consistently late zeolites.

Zeolite reactions in nontuffaceous rocks in saline, alkaline lakes

Zeolites, particularly those in nontuffaceous rocks, can form by reactions other than those previously discussed. Analcime in nontuffaceous lacustrine rocks apparently formed during diagenesis by reaction of saline, alkaline pore water with plagioclase, montmorillonite, kaolinite, and aluminosilicate gels (Hay and Moiola, 1963; Hay, 1966; Pipkin, 1967; Hay, 1970; Surdam and Eugster, 1976). Analcime in the Triassic Lockatong Formation of New Jersey and Pennsylvania probably formed at an early stage of diagenesis from a colloidal precursor or aluminosilicate mineral (Van Houten, 1965). Studies of the lithofacies at Olduvai Gorge by Hay (1970) showed that both analcime and clinoptilolite in lacustrine mudstones formed from the reaction of clay minerals with saline, alkaline brines. Surdam and Eugster (1976) studying the sediments at Lake Magadi, demonstrated that some analcime formed from a hydrous sodium aluminosilicate gel.

ZEOLITES IN HOLOCENE SALINE, ALKALINE-LAKE DEPOSITS

Authigenic minerals in Holocene lacustrine sediments can be understood by considering the following: (1) composition of the starting materials; (2) composition of the solutions interacting with them; (3) hydrologic conditions; and (4) kinetic factors. Holocene deposits such as the sediments in the Lake Magadi region, Kenya, or Teels Marsh, Nevada, provide unique opportunity to study authigenic minerals and to determine the processes responsible for their formation.

Bulk composition

Surdam and Eugster (1976) showed at Lake Magadi that erionite formed from trachytic glass chiefly by the addition of water (Table 4). At Teels Marsh, Nevada, however, the reaction of rhyolitic glass to phillipsite involves not only the addition of water, but a loss of SiO_2 and a gain of sodium and potassium (Table 5 and Fig. 19). The released SiO_2 at Teels Marsh forms searlesite, and the saline, alkaline pore water is the source of sodium and potassium (Table 6).

TABLE 4. Chemical analyses of trachytic glass and erionite from a modern mud (644) of Lake Magadi*

	Chemical analyses (in weight percent)			Barth standard cell		
Oxide	Glass	Average Trachyte	Erionite	Element	Trachyte	Erionite†
SiO_2	59.2	62.7	60.0	Si	58.7	51.7
Al_2O_3	n.d.	13.7	14.8	Al	15.0	15.0
Fe_2O_3	n.d.	8.4	n.d.	Fe	5.9	n.d.
MgO	<0.5	0.5	0.1	Mg	0.7	0.1
CaO	<1.0	1.3	0.6	Ca	1.3	0.6
Na_2O	6.1	6.3	6.5	Na	11.5	10.9
K_2O	5.0	5.4	4.9	K	6.4	5.4
MnO	n.d.	0.3	n.d.	Mn	0.2	n.d.
Σ Oxides	–	98.6	–	Σ Oxygen	160	–

*After Surdam and Eugster (1976).
†Calculated with respect to Al content of trachyte Barth standard cell.

TABLE 5. Chemical analyses of rhyolite glass and phillipsite from Holocene tuffs at Teels Marsh, Nevada

Oxide	Chemical analyses (in weight percent) Rhyolite glass*	Phillipsite*	Element	Barth standard cell Rhyolite glass	Phillipsite†
SiO_2	77.2	56.7	Si	66.6	41.6
Al_2O_3	12.6	15.3	Al	12.8	12.8
Fe_2O_3	1.2	1.2	Fe	0.7	0.7
CaO	1.2	0.1	Ca	1.1	0.1
Na_2O	3.5	5.5	Na	5.7	7.6
K_2O	4.2	6.2	K	4.7	5.6
Σ Oxides	99.2		Σ Oxygen	160	–

*Microprobe analysis.
†Calculated with respect to the Al content of rhyolite Barth standard cell.

TABLE 6. Chemical analyses of brines from Teels Marsh, Nevada*

Constituents in ppm	A	1 (3)	2 (3)	3 (4)	4 (4)	5 (2)
SiO_2	71	65	65	60	65	85
Al	1	0	1	0	1	0
Na	308	118,000	129,000	118,000	123,000	122,000
K	8	3030	3450	3210	3300	3410
Ca	13	0	0	0	0	0
Mg	tr	tr	tr	tr	tr	tr
pH	9.0	9.5	9.5	9.5	9.5	9.5
	6 (4)	7 (1)	8 (3)	9 (4)	10 (3.5)	12 (3.5)
SiO_2	70	70	80	70	70	65
Al	0	1	0	2	0	0
Na	115,000	106,000	124,000	121,000	114,000	124,000
K	3160	3030	3260	3090	3070	3180
Ca	0	0	0	0	0	0
Mg	tr	tr	tr	tr	tr	tr
pH	9.4	9.5	9.4	9.6	9.6	9.5
	13 (3)	RLH4 (12)	RLH5 (7–10)			
SiO_2	80	128	122			
Al	2	n.d.	n.d.			
Na	116,000	64,600	11,500			
K	3200	2630	625			
Ca	0	n.d.	n.d.			
Mg	tr	n.d.	n.d.			
pH	9.5	9.4	8.9			

*Samples A to 13 are from this study.
Samples RLH4 and RLH5 are from Hay (1966, Table 6).
Number in parenthesis is depth of sample in feet.

Importance of pore water composition

Erionite is the most common alteration product of glass in the Lake Magadi region (Surdam and Eugster, 1976). This distinguishes the Lake Magadi region from other well-studied zeolite deposits such as the tuffaceous rocks of the Lake Tecopa, Barstow, and Wikieup areas (Sheppard and Gude, 1968, 1969, 1973). Why is erionite so much more abundant at Lake Magadi? To answer this question, consider the following observations by Surdam and Eugster (1976):

1. The trachytic glass, which is the parent material at Lake Magadi, is rich in alkalis and poor in alkaline earths.
2. There is an abundance of silica and silicate phases in the sediment, such as magadiite and hydrous sodium aluminosilicate gels, indicating high activities of SiO_2 and Na_2O.
3. The inflow waters to the Magadi basin and the brines in the basin are rich in sodium and silica and are chiefly devoid of alkaline earths.

These observations suggest that erionite is the preferred initial zeolite in an environment rich in sodium and silica and poor in alkaline earths, especially calcium. Surdam and Eugster (1976) checked this conclusion further by considering the compositional range of the other common alkalic zeolite, such as phillipsite and clinoptilolite. Thirteen acceptable published analyses of erionite, phillipsite, and clinoptilolite were considered and are given in Table 7. The criterion for choosing these analyses and rejecting others was the $Al_2O_3 + Fe_2O_3$ to $MgO + CaO + Na_2O + K_2O$ balance. If the balance was not within 5%, the analyses were rejected. The compositional relationships of phillipsite, erionite, and

TABLE 7. Chemical composition of phillipsites, erionites, and clinoptilolites*

Phillipsite	1		2		3		4	
	wt%	moles	wt%	moles	wt%	moles	wt%	moles
SiO_2	54.16	0.9015	56.34	0.9377	53.95	0.9812	59.69	0.9935
Al_2O_3	17.00	0.1660	15.22	0.1493	15.14	0.1485	14.70	0.1442
Fe_2O_3	1.35	0.0085	1.28	0.0080	0.92	0.0058	0.67	0.0042
MgO	0.58	0.0142	0.09	0.0022	0.62	0.0154	0.47	0.0116
CaO	0.70	0.0125	1.80	0.0321	1.37	0.0240	0.84	0.0150
Na_2O	4.92	0.0794	5.34	0.0861	3.72	0.0620	4.19	0.0676
K_2O	6.01	0.0638	3.53	0.0375	5.18	0.0550	5.25	0.0557

Erionite	1		2		3		4	
SiO_2	61.40	1.0220	57.24	0.9527	60.39	1.0052	59.51	0.9903
Al_2O_3	13.26	0.1301	13.93	0.1366	13.32	0.1307	14.20	0.1393
Fe_2O_3	1.48	0.0093	1.95	0.0122	1.31	0.0082	0.73	0.0046
MgO	0.23	0.0057	0.15	0.0037	0.49	0.0122	0.14	0.0035
CaO	0.57	0.0102	0.00	0.0000	1.30	0.0232	0.01	0.0002
Na_2O	4.21	0.0679	6.24	0.1007	3.48	0.0561	5.92	0.0955
K_2O	5.42	0.575	4.10	0.0435	4.33	0.0460	3.64	0.0386

Clinoptilolite	1		2		3		4		5	
SiO_2	66.40	1.1050	65.30	1.0867	65.21	1.0852	64.70	1.0763	63.69	1.0599
Al_2O_3	11.64	0.1141	11.47	0.1125	12.61	0.1236	12.43	0.1219	12.47	0.1223
Fe_2O_3	0.47	0.0029	1.36	0.0085	0.40	0.0025	0.44	0.0028	0.62	0.0039
MgO	0.86	0.0213	1.22	0.0303	1.09	0.0270	0.34	0.0084	1.42	0.0352
CaO	2.55	0.0455	1.80	0.0321	1.84	0.0328	1.26	0.0223	2.25	0.0401
Na_2O	1.85	0.0298	1.76	0.0284	1.76	0.0284	4.32	0.0697	2.46	0.0397
K_2O	1.93	0.0205	2.49	0.0264	3.79	0.0402	2.28	0.0242	1.80	0.0191

*After Surdam and Eugster (1976).

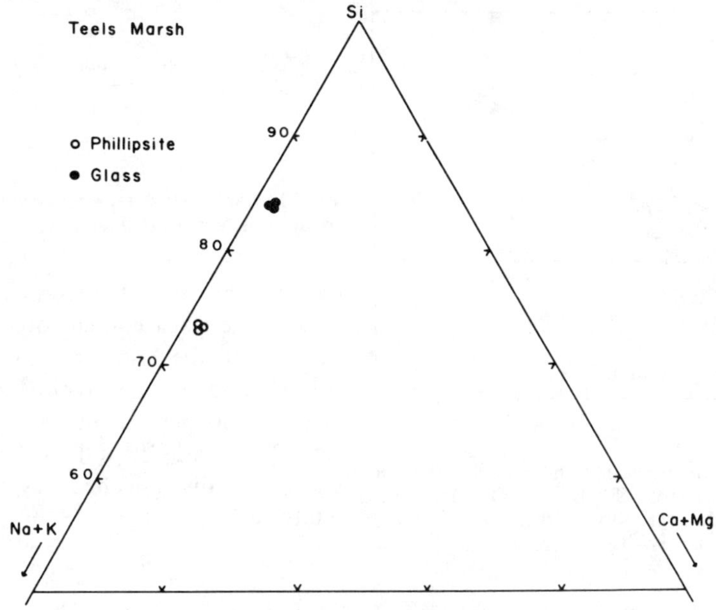

Fig. 19. Diagram of the compositional changes that occur when rhyolitic glass is altered to phillipsite at Teels Marsh, Nevada.

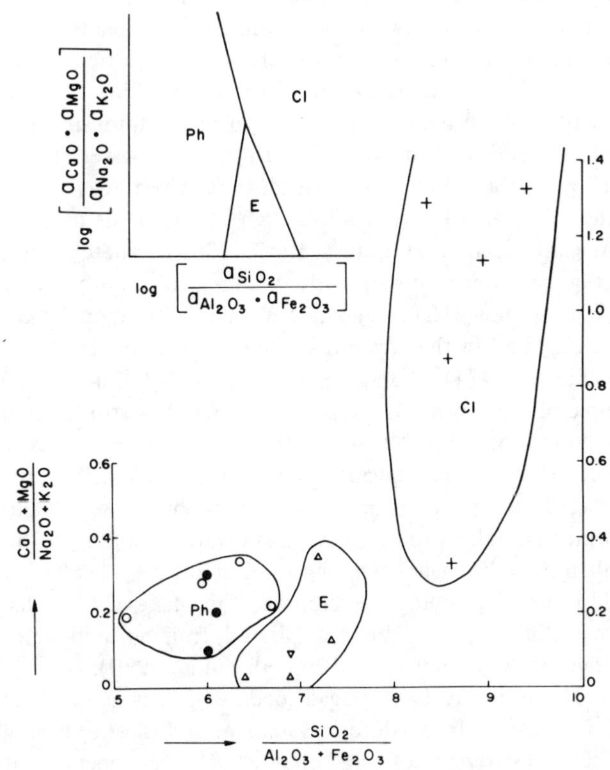

Fig. 20. Alkalic, silicic zeolites from Table 7 plotted on a mole fraction diagram. Circles are phillipsite (Ph), triangles are erionite (E) and crosses are clinoptilolite (Cl). Black circles are phillipsite from Teels Marsh, Nevada. In the erionite field, the triangle with the point down is the typical composition of erionite from the High Magadi Beds, Kenya. Insert shows the schematic activity diagram constructed from the mole fraction plot. (Modified from Surdam and Eugster, 1976.)

clinoptilolite are illustrated in Fig. 20, a diagram of $SiO_2/(Al_2O_3 + Fe_2O_3)$ versus $(CaO + MgO)/(Na_2O + K_2O)$. This diagram also emphasizes the important parameters of silica, alkali, and alkaline-earth activities. Based on the selected analyses, the compositional ranges of the three zeolites are clearly separated, with only minor overlap. Phillipsite occupies the low silica and low alkaline-earth region. Erionite occupies an intermediate silica and very low alkaline-earth area, and clinoptilolite occupies the high silica and high alkaline-earth area (Surdam and Eugster, 1976).

Using the techniques of Korzhinskii (1959), an activity diagram was constructed from the mole-fraction plot (insert of Fig. 20). The relationships show that erionite is expected to form in an environment characterized by high silica, high sodium, and extremely low calcium activities. These conditions are compatible with the observations summarized above and explain the predominance of erionite in the Magadi basin (Surdam and Eugster, 1976). The alkaline brines of Lake Magadi are free of calcium and are ideal agents for the hydration and solution of trachytic glass to form erionite.

The more siliceous rhyolitic glass at Teels Marsh reacts with saline, alkaline brines to form phillipsite, a less siliceous zeolite (see Table 7). The reason for this is found in the chemistry of the pore water.

Boron effect

The black dots in Fig. 20 represent the phillipsite at Teels Marsh and show that its crystallization should be

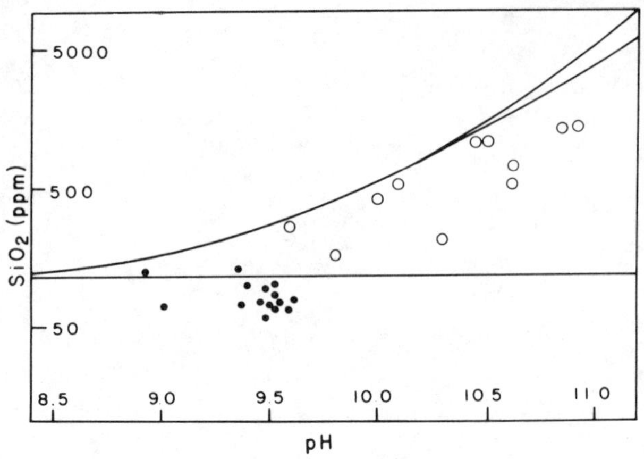

Fig. 21. Silica in solution versus pH. The three solid curves are (from low to high SiO_2) for $Si(OH)_4$, $SiO(OH)_3^-$, and $SiO_2(OH)_2^{--}$. The black dots are brines from Teels Marsh, Nevada, and the open circles are brines from Lake Magadi, Kenya.

favored by a solution low in silica and alkaline earths and high in alkalines. The brines at Teels Marsh are low alkaline earth brines because they average 0.13 meq./l calcium and 0.20 meq./l magnesium (Smith, 1974). The brines at Teels Marsh are also low-silica brines (Fig. 21). A surprising fact is that these alkaline brines (pH $>$ 9.5), even with the zeolitization of the rhyolitic glass releasing silica, are low in silica. The explanation is related to the boron content of the natural waters in the Teels Marsh basin (Fig. 22). The boron content of the brines is high

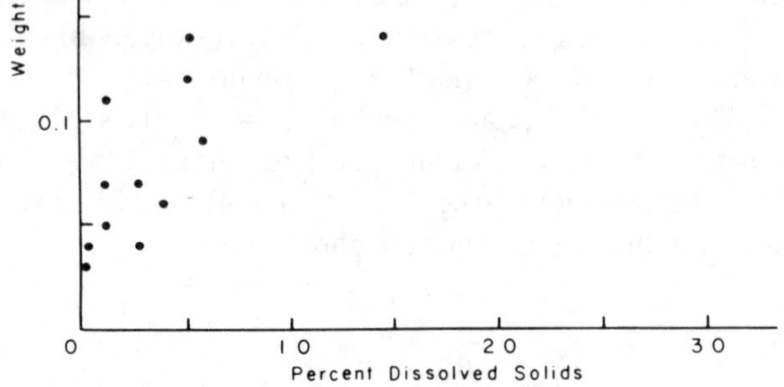

Fig. 22. Boron (B_2O_3) in solution versus percent dissolved solids (salinity) for brines from Teels Marsh, Nevada.

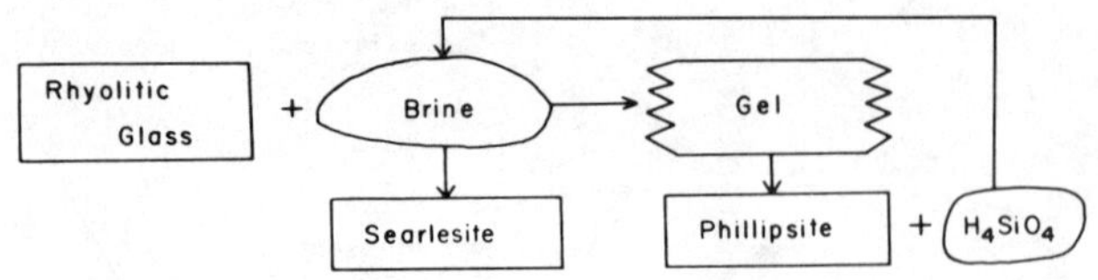

Fig. 23. Schematic representation of authigenic reactions in modern tuffaceous sediments at Teels Marsh, Nevada.

enough to allow the formation of searlesite which buffers the silica content of the brines at much lower values than the brines at Lake Magadi (Fig. 21). Thus, the alteration of the rhyolitic glass to phillipsite at Teels Marsh is coupled to the reaction that forms searlesite (Fig. 23). Searlesite at Teels Marsh is ubiquitous, and it occurs in the ash beds as well as in the adjacent mudstones.

Hydrous sodium silicates

Unlike the natural waters in the Teels Marsh hydrographic basin, the waters in the Lake Magadi region are very low in boron and do not lead to the formation of searlesite but evolve into silica-rich brines (Fig. 21). This sodium- and silica-rich, alkaline environment leads to the formation of hydrous sodium silicates, i.e. magadiite, and ultimately to chert (Eugster, 1967, 1969). Magadiite and magadi-type chert were first described in the Pleistocene High Magadi beds of Lake Magadi, Kenya (Eugster, 1967, 1969). These cherts (Fig. 24) were subsequently recognized in other east African lakes (Hay, 1968), in Wyoming (Surdam, *et al.*, 1972), and in the western United States (Sheppard and Gude, 1974). Magadiite ($NaSi_7O_{13}(OH)_3 \cdot 3H_2O$) precipitates from an alkaline lake that has stored large amounts of silica in solution (e.g. Lake Magadi, Jones *et al.*, 1967). The specific mechanism responsible for magadiite precipitation is unknown, but a seasonally stratified lake produced by fresh-water flooding of an alkaline lake has been suggested (Eugster, 1969). In addition to being precipitated in lake-wide beds, magadiite can also be precipitated from ground-water reactions (Rooney *et al.*, 1969; Maglione, 1970).

Magadiite at Lake Magadi occurs in beds as much as 2 feet thick. It is white, crystalline, and plastic, having the consistency of soft putty. The conversion of magadiite to chert has been observed in place (Eugster, 1969, Fig. 5). The conversion occurred during the depositional interval for magadiite as well as up to the present time. The conversion involves a loss of sodium and water and a reduction in volume of at least 25%. The effect the conversion has on the activity of silica at Lake Magadi seems to affect directly the erionite to analcime reaction by reducing the activity of silica and

Fig. 24. Photograph of typical magadi-type chert. Note reticulation on surface of chert.

increasing the activity of sodium.

The transformation of erionite to analcime involves the addition of sodium and the loss of silica and potassium (equation (3) above). Lowering the activity of silica and increasing the activity of sodium is also associated with the reaction of magadiite to quartz (Bricker, 1969). There is evidence in the Lake Magadi region that the erionite to analcime reaction is coupled to the magadiite–chert conversion. Samples containing analcime as a major phase rarely contain even a trace of magadiite (Surdam and Eugster, 1976).

Two mechanisms have been suggested for the magadiite–chert conversion. O'Neill and Hay (1973) used isotopic evidence to show that quartz may crystallize spontaneously from magadiite in the presence of sodium-rich brines. Eugster (1969) observed the *in situ* conversion of magadiite to chert in outcrops and shallow pits, and he attributed the conversion to the removal of sodium by dilute ground waters. Both mechanisms are operable, as Fig. 25 demonstrates; the natural waters in the Lake Magadi region plot in the quartz-stability field.

TABLE 8. Chemical composition of anhydrous gels from the Lake Magadi area, Kenya, in weight percent*

	M47H*	M48J	M48H	M62H	M14J
SiO_2	53.75	40.14	39.21	46.73	51.50
TiO_2	1.03	1.29	0.81	0.75	1.26
Al_2O_3	11.30	18.00	21.22	19.03	10.65
Fe_2O_3	9.61	11.50	6.38	6.72	9.76
MgO	1.62	2.92	1.40	1.23	2.66
CaO	1.96	4.47	4.75	3.07	4.20
Na_2O	15.67	18.24	22.54	18.50	13.98
K_2O	5.06	3.44	3.69	3.96	5.99
			H_2O^-		
	90–93	52–63	52–63	64–72	72–77
		Loss of ignition to 900°C (purified samples)			
	61.63	45.94	2.34	2.28	n.d.

*From Eugster and Jones (1968).

Gel to analcime reaction

In addition to forming from a zeolite precursor, analcime can also form from a hydrous sodium aluminosilicate gel as suggested by Eugster and Jones (1968). Their data on the chemical composition of some typical Lake Magadi gels are given in Table 8. Recent field work has demonstrated that these gels are more widespread than previously recognized, and they occur

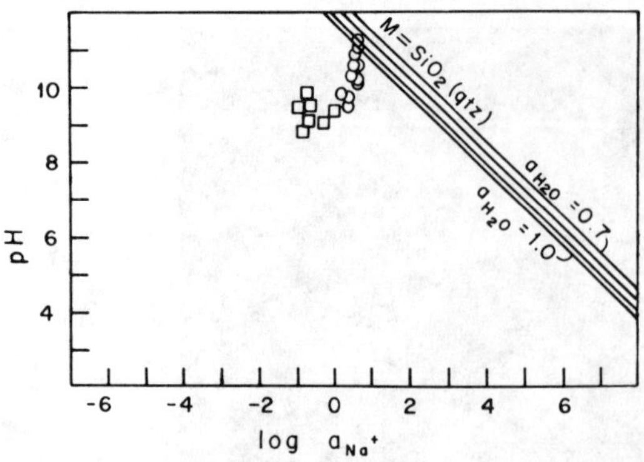

Fig. 25. Conversion boundaries for magadiite (M) to quartz in terms of pH and a_{Na^+}. The effect of a_{H_2O} changes is indicated for the M = SiO_2 boundary for values of 1.0, 0.9, 0.8, and 0.7. Squares are hot springs and circles are brines from the Lake Magadi region, Kenya. (Modified from Eugster, 1969.)

throughout the Magadi basin (Surdam and Eugster, 1976).

The gels are at or near the orifices of most of the hot springs and at the shores of Lake Magadi and Little Magadi, where alkaline brines are in contact with trachytic debris (Surdam and Eugster, 1976). At the north end of Little Magadi, gel accumulations as much as 30 cm thick have been observed. A reaction responsible for these gels has been suggested by Eugster and Jones (1968). Trachytic material reacts with alkaline solutions according to:

$$\text{trachyte} + Na^+ + H_2O \rightarrow \text{Na-Al-Si gel} + SiO_2,$$

assuming that aluminum is conserved during the reaction. Observations of the gel accumulations north of Little Magadi during four field seasons suggest that the gels are washed into the lake yearly, particularly during the rainy seasons (Surdam and Eugster, 1976). Therefore, it is not surprising that similar gels are common in the drill-core samples taken at Lake Magadi (Surdam and Eugster, 1976). The gels in these core samples are commonly associated with amorphous silica.

There also seems to be an antithetical relationship between the gel and the analcime. In samples of core F6 from Lake Magadi, the formation of analcime from gel was demonstrated (Fig. 26). This reaction may account for the abundance of analcime in the deeper portions of the core; however, the analcime distribution is not strictly controlled by depth. Erionite may also have formed locally from gels in the core samples, but Surdam and Eugster (1976) have no direct evidence for this reaction.

Crystallization of Magadi gels under hydrothermal (350°C and 2 kb pressure) conditions yielded analcime (Eugster and Jones, 1968). In order to study the crystallization products at lower temperatures and

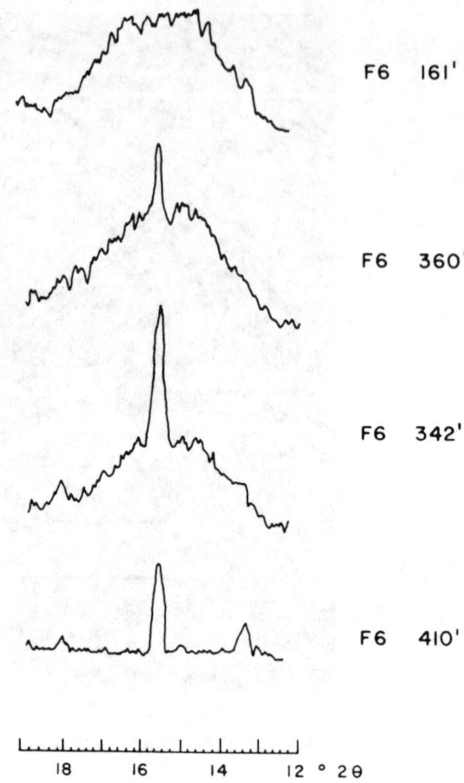

Fig. 26. Diffractometer patterns over the interval 12–19° (CuKα radiation) of samples from drill core F6 at Lake Magadi, Kenya. Upper sample (161 ft) is gel-rich, whereas the lower sample (410 ft) is gel-free. Intermediate samples contain gelatinous material reacting to analcime in varying proportions (from Surdam and Eugster, 1976).

pressures, purified Magadi gels were placed in Na_2CO_3 solutions of varying pH and heated at 1 atm pressure at 60°C for 5 weeks. These experiments yielded magadiite + gel (Surdam and Eugster, 1976). Similar experiments made with gels in Na_2CO_3 solutions at 80°C for 2 weeks yielded magadiite + erionite. In the latter experiments, erionite and magadiite formed in all solutions that originally had a pH of 9.0 or greater. When K_2CO_3 solutions were substituted for the Na-rich solutions and all other conditions kept constant, the gels still yielded erionite + magadiite. No silicate phases formed in Ca-rich solutions.

Potassium feldspar

Most surprising is the lack of authigenic potassium feldspar in the Magadi or Teels Marsh sediments. Potassium feldspar is a common alteration product in the Green River Formation as well as many Pleistocene and older saline alkaline lake deposits (Hay, 1966;

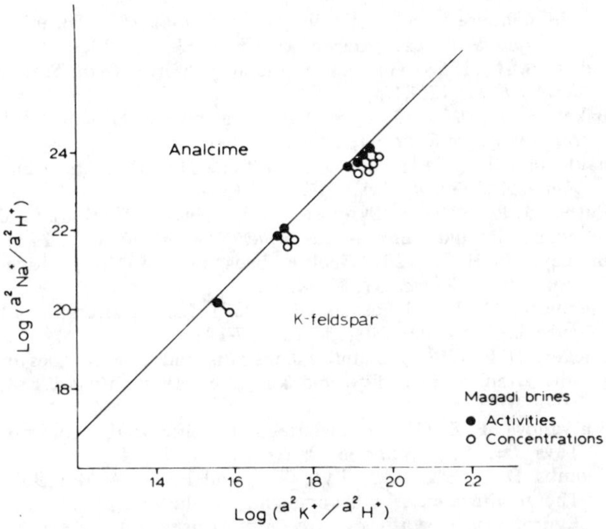

Fig. 27. Stability of analcime and potassium feldspar in the system K_2O–Na_2O–Al_2O_3–SiO_2–H_2O at 25°C and 1 atm total pressure (after Garrels and Mackenzie, 1971). It is assumed that quartz and water are ubiquitous. Magadi brines from Eugster (1970) are plotted on the diagram; solid circles, activities; open circles, concentrations.

Sheppard and Gude, 1968, 1969; and Hay, 1970). In these deposits, both the reaction zeolite to potassium feldspar and analcime to potassium K-feldspar has been documented (Sheppard and Gude, 1963; Iijima and Hay, 1968). An insight into this problem is gained by comparing the solution compositions in equilibrium with analcime and potassium feldspar to the compositions of the Magadi Lake brines. Figure 27 shows the stabilities of analcime and potassium feldspar with respect to the activities of Na^+, K^+, and H^+ (at 25°C and 1 atm pressure), as well as the activities of those lake brines for which field pH and potassium analyses are available. The concentration values were recalculated approximately to activities by using activity coefficients of the NaCl–KCl–H_2O system. Activities for the lake brines lie very close to the analcime–potassium feldspar boundary, indicating that authigenic K-feldspar could form either by an addition of potassium or perhaps by a rise in temperature.

Reaction mechanism

In modern environments where authigenic silicates are forming, the Si/Al ratios of the low-temperature zeolitic alteration products of silicic glass are related to the alkalinity of the waters responsible for that alteration (Hay, 1966). Specifically, more siliceous zeolites, such as clinoptilolite and mordenite, are the most common alteration products of silicic glass in marine and fresh-water environments, whereas less siliceous phillipsite and erionite are the dominant alteration products in saline, alkaline lakes (Hay, 1966). Mariner and Surdam (1970) suggested that the relation between pH and Si/Al ratio of the zeolite is a result of the reaction between silicic glass and alkaline solution whereby (i) a gel forms whose Si/Al ratio is controlled by the Si/Al ratio of the solution and (ii) a zeolite forms from the gel whose Si/Al ratio is controlled by the composition of the gel. Also, they demonstrated experimentally that the solubility of rhyolitic glass increases with increasing alkalinity.

Surdam and Mariner (1971) tested this reaction mechanism in modern, saline, alkaline lakes and found a gelatinous phase in the Holocene tuffs at Teels Marsh, Nevada. The gel is associated with rhyolitic glass and phillipsite in a thin ash bed that may be no older than 600 years (Surdam and Mariner, 1971). After separating this gel, Surdam and Mariner (1971) were able to crystallize it in sodium carbonate solutions to phillipsite. This work strongly suggests that some natural zeolites form as a result of a glass + H_2O → gel → zeolite or glass + H_2O → gel → solution → zeolite reaction. Laboratory and commercial synthesis of zeolites is commonly accomplished from gels (Zhdanov, 1971).

The data for the natural phillipsite in the thin ash beds at Teels Marsh fall on the experimentally derived curve for the relationship between pH and atoms of Si in phillipsite (Fig. 28). Mariner and Surdam (1970)

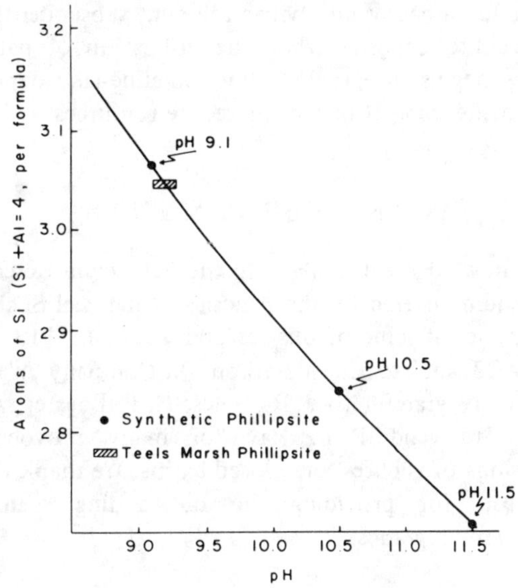

Fig. 28. A plot of the number of atoms of Si (on the assumption that S + Al = 4) in synthetic phillipsite (black dots) as a function of pH. Numbers of atoms of Si in natural phillipsite from Teels Marsh, Nevada (shaded box), are plotted for comparison. (Modified from Mariner and Surdam, 1970.)

observed no gels in the pH range of 7–9, where siliceous zeolites such as mordenite and clinoptilolite should form. This lack of gels can be explained by the relatively low solubility of Al in the pH range of 7–9. In experiments of longer duration or in a natural environment, the kinetic effect should be reduced. However, if the model proposed by Mariner and Surdam is correct, and if formation of the gel is dependent on the charged aluminum species in solution, the formation of the more siliceous zeolites should be more sluggish.

ECONOMIC POTENTIAL

Saline, alkaline lakes form in closed hydrographic basins and are a product of a special geological setting. There is a high probability that this special geological setting, either a block faulted or rifted terrain, will also be characterized by volcanism and saline, alkaline brines. This combination inevitably leads to the formation of zeolites.

An ideal setting for the formation of a commercial zeolite deposit is a saline, alkaline lake into which much volcanic glass was deposited. The possibility of a commercial zeolite deposit is increased by a high content of glass and low content of crystal and rock fragments in the parent ash bed. In either a playa-lake complex or a rift system, the fluvial processes are minimized, thereby effectively filtering much of the clastic debris from the saline, alkaline lake. These processes provide an ideal setting for a nearly pure vitric ash and, subsequently, a pure zeolite deposit. When the utilization of natural zeolites begins in earnest, saline, alkaline-lake deposits will provide some of our major zeolite resources.

ACKNOWLEDGMENTS

Acknowledgment is made to the Petroleum Research Fund, administered by the American Chemical Society, for support of some of this research (Surdam PRF No. 6678-AC2) and to the Marathon Oil Company. We are particularly grateful to J. R. Boles, H. P. Eugster, A. J. Gude, 3rd, and R. L. Hay for many provocative discussions of zeolites and closed basins. We thank F. A. Mumpton for providing the outstanding scanning electron micrographs.

REFERENCES

Andreis, R. R. and Rodriquez, A. M. I. (1968) Clinoptilolite de la formazion la Matilde (Upper Jurassic), Provincia de Santa Cruz: *Revista Museo la Plata, Secc. Geol.* **6**, 221-234.

Ataman, G. and Beseme, P. (1972) Découverte de l'analcime sédimentaire en Anatolie du nord-ouest (Turquie): minéralogie, genèse, paragenèses: *Chem. Geol.* **9**, 203-225.

Baker, B. H. (1958) Geology of the Magadi area: *Geol. Survey Kenya Rept.* **42**, 81 pp.

Baker, B. H. (1963) Geology of the area south of Magadi: *Geol. Survey Kenya Rept.* **61**, 27 pp.

Baldwin, J. L. (1944) Tupungate oil field, Mendoza, Argentina: *Amer. Ass. Petrol. Geol. Bull.* **28**, 1455-1484.

Boles, J. R. (1971) Synthesis of analcime from natural heulandite and clinoptilolite: *Amer. Mineral.* **56**, 1724-1734.

Bradley, W. H. (1928) Zeolite beds in the Green River Formation: *Science* **67**, 73-74.

Bramlette, M. N. and Posnjak, E. (1933) Zeolitic alteration of pyroclastics: *Amer. Mineral.* **18**, 167-171.

Bricker, O. P. (1969) Stability constants and free energies of formation of magadiite and kenyaite: *Amer. Mineral.* **54**, 1026-1033.

Bur'yanova, E. Z. (1960) Analcite- and zeolite-bearing rocks of Tuva: *Izv. Akad. Nauk SSSR, Ser. Geol.* **6**, 74-84.

Coombs, D. S., Ellis, A. J., Fyfe, W. S. and Taylor, A. M. (1959) The zeolite facies, with comments on the interpretation of hydrothermal synthesis: *Geochim. Cosmochim. Acta* **17**, 53-107.

Deffeyes, K. S. (1959) Zeolites in sedimentary rocks: *J. Sediment. Petrol.* **29**, 602-609.

Drever, J. I. (1972) Relations among pH, carbon dioxide pressure, alkalinity, and calcium concentration in waters saturated with respect to calcite at 25°C and one atmosphere total pressure: *Contr. Geol.* **11**, 41-42.

Echle, Wolfram (1974) Zur Mineralogie and Petrogenese jungtertiarer tuffitischer Sedimente in Neogen-Becken nordlich Mihaliccik (Westanatolien, Turkei): *Neues Jb. Mineral., Abh.* **121**, 43-84.

Echle, Wolfram (1975) Zusammensetzung und Entstehung sedimentarer Analcime in juntertiarer pyroklastischen Gesteinen nordlich Mihaliecik, Westanatolien, Turkei: *Neues Jb. Mineral., Abh.* **124**, 128-146.

Eugster, H. P. (1967) Hydrous sodium silicates from Lake Magadi, Kenya: Precursors of bedded chert: *Science* **157**, 1177-1180.

Eugster, H. P. (1969) Inorganic bedded cherts from the Magadi area, Kenya: *Contr. Mineral. Petrol.* **22**, 1-31.

Eugster, H. P. (1970) Chemistry and origin of the brines of Lake Magadi, Kenya: *Mineral. Soc. Amer. Spec. Paper* **3**, 215-235.

Eugster, H. P. and Hardie, L. A. (1975) Sedimentation in an ancient playa-lake complex: the Wilkins Peak Member of the Green River Formation of Wyoming: *Geol. Soc. Amer. Bull.* **86**, 319-334.

Eugster, H. P. and Jones, B. F. (1968) Gels composed of sodium-aluminum silicate, Lake Magadi, Kenya: *Science* **161**, 160-163.

Eugster, H. P. and Surdam R. C. (1973) Depositional environment of the Green River Formation of Wyoming: a preliminary report: *Geol. Soc. Amer. Bull.* **84**, 1115-1120.

Fuchs, V. E. (1939) The geological history of the Lake Rudolf basin Kenya colony: *Royal Soc. London Philos. Trans.* **229**, 219-274.

Fuller, A. O. and Moir, G. J. (1967) Analcite in the upper Karroo sediments, Herschel District, South Africa: *J. Sediment. Petrol.* **37**, 1236-1237.

Garrels, R. M. and MacKenzie, F. T. (1967) Origin of the chemical composition of some springs and lakes: *Advances in Chemistry Series: Amer. Chem. Soc.*, pp. 222-242.

Hardie, L. A. (1968) The origin of the Recent non-marine evaporite deposit of Saline Valley, Inyo County, California: *Geochim. Cosmochim. Acta* **32**, 1279-1301.

Hardie, L. A. and Eugster, H. P. (1970) The evolution of closed-basin brines: *Mineral. Soc. America Spec. Paper* **3**, 273-290.

Harding, S. T. (1965) Recent variations in the water supply of the Great Basin: *Calif. Univ. Water Resources Center Archives, Ser. Rept.* **16**.

Hay, R. L. (1966) Zeolites and zeolitic reactions in sedimentary rocks: *Geol. Soc. Amer. Spec. Paper* **85**, 130 pp.

Hay, R. L. (1968) Chert and its sodium-silicate precursors in sodium-carbonate lakes of East Africa: *Contr. Mineral. & Petrol.* **17**, 255-274.

Hay, R. L. (1970) Silicate reactions in three lithofacies of a semi-arid basin. Olduvai Gorge, Tanzania: *Mineral. Soc. Amer. Spec. Paper* **3**, 237-255.

Hay, R. L. and Moiola, R. J. (1963) Authigenic silicate minerals in Searles Lake, California: *Sedimentology* **2**, 312-332.

Hess, P. C. (1966) Phase equilibria of some minerals in the K_2O–Na_2O–Al_2O_3–SiO_2–H_2O system at 25°C and 1 atmosphere: *Amer. J. Sci.* **264**, No. 4, 289-309.

Hunt, C. B. (1960) The Death Valley salt pan, a study of evaporite: *U.S. Geol. Survey Prof. Paper* **400-B**, B456-B457.

Iijima, Azuma and Hay, R. L. (1968) Analcime composition in the Green River Formation of Wyoming: *Amer. Mineral.* **53**, 184-200.

Jones, B. F. (1965) The hydrology and mineralogy of Deep Springs Lake, Inyo County, California: *U.S. Geol. Survey Prof. Paper* **502-A**, 56 pp.

Jones, B. F. (1966) Geochemical evolution of closed basin waters in the western Great Basin: *Symp. Salt, 2nd, Ohio Geol. Soc. Cleveland, Ohio 1*, pp. 181-200.

Jones, B. F., Eugster, H. P. and Rettig, S. L. (1977) Hydrochemistry of the Lake Magadi Basin, Kenya: *Geochim. Cosmochim. Acta* **41**, 53-72.

Jones, B. F., Rettig, S. L. and Eugster, H. P. (1967) Silicate in alkaline brines: *Science* **158**, 1310-1314.

Joulia, Francois; Bonifas, Marthe; Camez, Thérèse; Millot, Georges and Weid, René (1959) *Découverte d'un important niveau d'analcimolite qresèuse dans le Continental intercalaire de l'ouest de l'Air (Sahara central)*: Dakar Service de Géologie et de Prospection Minière, 40 pp.

Korzhinskii, D. S. (1959) *Physical-chemical Basis of the Analysis of the Paragenesis of Minerals*: Consultants Bureau, New York, 142 pp.

Kossovskaya, A. G. (1975) Genetic types of zeolites in stratified rocks: *Lithology Mineral. Res.* **10**, Mar./Apr., 1975.

Langbein, W. B. (1961) Salinity and hydrology of closed lakes: *U.S. Geol. Survey Prof. Paper* **412**, 20 pp.

Linkletter, G. O. (1974) Authigenic phillipsite in Antarctic lacustrine sediments: *Geol. Soc. Amer. Abstracts with Programs* **6**, 206-207.

Loughnan, F. C. (1966) Analcite in the Newcastle Coal Measure sediments of the Sydney Basin, Australia: *Amer. Mineral.* **51**, 486-494.

McKie, Duncan (1958) Mineralogical notes: *Geol. Survey Tanganyika Records* **6**, 87-93.

Maglione, G. (1970) La magadiite silicate soduque de neoformation des facies. evaporitiques du Kanem (Littoral Nord-est du Lac Tchad): *Bull. Serv. Carte Geol. Alsace Lorraine* **23**, 177-189.

Mariner, R. H. (1971) Experimental evaluation of authigenic mineral reactions in the Pliocene Moonstone Formation: University of Wyoming Ph.D. thesis, 133 pp.

Mariner, R. H. and Surdam, R. C. (1970) Alkalinity and formation of zeolites in saline alkaline lakes: *Science* **170**, 977-980.

Mason, Brian and Sand, L. B. (1960) Clinoptilolite from Patagonia, the relationship between clinoptilolite and heulandite: *Amer. Mineral.* **45**, 341-350.

Mumpton, F. A. (1973) Scanning electron microscopy and the origin at sedimentary zeolites: *Molecular Sieves*, Proc. 3rd Int. Cont. Mol. Sieves, Uytterhaoeven, J. B., ed.: Leuven Univ. Press, pp. 56-61.

Mumpton, F. A. (1975) Commercial utilization of natural zeolites: In: *Industrial Minerals and Rocks*, 4th ed., Leford, S. J., ed.: Amer. Inst. Mining, Metall., and Petroleum Engineers, Inc., New York, 1975, pp. 1262-1270.

Mumpton, F. A. and Ormsby, W. C. (1976) Morphology of zeolites in sedimentary rocks by scanning electron microscopy: *Clays and Clay Minerals* **24**, 1-23.

Munson, R. A. and Sheppard, R. A. (1974) Natural zeolites: their properties, occurrences, and uses: *Mineral Sci. Engng* **6**, Jan. 1974, 19-34.

O'Neil, J. R. and Hay, R. L. (1973) O^{18}/O^{16} ratios in cherts associated with the saline lake deposits of East Africa: *Earth and Planetary Sci. Letters* **19**, 257-266.

Pacquet, Andre (1968) Analcime et argiles diagenetiques dans les formations sedimentaires de la reion d'Agades (Republique da Niger): *Memories Services de la Carte Geologique d'Alsace et de Lorraine* **27**, 221 pp.

Papke, K. G. (1972) Erionite and other associated zeolites in Nevada: *Nev. Bur. Mines & Geol. Bull.* **79**, 32 pp.

Pipkin, B. W. (1967) Mineralogy of 140-foot core from Wilcox Playa, Cochise Arizona: *Amer. Ass. Petroleum Geologists Bull.* **51**, 478 pp.

Rooney, T. P., Jones, B. F. and Neal, J. T. (1969) Magadiite from Alkali Lake, Oregon: *Amer. Mineral.* **54**, 1034-1043.

Ross, C. S. (1928) Sedimentary analcite: *Amer. Mineral.* **13**, 195-197.

Russel, I. C. (1896) Present and extinct lakes of Nevada: In: Natl. Geog. Soc., *The Physiography of the United States*, American Book Co., New York.

Senderov, E. E. (1963) Crystallization of mordenite under hydrothermal conditions: *Geochemistry* **9**, 848-859.

Sheppard, R. A. (1969) Diagenesis of tuffs in the Barstow Formation, Mud Hills, San Bernardino County, California: *U.S. Geol. Survey Prof. Paper* **634**, 35 pp.

Sheppard, R. A. (1971) Zeolites in sedimentary deposits of the United States – a review: In: Gould, R. F., ed., *Molecular Sieve Zeolites – I*: Amer. Chem. Soc., *Advances in Chemistry Ser.* **101**, 279-310.

Sheppard, R. A. (1973) Zeolites in sedimentary rocks: *U.S. Geol. Survey Prof. Paper* **820**, 689-695.

Sheppard, R. A. (1973) Zeolites and associated authigenic silicate minerals in tufaceous rocks of the Big Sandy Formation, Mohave County, Arizona: *U.S. Geol. Survey Prof. Paper* **830**, 36 pp.

Sheppard, R. A. (1974) Chert derived from magadiite in a lacustrine deposit near Rome, Malheur County, Oregon: *J. Res. U.S. Geol. Survey* **2**, 625-630.

Sheppard, R. A. (1975) Zeolites in sedimentary rocks: In: *Industrial Minerals and Rocks*, 4th ed., Leford, S. J., ed.: Am. Inst. Mining, Metall., and Petroleum Engineers, Inc., New York, pp. 1257-1262.

Sheppard, R. A. and Gude, A. J., 3rd (1968) Distribution and genesis of authigenic silicate minerals in tuffs of Pleistocene Lake Tecopa, Inyo County, California: *U.S. Geol. Survey Prof. Paper* **597**, 38 pp.

Smale, David (1968) The occurrence of clinoptilolite in pan sediments in the Nata area, Northern Botswana: *Geol. Soc. South Africa Trans.* **71**, 147-152.

Smith, C. L. (1974) Chemical controls on weathering and trace metal distribution at Teels Marsh, Nevada: University of Wyoming, Ph.D. thesis, 96 pp.

Smith, C. L. and Drever, J. I. (1976) Controls on the chemistry of springs at Teels Marsh, Mineral County, Nevada: *Geochim. Cosmochim. Acta* (in press).

Smith, G. I. (1966) Geology of Searles Lake: In: Rau, J. L., ed., *Second Symposium on Salt, Northern Ohio Geol. Soc.* **1**, 167-180.

Sturm, W. and Morgan, J. J. (1970) *Aquatic Chemistry*: New York, John Wiley, 583 pp.

Surdam, R. C. (1972) Economic potential of zeolite-rich sedimentary rocks in Wyoming: *Wyo. Geol. Ass. Sci. Bull.*, Mar. 1972, pp. 5-8.

Surdam, R. C. and Eugster, H. P. (1976) Mineral reactions in the sedimentary deposits of the Lake Magadi region, Kenya: *Geol. Soc. Amer. Bull.* (in press).

Surdam, R. C., Eugster, H. P. and Mariner, R. H. (1972) Magadi-type chert in Jurassic and Eocene to Pleistocene rocks, Wyoming: *Geol. Soc. Amer. Bull.* **83**, 2261-2266.

Surdam, R. C. and Mariner, R. H. (1971) The genesis of phillipsite in Recent tuffs at Teels Marsh, Nevada: *Geol. Soc. Amer. Abs. with Programs* 3,

Surdam, R. C. and Parker, R. B. (1972) Authigenic aluminosilicate minerals in the tuffaceous rocks of the Green River Formation, Wyoming: *Geol. Soc. Amer. Bull.* **83**, No. 3, 689-700.

Teruggi, M. E. (1964) A new and important occurrence of sedimentary analcime: *J. Sediment. Petrol.* **34**, 761-767.

Teruggi, M. E. and Andreis, R. R. (1963) Revision de las zeolitas con especial referencia a su importancia sedmentologica: *Revista de la Asociacion Geologica Argentina* **18**, 73-95.

Vanderstappen, R. and Verbeek, T. (1959) Presence d'analcime d'origine sedimentaire dans le Mesozoique du Bassin du Congo: *Soc. Belge Geologie, Paleontologie, et Hydrologie Bull.* **68**, 417-421.

Vanderstappen, R. and Verbeek, T. (1964) Analcime et mineraux des formations geologiques de la curvet congolaise (Republique du Congo): *Musee Royal de l'Afrique Centrale, Belgique Annales,* Ser. in-8°, *Sciences geologique* **47**, 88 pp.

Van Houten, F. B. (1964) Tertiary geology of the Beaver Rim area, Fremont and Natrona counties, Wyoming: *U.S. Geol. Survey Bull.* **1164**, 99 pp.

Van Houten, F. B. (1965) Composition of Triassic Lockatong and associated formations of Newark group, central New Jersey and adjacent Pennsylvania: *Amer. J. Sci.* **263**, 825-863.

Vernet, J. P. (1961) Concerning the association montmorillonite–analcime in the series of Stanleyville, Congo: *J. Sediment. Petrol.* **31**, 293-295.

Zhdanov, S. P. (1971) Some problems of zeolite crystallization: In: Gould, R. F., ed., *Molecular Sieve Zeolites* – I. Am. Chem. Soc., *Advances in Chemistry Ser.* **101**, 20-43.

GEOLOGICAL OCCURRENCES OF ZEOLITE IN MARINE ENVIRONMENTS

AZUMA IIJIMA

Geological Institute, University of Tokyo, Japan

Abstract – Twenty-two species of zeolites have been found in marine deposits of Quaternary through Devonian age which are now on land areas. Fresh-water non-marine deposits contain the same species plus faujasite. Geological occurrences of these zeolites are classified into five genetic types: burial diagenesis, contact metamorphism, hydrothermal alteration, submarine hydrothermal alteration, and percolating groundwater alteration. Rock-interstitial water interaction is prevalent during the zeolitization, and migration of chemical components occurs to some degree in all of the types.

Zeolite deposits of burial diagenesis origin in a thick sedimentary column of marine deposits are characterized by regional vertical zoning. In silicic tuff sequences, an unaltered glass zone without zeolites (Zone I) is successively followed by an alkali clinoptilolite–mordenite zone (Zone II), an analcime zone (Zone III), and an albite zone (Zone IV) in silica-saturated environments. Zone III is subdivided into analcime–heulandite (IIIa) and analcime–laumontite (IIIb) subzones, though the occurrences of calcium zeolites are generally local. The downward succession of authigenic silicates represents decreasing hydration. Both field and laboratory evidence prove that the alkali zeolites–albite reaction series is dependent upon temperature. Subsurface data obtained from deep wells show that it is 84–91°C at the boundary Zone II/Zone III and 120–125°C at Zone III/Zone IV. Na^+ is the predominant cation of interstitial water in marine deposits and alkali-saline lake deposits probably causes drastic lowering of the equilibrium temperature of the alkali zeolites–albite reaction series. For example, the reaction analcime + SiO_2 → albite + H_2O occurs at a P_{H_2O} less than 2 kb at 180–190°C in pure water; 120–125°C in interstitial water of 1.5–2.5 × 10^4 ppm Na^+ in marine deposits; and 55–65°C in water of 10^5 ppm Na^+ in alkali-saline lake deposits.

Compositional changes of zeolites within diagenetic zones are still not fully understood. In the heulandite group, high-silica, alkali-rich clinoptilolite in Zone II is replaced by high-silica heulandite in Zone III, probably via intermediate Ca-rich clinoptilolite. In the mordenite group, Na-rich varieties in Zone II tends to be replaced by Ca-rich varieties in Zone II. Systematic changes in analcime composition have not been recognized in Zone III. Albite derived from analcime seems to be richer in Ca and K than that from plagioclase grains.

Statistics of 812 deep-sea zeolite samples collected from Initial Reports of the Deep Sea Drilling Project, Volumes 1 through 32 except 30, reveal that phillipsite concentrates in pelagic brown clay and calcareous sediments of younger age in the Pacific proper, while clinoptilolite concentrates in terrigenous mud, silicic volcanic sediments, and calcareous sediments in the Pacific margin, and in the Atlantic and the Indian Oceans. Most phillipsite may be derived from mafic glass, whereas most clinoptilolite formed from silicic glass or from mafic glass, with excess free silica being of biogenic or weathering origin. This hypothesis can easily explain why phillipsite does not occur in the silicic tuff sequences of marine deposits which are now on land. Temperature seems to be insufficient to promote the reaction clinoptilolite–analcime in deep-sea sediments at a sub-bottom depth less than 1000 m.

INTRODUCTION

As early as the end of the nineteenth century, zeolites were found by Murray and Renard (1891) from the deep-sea sediments collected by the *Challenger* Expedition. Before 1950, however, their occurrences in marine deposits were restricted to a few bentonite deposits (Kerr, 1931; Ota and Sudo, 1949), some silicic and palagonitic tuffs (Sargent, 1925; Bramlette and Posnjak, 1933; Raw and Matley, 1943), and terrigenous rocks (Milton, 1936; Rengarten, 1945; Gilbert and McAndrews, 1948). The classic work by Coombs (1954) on the burial metamorphism of Triassic geosyncline deposits of New Zealand and the overview paper by Coombs *et al.* (1959) on the metamorphic "zeolite facies" stimulated studies on zeolites in sediments and sedimentary rocks. Since then, research on zeolites in marine deposits has rapidly increased in many parts of the world; e.g., Australia, Bulgaria, Italy, Japan, the U.S.A., and the U.S.S.R. as well as New Zealand. Studies on sedimentary zeolites in alkali-saline lake deposits of the western United States and East Africa by Hay, Sheppard and Gude, Surdam, etc., have offered important information relevant to the studies on marine zeolites (Iijima, 1975). The occurrences of natural zeolites in sedimentary rocks were reviewed by Deffeyes (1959), Iijima and Utada (1966, 1972), Hay (1966), Sheppard (1971, 1973), Kossovskaya (1974), and several others.

Zeolite occurrences in fresh-water, non-marine

deposits are almost the same as those in marine deposits on land (Hay, 1966). In the latter, different zeolite suites are formed in silica-undersaturated and silica-saturated environments, the latter of which are far more common and significant. Both calcium- and alkali-zeolites react in silica-saturated environments. The calcium-zeolite reaction series have been thoroughly reviewed by Seki (1969) and Coombs (1971) from the viewpoint of the metamorphic "zeolite facies". In this paper the emphases is therefore on the alkali-zeolite reaction series which are important both in "burial diagenesis" of thick sedimentary columns and in alkali-saline lake deposits.

The knowledge of deep-sea zeolites has rapidly increased with the progress of the Deep Sea Drilling Project (DSDP) by the *Glomar Challenger*, which is now succeeded by the International Project of Ocean Drilling (IPOD). Zeolite occurrences in deep-sea sediments as old as Late Jurassic age and to a sub-bottom depth of more than 1 km are now recognized. Most deep-sea zeolites are alkali-rich varieties: clinoptilolite and phillipsite predominate along with a small amount of analcime. Why are calcium zeolites rare in deep-sea sediments? Why is phillipsite abundant in deep-sea sediments and alkali-saline deposits and not found in silicic tuffs of the marine deposits on land? Does phillipsite transform into clinoptilolite or analcime in deep-sea sediments? Does the clinoptilolite-analcime reaction take place in deep-sea sediments as it does in the marine deposits on land? These questions will be discussed in this paper after compiling occurrences of zeolites in the marine deposits and deep-sea sediments.

OCCURRENCES OF ZEOLITES IN MARINE DEPOSITS

Twenty-two species of the zeolite group have been reported from the marine sediments and sedimentary rocks which are now on land. The occurrences of

TABLE 1. Zeolites in marine and fresh-water non-marine deposits*

Zeolite species	Dominant cation	Geological age of host rocks	Abundance
Faujasite	Ca, Na	Quaternary	Rare
Clinoptilolite	Na, K, Ca	Quat. ~ Carbon.	Abundant
Mordenite	Na, Ca, K	Quat. ~ Carbon.	Abundant
Ferrierite	K, Na, Mg	Miocene	Rare
Erionite	K, Na, Ca	Quat. ~ Eoc.	Rare
Chabazite	Na, K, Ca	Quat. ~ Mio.	Common
Stilbite	Ca, Na	Quat. ~ Jura.	Common
Heulandite	Ca, Na	Plio. ~ Carbon.	Abundant
Phillipsite	K, Na, Ca	Quat. ~ Carbon.	Rare
Harmotome	Ba	Neogene	Rare
Levyn	Ca	Pliocene	Rare
Epistilbite	Ca	Miocene	Rare
Garronite	Ca, Na	Miocene	Rare
Gismondine	Ca, Na, K	Quat. ~ Carbon.	Rare
Laumontite	Ca	Plio. ~ Devon.	Abundant
Mesolite	Ca, Na	Neogene	Rare
Scolecite	Ca	Mio. ~ Perm.	Rare
Yugawaralite	Ca	Miocene	Rare
Thomsonite	Ca, Na	Quat. ~ Jura.	Common
Gonnardite	Na, Ca	Quat. ~ Mio.	Rare
Natrolite	Na	Quat. ~ Perm.	Common
Analcime	Na	Quat. ~ Carbon.	Abundant
Wairakite	Ca	Quat. ~ Cret.	Common

*Sources of data are Hay (1966), Iijima and Utada (1966, 1972), and Sheppard (1971) with the following supplements: Utada *et al.* (1974) on ferrierite; Iijima and Harada (1968) on faujasite, natrolite, gonnardite, and gismondine; Harada *et al.* (1968) on mesolite and scolecite. Species are arranged in decreasing order of $H_2O:Al_2O_3$ ratios.

zeolites in fresh-water, non-marine deposits are not different from those in the marine deposits (Hay, 1966; Iijima and Utada, 1966), except for faujasite which has been reported in Pleistocene mafic palagonite tuffs of Iceland (Peacock, 1926) and Oahu, Hawaii (Iijima and Harada, 1968). The zeolites in both marine and fresh-water, non-marine deposits are listed in Table 1. Zeolite species in volcanic rocks interbedded with those deposits are almost the same, though they commonly have different chemical composition as shown by Sheppard (1971). Zeolites filling veins and joints are not discussed in this paper.

Geologic age and abundance

Most zeolite species occur predominantly in marine and fresh-water non-marine deposits of Cenozoic and Mesozoic ages. The oldest zeolite-containing deposits (laumontite) are Devonian in age in Siberia (Buryanova, 1960) and in New South Wales (Packham and Crook, 1960); they are Jurassic in age in Japan (Fukuda, 1972). Laumontite has also been reported in Pre-cambrian volcanic sediments in Georgia, U.S.A. (Ross, 1958).

Genetic types of zeolitization

The occurrences of zeolites in marine deposits on land are classified into five genetic types, as illustrated schematically in Fig. 1:

Burial diagenesis.
Contact metamorphism.
Hydrothermal alteration.
Submarine hydrothermal alteration.
Percolating groundwater alteration.

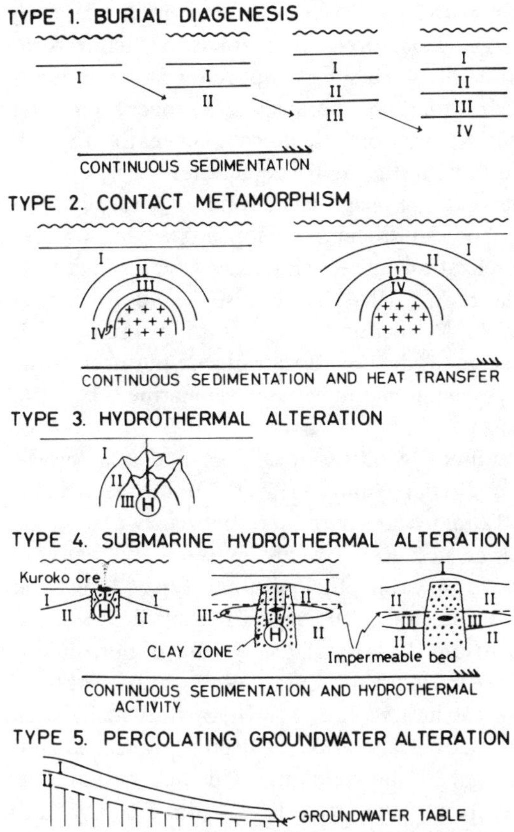

Fig. 1. Five genetic types of the occurrences of zeolites in marine and non-marine fresh-water sediments.

Burial diagenesis. Zeolites are formed as sediments are successively buried by overlying, younger sediments. As a result, zeolites occur in a progressive zonal arrangement in a thick sedimentary column (Iijima and Utada, 1971). This phenomenon is very similar to or the same as the burial metamorphism of Coombs (1954, 1971); however, the term burial diagenesis is preferable because alkali zeolites such as clinoptilolite, mordenite, and analcime are almost always present in the marine deposits and Coombs (1971) excluded these alkali zeolites from the metamorphic régime.

Contact metamorphism. Zeolites are formed in contact aureoles of granite intrusives, resulting in a zonal arrangement around the intrusive body. This type of occurrence seems to be limited to younger orogenic belts where relatively small-scale intrusions occur at shallow depths below the surface or sea-floor into incompletely consolidated, Tertiary volcanic sediments, e.g., in the Green Tuff region of Japan (Seki *et al.*, 1969; Motojuku Rock Alteration Research Subgroup, 1970; Utada, 1973, 1974) and Bulgaria (Kostov, 1969). Calcium zeolites, such as stilbite, heulandite, mordenite, laumontite, wairakite, and yugawaralite, are characteristic in this type.

Hydrothermal alteration. Zeolites are formed by hydrothermal alteration in active geothermal areas such as Wairakei in New Zealand (Steiner, 1953), Onikobe (Seki *et al.*, 1969), and Otake (Hayashi, 1973) in Japan, Yellowstone (Honda and Muffler, 1970), etc. Calcium zeolites, such as mordenite, laumontite, and wairakite, are characteristic of this type.

Submarine hydrothermal alteration. Zeolites are formed by submarine hydrothermal alteration which is commonly related to the Kuroko Pb–Zn–Cu sulfide-ore mineralization, e.g., in the Odate (Iijima, 1974) and Aizu (Utada *et al.*, 1974) districts in the Green Tuff region of northeast Honshu. A lenticular analcime zone is characteristic, which overlies and grades laterally into the argillaceous ore-bearing zones. Analcime formed by reaction of precursor clinoptilolite and mordenite with Na-rich solution permeating penecontemporaneous, porous, pumiceous sediments beneath the Miocene sea-floor from the argillaceous zones.

Ferrierite, a magnesium zeolite, is locally associated with mordenite and clinoptilolite in the Kuroko mineralization areas (Hayakawa and Suzuki, 1970; Utada *et al.*, 1974). It is also found in veins in the Kuroishi Mg-bentonite deposit of Aomori which is an alteration product of Miocene dacite pumiceous tuffs due to penecontemporaneous submarine hydrothermal activity.

Laumontite occurs locally in a deeper level of the Odate district (Iijima, 1974). To the east, laumontite and wairakite are common in the Hanawa mining district (Seki *et al.*, 1968) where gypsum and anhydrite are abundant with the Kuroko-type ore deposits. Laumontite forms an irregularly shaped zone around the gypsum ore deposits of the Wanibuchi mine in Shimane, southeast Honshu (Miura, 1969; Yanagimoto, 1970). These calcium zeolites may have formed by submarine hydrothermal alteration at deeper levels below the sea floor than the sodium and magnesium zeolites. Alternatively, they may have formed by hydrothermal alteration on land at a later stage.

Percolating groundwater alteration. Zeolites are formed by reaction of volcanic glass with percolating groundwater through porous tuffs. Zeolitization occurs at a depth of 200 m below the surface in silicic tuffs of the John Day Formation of Oregon (Hay, 1963) and at a depth of about 5–50 m in ultramafic palagonite tuffs of the Honolulu Group on Oahu Island, Hawaii (Hay and Iijima, 1968 a, b).

Rock-interstitial water reactions take place during zeolitization. In addition to alkalis and alkali earths, silica and alumina are also dissolved from glass during palagonitization of ultramafic tuffs of Oahu; most of the elements are precipitated as zeolites and carbonates in cement, but some are leached from the system (Hay and Iijima, 1968b). The percolating groundwater type is really an open-system as classified by Munson and Sheppard (1974). In other types also, significant migration of chemical components occurs on various scales (Utada, 1965, 1971; Utada *et al.*, 1967, 1974; Iijima, 1974; Iijima and Utada, 1967).

Silica-saturated and silica-undersaturated environments in zeolitic reactions

The silica content of zeolites is generally related to the availability of free silica in a reaction system (Coombs *et al.*, 1959). The chemical environments under which zeolites form can be practically classified into silica-saturated and silica-undersaturated environments, because the varieties and chemical composition of zeolites are quite different between reactions associated with and without excess free silica.

Silica-saturated environments. Figure 2 shows a flow sheet of the zeolite reaction series progressing in a thick column of marine sediments due to burial diagenesis. Overburden and temperature generally increase toward the bottom of the sheet. Clinoptilolite, mordenite, analcime, and albite comprise the alkali zeolites–alkali feldspar reaction series. Ca-clinoptilolite, heulandite, Ca-mordenite, and laumontite comprise the calcium zeolite series. Laumontite and analcime are partly formed from montmorillonite (Murata and Whiteley, 1973; Iijima, 1974). Heulandite and laumontite commonly replace plagioclase (Utada, 1965).

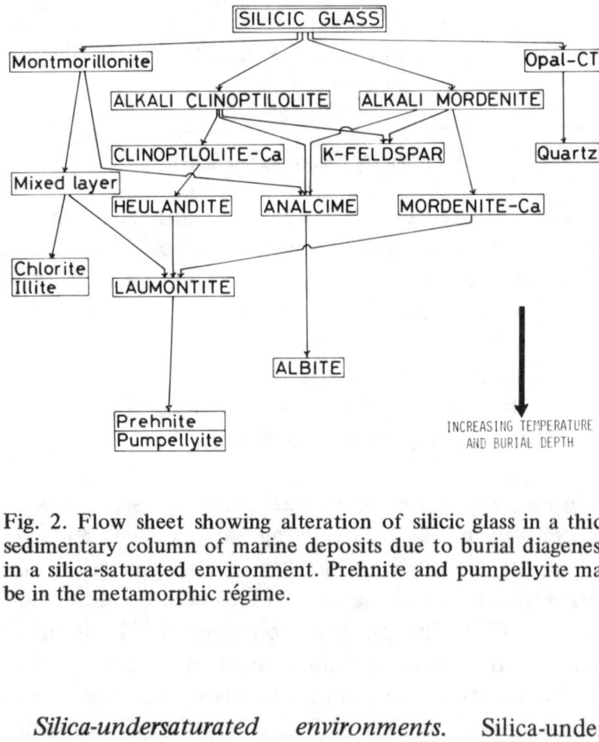

Fig. 2. Flow sheet showing alteration of silicic glass in a thick sedimentary column of marine deposits due to burial diagenesis in a silica-saturated environment. Prehnite and pumpellyite may be in the metamorphic régime.

Silica-undersaturated environments. Silica-undersaturated environments are generally found in mafic and ultra-mafic tuffs and lavas, though their distribution is local. In palagonite tuffs of the Pleistocene Honolulu Group of Oahu, Hawaii, the following sequence of formation of zeolites is recognized: phillipsite → chabazite → gonnardite → natrolite → analcime. These zeolites precipitated from percolating groundwater which dissolved sideromelane at low temperature and pressure (Hay and Iijima, 1968 a, b).

In thick successions of basalt lavas, many varieties of zeolites with relatively low silica content are found in amygdules without free silica (Walker, 1951, 1959, 1960 a, b; Shimazu, 1968; Mizuno, 1974). The zeolites occur in vertical zones. In the Tertiary lavas of eastern Iceland, for example, a chabazite-thomsonite zone in the upper part of the lavas is underlain by an analcime zone which in turn overlies a mesolite-scolecite zone (Walker, 1960b). Walker considers that such zoning reflects temperature distribution within the lava succession.

ZEOLITIC BURIAL DIAGENESIS

Zeolites in marine deposits which are now on land have been largely formed by burial diagenesis unless local heat sources, such as igneous intrusions or hydrothermal activity, exist. Burial diagenesis is characterized by regional occurrences and vertically zoned arrangements of zeolites and related silicate minerals if the deposits contain the proper raw material throughout the sedimentary basins (Iijima and Utada, 1966; Hay, 1966). Volcanic glass is generally the raw material. So-called non-tuffaceous sediments containing zeolites have often been shown to have included volcanic material (Kossovskaya, 1974). Silicic glass, above all, is most important because of its vast quantity and wide distribution in tuffs and tuffaceous sediments. Consequently, silica-saturated environments prevail in zeolitic burial diagenesis of the marine deposits. The following discussion is therefore focused on the alkali zeolites which have been excluded by Coombs (1971) from the metamorphic zeolite facies and which are important as natural zeolite resources.

Zeolite zoning under silica-saturated environments

In silicic volcanic sediments interbedded in a thick geosynclinal column of marine and freshwater deposits, authigenic zeolites and related silicates are distributed in a vertical zonal arrangement due to burial diagenesis. The zones are distinguished mineralogically on the basis of the reaction series of silicic glass-alkali zeolites-albite, as shown in Fig. 3. Zone I is characterized by an absence of zeolites and the presence of silicic glass which is partly altered to montmorillonite and opal-A or opal-CT defined by Jones and Segnet (1975). Zone II is characterized by the reaction of silicic glass with interstitial water to form alkali clinoptilolite, alkali mordenite, opal-CT, and montmorillonite. Zone III is characterized by the transformation of clinoptilolite and mordenite into analcime. Clinoptilolite changes partly to heulandite in the upper part (IIIa), and the heulandite itself to laumontite in the lower part (IIIb) of this zone. Ca-mordenite derived from the alkali mordenite coexists with the analcime in places. Plagioclase grains locally undergo extensive laumontization and faint albitization in Zone IIIb. Zone IV is characterized by transformation of analcime into albite. Plagioclase grains also undergo extensive albitization.

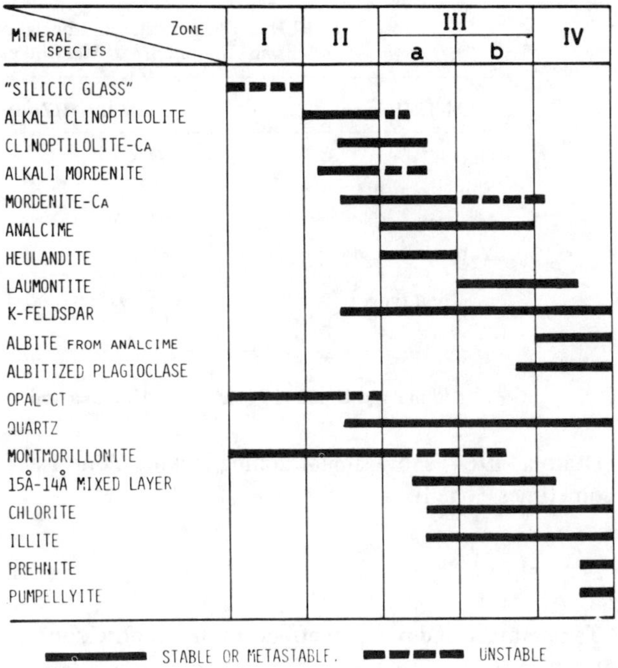

Fig. 3. Schematic diagram showing zonal distribution of authigenic zeolites and silicates in silicic volcanic sediments in a thick column of marine deposits due to burial diagenesis. The zoning is based on the reaction series, silicic glass → alkali clinoptilolite and mordenite → analcime → albite. Zone III is subdivided into two subzones on the basis of the reaction, heulandite → laumontite. Prehnite and pumpellyite may be in the metamorphic régime.

This downward succession of authigenic minerals represents as a whole, decreasing hydration with depth. The precursor zeolites commonly persist as relics, especially in younger Tertiary sediments (Iijima, 1975); e.g., clinoptilolite and mordenite are associated with analcime in Zone III. This phenomenon may be due to a sluggish rate of reaction compared with the rate of sedimentation, or to low permeability of host rocks which prevents rock-interstitial water interaction, or both. The zoning is well documented in both subsurface and subaerial strata of Cenozoic and Cretaceous age throughout the Japanese Islands (Yoshimura, 1961; Utada, 1970; Iijima and Utada, 1971; Iijima, 1975; Owa, 1975; etc.), Triassic sediments in New Zealand (Coombs, 1954; Boles and Coombs, 1975), Upper Paleozoic sediments in New South Wales (Packham and Crook, 1960), and Cretaceous sediments in Puerto Rico

TABLE 2. Temperature and burial depth at boundaries Zone II/Zone III and Zone III/Zone IV in MITI holes in the Niigata and Yabase oilfields

MITI Hole	Zone II/Zone III		Zone III/Zone IV	
Obuchi*	84°C	2900 m	120°C	4160 m
Shimoigarashi*	87°	3500	124°	4480
Masugata*	91°	3490	124°	4500
Yabase oil field†	88°	1700	120°	2500
Total range	84° to 91°	1700 to 3500	120° to 124°	2500 to 4500

*Iijima and Utada (1971). †Huzioka and Yoshikawa (1969).

(Otárola, 1964) show similar zoning lacking Zone I and sometimes Zone II.

Temperature and pressure effects on the zeolite zoning

Direct field evidences. Direct evidences are obtained from deep wells penetrating gently dipping, undeformed strata in which burial diagenesis has been progressing since the beginning of sedimentation. Table 2 gives the data of temperature and burial depth at the top of a zeolite zone, after Iijima and Utada (1971). These values were obtained from four MITI holes which were promoted by the Ministry of Trading and Industry of Japan and which were drilled into Pleistocene and Neogene marine deposits in the Niigata and Akita oilfields in northern Honshu. The transition temperatures in each hole from Zone II to Zone III and from Zone III to Zone IV are in good accord with one another in the different holes. By contrast the burial depth at the transition are variable. The observed fluid pressures including P_{H_2O} are not abnormal. Consequently, the clinoptilolite and/or mordenite-analcime and the analcime-albite reactions are principally influenced by temperature, which is not drastically shifted by P_{total}, P_{H_2O} or the difference between them.

Zeolite zoning in four MITI holes in Hokkaido is not incompatible with the above values as illustrated in Fig. 4. In the MITI-Hamayuchi hole the transitional temperature shows the same range as the holes in the Niigata and Akita oilfields. In the other holes, the zone boundaries occur at lower temperature than in the Hamayuchi hole. This is probably caused by a late uplift and subsequent removal of about 1 to 1.5 km of overburden (Iijima, 1975). In the Kanto Basin and Akita oilfield, Utada (1971) reports the top of the albite zone to occur at a temperature of 70°C, but he does not consider the analcime-albite reaction to have occurred at such a low temperature (personal communication).

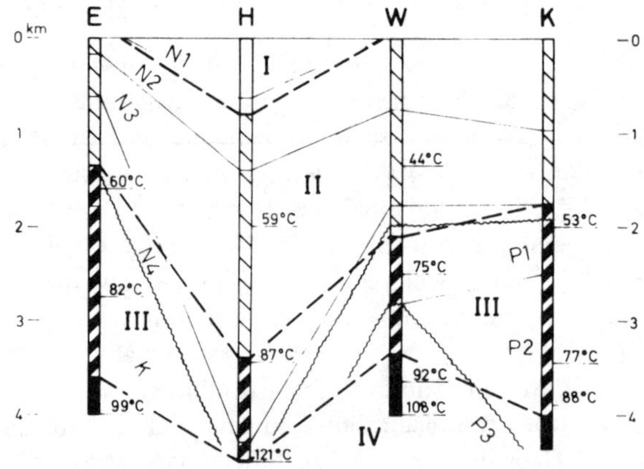

Fig. 4. Zeolite zoning in four MITI holes drilling into folded strata of Tertiary and Cretaceous sediments in the central-axial part of Hokkaido. Temperature indicated is the bottom hole temperature. Uplift and subsequent erosion of overburden in all holes except the Hamayuchi may have caused the lowering of temperature. Abbreviations – E: MITI-Enbetsu, H: MITI-Hamayuchi, W: MITI-Wakkanai, K: MITI-Karumai, N1~4: Neogene formations, Pl~3: Paleogene formations, and K: Upper Cretaceous formations.

Indirect field evidences. The degree of coalification is a function of temperature and duration of heating time: the maximum temperature to which coal has been heated has the greatest effect on increasing coal rank (Teichmüller and Teichmüller, 1966). Correlation between the metamorphic mineral phase and the coal rank were tried by Kisch (1969, 1974). In Japanese Paleogene coal measures, a clear relationship has been found between the zeolite zoning and the degree of coalification (Shimoyama and Iijima, 1976). Lignite and sub-bituminous coal belong to Zone II (clinoptilolite-mordenite), and coking bituminous coal to Zone III (analcime). The coal ranks were defined by the mean values of reflectance of vitrinite supplemented with the O/C and H/C ratios of pure coal. The degree of coalification is not influenced by the burial depths in a range of approximately 1.5-6 km (Fig. 5), nor by the degree of deformation ranging from gentle tilting to overturned folding. Shimoyama and Iijima (1976)

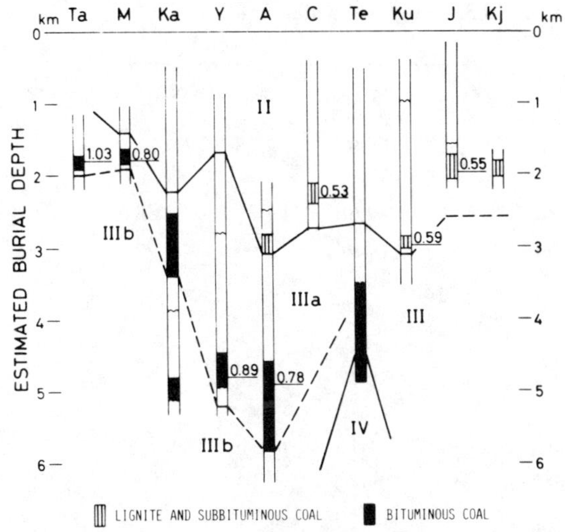

Fig. 5. Schematic diagram showing relationship among zeolite zoning, degree of coalification, and estimated burial depth of the Tertiary coal measures in Japan. Values indicated are the mean reflectance of vitrinite in coal seams. (Modified Fig. 3 of Shimoyama and Iijima, 1976.) Abbreviations – Ta: Takashima, M: Miike, Ka: Karatsu, Y: Yubari, A: Ashibetsu, C: Chikubetsu, Te: Tempoku, K: Kushiro, J: Joban, and Kj: Kuji.

concluded that both zeolite zoning and coalification are a function of the geothermal gradient.

Thermodynamical and experimental data. The equilibria of the reaction, analcime + quartz = albite + H_2O, were calculated thermodynamically by Campbell and Fyfe (1965) and refined by Thompson (1971), based on a hydrothermal experiment by the reaction-rate method. It is remarkable that the equilibrium reaction temperature is constant at 180–190°C at a P_{H_2O} less than approximately 2 kb, a typical pressure during zeolitic burial diagenesis. This is compatible with the above-stated field evidences, though the temperature is considerably higher than that in the marine deposits (*vide infra*).

The above-stated evidences lead to the conclusion that the clinoptilolite and/or mordenite–analcime and the analcime–albite reactions are essentially controlled by temperature under silica-saturated environments in marine deposits on land.

Interstitial water in marine deposits on land

It is difficult to know the true chemical composition of interstitial water in zeolitic rocks because of the difficulty in extracting it from the rocks without change of the composition. However, we are probably able to estimate it based upon certain assumptions.

Weaver and Beck (1971) studied the interstitial water of a marine Neogene mudstone of the Chevron cores beneath the Mississippi Delta where sedimentation has taken place since Miocene time. The interstitial water is characterized by dominant Na^+ and HCO_3^- (including CO_3^{--} and CO_2) and moderately high pH, 8–9. It is remarkable that the Na content is almost constant at a range of 1.5–2.5 × 10^4 ppm over the whole observed burial depth, ranging from 1.2 km to 4.8 km, excepting for local high concentration of about 3 × 10^4 ppm at a horizon of unusually high fluid pressure. The K and Mg contents increase, while the Ca content decreases gradually with increasing burial depth. The abundances of cations are Na >> K > Mg > Ca, whereas those in ocean water are Na >> Mg > Ca ≒ K. The Cl content declines gradually with increasing burial depth from about 10,000 ppm at a depth of 1.2 km to about 2000 ppm at a depth of 4.8 km. The content of Cl is much less than that of normal sea water.

The Cl content in interstitial water of the marine Neogene mudstone and the interbedded, zeolitized, fine-grained, silicic tuff of the MITI cores beneath Niigata Plain was measured (JPDC, 1968). The Niigata Plain has, like the Mississippi Delta, also been the site of sedimentation since Miocene time. The Cl content decreases gradually with increasing burial depth from about 12,000 ppm at a depth of 1 km to about 500 ppm at a depth of 4 km (Iijima and Utada, 1971). At a burial depth less than 1 km, it decreases rapidly toward the surface, probably due to the dilution by meteoric water. The Cl content at the surface is extrapolated to be that of normal, undiluted sea water.

The behavior of chlorine in the interstitial water of mudstone and the geological situations of the MITI and the Chevron cores are quite similar to each other. Also, the Cl contents in both mudstone and zeolitized silicic tuff fall into the same range (Iijima, 1975). Consequently, we may consider that the interstitial water in the zeolitized tuff of the MITI cores might be similar in composition to that in the mudstone of the Chevron cores. We may further generalize this consideration to folded marine deposits during zeolitic burial diagenesis.

Some mineralogical notes

Clinoptilolite and heulandite. Boles (1971) proposed that minerals of the heulandite group having a Si/Al ratio 4.0 or greater are termed clinoptilolite; those with a Si/Al ratio less than 4.0 are termed heulandite. Filizova *et al.* (1972) classified the group into heulandite, Ca-clinoptilolite, and clinoptilolite based on the Si content and thermal behavior. Clinoptilolite represents Na-, K-, and Ca-rich types (Minato and Utada, 1971; Sheppard, 1971), although Mason and Sand (1960) have defined it as the alkali-rich varieties. Heulandite is

TABLE 3. Chemical composition of some authigenic zeolites and albite of burial diagenesis origin

	(1)*	(2)*	(3)†	(4)†	(5)†	(6)†	(7)†	(8)†
SiO_2	66.00	59.65	68.07	60.10	69.60	68.91	68.78	69.21
TiO_2	0.04	0.06	–	–	–	–	–	–
Al_2O_3	11.67	14.06	13.01	19.80	18.90	19.66	19.75	19.88
Fe_2O_3	0.16	0.53	0.01	0.00	0.02	0.02	0.00	0.04
FeO	0.13	0.20	–	–	–	–	–	–
MnO	0.08	0.03	–	–	–	–	–	–
MgO	0.47	0.30	–	–	–	–	–	–
CaO	1.78	5.27	3.73	0.04	0.16	0.18	0.00	0.04
SrO	0.42	1.74	–	–	–	–	–	–
Na_2O	1.54	1.18	0.36	11.50	11.24	11.13	11.43	10.80
K_2O	3.08	1.14	0.31	0.02	0.08	0.10	0.04	0.03
$H_2O(+)$	8.96	11.16	14.51	8.54	–	–	–	–
$H_2O(-)$	5.64	4.65						
P_2O_5	0.03	0.03	–	–	–	–	–	–
	100.00	100.00	100.00	100.00	100.00	100.00	100.00	100.00
Mean ref. index	1.478 ±0.002	1.493 ±0.002						

*Wet analyses by H. Haramura, except for SrO which is analysed by the atomic absorption method.

†X-ray microprobe analyses by R. Matsumoto using the method of Bence and Albee (1968) as the correction factor.

(1) K-rich clinoptilolite in rhyolitic vitric tuff of the Hakobuchi Group of Late Cretaceous; Chikubetsu coal mine abandoned, Haboro Town, Rumoi, Hokkaido (after Iijima, 1971; but SrO is corrected).

(2) Si-rich heulandite in rhyolitic vitric tuff in the lowermost part of the Zone IIIa of the Hakobuchi Group; Tanzangawa, Ashibetsu City, Ishikari, Hokkaido (after Owa and Iijima, 1973).

(3) Ca-rich mordenite in rhyolitic tuff in the uppermost part of the Zone IV of the Fukuyama Formation of Lower Miocene; Shirafugawa, Fukushima Town, Oshima, Hokkaido.

(4) Analcime in rhyolitic vitric tuff in the lowermost part of the Zone III of the Upper Yezo Group of Late Cretaceous; 3766 m deep of the MITI–Enbetsu Hole, in the Tempoku oilfield, Hokkaido.

(5) Albite replacing analcime in rhyolitic vitric tuff in the Zone IV of the Izumi Group; east of Matsuyama, Ehime, Shikoku.

(6) Albite replacing analcime in the same specimen as (3).

(7) Albitized plagioclase in dacitic tuff breccia in the upper part of the Zone IV of the Nanatani Formation; 4488.6 m deep of the MITI–Obuchi Hole in the Niigata oilfield, northeast Honshu.

(8) Albitized plagioclase in rhyolitic sandstone in the lower part of the Zone III of the Upper Yezo Group of Late Cretaceous; 3746 m deep of the MITI–Enbetsu Hole in the Tonpoku oilfield, Hokkaido.

usually rich in Ca except for K-rich heulandite reported by Alietti (1972). Clinoptilolite with the higher thermal stability as defined by Mumpton (1960) belongs to the alkali-rich varieties with high silica content (Minato and Utada, 1971; Filizova *et al.*, 1972).

In marine deposits which are now on land, Na- and K-rich clinoptilolite prevails in altered silicic glassy tuffs of the Zone II. Ca-rich clinoptilolite occurs in mafic volcanics (Sheppard, 1971); it also appears to be common in altered silicic tuffs in the lower part of the Zone II and in the upper part of the Zone III in the Neogene Green Tuff formation in Japan. Heulandite prevails in altered silicic and mafic tuffs of the Zone III. Some heulandite in altered silicic tuffs contains an appreciable amount of Sr (Di Paola, 1965; Owa and Iijima, 1973). Heulandite in altered silicic tuff of the

Late Cretaceous Hakobuchi group in Hokkaido represents the high-silica type of Boles classification even in the lowermost part of the Zone IIIa, just before replacement by laumontite (Owa, 1975). This heulandite was probably derived from a K-rich clinoptilolite with high silica content as shown in Table 3.

The Si/Al ratio of clinoptilolite correlates with that of the original material, and the dominant cations may be controlled by the chemical composition of interstitial water as suggested by Sheppard (1971). In silica-saturated environments, the replacement of clinoptilolite by heulandite may principally be affected by increasing temperature as evidenced by the vertical distribution of both zeolites. The Si/Al ratio of heulandite, however, seems to be strongly influenced by that of the precursor clinoptilolite.

Mordenite. Mordenite has a wide range of chemical composition. According to the hydrothermal synthesis by Nakajima (1973), Ca-mordenite tends to be more thermally stable than Na-mordenite in a silica-saturated environment. Seki (1973) concludes on the basis of X-ray diffraction analyses that Na- and Ca-rich mordenite occurs in the upper and lower parts, respectively, of his mordenite zone including burial diagenesis, hydrothermal alteration and contact metamorphism. Na-rich mordenite occurs dominantly in altered silicic glassy tuffs of the Zone II. Ca-rich mordenite is rarely found even in albitized silicic tuff at the top of the Zone IV in the Fukushima district, Hokkaido, as shown in Table 3. However, more data are needed in order to solve the problem of whether the Ca/Na ratio in mordenite is a function of temperature or not.

Analcime. Analcime in altered silicic tuffs ranges from about 1.95 to 2.93 in Si/Al ratio (Iijima and Hay, 1968; Sheppard, 1971). The Si/Al ratio of analcime is influenced by temperature according to hydrothermal synthesis (Saha, 1959). However, it is affected by the salinity of interstitial water (Iijima and Hay, 1968) and by the Si/Al ratio of precursor zeolites (Sheppard and Gude, 1969) in alkali-saline lake deposits.

Nakajima and Koizumi (1966) considered that the Si/Al ratio of analcime from the Izumi group in the Izumi Mountains, Central Honshu, decreases with the increase of burial depth. However, re-examination reveals that the composition of analcime from the Izumi group and from the Cretaceous–Tertiary deposits in Central Hokkaido does not change systematically as a function of burial depth, and therefore, temperature, as illustrated in Fig. 6 (Iijima, 1975). Furthermore, analcime of burial diagenetic origin in silicic tuffs of the Japanese marine deposits is largely intermediate in Si

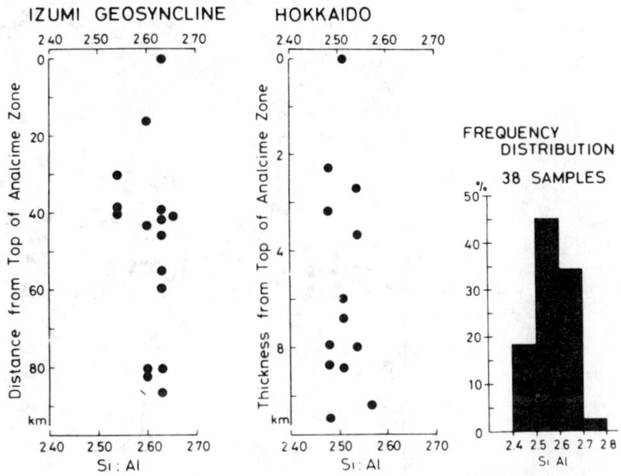

Fig. 6. The Si:Al ratio of analcime in silicic tuffs interbedded with Tertiary and Cretaceous marine deposits in Japan. The analcime is intermediate in composition, which seems not to be a function of burial depth. The ratio is obtained from X-ray diffraction diagrams after the calibration curve of Coombs and Whetten (1967).

content. This is consistent with the result by Coombs and Whetten (1967). These facts strongly suggest that the composition of analcime is principally controlled by the composition of precursor zeolites, clinoptilolite and mordenite.

An X-ray microprobe analyses of analcime samples in altered silicic tuffs from the Tertiary and Cretaceous marine deposits of Japan reveals that sodium is the exclusively dominant cation throughout Zone III and that Ca does not substitute for Na with increasing burial depth (Iijima, 1975; Owa, 1975).

Albite. Authigenic albite of burial diagenetic origin is slightly different in composition depending on the nature of its precursor mineral. In Tertiary and Cretaceous silicic tuffs in Japan, the albite replacing plagioclase is nearly ideal in composition, whereas that replacing analcime contains appreciable amounts of Ca and K as shown in Table 3. This suggests that the composition of albite is not only the function of temperature but also affected by composition of precursors.

ZEOLITES IN DEEP-SEA SEDIMENTS

Since phillipsite and analcime were discovered by Murray and Renard (1891) in the deep-sea sediments collected by the *Challenger* Expedition, zeolites have been reported from many locations (Arrhenius, 1963; Bonatti, 1963; Goldberg, 1961; Zen, 1959, etc.). Our knowledge on deep-sea zeolites, however, has rapidly

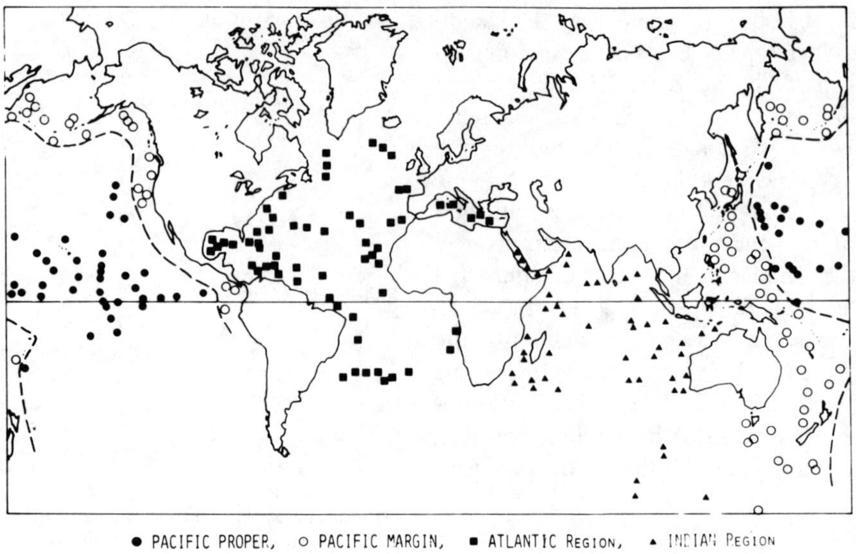

Fig. 7. Division of the DSDP sites into oceanic regions. Data are collected from the DSDP Initial Reports, Vols. 1 through 32 except 30.

increased with a large number of deep holes penetrating the sub-bottom sediments by the *Glomar Challenger* under the Deep Sea Drilling Project (DSDP). The following compilations were made principally from the results of the DSDP Legs 1 through 32 except 30 (Cook *et al.*, 1971; Cook and Zemmels, 1971, 1972, 1973; Cook *et al.*, 1974 a, b, c, 1975 a, b; Fan and Zemmels, 1972; Fan and Rex, 1972; Fan *et al.*, 1973; Matti *et al.*, 1973 a, b, 1974 a, b, c, d; Rex, 1969, 1970 a, b; Rex and Murray, 1970 a, b; Rex *et al.*, 1971; Zemmels, 1973; Zemmels *et al.*, 1972; Zemmels and Cook, 1973 a, b, c, d, 1975) and Leg 40 (Matsumoto *et al.*, 1976).

The world deep-sea area is divided into the Atlantic,

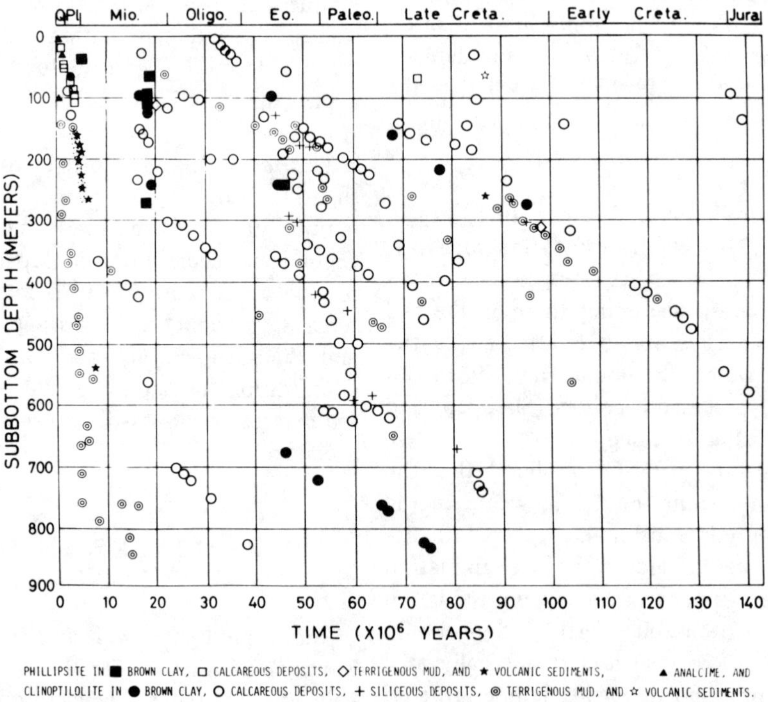

Fig. 8. Diagram showing the occurrences of zeolites in the DSDP curves in the Atlantic region. Data are collected from the DSDP Initial Reports, Vols. 1–4, 10–15. Statistics of deep-sea zeolites are based on this and similar diagrams in other regions.

Indian, and Pacific regions, the last of which is further subdivided into the Pacific Proper and Pacific Margin, as illustrated in Fig. 7. Zeolites in marginal seas are included in the connecting oceans; e.g. the Mediterranean Sea belongs to the Atlantic, the Red Sea to the Indian, and so on. Diagrams showing the relationship among zeolite species, lithologic types of zeolitic sediments, sub-bottom depth, and age of the sediments were made for each of the regions. The diagram of the Atlantic region is shown in Fig. 8. All statistics in the following sections are based on these diagrams.

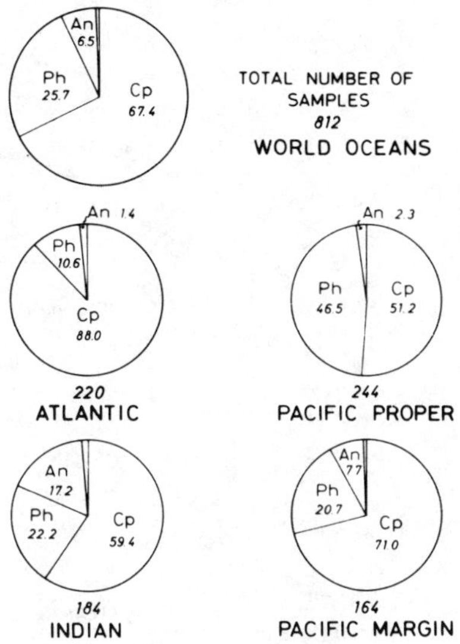

Fig. 9. Number percentage of zeolite samples in the world oceans. Data are collected from the DSDP Initial Reports, Vols. 1 through 32 except 30. Abbreviations – An: analcime, Cp: clinoptilolite, Ph: phillipsite.

Distribution and statistics of deep-sea zeolites in DSDP holes

The distribution of zeolites in the several deep-sea regions is shown in Fig. 9, which has been prepared from X-ray diffraction data on 812 selected samples reported in the Initial Reports of DSDP, Vols. 1 through 32 except 30. Analcime, clinoptilolite–heulandite, erionite, harmotome, mordenite, natrolite, and phillipsite have been found in deep-sea sediments. Approximately two-thirds of all occurrences are clinoptilolite, and about one-fourth are phillipsite. Analcime represents less than one-tenth. Erionite and mordenite are very rare. Additionally, natrolite occurs locally in Quaternary palagonitic mud of the Pacific and the Indian oceans (Arrhenius, 1963; Bonatti, 1963).

The distribution of deep-sea zeolites is distinct in each of the oceanic regions. The Atlantic region is characterized by predominant clinoptilolite, which amounts to 88%, and little phillipsite. In contrast, phillipsite is equally as abundant as clinoptilolite in the Pacific Proper, an area separated from the adjacent continents and island arcs by deep trenches except for the border of North America and Antarctic continents. In the latter areas, the edges of terrigenous turbidite bodies are drawn as the borderline. In the Pacific Margin, on the contrary, clinoptilolite overwhelmingly predominates over phillipsite as in the Atlantic region. Nevertheless, phillipsite is concentrated in pelagic sediments of such marginal seas that intervene between island arcs, such as the Philippine Sea and the Solomon Sea. The Indian region contains dominant clinoptilolite and common phillipsite and analcime, the last of which, above all, represents one-sixth of the total zeolite samples. This value for analcime compares with 7.7% in the Pacific Margin and with much lower amounts in the Atlantic and Pacific regions.

TABLE 4. Lithological types of deep-sea sediments containing zeolites

PELAGIC SEDIMENTS
Brown zeolitic clay
Siliceous deposits
Radiolaria ooze, diatom ooze, porcelanite, and chert
Calcareous deposits
Nannoplankton ooze, foraminifer ooze, chalk, marl, and limestone
TERRIGENOUS SEDIMENTS
Clay, claystone, mud, mudstone, sand, sandstone, and sapropelic shale
VOLCANIC SEDIMENTS
Ash, tuff, volcanic sand, volcanic sandstone, volcanic breccia, and tuffaceous sediments

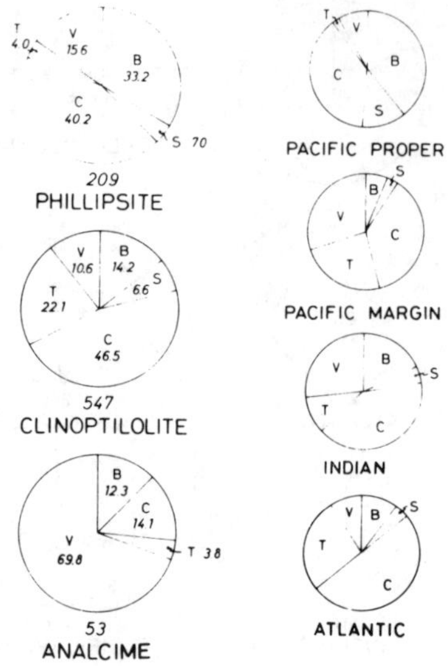

Fig. 10. Number percentage of zeolite samples in lithological types. Data are collected from the DSDP Initial Reports, Vols. 1 through 32 except 30. Abbreviations – B: Brown clay, S: Siliceous sediments, C: Calcareous sediments, T: Terrigenous mud, and V: Volcanic sediments.

Lithological types of zeolite-containing sediments

Deep-sea zeolites occur in pelagic, terrigenous, and volcanic sediments, as shown in Table 4. The statistics of lithologic types of zeolite-containing sediments are illustrated in Fig. 10 where lithologies are grouped into five types: brown clay, siliceous deposits, calcareous deposits, terrigenous sediments, and volcanic sediments.

Approximately 80% of the 209 phillipsite occurrences are in pelagic sediments, especially calcareous deposits and brown clay, most of which contain varying amounts of volcanic glass or palagonite. Phillipsite is not uncommon in basaltic volcanic sediments, but is rather rare in siliceous deposits and terrigenous mud. Two-thirds of the 547 clinoptilolite occurrences are in pelagic sediments, mainly in calcareous deposits. Its abundance in terrigenous mud and rarity in brown clay are contrasted with the rare occurrence of phillipsite in the former and abundance in the latter. Analcime is present largely in basaltic volcanic sediments, and its distribution is local. Erionite, mordenite, and natrolite rarely occur in basaltic sediments or in palagonitic mud.

The relationship between lithologic types and occurrences of deep-sea zeolites can explain the distinct distribution in different oceanic regions, as stated in the preceding section. Clinoptilolite is predominant in the Atlantic region where calcareous deposits and terrigenous mud prevail (Fig. 10). Phillipsite is prominent in the Pacific Proper where brown clay is dominant. In the Pacific Margin, where terrigenous mud is much more common than in the Pacific Proper, clinoptilolite occurrences increase correspondingly. Analcime is significant in the Indian region and in the Pacific Margin where basaltic volcanic sediments are significant.

Geological age

Zeolites have been recognized in deep-sea sediments of Recent through Late Jurassic age, as shown in the time frequency diagram of Fig. 11 which was also prepared from the Initial Reports of DSDP, Vols. 1 through 32 except 30. The frequencies are represented by the percentages of zeolite samples in increments of 10^7 years calculated from the time versus sub-bottom-depth diagrams of each oceanic region, one of which is shown in Fig. 8. Deep-sea zeolite occurrences tend to decrease gradually with increasing age. Three reasons are offered to explain such a distribution. The first is that the older the layer, the more difficult it is to obtain samples, and therefore fewer samples of older rocks are available. This is the most likely factor for Cretaceous samples. The second is that alteration of zeolites has taken place in older layers. The third is that reworking due to regional subaqueous erosion has taken place, as suggested from biostratigraphic gaps described by Kennett *et al.* (1975).

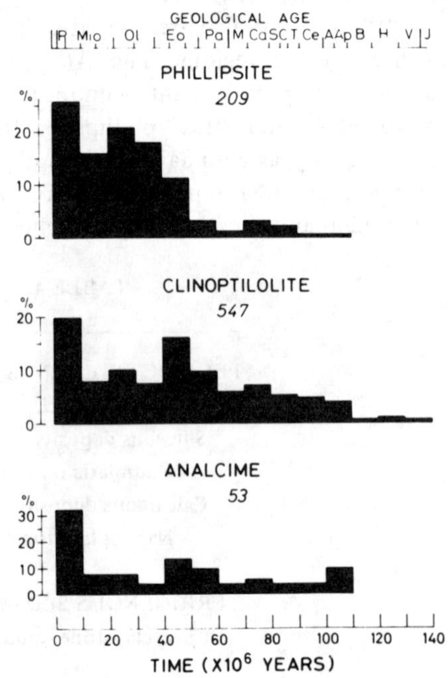

Fig. 11. Time frequency distribution of deep-sea zeolites. Data are collected from the DSDP Initial Reports, Vols. 1 through 32 except 30.

The number of phillipsite samples decrease explicitly with increasing age. Few phillipsites are found in Early Eocene through Turonian, and this mineral is rare in Cenomanian through Aptian. The distribution of clinoptilolite is similar, although a prominent mode occurs in Eocene. Analcime is rather evenly distributed except for a few prominent occurrences in the most recent 10^7 years. Also phillipsite and analcime are not found in pre-Barremian sediments but clinoptilolite has been found in sediments as old as Late Jurassic. Perhaps this could be explained by the first of the above-stated reasons, especially for analcime. Because the number of samples containing phillipsite and analcime is much smaller than those containing clinoptilolite, it has been said that phillipsite is concentrated in Quaternary and Younger Tertiary sediments, whereas clinoptilolite is more common in Older Tertiary and Late Cretaceous sediments (von Rad and Rösch, 1972; Rex and Fan, 1974). The age distribution of phillipsite in the Atlantic region and those of clinoptilolite in the Pacific Proper seem to support this pattern. However, phillipsite is most abundant in Oligocene sediments in the Pacific Proper, and clinoptilolite is rich in post-Late Miocene sediments in the Indian and the Atlantic regions. It is therefore difficult to generalize that phillipsite is concentrated in Quaternary and late Tertiary sediments and that clinoptilolite is concentrated in early Tertiary and Late Cretaceous sediments.

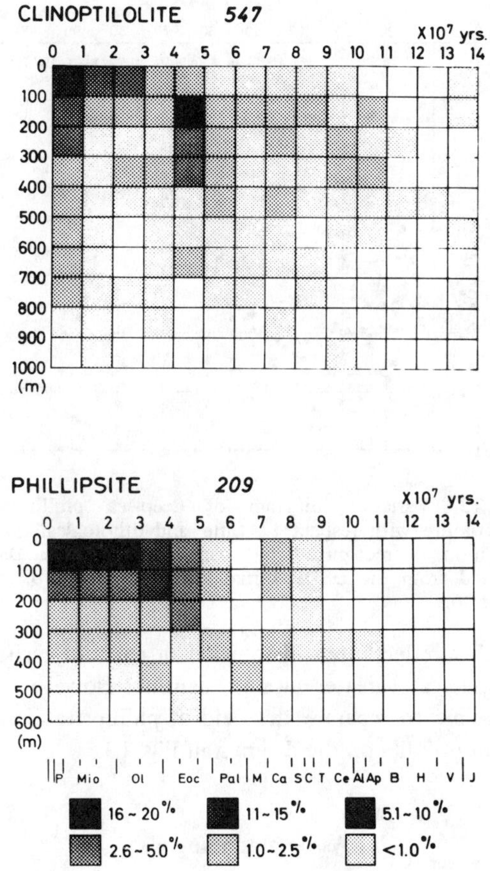

Fig. 12. Density charts showing frequency distribution of deep-sea zeolites by time and by sub-bottom depth. Data are collected from the DSDP Initial Reports, Vols. 1 through 32 except 30.

Sub-bottom depth

Zeolites occur throughout the whole range of sub-bottom depths of the DSDP holes in Legs 1 through 32 except 30, the deepest hole of which penetrated 1300 m below the ocean bottom at Site 222 (Whitmarsh *et al.*, 1974). The density charts in Fig. 12 represent the frequency of occurrences of clinoptilolite and phillipsite samples for every 100-m interval of sub-bottom depth in every period of 10^7 years.

Clinoptilolite is widely distributed in every depth interval as deep as 1000 m in post-Late Jurassic sediments. Its occurrences seem to be more common at shallower depths and in younger sediments. In sediments with an age less than 5×10^7 years (Early Eocene), the frequencies amount to about 18% of the total clinoptilolite samples at a depth range of 0–100 m, and about 31% at a depth range of 0–200 m. Such slight concentration, however, might be due to the sampling deviation discussed in the preceding section.

Phillipsite occurs in post-Barremian sediments at a sub-bottom depth of less than 600 m. Its abundance in shallower and younger sediments is very conspicuous. In sediments with an age less than 5×10^7 years, the frequency of abundance amounts to about 53% of the total phillipsite samples at a depth range of 0–100 m, and about 73% at a depth range of 0–200 m. Indeed, these figures are approximately 3 and 2.4 times, respectively, those of clinoptilolite. An insufficient number of analcime-containing samples precludes a meaningful characterization of their distribution.

Water depth

The relationship among water and sub-bottom depth, geological time, and lithological types of deep-sea zeolite-containing samples is shown in Fig. 13, prepared from the DSDP Initial Reports, Vols. 1 through 32 except 30. Deep-sea zeolites are present at a depth of water plus sub-bottom depth from approximately 1.2 km to 6.3 km. In general, the zeolites contained in calcareous sediments occur at shallower depths than those in non-calcareous sediments. This occurrence

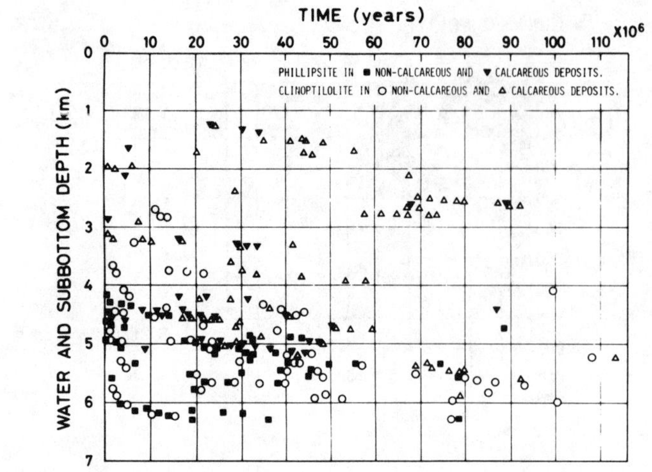

Fig. 13. Variation diagram of deep-sea phillipsite and clinoptilolite with respect to time and lithological types. No distinction is recognized between both zeolites. Data are collected from the DSDP Initial Reports, Vols. 1 through 32 except 30.

probably reflects that dissolution of calcium carbonates increases at greater water depths. However, it is impossible to separate the field of phillipsite from that of clinoptilolite on the diagram in Fig. 13.

Associated minerals

Deep-sea zeolites are associated with various minerals. Smectite is closely associated with phillipsite and clinoptilolite as pointed out by many researchers (Zemmels *et al.*, 1972; Rex and Fan, 1974). On the one hand, smectite is a main alteration product of volcanic glass in ash beds and pumice grains in deep-sea sediments (Andrews and Ovenshine, 1975; Lisitzin, 1972). On the other hand, it is derived from weathered detritus, especially in tropical zone (Lisitzin, 1972). In the Pacific Proper, volcanic glass is probably a principal source of smectite in pelagic sediments. *Sepiolite* and *palygorskite* are commonly found with clinoptilolite especially in the Atlantic region (Bonatti and Joensum, 1968; Zemmels *et al.*, 1972). However, the genetical relation of the fibrous Mg silicates to clinoptilolite is still a problem.

Opal-CT is a main form of the silica of deep-sea "chert" originating from biogenic amorphous silica such as diatom frustules and radiolaria tests (Jones and Segnit, 1975; Florke *et al.*, 1975). Clinoptilolite is commonly associated with "cristobalite" and "tridymite" (X-ray mineralogy tables in the DSDP Initial Reports, Vols. 1 through 32 except 30), which are probably Opal-CT. In the clinoptilolite-opal-CT assemblage, there is a rough correlation between the sub-bottom depth and the age of sediments, that is its first appearance seems to become shallower as the age of sediments is older. The assemblage appears at a depth range of 400–500 m in sediments less than 2×10^7 years old, at a depth range of 300–400 m in sediments $2\text{-}4 \times 10^7$ years old, at a depth range of 100–200 m in sediments of $3\text{-}4 \times 10^7$ years old, and at a depth less than 100 m in sediments more than 4×10^7 years old. It is remarkable that phillipsite and analcime are rarely associated with opal-CT. Authigenic chalcedonic quartz

TABLE 5. Chemical composition of interstitial water in the DSDP cores selected from legs 5, 6, 9, 11, 12, and 14

	Phillipsite-containing sediments	Clinoptilolite-containing sediments	Non-zeolitic sediments	Surface sea water
Na^+(g/l)	10.5 ~ 11.3	9.5 ~ 15.2	9.2 ~ 15.7	10.0 ~ 11.5
K^+(g/l)	0.33 ~ 0.48	0.19 ~ 0.56	0.12 ~ 0.53	0.37 ~ 0.41
Ca^{++}(g/l)	0.31 ~ 0.60	0.17 ~ 2.51	0.10 ~ 3.20	0.39 ~ 0.44
Cl^-(g/l)	19.2 ~ 19.6	18.15 ~ 27.94	18.67 ~ 25.28	18.1 ~ 20.76
Si (ppm)	5.1 ~ 21.0	5.5 ~ 42.0	2.6 ~ 45.0	0.4 ~ 4.2
Na/K	22.7 ~ 32.1	19.4 ~ 54.0		
Na+K/Ca	18.2 ~ 36.4	6.3 ~ 51.8		
pH	6.9 ~ 7.8	6.7 ~ 8.1	7.5 ~ 7.9	
Geological age	Quat. ~ Paleocene	Quat. ~ Late Cret.	Quat. ~ Late Cret.	Recent
Sub-bottom depth	2 ~ 120 m	8 ~ 660 m	2 ~ 686 m	
Total number of samples	17	36	87	5

may locally coexist with clinoptilolite in Upper Cretaceous siliceous sediments at a depth of more than 400 m.

Calcium carbonates are commonly associated with deep-sea zeolites. They occur in the form of fossil remains and chemical precipitates which probably originated from biogenic carbonates. As pointed out above, both clinoptilolite and phillipsite occur commonly in calcareous deposits.

Interstitial water of zeolite-containing sediments

The chemical composition of interstitial water of sediments may control the varieties of zeolites contained in them. The content of Na^+, K^+, Ca^+, Cl^-, and Si of interstitial water of the DSDP core sediments and surface sea water are summarized in Table 5 prepared from the DSDP Initial Reports, Vols. 4, 5, 6, 9, 11, 12, and 14 (Manheim *et al.*, 1970; Manheim and Sayles, 1971; Manheim *et al.*, 1972; Sayles *et al.*, 1971, 1972 a, b; Waterman *et al.*, 1972). Phillipsite samples occur in a sub-bottom depth range of 2–120 m, while clinoptilolite occurs in the range of 8–660 m. The interstitial water of phillipsite-containing sediments is considerably narrower in composition than that of clinoptilolite-containing sediments. Both waters, however, are similar to one another in composition in the same range of sub-bottom depth less than 120 m (Table 5). It is remarkable, furthermore, that the interstitial water of non-zeolitic sediments is not very different in composition from that of zeolitic sediments. The content and state of aluminum in interstitial water are probably distinct in both type of sediments, though no data are available.

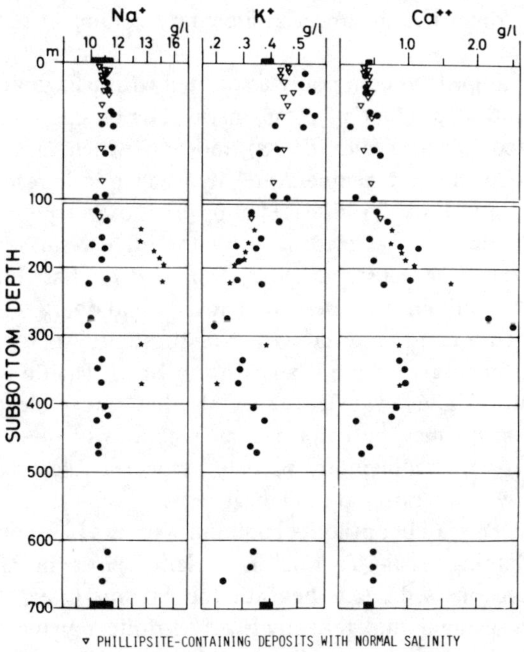

Fig. 14. Diagram showing relationship between cation content of interstitial water of zeolite-containing, deep-sea sediments and sub-bottom depth. The composition of surface sea water is indicated by solid bars on the scale. Data are collected from the DSDP Initial Reports, Vols. 4, 5, 6, 9, 11, 12, and 14.

The vertical change of the content of principal cations in interstitial water of zeolite-containing sediments within the sedimentary column is shown in Fig. 14. The Na^+ content is 9.5–15.2 g/l; it is significant that most fall in the narrow range of 10.0–11.5 g/l, the same as surface sea water, throughout the whole range of sub-bottom depths observed. Interstitial water of high salinity occurs locally in sediments of continental rises off west Africa (Site 140) and northeast Brazil (Site 144).

The K^+ content ranges from 0.19–0.56 g/l. In zeolitic sediments from a sub-bottom depth less than approximately 120 m, it is slightly greater than surface sea water. In some calcareous ooze and chalk containing clinoptilolite it is higher than in phillipsite-containing sediments. The K^+ content tends to decrease gradually to a depth of about 300 m, below which it is rather constant in a range of 0.14–0 42 g/l.

The Ca^+ content ranges from 0.17–2.51 g/l. This is considerably different from the Ca^+ content of surface sea water, except where the sub-bottom depth is less than approximately 25 m. In a sub-bottom depth range of 2–120 m, the Ca^+ content is similar in both phillipsite- and clinoptilolite-containing sediments. At depths of more than 120 m, clinoptilolite occurs commonly in interstitial water rich in Ca^+. The Si content of interstitial water of zeolitic sediments ranges from 5–42 ppm for both phillipsite- and clinoptilolite-containing sediments as deep as about 50 m of sub-bottom depth. At depths of more than 50 m, clinoptilolite commonly occurs in sediments containing high-silica interstitial water. The Cl^- content varies from 18.15–27.94 g/l, almost the same range of surface sea water. It tends to decrease with increasing sub-bottom depth, except for high chlorinity in the sediments of continental rises at the DSDP Sites 140 and 144.

The chemical composition of interstitial water in the DSDP cores is represented in the [Na/K]- and [(Na+K)/Ca]-SiO_2 diagrams in Fig. 15. The phillipsite field is more restricted than the clinoptilolite field which includes it. The interstitial water of sediments rich in phillipsite has a compositional range of Na/K = 22.8–25.2, (Na+K)/Ca = 25.2–36.5, and the Si = 5–18 ppm. The interstitial water of sediments rich in clinoptilolite has a compositional range of Na/K =

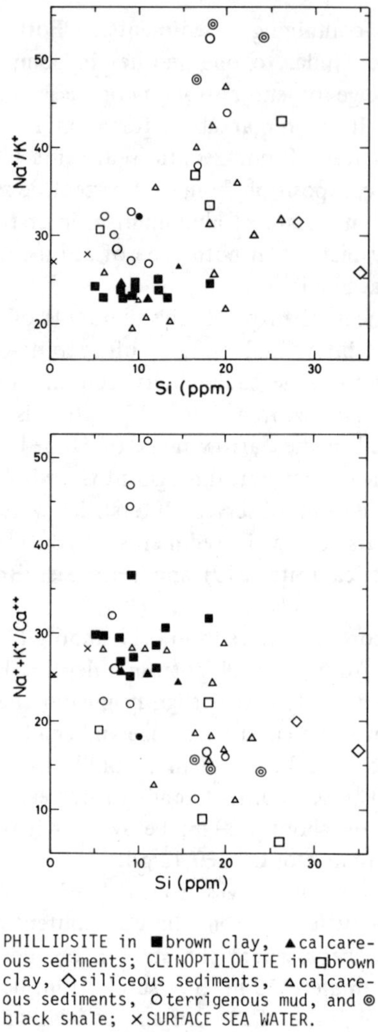

Fig. 15. Variation diagrams of composition of interstitial water in deep-sea sediments containing phillipsite and clinoptilolite with respect to the Na/K and Na+K/Ca ratios and the Si content. Smaller symbols represent the sediments which do not contain enough zeolites to be detected by X-ray diffraction of bulk samples. Data are collected from the DSDP Initial Reports, Vols. 4, 5, 6, 9, 11, 12, and 14.

TABLE 6. Diagenetic origin of phillipsite and clinoptilolite in deep-sea sediments

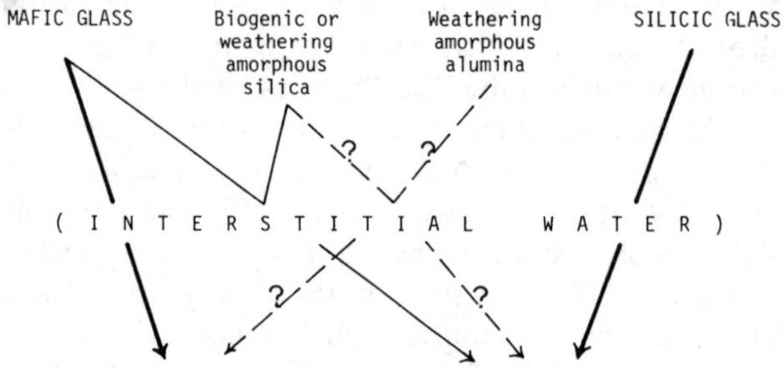

25.6–54.0, (Na+K)/Ca = 6.4–26.0, and Si = 7.4–35.0 ppm. It is probable that the chemical environment which favors clinoptilolite has higher Si/Al and lower (Na+K)/Ca ratios, while that which favors phillipsite has lower Si/Al and higher (Na+K)/Ca ratios, and above all, more K^+. The environment favorable to phillipsite prevails in shallow depths of burial, usually less than 120 m. On the contrary, that favorable to clinoptilolite prevails in deeper depths of burial. As implied above, water depth seems not to influence the chemical environments.

ORIGINS OF DEEP-SEA ZEOLITES

Three origins of deep-sea zeolites

Three origins of deep-sea zeolites may be considered: diagenetic, hydrothermal, and metamorphic. Some zeolites are probably formed by submarine hydrothermal alteration as suggested by Hathaway and Sachs (1965). Some are probably formed by thermal metamorphism at the contacts with intrusive basalts. Local occurrences of mordenite, erionite, and analcime suggest such origins. Most phillipsite and clinoptilolite are probably of a diagenetic origin, because of their regional occurrence and dominant appearance in younger and somewhat shallower sedimentary columns, as stated above.

Phillipsite is commonly associated with palagonite, an alteration product of clear, mafic glass or sideromelane (Peacock and Fuller, 1928), most of which may have formed by low-temperature hydrochemical reactions (Hay and Iijima, 1968b). Phillipsite is not in equilibrium with sea water as evidenced by the etched surfaces of crystals (Morgenstein, 1967; Rex, 1967). Based on a study on Th isotopes, Bernat and Goldberg (1969) proved a continuous growth of phillipsite crystals within a sedimentary column beneath the North Pacific ocean floor. These evidences suggest that most deep-sea phillipsite may have formed by reactions of interstitial water with sedimentary particles, especially glass shards, at low temperature during diagenesis.

Deep-sea clinoptilolite commonly occurs in sediments containing opal-CT. Coalified plant debris in Upper Cretaceous sediments beneath the Atlantic Ocean floor show a maximum reflectivity of vitrinite ranging from 0.35–0.57%, corresponding to lignite to sub-bituminous B-coal (Berger and von Rad, 1972). Alkali clinoptilolite is stable as high as 84–91°C with excess opal-CT ("low cristobalite") in Tertiary and Cretaceous marine deposits on land with a burial depth as much as 3.5 km (Iijima and Utada, 1971; Iijima, 1975). The clinoptilolite-mordenite zone corresponds to a degree of coalification of lignite and sub-bituminous coal (Shimoyama and Iijima, 1976). Consequently, the physicochemical condition of the formation of clinoptilolite seems not to be different between the marine deposits on land and deep-sea sediments.

Origin of phillipsite and clinoptilolite

What is the difference in origin between phillipsite and clinoptilolite? Why is not phillipsite found in silicic tuffs interbedded with the marine deposits which are now on land, whereas it is common in both deep-sea and alkali-saline deposits?

Chemical analyses of deep-sea phillipsite reveal that the Si/Al ratio is 2–3 which is intermediate between that in mafic volcanics and silicic tuffs of alkali-saline lake deposits. Alkali cations are also dominant and K predominates over Na (Sheppard and Gude, 1970). Alkali clinoptilolite usually has a Si/Al ratio of 5–7 (Minato and Utada, 1971; Sheppard, 1971; Boles, 1972; Filizova *et al.*, 1972). Consequently, phillipsite should be formed by a chemical environment with a lower activity of silica and higher activities of alumina and potassium. Such an environment fits with that described for pelagic sediments beneath the floor with a sub-bottom depth less than about 120 m, as stated above. Moreover, alumina may be readily available by the palagonitization of mafic glass (Hay and Iijima, 1968 a, b; Iijima and Harada, 1968). In fact, phillipsite is commonly found in basaltic volcanic sediments and palagonitic brown clay of younger geologic age (Bonatti, 1963; Alexanderssen, 1970; Rex, 1967; Cook and Zemmels, 1972). It also occurs in pores of radiolaria tests (Arrhenius, 1963) and in cement of chalk (Zemmels *et al.*, 1972). It is rare in younger sediments of the Atlantic and the Pacific Margin where large amounts of silicic volcanic materials and fine particles of weathered detritus have been derived from the neighboring continents and island arcs, as stated above. Clinoptilolite, not phillipsite, is dominant here. It is remarkable that no phillipsite has been recorded from Quaternary and Neogene silicic tuffs and tuffaceous sediments in the Pacific Margin, e.g., the DSDP Hole 84 off Panama (Cook and Zemmels, 1972). The moderately high silica content of deep-sea phillipsite, as compared with phillipsite in mafic volcanics on land (Sheppard and Gude, 1970), may be due principally to the substitution (K,Na)Si ⇌ CaAl (Hay, 1964) in interstitial water rich in alkalis beneath the ocean floor. High-silica phillipsite forms by reaction of silicic glass with the highly alkaline waters of saline lake deposits (Hay, 1964; Sheppard, 1971).

The high Si/Al ratio in clinoptilolite strongly suggests that a favorable chemical environment for its formation is one with high silica and low alumina activities. Such an environment may be easily achieved by dissolution of silicic glass. Palagonitization of mafic glass may be favored if excess soluble silica is supplied from other sources, e.g., dissolution of biogenic silica or amorphous silica from weathering. Euhedral crystals of clinoptilolite occur in pores of radiolaria tests in sediments which show no contamination with volcanic ash (von Rad and Rösch, 1972). In this case, biogenic silica may have supplied silica for the clinoptilolite, but is the source of the alumina for it.

The uppermost occurrence of clinoptilolite in a deep-sea sedimentary column seems to be much shallower than in the marine deposits which are now on land. On the one hand, clinoptilolite was found at a sub-bottom depth of 0.05–7.57 m in Pleistocene volcanic, montmorillonite-rich, radiolarian-nannofossil ooze from the DSDP Hole 84 (Cook and Zemmels, 1972). On the other hand, it is found only below a burial depth of 800 m in Pleistocene tuffaceous mudstone from the MITI-Obuchi Hole in Japan (Iijima and Utada, 1971). Clinoptilolite is common in deep-sea sediments at sub-bottom depths less than 100 m, whereas in marine deposits it is restricted to burial depths of more than a few hundred meters. Are such shallow occurrences of the deep-sea clinoptilolite caused by submarine erosion or a reworking origin? Or are they either affected by the composition of interstitial water or large water depth? Zeolites are decomposed so easily by meteoric water that they are never seen as fine, detrital particles of a terrestrial origin. In marine environments, however, reworked zeolite grains are not uncommon. For example, zeolitized radiolarian tests of Cretaceous age are sometimes recognized in the Tertiary sediments above a submarine hiatus (Kennett *et al.*, 1975). The occurrences of clinoptilolite in the Blake–Bahama Outer Ridge section in the northwest Atlantic (Zemmels, Cook, and Hathaway, 1972; Paulus, 1972) are very suggestive, as shown in Fig. 16. The distribution pattern is not concordant with the present sea floor but with the stratigraphic sequence, suggesting submarine erosion after the formation of clinoptilolite due to burial diagenesis. This contrasts with the occurrences of authigenic carbonates which is affected by a methane-rich zone, which is concordant with the present submarine configuration as shown by Lancelot and Ewing (1972).

Alteration of deep-sea zeolites

Phillipsite tends to be abundant in younger pelagic sediments at shallower sub-bottom depths, while clinoptilolite is widely distributed in sediments of Quaternary through Late Jurassic age, as stated above. Is phillipsite altered to clinoptilolite or to other minerals in older or deep-buried sediments? Kossovskaya (1974) considers that prismatic crystals of clinoptilolite are pseudomorphic after phillipsite and that tabular crystals are primary. Nevertheless, both habits exist in the primary clinoptilolite (G. N. Kirov, personal communication). In alkali-saline lake deposits, phillipsite coexists with clinoptilolite; but their interaction has not been recognized. Both zeolites change to analcime and K-feldspar by reaction with alkali-saline water (Hay, 1966; Sheppard and Gude, 1968, 1969). Authigenic K-feldspar occurs in the form of euhedral crystals in deep-sea sediments (von Rad and Rösch, 1972). Some may have originated from phillipsite and possible clinoptilolite.

Analcime occurs locally in basaltic sediments and palagonitic mud. The occurrence of deep-sea analcime is quite different from the analcime zone of the marine deposits on land. In silicic tuffs of the marine deposits, alkali clinoptilolite and mordenite change into analcime at 84–91°C in a quartz-saturated environment due to burial diagenesis (Iijima and Utada, 1971; Iijima, 1975). In the Pleistocene basanite–palagonite tuffs of Koko Crater on Oahu Island, Hawaii, analcime is precipitated in pores from percolating cold water which reacts with phillipsite and chabazite (Hay and Iijima, 1968a; Iijima and Harada, 1968). However, it is not clear whether the deep-sea analcime is precipitated from cold interstitial water or formed by hydrothermal alteration.

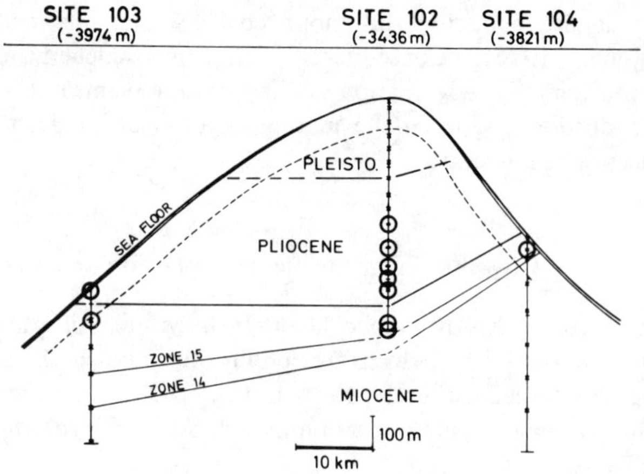

Fig. 16. Cross-section of the Blake–Bahama Outer Ridge (modified and simplified Fig. 3 of Paulus, 1972), showing that the occurrences of clinoptilolite marked in circles are controlled by stratigraphy. The broken line running parallel to the sea floor represents the top of methane-rich sediments.

EFFECT OF PORE WATER ON THE CLINOPTILOLITE–ANALCIME–ALBITE REACTION SERIES

Both field and synthetic data, as stated above, have proved that the analcime–albite equilibria are essentially affected by temperature at a P_{H_2O} less than approximately 2 kb in silica-saturated environments. The alkali clinoptilolite–analcime equilibria might also be influenced by temperature at lower P_{H_2O} as suggested by the MITI drilling data. Nevertheless, these reactions take place at different temperature in various environments.

Analcime–albite equilibria

The analcime–albite reaction takes place at 180–190°C at a P_{H_2O} less than 2 kb (Campbell and Fyfe, 1965; Thompson, 1971). In nature, analcime starts to react with quartz to form albite at 120–125°C at a burial depth ranging from 2.5 km to 4.5 km in silicic tuffs of the Miocene marine deposits (Iijima and Utada, 1971; Iijima, 1975). The difference between P_{total} and P_{H_2O} does not cause any drastic change of temperature. The same reaction occurs at 55–65°C at a burial depth of about 2 km in silicic tuffs of the Eocene Green River Formation (Iijima and Hay, 1968) which is the only known alkali-saline lake deposits containing authigenic albite derived from analcime. What causes the lowering of the reaction temperature?

Ions in interstitial water probably play a decisive role in the lowering of reaction temperature. It is not reasonable that the concentrations of K, Ca, Mg, and Cl affect the reaction temperature, because they change only gradually with burial depth in thick columns of marine deposits as stated above. Na^+, HCO_3^-, and CO_3^{--} are dominant ions in interstitial water of both marine and alkali-saline deposits. The concentration of Na^+ is probably one of the essential factors in controlling the reaction temperature. As shown in Table 7, temperature is lowering with increasing Na^+ concentration. Hydrothermal experiments (Campbell and Fyfe, 1965; Nakajima, 1973) support this hypothesis. Lowering of a H_2O activity through increase of salinity also is a factor (Hay, 1966).

The relationship between the concentration of Na^+ in pore water and the equilibrium temperature at a P_{H_2O} less than 2 kb is illustrated on a schematic diagram in Fig. 17. The smooth curve connecting the thermodynamic equilibria, the marine deposits, and the alkali-saline lake deposits probably represents the analcime–albite equilibria at the lower P_{H_2O}. In the Otake geothermal area in Kyushu, albite is formed by reaction of plagioclase with weakly alkaline hot water of

TABLE 7. The Na^+ content in interstitial water and the equilibrium or reaction temperature of the formation of albite from analcime

	Na^+ content (ppm)	Temperature (°C)
Alkali-saline lake deposits of the Green River Formation*	10^5	55° ~ 65°
Marine deposits of the Tertiary and Upper Cretaceous in the MITI holes†	$1.5 \sim 2.5 \times 10^4$	120° ~ 124°
Albitization of plagioclase in hydrothermal water in the Otake geothermal area‡	$0.8 \sim 1.5 \times 10^3$	150° ~ 240°
Thermodynamical calculation and hydrothermal experiment §	Nil or trace	180° ~ 190°

*Na^+ content: Surdam and Parker (1972), Temperature: Iijima and Hay (1967).

†Iijima and Utada (1971) and Iijima (1975).

‡Hayashi (1973).

§Campbell and Fyfe (1965) and Thompson (1971).

$0.8-1.5 \times 10^3$ ppm Na^+ at 150–240°C (Hayashi, 1973). These figures are not inconsistent with the equilibrium curve, reflecting the fact that albitization of plagioclase occurs in deposits exhibiting the analcime–albite reaction.

Alkali clinoptilolite–analcime equilibria

The alkali clinoptilolite–analcime reaction begins at 84–91°C at a burial depth of 1.7 km to 3.5 km in silicic tuffs of the marine deposits of the MITI holes (Iijima and Utada, 1971; Iijima, 1975). This suggests that the reaction temperature is not affected drastically by a difference in burial depth. The reaction occurs at normal temperature and pressures in silicic tuffs in alkali-saline lake deposits of the western part of the United States (Hay, 1966; Sheppard and Gude, 1968, 1969; Sheppard, 1971). In the laboratory Boles (1971) transformed clinoptilolite into analcime in a Na_2CO_3–NaOH solution of 4600 ppm Na^+ at 100°C.

The discrepancy in the reaction temperature is interpreted to be mainly caused by the concentration of Na^+ in interstitial water; that is, the temperature of the analcime–albite reaction is lowering with increased concentration of Na^+. The clinoptilolite–analcime reaction is also illustrated on the temperature versus Na^+-concentration diagram in Fig. 17.

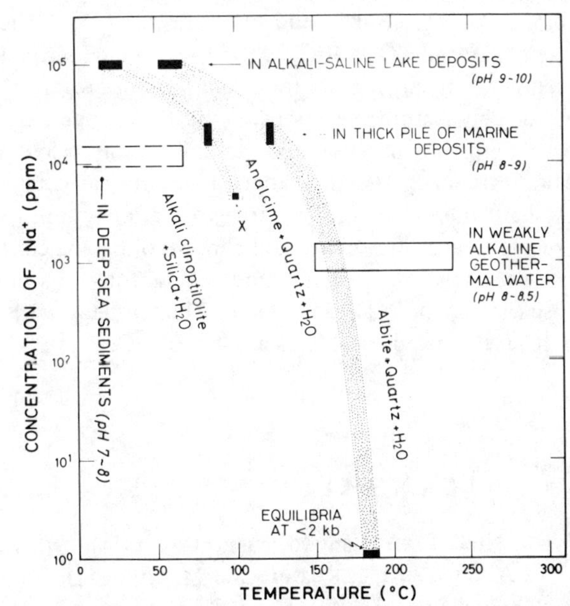

Fig. 17. Diagram showing equilibria of the reaction series, alkali clinoptilolite → analcime → albite at a P_{H_2O} less than approximately 2 kb in a silica-saturated environment with respect to temperature and Na concentration of the interstitial water (modified Fig. 7 of Iijima, 1975).

Clinoptilolite is abundant in deep-sea sediments, while analcime is localized and less abundant as already stated. Clinoptilolite is commonly associated with

Opal-CT in the deeper or older parts of the sedimentary columns. Most clinoptilolite is probably not subjected to temperature greater than 60°C at a sub-bottom depth less than 1 km except for that in contact with intrusive basalt and in a hydrothermal area. Plant debris contained in the Upper Cretaceous sediments of the Atlantic has the ranks of lignite to sub-bituminous B-coal (Berger and von Rad, 1972) which correlates with Zone II (clinoptilolite–mordenite zone) in the Paleogene coal measures in Japan. The Na^+ concentration of interstitial water in deep-sea fine-grained sediments is almost constant at a range of 1.0–1.5 × 10^4 ppm regardless of sub-bottom depth as stated above. This is consistent with the interstitial water in marine mudstone on land. The equilibrium diagram suggests that it is too cold to promote the clinoptilolite–analcime reaction in the deep-sea sediments as shown in Fig. 17.

Remarks

In the heulandite–laumontite reaction in silica-saturated environments, the equilibrium temperature may be lowered by the difference between P_{total} and P_{fluid} and the nature of the ions in stratal water (Coombs, 1971; Boles and Coombs, 1975). The concentrations of CO_2, HCO_3^-, and CO_3^{--} might affect the temperature much more for calcium zeolite reactions than for alkali zeolite reactions, if the Ca^{++} concentration is critical, because the solubility of calcium carbonate is much less than that of sodium carbonate. The heulandite–laumontite transformation appears between the analcime–albite and clinoptilolite–analcime transformations in Tertiary and Cretaceous marine deposits in Japan, as already stated. The temperature is, therefore, estimated roughly as around 100°C.

ACKNOWLEDGMENTS

I benefited from many discussions with Drs. R. L. Hay, R. A. Sheppard, I. Kostov, and M. Utada. Dr. Hay read the manuscript. Miss Yuki Komoto typed the manuscript.

REFERENCES

Alexanderssen, T. (1970) The sedimentary xenoliths from Surtsey: marine sediments lithified on the sea-floor: *Surtsey Research Progress Report*, pp. 1-6.

Alietti, A. (1972) Polymorphism, and crystal-chemistry of heulandites and clinoptilolites: *Amer. Mineral.* **57**, 1437-1451.

Andrews, P. B. and Ovenshine, A. T. (1975) Terrigenous and clay facies: deposits of the early phase of ocean basin evolution: *Initial Reports of the Deep Sea Drilling Project*, Vol. **29**, Washington (U.S. Government Printing Office), pp. 1049-1063.

Arrhenius, B. O. S. (1963) Pelagic sediments: *The Sea* **3**, 655-727.

Bence, A. E. and Albee, A. L. (1968) Empirical correction factors for the electron microanalysis of silicates and oxides: *J. Geol.* **76**, 382-403.

Berger, W. H. and von Rad, U. (1972) Cretaceous and Cenozoic sediments from the Atlantic Ocean: *Initial Reports of the Deep Sea Drilling Project*, Vol. **14**, Washington (U.S. Government Printing Office), pp. 788-954.

Bernat, M. and Goldberg, E. D. (1969) Thorium isotopes in the marine environment: *Earth Planet. Sci. Letters* **5**, 308-312.

Boles, J. R. (1971) Synthesis of analcime from natural heulandite and clinoptilolite: *Amer. Mineral.* **56**, 1724-1734.

Boles, J. R. (1972) Composition, optical properties, cell dimensions, and thermal stability of some heulandite group zeolites: *Amer. Mineral.* **57**, 1463-1493.

Boles, J. R. and Coombs, D. S. (1975) Mineral reactions in zeolitic Triassic tuff, Hokonui Hills, New Zealand: *Geol. Soc. Amer. Bull.* **86**, 63-173.

Bonatti, E. (1963) Zeolites in Pacific pelagic sediments: *N.Y. Acad. Sci. Trans.*, Ser. II, **25**, 938-948.

Bonatti, E. and Joensum (1968) Palygorskite from Atlantic deep sea sediments: *Amer. Mineral.* **53**, 975-983.

Bramlette, M. N. and Posnjak, E. (1933) Zeolitic alteration of pyroclastics: *Amer. Mineral.* **18**, 167-171.

Buryanova, E. Z. (1960) Analcime- and zeolite-bearing rocks of Tuva: *Akad. Nauk SSSR, Izvestiya*, Geol. Series, no. 6, 74-84.

Campbell, A. S. and Fyfe, W. S. (1965) Analcime–albite equilibria: *Amer. J. Sci.* **263**, 807-816.

Cook, H. E., Rex, R. W., Eklund, W. A. and Murray, B. (1971) X-ray mineralogy studies, Leg 7: *Initial Reports of the Deep Sea Drilling Project*, Vol. **7**, part 2, Washington (U.S. Government Printing Office), pp. 845-1757.

Cook, H. E. and Zemmels, I. (1971) X-ray mineralogy studies, Leg 8: *Initial Reports of the Deep Sea Drilling Project*, Vol. 8, Washington (U.S. Government Printing office), pp. 901-950.

Cook, H. E. and Zemmels, I. (1972) X-ray mineralogy studies, Leg 9, DSDP: *Initial Reports of the Deep Sea Drilling Project*, Vol. **9**, Washington (U.S. Government Printing Office), pp. 707-778.

Cook, H. E. and Zemmels, I. (1973) X-ray mineralogy studies of Leg 10 cores in the Gulf of Mexico: *Initial Reports of the Deep Sea Drilling Project*, Vol. **10**, Washington (U.S. Government Printing Office), pp. 337-374.

Cook, H. E., Zemmels, I. and Matti, J. C. (1974a) X-ray mineralogy data, Southern Indian Ocean – Leg 26 DSDP: *Initial Reports of the Deep Sea Drilling Project*, Vol. **26**, Washington (U.S. Government Printing Office), pp. 573-592.

Cook, H. E., Zemmels, I. and Matti, I. (1974b) X-ray mineralogy data, Eastern Indian Ocean – Leg 27, Deep Sea Drilling Project: *Initial Reports of the Deep Sea Drilling Project*, Vol. **27**, Washington (U.S. Government Printing Office), pp. 535-548

Cook, H. E., Zemmels, I. and Matti, J. C. (1974c) X-ray mineralogy data, Campbell Plateau and South Tasman Sea; Leg 29, DSDP: *Initial Reports of the Deep Sea Drilling Project*, Vol. **29**, Washington (U.S. Government Printing Office), pp. 1173-1186.

Cook, H. E., Zemmels, I. and Matti, J. C. (1975a) X-ray mineralogy data, Austral–Antarctic Region, Leg 28, Deep Sea Drilling Project: *Initial Reports of the Deep Sea Drilling Project*, Vol. **28**, Washington (U.S. Government Printing Office), pp. 981-998.

Cook, H. E., Zemmels, I. and Matti, J. C. (1975b) X-ray mineralogy data, Far Western Pacific, Leg 31, Deep Sea Drilling Project: *Initial Reports of the Deep Sea Drilling Project*, Vol. **31**, Washington (U.S. Government Printing Office), pp. 883-896.

Coombs, D. S. (1954) The nature and alteration of some Triassic sediments from Southland, New Zealand: *Roy. Soc. New Zealand Trans.* 82, 65-109.

Coombs, D. S. (1971) Present status of the zeolite facies: *Molecular Sieve Zeolites – I, Advances in Chem. Series* **101**, pp. 317-327.

Coombs, D. S. and Whetten, J. T. (1967) Composition of analcime from sedimentary and burial metamorphic rocks: *Geol. Soc. Amer. Bull.* 78, 259-268.

Coombs, D. S., Ellis, A. J., Fyfe, W. S. and Taylor, A. M. (1959) The zeolite facies, with comments on the interpretation of hydrothermal syntheses: *Geochim. Cosmochim. Acta* **17**, 53-107.

Deffeyes, K. S. (1959) Zeolites in sedimentary rocks: *J. Sediment. Petrol.* **29**, 602-609.

Fan, P. and Rex, R. W. (1972) X-ray mineralogy studies – Leg 14: *Initial Reports of the Deep Sea Drilling Project*, Vol. **14**, Washington (U.S. Government Printing Office), pp. 677-726.

Fan, P. and Zemmels, I. (1972) X-ray mineralogy studies: *Initial Reports of the Deep Sea Drilling Project*, Vol. **12**, Washington (U.S. Government Printing Office), pp. 1127-1154.

Fan, P., Rex, R. W., Cook, H. E. and Zemmels, I. (1973) X-ray mineralogy of the Caribbean Sea – Leg 15: *Initial Reports of the Deep Sea Drilling Project*, Vol. **15**, Washington (U.S. Government Printing Office), pp. 847-922.

Filizova, L. D., Kirov, G. N. and Vassilenko, V. B. (1972) Chemical composition, properties and nomenclature of the heulandite group of zeolites: *Compte rendus de l'Acad. bulgare des Sciences* **25**, 1081-1084.

Flörke, O. W., Jones, J. B. and Segnit, E. R. (1975) Opal-CT crystals: *Neues Jb. Miner. Mh.*, pp. 369-377.

Fukuda, T. (1972) Geology of the Jurassic Sôma Group – especially the petrographical study of the Koike Limestone: Graduate Thesis at Geol. Inst., Univ. Tokyo, pp. 1-83.

Gilbert, C. M. and McAndrews, M. G. (1948) Authigenic heulandite in sandstone, Santa Cruz, California: *J. Sediment. Petrol.* **18**, 91-99.

Goldberg, E. D. (1961) Chemical and mineralogical aspects of deep-sea sediments: *Phys. Chem. Earth* **4**, 281-302.

Harada, K., Hara, M. and Nakao, K. (1968) Mineralogical notes on mesolite and scolecite from Japan: *Mineral. J.* **5** (5), 309-320.

Hathaway, J. C. and Sachs, P. L. (1965) Sepiolite and clinoptilolite from the Mid-Atlantic Ridge: *Amer. Mineral.* **50**, 852-867.

Hay, R. L. (1963) Stratigraphy and zeolitic diagenesis of the John Day Formation of Oregon: *Univ. Calif. Publ. in Geol. Sci.* **42**, 199-262.

Hay, R. L. (1964) Phillipsite of saline lakes and soils: *Amer. Mineral.* **49**, 1366-1387.

Hay, R. L. (1966) Zeolites and zeolitic reactions in sedimentary rocks: *Geol. Soc. Amer. Special Papers* **85**, 1-125.

Hay, R. L. and Iijima, A. (1968a) Petrology of palagonite tuffs of Koko Craters, Oahu, Hawaii: *Contr. Mineral. Petrol.* **17**, 141-154.

Hay, R. L. and Iijima, A. (1968b) Nature and origin of palagonite tuffs of the Honolulu Group on Oahu, Hawaii: *Geol. Soc. Amer. Mem.* **116**, 331-376.

Hayakawa, N. and Suzuki, S. (1970) Zeolitized tuffs and occurrence of ferrierite in Tadami-machi, Fukushima Prefecture, Japan: *Mining Geol.* **20**, 295-305 (in Japanese).

Hayashi, M. (1973) Hydrothermal alteration in the Otake geothermal area, Kyushu: *J. Japan Geothermal Energy Ass.* **10**, 9-46.

Heath, G. R. (1969) Mineralogy of Cenozoic deep-sea sediments from the equatorial Pacific Ocean: *Geol. Soc. Amer. Bull.* **80**, 1997-2018.

Honda, S. and Muffler, L. J. P. (1970) Hydrothermal alteration in core from research drill hole Y-1, Upper Geyer Basin, Yellowstone National Park, Wyoming: *Amer. Mineral.* **55**, 1714-1737.

Huzioka, K. and Yoshikawa, T. (1969) Zeolitic alteration of vitric tuffs in Akita oil field: *J. Japan. Ass. Petrol. Technologist* **34**, 1-10 (in Japanese).

Iijima, A. (1971) Composition and origin of clinoptilolite in the Nakanosawa Tuff of Rumoi, Hokkaido: *Molecular Sieve Zeolites – I, Advances in Chem. Ser.* **101**, 334-341.

Iijima, A. (1974) Clay and zeolitic alteration zones surrounding Kuroko deposits in the Hokuroku District, northern Akita, as submarine hydrothermal-diagenetic alteration products: *Mining Geol. Special Issue* **6**, 267-289.

Iijima, A. (1975) Effect of pore water to clinoptilolite analcime–albite reaction series: *J. Fac. Sci., Univ. Tokyo*, Sec: II **19**, 133-147.

Iijima, A. and Harada, K. (1968) Authigenic zeolites in zeolitic palagonite tuffs on Oahu, Hawaii: *Amer. Mineral.* **54**, 182-197.

Iijima, A. and Hay, R. L. (1968) Analcime composition in tuffs of the Green River Formation of Wyoming: *Amer. Mineral.* **53**, 184-200.

Iijima, A. and Utada, M. (1966) Zeolites in sedimentary rocks, with reference to the depositional environments and zonal distribution: *Sedimentology* **7**, 327-357.

Iijima, A. and Utada, M. (1967) Zeolitic alteration of vitric tuffs – a diagenetic change of volcanic glass: *Jubilee Papers Prof. Y. Sasa's 60th Birthday*, pp. 107-121.

Iijima, A. and Utada, M. (1971) Present-day zeolitic diagenesis of the Neogene geosynclinal deposits in the Niigata oil field, Japan: *Molecular Sieve Zeolites – I, Advances in Chem. Ser.* **101**, 342-349.

Iijima, A. and Utada, M. (1972) A critical review on the occurrence of zeolites in sedimentary rocks in Japan: *Japan. J. Geol. Geogr.* **42**, 61-84.

Jones, J. B. and Segnit, E. R. (1975) Nomenclature and structure of natural disordered (opaline) silica: *Contrib. Mineral. Petrol.* **51**, 231-234.

JPDC (1968) Annual report on exploration of natural gas and oil 1967; Core tests of the MITI-Obuchi, -Shimoigarashi, and -Nishimeoki Holes in 1967: Japan Petroleum Development Corporation, p. 46 (in Japanese).

Kennett, J. P. *et al.* (1975) Cenozoic paleoceanography in the Southwest Pacific Ocean, Antarctic glaciation, and the development of the Circum-Antarctic current: *Initial Reports of the Deep Sea Drilling Project*, Vol. **29**, Washington (U.S. Government Printing Office), pp. 1155-1169.

Kerr, P. F. (1931) Bentonite from Ventura, California: *Econ. Geology* **26**, 153-168.

Kisch, H. J. (1969) Coal-rank and burial-metamorphic mineral facies: *Advances in Organic Geochim.* **1968**, 407-425.

Kisch, H. J. (1974) Anthracite and meta-anthracite coal ranks associated with "anchimetamorphism" and "very-low-stage" metamorphism. 1: *Koninkl. Nederl. Akademie van Wetenschappen*, Ser. B, 77, 81-118.

Kossovskaya, A. G. (1974) Genetical types of zeolites in stratigraphical formation: *Litol. i Polez. Iskop., Akad. Nauk S.S.S.R.* **2**, 23-44 (in Russian).

Kostov, I. (1969) Zoning in the development of the volcanogenic zeolites: *Neues Jb. Mineral., Abh.* **111**, 264-278.

Lancelot, Y. and Ewing, J. I. (1972) Correlation of natural gas zonation and carbonate diagenesis in Tertiary sediments from the north-west Atlantic: *Initial Reports of the Deep Sea Drilling Project*, Vol. **11**, Washington (U.S. Government Printing Office), pp. 791-799.

Lisitzin, A. P. (1972) Sedimentation in the world ocean: *Soc. Econ. Paleont. Mineral. Special Pub.* **17**, 1-218.

Manheim, F. T., Chan, K. M. and Sayles, F. L. (1970) Interstitial water studies on small core samples, Deep Sea Drilling Project, Leg 5: *Initial Reports of the Deep Sea Drilling Project*, Vol. **5**, Washington (U.S. Government Printing Office), pp. 501-512.

Manheim, F. T. and Sayles, F. L. (1971) Interstitial water studies on small core samples, Deep Sea Drilling Project, Leg 6: *Initial Reports of the Deep Sea Drilling Project*, Vol. **6**, Washington (U.S. Government Printing Office), pp. 811-822.

Manheim, F. T., Sayles, F. L. and Waterman, L. S. (1972) Interstitial water studies on small core samples, Deep Sea Drilling Project, Leg XII: *Initial Reports of the Deep Sea Drilling Project*, Vol. **12**, Washington (U.S. Government Printing Office), pp. 1193-1200.

Mason, B. and Sand, L. B. (1960) Clinoptilolite from Patagonia. The relationship between clinoptilolite and heulandite: *Amer. Mineral.* **45**, 341-350.

Matsumoto, R., Utada, M. and Kagami, H. (1976) Sedimentary petrology of DSDP cores from Angola Basin and Walvis Ridge, Leg 40: *Initial Reports of the Deep Sea Drilling Project*, Vol. **40** (in preparation).

Matti, J. C., Zemmels, I. and Cook, H. E. (1973a) X-ray mineralogy of sediments from the Far Western Pacific, Leg 20, DSDP: *Initial Reports of the Deep Sea Drilling Project*, Vol. **20**, Washington (U.S. Government Printing Office), pp. 323-334.

Matti, J. C., Zemmels, I. and Cook, H. E. (1973b) Mineralogy and mineralogic trends in sediments from the Tasman and Coral Seas, Leg 21, Deep Sea Drilling Project: *Initial Reports of the Deep Sea Drilling Project*, Vol. **21**, Washington (U.S. Government Printing Office), pp. 723-750.

Matti, J. C., Zemmels, I. and Cook, H. E. (1974a) X-ray mineralogy data, northeastern part of the Indian Ocean, Leg 22, DSDP: *Initial Reports of the Deep Sea Drilling Project*, Vol. **22**, Washington (U.S. Government Printing Office), pp. 693-710.

Matti, J. C., Zemmels, I. and Cook, H. E. (1974b) X-ray mineralogy data: *Initial Reports of the Deep Sea Drilling Project*, Vol. **23**, Washington (U.S. Government Printing Office), pp. 1137-1156.

Matti, J. C., Zemmels, I. and Cook, H. E. (1974c) X-ray mineralogy data, Western Indian Ocean – Leg 24, Deep Sea Drilling Project: *Initial Reports of the Deep Sea Drilling Project*, Vol. **24**, Washington (U.S. Government Printing Office), pp. 811-826.

Matti, J. C., Zemmels, I. and Cook, H. E. (1974d) X-ray mineralogy data, Western Indian Ocean – Leg 25, Deep Sea Drilling Project: *Initial Reports of the Deep Sea Drilling Project*, Vol. **25**, Washington (U.S. Government Printing Office), pp. 843-862.

Milton, C. (1936) A foraminiferal analcite shale from Texas: *Washington Acad. Sci. J.* **26**, 1-386.

Minato, H. and Utada, M. (1971) Clinoptilolite from Japan: *Molecular Sieve Zeolites – I, Advances in Chem. Ser.* **101**, 311-316.

Miura, K. (1969) On the stratigraphy and alteration of the Neogene at Shimane Peninsula and its vicinities: Pub. Symp., *Problems on Green Tuff*, Geol. Soc. Japan, pp. 225-230 (in Japanese).

Mizuno, M. (1974) Zeolitic alteration of Neogene acidic and basic volcanics in the West Part of the Nisaka Mountains, Central Japan: Master Thesis at Geol. Inst., Univ. Tokyo, pp. 1-49.

Morgenstein, M. (1967) Authigenic cementation of scoriaceous deep-sea sediments west of the Society Ridge, South Pacific: *Sedimentology* **9**, 105-118.

Motojuku Rock Alteration Research Subgroup (1970) Study on rock alteration of the Motojuku volcanic rocks: *Chidanken Senpo* **16**, 65-78 (in Japanese).

Mumpton, F. A. (1960) Clinoptilolite redefined: *Amer. Mineral.* **45**, 351-369.

Munson, R. A. and Sheppard, R. A. (1974) Natural zeolites: their properties, occurrences, and uses: *Minerals Sci. Engng* **6**, 19-34.

Murata, K. J. and Whiteley, K. R. (1973) Zeolites in the Miocene Briones Sandstone and related formations of the Central Coast Ranges, California: *J. Res. U.S. Geol. Survey* **1**, 255-265.

Murray, J. and Renard, A. F. (1891) Report on deep-sea deposits: *Report on the Scientific Results of the Voyage of "H.M.S. Challenger" During the Years 1873-76*, Neill & Co., Edinburgh, pp. 1-520.

Nakajima, W. (1973) Stability relations of mordenite and related minerals in the system $Na_2Al_2Si_{10}O_{24}$–$CaAl_2Si_{10}O_{24}$–H_2O: *Geol. Soc. Japan Memoir* **9**, 239-247 (in Japanese with English abstract).

Nakajima, W. and Koizumi, M. (1966) On the chemical composition of analcime from the low-grade metamorphic rocks in Japan: *J. Geol. Soc. Japan* **72**, 517-521.

Ota, S. and Sudo, T. (1949) Studies on "Oya-ishi", Part II (Mineralogical composition): *J. Geol. Soc. Japan* **55**, 242-246 (in Japanese).

Otárola, G. (1964) Zeolites and related minerals in Cretaceous rocks of east-central Puerto Rico: *Amer. J. Sci.* **262**, 726-734.

Owa, I. (1975) Zeolitic diagenesis of the Cretaceous sediments in the central-axial part of Hokkaido, Japan: Master Thesis at the Geol. & Mineral. Inst., Hokkaido Univ., pp. 1-81 (in Japanese).

Owa, I. and Iijima, A. (1973) Si-rich heulandite in Cretaceous Hakobuchi group of Sorachi, Hokkaido: Abstracts of Papers *Ann. Meeting Clay Mineral. Soc. Japan.*, pp. 1-70 (in Japanese).

Packham, G. H. and Crook, K. A. W. (1960) The principle of diagenetic facies and some of its implications: *J. Geology* **68**, 392-407.

Di Paola, E. C. (1965) Heulandita autigenica en formaciones mesozoicas de Neuguen y Rio Negro: *Rev. de la Asoc. Geol. Argentina* **20**, 229-240.

Paulus, F. J. (1972) Leg 11 measurements of physical properties in sediments of the western north Atlantic and their relationship to sediment consolidation: *Initial Reports of the Deep Sea Drilling Project*, Vol. **11**, 667-722.

Peacock, M. A. (1926) The petrology of Iceland, Part I. The basic tuffs: *Royal Soc. Edinburgh Trans.* **55**, 53-76.

Peacock, M. A. and Fuller, R. E. (1928) Chlorophaeite, sideromelane, and palagonite from the Columbia River Plateau: *Amer. Mineral.* **13**, 360-382.

Von Rad, U. and Rösch, H. (1972) Mineralogy and origin of clay minerals, silica and authigenic silicates in Leg 14 sediments: *Initial Reports of the Deep Sea Drilling Project*, Vol. **14**, 727-751.

Raw, F. and Matley, C. A. (1943) Some altered palagonite tuffs from Jamaica and the origin and history of their chlorites: *J. Geology* **51**, 215-243.

Rengarten, N. V. (1945) A zeolite from the mordenite group in the Upper Cretaceous and Paleocene marine deposits of the eastern slope of the Urals: *Dokl. Akad. Nauk SSSR* **48**, 591-593 (English abstract in *Chem. Abstracts*, 1946, **40**, 4625).

Rex, R. W. (1967) Authigenic silicates formed from basaltic glass by more than 60 million years, contact with sea water, Sylvania Guyot, Marshall Islands: *Clays Clay Minerals, Proc. Nat. Conf.* **15** (1966), 195-203.

Rex, R. W. (1969) X-ray mineralogy studies – Leg 1: *Initial Reports of the Deep Sea Drilling Project*, Vol. **1**, Washington (U.S. Government Printing Office), pp. 354-367.

Rex, R. W. (1970a) X-ray mineralogy studies – Leg 2: *Initial Reports of the Deep Sea Drilling Project*, Vol. **2**, Washington (U.S. Government Printing Office), pp. 329-346.

Rex, R. W. (1970b) X-ray mineralogy studies – Leg 3: *Initial Reports of the Deep Sea Drilling Project*, Vol. **3**, Washington (U.S. Government Printing Office), pp. 509-582.

Rex, R. W. and Murray, B. (1970a) X-ray mineralogy studies,

Leg 4: *Initial Reports of the Deep Sea Drilling Project*, Vol. **4**, Washington (U.S. Government Printing Office), pp. 325-370.

Rex, R. W. and Murray, B. (1970b) X-ray mineralogy studies: *Initial Reports of the Deep Sea Drilling Project*, Vol. **5**, Washington (U.S. Government Printing Office), pp. 441-484.

Rex, R. W., Eklund, W. A. and Jamieson, I. M. (1971) X-ray mineralogy studies – Leg 6: *Initial Reports of the Deep Sea Drilling Project*, Vol. **6**, Washington (U.S. Government Printing Office), pp. 753-810.

Rex, R. W. and Fan, P. F. (1974) Zeolites in deep sea sediments: US–Japan Cooperative Science Seminar on Sedimentary Zeolites: unpublished abstract.

Ross, C. S. (1958) Welded tuff from deep-well cores from Clinch County, Georgia: *Amer. Mineral.* **43**, 537-543.

Saha, P. (1959) Geochemical and X-ray investigation of natural and synthetic analcites: *Amer. Mineral.* **44**, 300-313.

Sargent, H. C. (1925) An analcite-bearing tuff in the Carboniferous limestone of Derbyshire: *Geol. Mag.* **62**, 462-463.

Sayles, F. L., Manheim, F. T. and Chan, K. M. (1970) Interstitial water studies on small core samples, Leg 4: *Initial Reports of the Deep Sea Drilling Project*, Vol. **4**, Washington (U.S. Government Printing Office), pp. 401-414.

Sayles, F. L., Manheim, F. T. and Waterman, L. S. (1972a) Interstitial water studies on small core samples, Leg 9, Deep Sea Drilling Project: *Initial Reports of the Deep Sea Drilling Project*, Vol. **9**, Washington (U.S. Government Printing Office), pp. 845-856.

Sayles, F. L., Manheim, F. T. and Waterman, L. S. (1972b) Interstitial water studies on small core samples, Leg XI: *Initial Reports of the Deep Sea Drilling Project*, Vol. **11**, Washington (U.S. Government Printing Office), pp. 997-1008.

Seki, Y. (1969) Facies series in low-grade metamorphism: *J. Geol. Soc. Japan* **75**, 255-266.

Seki, Y. (1973) Ionic substitution and stability of mordenite: *J. Geol. Soc. Japan* **79**, 669-676.

Seki, Y., Takayasu, T., Nakajima, M. and Onuki, H. (1968) Wairakite from Hanawa mining district, northern Japan: *J. Japan. Ass. Mineralog. Econ. Geolog.* **59**, 236-246.

Seki, Y., Oki, Y., Matsuda, T., Mikami, K. and Okumura, K. (1969) Metamorphism in the Tanzawa Mountains, Central Japan: *J. Japan. Ass. Mineral. Petrol. Econ. Geol.* **61**, 1-24, 50-75.

Seki, Y. *et al.* (1969) Zeolite distribution in the Katayama geothermal area, Onikobe, Japan: *J. Geol. Geog. Japan* **40**, 63-79.

Seki, Y., Oki, Y., Odaka, S. and Ozawa, K. (1972) Stability of mordenite in zeolite facies metamorphism of the Oyama – Isehara district, East Tanzawa Mountains, Central Japan: *J. Geol. Soc. Japan* **78**, 145-160.

Sheppard, R. A. (1971) Zeolites in sedimentary deposits of the United States – A review: *Molecular Sieve Zeolites – I, Advances in Chem. Ser.* **101**, 279-310.

Sheppard, R. A. (1973) Zeolites in sedimentary rocks: *U.S. Geol. Survey Prof. Paper* **820**, 689-695.

Sheppard, R. A. and Gude, A. J. 3rd (1968) Distribution and genesis of authigenic silicate minerals in tuffs of Pleistocene, Lake Tecopa, Inyo County, California: *U.S. Geol. Surv. Prof. Paper* **597**, 1-38.

Sheppard R. A. and Gude, A. J. 3rd (1969) Diagenesis of tuffs in the Barstow Formation, Mud HIlls, San Bernardino County, California: *U.S. Geol. Surv. Prof. Paper* **634**, 1-35.

Sheppard, R. A. and Gude, A. J. 3rd (1970) Chemical composition and physical properties of phillipsite from the Pacific and Indian Oceans: *Amer. Mineral.* **55**, 2053-2062.

Shimazu, M. (1968) The alteration of basic volcanic rocks: Pub. Symp. *The Alteration of Green Tuff Region*, Geol. Soc. Japan, pp. 17-28 (in Japanese).

Shimoyama, T. and Iijima, A. (1976) Influence of temperature on coalification of Tertiary coal in Japan (summary): *Ass. Amer. Petrol. Geol. Memoir* **25**, 15-22.

Steiner, A. (1953) Hydrothermal rock alteration at Wairakei, New Zealand: *Econ. Geol.* **48**, 1-13.

Surdam, R. C. and Parker, R. D. (1972) Authigenic aluminosilicate minerals in the tuffaceous rocks of the Green River Formation, Wyoming: *Geol. Soc. Amer. Bull.* **83**, 689-700.

Teichmüller, M. and Teichmüller, R. (1966) Geological causes of coalification: *Coal Sciences*, pp. 133-155.

Thompson, A. B. (1971) Analcite–albite equilibria at low temperatures: *Amer. J. Sci.* **271**, 79-92.

Utada, M. (1965) Zonal distribution of authigenic zeolites in the Tertiary pyroclastic rocks in Mogami District, Yamagata Prefecture: *Sci. Papers Coll. Gen. Educ. Univ. Tokyo* **15**, 173-216.

Utada, M. (1970) Occurrence and distribution of authigenic zeolites in the Neogene pyroclastic rocks in Japan: *Sci. Papers Coll. Gen. Educ. Univ. Tokyo* **20**, 191-262.

Utada, M. (1971) Zeolitic zoning of the Neogene pyroclastic rocks in Japan: *Sci. Papers Coll. Gen. Educ. Univ. Tokyo* **21**, 189-221.

Utada, M. (1973) The types of alteration in the Neogene sediment relating to the intrusion of volcano-plutonic complexes in Japan: *Sci. Papers Coll. Gen. Educ. Univ. Tokyo* **23**, 167-216.

Utada, M. (1974) An interpretation of the types of alteration in the Neogene sediments relating to the intrusion of volcano-plutonic complexes, correlating to the calculated temperature patterns: *Sci. Papers Coll. Gen. Educ. Univ. Tokyo* **24**, 65-77.

Utada, M., Minato, H. and Hattori, N. (1967) Migration of chemical components of vitric tuff accompanying zeolitization and formation of calcareous nodules: *Sci. Papers Coll. Gen. Educ. Univ. Tokyo* **17**, 121-140.

Utada, M., Minato, H., Isikawa, T. and Yoshizaki, Y. (1974) The alteration zone surrounding Kuroko-type deposits in the Nishi-Aizu District, Fukushima Prefecture, especially the analcime zone as an indicator of exploration of the ore deposits: *Mining Geol., Spec. Issue* **6**, 291-301.

Walker, G. P. L. (1951) The amygdale minerals in the Tertiary lavas of Ireland. I. The distribution of chabazite habits and zeolites in the Garron plateau area, County Antrim: *Mineral. Mag.* **29**, 773.

Walker, G. P. L. (1959) The amygdale minerals in the Tertiary lavas of Ireland. II. The distribution of gmelinite: *Mineral. Mag.* **32**, 202.

Walker, G. P. L. (1960a) The amygdale minerals in the Tertiary lavas of Ireland. III. Regional distribution: *Mineral. Mag.* **32**, 503.

Walker, G. P. L. (1960b) Zeolite zones and dike distribution in relation to the structure of the basalts of eastern Iceland: *J. Geol.* **68**, 515.

Waterman, L. S., Sayles, F. L. and Manheim, F. T. (1972) Interstitial water studies on small core samples, Leg XIV: *Initial Reports of the Deep Sea Drilling Project*, Vol. **14**, Washington (U.S. Government Printing Office), pp. 753-762.

Weaver, C. E. and Beck, K. C. (1971) Clay water diagenesis during burial: How mud becomes gneiss: *Geol. Soc. Amer. Spec. Papers* **134**, 1-96.

Whitmarsh, R. B., Weser, O. E., Ross, D. A. *et al.* (1974) *Initial Reports of the Deep Sea Drilling Project*, Vol. **23**, Washington (U.S. Government Printing Office), pp. 1-180.

Yanagimoto, Y. (1970) Geology of the western part of Shimane Peninsula, Shimane Prefecture: Graduate Thesis at Geol. Inst., Univ. Tokyo, pp. 1-56.

Yoshimura, T. (1961) Zeolites in the Miocene pyroclastic rocks in the Oshima–Fukushima district, southwestern Hokkaido: *J. Geol. Soc. Japan* **67**, 578-583 (in Japanese).

Zemmels, I. (1973) X-ray mineralogy studies – Leg 16: *Initial Reports of the Deep Sea Drilling Project*, Vol. **16**, Washington (U.S. Government Printing Office), pp. 529-572.

Zemmels, I., Cook, H. E. and Hathaway, J. C. (1972) X-ray mineralogy studies, Leg XI: *Initial Reports of the Deep Sea Drilling Project*, Vol. **11**, Washington (U.S. Government Printing Office), pp. 729-790.

Zemmels, I. and Cook, H. E. (1973a) X-ray mineralogy studies of selected samples from the sea floor of the Northeast Atlantic and Mediterranean Sea: *Initial Reports of the Deep Sea Drilling Project*, Vol. **13**, Part 2, Washington (U.S. Government Printing Office), pp. 606-665.

Zemmels, I. and Cook H. E. (1973b) X-ray mineralogy of sediments from the Central Pacific Ocean: *Initial Reports of the Deep Sea Drilling Project*, Vol. **17**, Washington (U.S. Government Printing Office), pp. 517-560.

Zemmels, I. and Cook, H. E. (1973c) X-ray mineralogy of sediments from the North-east Pacific and Gulf of Alaska – Leg 18: *Initial Reports of the Deep Sea Drilling Project*, Vol. **18**, Washington (U.S. Government Printing Office), pp. 1015-1060.

Zemmels, I. and Cook, H. E. (1973d) X-ray mineralogy of sediments from the Northern Pacific and the Bering Sea – Leg 19: *Initial Reports of the Deep Sea Drilling Project*, Vol. **19**, Washington (U.S. Government Printing Office), pp. 667-698.

Zemmels, I. and Cook, H. E. (1975) X-ray mineralogy data from the Northwest Pacific, Leg 32, Deep Sea Drilling Project: *Initial Reports of the Deep Sea Drilling Project*, Vol. **32**, Washington (U.S. Government Printing Office), pp. 547-558.

Zen, E-An (1959) Mineralogy and petrography of marine bottom sediment samples off the coast of Peru and Chile: *J. Sediment. Petrol.* **29**, 513-539.

ZEOLITES IN PELAGIC SEDIMENTS OF THE ATLANTIC, PACIFIC, AND INDIAN OCEANS

M. KASTNER and S. A. STONECIPHER

Scripps Institution of Oceanography, La Jolla, California, 92093, U.S.A.

Abstract – Quantitatively, zeolites are among the most important authigenic minerals in pelagic sediments that form during and after burial. The horizontal and vertical distribution and the chemical composition of the various zeolites depend primarily on the following factors: (1) composition of precursors, (2) composition of host sediments and interstitial waters, (3) pH, (4) temperature, (5) thermodynamic stability, (6) hydrodynamics of the system, and (7) time. Zeolites extract major and trace elements from sea water and interstitial waters and redistribute them within the sedimentary column. Phillipsite and clinoptilolite predominate. Other less abundant zeolites are: analcime, chabazite, erionite, gmelinite, laumontite, thaumasite, thomsonite, and natrolite.

Phillipsite predominates in volcanogenic and clayey sediments (sedimentation rates of <5 m/10^6 years) of Recent to Eocene age. It starts to form at the sediment–water interface and usually is not found in great abundance below 150 m depth; below about 500 m depth, it is rarely present. The Si/Al ratio varies between 2.4 to 2.8, and K is in excess of Na and Ca. The most common precursor is volcanic (basaltic) material, in particular basaltic glass and palagonite. Harmotome is also present.

Clinoptilolite predominates in calcareous and clayey sediments (sedimentation rates >5 m/10^6 years) of Miocene to Cretaceous age. It persists even below 850 m depth. The Si/Al ratio varies between 4.2 to 5.2, with over 80% of clinoptilolites analyzed between 4.5 to 5.0. Alkalis are in excess of alkaline earths cations. Volcanic (basaltic to rhyolitic) precursors are important but do not seem to be essential.

Neither phillipsite nor clinoptilolite are suitable for K–Ar dating. All of the more calcic zeolites are always associated with basaltic material. Their formation might be controlled by local short- or long-lived elevated temperatures due to hydrothermal activity or high heat flow.

INTRODUCTION

Quantitatively, zeolites are among the most important authigenic minerals in pelagic sediments that form during and after burial. Their composition depends mainly on the chemistry of the precursors, chemistry of the solutions, composition of host sediments, and temperature. Zeolites are also among the most common authigenic aluminosilicates that form in sediments of non-marine environments, primarily as alteration products of volcanic glass in saline, alkaline lakes and in fresh-water deposits (Deffeyes, 1959; Hay, 1966; Goodwin and Surdam 1967; Eugster and Jones, 1968; Hay and Iijima, 1968; Iijima and Harada, 1968; Sheppard and Gude, 1968, 1969; Hay, 1970) and as metamorphic minerals of the zeolite facies (Coombs *et al.*, 1959).

Investigations of authigenic minerals in deep-sea sediments are extremely important because of their effect on the marine geochemical cycles of many elements. They could also elucidate the physico-chemical processes that have and/or are operating on the ocean floor and within the sediments. Zeolites gradually disappear with geologic age (Hay, 1966 and references therein), which suggests that they are either thermo-dynamically unstable phases in most sedimentary environments or that they are thermodynamically stable under very restricted physico-chemical conditions. The relative importance of temperature and pressure versus chemical variables such as cation activities on the stability of various zeolites is being debated (e.g., Boles and Coombs, 1975). In general, the less hydrated zeolites form at higher temperatures and pressures, but an increase in the activity of silica, for example, tends to increase their dehydration temperatures (Coombs *et al.*, 1959; Miyashiro, 1960; Hay, 1966).

Volcanic material, in particular volcanic glass (silicic as well as basaltic) and palagonite, is the most commonly reported precursor of sedimentary zeolites (e.g. Murray and Renard, 1891; Deffeyes, 1959; Bonatti, 1963, 1965; Hathaway and Sachs, 1966; Hay, 1966; Rex, 1967; Sheppard and Gude, 1968, 1969; Sheppard *et al.*, 1970; Kolla and Biscaye, 1973). Gels are commonly used as starting materials in zeolite synthesis, but the only report of the occurrence of natural gels as precursors of zeolites is by Eugster and Jones (1968) in Lake Magadi, Kenya. The Si/Al ratio of these gels is similar to the amorphous marine muds that were derived from weathered basalt in Hawaii (Moberly, 1963). Minute lath-shaped crystals enriched in K in an amorphous cement of pyroclastic sediments from Surtsey, were described by Alexanderson (1972). These crystals are conceivably a zeolite. Direct alteration of amorphous clay to clinoptilolite was reported, for example, by Hay (1970) in Pleistocene deposits of Olduvai Gorge, Tanzania. Other similar reports are by Brown *et al.* (1969) and Estéoule *et al.* (1971). The occurrence of zeolites as cement or cavity filling in sedimentary rocks (Murray and Renard, 1891; Morgenstein, 1967; Garrison *et al.*, 1973) indicates that they also form as primary chemical precipitates from solution.

The present paper is divided into three parts; a critical review of the literature by one of us (M. K.), a discussion of the distribution of phillipsite and clinoptilolite in deep-sea sediments by the other author (S. A. S.), and a general discussion by both authors.

A CRITICAL REVIEW OF THE LITERATURE

Phillipsite and clinoptilolite are the two most abundant zeolites in pelagic sediments. Phillipsite was first recognized by Murray and Renard (1891) as an important diagenetic mineral in pelagic sediments, associated with volcanic material, in the Central Pacific and Indian Oceans. Its occurrence in the Atlantic Ocean was reported by Young (1939). The distribution, mineralogy, chemistry, and mode of origin of phillipsite were subsequently studied by Goldberg and Arrhenius (1958), Goldberg (1961), Arrhenius (1963), Bonatti (1963), Biscaye (1965), Peterson *et al.* (1970), Czyscinski (1973), and references therein. Harmotome, the Ba analogue of phillipsite, was also identified in pelagic sediments of the Pacific Ocean by Murray and Renard (1891), Goldberg and Arrhenius (1958), Arrhenius (1963), Bonatti (1963), Arrhenius and Bonatti (1965), and Morgenstein (1967), but it is much less abundant than phillipsite. The importance of clinoptilolite as a diagenetic zeolite in the Atlantic Ocean was recognized rather recently (Biscaye, 1965; Hathaway and Sachs, 1965; Turekian, 1965; Peterson *et al.*, 1970; Berger and von Rad, 1972). Clinoptilolite also occurs in the Pacific and Indian Oceans (for example, Weaver, 1968; Cook and Zemmels, 1972). Recently, Kolla and Biscaye (1973), Cronan (1974), and Stonecipher (1976) summarized the occurrences and distribution of phillipsite and clinoptilolite in the Indian, Pacific, and Atlantic Oceans. Insight into the factors that control the formation of these zeolites is provided by the above and the present syntheses, as described below.

Based on presently published data, analcime is the next most abundant zeolite in deep-sea sediments associated with volcanic material in the Pacific and Indian Oceans (Murray and Renard, 1891; Griffin *et al.*, 1972; Bass *et al.*, 1973). Analcime commonly coexists with phillipsite and/or clinoptilolite. Occurrences of the following more Ca- and Na-rich phases in pelagic sediments are quantitatively not important: chabazite (Murray and Renard, 1891; Bass *et al.*, 1973); erionite (Rex *et al.*, 1971); laumontite (Bass *et al.*, 1973; Drever, 1976); gmelinite, natrolite and thomsonite (Bass *et al.*, 1973), all in the Pacific Ocean, and thaumasite (Walker, 1972) from the North Atlantic. All these phases are associated with volcanic material.

The most abundant deep-sea zeolites, phillipsite and clinoptilolite, commonly constitute a major portion of the sediment, up to 70–80% by weight (Czyscinski, 1973). Depending on their site of formation, near the sediment/water interface or within the sediment column, they extract major, minor, and trace elements (Si, Al, K, Na, Ca, Sr, Ba, Ti, P, Th, U, and Ra) and H_2O from sea water and interstitial waters. Upon their dissolution or transformation to another less hydrous aluminosilicate phase, they reintroduce water into interstitial waters and some fractions of the major, minor, and trace elements that they originally extracted from sea water and interstitial waters.

$\delta^{18}O$ analyses of interstitial water from DSDP cores show large and rather systematic $\delta^{18}O$ depletion with depth (Lawrence, 1973; Lawrence *et al.*, 1975). This is due to interstitial water interaction with sediments, volcanic ash, and basalt and to the formation of diagenetic minerals. The value of $\alpha_{\text{phillipsite}-H_2O}$ of approximately 1.034 (Savin and Epstein, 1970a) which is greater than any fractionation factor of clay-H_2O systems (Savin and Epstein 1970b) suggests formation under isotopic equilibrium conditions in the marine environment. Lawrence *et al.* (1975) showed that the alteration of basalt or volcanic ash to zeolites involves the largest interstitial water shift in $\delta^{18}O$. O'Neil and Taylor (1967) demonstrated that the tendency to concentrate ^{18}O in aluminosilicates is directly correlated with the relative number of Al–O and Si–O bonds. Accordingly, because of the significantly greater Si/Al ratio of clinoptilolite relative to phillipsite, $\alpha_{\text{clinoptilolite}-H_2O}$ should be even greater than the $\alpha_{\text{phillipsite}-H_2O}$. The quantitative importance of water release by the dissolution or transformation of hydrous zeolites to less hydrous or to anhydrous aluminosilicate phase on interstitial water $\delta^{18}O$ values and concentration profiles is, as yet, unknown.

Phillipsite

Occurrence. Phillipsite crystallizes in the monoclinic system (P_{2_1} or $P_{2_1/m}$) with no evidence of Al/Si ordering

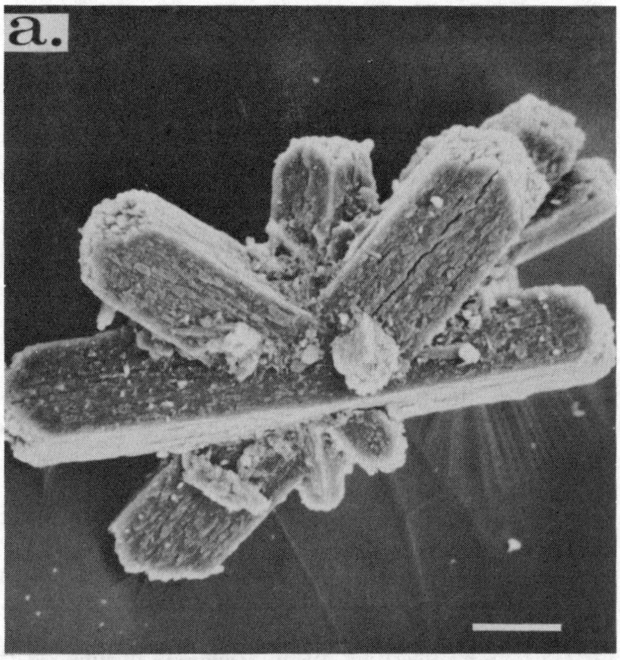

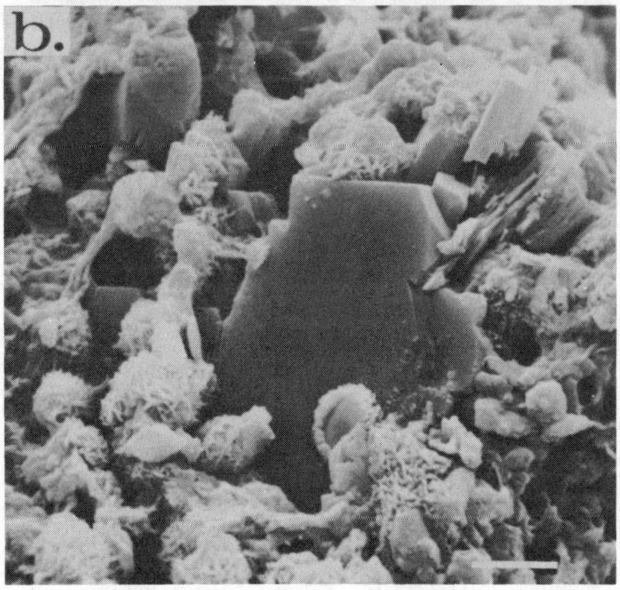

Fig. 1a. Complex sector-twinned phillipsite crystals from radiolarian clay, SCAN 16P, 200 cm; 16°25′N, 164°24′W. White line in lower right represents 25 μm.

Fig. 1b. Clinoptilolite crystals in a silicified limestone DSDP 116-27-1, 850 m; 57°30′N, 15°55′W. Calcite has been dissolved to reveal clinoptilolite, opal-CT lepispheres and clay minerals. White line in lower right represents 5 μm.

(Rinaldi *et al.*, 1974). It forms colorless to yellowish elongated prismatic crystals between 8–250 μm long, ball-shaped, and complex sector-twinned crystals (Murray and Renard, 1891; Sheppard *et al.*, 1970) as shown in Fig. 1a. They commonly contain numerous inclusions indicating rapid growth; the yellowish color of some phillipsite crystals is mainly due to Fe-oxide or hydroxide inclusions. Bonatti (1963) suggested that the well-shaped crystals indicate *in situ* formation. The ideal composition of phillipsite is (½Ca, Na, K)$_5$ $Al_5Si_{11}O_{32}\cdot 10H_2O$ (Steinfink, 1962). Mean indices of refraction and unit-cell dimensions of deep-sea phillipsites are given in Table 1. According to Hay (1966), the refractive indices of phillipsite vary inversely with SiO_2 content. Sheppard *et al.* (1970) did not observe a similar correlation in deep-sea phillipsites. Instead, they suggested that Ba increases and Na decreases the refractive indices of phillipsite and also concluded that variations in the chemistry of deep-sea phillipsites do not appreciably affect their cell dimensions.

TABLE 1. Mean index of refraction and unit cell dimensions of deep-sea phillipsites and clinoptilolites

	Mean index of refraction	*a* (Å)	*b* (Å)	*c* (Å)	$\bar{V}$ (Å)
Phillipsite [(a)]					
1	1.482	10.003	14.156	14.260	2019.4
2	1.481	9.962	14.169	14.243	2010.3
5	1.485	9.928	14.141	14.221	1996.6
9	1.483	9.975	14.182	14.274	2019.4
10	1.484	9.981	14.175	14.198	2008.7
11	1.486	9.960	14.133	14.208	1999.9
14	1.477	9.974	14.174	14.307	2022.6
15	1.478	9.970	14.165	14.264	2014.4
17	1.480	9.980	14.161	14.248	2013.6
18	1.485	10.005	14.182	14.282	2026.7
21	1.486	9.946	14.125	14.260	2003.3
22	1.483	9.974	14.124	14.290	2013.3
(b)	–	9.960	14.250	14.250	–
Clinoptilolite [(c)]					
15	1.484	17.649	17.902	7.407	2098.7

(a) After Sheppard *et al.* (1970). Sample designation is the same as in the original paper. They assumed orthorhombic symmetry after Steinfink (1962).
(b) Steinfink's (1962) orthorhombic cell.
(c) After Boles (1972). DSDP 5-3-1; 25°51.5′N, 92°11′W. β = 116°16.4′. Sample designation is the same as in the original paper.

As mentioned already, phillipsite is more abundant in the Pacific and Indian Oceans than in the Atlantic Ocean (Murray and Renard, 1891; Kolla and Biscaye, 1973; Stonecipher, 1976) and occurs mainly in Miocene sediments and younger (Peterson *et al.*, 1970; Kolla and Biscaye, 1973; Stonecipher, 1976). Murray and Renard (1891) observed that phillipsite is most common in red clays, less common in radiolarian oozes, and even less common in globigerina oozes. Goldberg (1961), however, described concentrations of up to 40% phillipsite (on a $CaCO_3$-free basis) in a Pacific core with 21% $CaCO_3$. Although common in red-brown clays in the Indian Ocean, Kolla and Biscaye (1973) reported phillipsite in calcareous oozes also off the Ninety East Ridge and the Mid-Indian Ridge. Stonecipher's (1976) statistical analysis shows the following sequence of decreasing phillipsite abundance as a function of lithology: clayey > volcanic > calcareous > siliceous. Consequently, a negative correlation between sedimentation rates and phillipsite abundance is manifested, as shown by Stonecipher (1976); contrasting with Bonatti's (1963) conclusion that such a correlation does not exist. Phillipsite was also reported as cement (Morgenstein, 1967), as cavity filling of tuff, Sylvania Guyot, Pacific Ocean (Rex, 1967), as fracture filling in basalts of Leg 17 (Bass *et al.*, 1973), and in dredged basaltic boulders from the vicinity of the Mid-Atlantic Ridge (Garrison *et al.*, 1973). In the latter dredge hauls, phillipsite was observed to replace fragments of volcanic debris and to form nearly 50% of the rock. The replacement of plagioclase by phillipsite was reported by Bass *et al.* (1973). Murray and Renard (1891), Bonatti (1963), and Bonatti and Nayudu (1965) reported phillipsite in the interior of Mn modules.

In several cores, phillipsite appears to have nucleated rapidly and to have grown at the sediment/water interface. Its growth appears continuous within the sediment column to form larger and more commonly twinned crystals (Bernat and Goldberg, 1969; Bernat *et al.*, 1970; Czyscinski, 1973). In Pacific Ocean cores, Bernat *et al.* (1970) showed that the Th and U concentrations decrease, while phillipsite concentration increases with depth. Also the larger phillipsite crystals have lower $^{40}K/^{40}Ar$ ratios than the smaller crystals. They concluded that phillipsite continues to grow for $>10^6$ y, and that deep-sea phillipsites are not suitable for K/Ar dating, confirming the results of Dymond (1966). Similarly, in several cores of brown clays with siliceous organisms, in the Central Indian Ocean and equatorial Pacific, Czyscinski (1973) observed an increase in phillipsite concentration and siliceous organisms dissolution with depth. According to him, the growth process of phillipsite crystals can go to completion in just 1.5×10^5 y. Both Bernat *et al.* (1970) and Czyscinski (1973) observed a positive correlation between phillipsite abundance and extent of siliceous organisms dissolution, in support of Rex's (1967) suggestion that phillipsite forms by interaction with solutions higher in silica concentration than sea water. Based on the above observations, it is suggested that the coexistence of large phillipsite crystals with well-preserved siliceous organisms at or close to the sediment/water interface indicates a reworked sediment.

At greater depths (in several cores already at about 4 m depth), phillipsite starts to dissolve, as indicated by etched crystal faces and by a decrease in phillipsite abundance as a function of depth; it is not found in great abundance below 150 m, and below about 500 m it is rarely present (*vide infra*). The few occurrences of phillipsite at greater depths are in volcanic ash sediments and in basalts of layer II.

In addition to the almost ubiquitous presence of basaltic material, in particular volcanic glass and palagonite, the most common coexisting authigenic minerals with phillipsite are smectite and Fe and Mn oxides and/or hydroxides (Murray and Renard, 1891; Bonatti, 1963; Stonecipher, 1976).

Chemistry. Chemical analyses of deep-sea phillipsites by Murray and Renard (1891), Goldberg (1961), Rex (1967), Sheppard *et al.* (1970), and Stonecipher (1976b) show that they are intermediate in their Si/Al ratio between phillipsites in mafic igneous rocks (that are also higher in alkaline earths) and phillipsites from silicic tuffs in saline lacustrine deposits (that are higher in alkalis and Na > K) and have high alkali/alkaline earths ratios. The latter high ratio is accounted for by the high Si/Al ratio (Goldberg, 1961). Generally, K > Na >> Ca, and Si, Al, K, and the H_2O show little variations, as shown in Table 2. The Si/Al ratio varies between 2.3 to 2.8. In addition to Ba that is present either in trace amounts (Goldberg and Arrhenius, 1958) or as a major element in harmotome, the following trace elements were reported in deep-sea phillipsites; Ra (Revelle *et al.*, 1955; Goldberg and Arrhenius, 1958); Ti (0.02%), and P (0.09%) in one Pacific phillipsite (Goldberg, 1961); 2.95–13.20 ppm Th and 0.25–0.69 ppm U (Bernat and Goldberg, 1969; Bernat *et al.*, 1970). Based on phase equilibria calculations in the $K_2O–Na_2O–Al_2O_3–SiO_2–H_2O$ system at 25°C and 1 atmosphere (Hess, 1966), phillipsite is a stable phase in oceanic sediments. His calculations, however, were not based on an average chemical formula for deep-sea phillipsites. Arrhenius (1963) suggested that both phillipsite and harmotome are stable between pH 7–8 to 9–10 in a K-hydrogen cation system such as sea water. Glaccum and Boström (1976) estimated free energies of formation for Na and K phillipsite with a Si/Al ratio of 2.33 (5337.6 ± 6.7 and 5382.4 ± 1.9 kcal/mole, respectively) and suggested that based on lack of reports that phillipsite undergoes further alteration, it is probably the thermodynamic stable end product of the alteration of volcanic matter at the sediment–sea water interface. But as mentioned above, at a depth of only a few meters phillipsite undergoes dissolution. Therefore, if Glaccum and Boström's (1976) suggestion that phillipsite is thermodynamically stable at P–T–X conditions of the deep sea floor, the rapid dissolution and decrease in abundance of phillipsite with depth implies a very restricted stability field.

Based on this implied very restricted stability field of phillipsite and the depth and age relations of phillipsite and clinoptilolite, it is suggested that phillipsite is not a thermodynamically stable phase in the deep sea environment. Instead, kinetic factors control its rapid nucleation and growth.

Origin. Basaltic material, in particular basaltic glass and/or palagonite, has been observed in most phillipsite-bearing cores and has been suggested as the ultimate precursor of phillipsite (Murray and Renard, 1891; Bonatti, 1963, 1965; Rex, 1967; Sheppard *et al.*, 1970; Kolla and Biscaye, 1973; Stonecipher, 1976). The main disagreement centers around the immediate precursor; can phillipsite crystallize from fresh anhydrous basaltic glass, or only from palagonite or smectite that formed from basaltic glass? No one seems to favor the fresh anhydrous basaltic glass alternative. Murray and Renard (1891), Bonatti (1963, 1965), Moore (1966), and Sheppard *et al.* (1970) suggested palagonite as the immediate precursor; Bonatti (1963) observed microscopically all stages between palagonite and phillipsite crystallization. Based on the Na/K ratios of basaltic glass (2–3) and palagonite (0.3), Moore (1966) concluded that the potassium content of deep-sea phillipsites has been inherited from palagonite and not from basaltic glass. Instead, in marine and non-marine environments, Arrhenius and Bonatti (1965) and Hay (1966), respectively, suggested that the formation of smectite is an essential prerequisite for phillipsite crystallization. Kastner (1976) in Site 323, Leg 35, observed that the alteration of basaltic glass to smectite preceded the formation of phillipsite. Kolla and Biscaye (1973) could not decide between the two last alternatives. Stonecipher (1976b) showed that in two phillipsite-bearing cores in the Atlantic and Pacific, even in the absence of visible basaltic material, the Ti/Al ratios of the bulk sediments suggest that basaltic material was present. Phillipsite with clinoptilolite was identified by Zemmels *et al.* (1975) in one core, Site 286, Coral Sea, in an andesitic conglomerate (Stoeser, 1975). The association of phillipsite with andesitic fragments and glass does not necessarily imply a genetic relation. Also, the observation that phillipsite is more abundant in clayey sediments relative to volcanic sediments (Stonecipher, 1976) does not negate the above suggestions that volcanic material is the ultimate precursor of most phillipsites in pelagic sediments.

It is interesting to note that in the marine environment, basaltic glass is the most common ultimate precursor of phillipsite, whereas in saline lacustrine deposits silicic, rhyolitic glass is the most common precursor (Hay, 1966; Sheppard and Gude, 1968, 1969).

TABLE 2. Chemical analyses and unit cell contents of deep-sea phillipsites

Sample	1	2	3	4	5	6
SiO_2	56.21	58.28	54.84	55.23	52.05	52.54
Al_2O_3	18.38	17.12	18.50	17.57	18.09	17.87
K_2O	7.17	6.72	8.37	7.02	7.30	6.45
Na_2O	2.35	4.72	3.58	4.30	5.51	5.33
CaO	3.17	0.27	1.25	0.51	0.10	0.79
MgO	0.19	–	–	0.64	0.08	0.56
FeO^a	0.42	0.35	–	–	–	–
H_2O^b	12.11	12.54	13.46	14.65	16.87	16.39
Totals	100	100	100	100	100	100
Number of cations on the basis of 32 oxygens						
Si	11.54	11.92	11.50	11.65	11.38	11.39
Al	4.45	4.14	4.57	4.37	4.66	4.56
K	1.88	1.76	2.24	1.89	2.04	1.78
Na	0.94	1.88	1.46	1.76	2.34	2.24
Ca	0.70	0.06	0.28	0.12	0.02	0.18
Mg	0.05	–	–	0.20	0.03	0.18
Si/Al	2.59	2.87	2.51	2.67	2.44	2.50
Na/K	0.50	1.07	0.65	0.93	1.15	1.26
$E_b{}^c$	2.70	9.28	6.85	1.74	4.28	–4.16

(1) DSDP 59.2-6-2; Paleocene dark brown clay sampled at 134 m; 11°47′N, 147°35′E; from Stonecipher (1976b).
(2) SCAN 16P 100; Eocene (?) brown clay sampled at 100 cm; 16°25′N, 164°24′W; same reference as (1).
(3) SCAN 16P 100; Eocene (?) brown clay sampled at 100 cm; same location and reference as (2).
(4) JYN V-8G; light brown clay sampled at 15–25 cm; 13°26′N, 152°17′W; from Sheppard *et al.* (1970).
(5) DODO 113D; manganese coated boulder; 23°16′S, 74°59′E; same reference as (4).
(6) DODO 232D; manganese coated boulder; 5°23′S, 97°29′E; same reference as (4).
(a) Total iron given as FeO. Inclusion impurity; not included in cell content calculations. Fe has already been subtracted from analyses by Sheppard *et al.* (1970).
(b) By difference for analyses by Stonecipher. Measured by weight loss at 1000°C in analyses by Sheppard *et al.* (1970).
(c) $E_b = \frac{Al - \Sigma_{cat}}{Al}$ where $\Sigma_{cat} = 2(Mg + Ca) + (Na + K)$.

Harmotome

Harmotome is isostructural with phillipsite, with Ba as the principal cation (Rinaldi *et al.*, 1974). It is distributed mainly in sediments of the Pacific Ocean in areas of slow sedimentation rates and commonly coexists with phillipsite (Murray and Renard, 1891; Goldberg and Arrhenius, 1958; Arrhenius, 1963; Bonatti, 1963; Hay, 1966; Cronan, 1974). Harmotome-rich scoriaceous basaltic fragments and shards were observed west of the Society Ridge, Pacific Ocean, by Morgenstein (1967). Harmotome crystals with palagonite centers indicate that harmotome is the alteration product of palagonite. According to Arrhenius and Bonatti (1965), harmotome appears to be limited to the early stages of interaction between sea water and volcanic glass; subsequently slow growth of phillipsite is quantitatively the most important mineral forming process.

Clinoptilolite

Occurrence. Clinoptilolite is a high-silica member of the heulandite group (Mumpton, 1960) and crystallizes in the monoclinic system (Cm) (Merkle and Slaughter, 1968). It forms fine-grained euhedral crystals (Fig. 1b), usually less than 45 μm long, and therefore it is difficult or impossible to observe with the petrographic microscope. Inclusions are much less common than in phillipsite, indicating slow growth rates. The ideal composition of clinoptilolite is $(Na, K)_4CaAl_6Si_{30}O_{72}\cdot 24H_2O$ (Sheppard, personal communication). By definition, the Si/Al ratio of clinoptilolite can vary between 4.0–5.25; such phases are length slow (Boles, 1972). Mean indices of refraction and unit-cell dimensions are given in Table 1. According to Boles (1972), the area of the *ac* plane decreases with increasing Si or divalent cation substitution, and *b* increases with increasing Mg substitution.

Most reports of deep-sea clinoptilolites (as the high-Si member of the heulandite group) are based on Mumpton's (1960) overnight heating test to 450°C and subsequent X-ray diffraction analysis. But Alietti (1972) showed that the distinction between clinoptilolite and heulandite must be based on heating tests to (a) 400°C overnight to show the lack of polymorphism and to (b) 550°C to show its thermal stability relative to heulandite. As a result, it is possible that some of the reported deep-sea clinoptilolite occurrences might have to be reclassified as heulandites.

Clinoptilolite is more abundant in the Atlantic Ocean than in the Pacific and Indian Oceans (Biscaye, 1965; Turekian, 1965; Kolla and Biscaye, 1973; Stonecipher, 1976). In the Indian Ocean, Kolla and Biscaye (1973) observed that clinoptilolite is restricted to Cretaceous to Eocene sediments close to continental land masses and that it is more abundant in indurated chalks. It is also commonly associated with chert and/or opal. They indicated that clinoptilolite is absent in sediments with abundant basaltic material. Peterson *et al.* (1970) reported clinoptilolite from the Central Atlantic Ocean only in pre-Miocene sediments. Stonecipher's (1976) statistical analysis shows that clinoptilolite occurs in sediments of Cretaceous to Miocene age, with a steady increase in abundance from Miocene to Cretaceous. Unlike phillipsite, clinoptilolite is least common in sediments of less than 100 m depth and is rather abundant between 700–1200 m depth. The following is a sequence of decreasing clinoptilolite abundance as a function of lithology: calcareous > clayey > volcanic > siliceous. This sequence suggests a positive correlation between sedimentation rates and clinoptilolite abundance, as indeed has been shown by Stonecipher (1976).

Clinoptilolite has been observed to occur as radiolarian casts (Berger and von Rad, 1972; Robinson *et al.*, 1974) and was reported to coexist with palygorskite, chert and/or opal, smectite, sepiolite, and rhyolitic glass or with basaltic glass and opal (A and CT)* (Hathaway and Sachs, 1965; Weaver, 1968; Cook and Zemmels, 1972; Kolla and Biscaye, 1973; Stonecipher, 1976; Couture, 1977). Association with volcanic material is less evident for clinoptilolite than for phillipsite.

Chemistry. Stonecipher (1976b) analyzed twenty-five deep-sea clinoptilolites; some of the analyses are given in Table 3. Their Si/Al ratio varies between 4.2 to 5.2, but 84% of the above analyzed clinoptilolites have Si/Al ratios between 4.5 and 5.0. Alkalis are in excess of alkaline earths cations, and in most samples K > Na. Clinoptilolites from carbonate-rich horizons are somewhat enriched in Ca. In most sedimentary clinoptilolites, however, Na predominates (Sheppard, 1971a).

Origin. Clinoptilolite, as a diagenetic alteration product of rhyolitic glass, has been reported from various sedimentary environments (Hay, 1966; Sheppard and Gude, 1968, 1969; Iijima, 1971). Sheppard and Gude (1969) suggested that in moderately alkaline and

* According to Jones and Segnit's (1971) classification, opal-A is a hydrous X-ray amorphous silica phase that shows a diffuse X-ray band centered at about 3.8–4.1 Å; opal-CT is a unidimensionally disordered three-layered structure of low cristobalite with two-layered low tridymite domains (Flörke, 1955). The X-ray diffraction pattern of opal-CT shows the low cristobalite broad peaks at about 4.1 Å and 2.5 Å, with a low tridymite peak at 4.3 Å.

TABLE 3. Chemical analyses and unit cell contents of deep-sea clinoptilolites

Sample	1	2	3	4	5	6
SiO_2	66.27	66.64	66.71	69.78	66.77	67.98
Al_2O_3	11.87	11.78	11.67	11.96	11.62	12.08
K_2O	6.16	6.06	4.70	5.44	2.56	6.42
Na_2O	2.01	2.29	0.75	1.24	0.44	1.25
CaO	0.45	0.53	2.50	1.14	3.40	0.63
MgO	0.28	0.30	0.35	0.59	0.96	0.62
FeO(a)	–	0.21	–	–	–	–
H_2O(b)	12.96	12.19	13.32	9.85	14.25	11.02
Totals	100	100	100	100	100	100
Number of cations on the basis of 72 oxygens						
Si	29.77	29.73	29.84	30.00	29.82	29.73
Al	6.28	6.19	6.15	6.06	6.12	6.23
K	3.53	3.45	2.68	2.98	1.45	3.58
Na	1.75	1.98	0.65	1.03	0.37	1.06
Ca	0.22	0.25	1.20	0.52	1.63	0.29
Mg	0.18	0.20	0.23	0.38	0.64	0.40
Si/Al	4.74	4.79	4.85	4.95	4.87	4.77
Na/K	0.49	0.57	0.02	0.34	0.02	0.30
E_b(c)	3.12	–2.37	–0.65	3.94	–3.80	3.00

(1) DSDP 105-14-cc; lower Cretaceous, black, carbonaceous clay sampled at 349 m; 34°54′N, 69°10′W; Stonecipher (1976b).
(2) DSDP 105-17-cc; lower Cretaceous, black, carbonaceous clay sampled at 405 m; same location and reference as (1).
(3) DSDP 105-21-cc; lower Cretaceous clay sampled at 457 m; same location and reference as (1).
(4) DSDP 9A-1-6; middle Eocene siliceous clay sampled at 682 m; 32°46′N, 59°12′W; same reference as (1).
(5) DSDP 163-27-1; upper Cretaceous nannochalk sampled at 271 m; 11°15′N, 150°17′W; same reference as (1).
(6) DSDP 59.2-5-1; Eocene brown clay in association with phillipsite, sampled at 127 m; 11°47′N, 147°35′E; same reference as (1).
(a) Total iron given as FeO. Inclusion impurity; not included in cell content calculations.
(b) By difference.
(c) $E_b = \frac{Al - \Sigma_{cat}}{Al}$ where $\Sigma_{cat} = 2(Mg + Ca) + (Na + K)$.

saline environments, clinoptilolite forms directly from silicic glass by solution precipitation. In the oceanic environment, Cook and Zemmels (1972) reported an occurrence of clinoptilolite associated with rhyolitic glass at Site 84, Leg 9, DSDP. But the presence of unaltered rhyolitic glass does not necessarily imply that the clinoptilolite formed from this glass; on the contrary, it might indicate that the rhyolitic glass is *not* the clinoptilolite precursor. In the oceans, silicic glass is much less common than basaltic glass; therefore, according to Cronan (1974), clinoptilolite is less abundant relative to phillipsite. Previous estimates of relative abundances of clinoptilolite and phillipsite may have been biased due to the difference in their distribution as a function of depth; in the sedimentary column, clinoptilolites occur at greater depths than phillipsites. More recent increased recoveries of sediments from greater depths indicate that they are about equally abundant in deep-sea sediments (*vide infra*). Peterson and Goldberg (1962) showed the possibility of local silicic volcanism in a silica-poor volcanic region, and rhyolite and obsidian were described by Daly (1925) on Ascension Island.

The abundance of clinoptilolite in smectite-rich sediments suggests a volcanic source. According to Bramlette and Posnjak (1933) smectite is the more complete alteration product of silicic glass relative to clinoptilolite.

Based on common observed assemblages, the following possible reactions for clinoptilolite formation were suggested:

(a) Basaltic glass + silica (mainly biogenic) → clinoptilolite (Weaver, 1968; Berger and von Rad, 1972; Cook and Zemmels, 1972; Hathaway and Sachs, 1972).
(b) Smectite + phillipsite + biogenic silica → clinoptilolite + palygorskite (Berger and von Rad, 1972; Couture, 1976).

In several Pacific cores between ∿10°N and ∿10°S phillipsite is present in the upper sections, but clinoptilolite predominates in the lower sections of the same core (Stonecipher, 1976). Phillipsite interaction with Si-rich interstitial waters might result in clinoptilolite formation. In Lake Magadi, Kenya, for example, phillipsite forms from trachyte tuff in the chert-free parts of the section, and clinoptilolite forms with magadiite and chert (Hay, 1968).

The growth of clinoptilolite crystals in the interior of radiolarians in a radiolarian mud with no detectable volcanic material suggested to Berger and von Rad (1972) that the clinoptilolite crystals precipitated from pore solutions.

Analcime

Analcime crystallizes in the isometric system (Ia3d) (Taylor, 1930). The ideal formula of analcime is $NaAlSi_2O_6 \cdot H_2O$. The Si/Al ratio of sedimentary analcimes varies between 2.0 and 2.9 (Hay, 1966), but no chemical analyses of deep-sea analcimes are available.

The occurrence of analcime in deep-sea sediments was first reported by Murray and Renard (1891). In deep-sea sediments, analcime is much less abundant than phillipsite or clinoptilolite. Basaltic components or glass are always present (Murray and Renard, 1891; Griffin *et al.*, 1972; Bass *et al.*, 1973). This observation does not prove that analcime in pelagic sediments forms directly from volcanic material, in particular, volcanic glass, in contrast to lake sediments in which most analcimes form from zeolitic precursors such as clinoptilolite, erionite, and phillipsite, when fresh glass is no longer present (Hay, 1966; Sheppard and Gude, 1968, 1969). Analcime has been described most often from volcaniclastic and calcareous sediments. The abundance of both clinoptilolite and analcime in calcareous sediments probably indicates a genetic relation between the two zeolites. The abundance of analcime increases with increasing age of the host sediments, but no correlation with burial depth has been observed (*vide infra*). Similarly, Hay (1966) observed that within silicic tuffs in alkaline saline lake deposits the amount of analcime clearly increases and that of phillipsite and clinoptilolite decreases as a function of age of the tuff, regardless of burial depth. On the basis of water content, analcime should be stable at higher temperatures than phillipsite or clinoptilolite (Coombs *et al.*, 1959; Hay, 1966).

Analcime in deep-sea sediments coexists with smectite, chlorite, celadonite, phillipsite, or clinoptilolite. Based on phase equilibria diagrams in the $K_2O-Na_2O-Al_2O_3-SiO_2$ system at 25°C and 1 atmosphere (Hess, 1966), analcime should be present only at $\log Na^+/H^+ \simeq 8$ under conditions of high Si activity. With increasing $\log Na^+/H^+$, the field of analcime expands to lower Si values, at the expense of montmorillonite. In sea water, $\log Na^+/H^+$ is about 7.2, and Si activity is low; therefore, analcime should not form in deep-sea sediments at or close to the sediment/water interface. Conditions for analcime formation may become more favorable in the sedimentary column, where Si activity of the interstitial water increases. A higher Na^+/K^+ ratio than in sea water may also increase its stability.

Other Zeolites

Only a few occurrences of more calcic zeolites were reported from deep-sea sediments. They are: chabazite, erionite, gmelinite, laumontite, thaumasite, and thomsonite (Merkle and Slaughter, 1968; Walker, 1972; Bass *et al.*, 1973; Drever, 1976). Occurrences of the sodic zeolite natrolite were reported by Nayadu (1964) and Bass *et al.* (1973). All these zeolites are associated with basaltic materials. Bass *et al.* (1973) observed that the calcic zeolites chabazite, gmelinite, and thomsonite replace plagioclase but not pyroxene. Thaumasite was described by Walker (1972) in the eastern North Atlantic as a cement of tuffaceous biocalcarenite and basalt and tuff breccia. Laumontite, common in zeolite facies metamorphic rocks (Coombs *et al.*, 1959), occurs as lath-shaped crystals, with Si/Al ratio of ∿2.5 and occurs with clinoptilolite in DSDP Site 323 (Drever, 1976). Based on the absence of epidote in this site, despite its coexistence with laumontite-rich rocks exposed nearby in Antarctica (Horn, 1968), Drever (1976) suggested that the laumontite is probably authigenic. However, due to the great difference in the densities of laumontite and epidote, they could have been separated during their transport from Antarctica. The authigenic origin of the laumontite from Site 323 could be confirmed on the basis of a comparative chemical study of both laumontites. Bass (personal communication) described a laumontite occurrence as veins in a moderately metamorphosed basalt dredged off Mussau Ridge by L. W. Kroenke.

Based on the very few reports of the above more Ca-rich zeolite occurrences in pelagic sediments, it is difficult to conclude whether they form at low temperature by basalt-sea water interaction, or at elevated temperatures as a result of local short- or long-lived hydrothermal activity, or of high heat flow.

The observation that several zeolites, such as mordenite, chabazite, and erionite, are common in sedimentary rocks (Hay, 1966; Sheppard, 1971b and references therein) but rare or absent in pelagic sediments might help to elucidate their mode of origin. For example, mordenite, a highly siliceous zeolite with Si/Al ratios similar to those of deep-sea clinoptilolites, but with Na in excess of K, has not been reported as yet from deep-sea sediments. The Na/K ratio of the solution might be an important factor that controls clinoptilolite versus mordenite formation. More experimental work is needed to determine whether kinetic or thermodynamic factors are responsible for the above observation.

Authigenic feldspars

Authigenic K-feldspars have been described from Atlantic and Pacific Oceans volcanic sediments and basalts, as euhedral or anhedral crystals replacing plagioclase or just calcic plagioclase cores (Mellis, 1952; Matthews, 1962, 1971; Bass *et al.*, 1973; Stewart *et al.*, 1973; Larson *et al.*, 1973; Kelts and McKenzie, 1976), as replacement of volcanic glass (Lancelot *et al.*, 1972), as discrete euhedral crystals in volcanic sediments (Donnelly and Nalli, 1973), in foraminifera chambers (Larson *et al.*, 1974), and in porcelanite with volcanic material (Kastner, 1976).

They were identified as intermediate microcline in basal volcaniclastic sediments of Sites 315A and 316 by Kelts and McKenzie (1976), as orthoclase replacement of plagioclase in dredged gabbro and tholeiite basalt from western and eastern North Atlantic by Mellis (1952) and Matthews (1962, 1971) respectively, and as sanidine replacing volcanic glass at Site 105 by Lancelot *et al.* (1972).

Except for numerous layers with up to 60% K-feldspar in basal volcaniclastic sediments of Sites 315A and 316 (Kelts and McKenzie, 1976), feldspars constitute a small fraction of the sediments or basalts.

Authigenic K-feldspars occur at all depths and are most common in upper Cretaceous volcanic sediments of the Pacific ocean, in particular in the coarser-grained sections of these sediments, as described by Kelts and McKenzie (1976). They implied that the distribution of the K-feldspars within the sediments is strongly affected by porosity.

Their euhedral habit, chemical purity, composition of inclusions, absence of cathodo-luminescence properties, and textural relations with the plagioclase crystals and volcanic glass that they replace were used as evidence for the diagenetic (Mellis, 1952; Matthews, 1962, 1971; Lancelot *et al.*, 1973; Donnelly and Nalli, 1973; Kastner, 1976; Kelts and McKenzie, 1976) or hydrothermal (Stewart *et al.*, 1973) origin of K-feldspars.

Illite and smectite are the most common associated clay minerals. Unlike tuffaceous deposits in alkaline saline lakes, where formation of zeolites such as clinoptilolite and analcime from volcanic glass precedes authigenic K-feldspar formation (e.g., Hay, 1966; Sheppard and Gude, 1968, 1969; and references therein) in the deep-sea environments, occurrence of the zeolites analcime and clinoptilolite with K-feldspar are rare and was reported only by Bass *et al.* (1973) and Kelts and McKenzie (1976), respectively. No obvious textural observations, suggesting a genetic relation between these zeolites and the K-feldspars, were reported.

The only authigenic albites around detrital albites were reported by Kastner and Siever (1972) as overgrowths in gypsum and dolomite-rich sediments associated with the evaporitic sections of Late Miocene to Early Pliocene in the Mediterranean Sea, Site 124. This occurrence of authigenic albite indicates the importance of chemical controls on the formation of authigenic feldspars.

DISTRIBUTION OF PHILLIPSITE AND CLINOPTILOLITE IN DEEP-SEA SEDIMENTS

Methods

Characterization of the sedimentary environments in which deep-sea zeolites occur can provide insight into the factors which control their formation. Spatial and temporal distributions and mineralogic associations in Indian Ocean sediments have been studied by Kolla and Biscaye (1973). Recently, Stonecipher (1976) carried out a similar investigation based on Deep Sea Drilling Project data through Leg 24.

All volumes of Initial Reports of the Deep Sea Drilling Project that describe sites in the Atlantic, Pacific, and Indian Oceans (1–9, 11, 12, 14, 16–27, 30–33, 35) were investigated for the following synthesis of all cores with horizons containing recognizable amounts of deep-sea zeolites. Each lithologic unit, as defined by the authors of that volume, which contained a zeolite was counted as one data unit. For each unit, information on variables such as geographic location, lithology, sedimentation rate, depth in the core, age of the sediment, and mineral assemblages was compiled and analyzed for interrelations as shown in Figs. 2 through 10.

Results

The geographic distribution of phillipsite and clinoptilolite in the Atlantic, Pacific, and Indian Oceans is given in Fig. 2. In the Atlantic, phillipsite is most commonly associated with sediments of Miocene age and younger that were deposited near sources of volcanic material, such as islands, seamounts, and the Mid-Atlantic Ridge. Clinoptilolite is more common than phillipsite and occurs in continental margin sediments and in zones of carbonate deposition on rises, plateaus, and the flanks of the Mid-Atlantic Ridge. Clinoptilolite also occurs in clayey sediments of Eocene to Cretaceous age in abyssal plain areas.

In the Pacific, phillipsite predominates in the central portions of circulatory gyres where biogenous sedimentation is minimal and in areas of volcaniclastic deposition associated with topographic highs. It also occurs in equatorial regions and other areas where sediments are rich in biogenic material. In these areas, phillipsite occurs in calcareous and siliceous oozes of Miocene age or younger, and commonly small amounts of volcanic material are also present. Clinoptilolite occurs in cores from continental margins, in equatorial regions, and in rises, plateaus, and seamounts where calcareous biogenous sediments predominate. It also occurs in areas of volcaniclastic deposition in marginal basins, near island-arc and seamount systems, and where abundant detrital material from sub-aerial volcanic terrains is being deposited. An interesting feature is the band of transitional cores which occurs between about 10°N and 10°S. Here phillipsite is most abundant in the upper sections of the core, whereas clinoptilolite

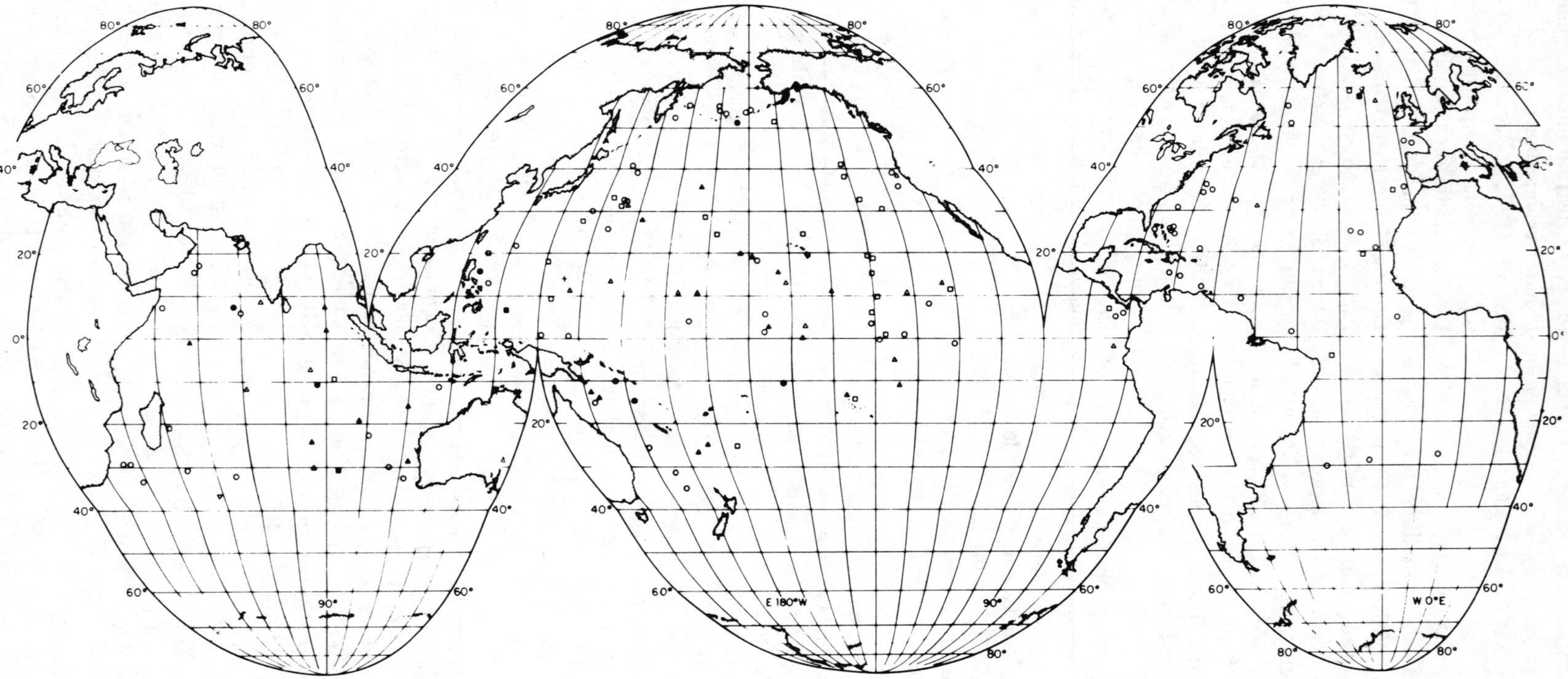

Fig. 2. Geographic location of all DSDP sites in which sediments containing zeolites occur. Note that these are not just surface sediment occurrences, but represent samples from all depths: □, cores that contain only phillipsite; ○, cores that contain only clinoptilolite; △, cores that contain phillipsite and clinoptilolite; ■, cores that contain phillipsite and analcime; ●, cores that contain clinoptilolite and analcime; ▲, cores that contain phillipsite, clinoptilolite, and analcime; ▽, cores that contain only analcime; X, cores that contain thaumasite; +, cores that contain erionite. In cores that contain more than one type of zeolite, the zeolites may occur together in the same horizon or separately in different horizons.

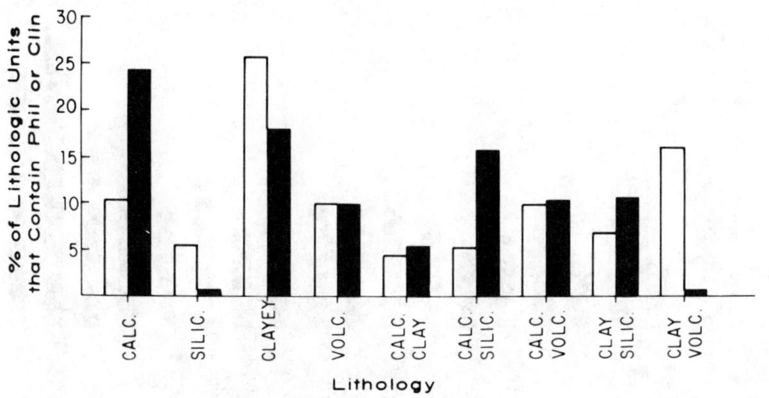

Fig. 3. Frequency of occurrence of phillipsite □ and clinoptilolite ■ in various lithologies. Calcareous sediments include oozes, chalks, and limestones; siliceous = radiolarian oozes; clayey = clays and mudstones; volcanic = hyaloclastites, ash layers, and volcaniclastic sandstones and mudstones; calc.-clay = clays interbedded or mixed with calcareous oozes; calc.-silic; = radiolarian/nanno oozes or chalk or limestone with chert; calc.-volc. = calcareous oozes interbedded or mixed with volcanic debris; clay-silic. = radiolarian clays or clays with chert; clay-volc. = clays bearing altered ash.

becomes dominant at depth. The transition is either sharp or gradational. A sharp transition is generally associated with a lithologic change from a clayey sediment above to calcareous ooze below. A gradational transition commonly occurs where there is no lithologic change.

In the Indian Ocean, phillipsite occurs in ash-bearing biogenous oozes of Miocene age or younger and in pelagic clays from the central portions of the basins. Clinoptilolite is found near the margins of the basins, and in carbonate sediments on topographic highs, and at the base of the transitional cores on the flanks of the ridges. It also occurs in clayey sediments of Cretaceous age from the central basinal areas.

Figure 3 shows the relation between the distribution of phillipsite and clinoptilolite and the various lithologies. Phillipsite occurs primarily in clayey and volcanogenic sediments, but it is also present in siliceous oozes of Eocene age or younger and in carbonate sediments younger than Miocene. Almost all of the sediments in which phillipsite occurs contain at least a small amount of basaltic material. Clinoptilolite occurs predominantly in calcareous oozes or in calcareous/siliceous sediments in which biogenous silica has transformed to opal-CT (Jones and Segnit, 1971) (porcelanite) or quartz (chert), where they contain greater than ∿10–15% clay minerals or volcanic components. Clinoptilolite occurs in clayey sediments, but only in those of Oligocene age or older. It also occurs in volcaniclastics and in turbidites composed of volcanic material interbedded and mixed with calcareous oozes and/or shallow water calcareous debris.

The differences in lithologic distributions of phillipsite and clinoptilolite are reflected in the average

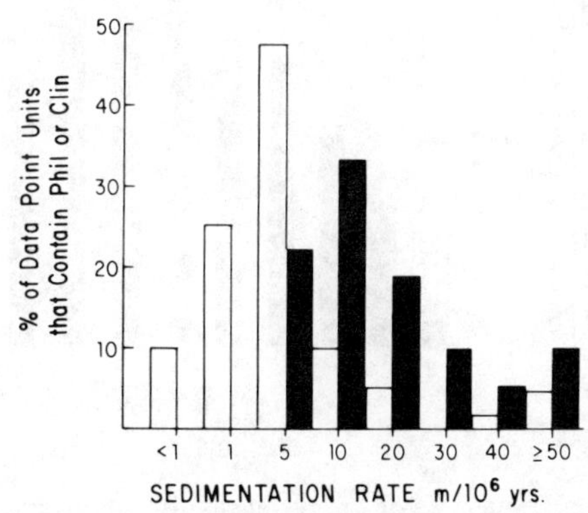

Fig. 4. Average sedimentation rates of sediments that contain phillipsite □ or clinoptilolite ■. Each data unit represents an interval of a core that contained a zeolite over which a constant sedimentation rate was maintained.

sedimentation rates for cores containing these minerals. Figure 4 demonstrates that phillipsite-bearing sediments are generally deposited at rates of $\lesssim 5$ m/10^6 y, whereas clinoptilolite-bearing sediments are generally deposited at rates of $\gtrsim 5$ m/10^6 y. Sediments of the volcaniclastic-turbidite facies (Appendix I) which commonly accumulate at rates of $\gtrsim 30$ m/10^6 y, are an exception to this observation.

Most phillipsite occurrences are concentrated in the 0–100 m depth range. A marked decrease in the number of phillipsite occurrences with increasing depth in the sediment is shown in Fig. 5. Clinoptilolite is least common in sediments of less than 100 m depth in the sediment column, but it seems to be nearly evenly distributed in sediments from 100–700 m in depth. In deeper horizons (700–∿1200 m) clinoptilolite is more common.

The distribution of phillipsite and clinoptilolite with

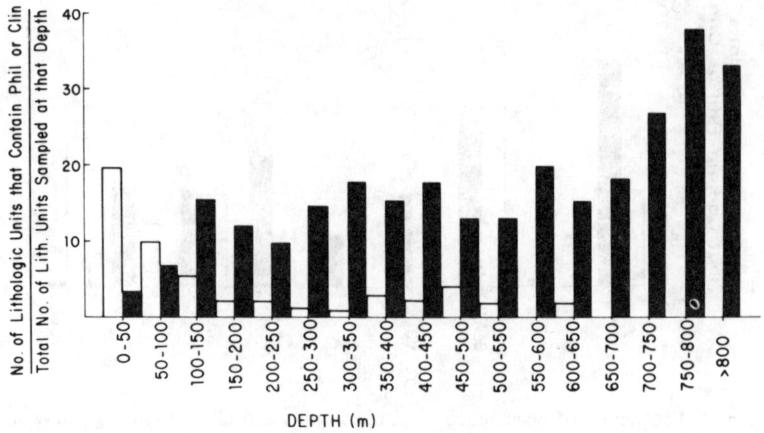

Fig. 5. Distribution of phillipsite □ and clinoptilolite ■ with respect to depth from the sediment/water interface.

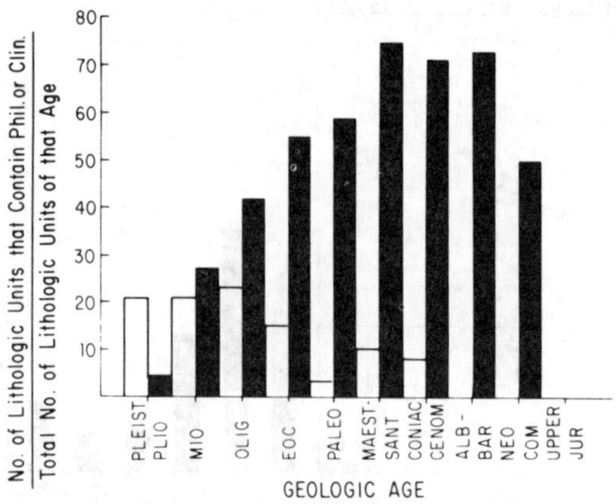

Fig. 6. Frequency of occurrence of phillipsite □ and clinoptilolite ■ in sediments of various geologic ages.

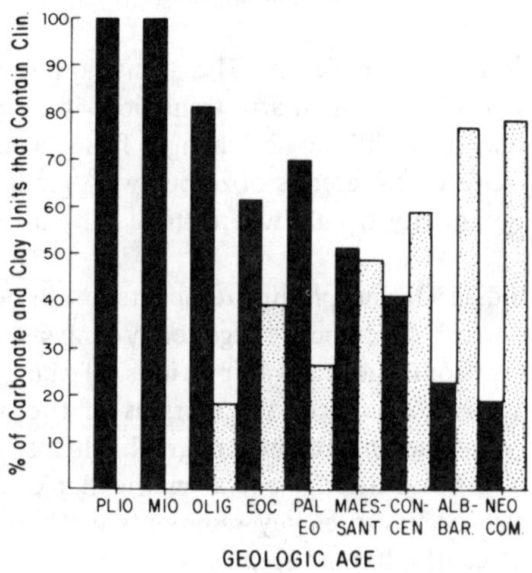

Fig. 7. Age distribution of clinoptilolite in clayey ▨ and carbonate ■ sediments.

respect to the age of the host sediment is presented in Fig. 6. Phillipsite is about equally common in sediments of Pleistocene to Oligocene age, whereas in sediments older than Oligocene, its frequency of occurrence decreases. All Cretaceous sediments that contain phillipsite are volcaniclastic sediments. The youngest sediments that contain clinoptilolite are of early Pliocene age. Some clinoptilolites were reported in Pleistocene sediments of DSDP cores and in less than 20 m depth (e.g., Sites 85, 131, 262). Some of these, however, most probably are redeposited clinoptilolites. With increasing age, a systematic increase in clinoptilolite abundance is shown, reaching a maximum in the mid to upper Cretaceous. The paucity of sediments of Neocomian age and older makes the determination of zeolite distribution patterns in these sediments difficult, but it seems that the abundance of clinoptilolite decreases in sediments older than Barremian. The age distribution of sediments that contain clinoptilolite is also a function of lithology, as seen in Fig. 7. Calcareous clinoptilolite-bearing sediments are generally significantly younger than clinoptilolite-bearing argillaceous sediments.

Minerals that commonly coexist with phillipsite or clinoptilolite are presented in Fig. 8. Quartz and plagioclase are almost ubiquitous and nearly equally prevalent with phillipsite and clinoptilolite. Iron oxides and barite are more commonly associated with phillipsite than with clinoptilolite, whereas calcite, K-feldspar, opal CT, pyrite, and siderite are more commonly associated with clinoptilolite.

The changes in clay mineral–zeolite assemblages as a function of geologic age are examined in Fig. 9. In sediments of Pleistocene to Eocene age, phillipsite seems to be most commonly associated with a detrital clay mineral assemblage of illite,* kaolinite, and smectite. Detailed X-ray diffraction and chemical analyses of phillipsite-bearing sediments (Stonecipher, 1976b) indicate the presence of altered basaltic volcanic material

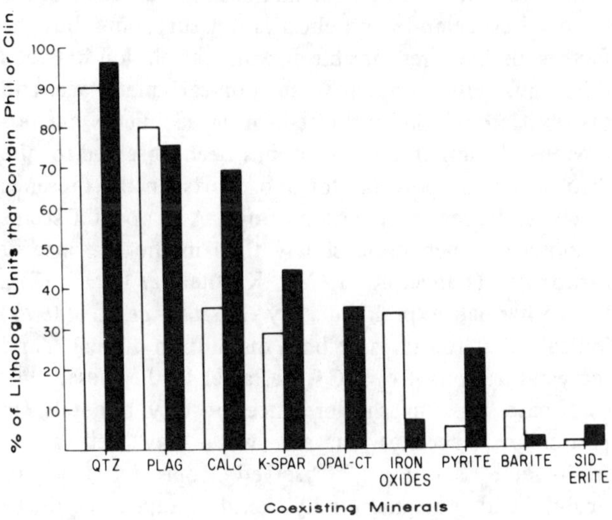

Fig. 8. Relative number of lithologic units in which phillipsite ☐ or clinoptilolite ■ coexist with each of the listed minerals.

even where no visible evidence of glass remains. Since it has been shown (for example: Arrhenius, 1963; Rex, 1967; Peterson and Griffin, 1964) that smectite is a common diagenetic product of the alteration of volcanic material in marine sediments, it is likely that at least part of the smectite in phillipsite-bearing sediments is authigenic. Phillipsite in Paleocene and upper Cretaceous sediments is associated with smectite-rich sediments. This correlates with the observation that most of these older sediments that contain phillipsite are volcaniclastic sediments. The assemblage phillipsite/palygorskite seems to be incompatible. In sediments of Pliocene to Eocene age, clinoptilolite is associated with smectite, illite, and kaolinite. In sediments of Eocene age and older, it occurs with palygorskite and illite, especially when associated with cherts. Pimm (1973), von Rad and Rösch (1972), and Berger and von Rad (1972) suggested that the association palygorskite/illite/chert is a diagenetic assemblage.

Analcime in deep-sea sediments is the next most abundant zeolite after phillipsite and clinoptilolite. Improved X-ray diffraction techniques have increased the number of occurrences reported in sediments of Phase III DSDP legs, and it may well be that analcime is actually more common in sediments of Phases I and II

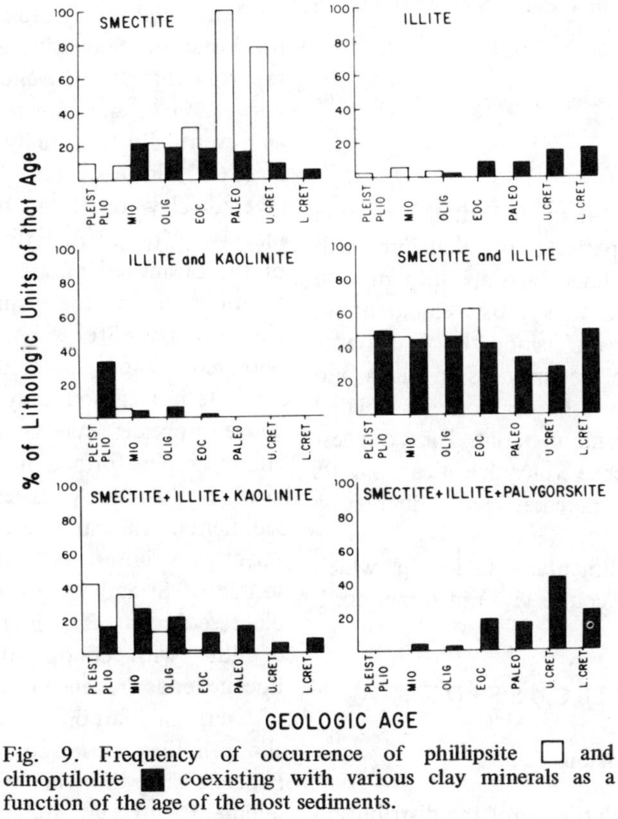

Fig. 9. Frequency of occurrence of phillipsite ☐ and clinoptilolite ■ coexisting with various clay minerals as a function of the age of the host sediments.

* Illite is used here to mean hydrous mica. See Yoder and Eugster (1955) for a discussion of the term.

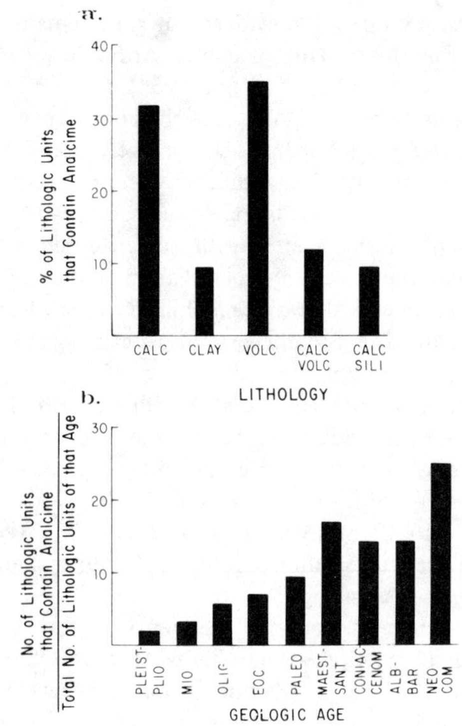

Fig. 10a. Frequency of occurrence of analcime in various sediment types. Except for calc. sediments, the lithologic classifications are the same as in Fig. 3. Calc. = ash-bearing carbonate oozes (<10% volcanic ash).

Fig. 10b. Distribution of analcime with respect to the age of the host sediment.

Deep Sea Drilling reports than indicated in these publications. Distribution patterns of analcime with respect to lithology and sediment age are presented in Figs. 10a and 10b. Analcime is most abundant in volcaniclastic sediments or in ash-bearing carbonate oozes. Other, less common occurrences in clays and nanno chalks with chert are also commonly associated with minor amounts of volcanic material. The youngest analcimes occur in Pleistocene volcaniclastics, and its frequency of occurrence increases with increasing geologic age.

A summary of the sedimentary facies in which clinoptilolite and phillipsite occur is given in Appendix I.

GENERAL DISCUSSION

Discussion

Correlation. The factors that control the distribution of zeolites in marine sediments are numerous, varied, and often difficult to isolate. However, differences in the physical and chemical characteristics of sediments associated with each zeolite and the chemistry of sea water and interstitial waters provide important information that helps to identify these factors. For example, because clinoptilolite is more siliceous than phillipsite, its association with opal-CT and authigenic quartz in the form of porcelanite and chert is not surprising; but the factors that are responsible for the correlation between lithology, carbonate-rich versus non-carbonate, age, and distribution of clinoptilolite-bearing sediments are less obvious. A similar correlation has been observed for the distribution of porcelanites and cherts. In the presence of clays, the recrystallization of opal-A to opal-CT seems to proceed much more slowly than in the presence of carbonate (Lancelot, 1973; Kastner *et al.*, 1977). Hydrothermal experiments by Kastner *et al.* (1977) indicate that the rates of both dissolution of opal-A and recrystallization of opal-CT are faster in the presence of carbonate than in the presence of clay minerals (in particular, smectite) in sea water, as well as in fresh-water solutions. Observed coprecipitation of opal-CT with a magnesium hydroxide complex requires a source of alkalinity (e.g., continuous dissolution of carbonate) and magnesium (sea water). In clayey sediments, particularly those rich in expandable clays, clay minerals compete with opal-CT for the available alkalinity and, as a result, decrease the rate of opal-CT precipitation. Similarly, Kastner and Bada (1974) also reported that the presence of carbonate accelerates the conversion of smectite to mixed layered smectite/illite and eventually to illite by at least an order of magnitude at 136°C. Since smectite is more siliceous than a mixed layered clay or illite, this conversion results in an increase in the Si/Al ratio of the interstitial fluids. Some of the Si and Al may be available for other diagenetic reactions such as the formation of zeolites, in particular clinoptilolite. Stonecipher (in preparation) has also observed a more rapid alteration of volcanic glass to smectite in a carbonate system. A combination of all the above processes may account for the observation that clinoptilolites formed in carbonate sediments that are younger than the youngest clinoptilolite-bearing clayey sediments. In carbonate-free sediments, the slower smectite to illite transformation may still prove to be a source of Si and Al for zeolite formation, but only in older sediments. The increasing number of associations of illite with clinoptilolite in sediments older than Eocene lends credence to this suggestion.

Even in carbonate-rich sediments, the youngest clinoptilolites observed in sediments are of Lower Pliocene age, but occur from 20–100 m below the sediment surface and not at the sediment/water interface. This indicates that clinoptilolite does not form by interaction of sediments with sea water, but rather with interstitial waters; it also suggests that burial depth is not the only critical factor for clinoptilolite formation. Lithology, as well, does not seem to be the only critical factor that controls clinoptilolite formation; clinoptilolite consistently occurs only in sediments older than Lower Pliocene, even in primarily carbonate ooze cores. Phillipsite, however, commonly occurs in post-Miocene carbonate sediments, even in the uppermost few centimeters of Pleistocene sediments. These observations suggest either a difference in the rates of nucleation or crystal growth between clinoptilolite and phillipsite, or probably a change in some chemical factors, such as Si activity, as a function of burial depth or time. High silica activities are often attained in interstitial waters of biogenic opal-rich sediments, but not near the sediment/water interface, where phillipsite predominates (e.g., Manheim and Sayles, 1974; Gieskes, 1975). The latter possibility is supported by the higher Si/Al ratio of clinoptilolites and by the dissolution experiments of Lerman *et al.* (1975). For any diffusion coefficient, concentration gradients of interstitial water depend on diagenetic reactions and sedimentation rates, as shown by Gieskes (1975). Therefore, depending on sedimentation rates, high silica concentrations could be attained in sediments of various lithologies, such as calcareous-volcanic turbidites and foram-nanno oozes with a minor amount of terrigenous clay minerals.

Kinetic vs. thermodynamic factors. Closely related to the question of rates of formation of the two zeolites is their relative thermodynamic stability and rates of dissolution. Phillipsite forms much earlier than clinoptilolite, but it also dissolves more rapidly. Commonly, the size, complexity of twinning, and perfection of crystal faces of phillipsite crystals increases or maintains a constant level to a depth of 1–50 m below the sediment surface. Below this depth, the crystals show the effects of dissolution, for example, etched crystal faces. The abundance of phillipsite also decreases with depth, and eventually this zeolite disappears (Fig. 5). Rex (1967) described similar evidence of solution and etching of phillipsite crystals from Sylvania Guyot. Direct evidence of clinoptilolite dissolution has been reported only once in upper Cretaceous sediments, at the boundary between carbonate and clayey sediments. Site 137, Atlantic Ocean by Berger and von Rad (1972), but this may be a reflection of the paucity of Lower Mesozoic sediments in the oceans. The oldest clinoptilolite known on land is in sediments of Triassic age (Boles and Coombe, 1975).

Inasmuch as only few cores of Upper Jurassic to Lower Cretaceous sediments were recovered by the DSDP, it is difficult to determine whether the scarcity of clinoptilolite in these older sediments is due to its dissolution with age, or less likely due to primary conditions that were unfavorable for zeolite formation in these few old samples. Several authors (e.g. Hay, 1966; Senderov, 1965) suggested that zeolites are thermodynamically unstable at surface temperatures and pressures. Hay (1966) based his conclusion on the observation that phillipsite and clinoptilolite (and also erionite, mordenite, and chabazite) are more common in Cenozoic sedimentary rocks than in pre-Cenozoic rocks, but analcime and possibly laumontite are the dominant zeolites in rocks of Mesozoic age, and albite and K-feldspar become increasingly abundant in rocks of Paleozoic age. More experimental work is needed to determine whether phillipsite and clinoptilolite are indeed thermodynamically unstable in the deep-sea environment, or whether they are thermodynamically stable, but have a relatively small stability field, as implied by Glaccum and Boström's (1976) suggestion that phillipsite is thermodynamically stable at P–T–X conditions of the deep-sea floor; thus, their dissolution or transformation to another aluminosilicate phase, such as authigenic feldspar, is due to changes in the original physico-chemical conditions.

Analcime has been identified in samples from all ages, and it is generally associated with basaltic volcanic material. It commonly occurs with phillipsite or clinoptilolite. Its abundance, however, does increase with increasing age, but no textural evidence that points towards its secondary origin after phillipsite or clinoptilolite has been forwarded. K-feldspars have been reported more often in Cretaceous sediments, but the diagenetic versus hydrothermal nature of these K-feldspars has not always been established. Assuming that they are all diagenetic, it is possible but, *a priori*, not intrinsic that similar to tuffaceous sediments in alkaline saline lake environments, there is a genetic relation between zeolites and K-feldspars in the deep-sea environment.

Precursors. Many investigators (e.g., Murray and Renard, 1891; Arrhenius, 1963; Bonatti, 1963) have noted the common association of phillipsite with basaltic glass and/or palagonite and have suggested them as precursors of deep-sea phillipsites. Stonecipher (1976b) substantiated these suggestions by demonstrating that in two cores from the Atlantic and Pacific Oceans, where no direct evidence of basaltic glass is available, chemical analyses of the bulk host sediment suggest that there was originally a basaltic volcanic component to that sediment. An important question is whether or not hydration of the volcanic glass is a necessary step prior to zeolite formation in sedimentary environments. Murray and Renard (1891), Deffeyes (1959), and Bonatti (1963, 1965), suggested that palagonitization precedes zeolite formation. At low temperatures, hydration of volcanic glass by the diffusion of water molecules into the glass is extremely slow. Scholze and Mulfinger (1959) indicated that the activation energy for water diffusion into glass decreases with the decrease in silica content of the glass, and Marshall (1961) estimated a diffusion coefficient of the order of 10^{-10} $cm^2/10^6$ years at 20°C for water diffusion into volcanic glass. But palagonitization of basaltic glass by dissolution and precipitation at low temperatures is rather fast, as suggested by Moore (1966) and by Hay and Iijima (1968). If hydration of volcanic glass by either diffusion of water molecules or dissolution–precipitation is an essential primary step prior to zeolite formation, at low temperatures basaltic glass will be more susceptible to palagonitization and zeolite formation than rhyolitic glass will be to hydration, and zeolite formation from volcanic glass will be favored in higher pH environments.

In the marine environment, basaltic glass is the most common phillipsite precursor, whereas in the alkaline, saline lake environment, rhyolitic glass is the most common precursor of phillipsite (Hay, 1966; Sheppard and Gude, 1968, 1969). Rhyolitic glass in the marine environment most commonly alters to clinoptilolite. Deep-sea phillipsites are also less silicic than phillipsites from alkaline saline lake deposits (Sheppard *et al.*, 1970; Stonecipher, 1976b). These observations could be explained on the basis of Mariner and Surdam's (1970) experiments, which showed that the solubility of rhyolitic glass increases, but the Si/Al ratio in the solution decreases, with increasing alkalinity. Accordingly, from the same rhyolitic glass precursor, Si-rich phillipsite will form in the more alkaline environment (alkaline saline lakes) and clinoptilolite will form in the less alkaline environment (sea water). Furthermore, phillipsite will form in the sea water environment from a less silicic volcanic glass precursor (basaltic glass) than in more alkaline environments.

A more difficult question concerns the precursors of clinoptilolite. Many authors (e.g., Hathaway and Sachs, 1965; Hay, 1966, and references therein) have suggested that the presence of intermediate to silicic volcanic glass is necessary for the formation of clinoptilolite. However, only about 20–25% of the deep-sea clinoptilolites occur in sediments which are predominantly volcanogenic. Perhaps another 40–60% occur in biogenous sediments and clays which contain trace to minor amounts of volcanic glass, both basaltic and silicic. But the remainder occurrences of clinoptilolite are in sediments which contain no visible evidence of volcanic glass. Some of these sediments are of Cretaceous age, and thus diagenetic processes may have removed all the visible evidence of volcanic glass. The chemical fingerprint of silicic to intermediate volcanic material is less obvious than that for basaltic material, and it is difficult to prove previous presence of silicic to intermediate volcanic material in these sediments. Thus, in many examples of deep-sea sediments containing clinoptilolite, a strong case for a volcanic precursor for clinoptilolite is difficult and unnecessary to make.

Arrhenius (1963) described zeolite growth within dissolving siliceous microfossils in sediments which contain no volcanic material, and suggested that zeolites in these sediments originated from Si derived from this biogenic source and from other elements dissolved in sea water, such as Al and various alkali and alkaline earth cations. Goodell (1965) and Berger and von Rad (1972) also concluded that dissolution of siliceous microfossils provided silica for the formation of clinoptilolite in the sediments they examined. Diagenetic alteration of detrital terrigenous clay-rich sediments to an assemblage of smectite, K-feldspar, and clinoptilolite was suggested by Donnelly and Wallace (1976) in three Antarctic cores (322, 323, and 325). Brown *et al.* (1969), Estéoule *et al.* (1971), and Curtis and Cornell (1972) also described diagenetic alteration of detrital terrigenous clays to clinoptilolite in marine sediments in England, France, and California, respectively; as did Hay (1970) for samples from Olduvai Gorge, Tanzania. These observations imply that clinoptilolite can also form from non-volcanic precursors.

Horizontal and vertical distribution. The distribution of phillipsite and clinoptilolite with respect to depth is primarily a reflection of the distribution of these zeolites with respect to age, lithology, and sedimentation rates. Phillipsite occurs primarily in younger sediments; clinoptilolite is more common in older sediments. Thus, phillipsite is more prevalent at shallow depths, and clinoptilolite is prevalent at greater depths. But the depth to the first or last appearance, or to the point of greatest abundance of either of these zeolites varies widely and is not systematic. The difference in average sedimentation rates for sediments that contain phillipsite and clinoptilolite seems to be a result of the dichotomous lithologic associations of these zeolites. As mentioned already, phillipsite most commonly occurs in clays, and clinoptilolite is most abundant in calcareous or calcareous/siliceous oozes. According to Heezen and Hollister (1971), the average calcareous ooze accumulates at a rate of 10–30 $m/10^6$ years, whereas siliceous oozes are deposited at rates of 2–10 $m/10^6$ years. Therefore, average sedimentation rate is a reflection of dominant lithologic association, but it also affects interstitial water concentration gradients (Gieskes, 1975).

It would seem that time and lithology are the two

most important variables in determining the distribution of deep-sea zeolites; time as a kinetic variable, and lithology as a variable that controls the availability of precursors and of carbonate. The synthesis of all the above factors which control the formation of deep-sea zeolites is, in effect, reflected in their geographic distribution (Fig. 2). If one re-examines this distribution with an understanding of these factors, the following observations may be made.

The Atlantic Ocean is smaller and shallower than the Pacific. There are many major rivers emptying sediments into the Atlantic but not into the Pacific Ocean. Due to differences in bottom water characteristics discussed by Berger (1970), an average Atlantic sediment has a higher carbonate content than an average Pacific sediment. Thus, in the Atlantic there is a proportionally larger area of sediments with significant carbonate and/or terrigenous detritus content than in the Pacific. These same two characteristics have been shown to affect the formation of clinoptilolite, and therefore one would expect clinoptilolite to dominate in the Atlantic, as it does.

Unlike the Atlantic, the Pacific Ocean has vast areas of sea floor which are below the Carbonate Compensation Depth (CCD). Pelagic clays that contain a low proportion of biogenous, especially calcareous, material are the dominant sedimentary facies in these regions. Widespread chains of basaltic seamounts have provided abundant volcanic material and ash, and phillipsite is the most common zeolite in these regions. In marginal areas and in equatorial regions, where sediments higher in carbonate contents and sedimentation rates occur, clinoptilolite is more abundant. Clinoptilolite is also common in marginal basins where silicic to intermediate volcaniclastic sediments have been deposited.

The Indian Ocean is more complex and shows elements of zeolite distribution patterns observed both in the Atlantic and Pacific Oceans.

The geographic distribution of deep-sea zeolites is better understood on the basis of Berger and Winterer's (1974) discussion of the facies relationships of sediments deposited on horizontally moving plates, which are also subsiding as they cool (Sclater *et al.*, 1971). The upper portions of mid-ocean ridges generally lie above the CCD. Thus, newly formed sea floor is covered with carbonate sediments, which are in turn covered with clays as that portion of crust subsides below the CCD. Plate subsidence and latitudinal displacements of plates may explain the distribution of transitional cores in the Atlantic, Pacific, and Indian Oceans. In the Pacific, for example, the northward component of Pacific plate motion has resulted in a displacement of the equatorial sediment bulge with time (Ewing *et al.*, 1968), with the total displacement being as much as 30° (Winterer, 1973). The vertical distribution of various lithologies which result from this plate motion may explain the transitional cores which occur in the Pacific Ocean. In Site 163, for example, the transition foram/nannò ooze → nanno ooze → clayey radiolarian ooze → radiolarian clay is observed with decreasing depth in the sediment. Clinoptilolite occurs in the calcareous oozes, but phillipsite dominates in the clays. In this case, the change from one zeolite to another is correlated with a change in lithology.

Site 163 has been below the CCD since the Eocene. Subordinate amounts of clinoptilolite occur with phillipsite in Lower Oligocene sediments, and its abundance relative to phillipsite increases downward to the point of transition to calcareous ooze, below which clinoptilolite is the only zeolite present. In Hole 164, however, the site has been below the CCD since the Campanian. Radiolarian clays and oozes with chert and porcelanite at greater depth are dominant throughout the sedimentary column. In this hole, beginning in the lower Oligocene, a transition from phillipsite to phillipsite + clinoptilolite to clinoptilolite occurs with increasing age. In this site the transition is gradational, but in Site 163, where a lithological change occurs, the transition is sharp.

CONCLUSIONS

(1) Quantitatively, zeolites are among the most important authigenic minerals in pelagic sediments. Phillipsite and clinoptilolites are the two most common zeolites in deep-sea sediments. Other zeolites which occur in deep-sea sediments are: analcime, chabazite, erionite, gmelinite, harmotome, laumontite, thomsonite, thaumasite, and natrolite, of which analcime is the most abundant zeolite.

(2) Phillipsite is associated with argillaceous, volcanic, and siliceous sediments; it occurs in areas and sediments of slow sedimentation rates, at shallow depths in the sediment, and in very young sediments. It occurs in sediments which show visible or chemical evidence of volcanic glass or palogonite. Phillipsite is most commonly associated with smectite, at least part of which is probably authigenic. Its apparent dissolution at depth may indicate that it is thermodynamically unstable in the marine environment.

(3) Clinoptilolite is associated with calcareous and opal-CT-rich (with or without opal-A) sediments; it occurs in areas and sediments of high sedimentation rates, and in sediments of lower Pliocene age and older. The youngest clinoptilolite-bearing carbonate sediments are significantly younger than the youngest clinoptilolite-bearing clayey sediments. Even in the presence of carbonate, clinoptilolite forms in older

sediments than does phillipsite. Clinoptilolite is mainly associated with primarily detrital smectites and illites in younger sediments and with palygorskite and illites (probably authigenic) in older sediments.

(4) The most common precursor for phillipsite seems to be palagonite. Clinoptilolite seems to be able to form from both volcanic and non-volcanic precursors. Possible precursors for clinoptilolite include: rhyolitic to andesitic glass; basaltic glass + silica (mainly biogenic); smectite + phillipsite + biogenic silica; and perhaps smectite + biogenic silica and biogenic silica + Al and K.

(5) Analcime and more Ca-rich zeolites are associated with basaltic volcanic material.

ACKNOWLEDGMENTS

We are grateful to Drs. J. M. Gieskes and J. R. Boles for their helpful suggestions and to Ms. E. Bromley for her help in data collection. This work was supported by a Scripps Industrial Associates Traineeship and grants from the U.C.S.D. Academic Senate and the International Nickel Company, Inc. to the Institute of Marine Resources, Scripps Institution of Oceanography, and NSF grant OCE76-02128.

REFERENCES

Alexanderson, T. (1972) The sedimentary xenoliths from Surtsey: turbidites indicating shelf growth: *Surtsey Res. Progress Report* **6** 101-116.

Alietti, A. (1972) Polymorphism and crystal-chemistry of heulandites and clinoptilolites: *Amer. Mineral.* **57**, 1448-1462.

Arrhenius, G. (1963) Pelagic sediments: *The Sea*, Vol. **3** (edited by Hill, M. N.), Wiley Interscience, N.Y., pp. 655-727.

Arrhenius, G. and Bonatti, E. (1965) Neptunism and vulcanism in the ocean: *Progress in Oceanography*, Vol. **3** (edited by Sears, M.), Pergamon Press, Oxford, pp. 7-22.

Bass, M. N., Moberly, R., Rhodes, J. M., Shih Chi-yu and Church, S. E. (1973) Volcanic rocks cored in the Central Pacific, Leg 17, Deep Sea Drilling Project: *Initial Reports of the Deep Sea Drilling Project*, Vol. **XVII** (edited by Winterer, E. L. *et al.*); U.S. Printing Office, Washington, D.C., pp. 429-503.

Berger, W. H. (1970) Biogenous deep-sea sediments; fractionation by deep-sea circulation: *Geol. Soc. of Amer. Bull.* **81**, 1385-1401.

Berger, W. H. and von Rad, U. (1972) Cretaceous and Cenozoic sediments from the Atlantic Ocean: *Initial Reports of the Deep Sea Drilling Project*, Vol. **XIV** (edited by Hayes, D. E. *et al.*), U.S. Printing Office, Washington, D.C., pp. 787-954.

Berger, W. H. and Winterer, E. L. (1974) Plate stratigraphy and the fluctuating carbonate line: *Spec. Publs. Int. Ass. Sediment.* **1**, 11-48.

Bernat, M. and Goldberg, E. D. (1969) Thorium isotopes in the marine environment: *Earth Planet. Sci. Letts.* **5**, 308-312.

Bernat, M., Bieri, R. H., Koide, M., Griffin, J. J. and Goldberg, E. D. (1970) Uranium, thorium, potassium and argon in marine phillipsites: *Geochim. Cosmochim. Acta* **34**, 1053-1071.

Biscaye, P. E. (1965) Mineralogy and sedimentation of recent deep-sea clay in the Atlantic Ocean and adjacent seas and oceans: *Geol. Soc. Amer. Bull.* **76**, 803-832.

Boles, J. R. (1972) Composition, optical properties, cell dimensions, and thermal stability of some heulandite group zeolites: *Amer. Mineral.* **57**, 1463-1493.

Boles, J. R. and Coombs, D. S. (1975) Mineral reactions in zeolite Triassic tuff, Hokonui Hills, New Zealand: *Geol. Soc. Amer. Bull.* **86**, 163-173.

Bonatti, E. (1963) Zeolites in Pacific pelagic sediments: *Trans. N.Y. Acad. Sci.* **25**, 938-948.

Bonatti, E. (1965) Palagonite, hyaloclastites and alteration of volcanic glass in the ocean: *Bull. Volcanol.* **29**, 257-269.

Bonatti, E. and Nayudu, Y. R. (1965) The origin of manganese nodules on the ocean floor: *Amer. J. Sci.* **263**, 17-39.

Bramlette, M. N. and Posnjak, E. (1933) Zeolitic alteration of pyroclasts: *Amer. Mineral.* **18**, 167-171.

Brown, G., Catt, J. A. and Weir, A. H. (1969) Zeolites of the clinoptilolite-heulandite type in sediments of southeast England: *Min. Mag.* **37**, 480-488.

Cook, H. E. and Zemmels, I. (1972) X-ray mineralogy studies – Leg 9: *Initial Reports of the Deep Sea Drilling Project*, Vol. **IX** (edited by Hays, J. D. *et al.*), U.S. Government Printing Office, Washington, D.C., pp. 707-778.

Coombs, D. S., Ellis, A. J., Fyfe, W. S. and Taylor, A. M. (1959) The zeolite facies, with comments on the interpretation of hydrothermal syntheses: *Geochim. Cosmochim. Acta* **17**, 53-107.

Couture, R. A. (1977) Composition and origin of palygorskite-rich and montmorillonite-rich zeolite-containing sediments from the Pacific Ocean: *Chem. Geol.* **19**, 113-130.

Cronan, D. S. (1974) Authigenic minerals in deep-sea sediments: *The Sea*, Vol. **5** (edited by Goldberg, E. D.), Wiley Interscience, N.Y., pp. 491-525.

Curtis, C. D. and Cornell, W. C. (1972) Unusual occurrence of clinoptilolite: Fresno County, California: *Geol. Soc. Amer. Bull.* **83**, 833-838.

Czyscinski, K. (1973) Authigenic phillipsite formation rates in the central Indian Ocean and the equatorial Pacific Ocean: *Deep-Sea Res.* **20**, 555-559.

Daly, R. A. (1925) The geology of Ascension Island: *Amer. Acad. Arts. Sci. Proc.* **60**, 3-80.

Deffeyes, K. S. (1959) Zeolites in sedimentary rocks: *J. Sed. Petrol.* **29**, 602-609.

Donnelly, T. and Nalli, G. (1973) Mineralogy and chemistry of Caribbean sediments: *Initial Reports of the Deep Sea Drilling Project*, Vol. XV (edited by Edgar, N. T. *et al.*), U.S. Government Printing Office, Washington, D.C., pp. 929-961.

Donnelly, T. W. and Wallace, J. L. (1976) Major and minor element chemistry of Antarctic clay-rich sediments: Sites 322, 323, and 325, DSDP Leg 35: *Initial Reports of the Deep Sea Drilling Project*, Vol. **XXXV** (edited by Hollister, C. D. *et al.*), U.S. Government Printing Office, Washington, D.C., pp. 427-446.

Drever, J. I. (1976) Chemical and mineralogical studies, Site 323: *Initial Reports of the Deep Sea Drilling Project*, Vol. **XXXV** (edited by Hollister, C. D., *et al.*), U.S. Government Printing Office, Washington, D.C., pp. 471-477.

Dymond, J. (1966) Potassium-argon geochronology of deep-sea sediments: *Science* **152**, 1239-1241.

Estéoule, J., Estéoule-Choux, J. and Louail, J. (1971) Su la présence de clinoptilolite dans les dépôts marno-calcaires du Crétacé superieur de l'Anjou: *C. R. Acad. Sci. Paris* **272**, Série D, 1569-1572.

Eugster, H. P. and Jones, B. F. (1968) Gels composed of sodium–aluminum silicate, Lake Magadi, Kenya: *Science* **161**, 160-163.

Ewing, J. I., Ewing, M. and Ludwig, W. J. (1968) North Pacific sediment layers measured by seismic profiling: *Amer. Geophys. Union Monogr.* **12**, 147-173.

Flörke, O. W. (1955) Zur Frage des "Hoch"-Cristobalit in Opalen, Bentoniten und Gläsern: *N. Jahrb. Mineral. Monat.* **10**, 217-223.

Garrison, R. E., Hein, J. R. and Anderson, T. F. (1973) Lithified carbonate sediment and zeolitic tuff in basalts, Mid-Atlantic Ridge: *Sedimentology* **20**, 399-410.

Gieskes, J. M. (1975) Chemistry of interstitial waters of marine sediments: *Ann. Rev. Earth and Plan. Sci.* **3**, 433-453.

Glaccum, R. and Boström, K. (1976) (Na, K)-phillipsite: its stability conditions and geochemical role in the deep sea: *Mar. Geol.* **21**, 47-58.

Goldberg, E. D. (1961) Chemical and mineralogical aspects of deep-sea sediments: *Physics and Chemistry of the Earth,* Vol. 4 (edited by Ahrens, L. H., Press, F. and Runcorn, S. K.), Pergamon Press, Oxford, pp. 281-302.

Goldberg, E. D. and Arrhenius, G. (1958) Chemistry of Pacific pelagic sediments: *Geochim. Cosmochim. Acta* **13**, 153-212.

Goodel, H. G. (1965) The marine geology of the Southern Ocean: *Contr. 11, Sedimentol. Res. Lab., Florida State Univ.,* 196 pp.

Goodwin, J. H. and Surdam, R. C. (1967) Zeolitization of tuffaceous rocks of the Green River Formation, Wyoming: *Science* **157**, 307-308.

Griffin, J. J., Koide, M., Höhndorf, A., Hawkins, J. W. and Goldberg, E. D. (1972) Sediments of the Lau Basin – rapidly accumulating volcanic deposits: *Deep-Sea Res.* **19**, 139-148.

Hathaway, J. C. and Sachs, P. L. (1965) Sepiolite and clinoptilolite from the Mid-Atlantic Ridge: *Amer. Mineral.* **50**, 852-867.

Hay, R. L. (1966) Zeolites and zeolitic reactions in sedimentary rocks: *Geol. Soc. Amer. Special Paper* **85**, 130 pp.

Hay, R. L. (1968) Chert and its sodium-silicate precursors in sodium-carbonate lakes of East Africa: *Contr. Mineral. Petrol.* **17**, 255-274.

Hay, R. L. (1970) Silicate reactions in three lithofacies of a semi-arid basin, Olduvai Gorge, Tanzania: *Mineral. Soc. Amer. Spec. Paper* **3**, 237-255.

Hay, R. L. and Iijima, A. (1968) Nature and origin of palagonite tuffs of the Honolulu Group on Oahu, Hawaii: *Geol. Soc. Amer. Mem.* **116**, 331-376.

Heezen, B. C. and Hollister, C. D. (1971) *The Face of the Deep*, Oxford Press, N.Y.

Hess, P. C. (1966) Phase equilibria of some minerals in the K_2O–Na_2O–Al_2O_3–SiO_2–H_2O system at 25°C and 1 atmosphere: *Amer. J. Sci.* **264**, 289-309.

Horn, R. R. (1968) Authigenic prehnite, laumontite, and chlorite in the lower Cretaceous sediments of south-eastern Alexander Island: *Brit. Antarct. Surv. Bull.* **18**, 1-10.

Iijima, A. (1971) Composition and origin of clinoptilolite in the Nakanosawa tuff of Rumoi, Hokkaido: *Adv. in Chem.* No. 101, *Molecular Sieve Zeolites – I*, pp. 334-341.

Iijima, A. and Harada, K. (1968) Authigenic zeolites in zeolitic palagonite tuffs on Oahu, Hawaii: *Amer. Mineral.* **54**, 182-197.

Jones, J. B. and Segnit, E. R. (1971) The nature of opal. I. Nomenclature and constituent phases: *J. Geol. Soc. Aust.* **18**, 57-68.

Kastner, M. (1976) Diagenesis of basal sediments and basalts of Sites 322 and 323, Leg 35, Bellingshausen Abyssal Plain: *Initial Reports of the Deep Sea Drilling Project*, Vol. **XXXV** (edited by Hollister, C. D. *et al.*), U.S. Government Printing Office, Washington, D.C., pp. 513-527.

Kastner, M. and Siever, R. (1972) Diagenesis in Mediterranean Sea core samples from Site 124–Balearic Rise and Site 125A–Ionian Basin: *Initial Reports of the Deep Sea Drilling Project*, Vol. **XIII** (edited by Ryan, W. B. F., *et al.*), U.S. Government Printing Office, Washington, D.C., pp. 721-726.

Kastner, M. and Bada, J. (1974) Kinetics of montmorillonite to illite conversion: *Trans. Amer. Geophys. Union* **55**, 456.

Kastner, M., Keene, J. B., and Gieskes, J. M. (1977) Diagenesis of siliceous oozes – I. Chemical controls on the rate of opal-A to opal-CT transformation – an experimental study: *Geochim. Cosmochim. Acta* **41**, 1041-1059.

Kelts, K. and McKenzie, J. A. (1976) Cretaceous volcanogenic sediments from the Line Island chain: diagenesis and formation of K-feldspar, DSDP Leg 33, Hole 315A and Site 316: *Initial Reports of the Deep Sea Drilling Project,* Vol. **XXXIII** (edited by Schlanger, S. D., *et al.*), U.S. Government Printing Office, Washington, D.C., pp. 789-803.

Kolla, V. and Biscaye, P. E. (1973) Deep-sea zeolites: Variations in space and time in sediments of the Indian Ocean: *Mar. Geol.* **15**, 11-17.

Lancelot, Y. (1973) Chert and silica diagenesis in sediments from the Central Pacific: *Initial Reports of the Deep Sea Drilling Project,* Vol. **XVII** (edited by Winterer, E. L. *et al.*), U.S. Government Printing Office, Washington, D.C., pp. 377-405.

Lancelot, Y., Hathaway, J. C. and Hollister, C. D. (1972) Lithology of sediments from the Western North Atlantic, Leg 11, Deep Sea Drilling Project: *Initial Reports of the Deep Sea Drilling Project,* Vol. **XI** (edited by Hollister, C. D. *et al.*), U.S. Government Printing Office, Washington, D.C., pp. 901-973.

Larson, R., Moberly, R., Bukry, D., Foreman, H. P., Garner, J. V., Keene, J., Lancelot, Y., Luterbacher, H., Marshall, M. and Matter, A. (1974) Leg 32, Deep Sea Drilling Project: *Geotimes* **18**, 14-18.

Lawrence, J. R. (1973) Interstitial water studies, Leg 15 – stable oxygen and carbon isotope variations in water, carbonates and silicates from the Venezuela Basin (Site 149) and the Ares Rise (Site 148): *Initial Reports of the Deep Sea Drilling Project,* Vol. **XX** (Edgar, N. T. *et al.*, eds.). U.S. Government Printing Office, Washington, D.C., pp. 891-901.

Lawrence, J. R., Gieskes, J. M. and Broecker, W. S. (1975) Oxygen isotope and cation composition of DSDP pore waters and the alteration of layer II basalt: *Earth Planet. Sci. Letts.* **27**, 1-10.

Lerman, A., Mackenzie, F. T. and Bricker, O. P. (1975) Rates of dissolution of aluminosilicates in seawater: *Earth Planet. Sci. Letts.* **25**, 82-88.

Manheim, F. T. and Sayles, F. L. (1974) Composition and origin of interstitial waters of marine sediments, based on deep-sea drill cores: *The Sea,* Vol. **5** (edited by Goldberg, E. D.), Wiley Interscience, N.Y., pp. 527-568.

Mariner, R. H. and Surdam, R. C. (1970) Alkalinity and formation of zeolites in saline alkaline lakes: *Science* **170**, 977-980.

Marshall, B. R. (1961) Devitrification of natural glass: *Geol. Soc. Amer. Bull.* **72**, 1493-1520.

Matthews, D. H. (1962) Altered lavas from the floor of the Eastern North Atlantic: *Nature* **194**, 368-369.

Matthews, D. H. (1971) Altered basalts from Swallow Bank, an abyssal hill in the Northeastern Atlantic and from a nearby seamount: *Phil. Trans. Roy. Soc. London Acad.* **268**, 551-571.

Mellis, O. (1952) Replacement of plagioclase by orthoclase in deep-sea deposits: *Nature* **169**, 624.

Merkle, A. B. and Slaughter, M. (1968) Determination and refinement of the structure of heulandite: *Amer. Mineral.* **53**, 1120-1138.

Miyashiro, A. (1960) Thermodynamics of reactions of rock-forming minerals with silica. III. Variation of reaction temperature with degree of silica saturation of environment: *J. Geol. Geogr. Japan* **31**, 107-111.

Moberly, R. M. (1963) Amorphous marine muds from tropically weathered basalt: *Amer. J. Sci.* **261**, 767-772.

Moore, J. G. (1966) Rate of palagonitization of submarine basalt adjacent to Hawaii: *U.S. Geol. Surv. Prof. Paper* **550-D**, pp. 163-171.

Morgenstein, M. M. (1967) Authigenic cementation of scoriaceous deep-sea sediments west of the Society Ridge, South Pacific: *Sedimentol.* **9**, 105-118.

Mumpton, F. A. (1960) Clinoptilolite redefined: *Amer. Mineral.* **45**, 351-369.

Murray, J. and Renard, A. F. (1891) *Report on the Scientific Results of the Voyage of "H.M.S. Challenger" during the years 1873-76; Deep Sea Deposits:* Johnson Reprint Co., London, 525 pp.

Nayadu, Y. R. (1964) Palagonite tuffs (hyaloclastites) and the products of post-eruptive processes: *Bull. Volcanol.* **27**, 1-20.

O'Neil, J. R. and Taylor, H. P., Jr. (1967) The oxygen isotope and cation exchange chemistry of feldspars: *Amer. Mineral.* **52**, 1414-1437.

Peterson, M. N. A. and Goldberg, E. D. (1962) Feldspar distributions in South Pacific pelagic sediments: *J. Geophys. Res.* **67**, 3477-3492.

Peterson, M. N. A. and Griffin, J. J. (1964) Volcanism and clay minerals in the Southeastern Pacific: *J. Mar. Res.* **22**, 13-21.

Peterson, M. N. A., Edgar, N. T., von der Borch, C. C. and Rex, R. W. (1970) Cruise leg summary and discussion: *Initial Reports of the Deep Sea Drilling Project,* Vol. **II** (edited by Peterson, M. N. A. *et al.*), U.S. Government Printing Office, Washington, D.C., pp. 413-427.

Pimm, A. C. (1973) Trace element determinations compared with X-ray diffraction results of brown clay in the Central Pacific: *Initial Reports of the Deep Sea Drilling Project,* Vol. **XVII** (edited by Winterer, E. L. *et al.*), U.S. Government Printing Office, Washington, D.C., pp. 511-514.

Revelle, R., Bramlette, M., Arrhenius, G. and Goldberg, E. D. (1955) Pelagic sediments of the Pacific: *Geol. Soc. Amer. Spec. Paper 62*, pp. 221-235.

Rex, R. W. (1967) Authigenic silicates formed from basaltic glass by more than 60 million years' contact with sea water, Sylvania Guyot, Marshall Islands: *Clays and Clay Minerals, Proc. Nat. Conf.* **15**, 195-203.

Rex, R. W., Eklund, W. A. and Jamieson, I. M. (1971) X-ray mineralogy studies – Leg 6: *Initial Reports of the Deep Sea Drilling Project,* Vol. **VI** (edited by Fischer, A. G. *et al.*), U.S. Government Printing Office, Washington, D.C., pp. 753-810.

Rinaldi, R., Pluth, J. J. and Smith, J. V. (1974) Refinement of crystal structure of phillipsite and harmotome: *Acta Cryst.* **B30**, 2426-2433.

Robinson, P. T., Thayer, P. A., Cook, P. J. and McKnight, B. T. (1974) Lithology of Mesozoic and Cenozoic sediments from the Indian Ocean, Leg 27: *Initial Reports of the Deep Sea Drilling Project,* Vol. **XXVII** (edited by Veevers, J. J. *et al.*), U.S. Government Printing Office, Washington, D.C., pp. 1001-1048.

Savin, S. M. and Epstein, S. (1970a) The oxygen and hydrogen isotope geochemistry of ocean sediments and shales: *Geochim. Cosmochim. Acta* **34**, 43-63.

Savin, S. M. and Epstein, S. (1970b) Oxygen and hydrogen isotope geochemistry of clay minerals: *Geochim. Cosmochim. Acta* **34**, 25-42.

Scholze, H. and Mulfinger, H. O. (1959) Der Einbauder Wassers in Gläsern: *Glastech. Ber.* **32**, 331-336.

Sclater, J. G., Anderson, R. N. and Bell, M. L. (1971) Elevation of ridges and evolution of the central eastern Pacific: *J. Geophys. Res.* **76**, 7888-7915.

Senderov, E. E. (1965) Features of the conditions of zeolite formation: *Geochim. Int.* **2**, 1143-1155.

Sheppard, R. A. (1971a) Clinoptilolite of possible economic value in sedimentary deposits of the conterminous United States: *U.S. Geol. Surv. Bull.* **1332-B**, 1-15.

Sheppard, R. A. (1971b) Zeolites in sedimentary deposits of the United States – a review: *Molecular Sieve Zeolites I, Adv. in Chem.* Series 101 (Gould, R. F., ed.), Amer. Chem. Soc., Washington, D.C., pp. 279-310.

Sheppard, R. A. and Gude, A. J., 3rd (1968) Distribution and genesis of authigenic silicate minerals in tuffs of Pleistocene Lake Tecopa, Inyo County, California: *U.S. Geol. Surv. Prof. Paper* **597**, 38 pp.

Sheppard, R. A. and Gude, A. J., 3rd (1969) Diagenesis of tuffs in the Barstow Formation, Mud Hills, San Bernardino County, California: *U.S. Geol. Surv. Prof. Paper* **634**, 35 pp.

Sheppard, R. A., Gude, A. J., 3rd and Griffin, J. J. (1970) Chemical composition and physical properties of phillipsite from the Pacific and Indian Oceans: *Amer. Mineral.* **55**, 1053-2062.

Steinfink, H. (1962) The crystal structure of zeolite, phillipsite: *Acta Crystallog.* **15**, 644-651.

Stewart, R. J., Natland, J. H. and Glassley, W. R. (1973) Petrology of volcanic rocks recovered on DSDP Leg 19 from the North Pacific Ocean and the Bering Sea: *Initial Reports of the Deep Sea Drilling Project,* Vol. **XIX** (edited by Creager, J. S. *et al.*), U.S. Government Printing Office, Washington, D.C., pp. 615-627.

Stoeser, D. B. (1975) Igneous rocks from Leg 30 of the Deep Sea Drilling Project: *Initial Reports of the Deep Sea Drilling Project,* Vol. **XXX** (edited by Andrews, J. E. *et al.*), U.S. Government Printing Office, Washington, D.C., pp. 401-414.

Stonecipher, S. A. (1976) Distribution of deep-sea phillipsite and clinoptilolite: *Chem. Geol.* **17**, 307-318.

Stonecipher, S. A. (1976b) Origin, distribution and diagenesis of deep sea clinoptilolite and phillipsite. *Ph.D. Thesis, Scripps Inst. Oceanogr.*, 154 pp.

Taylor, W. H. (1930) The Structure of analcite ($NaAlSi_2O_6 \cdot H_2O$): *Zeits. Krist.* **74**, 1-19.

Tracey, J. I., Jr. *et al.* (1971) Leg 8 Summary: *Initial Reports of the Deep Sea Drilling Project*, Vol. **XXX** (edited by Tracey, J. I., Jr. *et al.*), U.S. Government Printing Office, Washington, D.C., pp. 17-45.

Turekian, K. K. (1965) Some aspects of the geochemistry of marine sediments: *Chemical Oceanography* Vol. **2** (edited by Riley, J. P. and Skirrow, G.), Academic Press, London, pp. 81-126.

von Rad, U. and Rösch, H. (1972) Mineralogy and origin of clay minerals, silica and authigenic silicates in Leg 14 sediments: *Initial Reports of the Deep Sea Drilling Project,* Vol. **XIV** (edited by Hayes, D. E. *et al.*), U.S. Government Printing Office, Washington, D.C., pp. 727-739.

Walker, G. P. L. (1972) Petrology of the volcanic hyaloclastic sandstone – Site 115: *Initial Reports of the Deep Sea Drilling Project,* Vol. **XII** (Laughton, A. S. *et al.*, Eds.), U.S. Government Printing Office, Washington, D.C., pp. 365-366.

Weaver, C. E. (1968) Mineral facies in the Tertiary of the continental shelf and Blake Plateau: *Southeastern Geology* **9**, 57-63.

Winterer, E. L. (1973) Sedimentary facies and plate tectonics of the equatorial Pacific: *Am. Assoc. Pet. Geol. Bull.* **57**, 265-282.

Yoder, H. S. and Eugster, H. P. (1955) Synthetic and natural muscovites: *Geochim. Cosmochim. Acta* **7**, 225-280.

Young, J. A. (1939) Minerals from deep-sea cores surface deposits of Bermudian calcareous sediments: *Amer. J. Sci.* **237**, 798-810.

Zemmels, I., Cook, H. E. and Matti, J. C. (1975) X-ray mineralogy data, Tasman Sea and far western Pacific, Leg 30, Deep Sea Drilling Project: *Initial Reports of the Deep Sea Drilling Project*, Vol. **XXX** (edited by Andrews, J. E. *et al.*), U.S. Government Printing Office, Washington, D.C., pp. 603-616.

APPENDIX I
SEDIMENTARY FACIES IN WHICH CLINOPTILOLITE OCCURS

Calcareous oozes and chalks

Age range: Miocene to Eocene.

Common occurrence: on ridges, plateaus, rises, and in areas beneath the equatorial high productivity belt.

Common characteristics: contain more than $\sim$10% clay minerals or trace amounts of volcanic glass; occur as basal sediments of the Line Islands facies type

(Tracey *et al.*, 1971). In sediments where clinoptilolite is associated with phillipsite and volcanic glass, the sediments are generally of Oligocene age or older and were deposited close to the CCD or near the margins of the high productivity belt. Moderate to high sedimentation rates.

Exemplary DSDP sites:

Pacific – 44, 62, 63, (74),* (75), (159), 171, 206, 210, 287, [296], 317.

Atlantic – 17, [19], 98.

Indian – 212, 214 (Paleocene),† 219, (238), 245, 249, 253, 258, 259.

Calcareous/siliceous oozes and limestones with chert

Age range: Miocene to Upper Cretaceous.

Common occurrence: on plateaus, rises, platforms, and at margins of the equatorial high productivity belt.

Common characteristics: contain more than $\sim$10% clay minerals or trace amounts of volcanic glass; moderate to high sedimentation rates.

Exemplary DSDP sites:

Pacific – 47, 48, 72, 78, 157, 163, 165, 171, 173, 206, 208, 209, 210, 220, 221, 289, 305, 313, 315, 316.

Atlantic – 22, 98, 99, 105, 116-117, 143-144.

Indian – 236-238.

Calcareous clays, siliceous/calcareous clays, and hemipelagic muds

Age range: Miocene to Eocene.

Common occurrence: on ridges, in basal sediments of Eocene age or older, near continental margins.

Common characteristics: clays may represent completely altered ash-rich calcareous basal sediments, or may still contain some visible altered glass and phillipsite; hemipelagic muds are composed of primarily terrigenous material; moderate sedimentation rates.

Exemplary DSDP sites:

Pacific – 32, 77, 155, 162, 166 (Middle Cretaceous), 169, 192.

Atlantic – 101, 102-104, 105, 106, 111, 119, 142, 43-144 (Middle Cretaceous), 216 (Upper Cretaceous).

Clays and clays with chert or porecelanite

Age range: Oligocene to Lower Cretaceous.

Common occurrence: on abyssal plains..

Common characteristics: commonly occurs in cores with the following transitions with depth P($\leq$Oligocene) $\rightarrow$ P + C (Oligocene to Eocene) $\rightarrow$ C ($\geq$Paleocene) (P = phillipsite; C = clinoptilolite); may or may not contain visible volcanic glass; commonly clinoptilolite occurs as molds of radiolarians associated with cherts and porcelanites; low sedimentation rates.

Exemplary DSDP sites:

Pacific – 68, 163, 164, 166, 196, 207, 303, 304, 307, 314.

Atlantic – 8, 9, 28, 105, 118, 137, 138, 140.

Indian – 212, 215, 250, 256, 257, 258, 259, 260, 261.

Siliceous oozes and siliceous oozes with porcelanites

Age range: Oligocene to Eocene.

Common occurrence: at margins of equatorial high productivity belt.

Common characteristics: may or may not contain visible volcanic glass; always contain clay minerals and commonly also some nanno fossils; moderate sedimentation rates.

Exemplary DSDP sites:

Pacific – 65, 166, 291.

Atlantic – 13.

Turbidites and volcaniclastics

Age range: Pliocene to Upper Cretaceous.

Common occurrence: near seamounts or island-arcs; in basal sediments.

Common characteristics: reworked; interbedded with calcareous oozes; associated with phillipsite; high sedimentation rates.

Exemplary DSDP sites:

Pacific – 84, 158, 165, 166, 171, 184, 185, 186, 188, 189, 190, 191, 198, 205, 207, 285, 288, 289, 294, 313, 317, 318.

Atlantic – 113.

Indian – 214, 216, 224, 236, 238, 245, 246-247, 248, 249, 258.

---

* Parenthesis indicate sites where clinoptilolite is associated with phillipsite and volcanic glass. Brackets indicate sites where Line Islands facies is present.

† Ages of these sites fall outside of the general age range for that facies.

SEDIMENTARY FACIES IN WHICH PHILLIPSITE OCCURS

Pelagic clays

Age range: Pleistocene to Eocene.
Common occurrence: on abyssal plains in the center of circulatory gyres, away from topographic highs.
Common characteristics: contain altered volcanic material; low sedimentation rates.
Exemplary DSDP sites:
Pacific – 37, 38, 39, 40, 41, 45, 49, 50, 51, 52, 53, 59, 68, 75, 161, 170, 172, 204, 205, 285, 287, 307, 314.
Atlantic – 23.
Indian – 213, 215.

Radiolarian oozes

Age range: Pleistocene to Eocene.
Common occurrence: on abyssal plains close to the margins of the equatorial high productivity belt.
Common characteristics: contain altered volcanic glass and/or clays; phillipsite occurs with clays and amorphous material inside radiolarian tests; ferruginous; moderate sedimentation rates.
Exemplary DSDP sites:
Pacific – 59, 70, 73, 74, 161, 162, 166, 199.

Calcareous and calcareous/siliceous oozes

Age range: Miocene.
Common occurrence: near ridges and rises in areas beneath the equatorial high productivity belt.
Common characteristics: contain minor volcanic material; moderate sedimentation rates.
Exemplary DSDP sites:
Pacific – 47, 74 (Eocene),* 157, 158, 161 (Oligocene), 162 (Eocene), 285, 305, 310, 313, 316.
Indian – 235, 238, 254, 255.

Radiolarian clays or clays with porcelanites

Age range: Miocene to Eocene.
Common occurrence: on abyssal plains at the margins of the equatorial high productivity belt below the CCD.
Common characteristics: similar to pelagic clay facies; volcanic material present; low sedimentation rate.
Exemplary DSDP sites:
Pacific – 59, 68, 159, 160, 163, 164, 167, 314.

Calcareous clays

Age range: Miocene to Eocene.
Common occurrence: at margins of the equatorial high productivity belt.
Common characteristics: deposited near CCD; low sedimentation rate.
Exemplary DSDP sites:
Pacific – 159, 162.
Atlantic – 10.
Indian – 239, 260.

Turbidites

Age range: Pleistocene to Eocene.
Common occurrence: near volcanic islands, seamounts, or ridge crests.
Common characteristics: volcanic material is interbedded and mixed with calcareous oozes and/or shallow water calcareous debris; high sedimentation rate.
Exemplary DSDP sites:
Pacific – 165, 173, 199, 205, 286, 290, 293, 311, 315, 318.
Atlantic – 12, 114, 115.
Indian - 236, 238.

Volcaniclastics

Age range: Eocene to Cretaceous.
Common occurrence: near seamounts, ridgecrests, or other volcanic topographic highs.
Common characteristics: volcaniclastic textures; little or no carbonate material; high sedimentation rate.
Exemplary DSDP sites:
Pacific – 171, 196, 286, 313.
Atlantic – 136.
Indian – 216, 219.

---

* Ages of these examples fall outside the general age range for that facies.

CHEMISTRY AND DEEP-SEA PHILLIPSITE, CLINOPTILOLITE, AND HOST SEDIMENTS

SHARON A. STONECIPHER*

Scripps Institution of Oceanography, La Jolla, California 92093, U.S.A.

Abstract – Chemical analyses of zeolites from various deep-sea sedimentary environments and bulk sediment and less than 2μm fractions of the host sediments were performed by electron microprobe and X-ray fluorescence techniques, respectively. Comparison of these analyses provides insight into the origins of deep-sea phillipsite and clinoptilolite.

The phillipsite from a radiolarian clay is slightly more siliceous than that from a clay with volcanic glass. Bulk sediment analyses suggest that both host sediments contain some basaltic material. According to Böstrom (1970), Ti/Al ratios of >0.039 indicate basaltic volcanic contributions to the sediment. Both sediment types show Ti/Al ratios of this magnitude, but the ratios are higher in the clay where the volcanic material is still visible. The Ti/Al ratio and phillipsite abundance both decrease with increasing depth, while clinoptilolite abundance increases with increasing depth.

The clinoptilolite from a nanno chalk contains significantly more calcium than the clinoptilolite from a carbonaceous clay. X-ray fluorescence analyses for the bulk clay samples show constant and low (<0.030) Ti/Al ratios. Lack of sufficient material precluded analyses of the bulk carbonate material.

These results suggest that whereas a certain proportion of basaltic volcanic material is essential for the formation of phillipsite, clinoptilolite seems to be able to form in non-volcanic siliceous terrigenous sediments as well as in sediments with volcanic components.

INTRODUCTION

Phillipsite and clinoptilolite are the two most common authigenic zeolites in deep-sea sediments. The physico-chemical conditions for their formation are not well known, but seem to be controlled by the sedimentary environment in which the zeolites occur. An effort has been made to understand bulk chemical and mineralogical variations of different sedimentary environments and to see if and how these variations are reflected in the chemical composition of the zeolites which occur in each environment.

Previous work (Stonecipher, 1976; Kastner and Stonecipher, 1978) indicates that phillipsite and clinoptilolite each occur in certain characteristic environments. Phillipsite is commonly associated with younger clays, volcanogenic sediments, and siliceous oozes. Clinoptilolite occurs predominantly in carbonate sediments and in older clays. Four cores representing these basic environments have been selected for detailed analysis. Bulk samples and the less-than-2μm fractions of the host sediments have been analyzed by X-ray fluorescence technique. Zeolites separated from these cores have been analyzed using an electron microprobe. The chemical compositions of these zeolites and the sediments of which they form a part will be compared.

CORE DESCRIPTIONS

SCAN 16P

This is a 9-m Scripps Institution of Oceanography piston core located near the center (16°25′N, 164°24′W) of a roughly circular basin, about 10 degrees across, bounded by ridges and seamounts. The sediment consists primarily of a homogeneous, dark-brown zeolitic clay. A thin porcellanite layer at 436 cm was described by Keene and Kastner (1974). Other smaller porcellanite horizons occur between this layer and the bottom of the hole. Seismic profiles from the area indicate a strong reflector, probably chert, at 30–40 m. Rare, poorly preserved radiolarians were used to determine the age of the sediment to be Middle Eocene. Basement age in this area is Cretaceous. Calcareous microfossils are absent, but fish debris are common. Glass shards and other evidence of volcanic components are absent. Phillipsite predominates above 400 cm depth, with trace amounts of clinoptilolite; below 400 cm, phillipsite disappears and the abundance of clinoptilolite increases.

DSDP 59

This Deep Sea Drilling Project (DSDP) hole penetrated 135 m of sediment on the western edge of the

*Present address: Marathon Oil Co., P.O. Box 269, Littleton, Colorado 80160.

Marianna Basin, just east of the Marianna Trench at 11°47′N, 147°35′E. The sediment consists of interbedded siliceous oozes and zeolitic clays, both of which contain variable amounts of volcanic glass, palagonite, feldspars, and heavy minerals. There are chert chips in the lowest portion of the core. Sediment age ranges from Quaternary for the first diatom ooze to Cretaceous for the last clay above the chert. Basement age in this area is Jurassic. In sediments younger than Oligocene, only phillipsite is found, whereas phillipsite and clinoptilolite occur together in sediments of Oligocene age and older.

DSDP 105

This hole was drilled in the Atlantic Ocean on the edge of the lower continental rise at 34°54′N, 69°10′W. Here the upper section of the sediment column consists of Quaternary to Tertiary hemipelagic muds and Cretaceous clays. The hemipelagic muds contain abundant terrigenous material and some calcareous microfossils, but no zeolites. The clays range from Upper Cretaceous(?) multicolored clays to Lower Cretaceous black, highly carbonaceous clays. The multicolored clays are devoid of biogenous components, but contain abundant iron and manganese oxides. In the black clays, foraminifera and calcareous nannoplankton are rare to common, and rare radiolarians are poorly preserved and commonly replaced by pyrite. The carbonaceous clays overlie a zone of Lower Cretaceous-Upper Jurassic interbedded limestones and clayey limestones. The limestone is composed primarily of recrystallized nanno-calcite and generally contains no clay or plant debris. The clayey limestones contain 30–50% clay and abundant organic matter. This sequence is underlain by Upper Jurassic limestones and red clayey limestones which extend to basement. Clinoptilolite is the only zeolite present; it occurs only in the Cretaceous clays and the Lower Cretaceous clayey limestones, and is most common in the black clays.

DSDP 163

This hole penetrated 294 m of sediment and basaltic basement in an abyssal hill region of the central equatorial Pacific at 11°15′N, 150°17′W. In this hole, Oligocene radiolarian clays grade downward into Eocene clayey radiolarian oozes and thin chert beds in the lower part of the section. These are underlain by 20 m of ferruginous Maestrichtian clay with few radiolarians and some thin cherts. Below the ferruginous clays lie a series of Campanian nannofossil chalks interbedded with cherts which overlie the basement basalts. Phillipsite occurs associated with amorphous material and clays as infillings of radiolarians in the radiolarian oozes and as large twinned crystals in the radiolarian clays. Clinoptilolite occurs in the 2–62-μm fraction of the lowermost radiolarian clay in association with phillipsite and is the only zeolite present in the nanno chalks.

EXPERIMENTAL METHODS

Non-carbonate samples were washed with distilled water until flocculation no longer occurred. Part of the sediment was dried for bulk chemical analysis; part was separated into size fractions. The >62-μm fraction was removed by sieving. The remainder was separated into >2- and <2-μm fractions by settling and decantation. Individual zeolites were hand picked from the >62-μm fraction where possible and mounted in epoxy for electron microprobe analysis. Heavy liquids were used to separate clinoptilolite from the 2–62-μm fraction of some of the SCAN 16P and DSDP 59 samples, and from DSDP 163 acid-residues for electron microprobe analysis. Carbonate samples were treated with a 0.1 M acetic acid–0.1 M Na acetate solution, pH 5, and the residues were allowed to re-equilibrate with seawater before washing and separation into size fractions. X-ray fluorescence analyses of the bulk and <2-μm fractions were performed on glass pellets made by the method described in Clague (1974). Due to lack of material, bulk analyses of the acid residues for DSDP 163 and trace element analyses of all samples were not possible. Electron microprobe analyses were performed on an ARL microprobe used in conjunction with a Nuclear-Semiconductor Autotrace, energy-dispersive system at 15 kV, ∿.01 μm amp, with a 10–15-μm spot size and a 120 sec count-time.

RESULTS

X-ray diffraction analyses are presented in Table 1. In core SCAN 16P, the dominant clay is a randomly interlayered smectite/illite. Both the proportion of smectite/illite (S/I) and the percent of expandable layers in the S/I increase below 400 cm. Palygorskite has been detected in the lowermost sample. Phillipsite with minor clinoptilolite is dominant in the upper sections of the core, whereas below ∿400 cm clinoptilolite is more common. K-feldspar is abundant below ∿700 cm.

Hole 59 shows more evidence of a substantial volcanic component in the sediment. The percent expandable layers in the interlayered S/I varies widely (60–80%), and in several horizons it is high enough to suggest that some of the clay in that horizon is

TABLE 1. X-ray diffraction data*

Sample	Depth in sed.	Age	Lithology	>62 μm	2–62 μm	<2 μm†
SCAN 16P			brn clay; rads			
SCAN 0	0–2 cm	Eoc ? ↓	brn clay ↓; rads ↓	100% phil	qtz:ph:cl:I:S/I 1:3.5:0.2:1:3	qtz:S/I (63):I:k 1:11:3:1.2
SCAN 50	50–52 cm			100% phil	qtz:ph:cl:I:S/I 1:5:0.2:1:3	qtz:S/I (66):I:k 1:13:3:0.7
SCAN 100	100–102 cm	↓		100% phil	qtz:ph:cl:I:S/I:pl 1:2:8:0.13:1:3:0.1	qtz:S/I (66):I:k 1:13:1.6:0.8
SCAN 200	202–202 cm	↑		100% phil	qtz:ph:cl:I:S/I 1:5.7:0.14:1.6:3	qtz:S/I (61):I:k:pl 1:8:2:0.7:0.4
SCAN 300	300–302 cm			100% phil	qtz:ph:cl:I:S/I 1:7.8:0.4:0.7:3	qtz:S/I (58):I:k 1:25:3:0.6
SCAN 400	400–402 cm		rads ↓	ph:apat 1:1.6	qtz:cl:pl:S/I 1:2:0.5:23	S/I (73):I 25:1
SCAN 500	500–502 cm	?		100% apat	qtz:cl 1:7.5	S/I (76):I:k 54:3:1
SCAN 600	600–602 cm			k-fd:apat 1:2.6	qtz:cl 1:7	S/I (69):I 50:1
SCAN 700	700–702 cm			100% apat	qtz:cl:k-fd:I 1:8.8:1.4:16	qtz:S/I (72) 1:18
SCAN 800	800–802 cm	↓	↓	100% apat	qtz:cl:k-fd:I:pal 1:8:4.7:4.4:9.2	qtz:S/I (75):I:k:cl 1:37:4:1.5:0.7
DSDP 59						
59.1-3-2	54 m	Plio-Pleist ↓	rad clay ↓	qtz:ph:pl also opal-A 1:4.5:3	qtz:ph:pl:S/I 1:4.2:4.8:13	qtz:S/I (69):I:k:pl 1:6:1:0.7:0.3
59.1-3-3	57 m			qtz:ph also opal-A 1:7	qtz:ph:pl:S/I:I 1:2:2:3:1.8	qtz:S/I (65):I:k 1:9:2:1:0.6
59.2-1-1	90 m	Mioc ↓		ph:S/I also opal-A 1:0.8	qtz:ph:pl:S/I:I 1:2:9:15:1.7	qtz:S/I (60):pl 1:14:3
59.2-3-1	108 m		brn clay ↓	qtz:ph:pl 1:40:20	qtz:pl:S/I 1:1.7:13	qtz:S/I (75) 1:12.5
59.2-3-2	111 m	Olig ↓		qtz:ph:pl:cl 1:14:3.6:3.4	qtz:ph:pl:cl:S/I 1:8.7:8:2:9.5	qtz:S/I (80):pl 1:14:2
59.2-4-1	117 m			qtz:ph:pl 1:4:18	qtz:ph:pl:cl 1:8:6:1.4	qtz:S/I (61):pl 1:3.5:1.6
59.2-5-1	127 m	Eoc ↓		tr. ph and opal-A	qtz:cl:pl:I:S/I 1:1.7:0.8:4:9	qtz:S/I (75):I 1:4:2
59.2-6-2	134 m	Paleoc	↓	qtz:ph:pl:cl & opal-A 1:0.6:2:0.2	qtz:ph:pl:cl:S/I 1:4.6:3:0.2:9	qtz:S/I (68):k:pl 1:11:1:1

TABLE 1 (*cont.*)

Sample	Depth in sed.	Age	Lithology	>62 μm	2–62 μm	<2 μm†
DSDP 105						
105-5-2	244 m	U. Cret	varicolored clay	qtz:k-fd 1:1	qtz:cl:I:k:pl:k-fd 1:0.8:0.8:0.2:0.02:0.15	S/I (67):I:k 1:0.4:0.8
105-9-4	291 m	↓	↓	qtz:pl:cl 1:0.8:0.2	qtz:cl:I:S/I:pl 1:0.3:1:6.7:0.3	S/I (90):I:k 1:0.3:0.17
105-10-2	303 m	↓	black carbonaceous clay	qtz:mica:chl 1:3:0.15	qtz:mica:chl:pl:S/I 1:5:0.5:0.4:5	S/I (77):I:k 1:0.09:0.15
105-12-2	316 m	↓	↓	qtz:cl:cal 1:0.7:3.6	qtz:cl:I:cal 1:0.6:3.7:2	S/I (74):I:cal 1:0.27:0.3
105-13-cc	327 m	L. Cret	↓	qtz:cl:cal 1:18:1	qtz:cl:cal:pl:SI/:I 1:0.3:0.8:0.2:1.4:1.3	S/I (73):I:cal 1:0.1:0.2
105-14 cc	349 m	↓	↓	qtz:cl:pl 1:3:0.4	qtz:cl:pl:I:S/I 1:0.5:0.2:1.8:1.8	qtz:S/I (77):I:k 1:4.8:0.5:1
105-15-6	371 m	↓	↓	100% clin	qtz:cl:pl:I:S/I 1:0.3:0.2:1.2:2.3	S/I (78):I 1:0.3
105-16 cc	386 m	↓	↓	qtz:cl 1:7	qtz:cl:k-fd:pl:I 1:1:0.2:0.2:1.5	S/I (80):I:cal 1:0.5:0.2
105-19-3	434 m	↓	↓	–	qtz:cl:crist:k-fd:S/I:I 1:0.4:1:0.4:3.5:0.9	S/I (80):k-fd:cal 1:0.6:12
105-21-cc	457 m	↓	↓	–	qtz:cl:cal:pl:I:S/I 1:0.3:0.5:0.2:1:3	qtz:S/I (83):I:cal 1:23:3:1.6

TABLE 1 (*cont.*)

Sample	Depth in sed.	Age	Lithology	>62 μm	2–62 μm	<2 μm†
DSDP 163						
163-2-3	5 m	U. Olig ↓	rad clay ↓	qtz:ph also opal-A 1:4.7	qtz:ph:pl:S/I 1:3:1.3:10	qtz:S/I (69):I:pl 1:11:2:0.9
163-3-4	14 m	↓	clayey rad ooze ↓	opal-A	qtz:cl:ph:pl:I:S/I 1:0.5:1.5:1:1.5:5	qtz:S/I (79) 1:39
163-4-5	25 m	L. Olig ↓	↓	qtz:ph:pl 1:3.6:1	qtz:ph:pl:S/I 1:0.76:1.6:15	qtz:S/I (70):I:pl 1:47:2.7:1.5
163-1A-4	142 m	Eoc (?) ↓	brn clay ↓	–	qtz:ph:cl:pl 1:2.4:3.4:1	qtz:S/I (73):cl 1:10:0.4
163-15-1‡	162 m	U. Cret ↓	nanno chalk ↓	–	qtz:cl:pl:k-fd:pal:I 1:0.4:0.5:0.75:0.5:0.75	qtz:S/I (58):I:k-fd 1:9:1.2:1.9
163-16-5‡	177 m	↓	↓	–	qtz:cl:pl:k-fd:pal:I 1:0.35:0.43:0.43:0.8:1.3	qtz:S/I (56):I:pal 1:3:2:1.4
163-17-5‡	187 m	↓	↓	–	qtz:cl:pl:k-fd:I 1:0.3:0.6:0.4:1	qtz:S/I (57):I:pal:k-fd 1:5:2:1.7:0.8
163-19-5‡	204 m	↓	↓	–	qtz:cl:k-fd:I 1:0.6:0.5:1.5	qtz:S/I (57):I:pal 1:5:2.4:1
163-21-5‡	224 m	↓	↓	100% qtz	qtz:cl:pl:k-fd:pal:I:S/I 1:0.2:0.4:0.6:0.4:0.6:1.6	qtz:S/I (58):I:pal 1:5:2.5:1.4
163-23-5‡	241 m	↓	↓	–	qtz:crist:pal 1:17:1.5	qtz:S/I (40):crist:pal:I 1:3:17:6:2.7
163-27-1‡	271 m	↓	↓	–	qtz:cl:pl:k-fd:pal:I 1:0.7:0.7:0.6:1.6:1.6	qtz:S/I (50):I:pal 1:3:3.5:3

*Given as abundance ratio relative to quartz using the method of Cook *et al.* (1975), Qtz–quartz; ph–phillipsite; cl–clinoptilolite; I–illite; S/I–randomly interlayered smectite/illite; pl–plagioclase; k-fd–k-feldspar; crist–cristobalite; pal–palygorskite; cal–calcite; k–kaolinite; chl–chlorite; apat–apatite; brn–brown; rad–radiolaria.

†Number in parenthesis is the percent of expandable layers in the randomly interlayered smectite/illites. Illite is discrete hydromica separate from the S/I.

‡Acid residues.

authigenic (Perry *et al.*, 1976a; Drever, 1976; Perry *et al.*, 1976b). Plagioclase is abundant in the upper sections of the core, but its abundance decreases near the base of the hole, where clinoptilolite occurs.

Clinoptilolite is the only zeolite present in the Cretaceous clays of Site 105. The percent expandable layers in the mixed layered S/I is also high (up to 90%) and variable (67-90%), suggesting that some horizons may have contained some contributions from volcanic sources. Plagioclase concentrations are quite low, however, which is atypical of sediment with significant admixtures of basaltic material.

Variations in the mineral suites occurring in DSDP 163 demonstrate the differences in mineral assemblages commonly associated with pelagic clays and carbonate oozes. The upper portions of the hole are radiolarian clays, clayey radiolarian oozes, and clays. Mineralogically, they contain interlayered S/I clays with 65-75% expandable layers, phillipsite, plagioclase, and rare to common clinoptilolite. The lower section of the core consists of a fairly homogeneous nannofossil chalk. The non-carbonate residues of these chalks differ mineralogically from the clays above and display several compositional trends downhole. The percentage of expandable layers in the interlayered S/I clays drops markedly below the transition from clay to carbonate and continues to decrease further downhole. Illite is more common, and palygorskite is present, as is K-feldspar. Cristobalite occurs in Core 23 above a section of cherts. Clinoptilolite is the only zeolite present in the carbonate section.

X-ray fluorescence results are presented in Table 2A, B, and C. Two of the most salient features of these results are the variation in Ti/Al and Si/Al ratios from site to site and with increasing depth in each core. These trends are plotted in Figs. 1a-c. In SCAN 16P, the Ti/Al ratio decreases markedly below 400 cm, a point which coincides approximately with the first porcellanite layer and the last occurrence of phillipsite. The Si/Al ratio also increases rapidly at this point.

Similar, though less smooth, trends are characteristic of DSDP 59 (Fig. 1b). Here the Ti/Al ratio is overall much higher and is more variable. Ti/Al ratios generally decrease and Si/Al ratios increase towards the lower portions of the core where clinoptilolite is present. In hole 105, Ti/Al ratios are intermediate between those of DSDP 59 and the lower portions of SCAN 16P, except for sample 105-9-4, where the ratio is moderately high. Si/Al ratios are more variable, but tend to increase towards the base of the section.

A summary of selected chemical variations from core to core is given in Fig. 2 a and b. In Fig. 2a it can be seen from Ti and Al values (Böstrom *et al.*, 1972; Donnelly and Nalli, 1973) that sediments from hole 59 contain

TABLE 2-A. X-ray fluorescence analyses*

Sample	Depth in core	SiO_2	Al_2O_3	K_2O	CaO	FeO†	TiO_2	Si/Al	Ti/Al
SCAN 16P									
Bulk samples									
SCAN 0	0 cm	54.03	16.31	3.40	3.64	8.08	1.05	2.92	0.073
SCAN 50	50 cm	53.44	15.80	3.55	3.34	7.91	0.98	2.99	0.070
SCAN 100	100 cm	53.81	16.28	3.61	3.60	7.94	0.93	2.92	0.065
SCAN 200	200 cm	52.86	15.75	3.85	4.62	6.31	0.83	2.96	0.060
SCAN 300	300 cm	53.08	15.71	4.24	4.71	5.72	0.67	2.98	0.048
SCAN 400	400 cm	59.55	14.27	2.14	4.78	7.46	0.36	3.68	0.028
SCAN 500	500 cm	53.08	11.82	2.40	6.44	7.07	0.32	3.97	0.030
SCAN 600	600 cm	59.88	13.61	2.53	5.80	7.45	0.36	3.89	0.029
SCAN 700	700 cm	54.10	11.89	2.31	5.53	7.07	0.33	4.02	0.032
SCAN 800	800 cm	56.16	12.12	2.57	5.64	7.10	0.33	4.09	0.031
<2 μm fraction									
SCAN 0	0 cm	51.40	17.10	3.12	1.26	11.73	1.11	2.66	0.073
SCAN 50	50 cm	53.88	16.31	3.01	1.34	11.52	1.10	2.92	0.076
SCAN 100	100 cm	53.22	17.10	2.91	1.36	11.69	1.06	2.75	0.070
SCAN 200	200 cm	52.79	16.68	2.91	1.38	11.66	1.12	2.80	0.076
SCAN 300	300 cm	52.86	16.45	2.65	1.45	11.38	1.06	2.84	0.072
SCAN 500	500 cm	52.79	11.80	1.84	2.53	9.93	0.45	3.95	0.043
SCAN 600	600 cm	52.28	11.75	1.79	2.22	9.17	0.42	3.93	0.040

TABLE 2-B. X-ray fluorescence analyses*

Sample	Depth in core	SiO_2	Al_2O_3	K_2O	CaO	FeO†	TiO_2	Si/Al	Ti/Al
DSDP 59									
Bulk samples									
59.1-3-2	54 m	51.26	15.93	2.39	3.10	13.52	2.65	2.84	0.187
59.1-3-3	57 m	51.18	17.33	2.61	1.85	11.52	2.54	2.61	0.165
59.2-1-1	90 m	55.64	15.81	1.84	3.57	12.43	2.19	3.11	0.157
59.2-3-1	108 m	52.50	14.57	1.42	4.50	13.08	2.44	3.84	0.229
59.2-3-2	111 m	54.39	13.32	2.30	5.32	11.73	1.85	3.60	0.157
59.2-5-1	127 m	60.00	12.45	3.47	3.71	5.90	1.25	4.25	0.114
59.2-6-2	134 m	60.92	12.34	1.70	4.47	9.53	1.43	4.35	0.132
<2 μm fraction									
59.1-3-2	54 m	51.55	15.76	2.05	1.51	13.80	1.81	2.89	0.129
59.1-3-3	57 m	52.11	15.29	2.11	1.36	12.92	1.84	3.01	0.135
59.2-1-1	90 m	51.26	14.51	1.29	1.82	12.83	1.87	3.12	0.146
59.2-3-2	111 m	55.97	11.87	1.60	1.75	15.94	1.73	4.16	0.166
59.2-5-1	127 m	54.76	14.07	2.51	2.89	8.44	0.72	3.22	0.058

TABLE 2-C. X-ray fluorescence analyses*

Sample	Depth in core	SiO_2	Al_2O_3	K_2O	CaO	FeO†	TiO_2	Si/Al	Ti/Al
DSDP 105									
Bulk samples									
105-5-2	244 m	59.75	16.63	3.30	0.69	5.95	0.68	3.17	0.046
105-9-4	291 m	53.81	16.10	3.48	0.53	7.43	1.02	2.96	0.071
105-10-2	303 m	58.64	17.95	2.87	0.48	7.21	0.82	2.89	0.052
105-12-2	316 m	49.14	11.28	2.61	15.66	4.27	0.45	3.85	0.045
105-14-cc	349 m	62.48	14.98	2.85	0.44	5.98	0.70	3.69	0.053
105-16-cc	386 m	60.98	14.19	2.99	0.40	7.01	0.66	3.79	0.053
105-19-3	434 m	64.57	11.99	2.86	2.84	4.83	0.54	4.76	0.050
105-21-cc	457 m	62.16	14.64	2.59	4.98	6.74	0.66	3.90	0.052
<2 μm fraction									
105-5-2	244 m	56.30	17.15	3.10	0.71	9.21	0.68	2.90	0.045
105-9-4	291 m	57.92	18.92	2.34	0.16	7.41	0.61	2.71	0.036
105-10-2	303 m	58.12	19.87	2.46	0.47	8.44	0.64	2.58	0.036
105-21-cc	457 m	56.49	17.00	2.22	4.33	7.79	0.60	2.93	0.040

*Weight percent.
†Total iron given as FeO.

significant but variable amounts of basaltic volcanic material; that sediments from the upper 400 cm of SCAN 16P contain some basaltic material, but less than is found in hole 59; and that sediments in hole 105 may contain some volcanic material which must be more andesitic-trachytic in composition. Of note are the points representing the lower 400 cm of SCAN 16P, which fall below the line representing average terrigenous material (Clarke, 1924). Figure 2b indicates that not only do these samples contain less Ti but also more Si than average terrigenous material. Si/Al ratios for samples from hole 59 are high, reflecting the siliceous ooze component in them. With the exception of 105-9-4 and 105-10-2, hole 105 sediments also appear to be more siliceous than average terrigenous material.

Microprobe analyses of individual zeolite crystals are

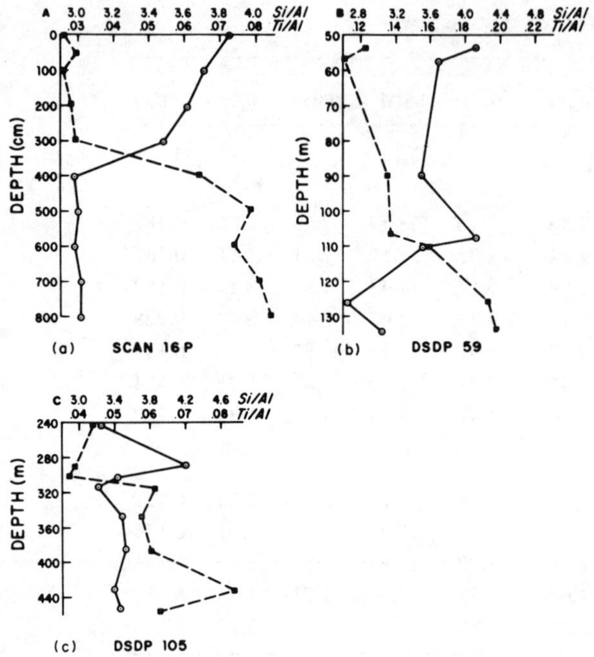

Fig. 1. Variations in Ti/Al and Si/Al ratios (elemental) of bulk samples of sediment containing phillipsite and/or clinoptilolite as a function of depth. (a) SCAN 16P; (b) DSDP 59; (c) DSDP 105. – – –□– – – Si/Al; ——○—— Ti/Al.

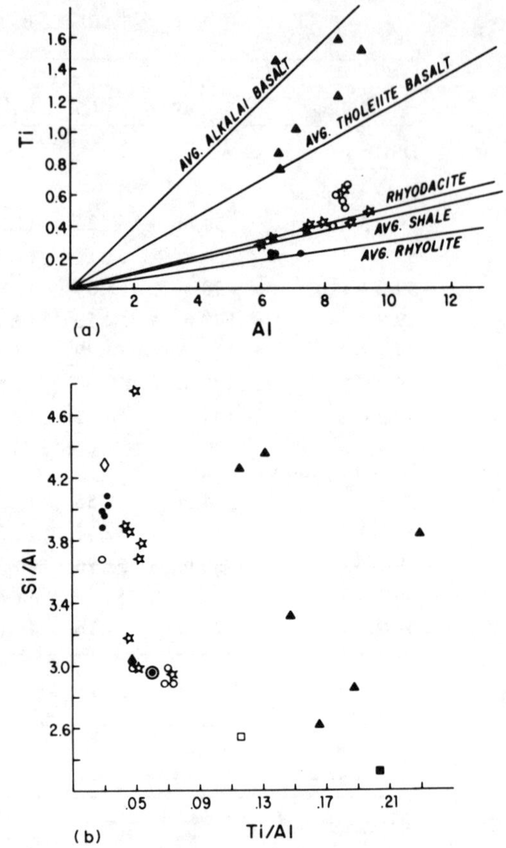

Fig. 2(a). Plot of Ti versus Al (weight % of the element) for bulk samples of sediments containing clinoptilolite and/or phillipsite. △ – DSDP 59; ○ – above 400 cm SCAN 16P; ● – below 400 cm SCAN 16P; ★ – DSDP 105. The plotted curves represent Ti/Al ratios for various average igneous and sedimentary rock types. Avg. alkali and tholeiite basalt – Engel *et al.* (1965); rhyodacite – Mauna Kea – Engel and Engel (1968); avg. shale – Clarke (1924); avg. rhyolite – Clark (1966). (b) Ti/Al versus Si/Al ratios (elemental) for bulk sediment samples and various average igneous and sedimentary rock types. Symbols and sources the same as for (a).

presented in Tables 3 and 4 for clinoptilolites and phillipsites, respectively. Perhaps the most important point indicated by the data is the very narrow compositional range for both zeolites (Figs. 3 a and b). Ninety-two percent of the phillipsites have Si/Al ratios between 2.5 and 2.8, and 84% of the clinoptilolites have Si/Al ratios between 4.5 and 5.0. With the exception of a few clinoptilolites from carbonate horizons, both zeolites contain predominantly monovalent cations, and in most samples potassium is in excess of sodium. Figure 4 shows the variation of the proportion of monovalent to divalent cations versus Si/Al ratios in phillipsites and clinoptilolites from the deep sea. Fields representing phillipsites and clinoptilolites from other geologic environments are also plotted (after Sheppard, 1971). Deep-sea clinoptilolites fall in the more siliceous portion of the field representing clinoptilolites from sedimentary rocks. Sodium is the dominant cation in most sedimentary clinoptilolites (Sheppard, 1971). Deep-sea clinoptilolites are predominantly potassic, except for clinoptilolites from carbonate horizons, which are enriched in calcium. Compared to heulandites and clinoptilolites from intermediate and basic igneous rocks and tuffs, deep-sea clinoptilolites are more siliceous and have a greater proportion of alkalis to alkaline earths. Phillipsites from deep-sea sediments have Si/Al ratios intermediate between phillipsites from mafic igneous rocks and phillipsites from saline lacustrine deposits. As with clinoptilolites, deep-sea phillipsites are dominantly potassic in contrast to lacustrine phillipsites, which generally have sodium as the dominant cation. Phillipsites from mafic igneous rocks commonly have a higher percentage of alkaline earths than deep-sea or lacustrine phillipsites. Neither phillipsite nor clinoptilolite shows any systematic compositional variation with depth over a depth interval containing a homogeneous sediment type. Slight differences in composition associated with the variations in bulk sediment chemistry between cores are detectable. The phillipsites from hole 59 are slightly less siliceous and contain a somewhat greater proportion of alkaline earths than do the phillipsites from SCAN 16P. Clinoptilolites show no definite trends in Si/Al ratios, but clinoptilolites from carbonates contain significantly more Ca than clinoptilolites from clays.

TABLE 3. Chemical analyses of clinoptilolites

Sample*	105-5-2	105-12-2	105-12-2	105-13-cc	105-14-cc	105-14-cc	105-15-6	105-15-6	105-16-cc
SiO_2	60.13	61.69	64.25	70.40	64.61	66.27	69.18	69.20	69.65
Al_2O_3	12.07	11.41	11.49	12.04	11.62	11.87	12.03	12.58	11.82
K_2O	4.50	4.57	6.26	5.30	5.92	6.16	7.97	8.10	6.95
Na_2O	1.76	0.54	0.76	1.69	1.60	2.01	1.01	1.25	1.43
CaO	0.89	1.85	0.95	1.09	0.45	0.45	0.57	0.52	0.60
MgO	0.83	0.71	0.41	0.27	0.26	0.28	0.28	0.21	–
FeO†	0.71	0.80	0.20	–	–	–	–	–	–
H_2O‡	19.11	18.43	15.68	9.21	15.54	12.96	8.96	8.14	9.55
Total	100	100	100	100	100	100	100	100	100
Number of cations on the basis of 72 oxygens (anhydrous cell)									
Si	29.07	29.47	29.77	30.05	29.82	29.77	29.87	29.67	30.09
Al	6.87	6.43	6.28	6.05	6.33	6.28	6.13	6.37	6.02
K	2.77	2.78	3.70	2.89	3.49	3.53	4.39	4.43	3.83
Na	1.65	0.50	0.68	1.40	1.43	1.75	0.84	1.04	1.19
Ca	0.46	0.94	0.47	0.50	0.22	0.22	0.26	0.24	0.28
Mg	0.59	0.50	0.28	0.17	0.18	0.18	0.18	0.13	–
Si/Al	4.23	4.59	4.74	4.96	4.71	4.74	4.88	4.66	5.00
Si+Al	35.94	35.90	36.05	36.10	36.15	36.05	36.00	36.04	36.11
Σ_{cat}	6.53	6.18	5.90	5.63	5.73	6.08	6.12	6.21	5.58
E_b§	4.93	3.78	5.83	6.84	9.39	3.12	0.08	2.36	7.19

Sample	105-17-cc	105-17-cc	105-19-3	105-19-3	105-19-3	105-21-cc	105-21-cc	163-17-5	163-19-5
SiO_2	66.06	66.64	68.79	70.13	68.99	69.20	66.71	66.82	68.51
Al_2O_3	11.57	11.78	11.64	12.59	11.85	12.52	11.67	11.92	11.84
K_2O	5.07	6.06	5.25	4.80	5.33	3.43	4.70	5.93	2.69
Na_2O	1.38	2.29	2.12	2.91	2.35	0.70	0.75	0.72	0.40
CaO	1.27	0.53	0.83	1.09	0.59	3.12	2.50	0.89	3.36
MgO	0.41	0.30	0.22	0.21	–	0.42	0.35	0.63	0.66
FeO	–	0.21	–	–	–	–	–	0.22	–
H_2O	14.24	12.19	11.17	8.27	10.89	10.61	13.32	12.87	12.54
Total	100	100	100	100	100	100	100	100	100
Number of cations on the basis of 72 oxygens									
Si	28.89	29.73	30.06	29.73	30.06	29.77	29.84	29.83	29.98
Al	6.17	6.19	5.99	6.29	6.09	6.35	6.15	6.27	6.10
K	2.92	3.45	2.93	2.59	2.96	1.88	2.68	3.38	1.47
Na	1.21	1.98	1.79	2.39	1.98	0.58	0.65	0.62	0.31
Ca	0.62	0.25	0.38	0.50	0.27	1.43	1.20	0.43	1.58
Mg	0.28	0.20	0.14	0.13	–	0.27	0.23	0.42	0.42
Si/Al	4.84	4.79	5.02	4.72	4.94	4.62	4.85	4.75	4.91
Si+Al	36.06	35.92	36.05	36.02	36.15	36.12	35.99	36.10	36.08
Σ_{cat}	5.93	6.34	5.78	6.24	5.49	5.88	6.19	5.70	5.78
E_b	3.90	–2.37	3.55	0.73	9.71	7.39	–0.65	9.18	5.20

TABLE 3 (*cont.*)

Sample	163-27-1	59.2-5-1 (2-62μ)	59.2-5-1 (2-62μ)	59.2-5-1 (2-62μ)	59.2-3-2	59.2-4-1	SCAN 16P 500 cm (2-6μ)
SiO_2	66.77	65.84	66.60	67.98	68.98	68.19	65.55
Al_2O_3	11.62	12.18	11.61	12.08	11.24	12.73	11.76
K_2O	2.56	6.94	6.35	6.42	2.79	3.47	4.67
Na_2O	0.44	0.90	1.11	1.25	2.61	0.96	1.23
CaO	3.40	0.45	0.43	0.63	2.12	2.81	1.79
MgO	0.96	0.90	0.57	0.62	–	0.45	0.49
FeO	–	0.54	0.42	–	–	–	–
H_2O	14.25	12.25	12.91	11.02	12.26	11.10	14.51
Total	100	100	100	100	100	100	100
Number of cations on the basis of 72 oxygens							
Si	29.82	29.52	29.87	29.73	30.21	29.58	29.74
Al	6.12	6.44	6.14	6.23	5.80	6.51	6.28
K	1.45	3.97	3.63	3.58	1.55	1.92	2.70
Na	0.37	0.78	0.96	1.06	2.22	0.81	1.08
Ca	1.63	0.22	0.21	0.29	0.99	1.30	0.87
Mg	0.64	0.50	0.38	0.40	–	0.29	0.33
Si/Al	4.87	4.59	4.87	4.77	5.20	4.54	4.73
Si+Al	35.94	35.96	36.01	35.96	36.01	36.09	36.02
Σ_{cat}	6.35	6.18	5.78	6.04	5.76	5.93	6.19
E_b	−3.80	3.90	5.80	3.00	0.60	8.90	1.50

*Each analysis represents a separate crystal from the stated horizon.

†Inclusion impurities – not included in cell contents. Total iron given as FeO.

‡By difference.

§ $E_b = \frac{Al - \Sigma_{cat}}{Al} \times 100$; $\Sigma_{cat} = 2(ca + Mg) + K + Na$.

Element abundances on basis of unit cell contents.

DISCUSSION

These four cores form a representative cross-section of the sedimentary environments in which deep-sea phillipsites and clinoptilolites occur. Examination of the combined chemical and mineralogical data indicates several general patterns of occurrence. The presence of basaltic volcanic material seems to be essential for the formation of phillipsite. Most occurrences of deep-sea phillipsites have been reported as directly associated with glass or other basaltic components as hole 59. In SCAN 16P. however, no direct evidence of volcanic

TABLE 4. Chemical analyses of phillipsites

Sample*	SCAN 16P O TW	SCAN 16P O TW	SCAN 16P 100 LTW	SCAN 16P 100 LTW	SCAN 16P 100 S	SCAN 16P 100 S	SCAN 16P 100 TW	SCAN 16P 100 TW	SCAN 16P 100 TW
SiO_2	54.10	56.65	58.28	55.54	56.32	57.61	53.57	55.09	54.84
Al_2O_3	17.47	17.89	17.12	17.27	17.52	17.99	17.48	17.65	18.50
K_2O	7.06	6.46	6.72	6.32	6.66	6.86	6.54	5.86	8.37
Na_2O	3.50	3.26	4.72	4.96	4.25	5.05	4.61	4.35	3.58
CaO	0.94	1.22	0.27	0.19	0.63	0.36	0.74	0.97	1.25
MgO	0.32	0.64	–	–	0.16	–	0.32	0.40	–
FeO†	0.87	0.90	0.35	0.46	0.60	0.53	0.72	0.98	–
H_2O‡	15.74	12.98	12.54	14.99	13.86	11.60	16.02	14.70	13.46
Totals	100	100	100	100	100	100	100	100	100
Number of cations on the basis of 32 oxygens									
Si	11.60	11.68	11.92	11.75	11.68	11.73	11.54	11.61	11.50
Al	4.41	4.34	4.14	4.31	4.28	4.32	4.44	4.38	4.57
K	1.93	1.70	1.76	1.78	1.76	1.78	1.79	1.57	2.24
Na	1.46	1.30	1.88	2.03	1.71	1.99	1.92	1.78	1.46
Ca	0.21	0.27	0.06	0.04	0.14	0.08	0.17	0.22	0.28
Mg	0.10	0.20	–	–	0.05	–	0.10	0.12	–
Si/Al	2.63	2.69	2.87	2.73	2.72	2.72	2.60	2.64	2.51
Si+Al	16.02	16.02	16.06	16.06	16.05	16.05	15.98	16.00	16.08
Σ_{cat}	4.02	3.94	3.76	3.89	3.86	3.93	4.26	4.04	4.26
E_b§	8.94	9.42	9.28	9.58	9.89	8.85	3.92	7.77	6.85

Sample	59.2-5-1	59.2-5-1	59.2-6-2	59.2-6-2
SiO_2	57.17	54.87	54.41	56.21
Al_2O_3	17.96	18.50	17.47	18.38
K_2O	6.41	7.02	7.14	7.17
Na_2O	4.00	3.60	2.19	2.35
CaO	0.86	1.17	2.37	3.17
MgO	0.46	0.55	0.44	0.19
FeO	0.52	0.82	–	0.42
H_2O	12.62	13.47	15.98	12.11
Totals	100	100	100	100
Number of cations on the basis of 32 oxygens				
Si	11.72	11.46	11.60	11.54
Al	4.33	4.55	4.39	4.45
K	1.67	1.87	1.94	1.88
Na	1.59	1.46	0.90	0.94
Ca	0.19	0.26	0.54	0.70
Mg	0.14	0.17	0.14	0.05
Si/Al	2.71	2.52	2.64	2.59
Si+Al	16.05	16.01	15.99	15.99
Σ_{cat}	3.93	4.20	4.21	4.33
E_b	9.00	7.70	4.00	2.70

***Each analysis represents a separate crystal from the stated horizon. TW–simply twinned; LTW–large, multiply twinned; S–single-untwinned.**

†Inclusion impurities – not included in cell contents. Total iron given as FeO.

‡By difference

$$\S E_b = \frac{Al - \Sigma_{cat}}{Al} \times 100 \quad \Sigma_{cat} = 2(Ca+Mg) + K + Na.$$

Elemental abundances based on unit cell-contents.

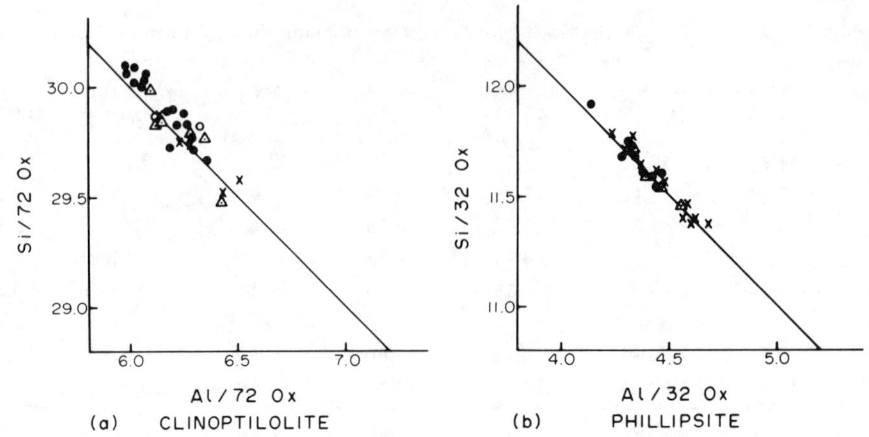

Fig. 3(a). Variation of Si vs. Al for clinoptilolites from various deep-sea environments. ○ – clinop. from deep-sea clays; △ – clinop. from horizons containing $CaCO_3$; X – clinop. from clays which also contain phillipsite. (b) Variation of Si vs. Al for deep-sea phillipsites. ○ – SCAN 16P; △ – DSDP 59. Also plotted are 12 analyses (x) from Sheppard *et al.* (1970) for comparison.

material was detected, but the bulk chemistry suggests a certain admixture of basaltic material. Mineralogical support for this analogy is supplied by the plagioclase concentrations which are high, both where glass can be observed and where its presence is only suggested by the bulk sediment chemistry.

The chemical environment necessary for the formation of clinoptilolite is more difficult to characterize. In hole 105, a volcanic origin for most of the Cretaceous clays is suggested by the absence of detrital clays and by structural characteristics such as numerous thin, sharply defined, ungraded beds which seem to be altered ash beds. However, the higher Si/Al ratios and low Ti/Al ratios indicated by the bulk chemical analyses and the lack of abundant plagioclase in these sediments suggest that the volcanic material was not basaltic. This would seem to support the suggestion by many authors (e.g., Hathaway and Sachs, 1965; Hay, 1966) that more silicic volcanic material is necessary for the formation of clinoptilolite. Data from SCAN 16P, however, do not definitely corroborate or refute this suggestion. It is possible that the sediment originally contained a certain proportion of rhyolitic material; the points representing samples below 400 cm lie near the line representing the Ti/Al ratio of an average rhyolite (Clark, 1966). However, the chemical fingerprint of rhyolitic material is

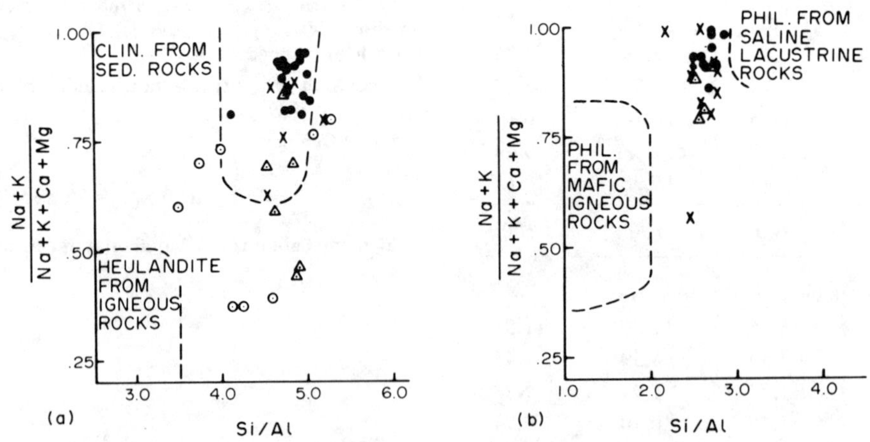

Fig. 4(a). Plot showing the proportion of monovalent to divalent cations versus the Si/Al ratios of clinoptilolites from various deep-sea environments. ○ – clinop. from deep-sea clays; △ – clinop. from horizons containing $CaCO_3$; X – clinop. from clays which also contain phillipsite. The bounded areas represent compositional fields for clinoptilolites and heulandites from other environments (after Sheppard, 1971). (b) Variation of Si/Al ratios versus the proportion of monovalent to divalent cations for deep-sea phillipsites. ○ – SCAN 16P; △ – DSDP 59. Also plotted are analyses (x) from Sheppard *et al.* (1970). Bounded areas represent compositional fields for phillipsites from other environments (after Sheppard, 1971).

not as clear cut as that for basaltic material. An equally possible explanation would be a sediment containing a greater proportion of smectites coupled with dilution with more biogenic silica. Smectites have a lower Ti/Al ratio than the average illite or kaolinite (Weaver and Pollard, 1973) and dilution with silica would account for the decrease in both Ti and Al. Absence of direct corroborative evidence, such as radiolarian molds or relict or altered glass shards, does not indicate which explanation is more accurate. More data from other similar environments is being collected to help resolve this problem.

Lack of chemical data makes the interpretation of hole 163 difficult; however, mineralogical data suggest several trends. The upper siliceous clay section of the core contains phillipsite with abundant plagioclase and interlayered S/I. This assemblage is analogous to hole 59 and similar to SCAN 16P, where siliceous clays with some basaltic material occur. The lower calcareous section presents an even more interesting puzzle. Extensive diagenesis of these sediments is indicated by several lines of evidence: the increase in the proportion of illite; the presence of palygorskite; the decrease in both the amount of and the percentage of expandable layers in the S/I interlayered clays; the abundance of K-feldspar. The question is, what was the parent material which diagenetically altered into this assemblage? Was it a mixture of terrigenous clays and siliceous microfossils thoroughly diluted with calcareous material, or was some type of volcanic material involved? Donnelly and Wallace (1976) have suggested that secondary alteration of terrigenous material may be responsible for similar diagenetic assemblages seen in Leg 35 sediments. More work on the chemical aspects of the non-carbonate portions of these sediments, especially the refractory elements Ti and Al, are warranted. Other questions, such as the thermodynamic relationship between the various phases in sediments which contain detrital clay assemblages and biogenic silica, need to be investigated. Stonecipher and Kastner (this volume) discuss a diagenetic trend based on observed mineralogic associations from many sites. Does this trend based on observed mineralogic data, i.e.

$$\text{S/I + opal A} \rightarrow \begin{matrix}\text{S/I + opal CT}\\ \text{+ clin}\end{matrix} \rightarrow \begin{matrix}\text{S/I + opal CT}\\ \text{+ clin + I}\end{matrix} \rightarrow$$

$$\begin{matrix}\text{S/I + opal CT}\\ \text{+ clin + I}\\ \text{+ K-feld + paly}\end{matrix} \rightarrow \begin{matrix}\text{qtz + I}\\ \text{+ K-feld + paly}\end{matrix}$$

represent a series of metastable assemblages as a function of time and/or burial depth, or does it reflect primary variations in bulk sediment chemistry? Are most diagenetic reactions observed in marine sediments isochemical or not? What is the mobility of aluminum in various marine diagenetic environments, and how does this affect diagenesis? More work needs to be done to answer some of these questions.

CONCLUSIONS

1. The presence of a certain proportion of basaltic volcanic material is essential for the formation of phillipsite. Clinoptilolite seems to be able to form in non-volcanic siliceous terrigenous sediments, as well as in sediments with volcanic components.

2. Deep-sea zeolites have relatively narrow ranges of chemical compositions. Ninety-two percent of the phillipsites have Si/Al ratios between 2.5 and 2.8, and 84% of the clinoptilolites have Si/Al ratios between 4.5 and 5.0. Almost all the deep-sea phillipsites and clinoptilolites have potassium as the dominant cation. Small bulk-sediment compositional variations may be reflected in minor changes in the composition of the zeolite.

3. Major differences in bulk sediment chemistry are generally accompanied by a change in the type of zeolite present. Phillipsite is commonly associated with basaltic glass in sediments with low Si/Al ratios and high Ti/Al ratios. Clinoptilolite is associated with siliceous volcanic material or terrigenous sediments with high Si/Al ratios and low Ti/Al ratios.

ACKNOWLEDGMENTS

I wish to thank Dr. Miriam Kastner for many helpful discussions and critical suggestions on this paper. I would also like to thank Dr. J. V. Smith for his advice and discussions on the electron microprobe analyses. The work was supported by a Scripps Industrial Associates Traineeship and UCSD Academic Grant 764-V.

REFERENCES

Böstrom, K. (1970) Submarine vulcanism as a source for iron: *E.P.S.L.* **9**, 348-354.

Böstrom, K., Joensuu, O., Valdes, S. and Riera, M. (1972) Geochemical history of South Atlantic Ocean sediments since Late Cretaceous: *Mar. Geol.* **12**, 85-121.

Clague, D. A. (1974) The Hawaiian Emperor Seamount Chain: its origin, petrology, and implications for plate tectonics: Ph.D. Dissertation, University of California, San Diego.

Clark, S. P. (1966) Composition of rocks. In: *Handbook of Physical Constants*, Clark, S. P., ed.: *Geol. Soc. Amer.*, pp. 1-6.

Clarke, F. W. (1924) Data of geochemistry: *U.S. Geol. Surv. Bull.* **770**.

Cook, H. E., Johnson, P. D., Matti, J. C. and Zemmels, I. (1975) Methods of sample preparation and X-ray diffraction data analysis, X-ray Mineralogy Laboratory, Deep Sea Drilling Project, Univ. of California, Riverside. In: *Initial Reports of the Deep Sea Drilling Project*, Hayes, D. E., Frakes, L. A. *et al.*, eds., **28**, 999-1007. U.S. Govt. Printing Office, Washington, D.C.

Donnelly, T. W. and Nalli, G. (1973) Mineralogy and chemistry of Caribbean sediments. In: *Initial Reports of the Deep Sea Drilling Project*, Edgar, N. T., Saunders, J. B. *et al.*, eds., **15**, 929-962. U.S. Govt. Printing Office.

Donnelly, T. W. and Wallace, J. L. (1976) Major-element chemistry of Antarctic clay-rich sediments: Sites 322, 323, and 325, Leg 35, DSDP. In: *Initial Reports of the Deep-Sea Drilling Project*, Worstell, P. *et al.*, eds., **35** U.S. Govt. Printing Office, Washington, D.C.

Drever, J. I. (1976) Chemical and mineralogical studies, site 323. In: *Initial Reports of the Deep Sea Drilling Project*, Worstell, P. *et al.*, eds., **35**, 471-478. U.S. Govt. Printing Office, Washington, D.C.

Engel, A. E. J., Engel, C. C. and Havens, R. G. (1965) Chemical characteristics of oceanic basalts and the upper mantle: *Bull. Geol. Soc. Amer.* **76**, 719-734.

Engel, A. E. J. and Engel, C. C. (1970) Mafic and ultramafic rocks. In: *The Sea*, Vol. 4, part 1, Maxwell, A. E., ed., pp. 465-519. Wiley-Interscience, New York.

Hathaway, J. C. and Sachs, P. L. (1965) Sepiolite and clinoptilolite from the Mid-Atlantic Ridge: *Amer. Mineral.* **50**, 852-867.

Hay, R. L. (1966) Zeolites and zeolitic reactions in sedimentary rocks: *Geol. Soc. Amer. Spec. Paper* **85**.

Kastner, M. and Stonecipher, S. A. (1978) Zeolites in pelagic sediments: this volume.

Keene, J. B. and Kastner, M. (1974) Clays and formation of deep-sea chert: *Nature* **249**, 754-755.

Perry, E. A., Jr., Gieskes, J. M. and Lawrence, J. R. (1976a) Mg, Ca, and O^{18}/O^{16} exchange in the sediment–pore water system, Hole 149, DSDP: *Geochim. Cosmochim. Acta* **40**, 413-423.

Perry, E. A., Jr., Beckles, E. C. and Newton, R. M. (1976b) Chemical and mineralogical studies, Sites 323 and 325. In: *Initial Reports of the Deep Sea Drilling Project*, Worstell, P. *et al.*, ed., **35**, 465-470. U.S. Govt. Printing Office, Washington, D.C.

Sheppard, R. A. (1971) Zeolites in sedimentary deposits of the United States – a review. In: *Molecular Sieve Zeolites.* I, *Adv. in Chem. Series* **101**, Gould, R. F., ed., pp. 279-310. Amer. Chem. Soc., Washington, D.C.

Sheppard, R. A., Gude, A. J., 3rd and Griffin, J. J. (1970) Chemical composition and physical properties of phillipsite from the Pacific and Indian Oceans: *Amer. Mineral.* **55**, 2053-2062.

Stonecipher, S. A. (1976) Origin, distribution, and diagenesis of phillipsite and clinoptilolite in deep-sea sediments: *Chem. Geol.* **17**, 307-318.

Weaver, C. E. and Pollard, L. D. (1973) *The Chemistry of Clay Minerals*, pp. 8-9; 58-59; 134. Elsevier, New York.

NATURE AND ORIGIN OF DEEP-SEA CLINOPTILOLITE

J. R. BOLES and W. S. WISE

Department of Geological Sciences, University of California, Santa Barbara, Santa Barbara, California 93106, U.S.A.

Abstract – Deep-sea clinoptilolites are relatively uniform in composition. Electron microprobe analyses of five samples show that Si/Al ratios vary between 4.1 and 4.6, although an alkali-rich heulandite from a lignite has a Si/Al = 3.6. Potassium generally predominates in the samples. K/Na ratios are between 0.95 and 1.56, and calcium is generally less than 0.6 ions/unit cell (Mg/Ca ratios are commonly >1.0).

Clinoptilolite occurs as euhedral crystals in all samples suggesting that it has crystallized from pore fluids rather than as a direct replacement of an earlier phase. Similar proportions of Na, K, and Ca in both deep-sea clinoptilolites and phillipsites suggest that the exchangeable cation content of these zeolites reflects pore-fluid composition. Compared with associated interstitial fluids, sodium is depleted by a factor of 80, calcium by 50, and magnesium by about 55 in the clinoptilolite, relative to potassium.

Clinoptilolite has been reported in deep-sea sediments as young as Pleistocene, but it is most abundant in Eocene and Cretaceous samples. Surprisingly, over 80% of the occurrences are in brown clays and microfossil-rich sediments rather than in ash beds. Clinoptilolite (?) has also been reported as a replacement of radiolarian tests. These observations suggest that volcanic glass is not a prerequisite for the formation of deep-sea clinoptilolite.

Phillipsite is the dominant zeolite in younger deep-sea sediments. We suggest that phillipsite forms as a metastable, silica-deficient phase in marine pore fluids and that it is eventually replaced by clinoptilolite. Present data indicate that time is a more important control on this reaction than depth of burial or silica concentrations in pore fluids. Some clinoptilolite may also have formed from the dissolution of siliceous microfossils and clay minerals in the absence of volcanic glass.

INTRODUCTION

Phillipsite has been regarded as the major zeolitic component of deep sea sediments since Murray and Renard (1891) recognized it in deposits on the Pacific Ocean floor. Clinoptilolite was not recognized in these sediments until the mid-1960s (Murata and Erd, 1964; Hathaway and Sachs, 1965). Since that time clinoptilolite has been found to be widespread in deep-sea sediments (e.g. see Stonecipher, 1976; and review by Kastner and Stonecipher, this volume). Aside from limited descriptions of a few DSDP drilling legs, there has been no overall summary of clinoptilolite occurrences in deep-sea sediments.* Rad and Rosch (1972) noted in Leg 14, what we will show to be a general rule, that clinoptilolite is most abundant in the older sediments (also see Stonecipher, 1976). In Leg 20, Okada and Tomita (1973) noted a vertical transition from phillipsite to clinoptilolite, which they attributed to burial diagenesis. Fan and others (1973, p. 850) reported that clinoptilolite "... usually shows a close correlation with rhyolite ashes ..." in marine environments. We present evidence in this paper that clinoptilolite is commonly found in apparently non-volcanogenic sediments.

The abundance of euhedral clinoptilolite crystals in marine sediments suggests that it is a diagenetic product. The composition and origin of this phase is, therefore, important to an understanding of the interaction between marine fluids and sediments at moderate pressures and low temperatures. To date no analyses of deep-sea clinoptilolites have been published. On the basis of physical properties and thermal stability Boles (1972) estimated that a deep-sea clinoptilolite from the Atlantic Ocean has a Si/Al ratio between 4.0 and 4.3 and that it is alkali-rich. Microprobe analyses of this sample and five others are presented in this paper. A comparison of the compositions of deep-sea phillipsites and clinoptilolites coupled with a synthesis of the reported DSDP occurrences of clinoptilolite suggests that there is a reaction relationship with phillipsite.

---

*After this paper was submitted, Stonecipher (1976) discussed the distribution of phillipsite and clinoptilolite in deep-sea sediments.

CHEMICAL COMPOSITION

Based on descriptions of cores in DSDP Preliminary Reports, Legs 1-33, twelve samples were requested from the Deep Sea Drilling Project. Of these only six samples were found to contain sufficient clinoptilolite for microprobe analysis (see Table 1). The clinoptilolite crystals range in size from 2 to 40 μm and exhibit euhedral crystal forms (Fig. 1). Even in these samples clinoptilolite crystals large enough for analysis were too sparse to be easily found in a grain mount. Therefore, the zeolite crystals were concentrated by obtaining a size fraction between 5 and 40 μm. Bulk samples were disaggregated by wet sieving in distilled water through a 61-μm sieve. The fraction smaller than 5 μm was removed by repeated centrifuging and the greater than 40-μm particles were removed in a settling column. Samples were dried at 40°C.

The zeolite concentrates were embedded in epoxy, ground to expose the centers of several crystals, polished, and coated with 250 Å carbon. The average exposed area was about 20 μm. Analyses were obtained with an ARL microprobe with an accelerating voltage of 15 kV for all elements, a sample current of 0.01 μamps, and a beam diameter of 5 μm. Each sample was analyzed for K, Ca, Na, Sr, Ba, Mg, Si, Fe, and Al, using feldspars and other silicate standards.

The count rates for K and Na declined significantly with successive 10-second counting periods, because the electron beam was driving the weakly bound positive ions out of the excitation volume of the sample. Because the crystals are too small to use large beam diameters or to move the spot, a sequence of counts from each crystal were obtained, and the counting rates extrapolated to zero time (see Boles, 1972, p. 1465). Corrections were made with a modified version of EMPADR 7 (Rucklidge and Gasparrini, University of Toronto, written communication, 1969).

The average of acceptable analyses of individual crystals in each sample are presented in Table 2. Because the crystals are so small, they were extensively dehydrated by the energy of the electron beam and the vacuum of the microprobe. Therefore, the totals in Table 2 indicate water contents about one-third the normal value.

Cell contents were calculated for the analyzed samples (Table 2), and these values are plotted in Figs. 2 and 3. In Fig. 2, a plot relating the silica content with the exchangeable cations, the deep-sea clinoptilolites are more alkali-rich than clinoptilolites from other occurrences, especially those from saline lacustine environments. Five of the samples cluster with Si/Al ratios above 4, while sample F has an Si/Al ratio of 3.57. The compositions of these samples differ mainly from those of deep-sea phillipsites only in silica content. The relative composition of exchangeable cations shown in Fig. 3 is remarkably uniform (and overlap the deep-sea phillipsite compositions as well), but are substantially different from any other reported clinoptilolites. The composition of sample E is within the range of that predicted from its physical properties (cf. Boles, 1972).

Samples A through E would be classified as clinoptilolites and sample F would be classified as a heulandite, using the classification scheme of Boles (1972) which is based on Si/Al ratio. Thermal stability tests indicate that samples A through E are clinoptilolites whereas sample F has an intermediate thermal stability with respect to most heulandite-clinoptilolites

TABLE 1. Location and lithology of analyzed clinoptilolite samples from the Deep Sea Drilling Project

Sample	DSDP drill site	Location	Water depth (m)	Core number	Section of core	Distance (cm) below top of section	Depth of sample below sea bed (m)	Lithology and age
A	259	East Indian Ocean	4696	10	3	22–24	89	zeolitic clay, Cretaceous
B	307	West Pacific Ocean	5696	5	core catcher		103	radiolarian-bearing altered volcanic ash, Cretaceous
C	214	Indian Ocean	1655	36	3	74–76	335.6	glauconitic carbonate sand, Paleocene
D	5	North Atlantic Ocean	5361	3	1	65–67	71.5	silty clay, Cretaceous
E	214	Indian Ocean	1655	36	3	54–56	335.4	glauconitic carbonate sand, Paleocene
F	214	Indian Ocean	1655	42	1	93-94	390.4	lignite, Paleocene

Fig. 1. Scanning electron micrograph of a clinoptilolite crystal from the zeolitic lignite, site 214, in the Indian Ocean.

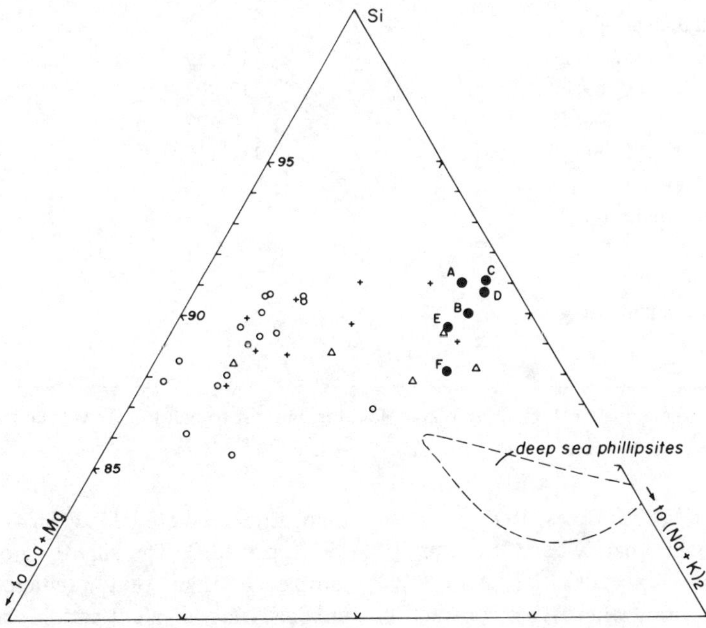

Fig. 2. Si – Ca + Mg – $(Na + K)_2$ atomic % plot of the compositions of the deep sea clinoptilolites (closed circles) given in Table 1. For comparison, the compositions of other clinoptilolites are those from saline lacustrine deposits (triangles) given by Sheppard and Gude (1968, 1969, and 1973); from marine tuff beds (open circles) of the Murihiku Supergroup, Hokonui Hills, New Zealand, from Boles (1972) and Boles and Coombs (1975); and from cavities in various volcanic rocks (crosses) from Wise and Tschnerich (1976) and unpublished data. The field of deep-sea phillipsites includes all analyses from Sheppard and Gude (1970).

TABLE 2. Chemical compositions, cell contents, and physical properties of six deep-sea clinoptilolites (see Table 1 for localities)

	(A)	(B)	(C)	(D)	(E)	(F)
SiO_2	72.40	71.58	71.83	71.11	72.75	67.80
Al_2O_3	13.17	13.71	14.27	14.22	15.03	16.12
Fe_2O_3	0.20	0.19	0.13	0.56	0.07	0.03
MgO	0.64	0.63	0.0	0.34	0.70	0.92
CaO	0.24	0.51	0.55	0.26	1.06	1.01
Na_2O	2.92	2.19	2.76	3.37	2.76	3.06
K_2O	4.93	3.16	5.88	5.09	5.83	5.76
Total*	94.50	91.97	95.42	94.95	98.20	94.70
Cell contents, based on 72 oxygens						
Si	29.71	29.35	29.38	29.21	28.99	28.18
Al	6.37	6.62	6.88	6.88	7.05	7.89
Fe	0.06	0.06	0.04	0.17	0.02	0.01
Mg	0.39	0.38	0	0.21	0.41	0.57
Ca	0.11	0.22	0.24	0.11	0.45	0.45
Na	2.32	2.19	2.19	2.69	2.13	2.46
K	2.58	3.16	3.07	2.67	2.96	3.06
Si/Al	4.66	4.43	4.27	4.25	4.11	3.57
Refractive Index (all ±0.002)						
β	1.480	1.480	1.482	1.484	1.480	1.486
Unit cell dimensions†						
a	17.67(1)	–	–	17.649	–	17.71(1)
b	17.89(2)	–	–	17.902	–	17.92(1)
c	7.408(5)	–	–	7.407	–	7.426(2)
β	116°9(4)′	–	–	116°16′	–	116°13(3)′
Thermal stability type†						
	C	C	C	C	C	I
Collapse temperature						
	–	–	–	–	–	272°C

*The remainder is H_2O; Sr and Ba looked for, but not found. Fe_2O_3 is total iron.
†See Boles (1972).

(cf. Mumpton, 1960; Alietti, 1972; Boles, 1972).

Boles (1972) has shown that when the sum of divalent cations (SD value) exceeds 1.87, contracted phase I and B form upon heating whereas no contraction occurs when the SD value is less than 1.00. Sample F has an SD value of 1.02 and shows a contracted phase when heated. This shows that thermal contraction can occur in heulandite group minerals with SD values as low as about 1.02. The temperature required for appearance of phase I in sample F is 272°C (2 hour heating, 1 hour cooling), which is similar to, but about 10°C higher than, that predicted for a Si/Al ratio of 3.57 (see Boles, 1972, p. 1489). The slightly greater thermal stability of sample F than that predicted may be due to its unusually low divalent cation content.

CLINOPTILOLITE OCCURRENCES

Correlations between clinoptilolite and phillipsite occurrences versus sediment age, sediment type, and

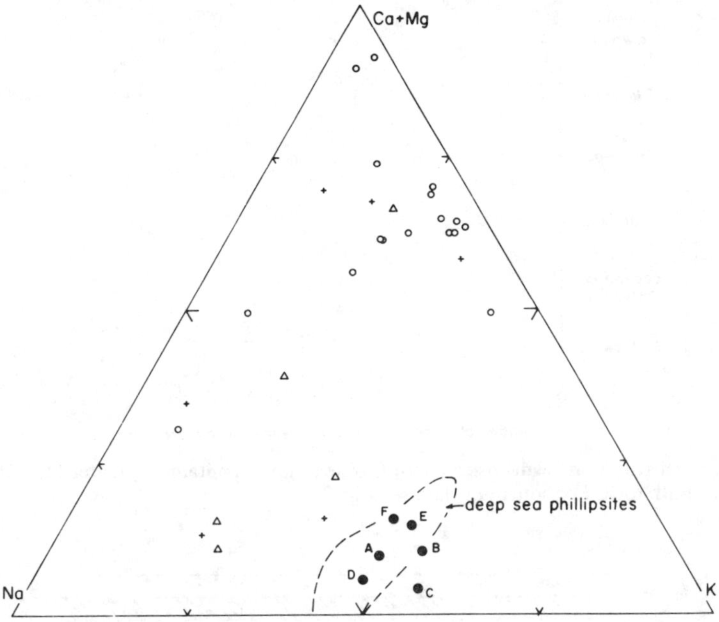

Fig. 3. Ca + Mg – Na – K atomic % plot of the compositions of the deep-sea clinoptilolites of Table 1, compared with the same clinoptilolites and phillipsites of Fig. 2. Note that about half the exchangeable cations in the deep-sea clinoptilolites is potassium, more than that of all the other samples.

depth of burial are shown in Figs. 4, 5, 6, and 7. We have not normalized these figures for the total number of recovered samples within each interval. Clinoptilolite occurs in all of the major ocean basins of the world, but these occurrences are most common in the older sediments. Figure 4 shows that there is almost a linear relationship between the frequency of clinoptilolite occurrences and geologic age. In addition to clinoptilolite being more common in older sediments, it is

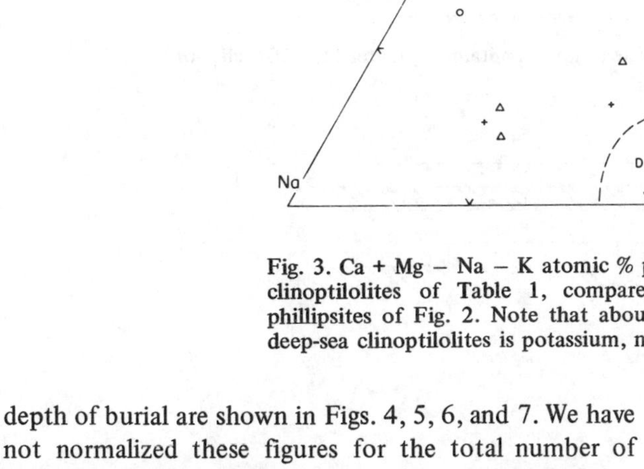

Fig. 4. Distribution of deep-sea clinoptilolite and phillipsite samples with respect to age of sediment. Compilation from bulk rock X-ray data of *Initial Reports of Deep Sea Drilling Project*, Legs 1 through 33. Total clinoptilolite occurrences = 584. Total phillipsite occurrences = 299. Eocene and Paleocene data have been combined owing to the paucity of data in Paleocene sediments.

frequently more abundant (Fig. 5). Clearly, there is trend for increasing quantities of clinoptilolite with increasing sediment age.

We had expected to see a strong correlation between the occurrence of clinoptilolite and the presence of volcanic ash in deep-sea sediments, based on this association in other sedimentary environments. Clinoptilolite is associated with volcanic ash beds at site 62, Leg 7; site 167, Leg 17; and site 253, Leg 26; however, at a number of other drill sites where volcanic ash beds are present clinoptilolite was not associated with the ash (e.g., Legs 11, 12, 14, 15, 24, 25, 28, and 32), but occurred at other horizons. Figure 6 shows the frequency with which clinoptilolite occurs in various sediment types. Surprisingly, over 70% of the occurrences are in pelagic clays, microfossil oozes, and chalks. Only 22% of the occurrences are in what DSDP investigators have described as "volcanogenic" sediments. Clinoptilolite replaces siliceous micro-fossil tests at drill site 119 of Leg 12, and at drill site 214 of Leg 22 clinoptilolite is in an ash-free lignite. These occurrences suggest that clinoptilolite may form in deep-sea sediments without a volcanic glass precursor. Many pelagic sediments contain at least traces of volcanic glass; and even in ash-free pelagic sediments, minor amounts of volcanic glass may have once been present but have subsequently been dissolved by reaction with sea water. Thus we cannot completely eliminate the possibility of a glass precursor for

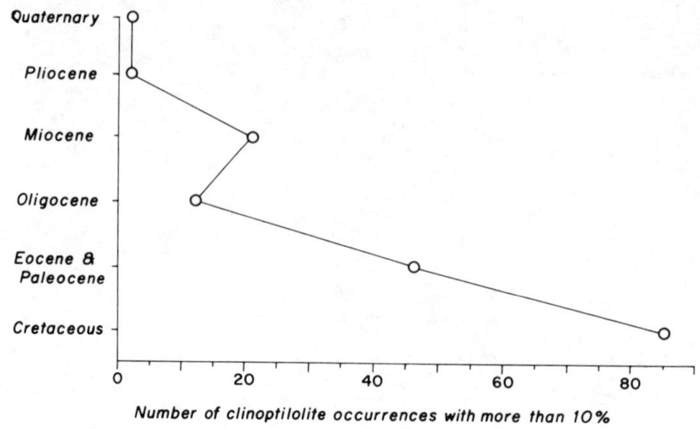

Fig. 5. Distribution of deep-sea clinoptilolite samples containing greater than 10% clinoptilolite in the bulk rock. For source of data see Fig. 3.

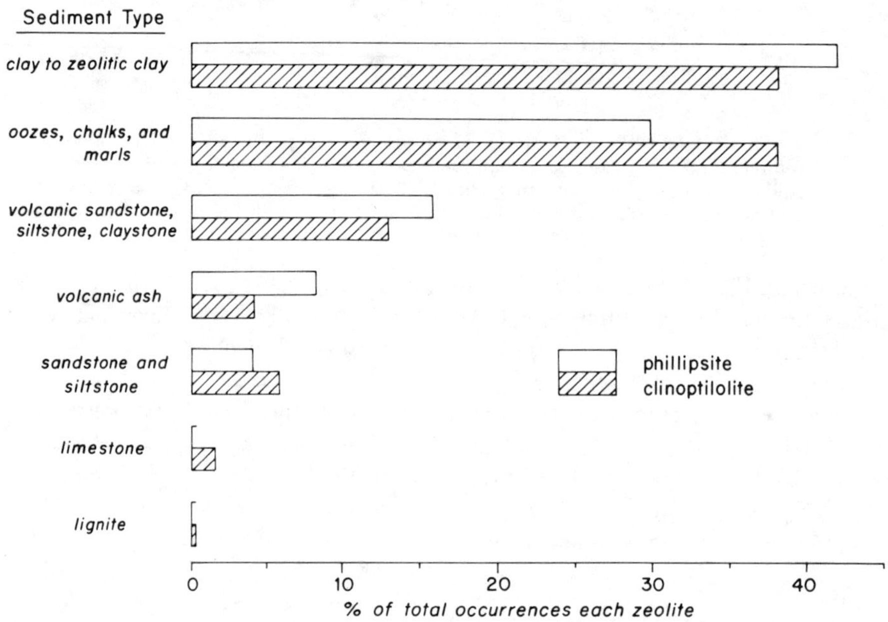

Fig. 6. Distribution of deep-sea clinoptilolite and phillipsites samples with respect to sediment type. For source of data see Fig. 3.

clinoptilolite in most occurrences. However, we contend that if clinoptilolite is forming only from glass, then the majority of the clinoptilolite should be associated with ash beds, but such is not the case.

RELATION OF INTERSTITIAL WATER TO CLINOPTILOLITE COMPOSITION

In rapidly deposited marine sediments from New Zealand, clinoptilolites have highly variable compositions which can be related to the composition of volcanic glass precursors (Boles and Coombs, 1975). In contrast, the compositions of deep-sea clinoptilolites studied here vary within narrow limits, particularly with respect to sodium, potassium, and calcium (Table 2 and Fig. 3). Even if these clinoptilolites have had ultimately a volcanic glass precursor, it seems unlikely that glasses over such a wide geographic area and from different horizons would have such uniform compositions. The fact that deep-sea clinoptilolites typically occur as euhedral crystals suggests crystallization directly from marine pore fluids.

We believe the uniform proportions of exchangeable cations in these clinoptilolites reflects the composition

TABLE 3. Composition of sea water and of interstitial fluids (in g/kg) associated with zeolitic sediments

		Sediments with clinoptilolite				Sediments with phillipsite	
	(1)	(2)	(3)	(4)	(5)	(6)	(7)
Na	10.6	10.7	10.6	9.9	9.8	10.60	10.79
K	0.38	0.32	0.24	0.27	0.22	0.47	0.47
Ca	0.40	1.19	1.34	0.71	1.57	0.46	0.32
Mg	1.27	0.80	0.82	1.16	1.00	1.27	1.32
Si	nd*	0.007	0.004	0.017	0.025	0.006	0.009
Cl	19.0	20.1	20.2	19.4	19.7	19.39	19.48
SO_4	2.65	1.90	1.64	2.30	2.24	2.02	2.31
HCO_3	0.14	nd	nd	0.07	0.05	0.09	–
pH	nd	7.8	6.7	7.0	6.7	7.5	7.5

*nd–not determined.

(1) "Average sea water" (Krauskopf, 1967).

(2) Interstitial water from grayish-olive-green silty sand, Paleocene, DSDP site 214, 364 m below sea bed (Manheim *et al.*, 1974).

(3) Interstitial water from grayish-black lignite, Paleocene(?), DSDP site 214, 400 m below sea bed (Manheim *et al.*, 1974).

(4) Interstitial water from zeolitic brown silty clay, Upper Cretaceous, DSDP site 137, 168 m below sea bed (Sayles and Manheim, 1975, table 6).

(5) Interstitial water from zeolitic clay, Upper Cretaceous, DSDP site 137, 225 m below sea bed (Sayles and Manheim, 1975, table 6).

(6) Interstitial water from brown zeolitic clay, undetermined age, DSDP site 51, 119 m below sea bed (Sayles and Manheim, 1975, table 2).

(7) Interstitial water from brown zeolitic clay, Miocene, DSDP site 50, 30 m below sea bed (Sayles and Manheim, 1975, table 2).

of the marine pore fluids in which they formed. Interstitial waters associated with clinoptilolite-bearing sediments have relatively uniform K/Na/Ca/Mg atomic ratios of about 1 to 63 to 4 to 6, respectively (Table 3). Deep-sea clinoptilolites analyzed here have average K/Na/Ca/Mg ratios of 1 to 0.80 to 0.08 to 0.11. This suggests that the clinoptilolite has preferentially fractionated potassium from the interstitial water relatively to sodium, calcium, and magnesium. Relative to potassium, sodium has been depleted by a factor of 80, calcium has been depleted by a factor of 50, and magnesium has been depleted by a factor of about 55. One wonders whether these depletion factors could be applied to other environments, for example, to estimate the proportions of Na, K, and Ca in saline-alkaline lake waters knowing the proportions of these cations in the lacustrine clinoptilolites.

Sample F of Table 2 has an anomalously low Si/Al ratio compared with other deep-sea clinoptilolites. In addition, the heulandite is found in a greyish-black lignite, an unusual lithology for deep-sea clinoptilolite occurrences (see Fig. 6). Siever (1962) noted that silica concentrations in natural peat waters are very low, even in the presence of quartz. Experimental runs for up to 2 years indicate that amorphous silica solubilities are reduced up to 10 times in natural peat waters; Siever attributes this to absorption of colloidal organic compounds on the silica surface. In accord with Siever's data, the silica content of the interstitial solution associated with sample F has a very low silica content (Table 3, analysis 3). We suggest that the abundance of organic material in this sample has maintained unusually low silica activities in the associated interstitial water. As a result, the clinoptilolite (variety heulandite) formed there has an unusually low silica content relative to other deep-sea samples.

ORIGIN OF DEEP-SEA CLINOPTILOLITE

We know of no conclusive *textural* evidence which demonstrates a precursor phase for deep-sea clinoptilolite (e.g., volcanic glass), aside from rare cases where clinoptilolite(?) has apparently replaced siliceous radiolarian tests (see Benson, 1972; Fan and Zemmels, 1972).

In contrast clinoptilolite commonly pseudomorphs glass shards in geosynclinal and lacustrine sediments (e.g., see Sheppard and Gude, 1973; Boles and Coombs, 1975). Nevertheless, the distribution of clinoptilolite relatively to phillipsite in deep-sea sediments suggests a reaction relation between these zeolites.

Figure 4 shows that phillipsite is most abundant in Miocene sediments and decreases in abundance in pre-Miocene sediments. If initial conditions and sediment types in the pre-Miocene were similar to those in younger sediments (i.e. glass compositions, ocean-water chemistry), as we believe they were, they why is phillipsite not more abundant in older sediments? We believe the phillipsite distribution pattern reflects its dissolution in pre-Miocene sediments. In accord with this suggestion isotope studies on relatively young phillipsite crystals (post-Miocene) have shown that the mineral continues to grow in the sedimentary column, in some cases for up to at least a million years (Bernat and Goldberg, 1969; Bernat and others, 1970). In contrast, Rex (1967) has shown that phillipsite formed from Eocene volcanic glass is highly corroded and suggests that the phillipsite was not in equilibrium with sea water.

Considering the fact that (1) clinoptilolite becomes progressively more abundant in older sediments whereas phillipsite becomes less common, (2) they both occur in similar sediment types, and (3) the two zeolites have similar compositions in deep-sea sediments, with respect to the nature of exchangeable cations, it is likely that phillipsite is reacting to form clinoptilolite.

The conversion of phillipsite to clinoptilolite involves mainly addition of silica and water, assuming conservation of alumina. Given an "average" deep-sea phillipsite with composition $Na_{2.40}K_{2.87}Ca_{0.45}Mg_{0.30}Al_{6.80}Si_{17.30}O_{48}16.6H_2O$ (Sheppard and Gude, 1970), with a density of 2.23 g/cm^3 (Deer and others, 1963) and an "average" deep-sea clinoptilolite with composition $Na_{2.40}K_{3.00}Ca_{0.30}Mg_{0.40}Al_{6.80}Si_{29.20}O_{72}20H_2O$ (Table 2, H_2O composition calculated from Boles, 1972, Fig. 2), with a density of 2.1 to 2.2 g/cm^3 (Deer and others, 1963), then 10 cm^3 of phillipsite would require about 3 cm^3 quartz and 1 cm^3 of H_2O to form 14 to 15 cm^3 of clinoptilolite.

Figure 7 shows that clinoptilolite occurs over a wide range of depths, that it is commonly found at less than 400 m, and that there is considerable overlap in the distribution of phillipsite and clinoptilolite with respect to depth of burial. It would appear from this that "depth of burial" or temperature is not a primary controlling factor on the proposed reaction, although the influence of temperature cannot be completely eliminated. Silica activity may be an important consideration as high silica activities would presumably favor clinoptilolite over phillipsite. However, Table 3 shows that silica activities associated with clinoptilolite and phillipsite bearing sediments are very similar.

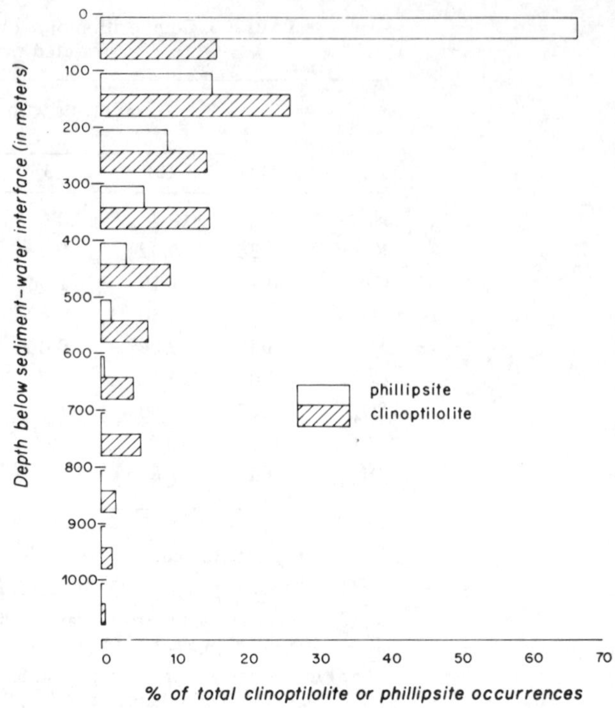

Fig. 7. Distribution of deep-sea clinoptilolite and phillipsite occurrences with respect to depth of burial. For source of data see Fig. 3.

The above considerations lead us to the conclusion that phillipsite forms metastably probably as a silica-deficient phase with respect to marine pore fluids. Phillipsite can be readily crystallized from volcanic glass in the laboratory (e.g., see Mariner and Surdam, 1970). It has been observed in a number of Pleistocene or younger non-marine saline–alkaline deposits (Hay, 1966, Table 4), but is rare in such deposits of Eocene and older ages. Sheppard and Gude (1969, 1973) have suggested that phillipsite forms before clinoptilolite in saline–alkaline lake deposits. These data suggest that phillipsite is a readily synthesized, early formed phase which does not persist. Furthermore, the distribution of phillipsite in saline-alkaline non-marine deposits is analogous to that in deep-sea marine sediments.

Although field and experimental evidence indicates that phillipsite is a common alteration product of volcanic glass and such a relationship has been shown in deep-sea deposits (e.g., see Rex, 1967), we are surprised by the fact that the most clinoptilolite–phillipsite occurrences are not associated with highly volcanogenic sediments (see Fig. 6). This may suggest that volcanic glass is not a prerequisite for phillipsite and clinoptilolite. Dissolution of siliceous microfossil may provide a ready source of silica in the absence of glass. Hurd

(1973) has shown that acid-cleaned opal tests can contain almost 0.77 wt.% Al, which would presumably be released and utilized for zeolite formation upon dissolution of the tests. Clay minerals (e.g., montmorillonite) may also be involved in such a reaction, possibly providing Al and other cations for zeolitization, but we have no direct evidence for this.

SUMMARY

The compilation of the occurrences of clinoptilolite in deep-sea sediments shows that it is most abundant in the older sediments (Figs. 4 and 5). Clinoptilolite abundance increases as phillipsite decreases with age (Fig. 4), but not with depth (Fig. 7). There is not a consistent relationship between the occurrence of clinoptilolite and ash layers in the sediments. We have also shown that a conversion of phillipsite to clinoptilolite requires only the addition of silica and water. From these observations we conclude that phillipsite forms as a silica-deficient, metastable phase early in diagenesis. With time, however, it dissolves in the pore fluids, and the more stable phase, clinoptilolite, crystallizes where sufficient silica is available.

REFERENCES

Alietti, A. (1972) Polymorphism, and crystal-chemistry of heulandites and clinoptilolites: *Amer. Mineral.* **57**, 1448-1462.

Benson, R. N. (1972) Radiolaria, Leg XII, Deep Sea Drilling Project: *Initial Repts. of Deep Sea Drilling Project*, Vol. **12**, Washington (U.S. Government Printing Office), pp. 1085-1113.

Bernat, M. and Goldberg, E. D. (1969) Thorium isotopes in the marine environment: *Earth Plant. Sci. Lett.* **5**, 308-312.

Bernat, M., Bieri, R. H., Koide, M., Griffin, J. J. and Goldberg, E. D. (1970) Uranium, thorium, potassium, and argon in marine phillipsites: *Geochim. et Cosmochim. Acta* **34**, 1053-1071.

Boles, J. R. (1972) Composition, optical properties, cell dimensions, and thermal stability of some heulandite group zeolites: *Amer. Mineral.* **57**, 1463-1493.

Boles, J. R. and Coombs, D. S. (1975) Mineral reactions in zeolitic Triassic tuff, Hokonui Hills, New Zealand: *Geol. Soc. Amer. Bull.* **86**, 163-173.

Deer, W. A., Howie, R. A. and Zussman, J. (1963) *Rock-forming Minerals*, Vol. **4**, *Framework Silicates*, Longmans, Green & Co. Publ., London, 435 pp.

Fan, P. and Zemmels, I. (1972) X-ray mineralogy studies – Leg 12: *Initial Repts. of Deep Sea Drilling Project*, Vol. **12**, Washington (U.S. Government Printing Office), pp. 1127-1154.

Fan, P., Rex, R. W., Cook, H. E. and Zemmels, I. (1973) X-ray mineralogy of the Caribbean Sea – Leg 15: *Initial Repts. of Deep Sea Drilling Project*, Vol. **15**, Washington (U.S. Government Printing Office), pp. 847-921.

Hathaway, J. C. and Sachs, P. L. (1965) Sepiolite and clinoptilolite from the mid-Atlantic ridge: *Amer. Mineral.* **50**, 852-867.

Hay, R. L. (1966) Zeolites and zeolitic reactions in sedimentary rocks: *Geol. Soc. Amer. Spec. Paper* **85**, 130 pp.

Hurd, D. C. (1973) Interactions of biogenic opal, sediment and seawater in the Central Equatorial Pacific: *Geochim. et Cosmochim. Acta* **37**, 2257-2282.

Krauskopf, K. (1967) *Introduction to Geochemistry*, McGraw-Hill Publ. Co., N.Y., 721 pp.

Manheim, F. T., Waterman, L. S. and Sayles, F. L. (1974) Interstitial water studies on small cores, Leg 22: *Initial Repts. of Deep Sea Drilling Project*, Vol. **22**, Washington (U.S. Government Printing Office), pp. 657-662.

Mariner, R. H. and Surdam, R. C. (1970) Alkalinity and formation of zeolites in saline–alkaline lakes: *Science* **170**, 977-979.

Mumpton, F. A. (1960) Clinoptilolite redefined: *Amer. Mineral.* **45**, 351-369.

Murata, K. J. and Erd, R. C. (1964) Composition of sediments in cores from the experimental Mohole Project (Guadelupe site): *J. Sediment. Petrol.* **34**, 633-655.

Murray, J. and Renard, A. F. (1891) *Deep Sea Deposits: Reports Challenger Expedition*, London.

Okada, H. and Tomita, K. (1973) Clay mineralogy of deep-sea sediments in the northwestern Pacific, DSDP, Leg 20: *Initial Repts. of Deep Sea Drilling Project*, Vol. **20**, Washington (U.S. Government Printing Office), pp. 335-357.

Rad, U. V. and Rosch, H. (1972) Mineralogy and origin of clay minerals, silica, and authigenic silicates in Leg 14 sediments: *Initial Repts. Deep Sea Drilling Project*, Vol. **14**, Washington (U.S. Government Printing Office), pp. 727-739.

Rex, R. W. (1967) Authigenic silicates formed from basaltic glass by more than 60 million years contact with sea water, Sylvania Guyot, Marshall Islands: *Clays and Clay Minerals, Proc. Nat. Conf.* (1966) **15**, 195-203.

Sayles, F. L. and Manheim, F. T. (1975) Interstitial solutions and diagenesis in deeply buried marine sediments: results from the Deep Sea Drilling Project: *Geochim. Cosmochim. Acta* **39**, 103-127.

Sheppard, R. A. and Gude, A. J., 3rd (1968) Distribution and genesis of authigenic silicate minerals in tuffs of Pleistocene Lake Tecopa, Inyo County, California: *U.S. Geol. Survey Prof. Paper* **597**, 38 pp.

Sheppard, R. A. and Gude, A. J., 3rd (1969) Diagenesis of tuffs in the Barstow Formation, Mud Hills, San Bernardino County, California: *U.S. Geol. Survey Prof. Paper* **634**, 35 pp.

Sheppard, R. A. and Gude, A. J., 3rd (1970) Chemical composition and physical properties of phillipsite from the Pacific and Indian Oceans: *Amer. Mineral.* **55**, 2053-2062.

Sheppard, R. A. and Gude, A. J., 3rd (1973) Zeolites and associated authigenic silicate minerals in tuffaceous rocks of the Big Sandy Formation, Mohave County, Arizona: *U.S. Geol. Surv. Prof. Paper* **830**, 36 pp.

Siever, R. (1962) Silica solubility, 0–200°C, and the diagenesis of siliceous sediments: *J. Geol.* **70**, 127-150.

Stonecipher, S. A. (1976) Origin, distribution and diagenesis of phillipsite and clinoptilolite in deep-sea sediments: *Chem. Geol.* **17**, 307-318.

Wise, W. S. and Tschnerich, R. W. (1976) Chemical composition of ferrierite: *Amer. Mineral.* **61**, 60-67.

GENERATION OF PHILLIPSITES BY PALAGONITIZATION OF BASALTIC GLASS IN SEA WATER AND THE ORIGIN OF K-RICH DEEP-SEA SEDIMENTS

JOSE HONNOREZ

Rosensteil School of Marine and Atmospheric Science, University of Miami, Miami, Florida 33149, U.S.A.

Abstract – Na > K-phillipsite with low Ca content and smectite* have been found to be the major secondary minerals resulting from the partial palagonitization* process which affected basaltic glass of Pliocene age from Palagonia, Sicily (Honnorez, 1967 and 1972). During this halmyrolysis* Ti and Fe were possibly accumulated in the residual glass ("palagonite"*), whereas Mg, Ca, K, Si, Al, and Na were leached in increasing order out of the glassy granules and precipitated as intergranular authigenic phillipsite and smectite. The chemical balance shows that Si, Ti, Al, and Fe concentrations in the total rock are unchanged, whereas half the initial amount of Ca is lost, and K, Na, and Mg are enriched by factors of 7, 2.4, and 2.2, respectively. It is suggested that sea water contributed to the enrichment of the authigenic cement with respect to Mg, Na, and K.

On the other hand, K > Na-phillipsite with no or very little Ca and smectite are the most ubiquitous authigenic silicates in deep-sea sediments from low sedimentation areas of the Pacific and Indian Oceans (Arrhenius, 1963; Bonatti, 1963, 1970; Sheppard *et al.*, 1970). These sediments are commonly abnormally rich with respect to K, Na, and Mn. Na > K and low Ca-phillipsite are therefore characteristic of the initial stage of the halmyrolysis of basaltic glass; these end-products accumulate K, Na, and Mg from sea water. The more mobile Na would thereafter be progressively displaced by K in the authigenic zeolite, and the end product of the final stage of the palagonitization (when no residual glass is left) is a K > Na and Ca-poor phillipsite. At this point the enrichment of the sediment in K would exceed by far that in Na and Mn.

I propose that the authigenic minerals (phillipsite and smectite) resulting from the halmyrolysis of the submarine basaltic glass behave as traps for sea water K, and to a lesser degree, Na. On the contrary, Ca is systematically released in sea water. This process probably does not regulate the alkali concentration in sea water but it affects the chemical composition of volcanogenic sediments and the pH of their interstitial waters.

INTRODUCTION

The first marine phillipsite was discovered by Murray and Renard (1891), who identified the zeolite in pelagic sediments dredged by H.M.S. *Challenger* from the floor of the Pacific Ocean. Phillipsite has since been found to be among the two most common authigenic zeolites in pelagic sediments from vast areas of the South Pacific and the Indian Oceans where it is almost always accompanied by smectites (Goldberg, 1961, 1962; Arrhenius, 1963; Bonatti, 1963, 1970; Arrhenius and Bonatti, 1965; Peterson and Griffin, 1964; Sheppard *et al.*, 1970). Phillipsite and a minor amount of smectite are also observed as the core of manganese nodules from the North Pacific (Nayudu, 1964) and the Indian Ocean (L. Dimitriev, pers. comm.).

On the other hand, both phillipsite and smectite are the most widespread authigenic minerals of the palagonitization process during which basaltic glass reacts at low temperature with sea or fresh water (Allen and Scheid, 1946; Bonatti, 1963, 1965, 1970; Nayudu, 1964; Bonatti and Nayudu, 1965; Honnorez, 1967 and 1972; Jakobsson, 1968; Hay and Ijima 1968 a and b; Ijima and Harada, 1969; Melson and Thompson, 1972). Hawkins and Roy (1963) concluded after their experimental study of the alteration of basaltic glasses by various aqueous solutions that "... the stable assemblage (resulting from the alteration of basaltic and andesitic glasses) at temperatures below 250°C is phillipsite–montmorillonite".

This paper discusses the origin of authigenic phillipsites and smectites as a result of the halmyrolysis of basaltic glasses generated by submarine eruptions.

MINERALOGICAL AND CHEMICAL VARIATIONS DURING PALAGONITIZATION

It is essential to distinguish between the initial, mature, and final stages of palagonitization if one intends

*The definition of the terms used in this paper are presented in an appendix at the end of the text.

TABLE 1. Chemical compositions recalculated on a water-free basis

	1	2	3	4	5
SiO_2	52.44	46.50	59.82	50.90	52.24
TiO_2	1.59	3.09	n.d.	1.71	1.56
Al_2O_3	16.17	12.90	24.36	15.41	13.13
Fe_2O_3	1.83	14.68	n.d.	7.22	9.82
FeO	8.03	3.20	n.d.	3.71	1.63
MnO	0.13	0.19	n.d.	0.12	0.16
MgO	7.21	8.13	n.d.	8.29	9.09
CaO	8.75	8.84	4.80	7.59	8.70
Na_2O	3.34	1.90	8.11	3.60	3.06
K_2O	0.28	0.28	3.25	1.07	0.33
P_2O_5	0.22	0.30	n.d.	.39	0.29
Total	100	100	100	100	100

1 = fresh glass in 4.
2 = altered residual glass in 4.
3 = zeolitic cement in 4.
4 = whole hyaloclastite from the inner part of the tunnel (not weathered).
5 = whole hyaloclastite from the entrance of the tunnel (weathered).

to follow the successive steps of this process and to compare the corresponding chemical and mineralogical changes. The three stages of the alteration of basaltic glass in sea water can be differentiated on the basis of the presence or absence of some types of glass and/or authigenic minerals as follows:

(1) The abundance of unaltered basaltic glass in a submarine volcaniclastic rock characterizes the initial stage of palagonitization; moreover, the altered residual glass ("palagonite") is not replaced by authigenic minerals which exclusively fill the spaces between the glass granules.

(2) Later on, all of the fresh glass is altered, and the residual glass begins to be replaced *in situ* by zeolites and smectites; the association of altered glass with both inter- and intragranular authigenic minerals is typical of rocks which have reached the mature stage of palagonitization.

(3) Finally, the material representing the ultimate stage of the alteration of the basaltic glass consists of an aggregate of authigenic minerals where inter- and intragranular phases cannot be distinguished because the contours of the parent glassy granules have been obliterated. At this stage, the residual glass has been replaced almost completely by secondary minerals.

The three stages of palagonitization can be divided by the following two major reactions:

(a) Fresh Glass + Sea Water →
"Palagonite" + Intergranular Authigenic Minerals.

(b) "Palagonite" + Sea Water →
Intragranular Authigenic Minerals.

The restricted case of the hydrothermal alteration of basaltic glass by subaerial acid fumeroles (Peacock and Fuller, 1928) will not be considered in the present article.

The initial stage of palagonitization

One recognizes that a given basaltic glassy rock remained at the initial stage of the palagonitization when unaltered glass is still present besides the residual altered glass ("palagonite") and the intergranular authigenic cement. Moreover, the "palagonite" is not yet replaced by authigenic minerals. In other words, during the initial stage of palagonitization, zeolites and smectites exclusively occupy pre-existing spaces between the glassy granules which now consist of cores of fresh glass surrounded by an outer zone of palagonitized glass.

Hyaloclastites of the Contrada Acqua Amara, near Palagonia (S.E. Sicily), offer an example of such an initial stage of palagonitization. The formation belongs to the submarine basaltic volcanic belt of Late Pliocene age which outcrops in the northern part of the Hyblean Mounts (Monti Iblei). The hyaloclastites are overlaid and protected by several tens of meters of closely packed pillow lavas which, in turn, are covered by the calcarenites of the Calabro-Sicilian transgression. I demonstrated (Honnorez, 1967 and 1972) that the halmyrolysis of the basaltic glass of the hyaloclastites was interrupted at an early stage by the emersion of the northern portion of the Hyblean Mounts during the Calabro-Sicilian and that, thereafter, the partly palagonitized hyaloclastites have been protected from subsequent weathering.

The material studied (Honnorez, 1967 and 1972) was sampled immediately after a 7.3 m-long tunnel was dug without success for collecting spring waters. The hyaloclastite of the first 80 cm from the entrance of the tunnel displays the typical yellowish-brown color. The weathered hyaloclastite is characterized by a higher oxidation coefficient* (0.847 vs. 0.637), a higher water content (H_2O+ = 10.03 vs. 8.51; H_2O- = 17.6 vs. 10.4) and a larger porosity (9.4% vs. 6.4%) than the hyaloclastite from deeper in the tunnel. Moreover, K_2O content of the weathered rock is much lower (0.33 vs. 1.08% on the water-free basis) and therefore closer to that of the initial fresh glass (see Table 1). This chemical difference reflects the fact that the intergranular cement of the weathered rock is mainly made up of chabazite, a more K-poor and Ca-rich zeolite than phillipsite.

The unweathered hyaloclastites from the inner

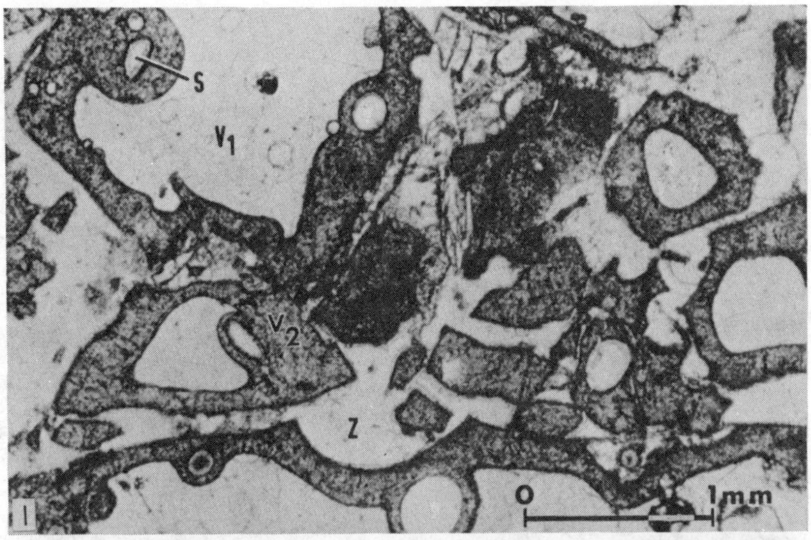

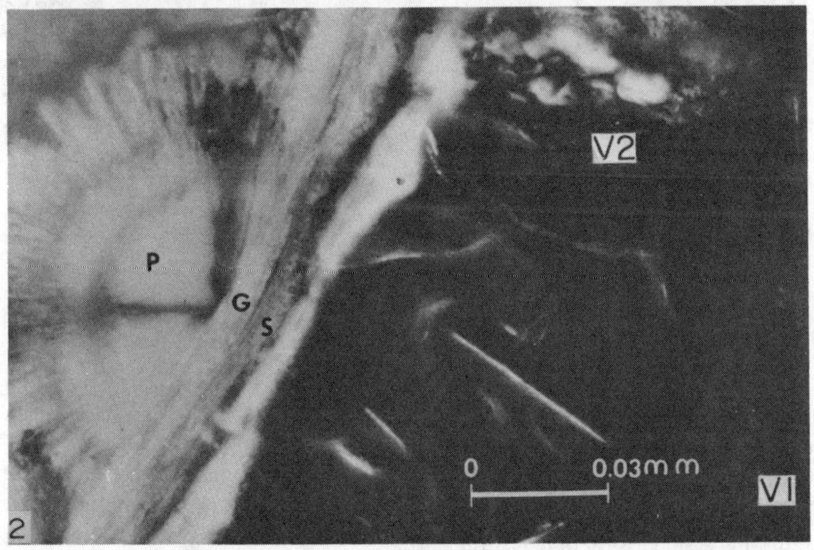

Fig. 1. Microphotograph (thin section) of a hyaloclastite from Palagonia (SE Sicily) illustrating the initial stage of palagonization. The glassy granules are made up of cores of unaltered basaltic glass (V1) surrounded by a rim of palagonitized glass (V2); they are cemented by intergranular zeolites (Z) among which phillipsite prevails. Thin smectite layers (S) coat the granule and vesicle surfaces.

Fig. 2. Microphotograph (thin section, crossed nicols) of a granule of a hyaloclastite from Palagonia: initial fresh isotropic basaltic glass in the lower right corner (V1), palagonitized glass (V2) in center displaying anistropy along microfissures and along the granule border, first smectite layer (S) overlain by a second gyrolite layer (G), and fibroradial phillipsite (P) of the intergranular cement.

portion of the tunnel (initially holohyaline pyroclastic rocks) are made up of 18% fresh glass, 45% palagonitized glass ("palagonite"), 27% intergranular zeolitic cement, and 8% saponite; 2% olivine microphenocrysts and rare calcite and opal are present in the glasses and the cement, respectively. The authigenic mineral sequence is the following: saponite, gyrolite, phillipsite, chabazite–gmelinite (when present), analcime and apophyllite (often absent), and the opal-calcite mixture (rare).

These hyaloclastites appear in thin sections to be made up of various kinds of glassy granules the diameter of which ranges from 0.08 to 1.2 mm. In addition to few microlitic lava fragments, the overwhelmingly most abundant granules are (Fig. 1):

either the largest and least vesicular ones; they are formed by a 0.3–8-mm diameter core of fresh glass surrounded by a zone of altered glass a few tenths of a millimeter thick;

or the smallest and most vesicular granules; these are

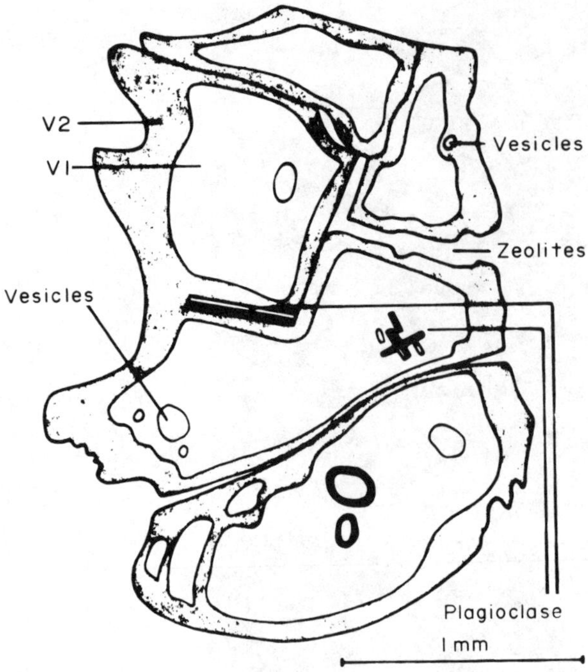

Fig. 3. Glassy granule of a hyaloclastite from Palagonia (V1 = fresh glass; V2 = palagonitized glass) showing that the halmyrolysis mainly took place along pre-existing fissures. The plagioclase microlites are unaltered.

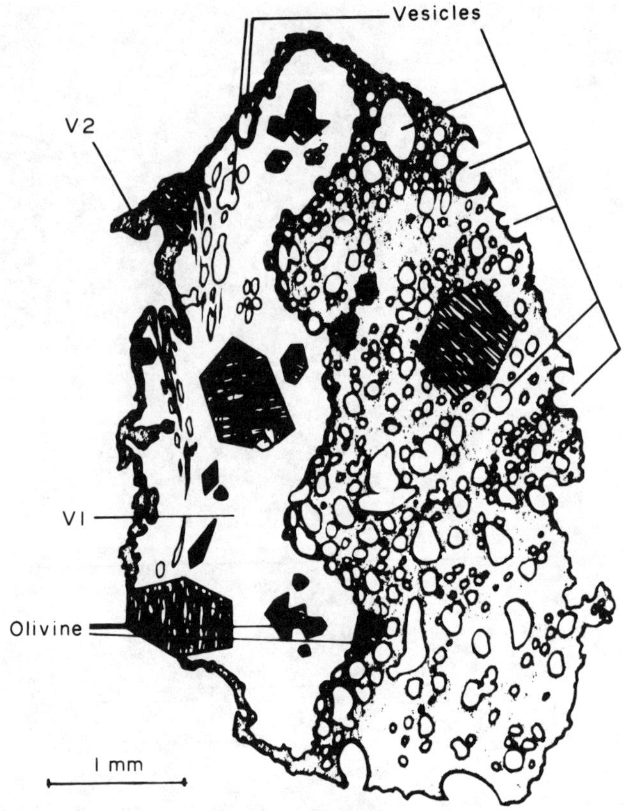

Fig. 4. Glassy granule of a hyaloclastite from Palagonia (V1 = fresh glass; V2 = palagonitized glass) illustrating the relationship between halmyrolysis intensity and/or rate and the vesicularity, and therefore the glass surface exposed to sea water reaction. The olivine microphenocrysts are unaltered.

almost always exclusively made up of altered glass.

The granules are coated by several thin (0.03–0.05 mm thick) layers of smectite usually followed by gyrolite (inversion and recurrence are observed), and they are cemented by an aggregate of authigenic zeolites among which phillipsite prevails (Fig. 2). Semi-quantitative emission spectrography shows that this smectite is essentially a silicate of magnesium with traces of Al and Ca; K and Ba are apparently not present in detectable amounts. I also checked by the same methods that the chemical composition of the gyrolite corresponds to that of a hydrated Ca silicate, but I also observed traces of Al. Finally, the presence of K and Ca was checked as well as that of Si, Al, and in the phillipsite. The vesicles are always lined and filled with the same authigenic minerals as those forming the intergranular phase, and they occur in the same sequence. The amounts of gyrolite present and those of other zeolites than phillipsite are negligible; therefore, these secondary minerals will no longer be referred to in the following pages.

The glass surrounding the vesicles is generally altered. The alteration increases as the granules become more vesicular (Fig. 4). I also observed that the palagonitization developed from the granule outlines toward the center, and that it progressed preferentially along cracks and fissures (Fig. 3). It was also noted that the few plagioclase, olivine, and augite phenocrysts present in the altered glass remained unaffected by the palagonitization. I concluded from these microscopic observations that: (1) the palagonitizing solution is exogenous; (2) the intensity and/or the rate of alteration was directly related to the "reactional surface area", i.e. the fundamental factor during the halmyrolysis of basaltic glass was the surface area in contact with sea water. Volcanic glasses forming hyaloclastites and hyalotuffs are the most strongly palagonitized because they have the largest surface exposed to the palagonitizing solutions. On the other hand, the glassy crusts of pillow lavas and lava flows of similar age and composition are not as strongly and as rapidly altered as the glass of the volcaniclastic rocks. In its turn, the alteration of the glassy rim of pillow lavas and lava flows proceeds at a different rate and degree than that of the inner, non-glassy parts of these lava units. Thompson (1973) showed that, for some elements, the effects of submarine weathering of basaltic glass were different from those of the halmyrolysis of non-glassy tholeiite. The Fe_2O_3, K_2O, H_2O total, and MgO variation trends (but not their intensity) are similar in both cases: the first three components have increased and the last one is always depleted in both the glassy margin and the crystallized core of the pillow lavas. On the other hand, CaO and FeO total are often leached from the glass rim while Na_2O and Al_2O_3 tend to be enriched in it, whereas this is not always the case for the non-glassy interior of the same pillows. The behavior of TiO_2, MnO, and SiO_2 are completely irregular according to Thompson (1973).

The chemical composition of the fresh initial glass, the altered residual glass, the zeolite cement (about 95% phillipsite), the bulk unweathered hyaloclastite and the bulk weathered hyaloclastite are presented in Table 1. The first three analyses were made on fractions isolated by heavy liquid technique and electromagnetic separation. They represent, therefore, average compositions of the three phases and not spot analyses which are obtained with microprobe electron technique. This is an important factor when one analyzes materials which are poorly or not crystallized, or which present local compositional variation.

The initial stage of the palagonitization process at Palagonia (Honnorez, 1967 and 1972) essentially consisted of a hydration and oxidation of the basaltic glass during which Fe and Ti were passively accumulated in the residual glass ("palagonite") whereas Mg, Ca, K, Si, Al, and Na (in increasing order) were leached from the glass granules and remobilized as authigenic minerals forming the intergranular cement (Table 1).

Jakobsson (1972) inferred similar chemical changes from a comparison of the chemical compositions of one fresh glass and the associated "palagonite" formed in 5 to 6 years at Surtsey. According to the preliminary microprobe analyses of Jakobsson (1972) all the elements, except Fe and Ti, are released during palagonitization so that the residual altered glass is depleted in Mg, Si, Al, Ca, K, and Na in increasing order and enriched in Fe, Ti, and H_2O. What Jakobsson studied was the very beginning of palagonitization of basaltic ashes generated by magmato-phreatic eruptions and deposited on land from November 1963 to April 1964. In September 1969 the glass which had fallen unaltered on the ground was already palagonitized; however, one does not know if the alteration had begun one year earlier since no sampling was made in 1968. The alteration took place in subaerial conditions; and the palagonitizing solutions were water vapors at not more than 100°C, probably derived from heated sea water, since thenardite (Na_2SO_4) was present in the ashes. Opal has been found since 1969, and the first zeolite (probably chabazite, but it was not identified by X-ray diffraction) was discovered in 1971. The palagonitization at Surtsey cannot properly be called a halmyrolysis because it does not take place at low temperature and on the sea floor but, with respect to the chemical changes, it is very similar to an alteration by cold sea water. The major difference is probably a kinetic one.

Hay and Iijima (1968 a and b) observed similar chemical trends when comparing the chemical compositions of fresh and altered glasses from hyaloclastites of Koko craters, Oahu (Hawaii): Ti was "enriched" in the palagonitized glass, while on the contrary about 25% of the SiO_2, 50% of the Al_2O_3 and MgO, and 75% or more of the CaO, Na_2O, and K_2O were leached from it; Fe behavior was "contradictory". Both fresh and sea waters were involved in the palagonitization of the basic glasses of Oahu, but it appears (Hay and Ijima, 1968 a and b) that the alteration took place mainly in subaerial conditions since the Koko craters resulted from littoral phreatic explosions and never were immersed.

At Palagonia, the chemical balance of palagonitization is measured from a graph (see Fig. 5) where we assume that the iron did not move during the initial stage of alteration ($FeO^* = FeO + MnO + Fe_3O_3$). One could equally well have chosen TiO_2 for that

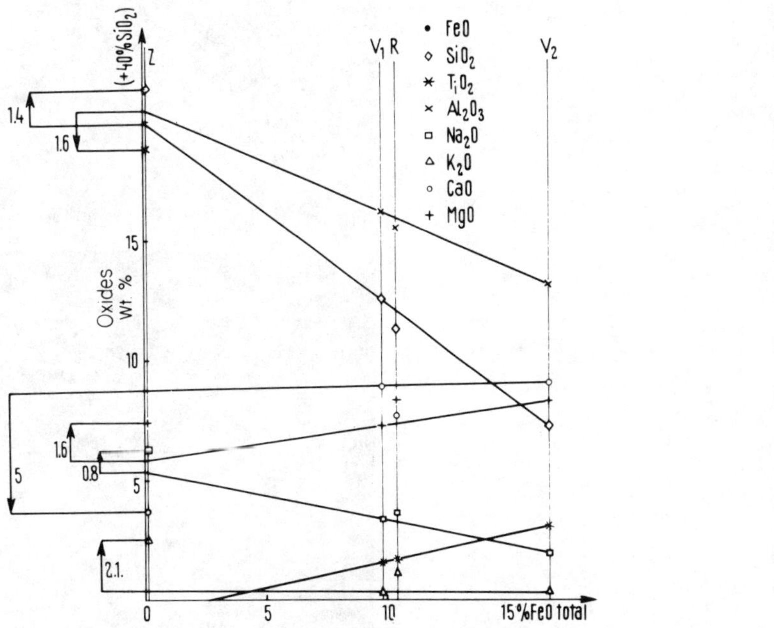

Fig. 5. The chemical budget of the initial stage of the palagonitization of the hyaloclastite from Palagonia. The various oxide weight percentages of the three major components of the hyaloclastite (V1 = initial unaltered glass; V2 = residual palagonitized glass; Z = zeolitic, mainly phillipsite, cement) are plotted on parallels to the ordinate corresponding to the respective total FeO* weight percentages of these three components (abscissa). R represents the chemical composition of the whole rock. One can see that the whole rock composition is very similar to that of the initial glass. The arrows pointing down and up show the gains and losses, respectively, during palagonitization when the final products are compared to the initial glass (from Honnorez, 1967 and 1972).

purpose. The bulk chemical balance (see Fig. 5) shows that the Si, Al, Fe, Ti, and P concentrations in the whole rock (i.e. fresh glass + altered glass + authigenic minerals) remained constant whereas half the Ca content was lost, and K, Na, and Mg were enriched by factors of 7, 2.2, and 2.4, respectively. It was concluded that the latter three elements were extracted from sea water which initially flowed through the spaces between the glassy granules. Therefore the authigenic minerals of the initial stage of the palagonitization precipitated from an intergranular solution where both ions from sea water and those released by the basaltic glass were present.

At Palagonia, the major authigenic minerals resulting from the initial phase of palagonitization are a more sodic than potassic phillipsite, which is relatively poor in Ca, and a magnesium-rich member of the smectite group, probably saponite. Iijima and Harada (1969) found that phillipsite and chabazite (i.e. the most K-rich zeolites) are the first and second authigenic minerals in sequential order and in abundance in the various palagonitized tuffs of Oahu Island (Hawaii). The chemical composition of both zeolites is a function of the original composition of the host rocks; for instance, phillipsite and chabazite in the melilite nephelinite tuff have higher Ca:Na ratios and lower Si:Al ratios than those in the more tholeiitic tuffs. Moreover, Iijima and Harada (1969) noticed that the Si:Al and Ca/(Na+K) ratios within a single crystal of phillipsite or chabazite roughly increases outward, that

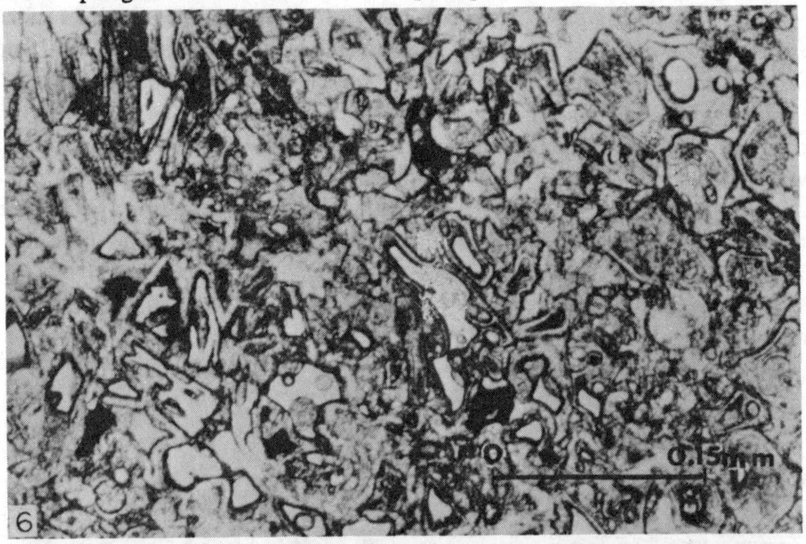

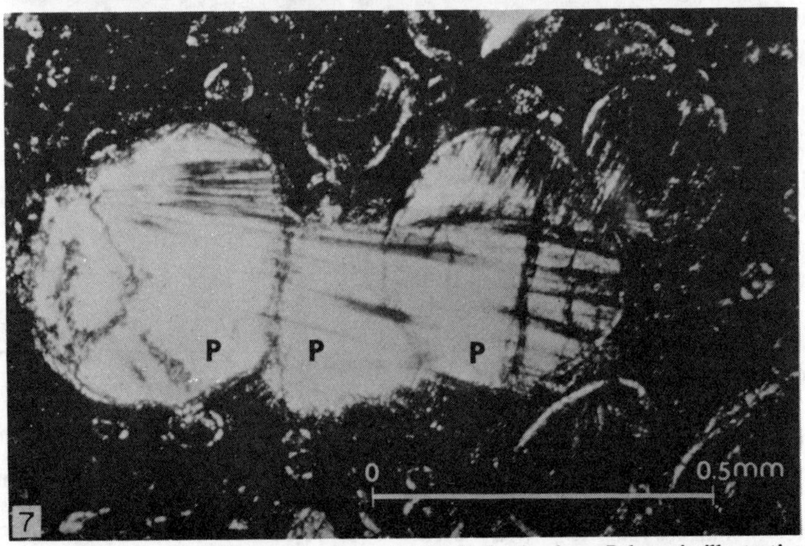

Fig. 6. Microphotograph (thin section) of a hyaloclastite from Palagonia illustrating the mature stages of palagonitization. The granules are exclusively made up of palagonitized glass (V2) and they are cemented by intergranular zeolites (Z).

Fig. 7. Microphotograph (thin section, crossed nicols) of a hyaloclastite from Palagonia at the mature stage of palagonitization. A bunch of phillipsite fibers filled two vesicles and replaced the glassy screen between them. Two alignments of tiny smectite and/or Fe–Mn oxide inclusions can be seen in the zeolite where the vesicle walls used to be.

is, in the growth direction. Melson and Thompson (1972) showed that the most abundant authigenic minerals generated during the incipient stage of the halmyrolysis of hyaloclastites and pillow-lava crusts dredged from the foot of the St. Peter and Paul Rocks are potassium-rich smectites (with more or less Mg and Fe), accompanied by analcime and minor phillipsite of undetermined composition.

In summary, the initial stage of palagonitization of basaltic glass in sea water is characterized in both cases by the association of a Na–K zeolite (phillipsite, or analcime and phillipsite) which is relatively Ca-poor, and a smectite (Mg-rich or K-rich).

The mature stage of palagonitization

The few remaining fresh-glass cores of the hyaloclastite granules are palagonitized at the same time as their altered residual glass margins begin to be replaced *in situ* by a mixture of authigenic minerals, essentially zeolites and possibly minor smectite. The hyaloclastites representing this stage typically consist of glassy granules entirely formed of altered residual glass, exclusive of fresh glass, which are cemented by intergranular zeolites (Fig. 6).

One also observes with the microscope that the altered glass in the areas of the granules farthest away from the center is partly replaced *in situ* by a whitish mosaic of fine zeolite crystals. Smectite was not observed within the borderlines of the zeolitized granules, but if the clay had been present among the zeolites as disseminated submicroscopic particles they could not have been detected. The intra-granular zeolite locally grades into larger fibroradial aggregates of phillipsite needles (see Fig. 8 a–c). Generally the plagioclase and augite microlites, and most often the olivine microphenocrysts, remain untouched by the zeolitization process. The pseudomorphosis of the glassy granules often respects the delicate texture of the thin multiple smectite and gyrolite layers covering the granule and vesicule surface. These fine coatings are the only boundaries separating the intergranular authigenic cement from the intragranular zeolites.

Figure 7 shows two close vesicles filled by one and the same bundle of phillipsite fibers. The zeolite probably started to crystallize from the left vesicle and expanded to the right one, replacing the glass screen between them. One notices across the phillipsite needles two curved alignments of fine dark inclusions corresponding to the old vesicle walls; the multiple smectite–gyrolite linings on the vesicle surface cannot now be seen.

To summarize, the mature stage of palagonitization is characterized by the completion of the alteration of the fresh glass into the residual "palagonite", and the beginning of the replacement *in situ* of the latter by phillipsite and minor smectite.

Final stage of palagonitization

When the palagonitization proceeds further, the authigenic minerals completely replace the altered glass so that the entire granule appears to be made up of zeolite (see Fig. 8d). The final stage of palagonitization is reached when the residual "palagonite" has been completely altered and replaced by authigenic minerals. During this stage the multiple smectite–gyrolite coatings underlining the granule contour have been almost completely obliterated. Hyaloclastites which have attained the final stage of palagonitization are now represented by an assemblage of phillipsite and smectite and Fe–Mn oxide aggregates, and only a few exceptional altered glass relics are still to be found.

Phillipsite is found over vast areas of Pacific sea-floor (Arrhenius, 1963; Bonatti, 1963, 1967; Peterson and Griffin, 1964) where it occurs with smectites (a nontronite according to Bonatti, 1963 and 1970) and iron-manganese oxides in pelagic sediments from regions which are characterized by widespread submarine basaltic volcanism and low sedimentation rates (less than 1 mm/year according to Glaccum and Bostrom, 1976). Sediments containing up to 50% of phillipsite have been recovered from the Pacific and Indian Oceans (Bonatti, 1963; Goldberg, 1962; Sheppard *et al.*, 1970). The phillipsite crystals are euhedral to subeuhedral and they range from 8 to 250 μm in length with an average size of about 10 to 20 μm (Fig. 9) (Sheppard *et al.*, 1970). Bonatti (1963) mentions "... that the phillipsite crystals are invariably accompanied by a great number of small grains of ... a material called palagonite" and that "... small crystallites of zeolites (are) growing in the palagonite material". Such frequent association of the phillipsite crystals with the "palagonite" strongly suggests that the zeolite is derived from the extreme alteration of small (from few to a couple of mm diameter) basaltic glass shards on the sea-floor. In fact, Bonatti (1963, 1965) describes all the steps of the substitution of palagonitized glass by zeolite and smectite, including the pseudomorphosis of glassy granules by a fibro-radial phillipsite within a cover of smectite and Fe–Mn oxides (see Fig. 10).

Nucleation of phillipsite crystals starts in the uppermost layer of the pelagic sediment column, and the crystals grow very slowly and over a long period of time (Bernat *et al.*, 1970; Czyscinski, 1973). The size of the phillipsite crystals does not change below the 1 to 2 m

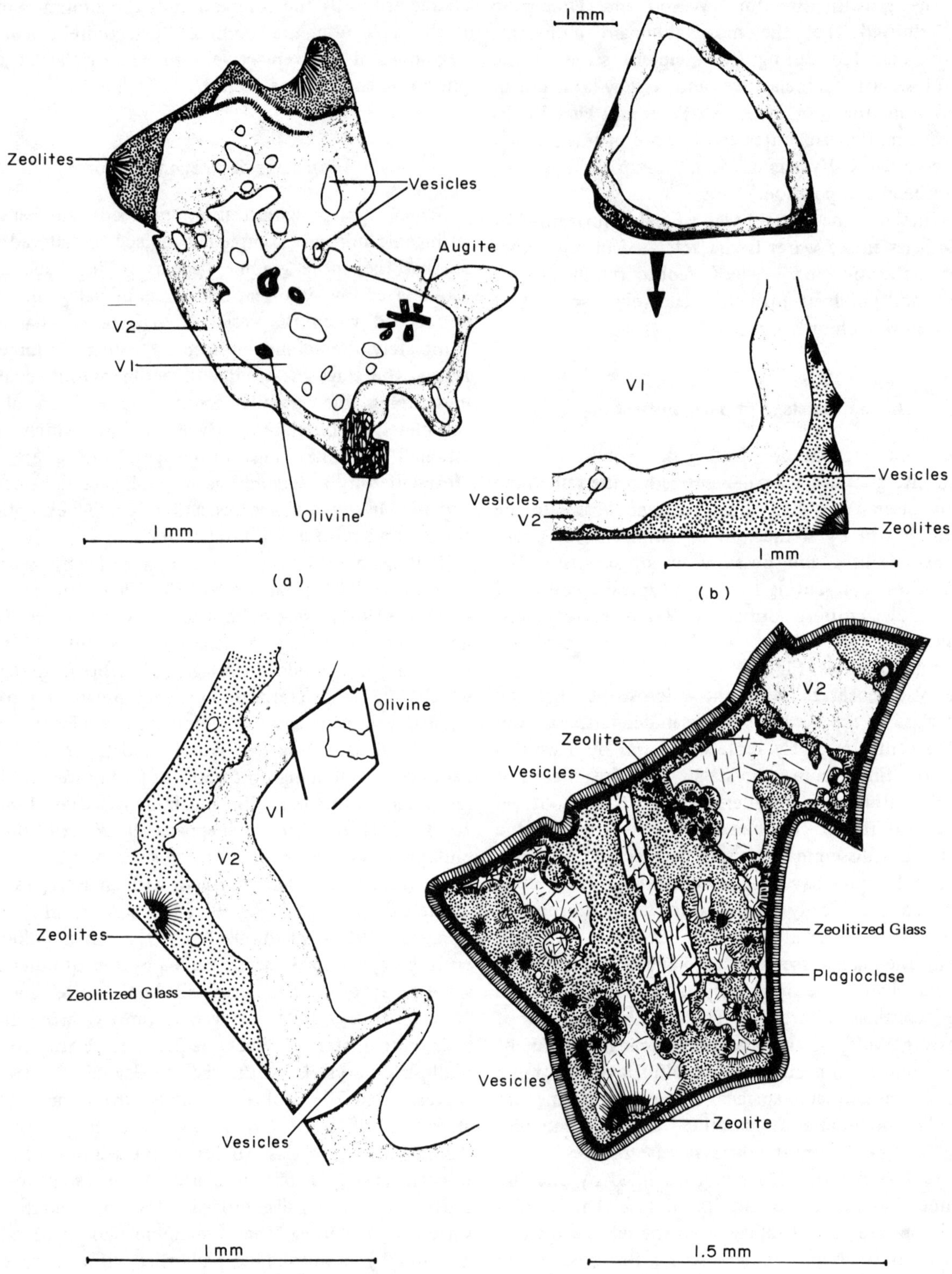

Fig. 8. Three examples (a,b,c) of incipient *in situ* zeolitization of glass granules in hyaloclastites from Palagonia (V1 = unaltered glass; V2 = palagonitized glass) and one example (d) of palagonitized granule almost completely replaced by phillipsite. The multiple smectite and gyrolite layers coating the granule and vesicle surfaces are shown only in the last case. The plagioclase, olivine, and augite microphenocrysts are unaltered.

AUTHIGENIC LAUMONTITE IN DEEP-SEA SEDIMENTS

CALVIN D. SANDS* and JAMES I. DREVER

Department of Geology, University of Wyoming, Laramie, Wyoming 82071, U.S.A.

Abstract – Laumontite was identified in sediments from a Deep Sea Drilling Project Core by X-ray powder diffraction of untreated and heated (350°C) heavy liquid separates. The laumontite-bearing sediments are from Site 323 located on the Bellingshausen Abyssal Plain and range in age from Cretaceous to Upper Miocene. Maximum burial depths range from 167 m for the youngest to 697 m for the oldest samples. Oxygen-isotope data indicate that the maximum temperature experienced by the deepest sample was not greater than 60°C.

Individual grains of laumontite were tentatively identified by scanning electron microscopy with X-ray energy spectrometry. Compared to the clinoptilolite, which makes up the bulk of the sample, the laumontite contains more Ca, less Na and K, and has a lower Si:Al ratio. The occurrence of a Ca-rich zeolite in these sediments may be related to the high concentrations of Ca^{+2} measured in the interstitial waters. Associated minerals indicate that the laumontite is not detrital.

The occurrence of laumontite in marine sediments which have not been subjected to elevated temperatures or pressures has important implications for the interpretation of mineral assemblages in zeolite-facies metamorphism.

INTRODUCTION

Laumontite is classified as a zeolite on the basis of its chemical, physical, and optical properties. Its crystal structure has been determined by Bartl and Fischer (1967). The ideal chemical formula is $CaAl_2Si_4O_{12}{\cdot}4H_2O$, but a more realistic formula, determined by Coombs (1952), is $Ca_x(Na, K)_yAl_{2x+y}Si_{24-(2x+y)}O_{48}{\cdot}16H_2O$. Sodium and potassium may substitute into the structure but calcium is usually the dominant cation. The Si:Al ratio also varies from occurrence to occurrence, but generally it is in the range 2.4–2.6:1.

Laumontite is monoclinic and generally has a fibrous or prismatic habit. It commonly occurs in veins and vesicles of igneous rocks and in low-grade metamorphic rocks where its presence is used to differentiate the first degree of metamorphism from diagenesis (Fyfe *et al.*, 1958). Its occurrence does not necessarily dictate a metamorphic reaction, however, and Hay (1966) states "... laumontite and quartz coexist in several deposits where maximum burial depths range from less than 3000 feet to 8000 feet, suggesting temperatures in the range of 50–100°C". Apparently laumontite formation can depend on other factors such as kinetics and solution chemistry in addition to pressure and temperature.

This study deals with the Deep Sea Drilling Project Site 33 located on the Bellingshausen Abyssal Plane (Fig. 1). Drever (1976) divides the core into two broad categories: (1) above 639 m derived predominantly from land; and (2) below 639 m derived predominantly from volcanic sources.

The predominantly land-derived sediments are primarily quartz, feldspars, micas, chlorite, and biogenic silica. The clay-size fraction (<2 μm) consists of illite, smectite, chlorite, and kaolinite in varying relative percentages. Small amounts of clinoptilolite and laumontite are also present in these sediments (Fig. 2).

Below 639 m an abrupt change in the mineralogy of the clay fraction occurs. Smectite, displaying no mixed layering (by the method of Reynolds and Hower, 1970), predominates with small amounts of illite, clinoptilolite, and quartz present. The silt-size fraction (>2 μm) is mostly quartz, feldspars, micas, and smectite. Calcite occurs in certain narrow intervals, and clinoptilolite occurs in the silt fraction from 669 m to about 702 m, where the basaltic basement is encountered.

TECHNIQUES AND RESULTS

Samples of core were enriched in the light fraction by heavy liquid separation using dibromomethane (specific gravity = 2.48). Sections of untreated core were placed

*Present address: American Stratigraphic Company, 524 W. Yellowstone, Casper, Wyoming 82601, U.S.A.

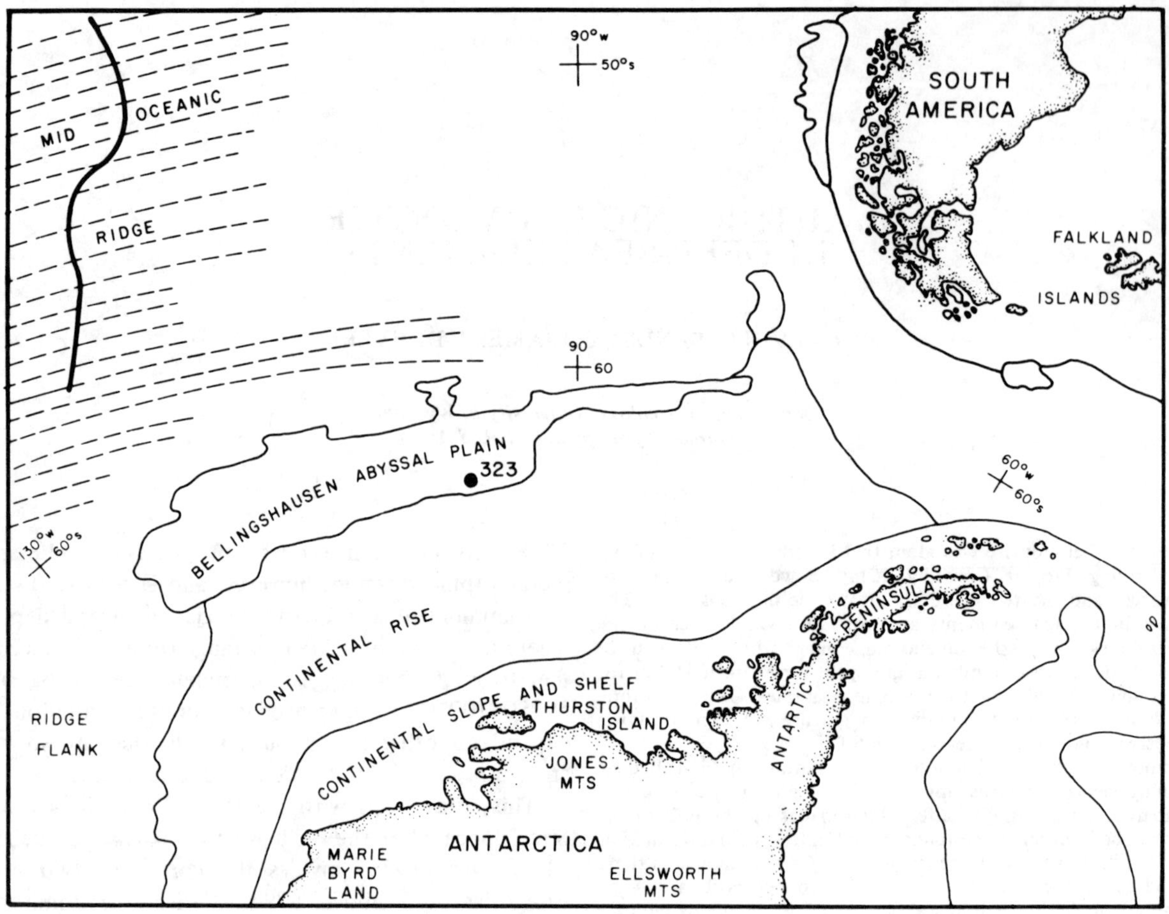

Fig. 1. Location of DSDP Leg 35, Site 323. Water depth at drill size was 5004 m (after Heezen and Tharp, 1972).

in centrifuge tubes containing dibromomethane and disaggregated using a sonic disruptor. The samples were centrifuged 30 minutes at 2000 rev/min, and the light fraction was filtered through a 0.45-μm Millipore filter. After this process was repeated two to three times, part of the sample was prepared for powder X-ray diffraction on a glass slide. Another portion of the sample was mounted onto a scanning electron microscope (SEM) stub with Duco cement and coated with gold or palladium.

X-ray diffraction was carried out on a General Electric XRD-5 at a goniometer speed of 2° 2θ per minute. Laumontite was tentatively identified by a strong peak at 9.6 Å and a weaker peak at 6.81 Å. These samples were heated at 350°C for 4 hours and then X-rayed again. Peaks at 10.8 Å, 8.6–8.8 Å, and 6.3 Å, corresponding to wairakite plus an unidentified phase (Furbish, 1965), replaced the previously described peaks (Fig. 3). The same heating and X-ray diffraction procedure was conducted on known laumontite from the Two Medicine Formation of Montana (Ogden, 1974). The peak changes observed were identical with those occurring in the deep-sea sediments. Several core samples in which clinoptilolite, a possible precursor of the laumontite, had been identified by X-ray diffraction of the bulk sediment were investigated by the heating and X-ray procedure. Laumontite was identified in samples 2-2(0-15) at 162.5 m depth, 4 cc at 322.5 m, 7-3(31-46) at 367.5 m, and one subsample of 18-4(10-11) from 697.5 m (Fig. 4).

Samples in which laumontite had been identified by X-ray diffraction techniques were observed by SEM, and semi-quantitative chemical analyses were made by an X-ray energy spectrometry (XES) system (Table 1) using analysed laumontite from the Briones Sandstone

TABLE 1. Semi-quantitative partial chemical analysis of laumontite from Site 323, DSDP

SiO_2	Al_2O_3	FeO_3	CaO	K_2O
48.3	17.3	4.4	9.2	4.1

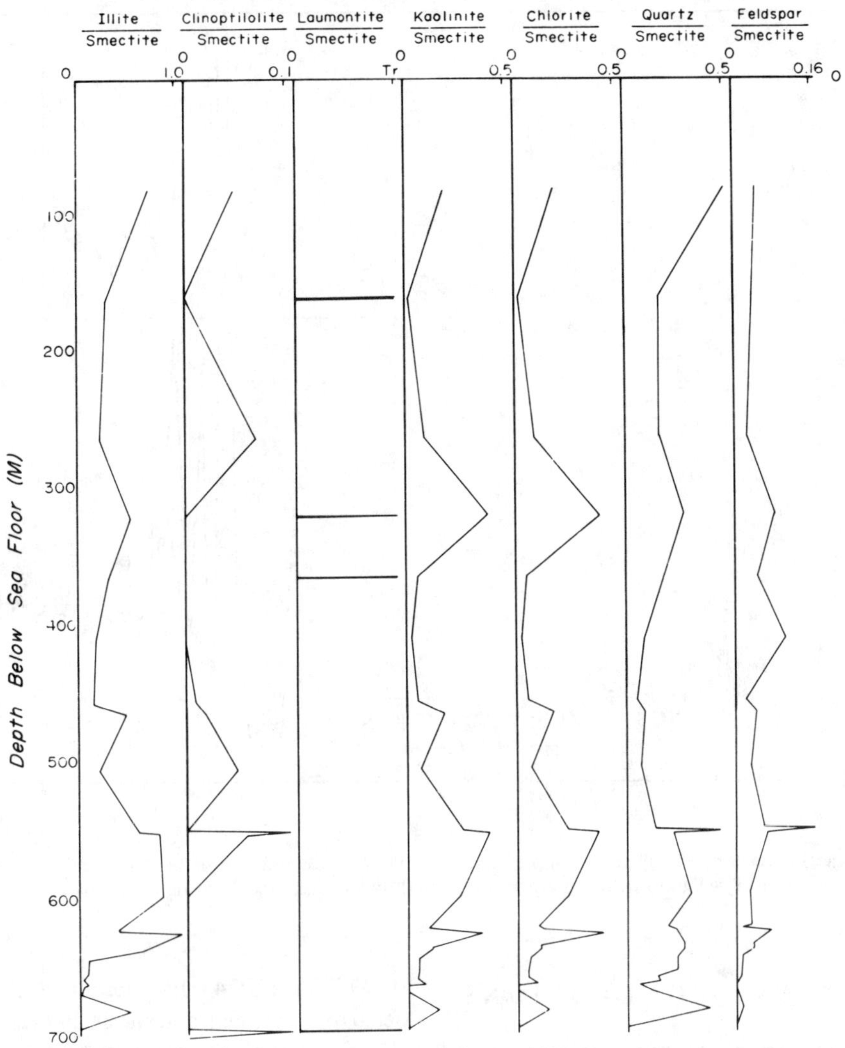

Fig. 2. Mineralogy of clay-size fraction expressed as ratio of X-ray diffraction peak heights to 17 Å smectite peak height.

(Murata and Whiteley, 1973) as a standard. Individual grains of laumontite were tentatively identified based on morphology and XES analysis (Fig. 5 A-C). Clinoptilolite in the samples (Fig. 5D) had a Si:Al X-ray peak-height ratio of 5.3:1, and potassium as the dominant cation. In contrast, the laumontite in these samples has a Si:Al peak-height ratio of 3.3:1 and a Ca:K peak-height ratio of about 2:1. Sodium was detectable, but no attempt was made to quantify the sodium concentration because sodium analyses by this method are very unreliable. The concentrations of calcium, sodium, and potassium in the same grain rules out the possibility of these grains being feldspar.

Heulandite, a silica-poor variety of clinoptilolite, frequently occurs with clinoptilolite in metamorphic terrains. Heulandite has a Si:Al ratio of approximately 3:1 and may have calcium, sodium, and potassium as cations. Mason and Sand (1960) report that heulandite and clinoptilolite can be distinguished optically. However, due to the small number and small size of the possible heulandite grains this method could not be employed. Another method for distinguishing between the two minerals was described by Mumpton (1960) in which X-ray diffraction slides were heated to 450°C overnight and X-rayed. The clinoptilolite diffraction pattern was unaffected, but heulandite became amorphous at that temperature. When this procedure was conducted on Site 323 samples no change occurred in the diffraction pattern indicating that no large amounts of heulandite were present.

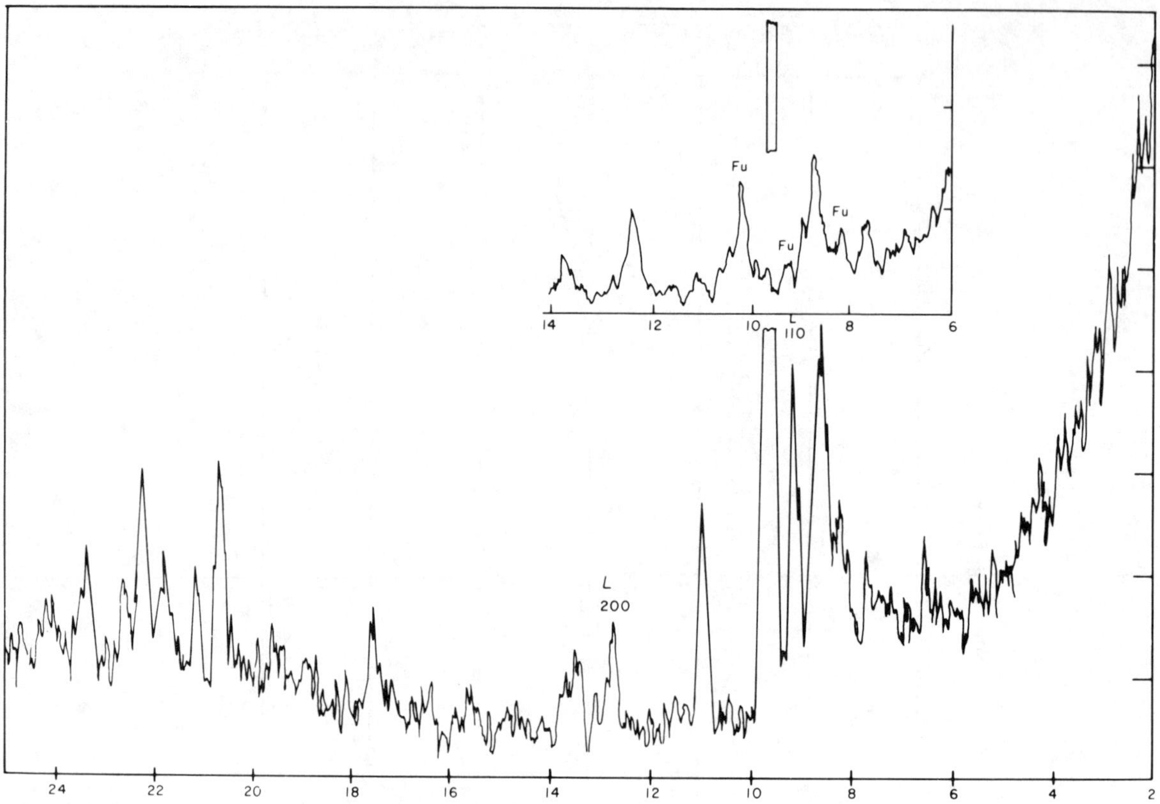

Fig. 3. X-ray diffraction pattern of unheated sample containing laumontite and sample heated to 350°C. Abscissa is degrees two theta Cu Kα radiation (L = laumontite, Fu = peaks reported by Furbish after heating laumontite).

DISCUSSION AND CONCLUSIONS

Gieskes (1976) reports a consistent increase in the Ca^{++} content and a decrease of Mg^{++} of the interstitial waters with depth at Site 323 (Fig. 6). Apparently at 450 ± 20 m reactions that absorb Mg^{++} and release Ca^{++} occur. Kastner (1976) suggested that these reactions are plagioclase (release of Ca^{++}) altering to smectite (sink for Mg^{++}) + chert + potassium feldspars (sink for K^{+}).

Other suggestions, made by Gieskes, to explain the gradients seen in Fig. 6 are that low-temperature reactions occur in the basaltic basement and supply a sink for Mg^{++} and source for Ca^{++}. Large O^{18} depletions in the interstitial water indicate that there is extensive interaction between the sediment pore waters and the underlying basalt (Lawrence *et al.*, 1976).

A definitive explanation of the reactions occurring in the sediments is not available. However, the net effect of these reactions is to create an environment with high calcium activity and low potassium and silica activities in the interstitial waters. That environment should favor the growth of a calcium-rich silicate-poor zeolite such as laumontite.

Winkler (1974) has described the zeolite facies as occurring between diagenesis and the greenschists facies. He places the laumontite subfacies at 200–300°C and 0.5–3 kilobars pressure. Anderson and Lawrence (1976), on the basis of 0^{18} analyses of the clays and authigenic carbonates, have set the maximum temperature at the base of Site 323 at 60°C. The deepest sample containing laumontite came from 698 m depth, and with an overlying water depth of 5 kilometers, the maximum pressure would be 640 bars. This is in the necessary pressure range postulated, but the temperature is much lower than that suggested necessary by Winkler.

Horne (1968) reported an occurrence of volcanically associated laumontite on Antarctica. Significant amounts of epidote occur with the laumontite on the Antarctic continent but no evidence of epidote was seen in the samples studied from Site 323. Laumontite and epidote vary considerably in specific gravity and therefore grains of the same physical size would not be hydraulically equivalent. However, if the DSDP laumontite was detrital from the locality discussed by Horne one might assume that epidote should also be present in trace amounts. From these data it can be concluded that the temperature is different than that postulated by

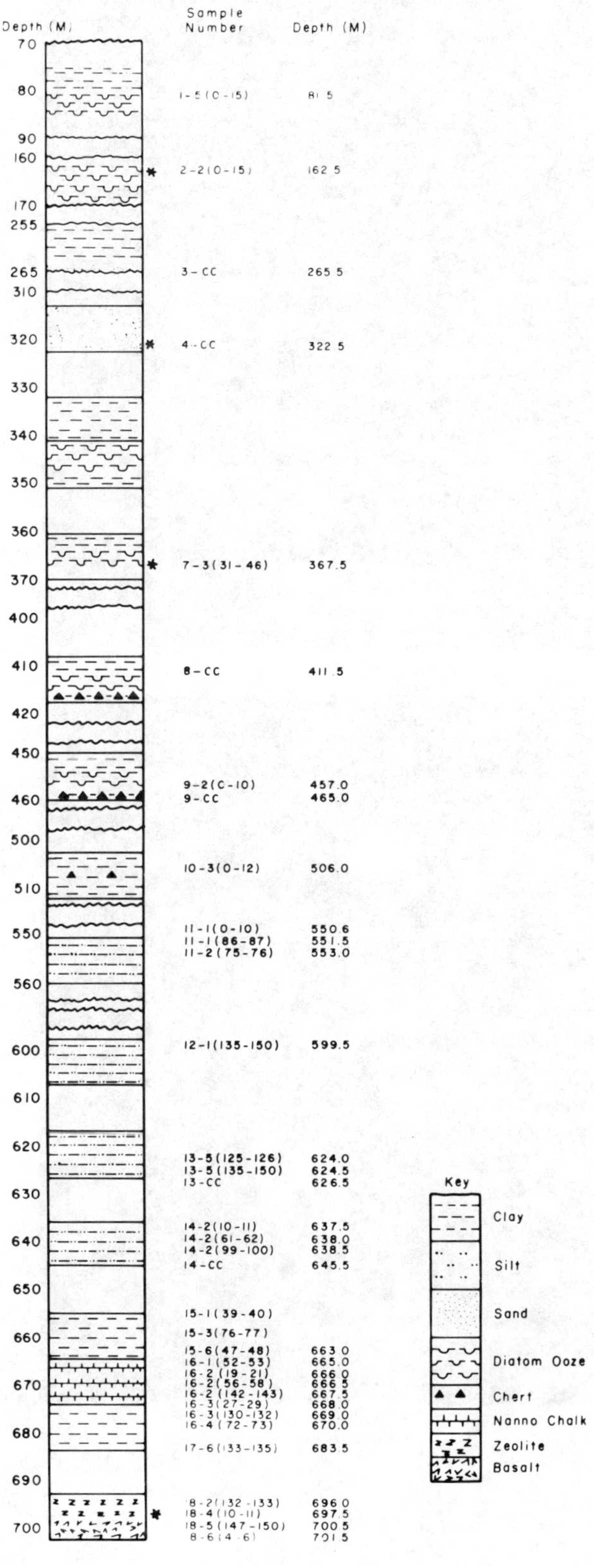

Fig. 4. Lithology, sample numbers, and sample depth at Site 323. Star signifies sample found to contain laumontite.

Winkler. Therefore, variables other than temperature are important in controlling laumontite formation.

Fyfe *et al.* (1958) originally defined the zeolite facies based on Coombs' (1954) observations in New Zealand. There, a comparatively homogeneous series of rocks had been subjected to pressure and temperature gradients which caused low-grade metamorphism to occur. Because of the small variation in original composition, pressure and temperature appeared to be the controlling factors.

Boles and Coombs (1975) report the occurrence of a laumontized tuff in the Triassic Murihiku Super-group of Southland, New Zealand. At this location heulandite has altered to laumontite in a zeolite facies reaction as defined by Coombs *et al.* (1959). The relationship of laumontite to other minerals in the tuff suggests that other factors, such as P_{H_2O} and ionic activities in interstitial waters, played a larger role in dictating the spatial distribution of laumontite than pressure and temperature.

ACKNOWLEDGMENTS

Dr. R. C. Surdam and Palmer R. Ogden, Jr. offered constructive discussions and suggestions during the study. This research was supported by National Science Foundation grant number GA-33503.

REFERENCES

Anderson, T. F. and Lawrence, J. R. (1976) Stable isotope investigations of sediments, basalts, and authigenic phases from leg 35 cores: *Initial Reports of the Deep Sea Drilling Project* **35** (edited by Hollister, C. *et al.*), Washington (U.S. Government Printing Office), pp. 497-505.

Bartl, H. and Fischer, K. F. (1967) Untersuchung der kristallstrucktur des zeolites laumontite: *Neues Jb. Miner., Mh.* **33**, 33-42.

Boles, J. R. and Coombs, D. S. (1975) Mineral reactions in zeolitic Triassic tuff. Hokonui Hills, New Zealand: *Bull. Geol. Soc. Amer.* **86**, 163-173.

Coombs, D. S. (1952) Cell size, optical properties and chemical composition of laumontite and leonhardite: *Amer. Mineral.* **37**, 812-830.

Coombs, D. S. (1954) The nature and alternation of some Triassic sediments from Southland, New Zealand: *Royal Soc. New Zealand Trans.* **82**, 65-109.

Coombs, D. S., Ellis, A. J., Fyfe, W. S. and Taylor, A. M. (1959) The zeolitic facies, with comments on the interpretation of hydrothermal synthesis: *Geochim. Cosmochim. Acta* **17**, 53-107.

Deer, W. A., Howie, R. A. and Zussman, J. (1963) *Rock-forming Minerals*, Vol. 4: Longmans, Green & Co. Ltd., London, p. 435.

Drever, J. I. (1976) Chemical and mineralogical studies, Site 323: *Initial Reports of the Deep Sea Drilling Project* **35** (edited by Hollister, C. *et al.*), Washington (U.S. Government Printing Office), pp. 471-477.

Furbish, W. J. (1965) Laumontite-leonhardite: *Southeastern Geology* **6**, 189-200.

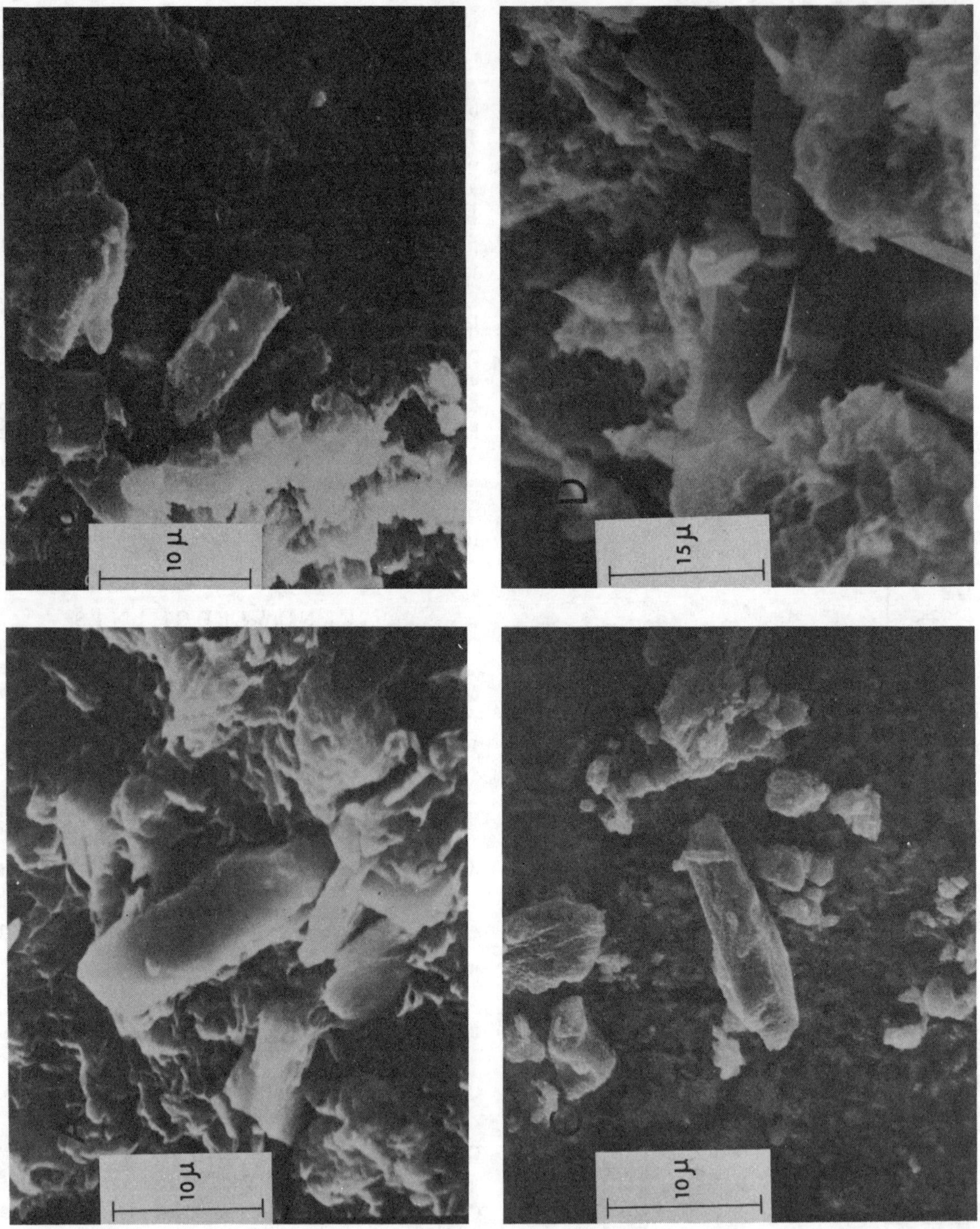
10μ
15μ
10μ
10μ

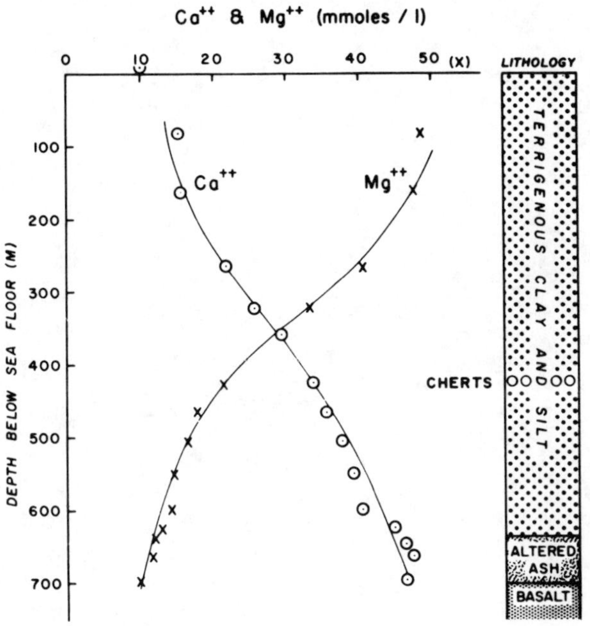

Fig. 6. Ca^{++} and Mg^{++} in Site 323 interstitial waters plotted against depth (from Lawrence *et al.*, 1976).

Fyfe, W. S., Turner, F. J. and Verhoogen, J. (1958) Metamorphic reactions and metamorphic facies: *Geol. Soc. Amer. Memoir* 73.

Gieskes, J. M. (1976) Interstitial water studies, leg 35: *Initial Reports of the Deep Sea Drilling Project* **35** (edited by Hollister, C. *et al.*), Washington (U.S. Government Printing Office), pp. 407-424.

Hay, R. L. (1966) Zeolites and zeolitic reactions in sedimentary rocks: *Geol. Soc. Amer. Special Paper* **85**, 130.

Heezen, B. C. and Tharp, M. (1972) Morphology of the sea floor: *in* folio 16, *Antarctic Map Folio Series*, Amer. Geog. Soc.

Horn, R. R. (1968) Authogenic prehnite, laumontite, and chlorite in the lower Cretaceous sediments of south-eastern Alexander Island: *Brit. Antarct. Surv. Bull.* **18**, 1-10.

Kastner, M. (1976) Diagenesis of basal sediments and basalts of Sites 322 and 323, leg 35, Bellingshausen Abyssal Plain: *Initial Reports of the Deep Sea Drilling Project* **35** (edited by Hollister, C. *et al.*), Washington (U.S. Government Printing Office).

Lawrence, J. R., Gieskes, J. M. and Anderson, T. F. (1976) Oxygen isotope material balance calculations, leg 35: *Initial Reports of the Deep Sea Drilling Project* **35** (edited by Hollister, C. *et al.*), Washington (U.S. Government Printing Office), pp. 513-527.

Mason, B. and Sand, L. B. (1960) Clinoptilolite from Patagonia; the relationship between clinoptilolite and heulandite: *Amer. Mineral.* **45**, 341-350.

Mumpton, F. A. (1960) Clinoptilolite redefined: *Amer. Mineral.* **45**, 351-369.

Murata, K. J. and Whiteley, K. R. (1973) Zeolites in the Miocene brines, sandstone and related formations of the Central Coast Ranges, California: *J. Research, U.S. Geol. Survey* **1** (3), 255-265.

Ogden, P. R., Jr. (1974) The mineralogy, petrology and petrogenesis of selected units of the Two Medicine Formation, Lewis and Clark County, Montana: M.S. thesis, Univ. of Florida.

Reynold, R. C. and Hower, J. (1970) The nature of interlayering in mixed-layer illite–montmorillonites: *Clays and Clay Minerals* **18**, 25-36.

Winkler, H. G. F. (1974) *Petrogenesis of Metamorphic Rocks*: Springer-Verlag.

ZEOLITE ZONES IN GEOTHERMAL AREAS IN ICELAND

HREFNA KRISTMANNSDÓTTIR

and

JENS TÓMASSON

Department of Natural Heat, National Energy Authority, Laugavegur 116, Reykjavík, Iceland

Abstract – Zeolites formed in basaltic flows and hyaloclastic rock sequences in Icelandic geothermal areas have been mapped and investigated. In the low-temperature geothermal areas (rock temperature <150°C in the uppermost 1 km) four distinct zeolite zones are found. The zeolite zones correspond to a rock-temperature interval. In the order of increased rock temperature the zones are (1) chabazite, (2) mesolite/scolecite, (3) stilbite, and (4) laumontite zones. Analcime is not confined to a distinct zone.

In most of the high-temperature geothermal areas (>200°C in the uppermost 1 km) zeolites are formed in the uppermost part of the upflow zone at rock temperatures below 250°C. At even higher rock temperatures analcime and wairakite are formed. The sequence of zeolites is normally not sharply defined in the upper levels of the high-temperature areas due to the sharply rising temperature gradients. Mordenite, heulandite, laumontite, and analcime are the zeolites most commonly found in the high-temperature geothermal areas. In a few of the high-temperature geothermal areas zeolites are almost lacking.

INTRODUCTION

Geothermal areas in Iceland have been investigated for the last 30 years, and deep drilling has been performed in some of the areas. Mineralogical investigation of the altered rocks has been greatly intensified over the last 5 years. The geothermal areas in Iceland are divided into two distinct groups (e.g., Palmason, 1974). The low-temperature areas are in Quaternary and Tertiary rock formations, and the high-temperature areas are in the recent volcanic zones (Fig. 1). In the low-temperature areas the temperature is below 150°C at 1 km depth, whereas in the high-temperature areas the temperature is higher than 200°C at this depth.

In all the high-temperature areas the dominant rock formations are basaltic hyaloclastites and basalt lavas. This is also the case for the low-temperature areas found in the Quaternary rock formation; however, in the low-temperature areas in rocks of Tertiary age the country rocks are mainly basalt lavas with sedimentary intercalations. The geothermal fluid is generally water of meteoric origin with a comparatively low content of dissolved solids (Arnórsson, 1974). However, in some geothermal areas the water is saline and contains a considerable amount of dissolved solids. The chemistry of geothermal water from several areas is shown in Table 1. The most common cause of saline water in Icelandic geothermal systems is direct mixing with sea water or an outwash of chlorine trapped in marine formations (Arnórsson and Tómasson, 1970). In the most extreme case, in the Reykjanes high-temperature geothermal area (Björnsson *et al.*, 1972; Tómasson and Kristmannsdóttir, 1972), the water has the same salinity as sea water. The pH of most geothermal water is high, 9 to 10 at 20°C in the meteoric originated water, and 7 to 8 in the saline water. The geothermal water of Leirá in Borgarfjöróur is special in that it is CO_2-rich.

Rocks in high-temperature areas are young and have never been deeply buried. The alteration caused by the circulation of the geothermal water overshadows completely any former deuteric or low-temperature alteration. In the low-temperature areas, the alteration minerals often reflect a complex thermal and geological history of the area. The effects of the present geothermal system are not as obvious as they are in the high-temperature geothermal areas.

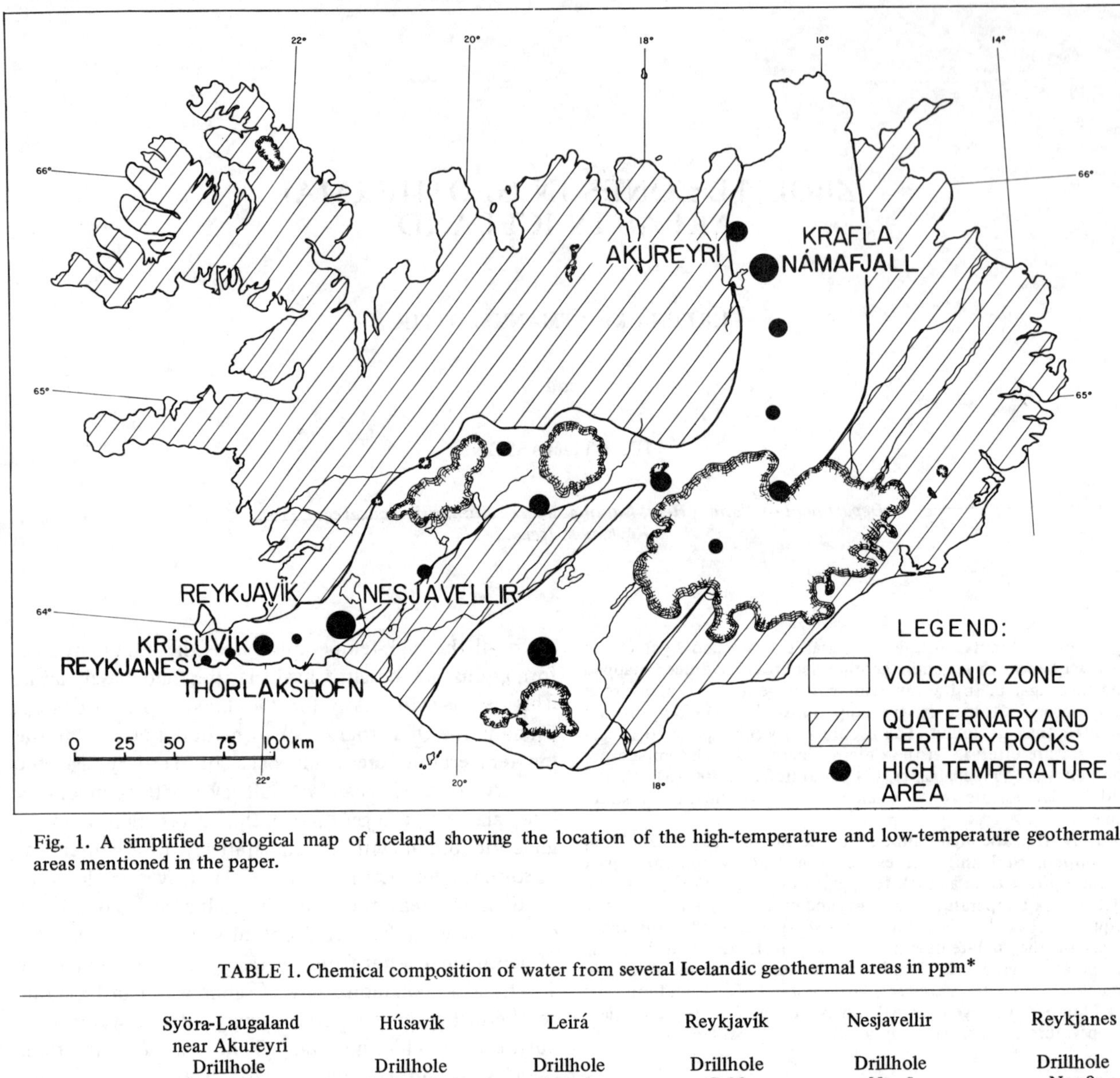

Fig. 1. A simplified geological map of Iceland showing the location of the high-temperature and low-temperature geothermal areas mentioned in the paper.

TABLE 1. Chemical composition of water from several Icelandic geothermal areas in ppm*

	Syöra-Laugaland near Akureyri Drillhole No. 5	Húsavík Drillhole No. 1	Leirá Drillhole No. 4	Reykjavík Drillhole G-23	Nesjavellir Drillhole No. 5	Reykjanes Drillhole No. 8
°C	89.5	94	87	100	261	270
pH/°C	9.81/20	7.4/20	8.10/20	9.60/20	7.69/261	6.27/270
SiO_2	106	80	149	130	559.6	592
Na^+	47.5	840	133.4	48	129.9	9854
K^+	1.2	18.7	16.1	1.2	24.2	1391
Ca^{++}	2.9	176	22.6	1.5	0.14	1531
Mg^{++}	0.15	10.6	1.20	0.24	0.05	1.15
CO_2(tot)	16.2	13	119.4	51	738.1	1427
$SO_4^=$	33.8	82	59.6	27	64.2	28.7
H_2S(tot)	<0.1	n.d.	0.26	n.d.	210	31.5
Cl^-	10.6	1633	135.7	19	5.8	18827
F^-	0.41	0.2	2.05	0.4	1.6	0.1

n.d.—not determined.

*National Energy Authority data and Arnórsson, 1974.

In the hydrothermally altered rocks zeolites are found mainly as minor constituents in small veins and amygdule fillings. The zeolites are also found as devitrification products of volcanic glass. However, in one of the Icelandic geothermal areas (Húsavík), zeolites have quantitatively replaced the original rocks.

ZEOLITIZATION IN GEOTHERMAL AREAS

Low-temperature areas

Low-temperature geothermal areas are found in Quaternary rocks which border active volcanic zones and also in Tertiary rock formations mainly in the northern and western parts of the country. Zeolites are apparently in all temperature regions up to 190°C in the deepest drillholes at the areas.

The geothermal areas within and in the neighborhood of Reykjavík are among the most intensively investigated low-temperature fields in the Quaternary rock formation. Here the rock formations are alternately hyaloclastites and basalt lavas in the uppermost 1000 m. Between 1000 m and below that depth to at least 2000 m basalt lavas are dominant. The maximum temperature obtained is 146°C at 2200 m depth in one of the areas, but in other areas the maximum temperatures vary from 100–120°C. Figure 2 shows a simplified geological section in a drillhole from one of the geothermal fields in Reykjavík, together with estimated rock temperatures and the distribution of secondary mineral species. In all of the Reykjavík geothermal fields the occurrence of zeolites is closely related to the prevailing rock temperature. The zeolite zones found are: an uppermost chabazite zone, a mesolite/scolecite zone, a stilbite zone, and a lowermost laumontite zone, listed as a function of increasing temperature and depth. Other alteration minerals are not directly related to the prevailing temperature. Relict alteration is found in an aureole around two nearby Quaternary volcanic centers. The intensity of the alteration in the different Reykjavík areas is governed by their relative distance to these volcanic centers. Epidote is always relic, and its retrograde transformation to prehnite has been observed. The clay minerals are partly relic and partly formed by

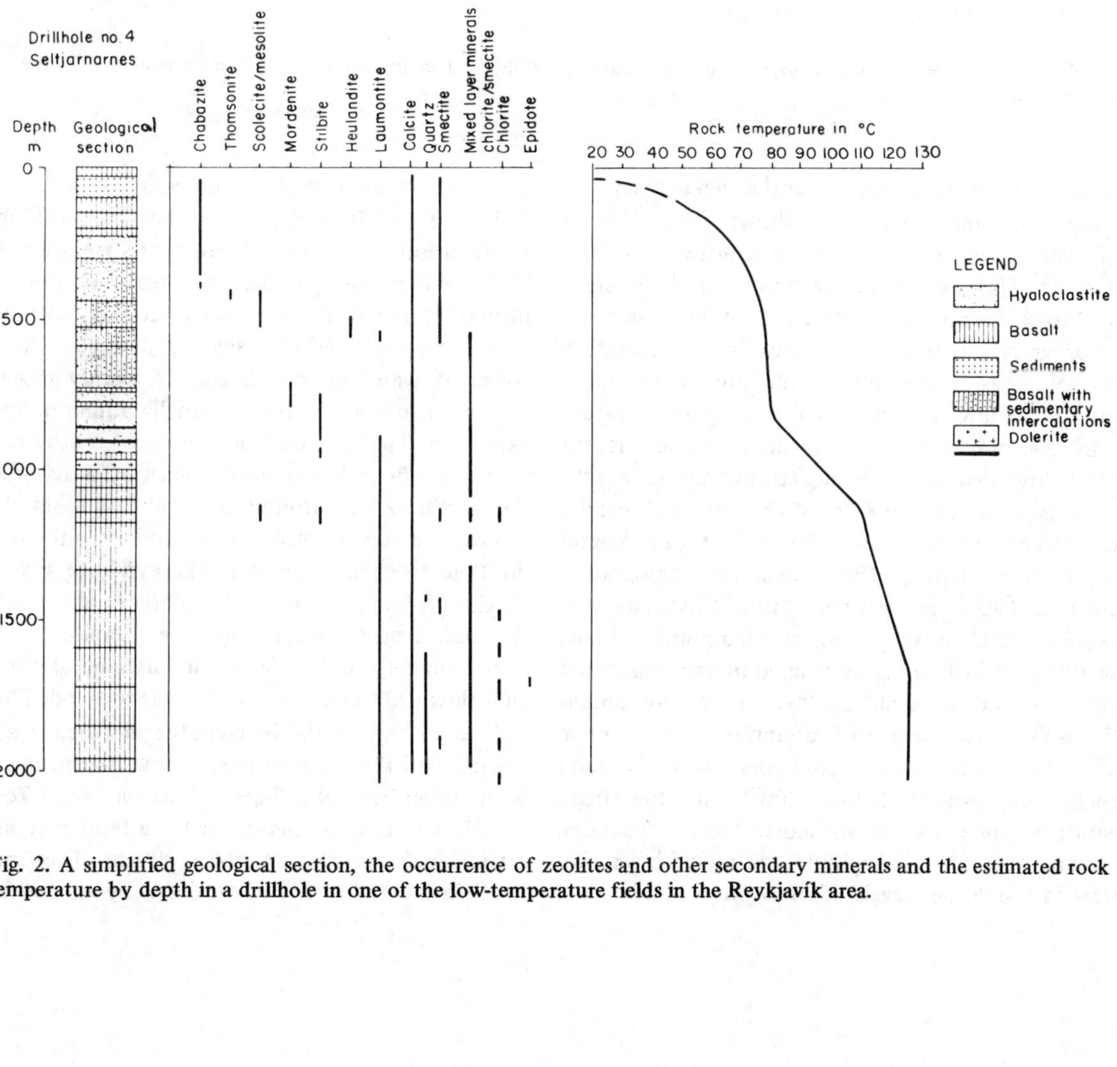

Fig. 2. A simplified geological section, the occurrence of zeolites and other secondary minerals and the estimated rock temperature by depth in a drillhole in one of the low-temperature fields in the Reykjavík area.

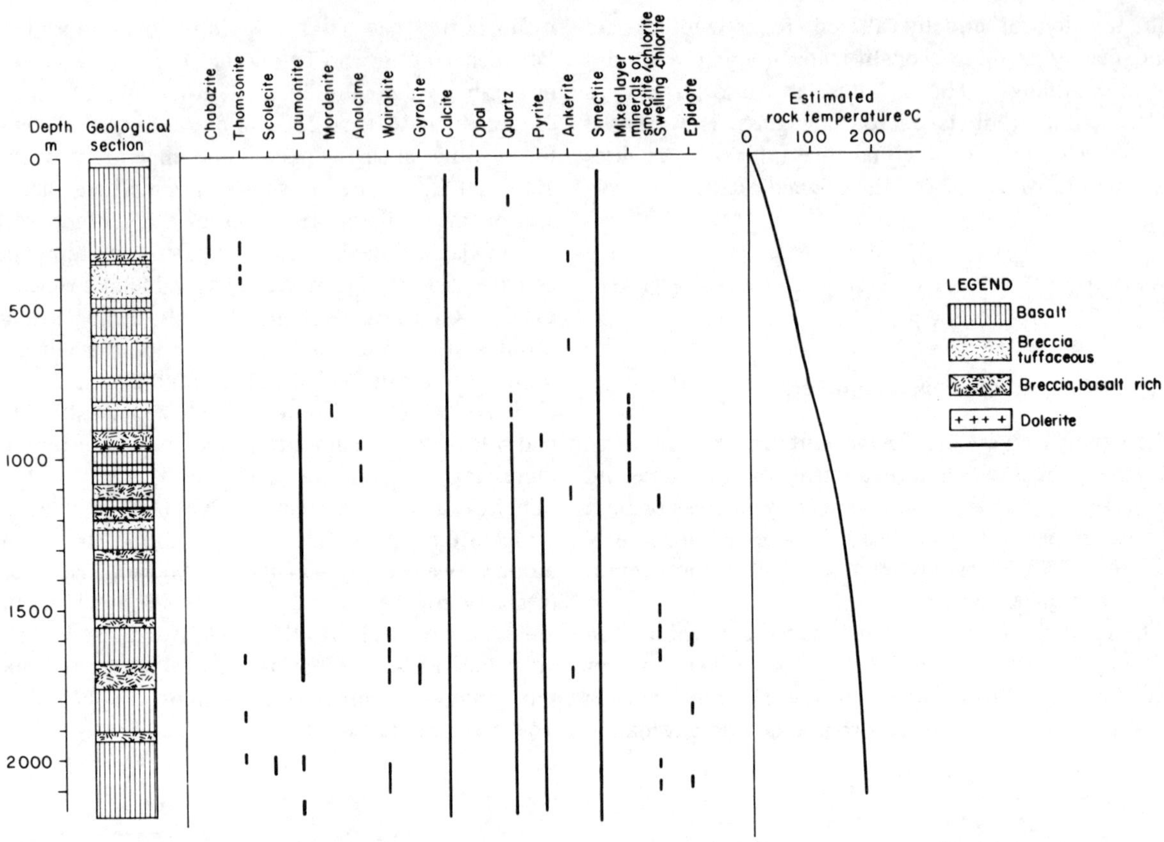

Fig. 3. A geological section, occurrence of secondary minerals, and estimated rock temperature in a drillhole near Thorlákshöfn.

the most recent low-temperature geothermal activity.

In Fig. 3 a similar diagram is shown for a 2187 m deep drillhole near Thorlákshöfn in southwest Iceland (see Fig. 1). This area is on the border of the western volcanic zone in southern Iceland and can be considered as a border-case between low- and high-temperature geothermal areas. According to the previously stated definition it is, however, classified as a low-temperature area. In the uppermost 300 m the rocks are recent basalts; below that depth hyaloclastites alternate with basalt lavas to about 1900 m, with dominant basalts below. Dolerite dikes intrude the other rocks, mainly below 1100 m depth. The maximum temperature obtained is 190°C at 2100 m depth. Chabazite and thomsonite occur above 450 m, and laumontite occurs below 800 m to 2100 m, accompanied by mordenite and analcime at 800–1000 m. Zeolites are nearly absent at 450–800 m. The degree of alteration is comparable to the uppermost smectite–zeolite zone, which is found at rock temperatures below 200°C in the high-temperature areas (Kristmannsdóttir and Tómasson, 1976). The occurrence of epidote and mixed-layer clay minerals in the deeper levels of the section indicate the beginning of the next alteration zone.

Walker (1960) mapped a well-defined zoning of the zeolitization of Tertiary formations in eastern Iceland. The zeolites in geothermal areas in the Tertiary formation show a very similar zoning, although the analcime zone is seldom sharply defined. The upflow zones of water in these areas are mainly governed by tectonic features and have normally a limited horizontal extension. The effect of circulation of geothermal water on the zeolitization is mostly seen by a disturbance of the zeolite zones around the main aquifers. Figure 4 shows the occurrence of secondary minerals in a drillhole from an area near Akureyri (see Fig. 1). The maximum temperature of the geothermal water is 96°C. A disturbance occurs in the zeolite zoning at approximately 620 m depth; at this same depth there is an inflow (23 l/s) of 94°C hot water to the drillhole.

The geology of the Húsavík thermal area in northeast Iceland and the geochemistry of the thermal water have been described elsewhere (Árnason and Tómasson, 1970). The area is transected by a fault system with a vertical displacement of 600–1000 m. The uppermost 1100 m are sediments intercalated by smaller amounts

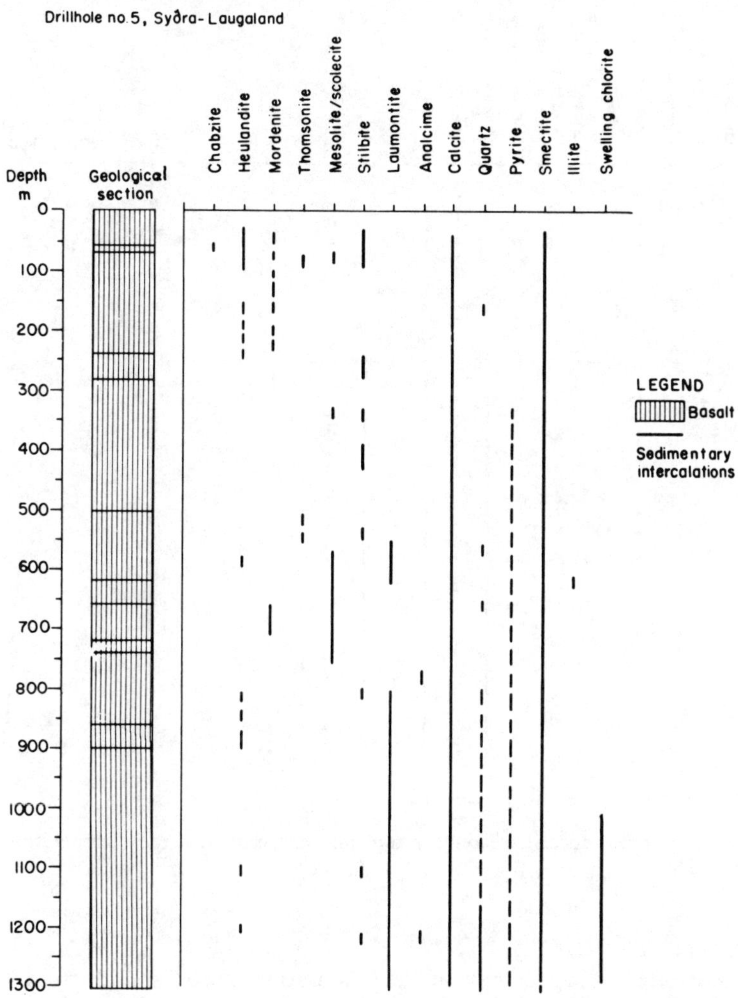

Fig. 4. A geological section, occurrence of secondary minerals, and estimated rock temperature in a drillhole in a low-temperature area near Akureyri.

of basalt lavas. The sediments are partly marine. Basalt lavas underlie the formation. The chlorine content of the water is from 400–1800 ppm, and the temperature of the geothermal water is from 45–100°C. The zeolites have nearly quantitatively replaced the original sedimentary rocks, especially the most fine grained. A small amount of primary pyroxene remains but plagioclase appears to be completely replaced. The appearance of the zeolite zones seems to be similar to that observed in other low-temperature areas. The mesolite/scolecite zone is quite clearly defined. The laumontite zone is also quite distinct in the sediments, but in the basalts it is not sharply defined. A separate stilbite zone seems to be lacking, but stilbite is found scattered below 400 m depth.

The geothermal area on Leirá, Borgarfjördur, in western Iceland contains CO_2-rich water, which is reflected by the paucity of secondary zeolites. Laumontite is scattered below 700 m at temperatures above 120°C, and analcime is present in only a few samples below 1000 m at temperatures of 130°C and higher.

High-temperature areas

A regular sequence in the appearance of zeolites is seldom found in the high-temperature areas. This is mainly due to the steep thermal gradient in the uppermost few hundred meters of the fields and also to

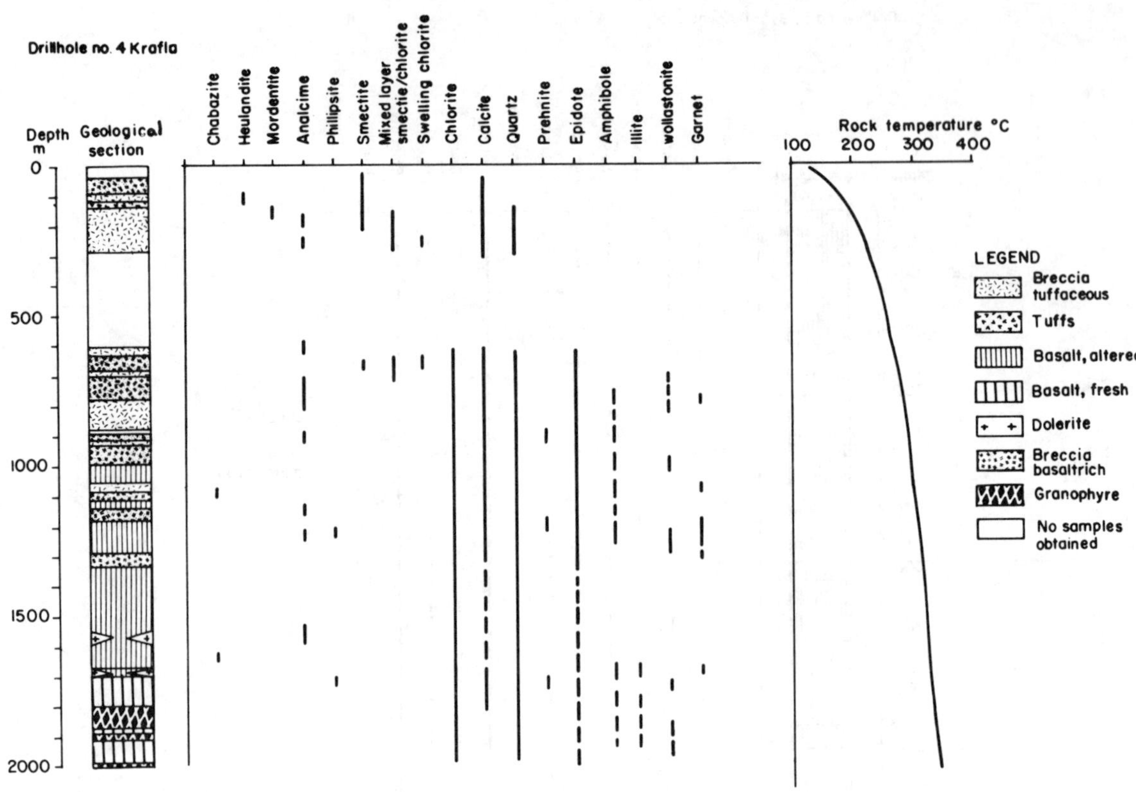

Fig. 5. A geological section, occurrence of secondary minerals, and estimated rock temperature in a drillhole in the Krafla high-temperature area.

inflow of cold groundwater into the upper part of the system.

Zeolites (except analcime/wairakite) occur mainly in the uppermost strata in the high-temperature areas at temperature up to 200–230°C and depths of 300–500 m. Analcime and wairakite are scattered in and below the zeolite zone up to rock temperatures as high as 300°C.

Zeolites have been observed in most of the high-temperature areas. However, they are scarce in some areas and completely lacking in one.

Figures 5 and 6 show the mineralogical data of two drillholes in the Krafla and Nesjavellir high-temperature areas (Fig. 1). A more detailed description of the Nesjavellir area has been published previously (Kristmannsdóttir and Tómasson, 1974). The investigation of the Krafla field is in progress, and only preliminary results from the first two shallow drillholes (1200 m) have been published (Kristmannsdóttir, 1975).

In the Krafla area all zeolites except analcime and wairakite are formed in the upper levels (200–300 m) at temperatures up to 230°C. Analcime and wairakite are scattered at greater depths and at temperatures up to 300°C. The magnitude and degree of alteration and the alteration zones found from the distribution of clay minerals, prehnite, epidote, and quartz are generally compatible with the prevailing temperature conditions, at least in the uppermost 1200 m of the profile. At greater depths, fresh impermeable dikes and smaller intrusions are more dominant in the section. Minerals which are stable at much lower temperatures (smectite, mixed-layer clay minerals, chabazite, phillipsite) coexist with high-temperature minerals formed at the contacts of the intrusions (wollastonite, andradite, hedenbergite). In some of the drillholes there is a fairly regular sequence of zeolites with depth: heulandite, mordenite, laumontite, and analcime/wairakite. Epistilbite is also found in the heulandite zone. Phillipsite occurs in the laumontite zone and also in the deeper intrusive rocks. The zeolitization found in the hydrothermally altered rocks in the Krafla area is typical of that in most other Icelandic high-temperature areas.

The Nesjavellir area is exceptional as it shows much less and a lower degree of hydrothermal alteration than other high-temperature areas with similar range of the rock temperature. In this geothermal area zeolites (other than analcime) are found at rock temperatures of 250°C and higher. The zeolites are commonly observed to 1000

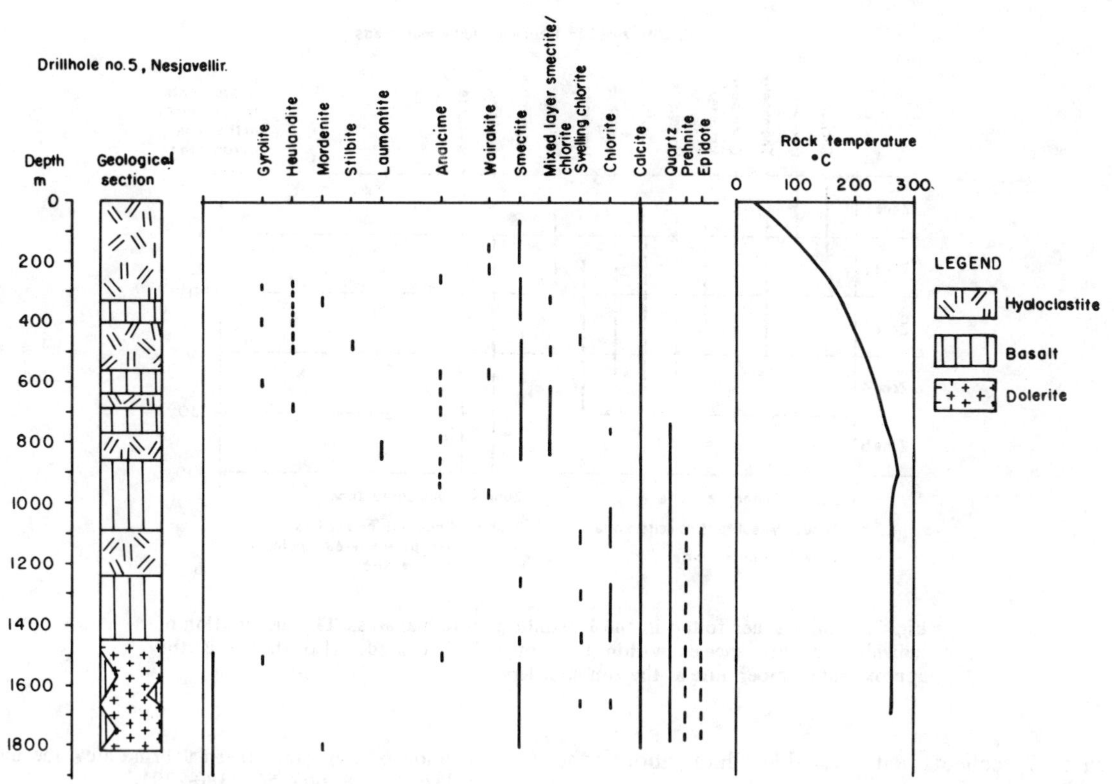

Fig. 6. A geological section, occurrence of secondary minerals, and estimated rock temperature in a drillhole in the high-temperature area at Nesjavellir.

m depth. Calcium silicates, such as gyrolite and reyerite, are found at the same depth as the zeolites. Zeolite zoning is not well defined, but it shows a trend similar to that found in Krafla. There is strong evidence for a recent displacement of this area (Kristmannsdóttir and Tómasson, 1974) that may explain the poor fit between the amount and degree of the alteration, the present secondary mineral paragenesis, and the prevailing temperatures.

Experimental studies by Liou (1970, 1971) of Ca-zeolites show the same sequence of zeolite species to be formed by increasing temperature as in the Icelandic geothermal areas. The equilibrium temperatures (at the same range of total pressure) for conversion from one Ca-zeolite to another are somewhat higher than those observed in the geothermal areas. Contaminants in the geothermal fluids will affect the equilibrium temperatures for formation of zeolites. In the experiments the aqueous fluid pressure is equal to total pressure, which is not necessarily the case in the geothermal areas. Lower aqueous fluid pressure and higher contents of dissolved solids in the geothermal fluids would tend to lower the temperatures of formation of the zeolites.

Studies of zeolites (e.g., Seki, 1966, 1973; Miyashiro and Shido, 1970) are in agreement with the range of stability observed for those species in the Icelandic geothermal areas.

CONCLUSIONS

The distribution of zeolites in Icelandic geothermal areas is characterized by the following factors: the original rocks are basaltic lavas and hyaloclastites in all different types of geothermal areas, except those in Tertiary settings where basalt lavas are predominant. In basaltic rocks the occurrence of individual zeolites is influenced by rock composition. For example, heulandite and stilbite are mostly found in tholeiites, while analcime is virtually absent. Successive zones characterized by chabazite, mesolite/scolecite, and laumontite are found in all areas. The permeability of the rocks, the age of the geothermal area and the rock formations, and the composition of the geothermal fluid are all factors that influence the degree and amount of the alteration. The rock temperature is, however, the most important single factor governing the degree of alteration.

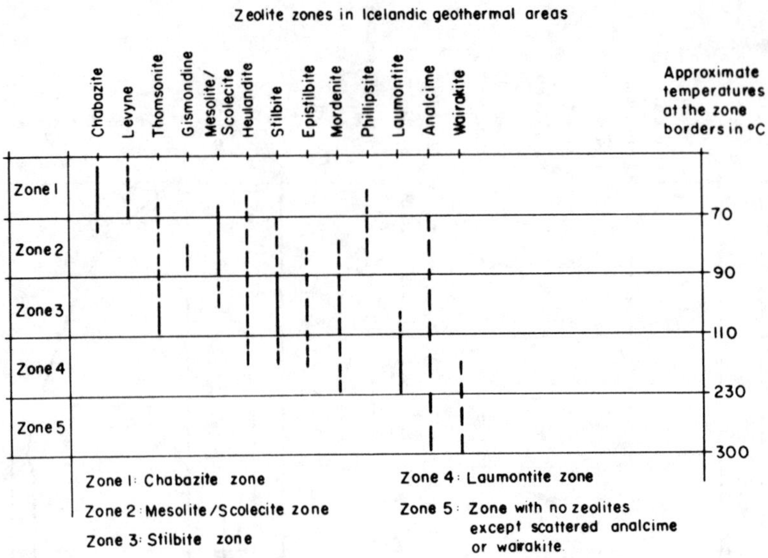

Fig. 7. Zeolite zones found in the Icelandic geothermal areas. The distribution of individual zeolite species within the zones is indicated. Also shown is the approximate temperature at the zone borders.

Figure 7 collects all available data about the occurrence of zeolites in the Icelandic geothermal areas and relates the characteristic zeolite zones to the temperature conditions at which they are formed. The occurrence of the zeolite species within the main zeolite zones in Icelandic geothermal areas is also shown in Fig. 7. In the figure the temperature range of each zeolite zone is indicated.

Experimental data for stability and formation for some of the leading zeolites (e.g., Liou, 1970, 1971; Seki 1966, 1973) are in fairly good agreement with the range of temperatures at which zeolites formed in the Icelandic geothermal areas.

REFERENCES

Árnason, B. and Tómasson, J. (1970) Deuterium and chloride in geothermal studies in Iceland: *Geothermics*, Spec. Issue 2, *U.N. Symposium on the Development and Util. of Geother. Res. Pisa, 1970* **2**, 1405-1415.

Arnórsson, S. (1974) The composition of thermal fluids in Iceland and geological features related to the thermal activity: In Kristjánsson, L., ed., *Geodynamics of Iceland and the North Atlantic Area*, pp. 307-323. Reidel Dordrecht (NATO Advanced Study Institute series).

Björnsson, S., Arnórsson, S. and Tómasson, J. (1972) Economic evaluation of Reykjanes thermal brine area, Iceland: *Amer. Ass. Petrol. Geol. Bull.* **56**, 2380-2391.

Kristmannsdóttir, H. and Tómasson, J. (1974) Nesjavellir. Hydrothermal alteration in a high-temperature area: *Proc. Int. Symposium on Water–Rock Interaction, Prague, 1974*, pp. 170-177.

Kristmannsdóttir, H. and Tómasson, J. (1976) Hydrothermal alteration in Icelandic geothermal fields: *Soc. Sci. Isl.* **5**, 171-180.

Kristmannsdóttir, H. (1975) Hydrothermal alteration of basaltic rocks in Icelandic geothermal areas: *Proc. 2nd U.N. Symposium on the Development and Use of Geothermal Resources, San Francisco* (in press).

Liou, J. G. (1970) Synthesis and stability relations of wairakite. $CaAl_2Si_4O_{12}2H_2O$: *Contr. Mineral. Petrol.* **27**, 259-282.

Liou, J. G. (1971) Stilbite-laumontite equilibrium: *Contr. Mineral. Petrol.* **31**, 171-177.

Miyashiro, A. and Shido, F. (1970) Progressive metamorphism in zeolite assemblages: *Lithos* **3**, 251-260.

Pálmason, G. (1974) Heat flow and hydrothermal activity in Iceland: In Kristjánsson, L., ed., *Geodynamics of Iceland and the North Atlantic Area*, pp. 297-306. *Reidel Dordrecht.* (NATO Advanced Study Institute series.)

Seki, Y. (1966) Wairakite in Japan II. *J. Japan Ass. Mineral. Petrol. Econ. Geol.* **56**, 30-39.

Seki, Y. (1973) Ionic substitution and stability of mordenite. *J. Geol. Soc. Japan* **79**, 669-676.

Tómasson, J. and Kristmannsdóttir, H. (1972) High temperature alteration minerals and thermal brines, Reykjanes, Iceland: *Contr. Mineral. Petrol.* **36**, 123-137.

Walker, G. P. L. (1960) Zeolite zones and dike distribution in relation to the structure of the basalts in eastern Iceland: *J. Geol.* **68**, 515-528.

OCCURRENCES AND USES OF ZEOLITES IN ITALY

RICCARDO SERSALE

Istituto di Chimica Applicata, Universita di Napoli, Naples, Italy

Abstract – An overview of Italian zeolites must treat both the great number of occurrences of igneous zeolites in well-developed crystals and the enormous deposits of pyroclastic and sedimentary rocks containing large amounts of disseminated phillipsite and chabazite. In view of their industrial utilization, the occurrence, production, and use potential of this latter type of deposit is emphasized in this report, and hypotheses are offered on their genesis.

The genesis of the zeolitic tuffs of the Campanian Region is related to the alteration of volcanics by groundwaters. The neoformation of a certain zeolite (chabazite or phillipsite) is a function of two factors: temperature and/or composition of pore water. The genesis of sedimentary rocks containing mainly phillipsite, occurring in Northern Italy, is attributed to a diagenetic process in a marine environment.

The most promising uses of Italian natural zeolites are (1) the manufacture of "pozzolanic" cements, by adding the pulverized tuff to clinker, (2) pollution abatement of natural or wastewaters by ion exchange, (3) the dehydration of industrial vapors and gases, and (4) the selective adsorption of SO_2 from stack gases.

INTRODUCTION

Italian zeolites are extremely varied and range from the well-formed crystals found in cavities of basic eruptive rocks (basalts and melaphyres) to the widespread occurrence of fine-grained chabazite and phillipsite found in altered pyroclastic and sedimentary rocks. Although there is a considerable variety of zeolites in the basic eruptive rocks, their concentration is very low, and they are mainly of interest from naturalistic and basic scientific viewpoints. In the pyroclastics and sedimentary rocks, the restricted number of zeolite varieties is amply compensated for by high zeolite concentrations in exceptionally large deposits, thus providing an economic incentive to study the possible utilization of such rocks either as-mined or after beneficiation.

HYDROTHERMAL ZEOLITES IN ERUPTIVE ROCKS

Numerous localities where hydrothermal zeolites occur are shown in Fig. 1. The Mesozoic basic volcanites in Trentino and Alto Adige which outcrop from Alpe di Siusi to Val di Fassa and on to Marmolada contain many zeolites (Alietti *et al.*, 1967a), commonly of museum crystal quality, including heulandite, stilbite, chabazite, analcime, natrolite, scolecite, and thomsonite. Zeolites which occur in the basic volcanites of Veneto, particularly those of Montecchio Maggiore, are analcime, heulandite, clinoptilolite, mordenite, and epistilbite. Ferrierite and a rare magnesian zeolite (Alietti *et al.*, 1967a) occur in porphyrite in the Albero Bassi (Vicenza) region in addition to other species. Natrolite and analcime are also abundant in the porphyrites of the valley of Cordevole and Val de Gares; mordenite and heulandite are less common. Numerous zeolites (heulandite, stilbite, chabazite, dachiardite, mordenite, and, probably, epistilbite) occur in the famous aplite-pegmatite veins of the Island of Elba. In the quarry of Capo di Bove in the Latial eruptive district occur almost colorless crystals of gismondine, as well as other zeolites, such as pseudophillipsite and herschelite. Phillipsite occurs in the leucitic lavas of the Latial volcano.

In the Campanian region, phillipsite is found in small cavities of the leucotephritic blocks of Monte Somma; other zeolites formed by Vesuvian volcanic action are thomsonite, scolecite, and analcime. In Sicily there are well-formed crystals of analcime, epistilbite, chabazite, herschelite, gmelinite, gonnardite, mesolite, natrolite, phillipsite, and thomsonite in the schists of the Peloritani Mountains and in the basic volcanites of the Mt. Etna and Mt. Judica regions, as well as in Val di

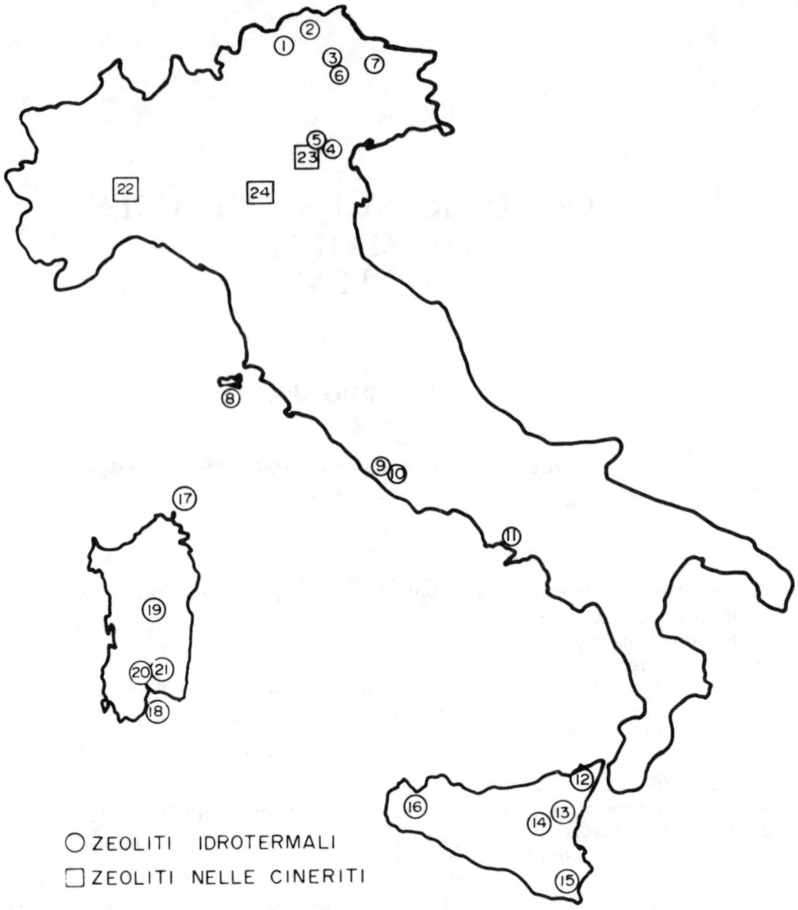

Fig. 1. Location map and areal distribution of hydrothermal zeolites (marked with circles) and of zeolitized cinerites (marked with squares).

Noto and western Sicily. In the pegmatitic cavities of granite on the Island of Maddalena in Sardinia, chabazite, stilbite, and scolecite occur. Also, in the small volcanoes of the Monastir zone, heulandite, stilbite, analcime, and gonnardite occur. Laumontite and harmotone have been found in the Sarrabus mines. In the fractures of an andesite rock, levynite and erionite have been discovered. Sodium stellerite occurs in an acid rock near Capo di Pula. Analcime, heulandite, stilbite, stellerite, sodium stellerite, chabazite, levynite, erionite, laumontite, mordenite, and mesolite have been detected in magmatic rocks.

ZEOLITES FORMED BY DIAGENESIS IN A CONTINENTAL ENVIRONMENT

In Italy zeolitized volcanic tuffs provide an inexhaustible supply of zeolites for industrial use (Sersale, 1972).

Tuff is widespread in the south central part of Italy, particularly in the volcanic areas of Campania, Latium, Monte Vulture (Potenza), and Roccamonfina (Caserta). These volcanic pyroclastics, largely vitreous, accumulated in every type of sedimentary basin. The silicate glasses present, because of their high free energy, are very reactive in aqueous conditions and transform by hydration, solubilization, and hydrolysis into clay minerals or zeolites. The most compact tuffs have a compressive strength of 30–60 kg/cm^2, which has enabled them to be used for building purposes since ancient times.

Products of the Campanian District

In the Campanian District there are two main types of alkali-trachytic tuffs: the older, locally known as *Campanian tuff* (Colella *et al.*, 1973), is found throughout Campania; the more recent tuffs, known as *Neapolitan yellow tuff*, cover the whole of the

Fig. 2. View of a deposit of Neapolitan yellow tuff.

Phlegrean Fields, including the Naples area.

The Campanian tuff formation is commonly zeolitized. Sometimes the zeolitized facies appear to be argillified in places.

The Neapolitan yellow tuff is a uniform straw color from top to bottom. The shape of the deposits may be seen in Fig. 2, which also gives some idea of the ore reserves. Commonly, the tuff is overlain by an amorphous, unwelded product known as *pozzolana*, which may be considered to be the unzeolitized part of the tuff. The chemical composition of Campania tuff (Di Girolamo, 1968) and Neapolitan yellow tuff (Sersale, 1958), along with their unzeolitized parts, are reported in Table 1. As seen from these data, zeolitization results

TABLE 1. Chemical compositions of Campanian and Neapolitan tuffs (wt. %)

	Campanian tuff	Unzeolitized equivalent	Neapolitan yellow tuff	Unzeolitized equivalent
SiO_2	52.64	56.12	54.68	57.0
$Al_2O_3(+TiO_2+Mn_3O_4)$	16.87	19.58	17.70	18.34
Fe_2O_3	3.63	2.68	3.82	2.20
FeO	0.51	1.79	0.29	2.14
MgO	1.67	1.41	0.95	0.97
CaO	4.50	3.62	3.66	3.15
K_2O	6.45	7.08	6.38	8.05
Na_2O	1.50	3.13	3.43	4.18
Cl_2	0.07	0.03	–	–
F_2	0.06	0.10	–	–
H_2O-	4.45	1.42	–	–
H_2O+	8.20	3.10	9.11	3.75
O/Cl_2-F_2	100.45– 0.04	100.06– 0.04	100.02	100.58
	100.41	100.02		

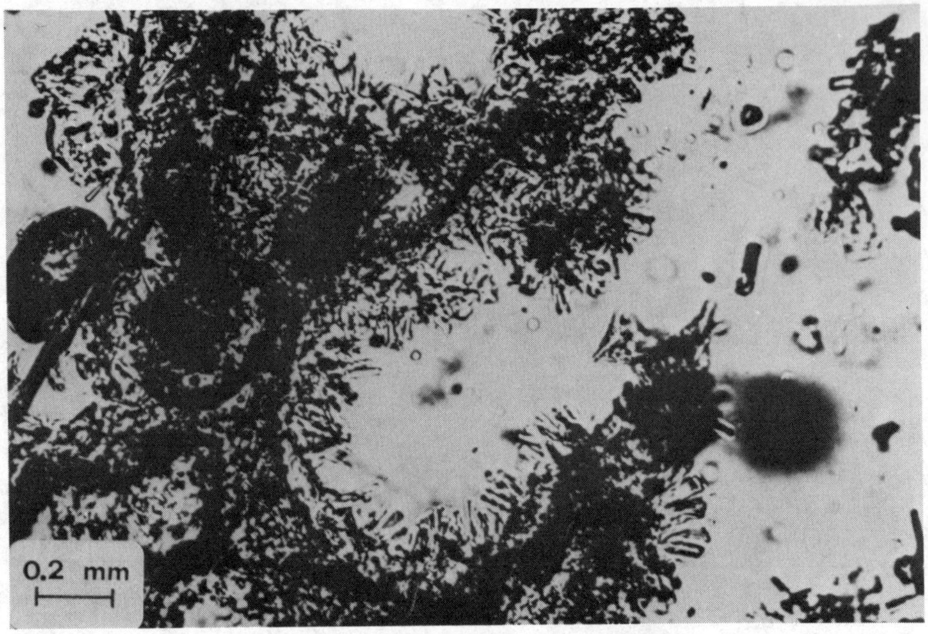

Fig. 3. Optical micrograph of Campanian tuff. Zeolite minerals in the small cavities of the rock. Nicols.

in an increase in water and lime and decreases in silica. alkalis, and alumina.

Principal deposits of Campanian tuff. Campanian tuff deposits normally lie in the fluvial-lacustrine basins; there are many outcrops where rivers have dissected the deposits. For the most part the tuff seems to be reworked and contains zones of rounded pumice fragments and of other non-volcanic detritus from the surrounding mountains. In many deposits a clearly sedimentary structure is to be seen.

Microscopic examination of thin sections of a typical tuff sample, collected in the Mercogliano (Avellino) area (Aiello *et al.*, 1974), reveals the rock to consist of a nearly isotropic matrix of cryptocrystalline zeolites which surround crystalline fragments of sanidine, biotite, and, to a lesser extent, pyroxene and plagioclase. Crystalline zeolites are also recognizable in the small cavities in the rock (Fig. 3).

From powder X-ray diffraction data, chabazite and phillipsite are the principal zeolites present in the tuff. The former is much more abundant than the latter, *phillipsite* making up only about 10% of the rock. Only a few samples from the different deposits were found to be without *phillipsite*. An approximate estimate of the chabazite content, made from different samples, is shown in Table 2 (Aiello *et al.*, 1974). Quantitative evaluations revealed that the highest percentages of zeolites were to be found in the deposits furthest from the source. This could be related to the higher amount of fine, vitreous material present and consequently its higher reactivity.

TABLE 2. Chabazite percentages in tuff samples collected from different quarries

Quarry location (see Fig. 3)	% Chabazite
Monte della Taglia (Cicciano)	55
Masseria delle Fosse (S.Felice a Cancello)	58
Monte Tagliato (Maddaloni)	62
Tufara (Mercogliano)	70
S.Caterina (Picarelli)	62
Starza (Altavilla Irpina)	64
Felette (Torre le Nocelle)	68
Madonna delle Grazie (Mirabella Eclano)	71
Cannito (Pastorano)	61
Campo Rosa (S.Agata dei Goti)	64

Principal deposits of Neapolitan yellow tuff. The occurrence and chemical features of the Campanian tuff and the more recent Neapolitan yellow tuff (2nd Phlegrean period) are very similar. However, in contrast to what has been observed with Campanian tuff, the peripheral deposits of the Neapolitan yellow tuff are the least zeolitized ones. In Fig. 4 one can see the distribution of the main deposits of Neapolitan yellow

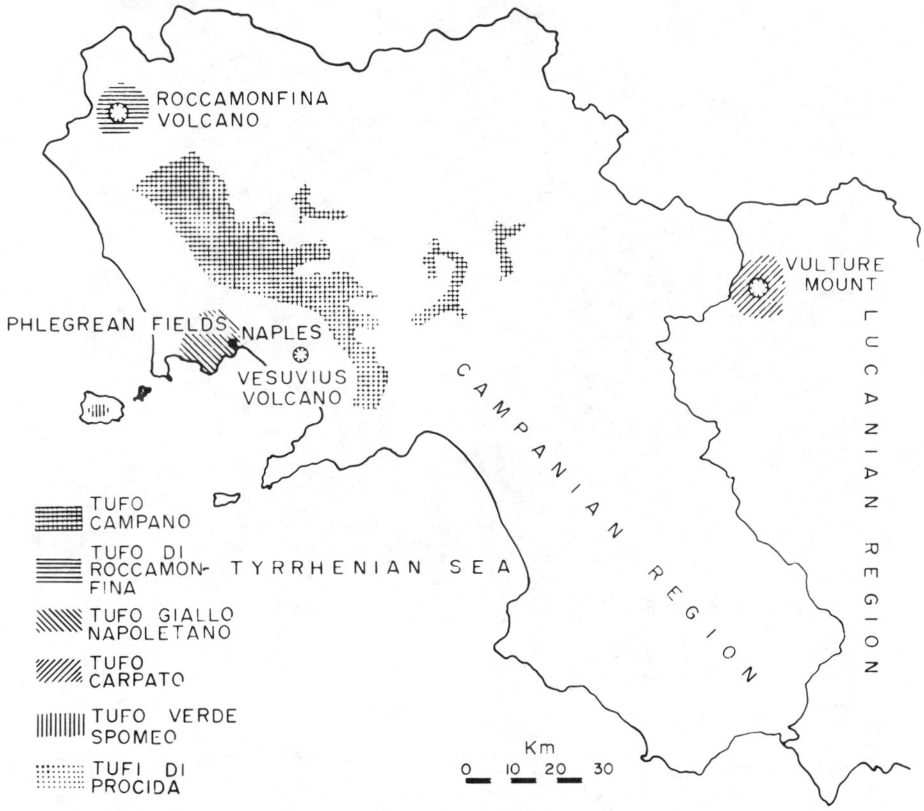

Fig. 4. Location map and areal distribution of the main deposits of zeolitic Campanian tuff, Neapolitan yellow tuff, and the Roccamonfina and Vulture zeolitic tuffs.

tuff. As with Campanian tuff, the ore reserves are very large.

The Neapolitan yellow tuff is highly homogeneous in appearance, rich in fibrous and bubbly fragments, and easily sawed and fabricated into dimension stones. Microscopically, the fresh rock contains unevenly distributed cavities, fragments of pumice, and crystals of plagioclase and a light-green pyroxene. The yellowish, nearly isotropic matrix which cements the aforementioned constituents was found by powder X-ray diffraction analysis to consist mostly of phillipsite and chabazite. Unlike the Campanian tuff, the Neapolitan yellow tuff is predominantly phillipsite, containing up to about 80% (Lenzi *et al.*, 1970) (Fig. 5).

Fig. 5. Scanning electron micrograph of Neapolitan yellow tuff. Prismatic crystals of phillipsite.

Deposits of the Latial District

The deposits of leucite tuff (Fornaseri *et al.*, 1963) of the Latial volcanoes are important both for their industrial potential and for the information they provide on zeolite paragenesis. The distribution of the principal deposits is shown in Fig. 6. They are as follows:

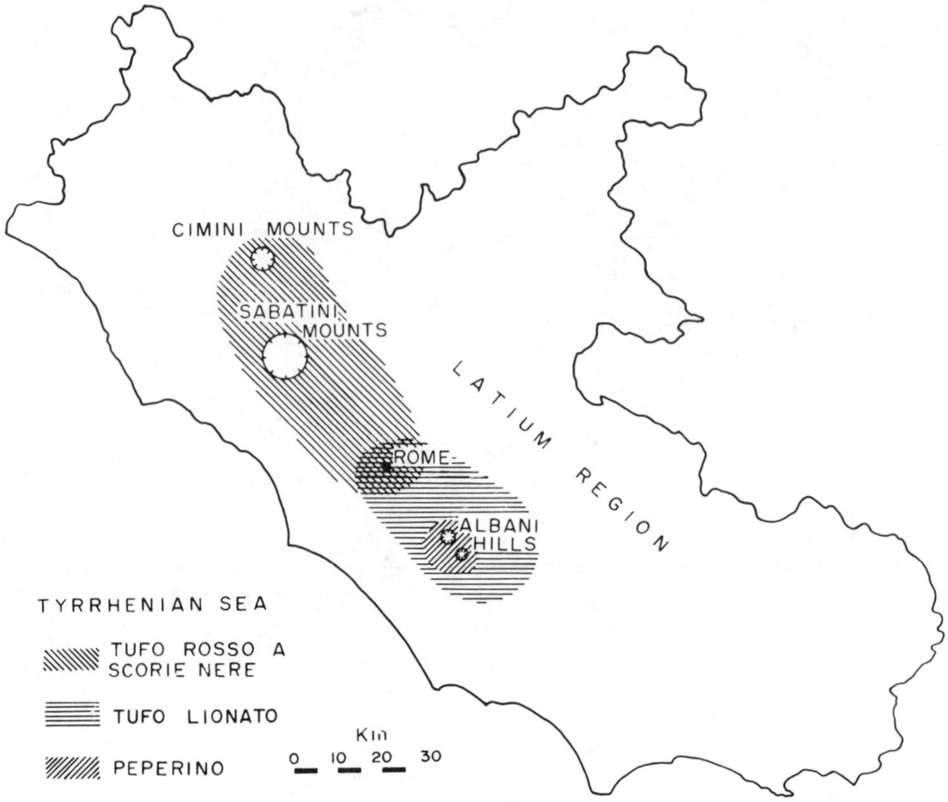

Fig. 6. Location map and areal distribution of the main deposits of zeolitic tuffs in the Latial district.

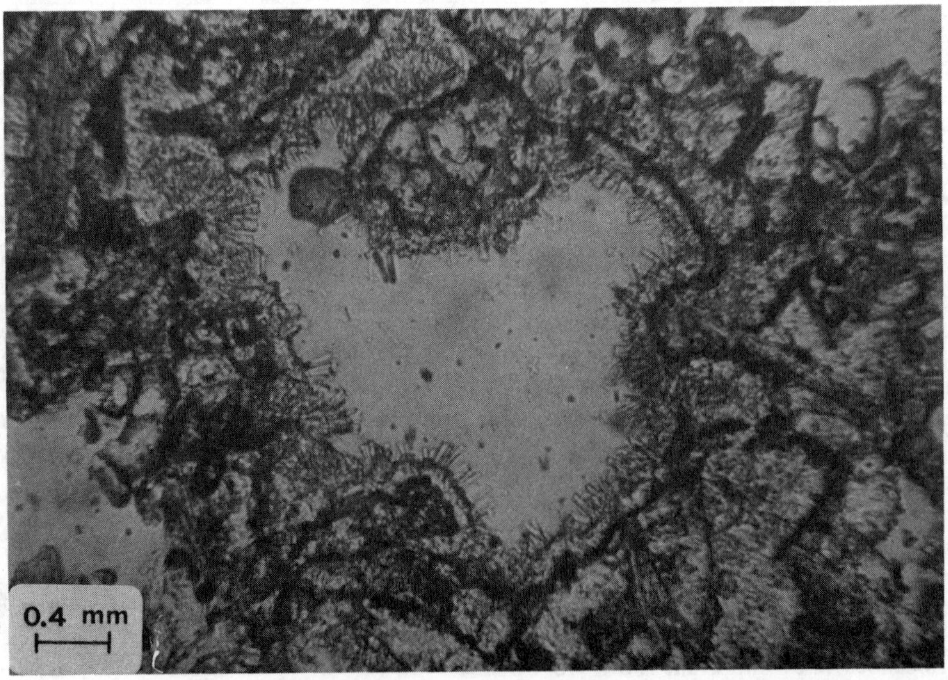

Fig. 7. Optical micrograph of "lionato" tuff. Nicols.

1. The *Lionato tuff* is a lithic, reddish-fawn tuff. It is grey or yellow in places and is particularly abundant in the region from Castel di Leva (Divino Amore) to Zagarolo (Sersale, 1959). The Lionato tuff originated from a violent explosion following the formation of black pozzolanas (Scherillo, 1933). Elevations of occurrences of this tuff generally range from about 20 m to 100 m above sea level. Some outcrops are between 260 m and 350 m above sea level. Optical examination of thin sections show the product to be a mixture of ash, scoriae, quite commonly rounded, and fragments of the earlier rocks. The rock cavities commonly are filled with crystals, mainly zeolites (Fig. 7). Dark-green pyroxene, mica, plagioclase, black leucitic inclusions, and calcareous inclusions are also present. Some calcite is of secondary origin. Chabazite is the predominant zeolite, although in some samples phillipsite predominates. An estimate of the chabazite content in most tuffs based on the water content is in the range 55 to 70%. Table 3 gives the chemical analyses of two Lionato tuffs, one zeolitized to chabazite, taken from the Via Prenestina; the other zeolitized to phillipsite, taken from Via Casilina, between Labico and Valmontone. Differential thermal analysis confirmed that modest quantities of clay minerals are present in the phillipsite-rich sample. The zeolitic Lionato tuff grades into argillitic tuff which contains no zeolites but which is similar in appearance.

2. The *peperino* tuff of Colli Albani is lithic, grey, and found in the area between Ariccia and Pomezia (Sersale, 1959). The Roman peperino is the name given to a product typical of the Latial volcano. It consists of grey ash with numerous inclusions, firmly cemented, and therefore prized as a building material since ancient times. It includes *typical peperino* (or *Lapis Albanus*) and *Lapis Gabinus*, the former being rich in isolated mineral crystals, with abundant inclusions of various kinds; the latter darker, quite rich in lapilli, but contains fewer inclusions and has a finer grain size.

Table 3 gives the chemical analyses of a sample of typical peperino taken from the Chigi Park in Ariccia and a sample of *Lapis Gabinus* taken from the south side of the Castiglione Valley. A high amount of fixed water and a small quantity of alkalis, especially potassium, are to be noted. Moreover, one should note the high proportion of calcium, part of which is visible to the eye as white carbonate inclusions. These deposits are mixtures of ash, leucite, mica, calcite, and pyroxene, and lava inclusions. Zeolite crystals, some with a fibrous-radial morphology, occur in cavities of the glassy fragments in the tuff. Zeolites and calcite are the cementing agents of the tuff. Powder X-ray diffraction analysis shows that phillipsite is the predominant zeolite in the matrix cement along with chabazite (var. herschelite). The Peperino tuff differs from the Lionato tuff in that peperino tuff contains more secondary calcite and no clay minerals.

3. The *red tuffs of the Sabatia and Cimina regions* are rich in pumice and black vitroclasts, respectively. A well-known feature of the Latial region is the large number of now-extinct volcanoes which once produced mainly leucitic lavas (Sersale, 1960a). It should first be pointed out that the Sabatini volcanoes which rim crater Lake Bracciano, which occupies the base of a crater, form the group nearest to Rome, north of the Tiber, in the Sabatia region (Scherillo, 1940). These volcanoes have created an extensive volcanic massif similar to that of the Latial Roccamonfina and Vesuvius volcanoes. Eruptional vents are numerous and widespread. Individual volcanic deposits, moreover, except for some explosive craters, are all relatively small and independent of each other, indicative of several short intermittent periods of activity. From this point of view, the Sabatia region, particularly the eastern half, is comparable to the Phlegrean Fields. The most significant outcrop of this tuff is found along Via Flaminia and initially follows the Tiber Valley from Grottarossa to Castel Giubileo to Prima Porta. It is also present by Via Cassia, close to the Isola Farnese and Veio, along Via Claudia near Galeria, near Cerveteri, and in the valleys between Via Cassia and Via Flaminia. In general, red tuffs with black pumices overlie tephryte-leucitic tuffs. Deep valleys with perpendicular walls dissect thick sequences of this tuff.

The tuff deposits of the Cimina region border those of the Sabatia region and appear to be a natural continuation of them. Two main eruptive centers are designated as the Cimino and Vico volcanoes, although these have been extinct for some time. The lithic tuffs with black vitroclasts erupted mainly from the Vico Volcano and are the result of a violent explosion which produced the large crater that is now a lake. The tuffs may be partly the result of Sabatini and Vulsini volcanic activity (Scherillo, 1940).

The Cimina tuffs are much more extensive than those of Sabatia and reach a maximum thickness of more than 70 m at the foot of the Vico volcano. Two distinct levels have been noted, separated by a seam of red tuff containing no vitroclasts or white pumice. According to Sabatini (1912), the tuffs overlie Pliocene conglomerate, Pliocene clay, and Eocene limestone.

Table 3 gives the chemical composition of a typical sample of red tuff with black pumice from the Sabatia region, taken from Via Flaminia, at Prima Porta (Rome), and a typical sample of red tuff with black vitroclasts from the Cimina region, taken from the amphitheater of Sutri (Viterbo). The town of Sutri is situated on a plateau consisting of this type of tuff, and the Sutri amphitheater (latter half of the first century B.C.) is entirely dug out of the tuff. The vitric inclusions in these

TABLE 3. Chemical compositions of tuffs, locally named "lionato" (wt. %)

	Chabazitic	Phillipsitic		Chabazitic	Phillipsitic	Chabazitic	Phillipsitic	Chabazitic		Phillipsitic		Chabazitic
	(1)	(2)		(3)	(4)	(5)	(6)	(7)		(8)		(9)
SiO_2	44.95	42.92	SiO_2	40.84	41.86	53.82	53.38	54.96	SiO_2	54.34	SiO_2	57.80
$Al_2O_3(+TiO_2+Mn_3O_4)$	20.49	22.53	$Al_2O_3(+TiO_2+Mn_3O_4)$	15.92	15.12	19.18	19.23	19.08	$Al_2O_3(+TiO_2+Mn_3O_4)$	19.65	$Al_2O_3(+TiO_2+Mn_3O_4+P_2O_5)$	18.62
Fe_2O_3	6.14	7.52	Fe_2O_3	7.11	6.78	3.91	4.03	2.66	Fe_2O_3	1.72	Fe_2O_3	2.42
FeO	0.17	0.25	FeO	2.91	1.13	0.41	0.73	0.36	FeO	1.12	FeO	0.32
CaO	10.32	8.00	CaO	13.56	13.78	4.08	4.04	4.20	CaO	3.47	CaO	3.09
MgO	2.81	2.37	MgO	4.54	3.35	1.34	1.43	1.09	MgO	1.03	MgO	1.00
Na_2O	1.75	1.31	Na_2O	1.80	2.02	1.50	1.97	3.89	K_2O	7.10	K_2O	7.07
K_2O	4.39	5.77	K_2O	4.75	5.90	5.42	7.40	3.93	Na_2O	4.15	Na_2O	5.34
H_2O+	9.51	10.00	CO_2	2.85	3.68	tr.	tr.	0.36	H_2O-	1.62	H_2O+	3.81
			H_2O+	5.82	5.99	9.85	8.02	9.73	H_2O+	4.68	CO_2	1.01
											Cl_2	0.21
											$-O/Cl_2$	0.04
	100.64	100.67		100.10	99.61	99.51	100.23	100.26		98.87		100.65

(1) Lionato chabazitic tuff.
(2) Lionato phillipsitic tuff.
(3) Typical Peperino tuff.
(4) Lapis Gabinus tuff.
(5) Tuff from Sabatia region.
(6) Tuff from Cimina region.
(7) Carpato tuff.
(8) Yellow tuff from Roccamonfina
(9) Green Epomeo tuff.

tuffs are better defined as vitroclasts rather than pumice, as they are always more compact than those in the tuffs of the Sabatia region.

The chemical compositions of the tuffs in the two regions are similar. Both have a high proportion of fixed water and significant quantities of potassium and calcium.

Powder X-ray diffraction analysis of material from the ground mass has shown chabazite to be the predominant zeolite (Lenzi, 1974). Based on the water content of the tuffs, the zeolite content (expressed as chabazite) is remarkably uniform and ranged from 40% to 45%.

Products of Vulture

This volcanic center was described in detail by De Lorenzo (1899). It consists of two distinct lava domes. The older one of Toppo S. Paolo is considerably altered and is hauynfonolitic. The other is hauynofiric and forms the platform on which Melfi is built. In addition, the deposit consists of the Vulture cone itself which opens westward onto a wide caldera whose southwest flanks were eccentrically rent by the violent explosion from the distoma crater which now contains the two lakes of Monticchio.

At the base of the whole Vulture volcanic deposit and lying directly on Eocene, Pliocene, and Pleistocene sediments are lithic tuffs (locally known as "carpati" tuffs) and unwelded products consisting of blocks, lapilli, pumices, and pozzolana. These tuffs surround the cone as a nearly continuous marginal band or fringe and occur from Foggiano through Monte Lapis, Serra Pirrone, Melfia, Macera Valley, Ripacandida, and Rionero, to the broad stream of Atella. The reserves of the deposits are very large; their distribution is shown in Fig. 4. Quarries in the past have supplied huge quantities of cut stone for ordinary building construction. At present the quarries are idle, owing to anti-seismic standards introduced to the region after the earthquake of 1930.

Table 3 lists the chemical analysis (Sersale, 1960b) of a tuff sample taken from the old "Severini" quarry at San Gillo (Melfi) on the slopes of Bicocca. The high amount of fixed water and the low amounts of alkalis and calcium are to be noted.

The tuff contains cavities about 2-3 mm in diameter throughout the mass, which, together with the almost total absence of pumices and vitroclasts, constitute its most striking feature. It is likely that the tuff drew its name "carpato" from these cavities which are similar to those found in the Apulian "carpato", a very porous type of sandy limestone. The walls of the cavities are lined with a whitish-yellow crumbly material, which

Fig. 8. Scanning electron micrograph of "carpato" tuff. Zeolite minerals covering the walls of the cavities.

under the microscope can be seen to consist of minute crystals with weak birefringence. X-ray diffraction analysis identifies these crystals as chabazite. Numerous transparent crystals of megascopic sanidine are also present.

Thin section examination of the tuff shows a yellow homogeneous mass consisting of a chabazite (with some analcime) cementing numerous sanidine inclusions, less abundant fragments of garnet, pyroxene, and amphibole. On the basis of the water content, the zeolite content of this tuff (Fig. 8), calculated as chabazite, is around 50%.

Products of Roccamonfina

The eruptive deposits of the Roccamonfina volcano, whose distribution is shown in Fig. 4, are the oldest in the entire Campanian district and overlie sedimentary deposits. Although these deposits vary in thickness, they have been zeolitized.

Table 3 gives the chemical composition of this tuff based on the analysis of a sample taken from Ventaroli (Aiello, 1962). In thin section it appears as a mixture of small vitroclasts and crystals in a slightly opaque matrix of about 25% chabazite (Fig. 9). Sanidine and pyroxene crystals are most abundant with lesser amounts of plagioclase and rare mica.

Product of Somma-Vesuvius

Only sporadic zones of zeolitized material are found in the pyroclastic products of Somma-Vesuvius, except

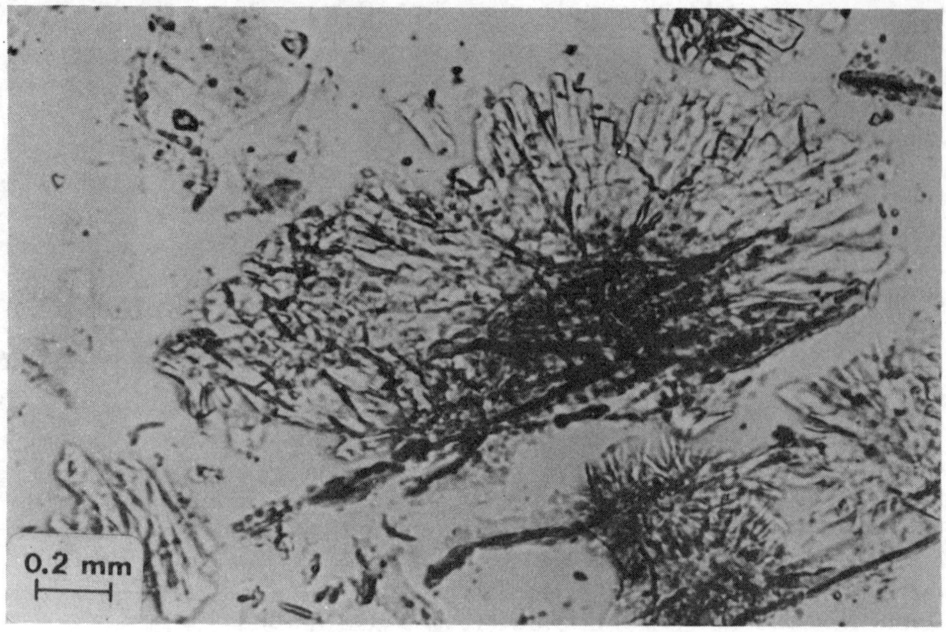

Fig. 9. Optical micrograph of yellow Roccamonfina tuff. Zeolite minerals on a glass fragment.

for the greyish-yellow lithic material covering Herculaneum (eruption in 79 A.D.). According to Franco (1976) this zone is about 10 m thick and contains chabazite, sometimes in macroscopic crystals.

Discussion of genesis

The zeolite deposits described in this section are the result of the diagenetic action of groundwaters on volcanic glass. Thus, the deposition should be assigned to the second group of the classification of Mumpton (1973). It seems likely that zeolitization takes place only after a portion of the glass constituents had dissolved (to a certain concentration of the soluble species) and remained in contact with the glass for a reasonable length of time. However, due to the lack of sufficient experimental data, it is difficult to explain the following phenomena:

- the coexistence of different zeolitic species in the same deposit, even if in different ratios;
- the greater occurrence of a particular zeolitic species;
- the presence of well-formed crystal zeolites in the larger cavities of the tuffs showing disseminated zeolites;
- the coexistence of clay and zeolitic minerals.

In attempting to explain these complexities, the following factors which substantially influence the zeolitization process, should be considered:

- the chemical composition and microstructure of the sedimentary material;
- the particle size distribution and permeability of the sedimentary material;
- the pH in the sedimentary environment;
- the ions present in solution;
- the age of the deposits;
- the pressure and temperature;
- the composition of the interstitial fluids.

Thus, the coexistence in a single deposit of one zeolitic species or another may be attributed to variations in the physical conditions present when the vitreous phase dissolved and to changes in the chemical composition of the contact solution, which inevitably occurred during the zeolitization process.

The appearance of particular zeolitic species, for example, chabazite and/or phillipsite, such as one finds in the deposits described above, may be accounted for by differences in the composition of the sedimentary starting material (especially the acidity of the glass), in the physical conditions, and in the composition of the contact solutions. The presence of well-formed crystals of zeolites in the cavities or cells of the vitroclasts suggests that they formed by the reaction of an aqueous solution percolating from the outside to the inside of the mass, with a consequent increase in the basicity owing to hydrolysis phenomena, or to leaching of the vitreous grains. Thus, localized attack on the vitreous phase with crystals forming in the intergranular spaces and cavities of the pumices and vitroclasts would occur only in those areas where factors combined to favor the growth: solution concentration, contact time, state of surface, and, eventually, temperature.

The coexistence of zeolites and clay minerals may be explained by supposing a previous argillification of the solid mass by diagenetic actions and later zeolitization of the clay minerals by the action of groundwater. The idea of zeolitization exclusively by groundwaters would agree, in fact, with the succession observed in some Latial tuffs (halloysite above and zeolites below), as well as in Campanian tuff deposits which generally show a seam of hydrated glass below, a zeolitized lithic nucleus in the middle, and a gradation into an argillified band above.

ZEOLITES FORMED BY DIAGENESIS IN AN EVAPORITIC ENVIRONMENT

Zeolitization in a transitional environment is characterized by the presence of clinoptilolite associated with sodium-montmorillonite in the silt-size fraction of sedimentary strata outcropping near Zovencedo (Colli Berici) (Alietti *et al.*, 1967). Clinoptilolite and sodium-montmorillonite are both known to have a high proportion of alkali elements and are commonly associated with each other as diagentic tuff products. Sodium-montmorillonite and clinoptilolite have not been found elsewhere in the Colli Berici region. The genesis of this formation may be attributed to diagenetic processes occurring in an evaporitic environment (saline lakes), comparable in pattern to the "Barstow Formation" (California). The deposition environment corresponds to the first group of Mumpton's classification.

ZEOLITES FORMED BY DIAGENESIS IN A MARINE ENVIRONMENT

The zeolite tuff formations of the islands of Ischia, Procida, and Vivara (Fig. 4) are particularly interesting, their depositional environment falling into Mumpton's third class.

Green tuff of the Island of Ischia (Epomeo)

The Ischia tuff is of great interest in the study of Ischian volcanic activity to which the island owes its existence. Such activity began before the Phlegrean eruptions and developed later on its own. Some has been of historical interest, such as the lava eruption at Arso in 1302. The distribution of the tuff is shown in Fig. 4. The formations are of considerable thickness as may be seen from the outcrops that reach as high as the crest of Epomeo at about 780 m. Table 3 lists the chemical

Fig. 10. Samples of green Epomeo tuff, Neapolitan yellow tuff and Campanian tuff.

Fig. 11. Optical micrograph of green Epomeo tuff. Glauconite in a pumice. Nicols.

composition of a sample from Monte Epomeo. It should be noted that this tuff has a lower water content than those previously discussed. A typical specimen is shown in Fig. 10.

Examination of a thin section of Epomeo tuff shows a mixture of pumice with numerous fragments of sanidine phenocrysts, sparse lamellae of black mica, plagioclase, and pyroxene. These constituents are cemented by a green interstitial matrix of analcime, glauconite, and some potassium feldspar (Sersale, 1959c) (Fig. 11). The complete lack of sorting of the constituents of the green Epomeo tuff, as well as the absence of marine fossils in the main body, suggests that it was derived from subareal-pyroclastic deposition. Subsequent tectonic movements dropped the deposit to below sea level and allowed a diagenetic zeolitization to take place (Rittmann, 1968). Successive tectonic movements uplifted the formation, so that the tuff is now clearly visible in the region of Mount Epomeo, from which it takes its name.

Tuffs of Procida and Vivara (Phlegrean Islands)

The islands of Procida and Vivara stand on a continental platform within the bay of Naples. Procida is 3 km long, 2 km at its widest, and consists essentially of clastic volcanites. The islet of Vivara is a crater, a portion of which is morphologically linked to Procida by a band of rearranged volcanites, which lie only a few meters above the sea. Zeolites are present as alteration products which cement lithic products (Cristofolini *et al.*, 1973). They also occur as well-crystallized grains filling bubbles as radial aggregates (Fig. 12). Commonly, such aggregates occur in the interstices between the shards. Also present are microcrystalline zeolite aggregates with a mosaic granular structure. Powder X-ray diffraction analysis of some fractions show that the zeolites are mainly chabazite and phillipsite. In the Vivara tuff, harmotone has been found (Fig. 13), and in samples, it is the only zeolite present (Porcelli, 1976).

Zeolites in the Cineritic levels of the Central-Northern Apennines

In the Italian sedimentary series, and particularly those from the Oligocene to the Pliocene, cineritic levels are often found, with thickness in meters, which have been zeolitized where conditions were favorable.

Some zeolite layers consisting of a whitish, fine-grained lithic material have been found recently in the extensive formations of greenish marls of the Oligocene–Miocene series of Garbagna (Alessandria) (Sersale *et al.*, 1963).

A cross-section on an arbitrary scale sketch of these layers is given in Fig. 14. It is seen that the zeolites constitute three definitive layers in the green marls; the lower is 1.2 m thick, a middle layer is 4.0 m thick and an upper layer is 1.7 m thick. The outcrop of these layers

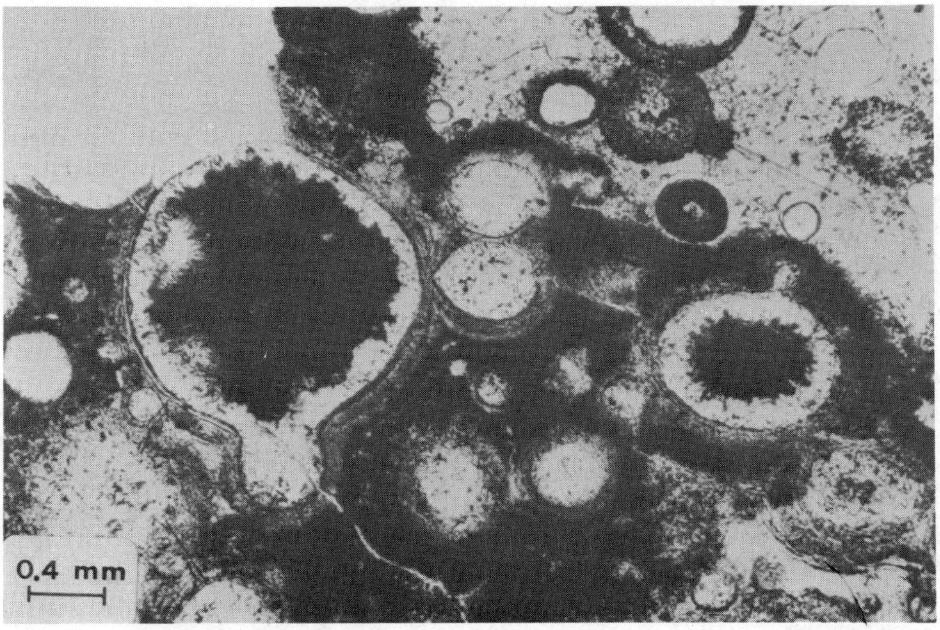

Fig. 12. Optical micrograph of Vivara tuff (Procida). Zeolite minerals in the small cavities of the rock. In the bubbles the glass is not zeolitized (in black).

Fig. 13. Scanning electron micrograph of Vivara tuff (Procida) showing harmotome crystals.

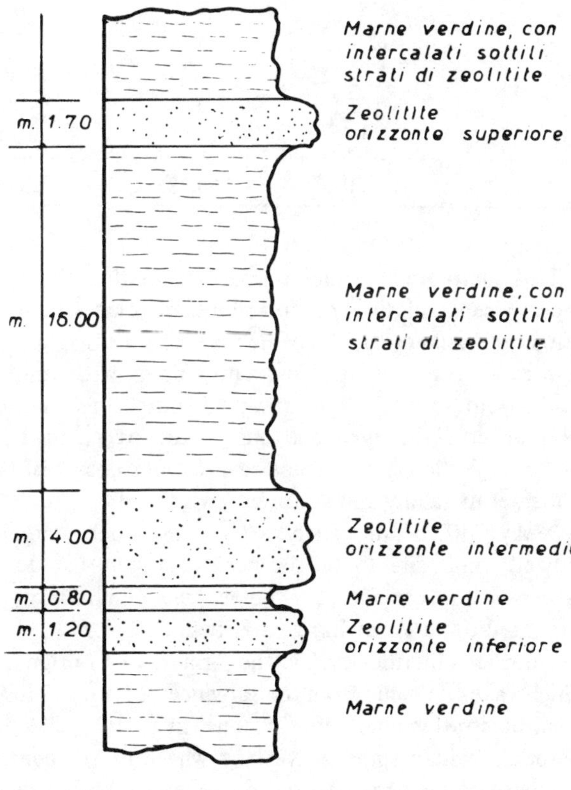

Fig. 14. Cross-sectional sketch (on an arbitrary scale) of Garbagna zeolitite layers in the green marls.

can be followed throughout the Garbagna basin and beyond; the ore reserves of these deposits are therefore very large.

The zeolitized material in all three layers is stratified and very compact. At the bottom in each layer the size of the fragments is greatest, the content of femic crystals is highest, and the marl–zeolite zones are always most distinct. This agrees with the hypothesis of a sedimentation of the volcanic material in calm waters.

The chemical analysis of samples from the three layers is given in Table 4.

TABLE 4. Chemical composition of zeolitic rocks from Garbagna (wt. %)

	Lower layer	Middle layer	Upper layer
SiO_2	48.00	54.28	50.73
Al_2O_3	15.96	17.21	17.65
Fe_2O_3	4.60	1.47	2.72
FeO	0.35	0.22	0.28
CaO	5.37	0.80	3.92
BaO	0.53	1.23	0.22
MgO	2.21	0.87	1.82
K_2O	2.70	2.70	2.39
Na_2O	2.32	6.07	2.74
H_2O+	8.50	6.67	8.27
H_2O-	8.72	8.34	9.36
CO_2	1.03	0.23	ass.
Cl_2	0.05	0.08	0.05
SO_3	0.17	0.11	0.11
	100.51	100.28	100.26

With petrographic microscope examination the rock appears as a mixture of crystalline fragments consolidated by a cementing mass which is nearly isotropic. The cavities in the rock are frequently lined with minute birefringent crystals consisting of numerous minute feldspar crystals, pyroxene and mica, fractured fragments of femic crystals, some calcitic microfossils of the globergerina family and stratified glass shards.

X-ray diffraction analysis of the three samples showed phillipsite to be the zeolitic mineral predominating in the cement. The percentage of phillipsite is estimated to be somewhat greater than 50%.

Another cineritic level is the Tripoli di Contignaco, which takes its name from the province of Parma. This is a litho-stratigraphic unit, consisting of siliceous materials, which spreads over the whole of the central Northern Apennines. According to some investigators (Gianello *et al.*, 1969), the lithic, zeolitic deposits of the Alexandrian Apennines correspond to the Tripoli di Contignaco, suggested by the presence of phillipsite in the latter's cineritic levels.

The presence of phillipsite in the lithic, zeolitized layers of the Alexandrian Apennines is due to the diagenitic transformation of the cineritic material in a marine environment as evidenced from the attitude of the beds in the extensive green-marl marine deposits and the presence of microfossils typical of this environment.

ZEOLITES IN THE IBLEAN JALOCLASTITES

This term refers to volcanic deposits consisting mainly of vitreous fragments altered to varying degrees in a subaqueous or at least in a water-rich environment (Cristofolini *et al.*, 1973). In the latter case the term refers to deposits previously known as "palagonitic tuffs". Thus, the depositional environment would fall into the fifth group in Mumpton's classification.

The northern part of the Iblean region (South Sicily) has vast outcrops of basic volcanites which extend over an area of about 350 km^2. The sedimentary series date from the Upper Miocene to the Pleistocene; in addition, there are less extensive outcrops of volcanics dating to the Cretaceous. A petrographic study of the Iblean volcanoclastics determined that the vitreous clastic materials are cemented by secondary zeolites and calcite. Zeolites are generally found in aggregates on the clay minerals which coat the surfaces of the vitreous grains. Cryptocrystalline, concentric growth bands indicate the existence of more than one growth stage of the zeolites. The zeolites were identified by powder X-ray diffraction analysis as phillipsite with a very small amount of chabazite. As regards reactions of glass in a subaqueous environment, the literature data indicate that they can cover a wide range of environmental conditions with respect to solutions of different composition and temperature. This is observed to be the case for the jaloclastitic deposits considering their attitudes and the variability of their depositional and environmental conditions in time following their emergence from the sea.

Thus, it is difficult to interpret the transformation of the glass subsequent to any possible early reaction with hydrothermal fluids or with slow circulating sea waters in the unwelded volcanite mass over long periods of time.

It is also difficult to determine the kinetics of the reactions, although it is seen in thin section that in the later stages the calcite crystals formed after the zeolites, and that in the early stages the hydration and extraction of elements from the glass occurred during the first transformational stage. The intermediate stage during which the zeolites of the cement and the recrystallizated clay minerals formed is not so clear. These phases are known to also grow in different environments and by different processes than those that form zeolites. However, one can postulate that the diffusion of ions from the glass is a significant factor in the formation of clay minerals owing to the recrystallization of the glass enriched in water and particularly impoverished in alkaline elements and calcium, and the process also provides selective removal of elements from the glass into the cavities where they become nutrients for zeolite growth – all part of a single process.

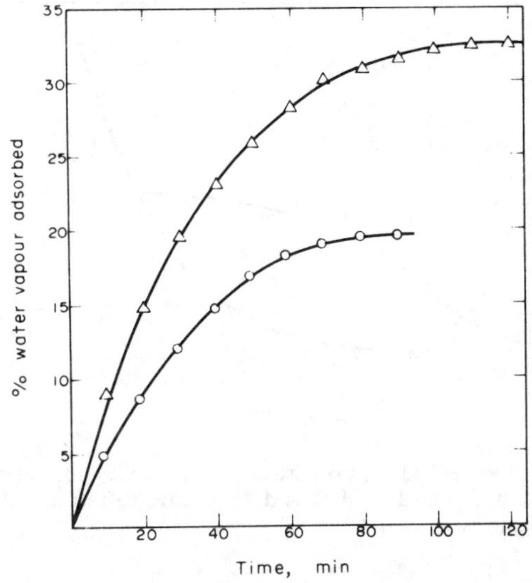

Fig. 15. Water-vapor adsorption by silica gel (marked with small triangles) and by chabazite tuff (marked with small circles) from humidified nitrogen flow.

UTILIZATION OF ZEOLITIC VOLCANIC TUFFS

The use of volcanic tuffs in important technological processes is still in the laboratory stage in Italy and has not yet reached the commercial stage. This situation is due to the fact that the discovery of such large uniform reserves of zeolites was made only recently, and research into their application is only beginning. Until now research has been concerned mainly with the chemical and structural aspects of the deposits. This paper, dealing with their mineral description, should therefore also be considered as background information leading towards their industrial utilization. At present, however, it is evident that they have a promising future as industrial minerals.

Beneficiating the tuffs to increase the zeolitic content has been proposed utilizing the heterogeneity of the rock and varying hardness of its constituents (Aiello *et al.*, 1972). A separation of the fractions was carried out by grinding the pulverized rock for varying intervals of time. As the zeolite fraction is the most friable, quite short grinding times favor its concentration. Grinding a sample of Campanian tuff from Mercogliano, first reduced to 2-3 mm particulates within a minute in a Fritsch Pulverisette mill separator, gave a 40% yield of 256 mesh powder containing more than 85% chabazite.

Samples of tuff enriched by this procedure were used to determine adsorption and cation-exchange capacities.

Volcanic tuffs as desiccants

The behavior of zeolite tuffs as desiccants was studied both in the liquid and gas phases (Colella *et al.*, 1973). The adsorption of water from the liquid phase was tested with water-dioxane solutions because of their complete miscibility and the fact that the tuff does not adsorb dioxane.

It was observed (Sersale, 1972) that the residual concentration of water in the system in equilibrium with the activated tuff powder was sufficiently low and that the tuff is competitive with 5A sieve.

Figure 15 gives results of adsorption from the gas phase and shows an adsorption-rate curve for water at 25°C under atmospheric pressure for a sample of chabazite tuff from Mercogliano using a humidified nitrogen flow. Silica gel was used as the reference. Under the same conditions the adsorption capacity of silica gel is 50% greater than that of tuff.

Figures 16 and 17 show the effect of temperature and gas pressure on adsorption properties. The adsorption isobar at 760 mmHg (Fig. 16) shows that tuff, even at a relatively high temperature (200°C), keeps its ability to adsorb discrete quantities of water vapor (5%), while silica gel is already unusable above 100°C. Figure 17 emphasizes the well-known fact that at 25°C silica gel adsorbs water vapor to an acceptable degree at pressures

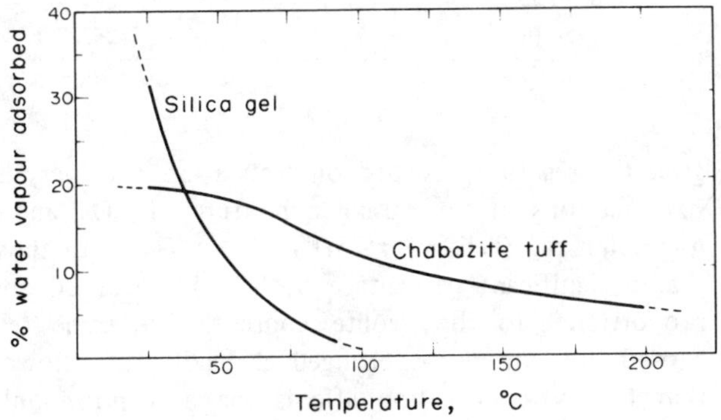

Fig. 16. Water adsorption at atmospheric pressure and 75% R.H.

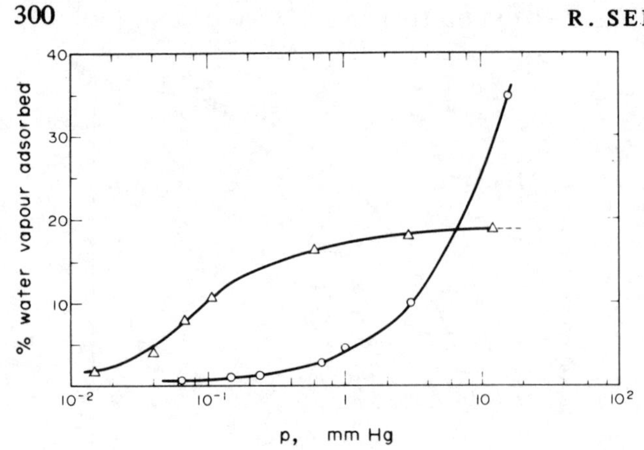

Fig. 17. Water adsorption isotherms at 25°C for silica gel (marked with small circles) and chabazitic tuff (marked with small triangles).

above 1 mmHg while tuff does so even at low gas pressures well below 10^{-1} mmHg. This comparison shows that tuff can be used to advantage as a drying substance in low humidity conditions and above normal temperatures (Colella *et al.*, 1973).

Volcanic tuffs as gas adsorbents

A study was made of the selective adsorption capacity of chabazite tuff from Mercogliano for SO_2 with a view to extending its use to atmospheric antipollution techniques (Aiello *et al.*, 1976). Table 5

TABLE 5. SO_2 adsorbed as a function of grain size by a chabazitic tuff from Mercogliano

Mesh size	Wt. % SO_2
35–50	17.27
50–80	18.94
80–100	20.17
>100	24.27

gives the results of adsorption tests at 35°C of varying size fractions of tuff, using a mixture of SO_2 and a non-polar gas (helium) at 5.18% SO_2. The data show that a significant percentage of SO_2 is adsorbed and proportional to the zeolite content. The same test carried out with ion-exchanged chabazite tuff showed that the type of cation affects the adsorption only slightly and that the adsorption capacity remains relatively high, even at temperatures above 100°C and with quite low concentrations of SO_2.

Volcanic tuffs as cation exchangers

The treatment of industrial effluent by ion-exchange techniques seems to offer the most promising opportunity for the use of the tuffs as ion exchangers, particularly at temperatures above normal. The exchange kinetics of these natural products are too low compared to organic exchangers for many uses.

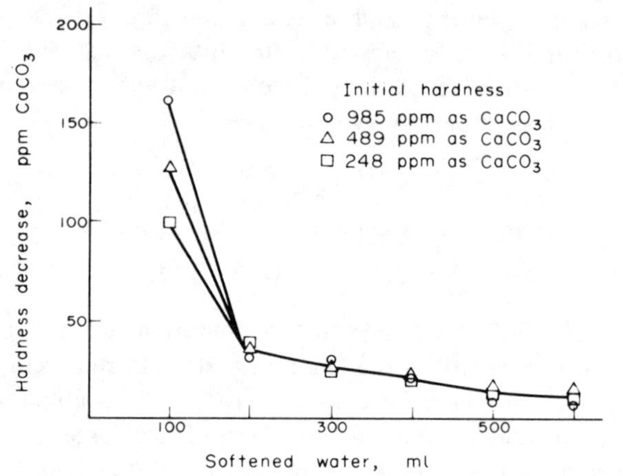

Fig. 18. Water-hardness decrease as a function of the amount of softened water.

Figure 18 shows the drop in hardness for three water samples with different initial hardness. The liquids were allowed to percolate at a constant velocity of 3 ml/min through a cylindrical bed of 5 g chabazite tuff with the ratio of diameter to height of the bed about 4. It was found that the largest decrease in hardness occurred during percolation of the first 100 ml of liquid after which the decrease became almost independent of the hardness. This is due to the low selectivity for calcium and the almost complete saturation of the bed under these conditions.

Figure 19 shows the decrease in hardness of 100 ml of water as a function of the initial hardness, and shows that the best results are reached where hardness values are initially the lowest.

The ion-exchange capacity of thin membranes of volcanic tuffs was also determined. Membranes of 170 mesh Neapolitan yellow tuff (first exchanged with the cation of the solution to be used) and polymethylacrylate powder were prepared by hot pressing (Drioli, 1967). A 1:1 ratio of tuff and polymethylacrylate was used. The membrane properties were then determined in 10^{-1} or 10^{-2} molar solutions of lithium, sodium, potassium, rubidium, cesium, magnesium, calcium, strontium, and barium chlorides at 25°C (Elia *et al.*, 1967). The exchange carried out on the powder can also be made on the already prepared membrane; however, the exchange then is not complete.

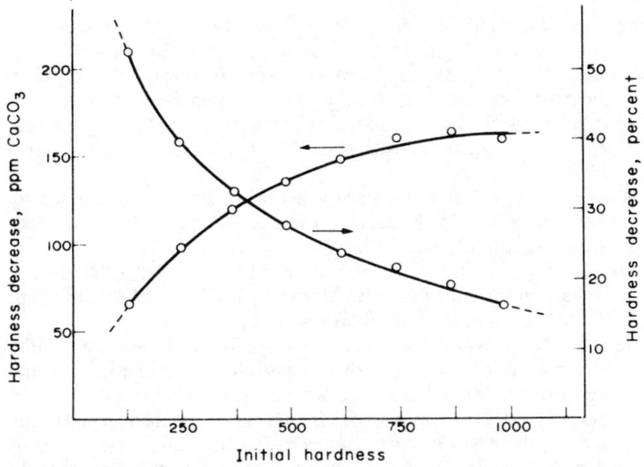

Fig. 19. Water-hardness decrease of 100 ml water volumes as a function of the initial hardness.

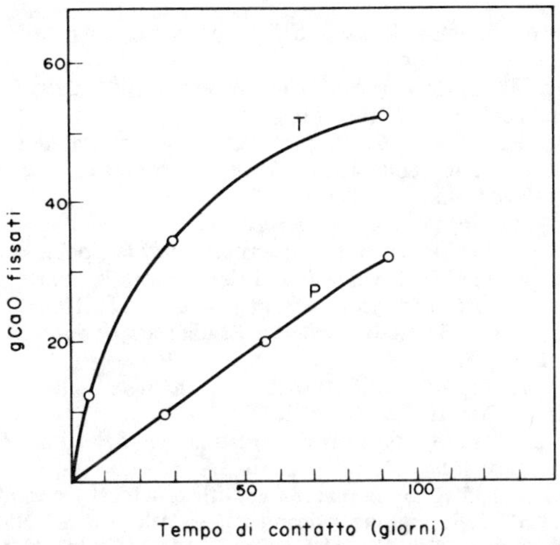

Fig. 20. Lime amount fixed by Neapolitan yellow tuff (T) and by its unzeolitized counterpart (pozzolana (P)), as a function of time.

The tests showed Neapolitan yellow tuff membranes to have good selectivity and that the selectivity tends to diminish as the electrolyte concentration increases. It was found that the zeolite matrix exchanges the alkaline earth and heavier alkaline metal cations preferentially, which indicates the possible use of zeolitic tuff membranes to concentrate and isolate radioactive species present in liquid waste from nuclear plants.

Volcanic tuffs as active additions to Portland clinker

Powdered volcanic tuffs have been found useful in the production of hydraulic cements resistant to chemical action such as that of sea water. Trass-Zemente in Germany mixes Portland clinker with Rhenish Trass (Sersale, 1959d), a volcanic tuff very similar chemically to Neapolitan yellow tuff. The product is well known and accepted. In Italy, the ready availability of true pozzolanas makes the use of tuff in hydraulic binders uneconomic, because the high grinding costs do not significantly increase the quality of the cement. However, tests have been made to evaluate the reactivity of powdered tuffs with lime. Figure 20 gives the kinetic curves for the fixing of lime by Neapolitan yellow tuff and its non-zeolitized counterpart, Phlegrean pozzolana, both reduced to the same specific surface areas. The curves were plotted of the amount of lime fixed in a given time at atmospheric temperature versus a given weight of solid in contact with the saturated calcium hydroxide solution. It may be noted that the fixing activity of Neapolitan yellow tuff ("pozzolanic" activity) is superior to that of pozzolana itself. Hydrated calcium silicate and aluminate are formed with both, but the zeolitic fraction shows greater reactivity to lime than does pozzolanic glass.

REFERENCES

Aiello, R. (1962) Studi petrografici sul settore meridionale del Vulcano di Roccamonfina: *Rend. Accad. Sci. Fis. Mat., Napoli* **29**, 241-258.

Aiello, R. and Colella, C. (1972) Zeolitic tuff as desiccant: *Chimica e Industria* **54**, 303-307.

Aiello, R. and Porcelli, C. (1974) Chabazitic tuff deposits of possible economic value in the Campanian region: *Rend. Accad. Sci. Fis. Mat., Napoli* **41**, 426-434.

Aiello, R. and Colella, C. (1976) Unpublished data.

Alietti, A., Galli, E., Gottardi, G., Guidetti Sorrivi, E., Passaglia, E. and Pederzolli Gottardi, L. (1967a) *Le Zeoliti in Italia.* Ist. Mineralogia Univ. Modena, pp. 245-314.

Alietti, A. and Ferrarese, G. (1967b) Clinoptilolite, Na-montmorillonite ed ossidi di manganese in una formazione sedimentaria a Zovencedo (Vicenza): *Min. e Petr. Acta* **13**, 139-145.

Colella, C., Aiello, R. and Sersale, R. (1973) Genesis, occurrence and properties of zeolitic tuff: *Rend. Soc. Ital. Min. Petr.* **29**, 439-451.

Cristofolini, R., Di Girolamo, P. and Stanzione, D. (1973) Caratteri genetici e mineralogici di jaloclastiti dell'Altopiano Ibleo (Sicilia) e dell'Isola di Procida (Campania): *Rend. Soc. It. Min. Petr.* **29**, 497-552.

De Lorenzo, G. (1899) Studio geologico del Monte Vulture: *Atti R. Acc. Sci. Fis. Mat., Napoli* **10**.

Di Girolamo, P. (1968) Petrografia dei tufi campani: il processo di pipernizzazione (tufo → tufo pipernoide → piperno): *Rend. Accad. Sci. Fis. Mat., Napoli* **35**, 329-394.

Drioli, E. and Elia, V. (1966) Comportamento elettrochimico di membrane scambiatrici inorganiche: Nota I: Membrane di tufo giallo napoletano: *Ric. Sci.* **36**, 835-840.

Elia, V., Caramazza, R. and Drioli, E. (1967) Comportamento elettrochimico di membrane scambiatrici inorganiche: Nota II: Influenza della granulometria: *Ric. Sci.* **37**, 413-417.

Fornaseri, M., Scherillo, A. and Ventriglia, U. (1963) *La Regione vulcanica dei Colli Albani.* Vulcano Laziale: Bardi, Roma.

Franco, E. (1976) Private communication.

Gianello, A. and Gottardi, G. (1969) Sulla zeolitizzazione del livello cineritico detto "Tripoli di Contignaco": *Min. Petr. Acta* **15**, 1-4.

Lenzi, G. and Pozzuoli, A. (1970) Sul tufo zeolitizzato di Chiaiano (Napoli) e sulla determinazione del suo contenuto phillipsitico: *Atti Accad. Sci. Fis. Mat.* **7**, 261-275.

Lenzi, G. and Passaglia, E. (1974) Fenomeni di zeolitizzazione

nelle formazioni vulcaniche della regione Sabatina: *Boll. Soc. Geol. It.* **93**, 623-645.
Mumpton, F. A. (1973) Worldwide deposits and utilization of natural zeolites: *Ind. Min.* 73, 1-11.
Nicotera, P. (1949) Contributo alla conoscenza del tufo trachitico della collina del Vomero (Napoli): *Boll. Soc. Naturalisti Napoli* **58**, 18-33.
Porcelli, C. (1976) Private communication.
Rittman, A. (1968) *I vulcani e la loro attività*: Cappelli, Milano.
Sabatini, V. (1912) I vulcani dell'Italia centrale: Parte II: I vulcani Cimini: *Mem. descr. Carta geol. d'It.* **15**, Roma.
Scherillo, A. (1933) Studi petrografici sulla regione Sabazia: *Per. Min.* **4**, 4-438.
Scherillo, A. (1940) I tufi litoidi a scorie nere della regione Sabazia e Cimina: *Per. Min.* **11**, 301-338.
Sersale, R. (1958) Genesi e costituzione del tufo giallo napoletano: *Rend. Accad. Sci. Fis. Mat., Napoli* **25**, 181-207.
Sersale, R. (1959a) Sulla natura zeolitica del tufo cosiddetto "lionato" della regione vulcanica dei Colli Albani: Nota I: *Rend. Accad. Sci. Fis. Mat., Napoli* **26**, 110-116: Nota II, 206-221.
Sersale, R. (1959b) Sulla natura zeolitica del "peperino" romano: *Rend. Accad. Sci. Fis. Mat., Napoli* **26**, 223-233.
Sersale, R. (1959c) Costituzione dei tufi verdi della regione flegrea: *Rend. Accad. Sci. Fis. Mat., Napoli* **26**, 355-373.
Sersale, R. (1959d) Analogie costituzionali fra il "trass" renano ed il tufo giallo napoletano: *Rend. Accad. Sci. Fis. Mat., Napoli* **26**, 117-125.
Sersale, R. (1960a) Sulla natura zeolitica dei tufi rossi a scorie nere delle regioni Sabazia e Cimina: *Rend. Accad. Sci. Fis. Mat., Napoli* **27**, 306-319.
Sersale, R. (1960b) Sulla natura zeolitica del tufo "carpato" della regione vulcanica del Monte Vulture: *Rend. Accad. Sci. Fis. Mat., Napoli* **28**, 543-556.
Sersale, R. (1972) Formazioni di tufo vulcanico dell'Italia Centro-Meridionale e loro possibile utilizzazione come zeoliti: *Atti Accad. Sci. Fis. Mat., Napoli* **8**, 1-11.
Sersale, R. (1975) Applicazioni industriali delle zeoliti sintetiche e naturali: *Atti Convegni Lincei* **5**, Bardi, Roma.
Sersale, R., Aiello, R. and Frigione, G. (1963) Sulla presenza di orizzonti zeolitici nella serie oligo-miocenica di Garbagna (Alessandria): *Atti Accad. Sci., Torino* **97**, 1-15.

COMPOSITION AND ORIGIN OF ANALCIME IN TAIWANITE, A GLASSY BASALTIC ROCK IN THE EASTERN COASTAL RANGE, TAIWAN

HUANN-JIH LO

Department of Geology, National Taiwan University, Taipei, Taiwan, Republic of China

Abstract – Analcime occurs as an alteration product of glass in a Tertiary glassy basaltic rock named taiwanite, in the Eastern Coastal Range, Taiwan. Two modes of occurrence of analcime were distinguished: it occurs as a replacement of, or as a fissure-filling in, the basaltic glass. The former is more common. That replacement analcime is indeed being formed by the reaction of palagonitic material and marine or fresh water is supported by the hydrothermal synthesis of analcime from palagonitic material of taiwanite in distilled water or NaCl solution at 330–345°C. Fissure-filling analcime was possibly formed by precipitation from interstitial fluid where the dissolved components were concentrated during the alteration of taiwanite. Chemical composition of analcimes from twenty rock samples ranges from $Na_{17.4}Al_{17.4}Si_{30.6}O_{96}{\cdot}nH_2O$ to $Na_{14.8}Al_{14.8}Si_{33.2}O_{96}{\cdot}nH_2O$ with an average composition of $Na_{15.9}Al_{15.9}Si_{32.1}O_{96}{\cdot}nH_2O$ as determined by X-ray methods. Analcime of the fissure-filling type, which is often associated with authigenic quartz, is poorer in silica than the replacement type.

INTRODUCTION

Taiwanite associated with dolerite occurs in the Lichi Formation of Miocene age in the Eastern Coastal Range, Taiwan. It is a glassy basaltic rock and is composed chiefly of glass with a few crystals of olivine and plagioclase (Juan *et al.*, 1953). The glass is partly altered to palagonite (Liou, 1974), whereas crystals of olivine and plagioclase are unaltered. Analcime and other zeolitic minerals, such as heulandite, natrolite, stilbite, thomsonite, laumontite, and prehnite, have been reported to occur as veins or veinlets in taiwanite (Juan *et al.*, 1965, 1967; Yen, 1967; Juan and Lo, 1971).

Analcime and other secondary minerals in taiwanite and in other igneous rocks in the Eastern Coastal Range have been considered to have formed by crystallization of hydrothermal solutions (Juan *et al.*, 1965; Yen, 1967; Juan and Lo, 1971) or by low-grade regional metamorphism (Yen, 1966, 1967). However, further studies of these secondary minerals are required for definite conclusions on their genesis; it is hoped that the present study will throw light on this problem.

INVESTIGATION METHODS

Analcimes in twenty rock samples of taiwanite collected at different sites near Kuanshan and Likiliki in the Eastern Coastal Range, Taiwan, were analyzed. Thin sections were prepared to examine the modes of occurrences of analcime and the nature of parent rock. Analcime concentrates for chemical analyses were obtained by magnetic separation after an initial crushing and grinding of each rock sample to 80 ~ 100 mesh. The $<2\ \mu m$ fraction, which is mainly a palagonitic material, was collected by a sedimentation method after gentle crushing and shaking in distilled water for 24 hours. Fresh glass was obtained by hand-picking from rock chips of the crushed sample followed by a magnetic separation of secondary minerals.

The composition of analcime was determined by the $\Delta 2\Theta$ method (Saha, 1959; Coombs and Whetten, 1967) using silicon as an internal standard ($\Delta 2\Theta = 2\Theta$ analcime$_{(639)}$$-2\Theta$ silicon$_{(331)}$; $CuK\alpha_1$ radiation). The analcime concentrate was mixed with silicon powder and analyzed on an X-ray diffractometer at a scanning rate of ¼° 2Θ/min. Hence, $\Delta 2\Theta$ is believed to be correct to ±0.02°. The chemical composition of $<2\ \mu m$ material (palagonite) was determined by atomic absorption.

Hydrothermal studies were carried out using a Tem-Pres HR-1B high temperature–high pressure unit. A powder of palagonite or fresh glass was sealed in a silver capsule with distilled water or a NaCl solution during hydrothermal treatment, and the mineral phases of the end product were identified by powder X-ray diffractometry.

GENESIS OF ANALCIME

Petrographic observations show that analcimes in the present study do not replace fresh glass, but that they crystallize around glass fragments which were replaced by palagonite. Two modes of occurrence of analcime were distinguished: it occurs as a replacement of, or as a fissure-filling, in the basaltic rock. Analcimes of these two types commonly occur together in a single sample; however, the replacement type is more common. The amount of fissure-filling analcime in each sample generally can be correlated with the quartz content, as they are commonly closely associated with each other.

Analcime of the replacement type is believed to have been formed by the reaction of palagonite and fresh or marine water. This inference is supported by the results of the hydrothermal study of taiwanite wherein analcime was synthesized at 300–345°C with different pressures by using palagonite or fresh glass of taiwanite in the presence of distilled water or NaCl solution (Table 1).

From Table 1 it can be noted that montmorillonite forms at low temperatures (below 250°C), whereas montmorillonite associated with analcime forms at higher temperatures from fresh glass; this indicates that montmorillonite forms before analcime. Consequently, analcime of the replacement type is believed to have been formed by the reaction between palagonite and fresh or marine water. This argument is also supported by the fact that analcime crystallizes only around palagonite as previously stated. Therefore, the transformation of palagonite to analcime can be idealized as a hydration reaction, but the actual chemical change must be more complex than this.

Analcime of the fissure-filling type is thought to have formed by precipitation from interstitial fluid. The chemical compositions of fresh glass and of two palagonites (TK401 and TK408) out of the twenty investigated are listed in Table 2. It is clearly indicated that the mechanism of the alteration of taiwanite involves both solution and hydration. Chemical components, such as SiO_2, Al_2O_3, CaO, Na_2O, and K_2O, were dissolved during the alteration and accumulated in the interstitial fluid, eventually to be precipitated as zeolites and other silicates. Among the precipitates, analcime, which crystallizes most easily (Coombs *et al.*, 1959; Saha, 1961), is the predominant mineral in the present case. That zeolites may be formed by precipitation from solution has already been noted by many previous workers (Deffeyes, 1959; Hay, 1963, 1966; Hay and Iijima, 1968) and has also been supported by experimental evidence (Mariner and Surdam, 1970). Recently, Surdam (1973) suggested that the chemical mechanism of the formation of precipitated zeolites involves (a) the formation of a gel from the solution and (b) the formation of a zeolite from the gel.

COMPOSITION OF ANALCIME

Analcime concentrates obtained by magnetic separation from each sample contain about 5–10% quartz and montmorillonite, which is the mineralogical composition of palagonite. These impurities are difficult to remove; therefore, the composition of analcime was estimated by Saha's X-ray diffraction method. The compositions of analcimes of twenty rock samples so

TABLE 1. Results of hydrothermal study on TK401 sample

Material	P_{H_2O} (psi)	T (°C)	D (days)	Medium	Products*	Intensity of analcime (211) reflection
Fresh glass	1 (atm)	80	7	distilled water	mont. (rare)	
Fresh glass	3000	110	7	distilled water	mont. (rare)	
Fresh glass	27,000	171	6	distilled water	mont.	
Fresh glass	32,000	244	12	distilled water	mont.	
Fresh glass	23,000	330	12	3% NaCl sol.	mont. + anal.	15
Fresh glass	23,000	330	12	5% NaCl sol.	mont. + anal.	30
Fresh glass	15,000	345	30	distilled water	mont. + anal.	7
Fresh glass	27,000	345	7	distilled water	mont. + anal.	12
Fresh glass	40,000	337	11	5% NaCl sol.	mont. + anal.	21
Palagonite	40,000	337	11	5% NaCl sol.	mont. + anal.	16
Palagonite	41,000	333	11	distilled water	mont. + anal.	18
Palagonite	41,000	333	11	3% NaCl sol.	mont. + anal.	20
Palagonite	41,000	333	11	5% NaCl sol.	mont. + anal.	34

*mont. = montmorillonite; anal. = analcime.

TABLE 2. Chemical composition of fresh glass and palagonite (wt. %)

	TK 401			TK 408		
	Fresh glass*	Palagonite	Gain (+) or loss (−)	Fresh glass*	Palagonite	Gain (+) or loss (−)
SiO_2	47.17	37.92	−9.25	48.63	37.65	−10.98
Al_2O_3	16.23	13.93	−2.30	15.53	13.50	−2.03
ΣFeO	10.47	11.09	+0.62	10.83	11.09	+0.26
MgO	9.47	10.88	+1.41	9.93	12.24	+2.31
CaO	11.35	5.07	−6.28	10.37	4.85	−5.52
Na_2O	2.77	0.83	−1.94	2.35	0.91	−1.44
K_2O	0.19	0.17	−0.03	0.20	0.13	−0.07
TiO_2	0.79	0.76	−0.03	0.85	0.79	−0.06
MnO	0.13	0.18	+0.05	0.14	0.23	+0.09
H_2O (total)	n.d.	19.14		n.d.	19.15	
Total	98.57	99.97		98.83	100.71	
Si/Al	2.58	2.41	3.57	2.78	2.47	4.78

*Electron microprobe data from Juan *et al.* (1976).

TABLE 3. Data on analcimes investigated

Sample No.	Δ2θ*	Composition calculated† Si	Na(+K+2Ca)	Al	Si/Al	Quartz‡
T510	1.66	30.6	17.4	17.4	1.76	Abundant
TK402	1.68	30.9	17.1	17.1	1.81	Abundant
T509	1.76	31.8	16.2	16.2	1.96	Abundant
K304	1.76	31.8	16.2	16.2	1.96	Moderate
TK406	1.76	31.8	16.2	16.2	1.96	Moderate
T504	1.77	32.1	15.9	15.9	2.02	Poor
T502	1.78	32.2	15.8	15.8	2.04	Poor
TK405	1.78	32.2	15.8	15.8	2.04	Poor
K312	1.78	32.2	15.8	15.8	2.04	Trace
K314	1.78	32.2	15.8	15.8	2.04	Trace
TK404	1.79	32.3	15.7	15.7	2.06	Poor
T501	1.79	32.3	15.7	15.7	2.06	Poor
TK410c	1.79	32.3	15.7	15.7	2.06	Poor
TK409	1.80	32.4	15.6	15.6	2.08	Poor
TK407	1.83	32.7	15.3	15.3	2.14	Trace
T507	1.84	32.8	15.2	15.2	2.16	Poor
K315	1.84	32.8	15.2	15.2	2.16	Poor
TK403	1.84	32.8	15.2	15.2	2.16	Trace
K313	1.85	32.9	15.1	15.1	2.18	Trace
TK401	1.87	33.2	14.8	14.8	2.24	Trace

* $\Delta 2\theta = 2\theta$ analcime (139)–2θ silicon (331).

† $\Delta 2\theta$ method of Saha (1959).

‡ Estimated from X-ray diffraction intensity.

obtained range from $Na_{17.4}Al_{17.4}Si_{30.6}O_{96}\cdot nH_2O$ to $Na_{14.8}Al_{14.8}Si_{33.2}O_{96}\cdot nH_2O$ (Table 3) with an overall average composition of $Na_{15.9}Al_{15.9}Si_{32.1}O_{96}\cdot nH_2O$.

Analcime in each sample generally gives a broad (639) diffraction peak suggesting a slight range in composition as a result of the composition difference between analcime of the fissure-filling type and that of the replacement type. In general, the Si/Al ratio of analcime in each sample decreases with increasing amount of the associated quartz (Table 3). Since the amount of analcime of the fissure-filling type in each sample is proportional to the quartz content, as stated before, it is inferred that analcime of the fissure-filling type is poorer in silica than that of the replacement type.

Assuming other factors to be constant, the Si/Al ratio of analcime will be mainly controlled by the Si/Al ratio of the parent material (Coombs and Whetten, 1967). Accordingly, analcime of the fissure-filling type, which was precipitated from interstitial fluids with high Si/Al ratios (3.57–4.78, see Table 2), should contain more silica than that of the replacement type, which was formed from palagonite with low Si/Al ratios (2.41–2.47, see Table 2). The formation of silica-poor analcime of the fissure-filling type from high Si/Al parent material is interpreted as being due to the polymerization of silica in the interstitial fluid to a semi-crystalline phase (Hay, 1964). Thus the activity of silica was greatly reduced, and it is not surprising that silica-poor analcime precipitated along with quartz in the present case.

DISCUSSION

The alteration of glass to palagonite is probably an important factor in the formation of analcime and other secondary minerals in taiwanite. Chemical changes during alteration not only provide materials for the formation of precipitated minerals, but also create a chemical environment, especially an alkaline pH induced by dissolved alkalis, suitable for the formation of zeolites.

Fieldes *et al.* (1956) concluded that palagonite from volcanic glass was formed under submarine conditions or under conditions created by impeded drainage similar to the formation of montmorillonite. Similar conditions can be assumed to have prevailed in the alteration of taiwanite, as suggested by the hydrothermal results of this study.

The poor crystallinity of montmorillonite indicates that the alteration of glass to palagonite may occur at low temperatures. Analcime–quartz associations in the studied samples suggest that the temperature of formation of analcime is below 190°C (Campbell and Fyfe, 1965). However, the hydrothermal data restrict the temperature of alteration of glass to montmorillonitic palagonite below 80°C (Table 1).

From field evidence, zeolites (such as heulandite, stilbite, and laumontite), prehnite, and other secondary minerals in taiwanite are considered to have formed in a manner similar to that of analcime of the fissure-filling type. The overlap of critical minerals such as heulandite, stilbite, laumontite, and prehnite, which are formed in the order of increasing temperature, is interpreted here in terms of ionic stability (Surdam, 1973).

The composition of analcime is mainly controlled either by the composition of the parent materials or by the chemical environment. Analcimes in siliceous tuffs of fresh or marine water are relatively siliceous, generally having Si/Al ratios between 2.4 to 2.7 (Saha, 1959; Wilkinson and Whetten, 1964; Coombs and Whetten, 1967). The Si/Al ratio is generally between 2.1 to 2.4 in analcimes of possible nontuffaceous origin in deposits of saline, alkaline lakes (Coombs and Whetten, 1967). The silica-poor nature of the studied analcimes, which have Si/Al ratios between 2.0 to 2.2, is believed to be inherited from the basaltic parent rock.

REFERENCES

Campbell, A. S. and Fyfe, W. S. (1965) Analcite–albite equilibria: *Amer. J. Sci.* **263**, 807-816.

Coombs, D. S., Ellis, A. J., Fyfe, W. S. and Taylor, A. M. (1959) The zeolite facies, with comments on the interpretation of hydrothermal syntheses: *Geochim. Cosmochim. Acta* **17**, 53-107.

Coombs, D. S. and Whetten, J. T. (1967) Composition of analcime from sedimentary and burial metamorphic rocks: *Geol. Soc. Amer. Bull.* **78**, 269-282.

Deffeyes, K. S. (1950) Zeolites in sedimentary rocks: *J. Sediment. Petrol.* **29**, 602-609.

Fieldes, M., Walker, I. and Williams, P. P. (1956) Clay mineralogy of New Zealand soils – III. Infrared absorption spectra of soil clays: *N.Z. J. Sci. Technol.* **38**, 31-43.

Hay, R. L. (1963) Stratigraphy and zeolitic diagenesis of the John Day Formation of Oregon: Univ. Calif. Publ. in *Geol. Sci.* **42**, 199-262.

Hay, R. L. (1964) Phillipsite of saline lakes and soils: *Amer. Mineral.* **49**, 1366-1387.

Hay, R. L. (1966) Zeolites and zeolitic reactions in sedimentary rocks: *Geol. Soc. Amer. Spec. Paper* **85**.

Hay, R. L. and Iijima, A. (1968) Petrology of palagonite tuffs of Koko Craters, Oahu, Hawaii: *Contr. Mineral. Petrol.* **17**, 141-154.

Juan, V. C., Liu, J. G. and Jahn, B. M. (1965) A preliminary study of minerals in the zeolite group in taiwanite from Taitung, Taiwan: *Proc. Geol. Soc. China* **8**, 85-90.

Juan, V. C. and Lo, H. J. (1971) The stability of natural laumontite and wairakite and their bearing on the zeolite facies: *Proc. Geol. Soc. China* **14**, 34-44.

Juan, V. C., Lo, H. J. and Chen, C. H. (1976) Crystallization of taiwanite at atmospheric pressure: *Proc. Geol. Soc. China* **19**, 87-97.

Juan, V. C., Tai, H. and Chang, F. H. (1953) Taiwanite, a new basaltic glassy rock of Eastern Coastal Range, Taiwan and its bearing on the parental magma type: *Acta Geologica Taiwanica* **5**, 1-25.

Juan, V. C., Youh, C. C. and Lo, H. J. (1967) Stability of prehnite: *Proc. Geol. Soc. China* **10**, 51-61.

Liou, J. G. (1974) Mineralogy and chemistry of glassy basalts, Coastal Range ophiolites, Taiwan: *Geol. Soc. Amer. Bull.* **85**, 1-10.

Mariner, R. H. and Surdam, R. C. (1970) Alkalinity and formation of zeolites in saline alkaline lakes: *Science* **170**, 977-980.

Saha, P. (1959) Geochemical and X-ray investigation of natural and synthetic analcites: *Amer. Mineral.* **44**, 300-313.

Saha, P. (1961) The system $NaAlSiO_4$ (nepheline) – $NaAlSi_3O_8$ (albite) – H_2O: *Amer. Mineral.* **46**, 859-884.

Surdam, R. C. (1973) Low-grade metamorphism of tuffaceous rocks in the Karmutsen Group, Vancouver Island, British Columbia: *Geol. Soc. Amer. Bull.* **84**, 1911-1922.

Wilkinson, J. F. G. and Whetten, J. T. (1964) Some analcime-bearing pyroclastic and sedimentary rocks from New South Wales: *J. Sediment. Petrol.* **34**, 543-553.

Yen, T. P. (1966) Occurrence of celadonite and zeolites in the pyroclastics of the Coastal Range, Eastern Taiwan: *Proc. Geol. Soc. China* **9**, 65-69.

Yen, T. P. (1967) A geologic consideration on the basic and ultrabasic plutonic rocks of the Coastal Range, Eastern Taiwan: *Proc. Geol. Soc. China* **10**, 25-39.

ZEOLITES IN TUFF BEDS OF THE MIOCENE WAITEMATA GROUP, AUCKLAND PROVINCE, NEW ZEALAND

T. SAMESHIMA

Geology Department, University of Auckland,
Auckland, New Zealand

Abstract – Analcime, clinoptilolite, chabazite, erionite, mordenite, and phillipsite are found in the Waitemata Group, an early Miocene marine sedimentary sequence, as alteration products of volcanic glass in volcaniclastic beds. The Si/Al + Fe^{+3} ratio of analcime from four localities, estimated from refractive indices and unit-cell dimensions, are in the silica-poor range (2.07–2.15) of "sedimentary analcime". A chemical analysis of clinoptilolite from North Cape showed a Si/Al+Fe^{+3} ratio of 3.85, and cell dimensions and refractive index data indicate that silica-poor and alkali-rich compositions may be characteristic of the Waitemata Group clinoptilolites. The Takapuna chabazite from the Auckland City region is the most silica-poor variety ever reported from sedimentary rocks. Its Si/Al+Fe^{+3} ratio is 2.35, and alkalis predominate over alkaline earths.

Erionite of the Timber Bay Formation on the Puketotara Peninsula, Kaipara, is the most silica-poor reported from a sedimentary environment; alkalis predominate over alkaline earths, and Si/Al+Fe^{+3} is 2.79. Mordenite is found as the major constituent mineral of an altered acidic tuff bed southeast of Whangarei. Minor amounts of phillipsite, chabazite, and clinoptilolite associated with calcite, quartz, chlorite, and smectite occur in the matrix of the "Parnell Grit", a submarine volcanic layer deposit in the Auckland City region.

The Waitemata Group zeolite data have been combined with data from the literature for heulandite–clinoptilolite, chabazite and erionite–offretite on three types of diagrams: Si:[Al+Fe^{+3}], [Mg+Ca]:Na:K and [Si+Na+K]:[Al+Mg+Ca]. On the last type of diagram for the heulandite–clinoptilolite series, thermally stable members of the series can be discriminated.

The North Cape clinoptilolite belongs to the C group of Boles (1972) with the lacustrine sedimentary origin minerals. Chabazite compositions plotted on a [Si+Na+K]:[Al+Mg+Ca] diagram are highly dispersed. This may be explained by post-formation cation exchanges. The Waitemata Group chabazite has excess Na which may be due to cation exchange with sea water.

INTRODUCTION

The Waitemata Group is an early Miocene sedimentary sequence in Auckland Province, North Island, New Zealand. According to Ballance (1974), the main part of the Group was deposited in a flysch basin of inter-arc character and crops out in a 100 × 200-km area. The Miocene sedimentary sequences in the North Cape region (Leitch, 1970) and in the Hokianga Harbor region are also regarded as the correlatives of the Waitemata Group (Fig. 1).

The Waitemata Group is poor in volcaniclastic materials in the southern half and richer in the northern half of the central flysch basin. Sparse intermediate and acidic tuffs are interbedded with the non-volcanic sandstones and siltstones in the southern half of the central flysch basin. Other significant facies are the

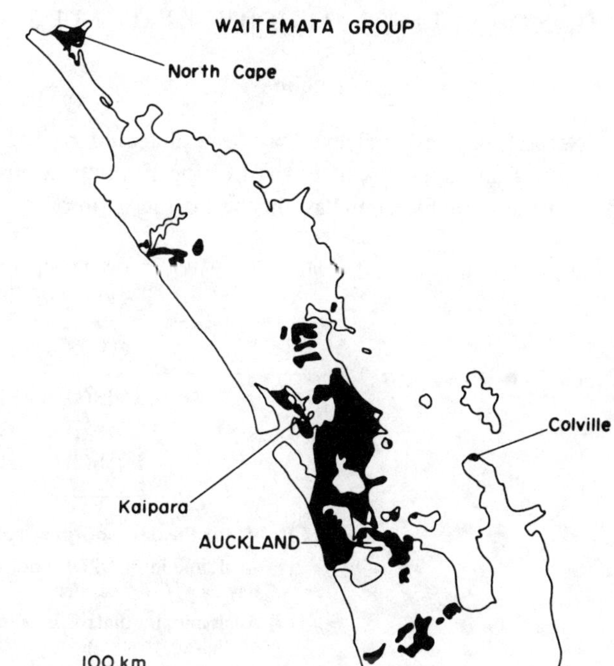

Fig. 1. Distribution of the Waitemata Group in the northern part of North Island, New Zealand.

volcanic mud-flow deposit or submarine lahar deposit, "Parnell grit", consisting of lithic and crystal debris of intermediate volcanics, and the "Albany conglomerate" which consists of andesite, hornblende diorite, and meta-greywacke pebbles in a clayey matrix. The northern half of the central flysch basin of the Group generally consists of interbedded dark-colored, tuffaceous sandstones and siltstones with a few interstratified acidic tuff beds. Volcanic glass shards in the volcaniclastic beds of the Group commonly are altered to zeolites such as clinoptilolite, analcime, chabazite, erionite, mordenite, and phillipsite.

Jones (1972) recognized clinoptilolite in acidic tuff from Pakaurangi Point, Hukatere Peninsula, Kaipara. Ballance and McCarthy (1975) reported chabazite in a tuff bed in Okahukura Peninsula, Kaipara.

Analcime, clinoptilolite, chabazite, erionite, mordenite, phillipsite, ferrierite, and harmotome are found in lacustrine tuff beds of the western United States; some of the species are also found in East Africa. Analcime, clinoptilolite, mordenite, and phillipsite are known from Cenozoic marine tuff beds in many localities throughout the world. Chabazite and erionite have not previously been reported from marine sediments except for a report of erionite in a deep-sea drilling core, east of Guam Island in the west Pacific (Rex *et al.*, 1971).

DESCRIPTION OF THE ZEOLITES

Analcime

Several percent analcime has been detected in the matrix of an andesitic grit in the Colville Formation on the east side of Fletcher Bay on the northern tip of the Coromandel Peninsula. The Colville Formation was correlated with the Waitemata Group by Skinner (1969) and is considered to be the eastern marginal facies of the Group (Ballance, 1974). Analcime is associated with quartz, andesine, calcite, chlorite, smectite, and trace amounts of clinoptilolite.

Analcime is also found in the matrix of the Albany conglomerate which crops out north of Wainui, 35 km north of Auckland. It constitutes up to 30% of the matrix and is associated with quartz, andesine, hornblende, calcite, chlorite, sericite, kaolinite, saponite, and montmorillonite. Analcime occurs extensively in the Kaipara region in tuffaceous sandstone.

Refractive index and cell dimension data for four analcimes are listed in Table 1. The inferred Si/Al+Fe^{+3} ratios of the analcimes, based on the cell dimensions (Saha, 1959), are between 2.07 and 2.16; these are in the silica-poor range of the "sedimentary analcime" of Coombs and Whetten (1967).

Clinoptilolite

Almost monomineralic clinoptilolite tuff beds are found in the Tom Bowling Formation exposed in sea cliffs at Wharekawa Pa (N.Z. Grid Ref. N1/491484), south-west of North Cape, where several 10-40-cm thick white tuff beds are interbedded with somewhat darker colored sandy tuff beds. The white tuff consists almost wholly of clinoptilolite with a minor amount of chlorite; the darker colored sandy tuff also contains up to 70% clinoptilolite.

Wet chemical analysis data and calculated cell contents of clinoptilolite separated from the white tuff are shown in Table 2, column 1. Cell dimensions and refractive index data of the mineral are shown in Table

TABLE 1. Cell-constant and refractive index data of analcime from the Waitemata Group

	(1)	(2)	(3)	(4)
a (Å)	13.713(4)	13.712(3)	13.707(6)	13.709(8)
Si/Al + Fe^{+3}	2.07	2.08	2.16	2.15
n	1.486(3)	1.485(3)	1.489(3)	1.486(3)

Errors for the data shown in parentheses.

(1) Analcime in matrix of andesitic grit, Colville Formation, Fletcher Bay east, Coromandel.

(2) Analcime in matrix of the Albany Conglomerate, 2 km north of Wainui, Auckland.

(3) Analcime in tuffaceous sandstone, Timber Bay Formation, north coast of the Puketotara Peninsula, Kaipara.

(4) Analcime in tuffaceous sandstone, Timber Bay Formation, north coast of the Puketotara Peninsula, Kaipara. Locality 0.5 km west of sample 3.

TABLE 2. Chemical composition (wt. %) and calculated unit-cell content for zeolites from the Waitemata Group

	(1) Clinoptilolite	(2) Chabazite	(3) Erionite
SiO_2	61.77	51.50	55.77
Al_2O_3	13.26	18.21	16.82
Fe_2O_3	0.59	0.56	0.19
FeO	–	–	0.14
MgO	0.51	1.10	1.51
CaO	0.89	0.55	1.13
SrO	0.03	–	0.01
BaO	0.10	–	0.10
Na_2O	4.19	6.12	4.77
K_2O	4.04	3.46	3.61
TiO_2	0.41	–	0.21
H_2O+	14.78	10.60	15.64
H_2O-	–	7.44	–
Total	100.57	99.53	99.90
Unit-cell contents on basis of 72 oxygens			
Si	28.50	25.36	26.35
Al	7.21	10.57	9.37
Fe^{+3}	0.20	0.21	0.07
Fe^{+2}	–	–	0.05
Mg	0.35	0.81	1.06
Ca	0.44	0.29	0.57
Sr	0.01	–	0.00
Ba	0.02	–	0.02
Na	3.75	5.85	4.37
K	2.38	2.18	2.18
$Si/Al + Fe^{+3}$	3.85	2.35	2.79

(1) Clinoptilolite from Tom Bowling Formation, 5 km southwest of North Cape. For cell dimension and refractive index see Table 3.

(2) Chabazite from Takapuna Beach, Auckland. For cell dimension and refractive index see Table 4.

(3) Erionite from Timber Bay Formation, north coast of the Puketotara Peninsula, Kaipara. For cell dimension and refractive indices see text.

3, column 1. This clinoptilolite is a high potash variety in which alkalis predominate over alkaline earths. It appears to be one of the most silica-poor variety ever reported from a sedimentary environment ($Si/Al+Fe^{+3}$ = 3.85). A sample of the clinoptilolite was heated at 450°C for 16 hours. Comparison of the X-ray diffraction data of the heated mineral with that of an unheated sample showed only a small intensity loss for the diffraction peaks in the heated sample, which indicates that the North Cape clinoptilolite is the C-type of Boles (1972).

Clinoptilolite was also found in tuff beds in the Kaipara region and from the Parnell Grit in Auckland. Cell dimensions and refractive indices of clinoptilolites from six localities in the Waitemata Group are listed in Table 3. Following the study of Boles (1972), these clinoptilolites also appear to be silica-poor varieties.

Chabazite

Thin tuff beds about 5 cm thick from Takapuna Beach (N.Z. Grid Ref. N42/303670) and Karaka Bay, both in the Auckland City area, were found to be

TABLE 3. Cell dimensions and refractive indices of clinoptilolite from the Waitemata Group

	(1)	(2)	(3)	(4)	(5)	(6)
a (Å)	17.69	17.72	17.70	17.62	17.70	17.68
b (Å)	17.92	18.00	17.97	17.93	18.00	17.93
c (Å)	7.40	7.41	7.40	7.41	7.40	7.39
β	116.3°	116.4°	116.3°	116.4°	116.3°	116.3°
ac (Å^2)	117.4	117.6	117.4	116.9	117.4	117.1
n	1.485	1.485	1.491	1.490	1.491	1.486

Errors for axis constants ±0.01, β±0.1, for refractive indices ±0.003.

(1) Clinoptilolite from Tom Bowling Formation, 5 km southwest of North Cape. For chemical composition see Table 2, column 1.

(2) Clinoptilolite from Okahukura Formation, 1 km north of Orongo Point, Okahukura Peninsula, Kaipara.

(3) Clinoptilolite from Okahukura Formation, 0.5 km north of Orongo Point, Okahukura Peninsula, Kaipara.

(4) Clinoptilolite associated with erionite and chabazite in a tuff bed of Timber Bay Formation, north coast of the Puketotara Peninsula, Kaipara.

(5) Clinoptilolite associated with chabazite in tuff bed of Timber Bay Formation, northwest coast of the Puketotara Peninsula, Kaipara.

(6) Clinoptilolite in the Parnell grit, associated with chabazite and phillipsite, Parnell, Auckland.

composed dominantly of chabazite with minor amounts of andesine, quartz, and chlorite. Wet chemical analysis data and calculated cell contents of the Takapuna chabazite are shown in Table 2, column 2. Cell dimensions and refractive indices of the mineral are shown in Table 4, column 1. The chabazite is the most silica-poor of reported sedimentary chabazite ($Si/Al+Fe^{+3}$ = 2.35); it is a potash-rich variety, and alkalis predominate over alkaline earths.

Chabazite is also found in tuff beds from the Kaipara region and from the Parnell Grit, Auckland. Cell dimensions and refractive indices of chabazite from five localities in the Waitemata Group are listed in Table 4. These chabazites have similar cell-volume sizes, 2445–2642 Å^3, which are intermediate between those from mafic igneous rocks (2470–2490 Å^3) and those

TABLE 4. Cell dimensions and refractive indices of chabazite from the Waitemata Group

	(1)	(2)	(3)	(4)	(5)
a (Å)	13.769(2)	13.771(5)	13.74(1)	13.802(3)	13.803(2)
c (Å)	14.997(9)	14.93(1)	14.937(4)	14.88(2)	14.89(1)
V (Å^3)	2462.2(15)	2454.4(16)	2444.9(18)	2454.7(33)	2459.7(16)
n	1.469(3)	1.470(3)	1.472(3)	1.480(3)	1.481(3)

Errors for the data shown in parenthesis.

(1) Chabazite in 5-cm-thick tuff bed at Takapuna Beach, Auckland. For chemical composition see Table 3, column 2.

(2) Chabazite in thin tuff bed at Karaka Bay, Auckland.

(3) Chabazite associated with erionite and clinoptilolite, Timber Bay Formation, north coast of the Puketotara Peninsula, Kaipara.

(4) Chabazite in tuff bed with clinoptilolite, Timber Bay Formation, northwest coast of the Puketotara Peninsula, Kaipara.

(5) Chabazite in Parnell grit, with clinoptilolite and phillipsite, Parnell, Auckland.

from lacustrine, tuffaceous sedimentary rocks (2410–2430 Å^3) following the study of Sheppard and Gude (1970).

Erionite

Erionite occurs in a 90 cm thick tuff bed of the Timber Bay Formation (Carter, 1971) on the northern coast of the Puketotara Peninsula, Kaipara (N.Z. Grid Ref. N28/832336). Microforaminifera in the tuff bed indicate a bathyal marine sedimentary environment and an early Miocene age. Within the tuff bed there are 3–5 cm thick, pinkish-brown zeolite layers predominantly composed of erionite with subordinate amounts of chabazite, clinoptilolite, and andesine. A 3 cm thick zeolite layer near the top of the tuff bed is composed almost wholly of erionite. The remainder of the tuff bed consists of 50–70% zeolite, 5% andesine, 5% quartz, and 20–40% beidellite.

Wet chemical analysis data and calculated cell contents of the Puketotara erionite are shown in Table 2, column 3. Hexagonal cell dimensions of the mineral are a = 13.250 ± 0.003 Å, c = 15.138 ± 0.01 Å, and V = 2301.5 ± 1.5 Å^3; it is uniaxial positive with ω = 1.469 ± 0.002, ϵ = 1.472 ± 0.002. The erionite is the most silica-poor variety among the reported sedimentary erionites (Si/Al+Fe^{+3} = 2.79), and alkalis predominate over alkaline earths. The mineral has the highest refractive indices and the largest cell volume of reported sedimentary erionites.

Mordenite

A 5 m thick siliceous tuff bed which outcrops southeast of Whangarei, and possibly belongs to the Waitemata Group, was found to be composed of mordenite with subordinate amounts of quartz and chlorite. As the tuff is fairly hard and tough, it is in part quarried for road aggregate.

Cell dimensions of the mordenite are, a = 18.12 ± 0.01 Å, b = 20.51 ± 0.02 Å, c = 7.50 ± 0.01 Å, V = 2787.2 ± 3.0 Å^3, and the mean refractive index is n = 1.478 ± 0.003.

Phillipsite

Parnell Grit crops out at a number of localities along the east coast between Auckland and Leigh. Several percent zeolite has been found in the Parnell Grit from the type locality, Parnell, Auckland City. The zeolite consists of subequal amounts of chabazite, clinoptilolite, and phillipsite; the zeolites are associated with quartz, andesine, and smectite. Clinoptilolite and chabazite in the Parnell Grit are described in Table 3 and Table 4, respectively.

The phillipsite has pseudo-orthorhombic cell dimensions a = 9.94 ± 0.01 Å, b = 14.24 ± 0.02 Å, c = 14.24 ± 0.02 Å, V = 2015.6 ± 2.8 Å^3, and a refractive index n = 1.499 ~ 1.505.

DISCUSSION

Origin

The extensive zeolitization found in the volcaniclastic materials of the Waitemata Group would most likely be due to hot-spring activity which accompanied the hydrothermal alteration at shallow burial depth during later geological times. The maximum thickness of the Waitemata Group sediments in the central flysch basin is roughly 1000 m, being thinner in the marginal zones (Ballance, 1974). As there is no evidence of a thick covering of younger sediments or a high degree of deformation which might have caused a deep burial condition, the burial depth of the Group was probably not greater than the maximum thickness of the sediment itself, i.e. 1000 m, and could have been much shallower. This depth is shallower than that reported by Iijima and Utada (1971) for the Cenozoic sediments of the Niigata oil field in Japan, where clinoptilolite occurs in tuffs buried deeper than 800 m, and analcime occurs at depths greater than 2900 m.

Several lines of evidence support the existence of the widespread hot-spring activity in the Waitemata Group, and include the occurrences of gypsum in silt beds, pyrite in tuff beds, regular mixed-layer clay minerals in tuff beds, and common veins of quartz, calcite, and gypsum. In addition, active hot springs in this region may be remnants of hot-spring activity which accompanied the early Miocene volcanism during the deposition of the Waitemata Group and which persisted into the Holocene (Sameshima, 1975).

Mineralogy

The Waitemata Group zeolitization is unique for the wide variety of individual zeolites and of assemblages. Clinoptilolite–chabazite–erionite and clinoptilolite–chabazite–phillipsite assemblages found in the Waitemata Group have also been reported from the lacustrine sediments of the Big Sandy Formation, Arizona, by Sheppard and Gude (1973) but have never been reported from a marine sediment.

Chemical compositions of the Waitemata Group zeolites, recalculated on the basis of 72 oxygen atoms per unit cell, are plotted on the diagrams Si:[Al+Fe^{+3}], [Mg+Ca+Sr+Ba]:Na:K, and [Si+Na+K]:[Al+Fe+Mg+Ca+Sr+Ba] with selected data from the literature (Figs. 2 to 10). The Si:[Al+Fe^{+3}] diagram for heulandite-clinoptilolite (Fig. 2) and for erionite-offretite (Fig. 4) show a fairly good concentration of the plotted points on the line Si+Al+Fe^{+3} = 36. In the case of the chabazite (Fig. 3), plotted points show an excellent fit to the line. This shows the high reliability of the recent chemical analysis, and also indicates for the chabazite that no Al or Fe^{+3} ions exist at the exchangeable cation site but only within the framework tetrahedra. For these three zeolites, the Waitemata Group minerals plot about mid-way along the lines.

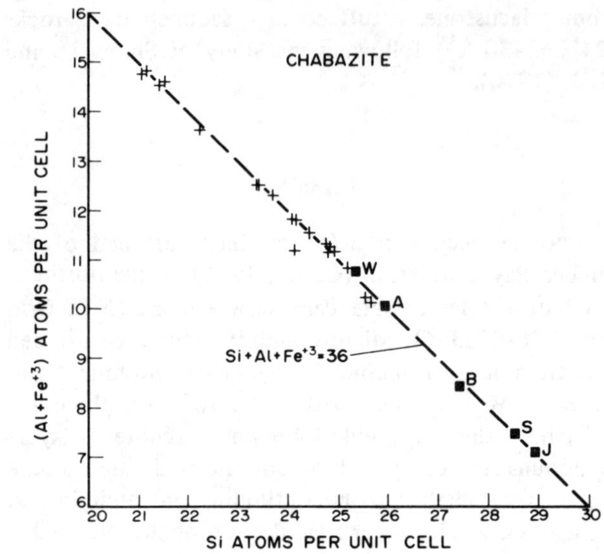

Fig. 3. Chabazite series Si:[Al+Fe^{+3}] diagram. + Basic volcanic rock origin (Passaglia, 1970; Nos. 1, 2, 3, 4, 5, 6, 7, 8, 9, 10, 11, 12, 13, 14, 16, 17, 18, 19, 25). ■ Cenozoic altered tuff origin. W = Waitemata Group, Takapuna Beach, Auckland, New Zealand, this paper. A = Olduvai Gorge, Tanzania (Hay, 1964). B = Big Sandy Formation, Mohave County, Arizona (Sheppard and Gude, 1973). J = John Day Formation, Oregon (Sheppard and Gude, 1970). S = Barstow Formation, California (Gude and Sheppard, 1966). Symbols, letters, and literature common to Figs. 6 and 9.

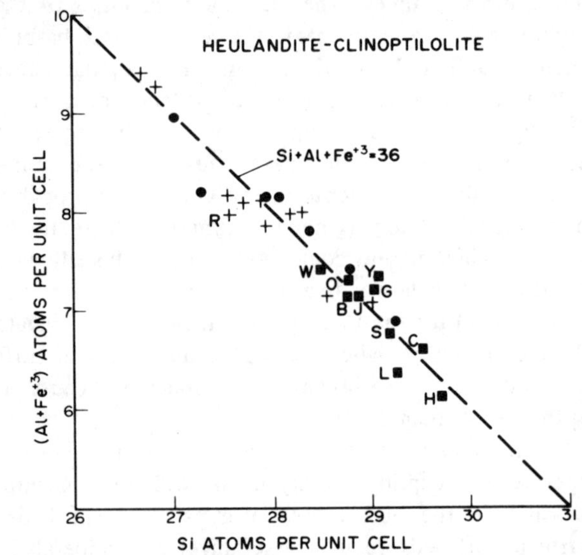

Fig. 2. Heulandite–clinoptilolite series Si:[Al+Fe^{+3}] diagram. + Basic volcanic rock origin (Alietti, 1972; Nos. 1, 3, 4, 5, 6, 7, 16, 17) (Boles, 1972; No. 1). R = unpublished new data from Tangihua volcanics, 6 km southeast of Cape Reinga, New Zealand. SiO_2 59.46, Al_2O_3 14.30, Fe_2O_3 0.40, MgO 0.16, CaO 7.46, SrO 0.10, BaO 0.34, Na_2O 1.98, K_2O 0.48, H_2O 15.64, TiO_2 0.09, Total 100.41. a = 17.71(1) Å, b = 17.92(1) Å, c = 7.42(1) Å, β = 116.4(1)°, n = 1.500(3). ● Triassic altered tuff of the Murihiku Supergroup origin. South Island, New Zealand (Boles, 1972; Nos. 2, 3, 4, 5, 7, 8, 11). ■ Cenozoic altered tuff origin. W = Waitemata Group, Tom Bowling Formation, North Cape, New Zealand, this paper. B = Big Sandy Formation, Mohave County, Arizona (Sheppard and Gude, 1973). J = John Day Formation, Oregon (Hay, 1963). H = Hector, California (Ames *et al.*, 1958). L = Lake Tecopa, Inyo County, California (Sheppard and Gude, 1968). O = Shizuma, Ota, Shimane, Japan (Minato and Utada, 1970). S = Barstow Formation, California (Sheppard and Gude, 1969a). Y = Yellowstone National Park, Wyoming (Honda and Muffler, 1970). C = Ricardo Formation, California (Sheppard *et al.*, 1965). G = Gem Hill Formation, California (Sheppard *et al.*, 1965). Symbols, letters, and literature common to Figs. 5 and 8.

The [Mg+Ca+Sr+Ba]:Na:K triangular diagram for heulandite-clinoptilolite (Fig. 5) shows the plotted points lie in the range K (0–40%), Na (0–80%), and [Mg+Ca+Sr+Ba] (10–90%). The Triassic Muruhiku Supergroup minerals (Boles, 1972) concentrate near the Mg+Ca corner. Clinoptilolites belonging to the thermal stability type-C (Boles, 1972) and corresponding to the clinoptilolite of Mumpton (1960) plot in a narrow area along the Na-K side. The diagram for chabazite (Fig. 6) shows plotted points in the range K (0–30%), Na (0–80%), and [Mg+Ca+Sr+Ba] (0–90%); for erionite-offretite (Fig. 7), K (20–60%), Na (0–70%), and [Mg+Ca+Sr+Ba] (0–70%). Judged from the diagram, potassium seems to be an essential constituent of the erionite-offretite series.

The [Si+Na+K]:[Al+Fe+Mg+Ca+Sr+Ba] diagram for heulandite-clinoptilolite (Fig. 8) separates two groups which correspond to the thermal stability group H + I and C of Boles (1972). The compositions of the H + I group $Ca_{31-n}(Na,K)_{n-26}Al_{36-n}Si_nO_{72}$ ($26 < n \leqslant 29$) cluster along the dashed line (Si+Na+K) + (Al+Mg+Ca) = 41, and compositions of the C group $Ca_{30-n}(Na,K)_{n-24}Al_{36-n}Si_nO_{72}$ ($28 \leqslant n \leqslant 30$), shown as the dashed line (Si+Na+K) + (Al+Mg+Ca) = 42 on the diagram.

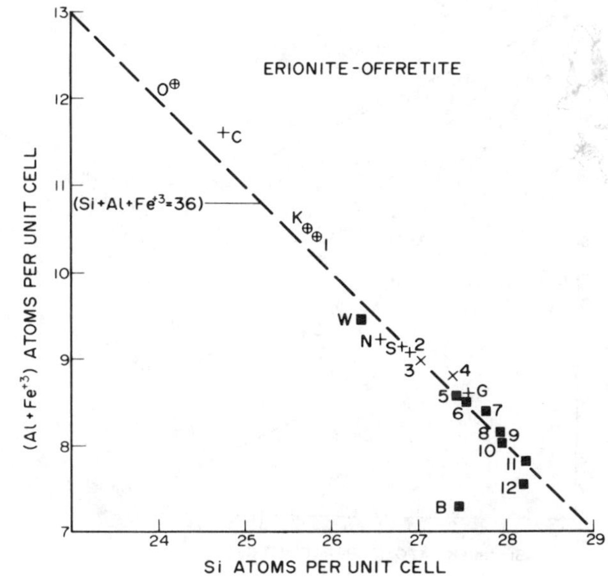

Fig. 4. Erionite–offretite series Si:[Al+Fe^{+3}] diagram. ⊕ Offretites, from basic volcanic rocks. 1 = Mount Simionse, France (Sheppard and Gude, 1969b). K = Sasbach, Kaiserstuhl, Germany (Rinaldi *et al.*, 1975). O = Beech Creek, Oregon (Sheppard *et al.*, 1974). + Erionite, basic volcanic rock origin. 2 = Maze, Niigata, Japan (Harada *et al.*, 1967). C = Chojabaru, Iki, Japan (Mizota *et al.*, 1974). G = Shurdo, Georgian SSR (Batiashvili and Gvakhariya, 1968). N = Nidym River, USSR (Belitskiy and Bukin, 1968). S = Sardinia, Italy (Passaglia *et al.*, 1974). × Erionite, rhyolitic welded tuff origin. 3 = Durkee, Oregon (Eakle, 1898). 4 = Durkee, Oregon (Staple and Gard, 1959). ■ Erionite, Cenozoic altered tuff origin. W = Waitemata Group, Puketotara Peninsula, Kaipara, New Zealand, this paper. B = Big Sandy Formation, Arizona (Sheppard and Gude, 1973). 5 = Lake Natron, Tanzania (Hay, 1966). 6 = Rome, Oregon (Eberly, 1964). 7 = Churchill County, Nevada (Sheppard and Gude, 1969b). 8 = Pine Creek, Nevada (Sheppard and Gude, 1969b). 9 = Jersey Valley Wash, Nevada (Sheppard and Gude, 1969b). 10 = Cady Mountains, California (Sheppard and Gude, 1969b). 11 = Crooked Creek, Oregon (Sheppard and Gude, 1969b). 12 = Tecopa, California (Sheppard and Gude, 1968). Symbols, letters, and literature common to Figs. 7 and 10.

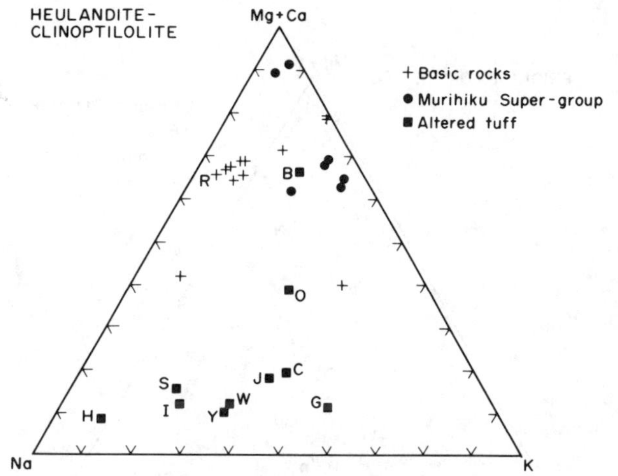

Fig. 5. Heulandite–clinoptilolite series [Mg+Ca]:Na:K diagram. Mg+Ca including Sr and Ba. For symbols, letters, and literature, see Fig. 2.

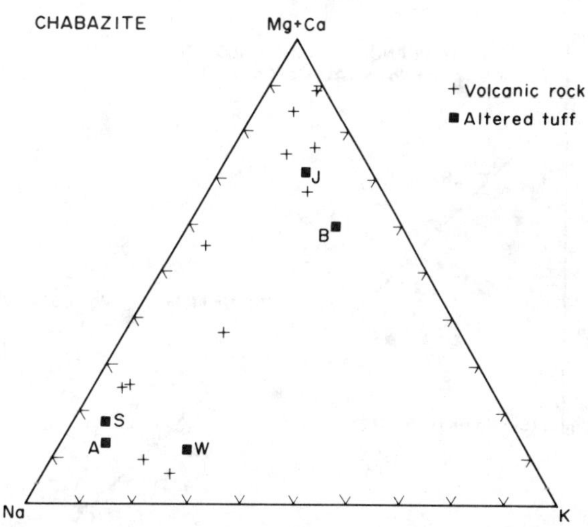

Fig. 6. Chabazite series [Mg+Ca]:Na:K diagram. Mg+Ca including Sr and Ba. For symbols, letters, and literature, see Fig. 3.

The diagram for chabazite shows plotted points dispersed over the theoretically bounded field (Fig. 9). The theoretical boundary drawn on the diagram is based on the substitution range $Al_{15}Si_{21}$ to Al_7Si_{29} without any restriction on the number of atoms of alkalis and alkaline earths except those required by total charge balance in the formula. The most reasonable chemical formula for chabazite deduced from the data is $Ca_{28-n}(Na,K)_{n-20}Al_{36-n}Si_nO_{72}$ where $21 \leqslant n \leqslant 29$. The formula assumes 8 atoms of alkalis and alkaline-earths instead of the usually assigned 6. The variation in alkalis and alkaline earths in the natural chabazite, as shown by the dispersion of the plotted points on the diagram, may be explained by cationic exchange which took place after the original crystallization. The high content of Na in the Waitemata Group chabazite may be due to a cation exchange with sea water, as the chabazite tuff bed at the Takapuna Beach is situated under the high-tide level.

The diagram for erionite–offretite series (Fig. 10) exhibits a similar trend to that of the heulandite–clinoptilolite series. The plotted points are not highly dispersed as is the case for chabazite; the offretite and erionites from basic volcanic rock and those from Cenozoic altered tuff plot in separate fields.

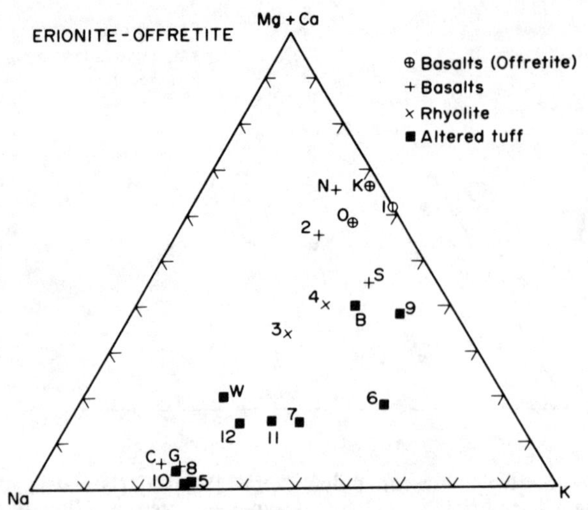

Fig. 7. Erionite–offretite series [Mg+Ca]:Na:K diagram. Mg+Ca including Sr and Ba. For symbols, letters, and literature, see Fig. 4.

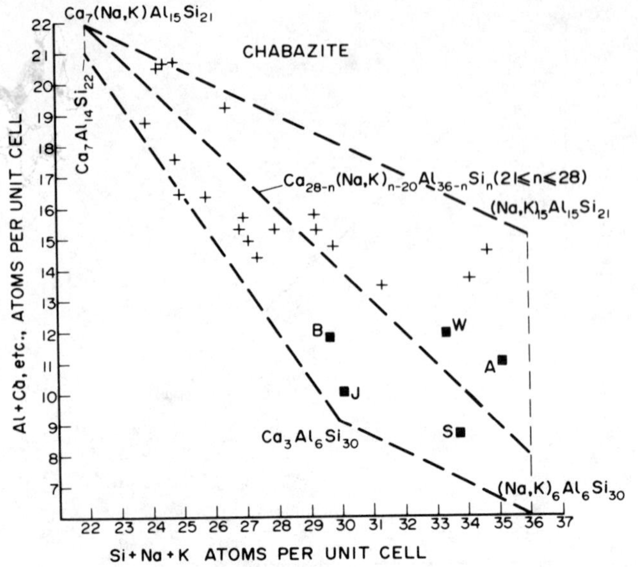

Fig. 9. Chabazite series [Si+Na+K]:[Al+Ca] diagram. Al+Ca including Fe^{+3}, Fe^{+2}, Mg, Sr, and Ba. For symbols, letters, and literature, see Fig. 3.

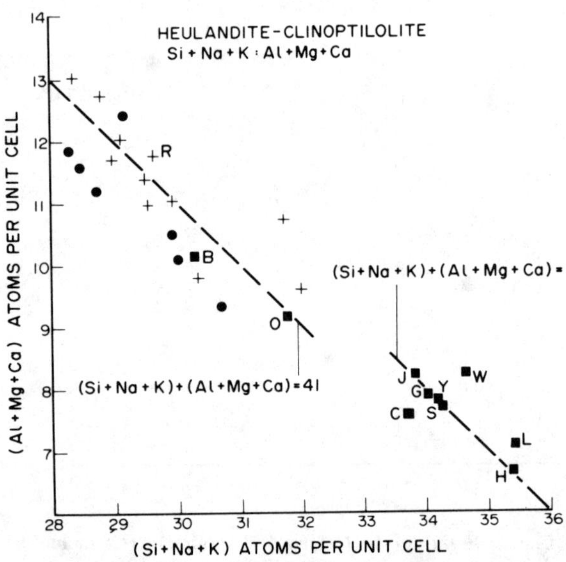

Fig. 8. Heulandite–clinoptilolite series [Si+Na+K]:[Al+Mg+Ca] diagram. Al+Mg+Ca including Fe^{+3}, Fe^{+2}, Sr, and Ba. For symbols, letters, and literature, see Fig. 2.

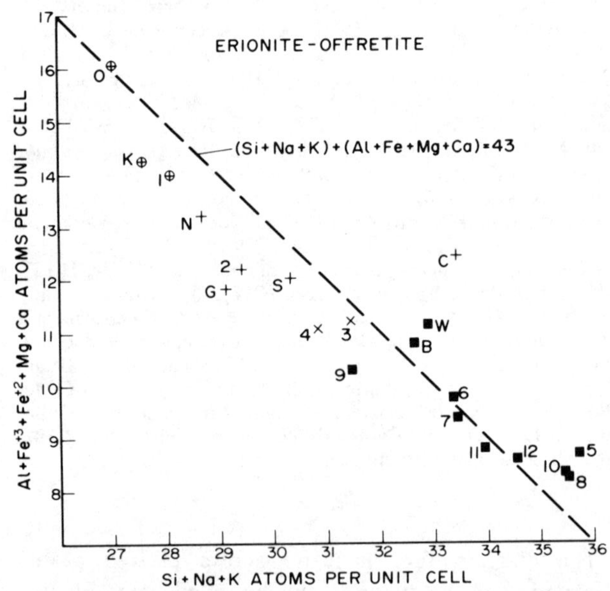

Fig. 10. Erionite–offretite series [Si+Na+K]:[Al+Fe^{+3}+Fe^{+2}+Mg+Ca] diagram. Sr and Ba are included in Ca. For symbols, letters, and literature, see Fig. 4.

ACKNOWLEDGMENTS

I am most grateful to Professor R. N. Brothers who made the facilities of the Department of Geology, University of Auckland, available to me. I am also indebted to Professor P. M. Black for criticisms of the manuscript.

REFERENCES

Alietti, A. (1972) Polymorphism and crystal-chemistry of heulandite and clinoptilolite: *Amer. Mineral.* **57**, 1448-1462.

Ames, L. L. Jr., Sand, L. B. and Goldich, S. S. (1958) A contribution on the Hector, California, bentonite deposit: *Econ. Geol.* **53**, 22-37.

Ballance, P. F. (1974) An inter-arc flysch basin in northern New

Zealand: Waitemata Group: *J. Geol.* **82**, 439-471.

Ballance, P. F. and McCarthy, J. A. (1975) Geology of Okahukura peninsula, Kaipara Harbour, New Zealand: *New Zealand J. Geol. Geophys.* **18**, 721-743.

Batiashvili, T. V. and Gvakhariya, G. V. (1968) Erionite found for the first time in Georgia: *Dokl. Akad. Nauk SSSR* **179**, 122-124. *Trans. Dokl. Akad. Nauk SSSR* **179**, 426-427 (1968).

Belitskiy, I. A. and Bukin, G. V. (1968) First find of erionite in the USSR: *Dokl. Akad. Nauk SSSR* **178**, 103-106. *Trans. Dokl. Akad. Nauk SSSR* **178**, 169-172 (1968).

Boles, J. R. (1972) Composition, optical properties, cell dimensions, and thermal stability of some heulandite-group zeolites: *Amer. Mineral.* **57**, 1463-1493.

Carter, L. (1971) Stratigraphy and sedimentology of the Waitemata Group, Puketotara Peninsula, Northland: *New Zealand J. Geol. Geophys.* **14**, 169-191.

Coombs, D. S. and Whetten, J. T. (1967) Composition of analcime from sedimentary and burial metamorphic rocks: *Geol. Soc. Amer. Bull.* 78, 269-282.

Eakle, A. S. (1898) Erionite a new zeolite: *Amer. J. Sci.* 4th ser. **6**, 66-68.

Eberly, P. E. Jr. (1964) Adsorption properties of naturally occurring erionite and its cation-exchanged forms: *Amer. Mineral.* **49**, 30-40.

Gude, A. J. 3rd and Sheppard, R. A. (1966) Silica-rich chabazite from the Barstow Formation San Bernardino County, Southern California: *Amer. Mineral.* **51**, 909-915.

Harada, K., Iwamoto, S. and Kihara, K. (1967) Erionite, phillipsite and gonnardite in the amygdales of altered basalt from Maze, Niigata Prefecture, Japan: *Amer. Mineral.* **52**, 1785-1794.

Hay, R. L. (1963) Stratigraphy and zeolitic diagenesis of the John Day Formation of Oregon: *Univ. Calif. Publ. Geol. Sci.* **42**, 199-262.

Hay, R. L. (1964) Phillipsite of saline lakes and soils: *Amer. Mineral.* **49**, 1366-1387.

Hay, R. L. (1966) Zeolite and zeolitic reaction in sedimentary rocks: *Geol. Soc. Amer. Spec. Paper* **85**, 130 pp.

Honda, S. and Muffler, L. J. P. (1970) Hydrothermal alteration in core from research drill hole Y-1, upper Geyser Basin, Yellowstone National Park, Wyoming: *Amer. Mineral.* **55**, 1714-1737.

Iijima, A. and Utada, M. (1971) Present-day zeolitic diagenesis of the Neogene geosynclinal deposits in the Niigata oil field, Japan: *Advances in Chemistry* Ser. No. 101, *Molecular Sieves Zeolites*, pp. 342-349.

Jones, B. G. (1972) Sedimentology of the Waitemata Group (Lower Miocene) at Pakaurangi Point, Kaipara, New Zealand: *J. Roy. Soc. New Zealand* **2**, 187-209.

Leitch, E. C. (1970) Contribution to the geology of northernmost New Zealand. II. The stratigraphy of the North Cape District: *Trans. Roy. Soc. New Zealand, Earth Sciences* **8**, 45-68.

Minato, H. and Utada, M. (1970) Clinoptilolite from Japan: Am. Chem. Soc. *2nd Int. Conf. Molecular Sieves Zeolite, Worcester, Mass.*, preprints 535-539.

Mizota, T., Shibuya, G., Shimazu, M. and Takeshita, Y. (1974) Mineralogical studies on levyne and erionite from Japan: *Mem. Geol. Soc. Japan* No. 11, 283-290.

Mumpton, F. A. (1960) Clinoptilolite redefined: *Amer. Mineral.* **45**, 351-369.

Passaglia, E. (1970) The crystal chemistry of chabazites: *Amer. Mineral.* **55**, 1278-1301.

Passaglia, E. (1975) The crystal chemistry of mordenites: *Contr. Mineral. Petrol.* **50**, 65-77.

Passaglia, E., Galli, E. and Rinaldi, R. (1974) Levynes and erionites from Sardinia, Italy: *Contr. Mineral. Petrol.* **43**, 253-259.

Rex, R. W., Eklund, W. A. and Jamieson, I. M. (1971) X-ray mineralogy studies Leg 6; in Fischer, A. G. *et al.*, eds., *Initial Report of the Deep Sea Drilling Project* **6** (U.S. Government Printing Office), Washington, pp. 691-708.

Rinaldi, R., Smith, J. V. and Jung, G. (1975) Chemistry and paragenesis of faujasite, phillipsite and offretite from Sasbach, Kaiserstuhl, Germany: *Neues Jb. Miner. Mh.* **10**, 433-443.

Saha, Prasenjit (1959) Geochemical and X-ray investigation of natural and synthetic analcites: *Amer. Mineral.* **44**, 300-313.

Sameshima, T. (1975) Silica indices of volcanoes in and around New Zealand with reference to volcanic zones in the North Island: *New Zealand J. Geol. Geophys.* **18**, 523-539.

Sheppard, R. A. and Gude, A. J. 3rd (1968) Distribution and genesis of authigenic silicate minerals in tuffs of Pleistocene Lake Tecopa, Inyo County, California: *U.S. Geol. Surv. Prof. Paper* **597**, 38 pp.

Sheppard, R. A. and Gude, A. J., 3rd (1969a) Diagenesis of tuffs in the Barstow Formation, Mud Hills, San Bernardino County, California: *U.S. Geol. Surv. Prof. Paper* **634**, 35 pp.

Sheppard, R. A. and Gude, A. J. 3rd (1969b) Chemical composition and physical properties of the related zeolite offretite and erionite: *Amer. Mineral.* **54**, 875-886.

Sheppard, R. A. and Gude, A. J. 3rd (1970) Calcic siliceous chabazite from the John Day Formation, Grant County, Oregon; in Geol. Surv. Research 1970: *U.S. Geol. Surv. Prof. Paper* **700D**, D176-D180.

Sheppard, R. A. and Gude, A. J. 3rd (1973) Zeolite and associated authigenic silicate minerals in tuffaceous rocks of the Big Sandy Formation, Mojave County, Arizona: *U.S. Geol. Surv. Prof. Paper* **830**, 36 pp.

Sheppard, R. A., Gude, A. J. 3rd and Munson, E. D. (1965) Chemical composition of diagenetic zeolites from tuffaceous rocks of the Mojave Desert and vicinity, California: *Amer. Mineral.* **50**, 244-249.

Sheppard, R. A., Gude, A. J. 3rd, Desborough, G. A. and White, J. S. Jr. (1974) Levyne offretite intergrowths from basalt near Beech Creek, Grant County, Oregon: *Amer. Mineral.* **59**, 837-842.

Skinner, D. N. B. (1969) Colville Formation – A new formation possibly correlative with the Waitemata Group: *New Zealand J. Geol. Geophys.* **12**, 349-360.

Staples, L. W. and Gard, J. A. (1959) The fibrous zeolite erionite; its occurrence, unit cell, structure: *Mineral. Mag.* **32**, 261-281.

BOWIE ZEOLITE DEPOSIT, COCHISE AND GRAHAM COUNTIES, ARIZONA

RICHARD A. SHEPPARD, ARTHUR J. GUDE, 3rd, and GARY M. EDSON

U.S. Geological Survey, Federal Center, Denver, Colorado 80225, U.S.A.

INTRODUCTION

The Bowie zeolite deposit is located in the San Simon Valley of southeastern Arizona, about 155 km east of Tucson. San Simon Valley is a northwest-trending intermontane basin in the Mexican Highland section of the Basin and Range physiographic province. The San Simon Valley is about 120 km long, extending from Rodeo, New Mexico, northward to Safford, Arizona. The valley is about 35 km wide at its central part near Bowie (Fig. 1).

San Simon River flows northward through the valley to join the Gila River near Safford. The San Simon Valley is bounded on the east by the Whitlock and Peloncillo Mountains and is bounded on the west by the Pinaleno, Dos Cabezas, and Chiricahua Mountains. The town nearest to the zeolite deposit is Bowie, located just north of Interstate Highway 10. Improved dirt roads provide access northward to the deposit from Bowie. Zeolite ore is shipped from Bowie on the Southern Pacific Railroad.

The Bowie zeolite deposit is significant from the standpoint of zeolite mining in the United States because there are only a few deposits being exploited specifically for zeolites. Zeolitic tuffs from other areas in the United States have been mined and quarried for many years, but their use as lightweight building stone and pozzolan is based on the gross properties of the rock rather than on the chemical and physical properties of the zeolites. Exploitation of the Bowie deposit is also significant because it demonstrates that beds of zeolitic tuff as thin as 15 cm can be economically mined.

The zeolites at the Bowie deposit were not positively identified until the late 1950s; however, Schwennesen (1917, p. 8) had described tuffs in lacustrine rocks that crop out in the lower dissected part of the San Simon Valley. Even earlier, Loew (1875, pp. 643-644) described and analyzed a zeolitic tuff from the San Simon Valley. The significance of Loew's report has only recently been recognized. Inasmuch as it is probably the earliest published record of a zeolite in a sedimentary deposit anywhere in the world, the following quote seems justified: "The volcanic tuffs have but seldom been the subject of close investigation; I therefore subjected a tuff of a peculiar character to analysis. On the plain extending from Fort Bowie to the Peloncillo Mountains, about four miles south of Whitlock's Cienega, are found stratified deposits of a yellowish, soft, porous material, in thickness from 12 to 13 feet, and in extent about one-sixth mile. The mass easily crumbled to powder between the fingers, had no similarity at all to clay, and the presence of very fine grit could be distinctly recognized." On the basis of chemical tests and a quantitative chemical analysis, Loew concluded "... that this tuff is a mixture of free particles of quartz, with a hydrous silicate, somewhat related to chabazite or stilbite". Although Loew did not report optical studies, he apparently was convinced of the zeolite content of the tuff because he later (p. 644) referred to the material as a "zeolitic deposit".

The first zeolite production from the Bowie deposit was by Union Carbide Corporation in 1961. Since then, at least four other companies have also mined the deposit on an intermittent basis, including W. R. Grace and Company, NRG Incorporated, Norton Company, and Filtrol Corporation. Although the deposit has been extensively drilled and studied by these companies, the results are chiefly unpublished (Sand and Regis, 1966; Regis and Sand, 1967; Mumpton, 1973). The Conservation Division of the U.S. Geological Survey drilled, sampled, and mapped the deposit in 1972 and 1973, but the results of these investigations are also not yet published.

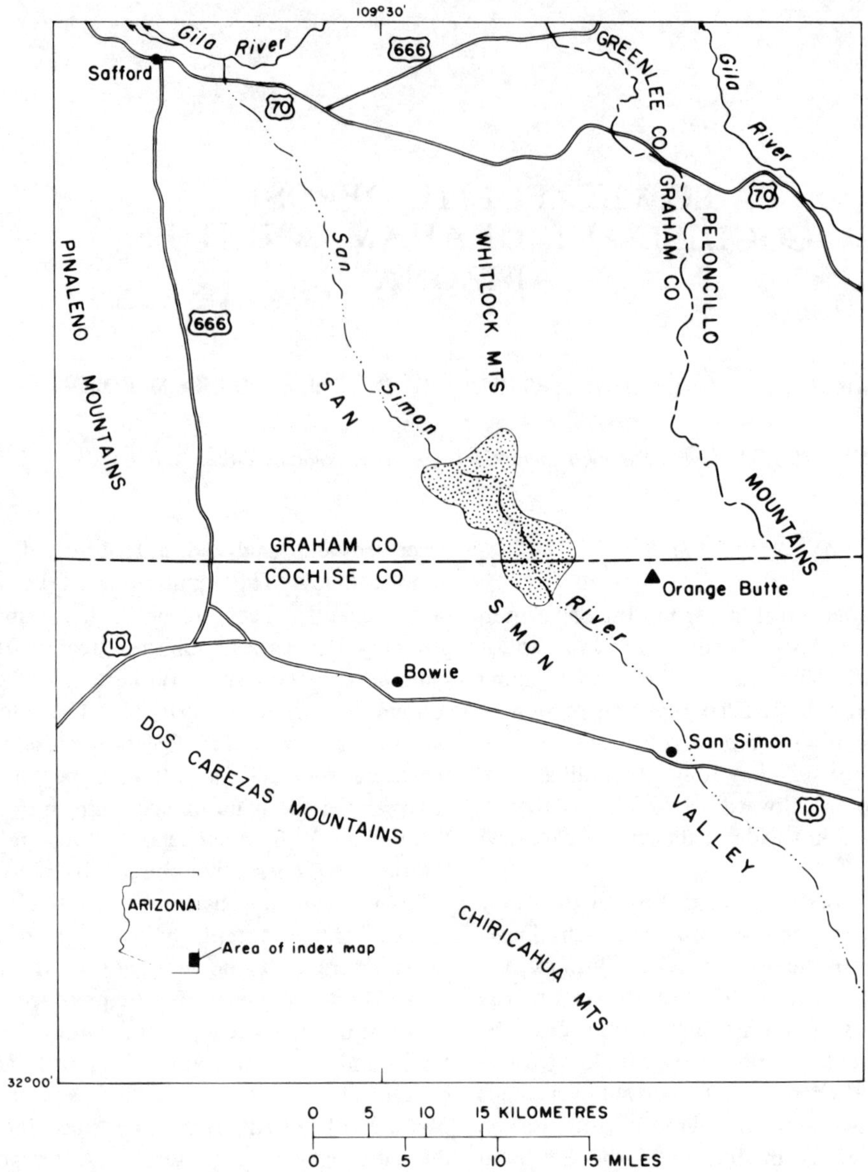

Fig. 1. Index map of southeastern Arizona, showing the Bowie zeolite deposit (shaded) in San Simon Valley.

GEOLOGIC SETTING

The mountain ranges bounding the San Simon Valley contain a wide variety of sedimentary, igneous, and metamorphic rocks. The Whitlock and Peloncillo Mountains are chiefly Tertiary and Quaternary volcanic rocks of rhyolitic to basaltic composition. The Pinaleno Mountains are Precambrian granite, granitic gneiss, and schist overlain by Tertiary volcanic rocks on the southern flank. The Dos Cabezas Mountains are also Precambrian granite, granitic gneiss, and schist as well as Paleozoic to Tertiary sedimentary rocks and Tertiary volcanic rocks. A similar variety of rocks makes up the Chiricahua Mountains; however, Tertiary volcanic rocks and Tertiary granitic intrusive rocks comprise most of the southern part of this range.

The nearly flat-lying fluviatile and lacustrine deposits of the San Simon Valley have a maximum thickness of at least 610 m (White, 1963) and are probably late Pliocene to Holocene in age (Knechtel, 1938). The

Fig. 2. Natural exposure of resistant marker tuff (M) at Field Trip Stop 4. The person at left is standing on a resistant siltstone bed that is stratigraphically beneath the marker tuff. View looking eastward across the San Simon River to the Peloncillo Mountains in background.

Fig. 3. Exposure of zeolitic marker tuff in a pit at Field Trip Stop 3. The ore bed (at hammer) is 13 cm thick and is overlain by thin-bedded tuffs that are 112 cm thick.

thickness and character of this valley fill is known chiefly from drillers' logs of water wells. Natural exposures of the sedimentary deposits are poor, and the maximum thickness exposed is commonly no more than several meters. A conspicuous zeolitic tuff, herein termed the "marker tuff", is part of a unit mapped as "older alluvial fill" by Cushman and Jones (1947). The older alluvial fill consists mainly of clay, silt, sandstone, and gravel, and Cushman and Jones concluded that some of the fine-grained detrital sediments of the San Simon Valley were deposited in a closed lake basin. The lower part of the older alluvial fill contains water under artesian pressure and is the source of most of the ground water in the San Simon Valley (White, 1963).

Although the detailed stratigraphy of the older alluvial fill is unknown, a hole drilled for the U.S. Geological Survey in the central part of the San Simon Valley penetrated the marker tuff and about 50 m of the underlying lacustrine claystone. Another zeolitic tuff (65 cm thick) was intersected by the drill hole about 23.5 m beneath the marker tuff. This stratigraphically lower tuff does not crop out in the valley. Several thin dolomite beds were also recognized in the lower 8 m of claystone penetrated by the drill hole. On the basis of drillers' logs, White (1963) estimated that the lacustrine claystone is as much as 180 m thick in the axial part of the San Simon Valley.

MARKER TUFF

The marker tuff is the most conspicuous and resistant unit in the lacustrine sediments (Fig. 2). Even so, natural exposures of the tuff are scarce and discontinuous. Data from the U.S. Geological Survey's drilling and mapping program (L. H. Godwin and P. C. Aguilar, written communication, 1975) suggest that the marker tuff originally was distributed over an area of about 85 km^2 in the central part of the basin. Its present areal distribution, however, is probably not more than 12 km^2. The present limited and discontinuous nature of the marker tuff is due chiefly to removal by erosion. Some of the tuff was eroded after the original glass altered to zeolites, as shown by rounded fragments of the zeolitic marker tuff which occur locally in unconsolidated sand and gravel of fluvial channel deposits. Furthermore, the tuff is still being eroded. Other areas of the tuff may have been eroded prior to zeolitization. Although difficult to prove, the discontinuous character of the marker tuff may be due, in part, to local areas of nondeposition of the original vitric ash.

Where observed in shallow pits and natural outcrops, the marker tuff is 22–125 cm thick and seems to thicken toward the southern part of the deposit. This tuff characteristically consists of a basal bed, 10–20 cm thick, overlain by a variable thickness of thin to very thin beds (Fig. 3). Only the basal bed is mined, and it is commonly termed the "ore bed". The color of the marker tuff is white, gray, yellow, or orange, but pastel

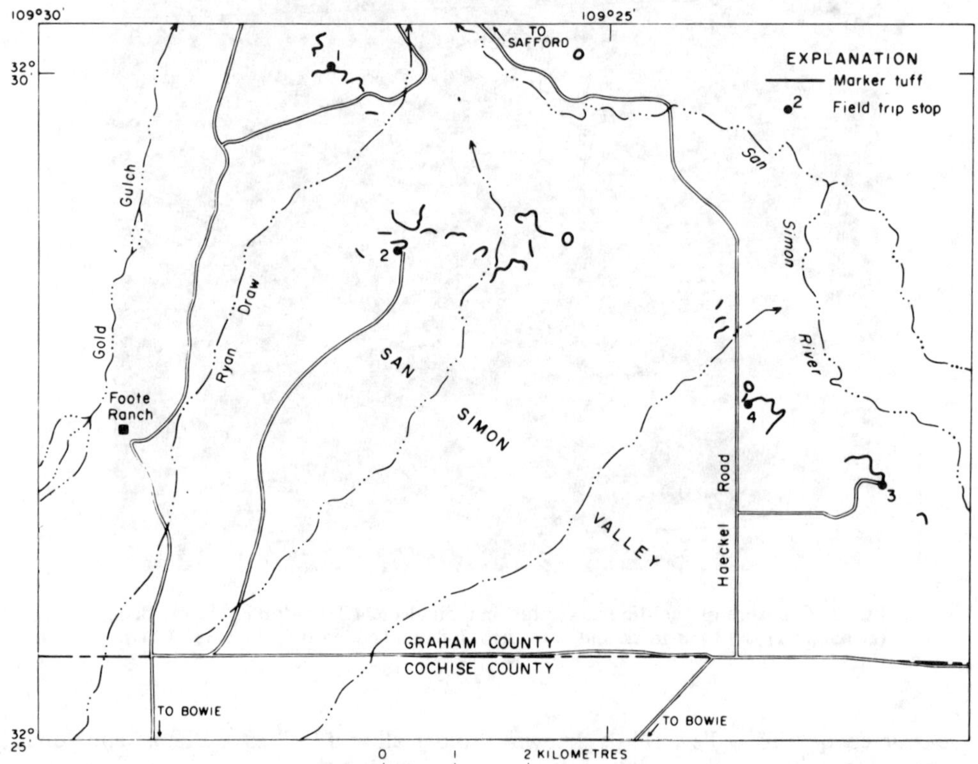

Fig. 4. Map showing the outcrop distribution of the marker tuff and the field trip stops at the northwestern and central parts of the Bowie zeolite deposit. The distribution of the marker tuff is modified from an unpublished map by L. H. Godwin and P. C. Aguilar (written communication, 1975).

shades of yellow seem to be most common. It has a dull or earthy luster and commonly breaks with a hackly or subconchoidal fracture. The original vitroclastic texture and sedimentary structures, such as ripple marks, are generally preserved in the zeolitic tuff. Convoluted structures are especially common in the basal bed, and ripple marks are especially common in the overlying thin beds.

MINERALOGY AND CHEMISTRY OF THE MARKER TUFF

The zeolites and associated authigenic silicate minerals in the marker tuff crystallized after the accumulation of the original volcanic ash in the lake. Alkaline and moderately saline pore water, trapped during sedimentation in a lacustrine environment, dissolved the original vitric material, thereby providing the chemical constituents necessary for the formation of these minerals. The zeolites, then, either precipitated directly from the solution provided by the dissolved glass shards or crystallized from an intermediate precipitate of aluminosilicate gel.

Parts of the marker tuff still locally contain relict glass shards. A particularly striking example is at Field Trip Stop 1 (Fig. 9), where most of the ore bed and much of the overlying thin-bedded tuffs consist of unaltered glass.

Those parts of the marker tuff that are chiefly fresh glass are gray and friable. The vitric particles are very fine to fine grained and are of two types: (1) platy bubble-wall shards formed from the walls of relatively large, broken bubbles, and (2) pumice shards that contain small elongated bubbles. Platy bubble-wall shards predominate in the marker tuff. The index of refraction of the unaltered glass is 1.506–1.512 and indicates a silicic composition. Even if relict glass were not recognized in the marker tuff, the generally well-preserved vitroclastic texture (Fig. 5) is convincing evidence that volcanic glass shards were the parent materials for the zeolites.

In addition to vitric material, the marker tuff contains minor crystal fragments of quartz, sodic plagioclase, hornblende, clinopyroxene, and zircon. These crystals, unlike the relatively unstable glass, are unaltered in the zeolitic tuff.

The mineralogy of the zeolitic marker tuff was determined from X-ray powder diffraction data of bulk samples, and some representative analyses are given in Table 1. The tuff commonly contains three zeolites, less commonly contains two zeolites, and rarely contains one or four zeolites. The most common zeolite assemblage is chabazite, erionite, and clinoptilolite. Small amounts of analcime rarely occur in the marker tuff at the southern part of the deposit. In addition to zeolites, the marker tuff commonly contains minor smectite, calcite, quartz, and plagioclase, and locally contains minor halite, gypsum, or thenardite.

The relative abundance of the three major zeolites in

Fig. 5. Scanning electron micrograph of the "ore bed" showing excellent preservation of vitroclastic texture. Tuff consists chiefly of chabazite. Micrograph by C. S. V. Barclay.

TABLE 1. Mineralogical composition of marker tuff and associated rocks at the Bowie zeolite deposit

Field Trip Stop (Fig. 4)	Sample	Stratigraphic Position of sample in cm above (+) or below (−) base of the marker tuff	X-ray diffraction analysis of bulk samples in parts of 10										
			Glass	Clay	Smec.	Anal.	Chab.	Cp	Er	Qtz	Plag.	Cal.	Other
1	F	+36, marker tuff, thin bed	4	–	1	–	2	–	2	1	–	–	–
	E	+28, marker tuff, thin bed	–	–	1	–	6	tr	1	–	–	2	–
	D	+10, marker tuff, ore bed	9	–	1	–	tr	–	–	–	–	–	–
	C	+1, marker tuff, ore bed	–	–	1	–	7	tr	–	1	1	–	–
	B	−5, sandstone	–	–	–	–	–	tr	–	2	1	7	–
	A	−75, siltstone	–	tr	2	–	–	1	–	4	3	–	–
2	F	+132, claystone	–	–	1	–	2	1	1	2	3	–	–
	E	+42, claystone	–	–	1	–	3	tr	2	1	3	–	–
	D	+20, marker tuff, thin bed	–	–	tr	–	7	–	3	–	–	–	–
	C	+9, marker tuff, ore bed	–	–	tr	–	9	tr	1	tr	tr	–	–
	B	+0.2, marker tuff, ore bed	–	–	tr	–	9	1	tr	–	–	–	–
	A	−3, claystone	–	tr	2	–	–	1	–	4	3	–	–
3	G	+165, claystone	–	2	2	1	–	–	–	3	2	–	–
	F	+122, marker tuff, thin bed	–	–	1	–	1	6	1	1	tr	–	–
	E	+65, marker tuff, thin bed	–	–	tr	–	–	2	7	–	–	–	1[3]
	D	+23, marker tuff, thin bed	–	–	tr	–	4	3	2	–	–	–	1[4]
	C	+8, marker tuff, ore bed	–	–	tr	–	6	2	2	–	–	–	–
	B	+5, marker tuff, ore bed	–	–	tr	1	5	3	1	–	–	–	
	A	−5, claystone	–	tr	4	1	–	–	–	1	1	2	1[3]
4	E	+60, marker tuff, thin bed	–	–	1	–	–	–	9	–	–	–	–
	D	+10, marker tuff, ore bed	–	–	tr	–	tr	4	5	–	–	1	–
	C	−2, siltstone	–	tr	tr	tr	–	1	tr	2	4	3	–
	B	−244, siltstone	–	1	1	2	–	–	–	3	2	1	–
	A	−260, siltstone	–	1	tr	1	–	–	–	5	3	–	–

Clay = illite/mica; Smec. = smectite; Anal. = analcime; Chab. = chabazite; Cp = clinoptilolite; Er = erionite; Qtz = quartz; Plag. = plagioclase; Cal. = calcite; [3] = gypsum; [4] = halite; tr = trace; – = looked for but not found.

the marker tuff is variable from bed to bed and from locality to locality within the same bed. We were unable to determine any systematic vertical or lateral variation of these zeolites in the marker tuff. Perhaps the most consistent mineralogical aspect of the marker tuff is the relatively high chabazite content of the basal ore bed. Although the chabazite content of the "ore bed" ranges from less than 5%, it is commonly in the range of 60–80%. Inasmuch as chabazite is the principal zeolite being sought at the Bowie deposit, it is understandable that only the "ore bed" is mined.

The zeolites in the marker tuff are generally finely crystalline, and examination under the petrographic microscope proved unsatisfactory. Using the scanning electron microscope, however, we were able to determine the crystal size, morphology, and grain relationships of the zeolites. Chabazite occurs as aggregates of minute rhombohedral crystals that form ragged thin plates (Figs. 6 and 7). The plates are 2–25 μm in their longest dimension and locally are arranged in radial groups. Individual chabazite crystals are commonly less than 0.1 μm to 0.8 μm in size, and this very small crystal size is probably responsible for the consistently low intensity and relatively broad character of peaks on X-ray diffractometer records. Erionite (Fig. 8) occurs as bundles of hexagonal rods. The bundles of erionite are 10–60 μm long, but most are 15–40 μm long. Clinoptilolite (Fig. 6) commonly occurs as well-formed tabular crystals that are 2–125 μm long. Most clinoptilolite, however, is 10–30 μm long. The relatively rare analcime occurs as well-formed trapezohedra that are as much as 40 μm across.

The sequence of formation of the authigenic silicate minerals in the marker tuff was studied chiefly by using the scanning electron microscope. Smectite was the earliest silicate mineral to form in the tuff, and

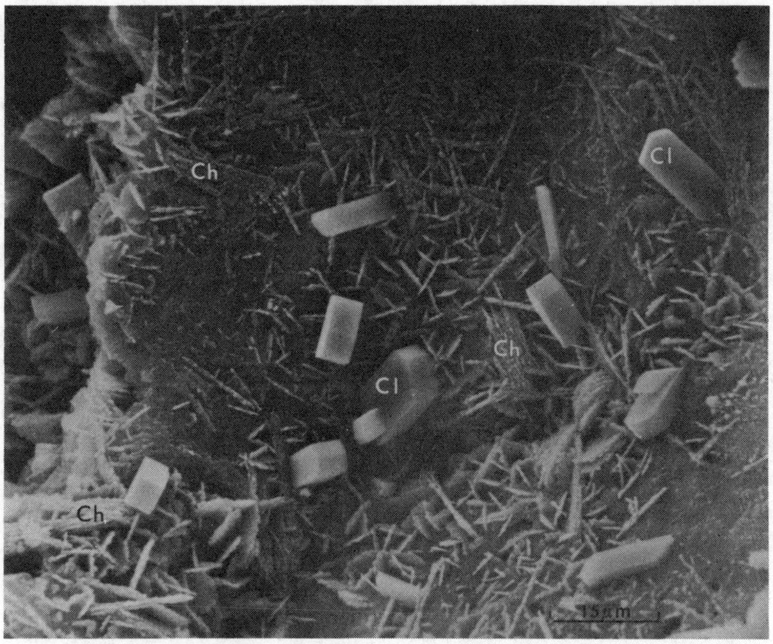

Fig. 6. Scanning electron micrograph of ore bed showing relatively large clinoptilolite (Cl) crystals perched on ragged plates of minute chabazite (Ch) crystals. Micrograph by C. S. V. Barclay.

Fig. 7. Scanning electron micrograph of "ore bed" showing the detail of the finely crystalline chabazite. Micrograph by C. S. V. Barclay.

commonly it is the material that has preserved the vitroclastic texture. The formation of some smectite, however, overlapped zeolite crystallization. Locally, an early-formed zeolite is coated with smectite, but later-formed zeolite is perched on the smectite coating. Chabazite seems to have been the earliest zeolite to crystallize in the marker tuff. Inasmuch as both clinoptilolite (Fig. 6) and erionite are perched on ragged aggregates of chabazite crystals and inasmuch as both project into minute open spaces, we believe these zeolites post-date the chabazite. No textural relationships were recognized to indicate the relative ages of the clinoptilolite and erionite, however. Analcime seems to have been the last zeolite to form. Other studies of

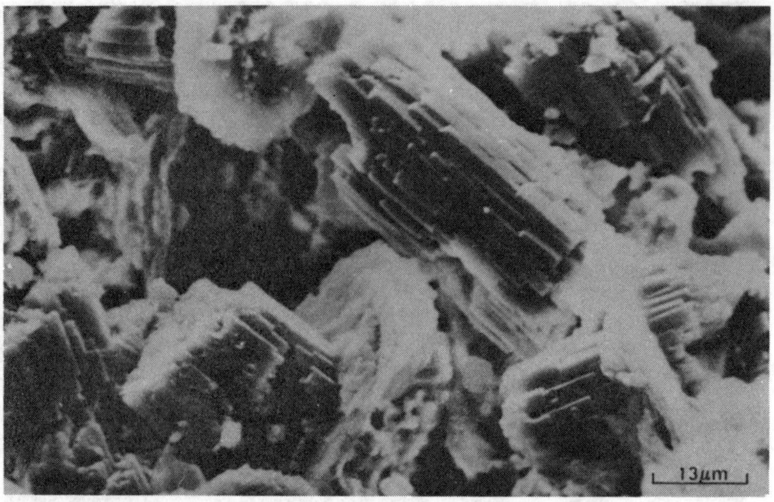

Fig. 8. Scanning electron micrograph of a thin-bedded tuff from the upper part of the marker tuff showing bundles of rods of erionite.

analcime in tuffs from alkaline, saline lacustrine deposits indicate that it generally formed by the reaction between early-formed zeolites and interstitial fluids (Sheppard and Gude, 1973). The Bowie analcime probably formed in a similar manner, but no textural evidence was recognized.

Chemical analyses of two bulk samples of the ore bed are given in Table 2. Both samples are mineralogically

TABLE 2. Chemical analyses of two ore-bed samples, in weight percent

	A	B
SiO_2	55.3	54.6
Al_2O_3	13.2	14.9
Fe_2O_3	3.40	2.28
MgO	0.98	0.60
CaO	3.25	0.22
Na_2O	2.85	6.67
K_2O	1.15	0.90
LOI	19.1	19.4
Total	99.23	99.57

A. Lab. No. D159477; locality: SW¼NW¼ sec. 13, T. 11 S., R. 28 E., Graham County, Arizona.

B. Lab. No. D161834; locality: NE¼NW¼ sec. 2, T. 12 S., R. 29 E., Cochise County, Arizona.

[SiO_2 and Al_2O_3 determined colorimetrically by G. T. Burrow; total Fe as Fe_2O_3, MgO, CaO, Na_2O, and K_2O determined by atomic absorption by Violet Merritt and Wayne Mountjoy; LOI determined at 900°C for 1 hour by G. D. Shipley.]

TABLE 3. Chemical analysis and composition of unit cell of chabazite

	A	B
Chemical analysis in weight percent		
SiO_2	51.9	52.1
Al_2O_3	12.0	14.2
Fe_2O_3	5.30	2.81
MgO	1.17	0.88
CaO	3.46	0.11
Na_2O	3.68	6.71
K_2O	0.38	0.68
H_2O	22.11	22.51
Composition of unit cell on the basis of O = 72		
Si	26.64	26.63
Al	7.26	8.56
Fe^{+3}	2.05	1.08
Mg	0.89	0.67
Ca	1.90	0.06
Na	3.66	6.65
K	0.25	0.44
H_2O	37.85	38.38
Si:Al+Fe^{+3}	2.86	2.76

A. Lab. No. D160295; separated from sample A of Table 2.

B. Lab. No. D161678; separated from sample B of Table 2.

[SiO_2 and Al_2O_3 determined colorimetrically by G. T. Burrow; total Fe as Fe_2O_3, MgO, CaO, Na_2O, and K_2O determined by atomic absorption by Violet Merritt and Wayne Mountjoy; H_2O calculated by difference.]

similar, consisting chiefly of chabazite with minor erionite, clinoptilolite, and smectite. Sample A was collected in the northern part of the deposit, near Field Trip Stop 2 (Fig. 4), and sample B was collected in the southern part of the deposit, about 3 km southeast of Field Trip Stop 3. The analyses are similar, except that sample A shows a much higher $CaO:Na_2O$ ratio than sample B. Additional unpublished analyses (C. S. V. Barclay and R. D. Hettinger, written communication, 1975) suggest that a decrease in the $CaO:Na_2O$ ratio from north to south is characteristic for the ore bed west of the San Simon River. Because chabazite is the major constituent of the ore bed, the same chemical trend is suggested for the chabazite.

Chemical analyses and the unit-cell compositions of chabazite separated from the two previously mentioned bulk samples are given in Table 3. These data show that chabazite from sample A has significantly higher calcium and lower sodium contents than chabazite from sample B. These data support the conclusion of Regis and Sand (1967) that there is a northward increase in the Ca:Na ratio of the chabazite. As shown in Table 3, both chabazites are siliceous, having $Si:Al+Fe^{+3}$ ratios of 2.86 and 2.76. Relatively small cell dimensions and low indices of refraction of the Bowie chabazite are consistent with the siliceous character (Sheppard, 1973).

UTILIZATION OF THE BOWIE ZEOLITE

The zeolitic tuff mined at the Bowie deposit has numerous industrial applications that are based chiefly on the physical and chemical properties of the chabazite; however, information on some uses or potential uses is not available because the companies mining the zeolite consider the data proprietary. A major use of the Bowie zeolite is in the drying and purification of gases. By appropriate pressure-swing adsorption processes, the zeolite is used to remove H_2S, CO_2, and H_2O from acidic natural gas streams. This application makes the natural gas acceptable to pipelines and public utilities. Bowie zeolite, marketed by the Linde Division of Union Carbide Corporation as AW-500, is claimed to be capable of removing HCl from reformer hydrogen streams, drying chlorine, purifying certain chlorinated hydrocarbons, drying and purifying certain fluorinated hydrocarbons, and purifying stack gases (Union Carbide Corporation, 1962). An experimental plant, operated by Reserve Synthetic Fuels Incorporated and utilizing the Bowie zeolite, began operation in early 1976 at the Palos Verdes landfill near Los Angeles, California. The zeolite purifies the methane produced by the decomposition of organic waste material in the landfill. Mumpton (1975) suggested that the Bowie zeolite could be similarly used to remove CO_2 and H_2S from methane gas produced during the anaerobic digestion of sludge at municipal sewage-treatment plants and of animal wastes from farms and feedlots. The Bowie zeolite can also be used for the recovery and storage of radioactive cesium. Once trapped in the zeolite, the cesium can be stored for ease of shipment or disposal, or it can be held for future recovery.

DESCRIPTION OF FIELD TRIP STOPS

The stops for the field trip at the Bowie zeolite deposit were chosen to show representative exposures of the marker tuff in both natural outcrops and in shallow pits over a northwest-southeast distance of about 10 km, west of the San Simon River. The four stops are located on Fig. 4, and the mineralogy of the marker tuff and enclosing sediments is given in Table 1.

Stop 1

The scraped area and adjacent natural outcrops to the north are on a mining claim of NRG Incorporated. These outcrops are the northwesternmost extent of the marker tuff. The marker tuff is 48 cm thick and consists of the chiefly unaltered "ore bed" (18 cm) and overlying thin-bedded tuffs (30 cm). The gray ore bed is mainly unaltered glass, but the basal 1-3 cm are yellow and consist chiefly of chabazite. Thin-bedded tuffs above the "ore bed" are 1 mm to 4 cm thick and consist mainly of varying amounts of relict glass, chabazite, and erionite. A gray calcareous sandstone (10 cm thick) underlies the marker tuff and is succeeded downward by about 1.5 m of brown claystone and siltstone that contain irregular, gray, calcareous concretions. About 3 m of brown siltstone, sandstone, and conglomerate overlie the marker tuff and are probably fluviatile and Quaternary in age.

Stop 2

This pit is mined by Union Carbide Corporation and exposes about 22 cm of the zeolitic marker tuff, most (18 cm) of which is the "ore bed" at the base. No relict glass has been found in the marker tuff; however, the original vitroclastic texture is well preserved, especially in the "ore bed" which consists mainly of chabazite. Thin-bedded tuff above the "ore bed" shows ripple marks and consists of chabazite and erionite. Brown claystone underlies the marker tuff and contains minor clinoptilolite. About 4 m of tuffaceous claystone and siltstone overlie the marker tuff and consist of 40-50% authigenic zeolites in addition to the detrital constituents. The zeolites are varying proportions of chabazite, clinoptilolite, and erionite. The claystone and siltstone overlying the marker tuff in this pit are probably lacustrine and are probably part of the same depositional unit as the marker tuff, in contrast to the sediments overlying the marker tuff at the previous stop.

Stop 3

This pit, operated by NRG Incorporated, exposes about 125 cm of the zeolitic marker tuff (Fig. 3). The basal orange and white "ore bed" is 13 cm thick and consists of major chabazite and minor clinoptilolite and erionite. Irregular segregations in the "ore bed" have a sugary texture and consist of relatively large clinoptilolite crystals that are as much as 125 μm long. The "ore bed" locally contains minor analcime and a white efflorescent coating of halite and thenardite. Gypsum is locally common at the top of the "ore bed". The overlying thin-bedded tuffs are yellowish-gray and consist of varying proportions of chabazite, clinoptilolite, and erionite. Clinoptilolite or erionite, rather than chabazite, commonly is the major zeolite in the thin-bedded tuffs. Green claystone, underlying and overlying the marker tuff, contains about 10% authigenic analcime.

Stop 4

Excellent natural exposures of the zeolitic marker tuff and a relatively thick section of the underlying rocks (Fig. 2) can be examined at this stop. The marker tuff is yellow, resistant, and about 65 cm thick. The basal "ore bed" is 20 cm thick; and, unlike most other exposures of the zeolitic "ore bed", at the locality examined it consists mainly of clinoptilolite and erionite with minor or no chabazite. Contorted laminations are common in the "ore bed", and ripple marks are characteristic of the overlying thin-bedded tuffs. About 5 m of brown claystone and siltstone crop out beneath the marker tuff. These rocks are commonly calcareous and contain calcareous concretions. All the sampled siltstones contain minor analcime, and the more resistant ones also contain calcite. A distinctive thin (3 cm), gray siltstone occurs beneath the base of the marker tuff. In addition to major contents of quartz, plagioclase, and calcite, this gray bed contains minor amounts of analcime, clinoptilolite, and erionite.

ACKNOWLEDGMENTS

Grateful appreciation is expressed to T. H. Eyde for sharing his knowledge of the Bowie zeolite deposit with us. L. H. Godwin and P. C. Aguilar supplied us unpublished field data, and C. S. V. Barclay and R. D. Hettinger allowed us access to unpublished chemical analyses of the marker tuff. T. L. Finnell is especially acknowledged for bringing to our attention the previously unrecognized pioneer work of Oscar Loew on the Bowie zeolite deposit.

EDITORIAL NOTE

A discovery of fossil tracks in the clay underlying the "ore" bed exposed in pits near Field Trip Stop 2 was reported by the Smithsonian Institution in 1976. The tracks are those of an extinct llama, with at least 10 individuals having been present. Also, 20 tracks of a Pleistocene camel (*Camelops* sp.) and one well-preserved pawprint of a large carnivore were noted in an area of more than 400 m^2.

REFERENCES

Cushman, R. L. and Jones, R. S. (1947) Geology and ground-water resources of the San Simon basin, Cochise and Graham Counties, Arizona, with a section on quality of water by J. D. Hem: U.S. Geol. Survey open-file report, 31 pp.

Knechtel, M. M. (1938) Geology and ground-water resources of the valley of Gila River and San Simon Creek, Graham County, Arizona: *U.S. Geol. Survey Water-Supply Paper* **796-F**, 181-222.

Loew, Oscar (1875) Report upon mineralogical, agricultural, and chemical conditions observed in portions of Colorado, New Mexico, and Arizona, in 1873: Pt. 6 of *U.S. Geog. and Geol. Explor. and Surveys West of 100th Meridian* (Wheeler) **3**, 569-661.

Mumpton, F. A. (1973) Worldwide deposits and utilisation of natural zeolites: *Indus. Minerals*, no. 73, 30-45.

Mumpton, F. A. (1975) Commercial utilization of natural zeolites: in Lefond, S. J., ed., *Industrial Minerals and Rocks*, 4th ed.; Amer. Inst. Mining, Metall., and Petroleum Engineers, Inc., New York, pp. 1262-1274.

Regis, A. J. and Sand, L. B. (1967) Lateral gradation of chabazite to herschelite in the San Simon basin [abs.]: *Clays and Clay Minerals* **27**, 193.

Sand, L. B. and Regis, A. J. (1966) An unusual zeolite assemblage, Bowie, Arizona: in Abstracts for 1965: *Geol. Soc. Amer. Spec. Paper* **87**, 145-146.

Schwennesen, A. T. (1917) Ground water in San Simon Valley, Arizona and New Mexico: *U.S. Geol. Survey Water-Supply Paper* **425**, 1-35.

Sheppard, R. A. (1973) Zeolites and zeolite ore from Union Carbide Corporation's EZ No. 225 placer mining claim, Graham County, Arizona: U.S. Geol. Survey open-file report, 38 pp.

Sheppard, R. A. and Gude, A. J. 3rd (1973) Zeolites and associated authigenic silicate minerals in tuffaceous rocks of the Big Sandy Formation, Mohave County, Arizona: *U.S. Geol. Survey Prof. Paper* **830**, 36 pp.

Union Carbide Corporation (1962) Linde molecular sieve type AW-500: *Linde Molecular Sieves Bull.* **F-1617**, 2 pp.

White, N. D. (1963) Analysis and evaluation of available hydrologic data for San Simon basin, Cochise and Graham Counties, Arizona: *U.S. Geol. Survey Water-Supply Paper* **1619-DD**, 33 pp.

EXPERIMENTS ON THE FORMATION OF ZEOLITES BY HYDROTHERMAL ALTERATION OF VOLCANIC GLASSES

H. HÖLLER and U. WIRSCHING

Institut für Technische Petrographie und Mineralogie der Technische Universität Graz, Austria

Abstract – Investigations of natural zeolites formed by hydrothermal alteration of volcanic glasses show that the following factors are influential: (1) temperature, (2) chemical composition and pH of the solutions, (3) chemical composition of the starting materials, and (4) existence of either a closed or open system during alteration. To investigate these factors, experiments were carried out holding parameters as close as possible to natural conditions.

In the case of a closed system, the chemical composition of the starting material is the most important factor. It determines the alteration rate and the zeolites that are formed at different temperatures and with different solutions. Therefore, basaltic and phonolitic glass react in a similar way because of their similar chemical compositions. Both glasses alter rapidly, and from both glasses chabazite, phillipsite, and analcime are formed within the temperature range 150° to 250°C. Rhyolitic glass, however, reacted much more slowly, and the zeolites mordenite and analcime formed. In closed systems, zeolites are formed predominantly at higher temperature (250°C), but with alkaline to weakly acid solutions (N/100 NaOH to N/1000 H_2SO_4).

An open system considerably influences the formation conditions of zeolites, as shown by experiments in a simulated open system using rhyolitic glass as a starting material. The cause is the possible addition and removal of material. In an open system zeolites are formed predominantly with alkaline solutions (N/100 NaOH), but at higher to lower temperatures (250°C to 150°C). Phillipsite may be formed transitorily, and mordenite may disappear with an increased addition of Na^+, meaning that the influence of the chemical composition of the starting material is reduced in an open system. Other important consequences of the open system are the change of the zeolite associations during the alteration and the non-uniform alteration within hydrothermal regions. Zonal arrangements may also be formed in open systems. These phenomena depend on the changing chemical composition of the reacting solutions caused by the reaction with the solid material, by the formation of minerals, and by the addition and removal of material.

INTRODUCTION

Zeolite occurrences formed by hydrothermal alteration of volcanic glass are characterized by several remarkable phenomena:

(1) Some zeolites are abundant, especially analcime.
(2) Different zeolites and zeolite assemblages are formed from the same starting material.
(3) Different degrees of alteration are to be found within a single occurrence.
(4) Zeolites occur in zonal arrangements.

Therefore, investigations on the formation of zeolites by hydrothermal alteration must answer the question – what are the factors that cause these phenomena?

The observation of natural occurrences show the following factors to be important: (1) temperature, (2) chemical composition and pH of the solutions, (3) chemical composition of the starting material, and (4) existence of an open or closed system during alteration. The observation of natural occurrences does not show, however, how and to what degree these factors influence zeolite formation. Moreover, it is not known how each of these factors influences the effect of the others. For these reasons, experiments were carried out recently to investigate the influence of each factor on zeolite formation (Höller *et al.*, 1974; Wirsching, 1975, 1976). Experimental conditions were held as close as possible to natural ones so that the results could be compared with natural alteration assemblages.

STARTING MATERIALS AND EXPERIMENTAL CONDITIONS

Starting materials were three volcanic glasses, the chemical compositions of which are listed in Table 1. Included were a basaltic glass from the East Styria volcanic area, Austria; a phonolitic glass from the Laach volcanic area, Eifel, Germany; and a rhyolitic glass from the Shoshone area, Inyo County, California. The basaltic and phonolitic glasses are hydrothermally altered to chabazite, phillipsite, and analcime in some areas (Flügel and Heritsch, 1968; Frechen, 1971; Höller *et al.*, 1974). The rhyolitic glass used in these experiments is fresh,

TABLE 1. Chemical composition of the volcanic glasses, used in the experiments (wt. %)

	Basaltic* Glass	Phonolitic† Glass	Rhyolitic‡ Glass
SiO_2	48.35	55.14	72.78
TiO_2	0,67	0.23	0.12
Al_2O_3	22.50	22.52	11.89
Fe_2O_3	0.79	1.18	0.55
FeO	5.42	0.97	0.99
MnO	0.13	0.33	0.04
MgO	2.65	0.35	0.22
CaO	5.27	0.98	0.55
Na_2O	8.05	10.02	3.03
K_2O	4.97	5.69	5.31
H_2O^+	0.99	2.38	3.86
H_2O^-	0.02		0.21
P_2O_5	0.50		0.01
CO_2			0.12
F			0.15
Total	100.31	99.79	99.84

*Schoklitsch (1935).
†Frechen (1971).
‡Sheppard and Gude (1968).

with only a minor amount of montmorillonite alteration, but it is known to have transformed into several different zeolites during diagenesis in a saline lake environment (Sheppard and Gude, 1968). At Wairakei, New Zealand, rhyolitic glass is known to have altered hydrothermally to montmorillonite and, at increasing depth, to mordenite (Coombs *et al.*, 1959; Ellis, 1960).

Experimental conditions

Temperature: 150°, 200°, 250°C.
Solution: (25 ml, 50 ml)
NaOH N/100, pH = ~ 10.5; NaOH N/1000, pH = ~ 8.
H_2O distilled, pH = ~ 5.
H_2SO_4 N/100, pH = ~ 2.
"Natural solution" formed by the reaction of distilled water with the volcanic glasses at 250°C for 3 days.
Pressure: Autogenous.
Solid weight: 0.5 g, 1.0 g finely ground.
Closed system: corresponding to the reaction in a closed bomb.
Open-system I: The reaction was stopped after a certain time and the solution filtered off and renewed. This was repeated several times. The time between changes of solution was 2, 4, and 8 days.
Open-system II: Three bombs worked in series (Table 5). The solution of bomb 1 was filtered off after a certain time of reaction and added to bomb 2, which contained fresh glass. After reaction in bomb 2, this solution was filtered off and added into bomb 3. Fresh solution was added to bomb 1 and treated in the same manner.

One difference from natural alteration conditions is that the starting materials were finely ground in the experiment. It may be assumed, however, that this only causes the alteration rate to be accelerated without influencing the kinds of minerals being formed or their formation range. Also, compared with nature, there is surplus of solution, which also tends to accelerate the alteration. The concentrations and pH values of the experimental solutions correspond to natural ones, as shown by a comparison with natural hydrothermal solutions (White and Waring, 1963; Raymahashay, 1968; Malkovský and Kačura, 1969; Miller *et al.*, 1975). Nevertheless, when comparing the experimental results with natural alteration products, one must not forget that pure NaOH solutions, etc., were used in the experiments.

The experiments were carried out in Pyrex glass bombs and in teflon-coated stainless-steel bombs. Products were identified by means of powder X-ray diffraction. The pH values of solutions were determined at room temperature by means of pH paper and a pH meter.

EXPERIMENTAL RESULTS

Closed-system experiments

Tables 2 and 3 show the zeolite formation during the alteration of the different volcanic glasses reacting with different solutions at different temperatures. Comparing the products from the three different glasses it is obvious that the *chemical composition of the starting material* has the greatest influence on the zeolites formed in a closed system. The chemical composition of the starting material determines (1) the alteration rate and (2) the kinds of zeolites being formed and their formation range. This is why the basaltic and phonolitic glass react in a similar way, due to their similar chemical compositions. Both glasses are characterized by rapid alteration, the basaltic one being faster. From both glasses the zeolites chabazite, phillipsite, and analcime formed (Table 2). Zeolites formed from the basaltic glass at lower temperatures and after shorter reaction times than from the phonolitic glass. Analcime is the dominant zeolite product within the temperature and solution

TABLE 2. Zeolite formation from the experimental alteration of basaltic and phonolitic glass in closed system*

Time	Basaltic Glass H_2O	Basaltic Glass N/100 NaOH	Basaltic Glass "nat. sol." (pH ~10.5)		Phonolitic Glass H_2O	Phonolitic Glass N/100 NaOH	Phonolitic Glass "nat. sol." (pH ~9.5)
(days)				**180°C**			
1	–	–	–		–	–	–
3	–	–	–		–	–	–
8	–	–	An		–	–	–
17	–	Ch,Ph	An		–	–	–
40	–	Ch,Ph	An		–	–	An
60	–	Ch,Ph			–	–	
(days)				**200°C**			
1	–	–	–		–	–	–
3	–	An	An		–	**Ch**,Ph,An	–
5	Ch,Ph,**An**	An	An		–	Ph,**An**	–
8	An	An	An		–	Ph,**An**	Ch
12	An	An	An		**Ch**,Ph,An	Ph,**An**,f?	Ch,Ph,**An**
17	An	An	An		Ch,Ph,**An**	Ph,**An**,f?	Ch,Ph,**An**
60					**An**,f,q?		Ph,**An**,f,q?
(hours)				**250°C**			
3	–	An	–				
13	An	An	An				
(days)							
1	An	An	An		–	An	–
3	An	An	An		An	An	**An**,f
8	An	An	An		**An**,f,q?	**An**,f?	**An**,f
17	An	An	An		**An**,f,q?	**An**,f,q?	**An**,f,q?

*The experiments were carried out with 0.5 g glass and 25 ml solution.

– = No mineral formation
Ch = Chabazite
Ph = Phillipsite
An = Analcime
f = Alkali feldspar
q = Quartz

Quantitatively prevailing minerals are printed in bold.

"nat. sol.": formed by the reaction of distilled water with the volcanic glass at 250°C for 3 days.

range investigated. The phonolitic glass also yields analcime, although at lower temperatures chabazite and phillipsite are formed transitorily. These zeolites are characterized by a higher water and a lower silica content.

Temperature and solution composition are obviously of little influence on the zeolite formation from these two glasses. The temperature influences the alteration rate only to a small degree. One must consider that the alteration rate is dependent on the dissolution rate of the glass and on the formation rate of the zeolites. Because the same zeolites formed from both glasses, the alteration rate depends on the dissolution rate alone, which on the other hand, increases with decreasing silica content of the glass. The zeolite formation itself is nearly independent of temperature, as shown by the fact that analcime is formed finally at all temperatures. The different zeolite associations observed at the different temperatures depends on the formation of transitory minerals and do not represent real stability boundaries of the zeolites.

Mariner and Surdam (1970) showed that the solubility of glass increases with increasing alkalinity of the solution; however, the influence of the kind of reacting solution on the actual zeolite formation is small and is effective only at low temperatures. The most influential parameter is the chemical composition of the basaltic and the phonolitic glass. Both contain all elements that are necessary for the formation of chabazite, phillipsite, and analcime. Therefore, the formation of these zeolites needs no additional elements from the solution.

The rhyolitic glass differs from the silica-poor glasses in having a slower alteration rate and in the nature of the zeolites formed (Table 3). The slower alteration rate depends mainly on a slower dissolution rate, which is caused by the higher silica content of the glass. The formation of the silica-rich zeolite, mordenite, also depends on the high activity of SiO_2 (Coombs *et al.*, 1959; Honda and Muffler, 1970). Furthermore, the chemical composition of the glass, especially the low alkali content, makes the chemical composition of the solution an important factor. The quantitative ratio of solution to glass is also important. The activity of alkalis necessary for the zeolite formation can be attained by reaction of the glass with a large amount of alkali-rich solution, although it also can be reached by the reaction of a small amount of alkali-free solution with a large amount of glass (Table 3). This points to the importance of porespace in the glass and to the availability of solution during the alteration in a closed system. Zeolites form from a rhyolitic glass chiefly at higher temperatures, however, in alkaline to weakly acidic solutions. At low temperatures, zeolite formation is observed with alkali-rich solutions only. With neutral solutions, a montmorillonite mineral is formed (Hemley, 1962; Hemley and Jones, 1964).

These experiments with different volcanic glasses show that the chemical composition of the starting material is the most important factor during the alteration in a closed system. It influences the mineral formation and determines to what extent temperature and solution composition are important. The less reactive the starting material and the less its chemical composition corresponds to that of the minerals being formed, the greater is the importance of temperature, composition, and the amount of reacting solution.

Open-system experiments

As can be concluded from results of the experiments in closed systems discussed above, the possibility of adding and removing material in an open system is important, especially if the starting material is slow to react and if its chemical composition does not correspond to that of the alteration products. Therefore, a series of experiments on the influence of an open system was carried out, using the rhyolitic glass as a starting material.

Table 4 shows the mineral formation during alteration in open-system I. It can be seen that an open system influences both the alteration rate and the kind of minerals being formed. The alteration rate is

TABLE 3. Mineral formation from the experimental alteration of rhyolitic glass in closed system

	Time (days)	N/100 NaOH	N/1000 NaOH	"nat. sol." (pH ~7)	H_2O			N/100 H_2SO_4
		(0.5 g/25 ml)	(0.5 g/25 ml)	(0.5 g/25 ml)		(0.5 g/25 ml)		(0.5 g/25 ml)
	40	(M, An)	mo			mo		–
150°C	60		mo			mo		(mo)
	84			–				
		(0.5 g/25 ml)	(0.5 g/25 ml)	(0.5 g/25 ml)		(0.5 g/25 ml)	(1 g/50 ml)	(1 g/ 25 ml)
	40	–	(M)			–	(mo)	mo
200°C	60	–				–		
	84			M, An, mo, f, q				
		(0.5 g/25 ml)	(0.5 g/25 ml)	(0.5 g/25 ml)	(1 g/25 ml)	(0.5 g/25 ml)	(1 g/50 ml)	(1 g/25 ml)
	20	(M), An, f, q		–		–		
250°C	40	M, An, f, q	**An**, f	–	An, f	(An), f	**f**, mo	An, **f**, mo
	60	M, An, f, q		M, (mo)		M, An, f		
	84			M, mo, f, q				

– = No mineral formation
M = Mordenite
An = Analcime
mo = Montmorillonite mineral
f = Alkali feldspar
q = Quartz

Traces of minerals are marked by ( ).

Quantitatively prevailing minerals are printed in bold.

TABLE 4. Mineral formation from the experimental alteration of rhyolitic glass in open-system I*

	Time (days)	N/100 NaOH	N/1000 NaOH	H_2O
	8	–	–	–
	16	–	–	–
	24	–	–	–
	32	Ph, (M)	–	–
150°C	40	Ph, (M)	–	–
	48	**Ph,** M, (An)	–	–
	56	**Ph,** M, (An)		(mo)
	64	**Ph,** M, An		mo
	72	Ph, M, **An**		mo
	80	Ph, M, **An**		
	4	–	–	–
	8	Ph	–	–
	12	Ph, (An)	–	–
	16	**Ph,** M, (An)	(mo)	(mo)
	20	Ph, M, An, (f)	(mo)	mo
	24	M, **An,** f	(mo)	mo
200°C	28	M, **An,** f	mo	mo
	32	(M), **An,** f	mo	mo
	36	(M), **An,** f	mo	mo
	40	**An,** f	mo	(f), mo
	44	**An,** f	mo	
	48	**An,** f		
	4	–	–	–
	8	M, **An,** f	–	–
	12	(M), **An,** f	(f, t)	–
	16	**An,** f	f, (t)	f, mo
250°C	20	**An,** f	f, (t)	f, mo
	24	**An,** f	M, An, f, (t)	f, mo
	28	**An,** f	An, f, mo	**f,** mo
	32	**An,** f	**f,** mo	f, mo, ka
	36	**An,** f	f, mo, ka	f, mo, **ka**
	40	**An,** f	f, mo, ka	

*1 g of glass in 50 ml of solution.

The solution was changed after 8, respectively, 4 days.

– = No mineral formation
Ph = Phillipsite
M = Mordenite
An = Analcime
f = Alkali feldspar
Mo = Montmorillonite mineral
ka = Kaolinite mineral
t = Montmorillonite or kaolinite mineral (not to be identified)

Traces of minerals are marked by ( ).

Quantitatively prevailing minerals are printed in bold.

considerably higher at all temperatures than in a closed system. This is caused by the change of solution, which accelerates the dissolution of the glass. It should also be noted that the zeolite formation range is extended to lower temperatures, although the formation range is restricted to alkaline solutions (N/100 NaOH). Both are caused by addition and removal of material; the continuous addition of fresh solution favors a faster dissolution of the glass. As a consequence, the minerals form more rapidly, even at lower temperatures. If the solutions are strongly alkaline, the addition of Na^+ favors the zeolite formation. If the solutions are weakly alkaline, neutral, or acidic (N/1000 NaOH to N/100 H_2SO_4), the removal of material dominates. By removal of surplus alkalis, clay minerals are formed (Grim, 1968). Zeolites are formed only transitorily and disappear with increasing alteration, i.e. increased removal of alkalis (Table 4, experiment with N/1000 NaOH at 250°C) (see also de Kimpe and Fripiat, 1968). This demonstrates the importance of permeability and percolation rates in natural systems.

Regarding the kinds of zeolites formed, two observations are important: (1) the formation of phillipsite, a phase which formed in closed system from SiO_2-poor glasses only and (2) the variation of zeolite associations during alteration. At 150°C and 200°C the silica-poor zeolites phillipsite and analcime are formed first. With increasing alteration the silica-rich zeolite mordenite forms, and phillipsite decreases or disappears completely. This order of succession points to an incongruent dissolution of the rhyolitic glass at the beginning of alteration. If the alteration is continued, the quantitative ratio of zeolites changes in favor of analcime, caused by the continuous addition of Na^+. At 200° and 250°C mordenite, an alkali-poor mineral compared with analcime, disappears with the addition of more Na^+. At this stage, alkali feldspar is the silica-rich mineral present along with analcime.

The above experimental arrangement imitates natural alteration within a certain restricted area only. Open-system experiments can also investigate the reactions of solutions formed by reaction with glass in one area with glass in an adjacent area. Thus experiments in open-system II were carried out (Table 5) with an aim to demonstrate how the variation of the solution composition caused by reaction with the volcanic glass influences the zeolite formation in the adjacent glass. Two results are to be emphasized:

(1) The alteration rate is different in the three bombs, corresponding to alteration in natural systems in different parts of the volcanic glass body.
(2) The zeolite associations being formed in the different bombs after a certain time of reaction are also different and vary with continued alteration.

The different alteration rates show that the solutions formed by reaction are less aggressive than the initial solution.

Considering the mineral formation in the bombs, the zeolite associations demonstrate in what way surplus elements are transported by the solution. For example, analcime formation does not start in bomb II until mordenite (an alkali-poor zeolite) no longer exists in bomb I. At that time surplus Na^+ exists in bomb I and is transported to bomb II by the solution, causing the formation of analcime. Correspondingly, no analcime is observed in bomb III while mordenite still exists in

TABLE 5. Zeolite formation from the experimental alteration of rhyolitic glass in open-system II*

N/100 NaOH →	Bomb I	solution →	Bomb II	solution →	Bomb III	solution →
Time (days)						
4	–		–		–	
8	Ph		–		–	
12	Ph, (An)		–		–	
16	**Ph**, M, (An)		–		–	
20	Ph, M, **An**, (f)		–		–	
24	M, **An** f		–		(M, f)	
28	M, **An** f		–		M, f	
32	(M), **An** f		M, f		**M**, f	
36	(M), **An** f		**M**, (An), f		**M**, f	
40	**An** f		M, An, f		**M**, f	
44	**An** f		M, **An**, f			
48	**An** f					

*Experiment with N/100 NaOH, at 200°C; change of solution after 4 days.

– = no mineral formation
Ph = Phillipsite
M = Mordenite
An = Analcime
f = Alkali feldspar

Traces of minerals are marked by ( ).
Quantitatively prevailing minerals are printed in bold.

bomb II. Therefore the Na^+ added to bomb II is used up by the alteration of mordenite to analcime (see also Abe and Aoki, 1975).

These experimental results show that: (1) In an open system, the possible addition and removal of material diminishes the influence of the chemical composition of the starting material on the mineral formation. This means that a single zeolite or a zeolite association does not ever characterize a particular starting material. (2) In open systems, different degrees of alteration and different zeolite associations can exist in adjacent areas after a certain time of reaction. (3) An open system shifts the boundary between the formation range of zeolites and that of clay minerals.

DISCUSSION

Many natural observations can now be explained with the aid of the experimental results.

1. The abundance of some zeolites, especially analcime, becomes comprehensible from the experiments. The results show that analcime can be formed from silica-poor and silica-rich glasses; i.e. its formation is not restricted to a glass of a particular composition. The experiments with distilled water show that the Na^+ content of the volcanic glasses is sufficient to form analcime; no special solutions are necessary. Moreover, an addition of Na^+ from the solutions significantly increases the formation of analcime. This is important because nearly all alkaline natural hydrothermal solutions contain predominantly Na^+. In addition, analcime is formed over a wide temperature range, also accounting for its abundance in nature.

The zeolites mordenite, chabazite, and phillipsite have a smaller formation range. The formation of mordenite from rhyolitic glass is restricted to conditions of high SiO_2-activity. Chabazite and phillipsite are formed predominantly at lower temperatures (Hoss and Roy, 1960) and from SiO_2-poor volcanic glasses. Their formation range can be extended to higher temperatures, but here they appear to be transitory minerals. Phillipsite can also occur during the incongruent dissolution of SiO_2-rich volcanic glasses which react with alkaline solutions.

2. Different zeolites can be formed from the same starting material because of (a) different temperature conditions, (b) reactions with solutions of different chemical composition, (c) reaction with different quantities of a certain solution, and (d) alteration in open system with different degrees of percolation and consequently different possibilities of addition and removal of material.

3. It is rather remarkable that different degrees of alteration are found within one occurrence; i.e. highly altered parts are observed adjacent to fresh starting material. The experiments in open-system II give a possible explanation. They show that the initial solution is more aggressive than the solution formed by the reaction with volcanic glass. The latter are charged with elements from the solid material and seem to be in chemical equilibrium with the solid material. Therefore the reaction with these solutions causes a slower mineral formation.

4. A zonal arrangement of zeolites can be caused by a temperature gradient but also by an open system. The experiments in open-system II demonstrate that different zeolite associations are formed in different bombs, i.e. in different zones caused by the continuous variation of the solution. These zones vary and move in the course of the alteration (Tsuzuki and Mizutani, 1971). This is illustrated by the formation of analcime which moves gradually from one bomb to the other, i.e. from one zone to the other in nature, with increasing alteration.

REFERENCES

Abe, H. and Aoki, M. (1975) Experiments on the hydrothermal alteration of mordenite rocks in sodium carbonate solution, with reference to analcimization around Kuroko Deposits: *Econ. Geol.* **70**, 770-780.

Coombs, D. S., Ellis, A. J., Fyfe, W. S. and Taylor, A. M. (1959) The zeolite facies, with comments on the interpretation of hydrothermal syntheses: *Geochim. Cosmochim. Acta* **17**, 53-107.

De Kimpe, C. R. and Fripiat, J. J. (1968) Kaolinite crystallization from H-exchanged zeolites: *Amer. Mineral.* **53**, 216-230.

Ellis, A. J. (1960) Mordenite synthesis in a natural hydrothermal solution: *Geochim. Cosmochim. Acta* **19**, 145-146.

Flügel, H. and Heritsch, H. (1968) Das Sterische Tertiär-Becken. *Samml. Geol. Führ.* **47**: Gebrüder Borntraeger, Berlin–Stuttgart.

Frechen, J. (1971) Siebengebirge am Rhein – Laacher Vulkangebiet – Maargebiet der Westeifel. *Samml. Geol. Führ.* **56**: Gebrüder Borntraeger, Berlin–Stuttgart.

Grim, R. E. (1968) *Clay Mineralogy*: McGraw-Hill Book Co., New York, 499 pp.

Hemley, J. J. (1962) Alteration studies in the system $Na_2O-Al_2O_3-SiO_2-H_2O$ and $K_2O-Al_2O_3-SiO_2-H_2O$: *Geol. Soc. Amer. Spec. Papers* **68**, 322.

Hemley, J. J. and Jones, W. R. (1964) Chemical aspects of hydrothermal alteration with emphasis on hydrogen metasomatism: *Econ. Geol.* **59**, 538-569.

Höller, H., Wirsching, U. and Fakhuri, M. (1974) Experimente zur Zeolithbildung durch hydrothermal Umwandlung: *Contr. Mineral. Petrol.* **46**, 49-60.

Honda, S. and Muffler, L. J. P. (1970) Hydrothermal alteration in core from research drillhole Y-1, Upper Geyser Basin, Yellowstone National Park, Wyoming: *Amer. Mineral.* **55**, 1714-1737.

Hoss, H. and Roy, R. (1960) Zeolite studies III: on natural phillipsite, gismondite, harmotome, chabazite, and gmelinite: *Beitr. Mineral. Petrol.* **7**, 389-408.

Malkovsky, M. and Kacura, G., eds. (1969) *Mineral and Thermal Waters of the World. A – Europe*: Academia, Prague.

Mariner, R. H. and Surdam, R. C. (1970) Alkalinity and

formation of zeolites in saline alkaline lakes: *Science* **170**, 977-980.

Miller, Th. P., Barnes, I. and Patton, W. W., Jr. (1975) Geologic setting and chemical characteristics of hot springs in West-Central Alaska: *J. Res. U.S. Geol. Surv.* **3**, 149-162.

Raymahashay, B. C. (1968) A geochemical study of rock alteration by hot springs in the Paint Pot Hill Area, Yellowstone Park: *Geochim. Cosmochim. Acta* **32**, 499-522.

Schoklitsch, K. (1935) Gläser und Nephelinakkumulationen vom Steinberg bei Feldbach: *Zentr. Mineral. Geol. Abt.* **A**, 368-382.

Sheppard, R. A. and Gude, A. J. (1968) Distribution and genesis of authigenic silicate minerals in tuffs of Pleistocene Lake Tecopa, Inyo County, California: *U.S. Geol. Surv. Prof. Paper* **597**, 1-38.

Tsuzuki, Y. and Mizutani, S. (1971) A study of rock alteration process based on kinetics of hydrothermal experiments: *Contrib. Mineral. Petrol.* **30**, 15-33.

White, D. E. and Waring, G. A. (1963) *Data of Geochemistry*, 6th ed.: U.S. Geol. Surv. Prof. Paper **440K**, 29pp.

Wirsching, U. (1975) Experimente zum Einfluß des Gesteinsglas Chemismus auf die Zeolithbildung durch hydrothermale Umwandlung: *Contr. Mineral. Petrol.* **49**, 117-124.

Wirsching, U. (1976) Experiments in hydrothermal alteration processes of rhyolitic glass in closed and "open" system: *Neues Jb. Miner.* **5**, 203-213.

HYDROTHERMAL SYNTHESIS OF CLINOPTILOLITE AND COMMENTS ON THE ASSEMBLAGE PHILLIPSITE—CLINOPTILOLITE—MORDENITE

D. B. HAWKINS

University of Alaska, Fairbanks, Alaska 99701, U.S.A.

and

R. A. SHEPPARD and A. J. GUDE, 3rd

U.S. Geological Survey, Denver, Colorado 80225, U.S.A.

Abstract – For the first time, clinoptilolite as a single phase has been reproducibly synthesized from volcanic ash. The conditions of synthesis were: volcanic ash (tuff A, Shoshone, California); 2.0 M carbonate solution, 1.0 M each in potassium and sodium; 1 kbar pressure; temperature range of 130–150°C; and runs of 3–15 days' duration. Composition of the synthetic clinoptilolite in weight percent is: SiO_2 65.2, Al_2O_3 13.2, CaO 0.1, Na_2O 1.1, K_2O 8.6, H_2O 11.8. Monoclinic cell parameters for the synthetic clinoptilolite are: a = 17.579 ± 0.01 Å, b = 17.884 ± 0.02 Å, c = 7.383 ± 0.005 Å, β = 116° 30.43′, and V = 2077 ± 2 Å³. The crystals are length slow and have a mean index of refraction of 1.465. Most crystals are less than 5 μm long, but some are as long as 20 μm. No change was noted in the (020) X-ray diffraction peak after heating for 24 hours at 650°C. Reduced intensity of this peak was evident after heating for 24 hours at 750°C.

Clinoptilolite forms over the compositional range from 2 M K_2CO_3 to 2 M Na_2CO_3 at 1 kbar and 145°C. It occurs as a single phase in the range of 3:2 K/Na to 2:3 K/Na. In more potassium-rich solutions, subordinate to dominant phillipsite is produced; in more sodium-rich solutions, mordenite, subordinate clinoptilolite, and minor phillipsite are formed. In 1:1 K/Na 2 M CO_3 solutions above 150°C, mordenite is formed. Phillipsite forms preferentially in potassium-rich solutions below 130°C. The mordenite–clinoptilolite boundary is shifted to 220°C at 1 kbar when 0.6 M mixed carbonate solution is used. Attempts to synthesize clinoptilolite at higher temperatures and lower carbonate concentrations were unsuccessful.

INTRODUCTION

Clinoptilolite, the alkali-rich and silica-rich member of the heulandite family of zeolites, is one of the most common natural zeolites. It occurs as an alteration product of volcanic ash in saline, alkaline lakes, as an authigenic silicate in marine deposits, and as a product of hydrothermal alteration of vitric tuffs in geothermal areas. In spite of its great abundance, true clinoptilolite has defied synthesis. This paper describes the first synthesis of clinoptilolite, presents a description of the synthetic phase, and presents speculations on the mode of formation of clinoptilolite in nature.

Background

Heulandite, the closely related alkaline earth zeolite, has been synthesized repeatedly in a variety of compositional forms (Hawkins, 1967). Ames (1963) synthesized a clinoptilolite-like or mordenite-like, lithium-bearing zeolite. L. B. Sand and D. T. Hayhurst (personal communication to Hawkins, 1975) synthesized clinoptilolite at Worcester Polytechnic Institute under hydrothermal conditions similar to those in this study but they apparently were unable to obtain clinoptilolite either as a major phase or as a single phase.

Sheppard and Gude (1968) dealing with the formation of clinoptilolite and other alkalic, silicic zeolites from rhyolitic tuffs in Pleistocene Lake Tecopa, Inyo County, California, provided a foundation for the synthesis of clinoptilolite. They recognized three diagenetic facies: fresh glass, zeolites, and potassium feldspar and searlesite. The zeolites are mainly phillipsite, clinoptilolite, and erionite, and trace to minor amounts of analcime and chabazite. The chemical composition of the Pleistocene lake varied from the margin basinward, as shown by the presence of unaltered glass at the margin of the lake, then zeolites, and finally potassium feldspar occupying the central basin. Evidence indicating the saline nature of the ancient lake is the presence of trona, halite, and molds of gaylussite and pirssonite. Experimental work (Freeth, 1923; Bury and Redd, 1933) suggested to Sheppard and Gude (1968) that a minimum salinity of 22% is indicated by trona, and 14 and 21% salinity by gaylussite and pirssonite, respectively.

With this information as background, the synthesis of clinoptilolite was attempted by the hydrothermal reaction of unaltered glass shards (tuff A) with 2 M (about 20%) alkali carbonate solutions.

EXPERIMENTAL TECHNIQUES

Reactants

Fresh, unaltered rhyolitic glass, tuff A of Sheppard and Gude (1968), was used as a solid reactant, and the composition is given in Table 1. Two-molar stock solutions of sodium carbonate and potassium carbonate were prepared by dissolving requisite quantities of reagent-grade Na_2CO_3 and K_2CO_3 in distilled water. The solutions were stored in polyethylene bottles, and working solutions were prepared from them.

Hydrothermal synthesis

The well-known technique of hydrothermal synthesis described by Roy and Tuttle (1956) was used. Twenty milligram quantities of glass were placed in gold capsules, having lengths of 2 cm and inside diameters of 2 mm. The capsules had been welded at one end. Twenty-five microliters of carbonate solution were added and the tubes welded shut by means of an arc welder. The capsules were weighed and placed in test-tube bombs. Experiments were run in duplicate, and as many as four capsules were placed in a single bomb. The bombs were sealed, and the desired pressure-temperature conditions were attained by means of a Tem Pres HR-4B Hydrothermal Unit. After an appropriate time, dictated by the experimental design, the bombs were quenched and the capsules weighed to test for leakage during the hydrothermal run. The contents of the capsule were transferred to a 45 μm millipore filter and washed three times with distilled

TABLE 1. Synthesis conditions, chemical composition, X-ray powder diffraction data, and cell parameters for synthetic clinoptilolite

Synthesis conditions: Initial synthesis was made with 0.20 mg tuff A and 25 μl 2 M mixed carbonate solution, 1 M each in sodium and potassium at 143°C, 1 kbar, and 10 days' duration. The material characterized below was prepared from 0.2 g tuff A, 250 μl 2 M CO_3, 1:1 Na:K, at 134°C, 1 kbar, and 15 days' duration.

Qualitative description of material analyzed: The material consisted chiefly of clinoptilolite. Optically, the major impurities are about 5% of a carbonate phase and about 10% unreacted glass shards. Very minor amounts of a phillipsite-like phase and even smaller amounts of "end-member" albite(?) were detected by X-ray powder diffraction.

Chemical composition

	Tuff A* (wt.%)	Clinoptilolite† (wt.%)		Unit-cell content (O = 72)	
SiO_2	72.78	65.2	(29)	Si	29.27
Al_2O_3	11.89	13.2	(18)	Al	6.98
Fe_2O_3	0.55	–	–		
FeO	0.99	–	–		
MgO	0.22	–	–		
CaO	0.55	0.1	(18)	Ca	0.05
Na_2O	3.03	1.1	(11)	Na	0.95
K_2O	5.31	8.6	(11)	K	4.92
H_2O+	3.86				
H_2O-	0.21	11.8		H_2O	17.67
Total	99.39	100.0		Si:Al	4.19

X-ray powder diffraction data

hkl	*d*(Å)	*I*
020	8.94	100
200	7.88	48
220	5.92	26
$\bar{3}11$	5.23	30
$\bar{1}31$	4.640	16
$\bar{4}01$	4.333	12
131	3.959	84
$\bar{4}21$	3.897	36
$\bar{3}12$	3.545	16
$\bar{2}22$	3.411	37
151	2.965	34

Monoclinic cell parameters are: a = 17.579±0.01 Å, b = 17.884±0.02 Å, c = 7.383±0.005 Å, β = 116° 30.43′, V = 2.077±2 Å^3.

*Analysis from Table 3, column 1, Sheppard and Gude (1968).

†Number in parentheses is the number of microprobe analyses made for the particular element. The reported value is the average of these analyses. Analyst: George A. Desborough.

water to remove soluble carbonates. The washed product was then mounted on a glass microscope slide for X-ray diffraction analysis.

The synthetic clinoptilolite described in Table 1 was prepared as above, except that 0.2 g of glass and 5 mm i.d. gold tubing were used to synthesize enough material for study.

SYNTHESIS RESULTS

Characterization of clinoptilolite

The chemical composition, X-ray powder diffraction data, and cell parameters for the synthetic clinoptilolite are given in Table 1.

Chemical composition. The chemical composition of the synthetic clinoptilolite was determined by electron-microprobe analysis because of the small amount of material available. An ARL model EMX-SM microprobe was used, and the electron beam was defocused to a diameter of 25 μm to eliminate decomposition effects. Matrix corrections of X-ray intensities were not made. Reference standards included albite, benitoite, three synthetic glasses, microcline, quartz, corundum, and anorthite. The water content was calculated by difference.

The analysis and unit-cell contents show that the clinoptilolite is alkali-rich, particularly potassium-rich. The Si:Al ratio of about 4.2 is within the range of natural clinoptilolites.

Scanning electron microscopy and optical data. Scanning electron micrographs (Fig. 1) show that the synthetic clinoptilolite occurs as thin monoclinic plates or sheaves of plates. Most crystals are less than 5 μm long, but some are as much as 20 μm in length. The relict glass shards associated with the clinoptilolite are conspicuously pitted. The small crystal size of the zeolite does not permit the determination of many optical parameters; however, the mean index of refraction is 1.465, and the crystals are length slow. Some relatively large crystals are optically zoned, and all zones are length slow.

X-ray powder diffraction data. Identification of the synthetic mineral phases was made by conventional X-ray powder diffractometry using Ni-filtered CuKα radiation. The monoclinic cell parameters and the d spacings were similarly derived from X-ray powder diffraction work. Five percent reagent-grade fluorite, recrystallized by calcining at 800°C, was used as an internal standard for the precision determination of the d spacings. The zeolite–fluorite mixture was then prepared as an unoriented collodion-membrane diffractometer mount and scanned at 1 degree 2θ per minute, using a diffracted-beam LiF crystal monochromator. The strip-chart record was made at 1 degree 2θ per inch on the blank side of the chart, and two continuous records, oscillated from 8 degrees 2θ to 60 degrees 2θ and back to 8 degrees 2θ, were made. Four charts obtained this way were then aligned by the fluorite pattern. The clinoptilolite reflections were visually "averaged", and a best-fit line was drawn

Fig. 1. Scanning electron micrograph of synthetic clinoptilolite showing the platy character of crystals.

through the corresponding peaks of the four charts. This carefully measured set of 2θ Kα_1 values provided the data to calculate the cell parameters and to index the *d* spacings (Table 1). Final cell dimensions were obtained by a least-squares refinement through use of the U.S. Geological Survey's FORTRAN IV Program W9214. The monoclinic cell parameters are: a = 17.579 ± 0.01 Å, b = 17.884 ± 0.02 Å, c = 7.383 ± 0.005 Å, β = 116° 30.43′, and V = 2077 ± 2 Å^3.

Thermal stability. The thermal stability (Mumpton, 1960; Alietti, 1972) of synthetic clinoptilolite was determined by heating the zeolite overnight at temperatures of 650°C and 750°C, cooling the material quickly to room temperature, and X-raying the product immediately. No change was observed in the (020) peak after heating at 650°C for 24 hours; however, it showed reduced intensity after heating at 750°C for 24 hours.

Other experimental results

In addition to the repeated synthesis of clinoptilolite as a single phase, several other hydrothermal experiments were performed to determine the effects of pressure, temperature, composition of the reactant solution, and nature of the solid reactant on the phases formed. Because of the large amount of experimental data, only a summary of the results is given here. The assemblages discussed below *do not* represent equilibrium assemblages but rather represent kinetic assemblages; i.e. phases that are formed in a particular temperature–pressure range on a specific duration of hydrothermal treatment. This phenomenon will be discussed in a subsequent part of this paper.

In 2 M or 4 M sodium carbonate solutions, the sequence mordenite, mordenite–analcime, mordenite–quartz–analcime, and quartz–albite is observed with increasing temperature at 1 kbar pressure for runs of 3–5 days' duration. Mordenite occurs from 100°C to about 180°C; mordenite–analcime (plus quartz at higher temperature) from 180°C to 220°C; analcime–quartz from 220°C to 280°C; and quartz–albite at temperatures greater than 280°C. Similar results have been reported by Abe and Aoki (1975).

In 2 M or 4 M potassium carbonate solutions, the assemblages phillipsite, phillipsite–clinoptilolite–mordenite, and potassium feldspar were produced with increasing temperature at 1 kbar pressure for runs of 3–5 days. The phillipsite occurs from 100°C to 125°C; phillipsite–clinoptilolite–mordenite occurs from 130°C to 145°C, with mordenite dominant at higher temperatures; and, finally, potassium feldspar occurs as the dominant phase at temperatures above 150°C.

The concentration of the carbonate solution has a pronounced effect on the hydrothermal reactions. Extensive reaction between glass and 2 M and 4 M carbonate solutions occurs at temperatures as low as 125°C. Similar reactions occur in 0.67 M carbonate solutions at temperatures near 200°C (Table 3). In 0.2 M and 0.02 M carbonate solutions, only slight reaction was observed in runs of 12 days for the temperature range of 225–300°C. No reaction occurred over the lower temperature range of these runs.

Zeolite formation was observed only with tuff A as a reactant. Coprecipitated gels of the approximate composition of tuff A were unreactive in mixed carbonate solutions over the temperature range of 125°C to 150°C at 1 kbar. Zeolite formation is apparently unaffected by pressure over the range of 0.2–1.3 kbar. A mixture of clinoptilolite, mordenite, and phillipsite was produced in a run from a 2 M mixed carbonate solution at 130°C, 0.2 kbar, and 14 days' duration.

The conditions under which clinoptilolite, with or without other zeolites, formed are summarized in Tables 2 to 4 and Fig. 2. A study of the kinetics of clinoptilolite crystallization is now under way, and preliminary results of the kinetic study are shown in Fig. 2.

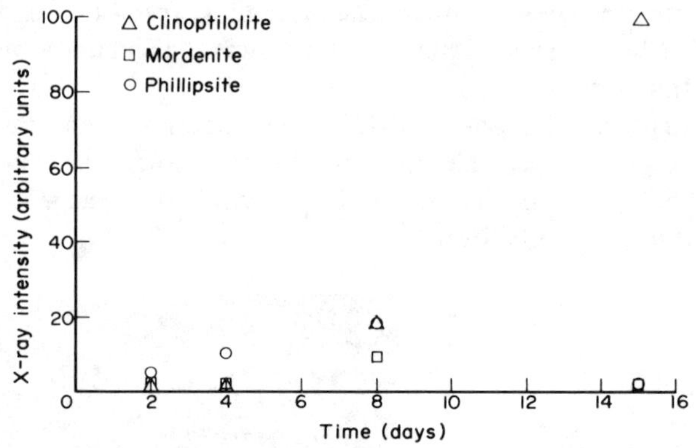

Fig. 2. Crystallization of clinoptilolite, mordenite, and phillipsite from tuff A in 2 M mixed carbonate solution, 1 M each in sodium and potassium, at 130°C and 1 kbar. Intensity indicates the degree of crystallization as determined from the mordenite (200) X-ray diffraction peak and the clinoptilolite and phillipsite (020) peaks.

DISCUSSION

Clinoptilolite synthesis

As shown by the data in Table 1, clinoptilolite as a major phase was synthesized hydrothermally. Clinoptilolite is favored over both phillipsite and mordenite by hydrothermal reaction in the temperature range of 130°C to 150°C in 2 M mixed carbonate solutions of about 1:1 potassium and sodium (Table 2).

TABLE 2. Variation in the relative amounts of zeolites formed near 145°C*

Temperature	Solutions	Products†
145 ± 1.5°C	2 M K_2CO_3	P > C = M
,,	2 M CO_3, 5:1, K:Na	C(?) slight reaction
,,	,, 3:1, ,,	M
,,	,, 3:2, ,,	C ≫ P > M
,,	,, 2:3, ,,	C ≫ P = M
,,	,, 1:3, ,,	C = M > P
140 ± 1.6°C	,, 1:5, ,,	C = M > P
,,	2 M Na_2CO_3	M > C > P

*Experimental conditions: 1 kbar pressure, 10-day runs, solutions range from 2 M Na_2CO_3 to 2 M K_2CO_3.

†Unreacted glass is in all products. C = clinoptilolite, M = mordenite, P = phillipsite.

This synthesis of clinoptilolite is a result of using silica-rich volcanic glass and, particularly, mixed potassium–sodium carbonate solutions as reactants in hydrothermal runs of nearly two weeks' duration. As discussed by Breck (1974, pp. 287 ff.), the presence of two different cations such as sodium and potassium in aluminosilicate gels has a profound effect on the nature of the zeolite phases produced by crystallization of these gels. Within this temperature–compositional region, pressure changes over the range 0.2–1.3 kbar do not affect the formation of clinoptilolite, phillipsite, or mordenite.

As expected, the concentration of the carbonate solution exerts a strong influence on the rate of formation of the zeolites from glass. Thus, for runs of 7 days' duration, clinoptilolite and mordenite form in 2 M mixed carbonate solution at 130°C. No reaction is observed for this temperature and duration in 0.67 M mixed carbonate solutions, and it is not until temperatures of 190°C to 225°C are reached that these zeolites form noticeably in 7-day runs (Table 3). Similar results for other zeolites have been reported by Höller *et al.* (1974). This change in reactivity is an effect of pH and ionic strength of the solutions. The rate of formation of these zeolites may be described approximately by a first-order reaction with respect to hydroxyl-ion concentration, as Sand (1974) suggested.

TABLE 3. Phases formed in 0.67 M mixed carbonate solution*

Temperature	Duration	Products†
174 ± 2.5°C	12 days	No reaction
197 ± 1.3°C	7 days	M ≥ C
216 ± 1.9°C	7 days	Q ≫ M
226 ± 2.3°C	12 days	C = M
225 ± 1.0°C	12 days	M > C
230 ± 2.0°C	7 days	M ≥ Q
246 ± 2.5°C	7 days	Q = M

*Experimental conditions: 1 kbar pressure, 7- to 12-day runs, solutions = 1 M each in Na and K.

†Unreacted glass is in all these runs. C = clinoptilolite, M = mordenite, Q = quartz.

Relations among clinoptilolite, mordenite, and phillipsite

Synthetic clinoptilolite is preferentially a potassium zeolite as shown in Table 1. Even though sodium and potassium were equally abundant in the reactant solution, the clinoptilolite that formed contained about 5 times as much potassium as sodium. Unfortunately, analyses of coexisting synthetic clinoptilolite, mordenite, and phillipsite are not available. Knowledge of the partitioning of major ions among these phases would be instructive. Phillipsite may form preferentially in potassium-rich solutions, and mordenite may form in sodium-rich solutions. However, as the data of Table 2 show, the effect of changes in the sodium to potassium ratio is not clear.

Clinoptilolite seems to be a lower temperature phase than mordenite. Although mordenite forms over the same temperature range as clinoptilolite, the converse is not true. Mordenite has a wide synthesis region. It is the dominant zeolite in 2 M mixed carbonate solutions in the range of 130°C to 150°C in this system in runs of long duration. Furthermore, clinoptilolite can easily be converted to mordenite hydrothermally in distilled water near 315°C and 1 kbar (Hawkins, 1967). The zeolite assemblages discussed here are kinetic assemblages and not equilibrium assemblages. In nature, the zeolites are probably unstable with respect to quartz and feldspars. The metastable persistence of the various zeolites is due to the slow rate of transformation of the zeolites at low temperature.

The crystallization of clinoptilolite and mordenite is slower than that of phillipsite. Phillipsite nucleates within 24 hours at 130°C. The nucleation periods at this temperature for mordenite and clinoptilolite are not known, but they are probably longer than that for phillipsite. As shown in the growth curve of Fig. 2, minor phillipsite forms after 2 days at 130°C, and after 4 days it is a major phase. Clinoptilolite and mordenite crystallize only in the 8-day runs at this temperature. Clinoptilolite is the dominant phase after 15 days, suggesting that clinoptilolite formation is favored by longer periods of hydrothermal treatment. A description of the phases in this 15-day run is given in Table 1. In Fig. 2 the data points for the 2-, 4-, and 8-day runs are

TABLE 4. Effect of time on the crystallization of various zeolites*

Temperature	Solution	Duration of run (hours)	Product†
143 ± 1.0°C	2 M CO_3, 2:3 K:Na	25	Minor P, slight reaction
”	” 3:2 ”	25	”
141 ± 2.0°C	” 2:3 ”	54	Minor P > M > C
”	” 3:2 ”	54	”
145 ± 1.5°C	” 2:3 ”	91	C > M = P
”	” 3:2 ”	91	P = C = M
144 ± 2.1°C	” 2:3 ”	195	C ≫ P = M
”	” 3:2 ”	195	C = M > P
143 ± 1.5°C‡	” 2:3 ”	24 (49)	P > M > C
	” 3:2 ”	24 (49)	Major P

*Experimental conditions: 1 kbar pressure, 145°C, 2 M carbonate solutions.

†Unreacted glass is in products. C = clinoptilolite, M = mordenite, P = phillipsite.

‡Reactants are products from the 25-hour run in this table. The products of the 25-hour run were filtered and washed, fresh carbonate solution was added, and the material was treated hydrothermally for an additional 24 hours. Values in parentheses are the total hours that this material was treated hydrothermally.

based on duplicate runs, whereas that for 15 days is for a single run only.

The early formation of phillipsite rather than of the more siliceous mordenite or clinoptilolite occurred because the ratio of aluminum to silicon is higher in strongly alkaline solutions than in the glass from which they were derived (Mariner and Surdam, 1970). The higher aluminum to silicon ratio favors the formation of phillipsite. Clinoptilolite and mordenite require a high silica activity which may occur at and be restricted to the boundary layer surrounding the glass shards, as discussed by Gogishvili *et al.* (1973). Possibly, clinoptilolite and mordenite form from the aluminum-poor, silica-rich rind of the glass shard, the rind having been produced by the preferential dissolution of aluminum. Phillipsite may form from the material initially leached from the glass, and it may persist or, depending on the temperature, dissolve once growth of mordenite and clinoptilolite begins. Thus, phillipsite seems to be unstable with respect to mordenite or clinoptilolite. Even though it forms first, it may dissolve upon prolonged treatment, during which time mordenite and clinoptilolite grow at its expense. A similar relation does not seem to exist between clinoptilolite and mordenite.

In the 15-day run plotted in Fig. 2, no mordenite was produced. The potassium selectivity of clinoptilolite requires that the mother liquor be enriched in sodium. Commonly, this sodium will go to form mordenite. However, as this and several other similar runs show, no mordenite was formed, and the sodium apparently remains in solution.

Mordenite, once formed, persists. This is clearly shown by the very common occurrence of mordenite with clinoptilolite throughout these experiments. Compositionally and structurally, mordenite and clinoptilolite are similar. Only a small free-energy difference should exist between the two, meaning that once both phases are present above some critical crystallite size, there should be very little driving force for one phase to grow at the expense of the other. Thus, the existence of clinoptilolite as a single phase in these experiments probably reflects the fact that even though both mordenite and clinoptilolite nucleated, the clinoptilolite crystallites were able to achieve a critical size more quickly than mordenite. Thus, the mordenite could not grow.

Chabazite, a minor phase in the Lake Tecopa tuffs, has been synthesized by Höller *et al.* (1974) and Wirsching (1975). The absence of chabazite in the experiments described here is probably due to the relatively high temperatures employed.

Natural zeolite assemblages

The zeolite assemblages formed in these hydrothermal experiments are similar to those found in nature except that zeolites from saline lakes and deep-sea sediments are characterized by an abundance of phillipsite and clinoptilolite and a relative scarcity of mordenite (Sheppard, 1971). Low-temperature zeolite assemblages from hydrothermal areas are characterized by clinoptilolite and mordenite and the absence of phillipsite (Honda and Muffler, 1970; Iijima, 1974). The results of our hydrothermal experiments suggest that in the low-temperature environment of saline lakes or the ocean floor, phillipsite should form first, followed by clinoptilolite. Mordenite formation may not be favored because of the low temperature. Where mordenite has formed under these low-temperature conditions, it is probably from solutions much enriched in sodium relative to potassium.

Clinoptilolite should grow at the expense of phillipsite. This growth rate will be controlled by the alkalinity of the solution, being relatively rapid in saline, alkaline lakes, and slower in the marine environments. The slow conversion of phillipsite to clinoptilolite under conditions of moderate alkalinity and relatively low temperature is probably unique to the marine environment, inasmuch as this is the only environment in which clinoptilolite, phillipsite, and fluid are likely to coexist for a sustained time to allow the reaction to approach completion.

In hydrothermal areas, clinoptilolite and mordenite should dominate the low-temperature (less than 150°C) mineral assemblage. Phillipsite should be absent from these regions for several reasons. First, even though phillipsite may form initially, it would be unstable with respect to clinoptilolite and mordenite. Second, phillipsite apparently requires strongly alkaline solutions for its formation, and these solutions seem to be absent from hydrothermal areas. Because clinoptilolite is a more hydrous mineral than mordenite, clinoptilolite should form preferentially under those conditions that tend to increase the chemical potential of water in the system; e.g., lower temperature, higher P_{H_2O}, and decreased salt concentration. Thus, at higher temperatures, clinoptilolite should become unstable with respect to mordenite.

ACKNOWLEDGMENTS

We thank Cheri Carver for her assistance with the hydrothermal experiments and George A. Desborough for the microprobe analyses. This work was supported in part by an NSF Institutional Grant to the University of Alaska.

REFERENCES

Abe, H. and Aoki, M. (1975) Experiments on the hydrothermal alteration of mordenite rocks in sodium carbonate solution, with reference to analcimization around Kuroko deposits: *Econ. Geol.* **70**, 770-780.

Alietti, A. (1972) Polymorphism and crystal chemistry of heulandites and clinoptilolite: *Amer. Miner.* **57**, 1448-1462.

Ames, L. L., Jr. (1963) Synthesis of a clinoptilolite-like zeolite: *Amer. Mineral.* **48**, 1374-1381.

Breck, D. W. (1974) *Zeolite Molecular Sieves, Structure, Chemistry, and Use*: Wiley-Interscience, Wiley & Sons, New York, 771 pp.

Bury, C. R. and Redd, R. (1933) The system sodium carbonate–calcium carbonate–water: *J. Chem. Soc. (London)*, 1160-1162.

Freeth, F. A. (1923) The system Na_2O–CO_2–NaCl–H_2O, considered as two four-component systems: *Roy. Soc. London Phil. Trans.*, ser. A, **223**, 35-87.

Gogishvili, V. G., Khundadze, A. G., Politova, Yu. V. and Urushadze, V. V. (1973) Significance of the boundary layer of solution in the mordenite–analcime relation: *Geochem. Int.*, pp. 1136-1144.

Hawkins, D. B. (1967) Zeolite Studies. I. Synthesis of some alkaline earth zeolites: *Mater. Res. Bull.* **2**, 951-958.

Höller, H., Wirsching, U. and Fakhuri, M. (1974) Experimente zur Zeolithbildung durch hydrothermale Umwandlung: *Contr. Miner. Petrol.* **46**, 49-60.

Honda, S. and Muffler, L. J. P. (1970) Hydrothermal alteration in core from research drill hole Y-1, Upper Geyser Basin, Yellowstone National Park, Wyoming: *Amer. Mineral.* **55**, 1714-1737.

Iijima, A. (1974) Clay and zeolitic alteration zones surrounding Kuroko deposits in the Hokuroku District, Northern Akita, as submarine hydrothermal-diagenetic alteration products: in *Geology of Kuroko Deposits*, S. Ishihara, K. Kanehira, A. Sasaki, T. Sato, and Y. Shimazaki, eds., Mining Geology Special Issue, no. 6, 267-289.

Mariner, R. H. and Surdam, R. C. (1970) Alkalinity and formation of zeolites in saline alkaline lakes: *Science* **170**, 977-980.

Mumpton, F. A. (1960) Clinoptilolite redefined: *Amer. Miner.* **45**, 351-369.

Roy, R. and Tuttle, O. F. (1956) Investigations under hydrothermal conditions: in *Physics and Chemistry of the Earth*, L. H. Ahrens, K. Rankama and S. K. Runcorn, eds., Pergamon Press, London.

Sand, L. B. (1974) Kinetics of phase transformation in zeolite formation: U.S.–Japan Cooperative Science Program Seminar on *The Occurrence, Origin, and Utilization of Zeolites in Sedimentary Rocks in the Circum-Pacific Region*, Menlo Park and Barstow, California, and Las Vegas, Nevada, July 1-7, 1974.

Sheppard, R. A. (1971) Zeolites in sedimentary deposits of the United States – A review: *Amer. Chem. Soc. Adv. Chem. Ser.* **101**, 279-310.

Sheppard, R. A. and Gude, A. J., 3rd (1968) Distribution and genesis of authigenic silicate minerals in tuffs of Pleistocene Lake Tecopa, Inyo County, California: *U.S. Geol. Surv. Prof. Paper* **597**, 38 pp.

Wirsching, U. (1975) Experimente zum Einfluss des Gesteinglas-Chemismus auf die Zeolithbildung durch hydrothermale Umwandlung: *Contr. Mineral. Petrol.* **49**, 117-124.

HYDRATION AS AN EARLY STAGE IN THE ZEOLITIZATION OF NATURAL GLASS

CARMINE COLELLA, ROSARIO AIELLO* and CESARE PORCELLI

Istituto di Chimica Applicata, Facoltà di Ingegneria, Università di Napoli, Italy

Abstract – Examination in thin section of a large number of samples of more or less zeolitized volcanics of the Campanian Region has made evident that glass hydration represents a preliminary stage of the zeolitization process.

The influence of the conditions and state of hydration of natural glasses has been studied by means of zeolite synthesis experiments starting from natural hydrated glasses and anhydrous natural glasses artificially hydrated. It has been pointed out that the hydration state of the glass may affect the successive zeolitization process both in the type of the newly formed zeolites and in the kinetics of the zeolite formation. The hydration state of the glass has been tentatively related to the possible formation of different zeolites in the same deposit.

INTRODUCTION

The hydration of glass has been thoroughly studied following the fundamental observations of Ross and Smith (1955). Natural glass can superficially absorb water, e.g., meteoric water, as shown by Friedman and Smith (1958), with consequent formation of a hydration rim. The rim is easily detectable under crossed polarized light because of its strain birefringence.

Friedman and Smith (1960), by measurements of hydration rims of obsidian artifacts of known age and origin, and Friedman *et al.* (1966), by the study of artificially hydrated glass, found that the diffusion rate of water into the glass follows the law $x^2 = kt$, where x is the depth of penetration of water in micrometers, k is a constant for a given temperature, related in a complex manner to the diffusion coefficient D defined by Fick's law, and t is the time in years.

Despite the large amount of data available on the physical chemistry of the glass hydration, not much attention has been given to the hydration phenomenon as a possible preliminary stage in zeolite formation.

Porcelli (1976), by a study in thin section of samples of more or less zeolitized volcanics from Southern Italy, found evidence that, at least in the diagenetic process of these materials, hydration of volcanic glass precedes zeolitization.

Such observations, together with the possibility that hydration conditions and therefore hydration state of the glass particles may differ within the limits of the same deposit and differently affect the zeolitization, might help to explain why different zeolites may occur in the same formation.

In this regard Aiello and Colella (1973), studying the formation of large-port zeolites from natural glass, reported that the rhyolitic obsidian and pumice (pyroclastic natural glass later hydrated in the depositional environment) show a different reactivity towards alkaline solutions.

The aim of this paper was to investigate the relationships between the hydration of glass and its zeolitization by examination of a large number of volcanics and by synthesis experiments starting from rhyolitic glasses (both anhydrous or naturally hydrated glass and artificially hydrated natural glass). From these investigations useful information on the influence of the state and the conditions of hydration on the zeolitization process therefore may be obtained.

EXPERIMENTAL

Natural materials examined in thin section were trachytic volcanics of the Campanian Region between about 200,000 and 3500 years in age (Porcelli, 1976) that were completely, partially, or unzeolitized. In order to preserve the original microtexture, especially in poorly or unlithified samples, thin sections were prepared at room temperature using Araldite diluted with xylene as the impregnating medium. Glasses employed for the synthesis experiments were rhyolitic obsidian (O) and pumice (P), from Rocche Rosse and

*Present address: Dipartimento di Chimica, Universita' Della Calabria, Cosenza, Italy.

TABLE 1. Chemical analyses of rhyolitic glasses (wt. %)

	Obsidian (Rocche Rosse, Lipari)	Pumice (Cava Grande, Lipari)
SiO_2	72.80	70.85
Al_2O_3	12.97	12.83
TiO_2	0.10	0.15
MnO	0.08	0.11
Fe_2O_3	0.69	1.02
FeO	1.40	1.35
CaO	1.26	0.83
MgO	0.60	0.55
Na_2O	4.37	4.46
K_2O	5.04	4.70
H_2O	0.20	3.70
	99.51	100.55

Cava Grande, respectively, Lipari Island, Italy. Their chemical analyses are reported in Table 1. The glass was ground to between 270 and 325 mesh.

Hydration of rhyolitic obsidian in the laboratory was carried out on 10 g samples placed in platinum crucibles. Runs were made in autoclaves at 270°C under autogeneous water vapor pressure for 98 and 186 hours. Table 2 reports the conditions of the hydration treatments and the resulting water content, evaluated by thermogravimetric (TG) analysis, both of the artificially hydrated obsidians (O_1 and O_2) and of the rhyolitic natural glasses (O and P). The hydrated glasses by powder X-ray diffraction analysis appeared to be completely amorphous. Infrared (IR) analyses were performed on glass specimens which were previously heated at 120°C for 2 hours, then rapidly dispersed in Nujol, spread out between flat NaCl plates, and scanned over the 4000–1000 cm^{-1} range.

Synthesis experiments were carried out by mixing weighed amounts of glass and NaOH solutions of variable concentrations in Teflon containers. These containers were placed in an air-thermostated oven at 80°C for a given time and rotated at 33 rpm. The ratio between the weight of the glass and the weight of the water contained in the alkaline solution is designated as the solid/liquid ratio (S/L).

TABLE 2. Hydration conditions and water content of treated products

Symbol	Glass	Hydration treatment: temperature (°C)	Hydration treatment: time (hr)	Water (%)
O	obsidian	–	–	0.20
O_1	obsidian	270	98	3.55
O_2	obsidian	270	186	4.95
P	pumice	–	–	3.70

Newly-formed phases were identified by powder X-ray diffraction analysis, and their yields in kinetics experiments were evaluated by the usual methods of quantitative X-ray diffraction analysis.

The following abbreviations have been used to indicate the products obtained: M = phillipsite; P_1,P_2 = pseudocubic and pseudoorthorhombic Na-P zeolite (isotypes of gismondine); X = faujasite; I = basic sodalite; U = no crystallization.

RESULTS

Optical microscopy of naturally hydrated glass

A common geological phenomenon in the pyroclastic formations of the Campanian Region is that of detectable gradations between zeolitized and non-zeolitized zones in the same deposit. The inner zones are normally zeolitized.

Thin-section examination of samples coming from the same formation, whose age is known by ^{14}C dating, provided the following data:

1. In non-zeolitized zones of the deposits the glass shards always show a hydration rim whose thickness is in agreement with that calculated from $x^2 = kt$ providing the value of k is known (Porcelli, 1976). Figure 1a shows a typical hydration rim around a crack in a glass fragment in the non-zeolitized zone of a deposit of Campanian tuff near Avellino. The small glass particles are completely hydrated and show no strain birefringence under crossed Nicols. If the non-zeolitized zone of the deposit is sufficiently thick, samples collected at various depths show different thickness of the hydration rim, probably due to the temperature gradient within the deposit.

2. In partially zeolitized zones of the deposits the peripheral part of the hydration rim appears to be transformed into zeolite. The thickness of the hydration rim, in fact, does not follow the above-mentioned law. Figures 1b and 1c, referring to samples of Campanian tuff from S. Agata dei Goti, show how newly-formed zeolite microcrystals grow, gradually replacing the hydrated glass.

3. In completely zeolitized zones of the deposits, glass shards which have not been zeolitized no longer show any hydration rim. This is totally substituted by an aggregate of zeolite microcrystals, as can be observed in Fig. 1d (sample of Campanian tuff from S. Agata dei Goti).

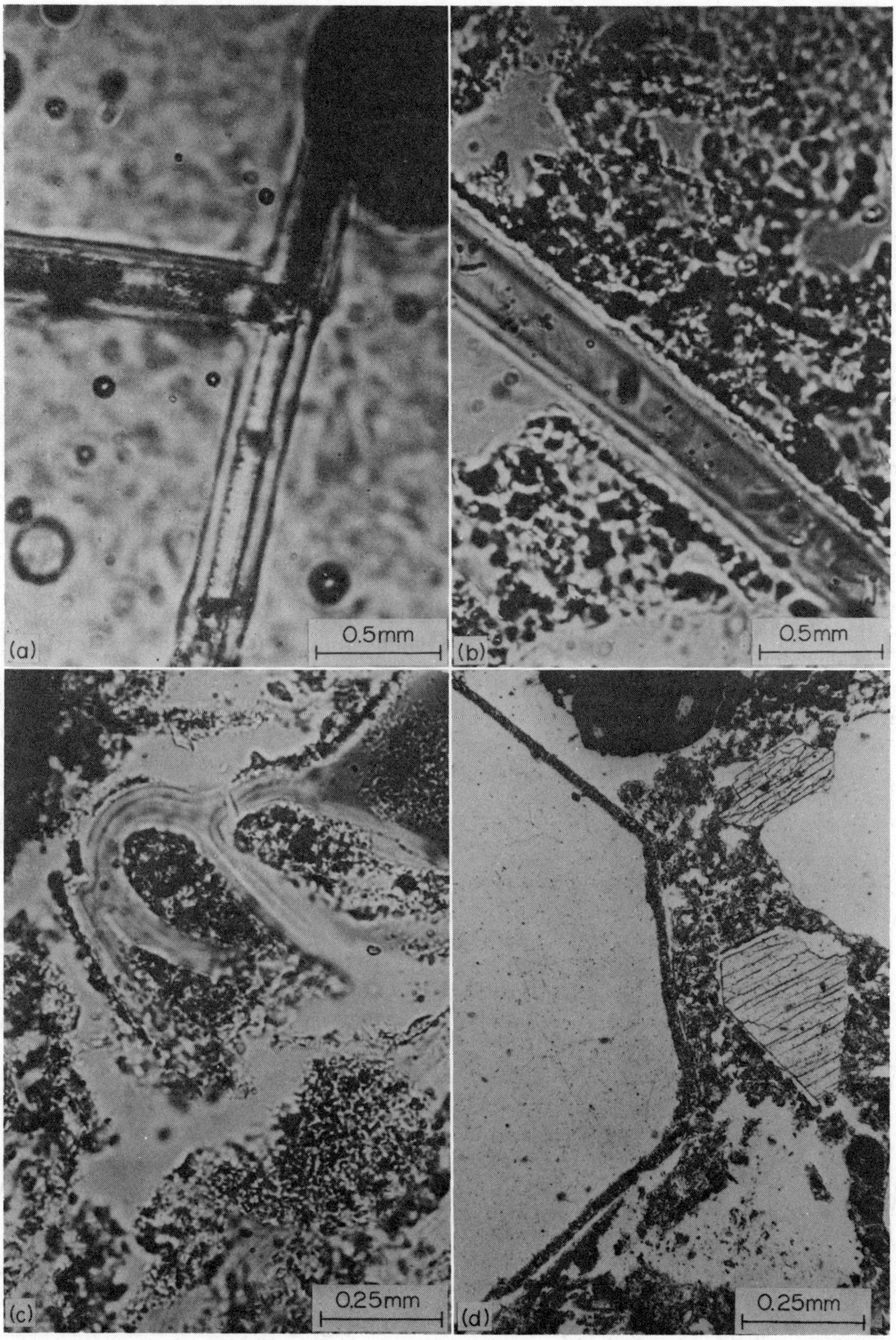

Fig. 1. Photomicrographs of thin sections of tuff samples from the Campanian Region. (a) Typical hydration rim around a crack in a glass fragment; (b) and (c) newly-formed zeolite microcrystals partially replacing the hydrated glass; (d) zeolite microcrystals completely replacing the hydration rim.

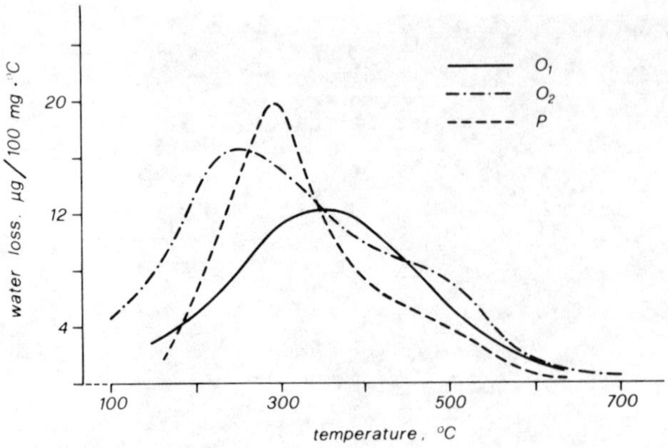

Fig. 2. Differential thermal gravimetric curves of hydrated rhyolitic glasses. Heating rate: 6.6°C min^{-1}.

Laboratory experiments on hydration and zeolitization of glass

Figure 2 gives the differential curves (DTG) obtained from TG analysis of samples O_1, O_2, and P. It can be seen that, while the curve of sample O_1 appears to be symmetrical, the other two curves (O_2 and P) show the presence of two dehydration effects which are evidently related to the loss of two different kinds of water. In order to investigate the position of the water and in which form it is present in the glass, IR spectra are given in Fig. 3.

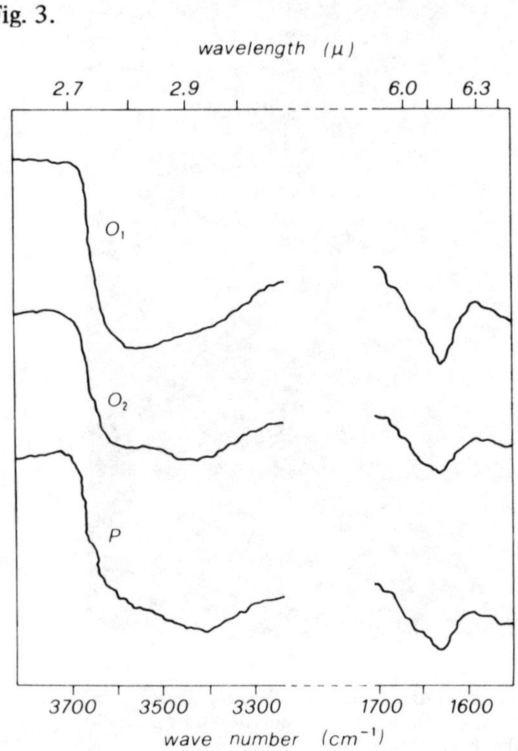

Fig. 3. Infrared spectra of hydrated rhyolitic glasses.

All spectra show bands at about 3700 and about 1650 cm^{-1}. They are attributed to the OH vibrational band of water molecules and to the usual bending vibration of water, respectively. The major difference between the spectra is the shape of the shoulder at about 3400 cm^{-1} which is much more pronounced in samples O_2 and P. This band, attributed to the presence of hydrogen bonded OH (Si–O–H---O$<$ in silicate glasses (Adams, 1961)), is more evident in the IR spectra of glasses O_2 and P after heating up to 400°C. The bands assigned to the presence of water molecules no longer exist, but the band at about 3400 cm^{-1} is still present. The spectrum of sample O_1, on the contrary, appears essentially flat in the zones at about 1650 cm^{-1} and at 3400–3700 cm^{-1}. Both kinds of analysis support the idea that in glasses O_2 and P water is present in both forms of H_2O and OH, while in glass O_1 water seems to be essentially present as H_2O. On the other hand, both possibilities have been reported in the literature (Drury *et al.*, 1962; Nasedkin, 1964).

The results of the synthesis experiments are summarized in Fig. 4. Crystallization fields are given for zeolites formed by hydrothermal reaction (at 80°C for 48 hours) of the glasses listed in Table 2 with NaOH solutions. The curves in Fig. 4 represent the limits of the conditions under which a certain phase grows. These limits are approximations and are sometimes not easily definable; therefore, a dotted line is given for zeolite X. The zeolites crystallizing from the four glasses are, with the exception of zeolite X, the same. The shapes of their

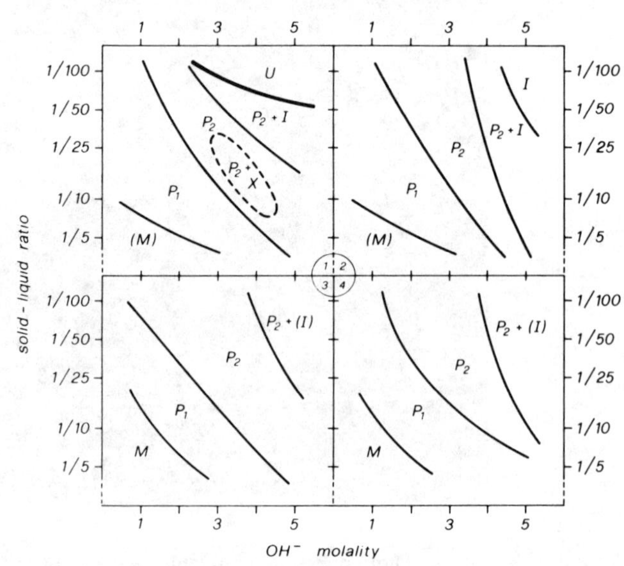

Fig. 4. Crystallization fields of the products obtained by reaction of the various rhyolitic glasses at 80°C for 48 hours, with NaOH solutions. Quadrant 1 = glass O; 2 = glass O_1; 3 = glass O_2; 4 = glass P. Ordinates in logarithmic scale. Abbreviations: M=phillipsite; P_1 and P_2=pseudocubic and pseudoorthorhombic Na–P zeolites; X=faujasite; I=basic sodalite; U=no crystallization.

crystallization fields, however, and sometimes their associations are rather different. It can be noticed in particular:

1. the crystallization fields of phases obtained from glasses O_2 and P are very similar, indirectly confirming the similarity of their hydration states;
2. phillipsite grows better from glasses O_2 and P than from glasses O and O_1;
3. basic sodalite grows better from glass O_1 than from glass O and has been obtained only in smaller amounts from glasses O_2 and P;
4. the presence of a zone of no crystallization (U) in the field of glass O shows that, at least at low solid/liquid ratios and higher alkalinities, this glass is less reactive than the others.

Finally, the reactivity of the various glasses in the presence of alkaline solutions was investigated concentrating on the crystallization kinetics of zeolite P_1 at high solid/liquid ratio and using alkalinities suitable for homogeneous comparison.

Figures 5 and 5a report the crystallization kinetics of zeolite P_1 from glasses O_1 and O (NaOH soln. 3 molal; S/L = 1/10) and from glasses O_2 and O (NaOH soln. 3.5 molal; S/L = 1/5), respectively.

It can be seen that there are no differences in kinetics in the growth of zeolite P_1 from glasses O_1 and O. Noticeable differences, especially at low reaction times, can be observed in the growth of the same zeolite from glasses O_2 and O. The higher reactivity of glass O_2 is confirmed by the values of dissolved silica in the contact solution during the initial stages of the reaction (Fig. 6b).

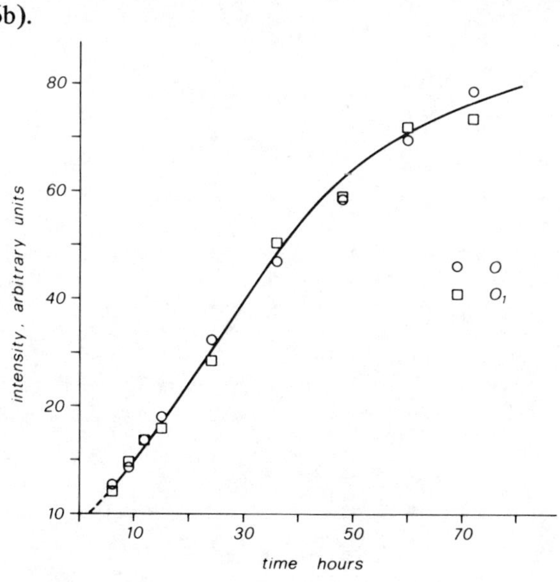

Fig. 5. Crystallization kinetics of zeolite P_1 by reaction of glasses O_1 and O, at 80°C, with 3 molal NaOH solution; S/L = 1/10. Intensity, in arbitrary units, indicates the degree of crystallization as determined by X-ray powder methods.

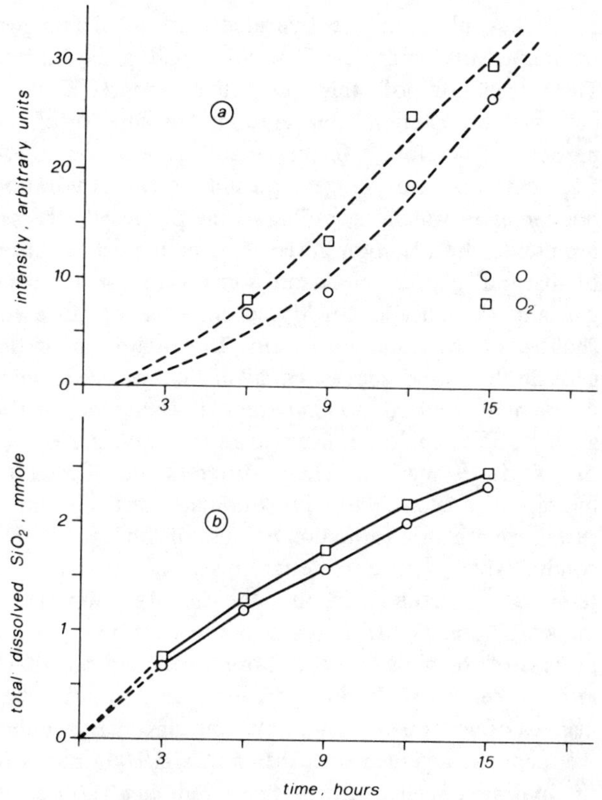

Fig. 6. Crystallization kinetics of zeolite P_1 (a) and total dissolved silica in the contact solution (b) by reaction of glasses O and O_2 at 80°C, with 3.5 molal NaOH solution; S/L = 1/5. Intensity, in arbitrary units, indicates the degree of crystallization as determined by X-ray powder methods.

DISCUSSION

From the interpretation of thin sections of samples from a large number of pyroclastic deposits of the Campanian region, a satisfactory mechanism for the zeolitization process of the volcanic glass can be proposed.

After the deposition of the pyroclastic material, the percolating water, essentially of meteoric origin and therefore practically neutral, comes in contact with the glass fragments and hydrates them by a mechanism which includes an initial stage of water diffusion into the glass followed by hydroxylation according to the scheme:

$$\text{–Si–O–Si–} + H_2O \rightarrow 2\text{Si–OH}$$

In the zones of the deposits in which ground water is present, the hydration process is followed by dissolution, particularly at the expense of the smallest, already completely hydrated particles. This dissolution determines the pH conditions in the pore water and also the concentration of the different ionic species suitable for the zeolite precipitation. The zeolitization process also takes place in the hydrated part of the largest fragments involving local dissolution and precipitation. The formation of this protective zeolitized layer prevents the solution from entering the interior of the particle. Therefore, further zeolitization of these fragments is no longer possible. The hydration phenomenon which, according to the proposed scheme, represents the common initial stage of the zeolitization of natural glasses, could in some cases at least be partially responsible for the occurrence of different zeolites in the same formation. The hydration of the glass in the same deposit could in fact proceed under different conditions, so that the hydration state of the glass particles could be different in the various zones of the same formation. These different conditions of hydration could occur, for example, vertically in a certain pyroclastic formation because of the low thermal conductivity of the material, resulting in a vertical temperature gradient in the deposit. As temperature markedly affects the kinetics of the depolymerization process of the glass (see the above-mentioned reaction), after a certain hydration period a range of different degrees of water penetrations into the glass may occur in the deposit, together with different OH/H_2O ratios in the hydrated layer of the particles. This can be observed in some deposits. On the other hand, laboratory synthesis experiments, even if performed under conditions of alkalinity, solid/liquid ratios, and temperature quite different from the natural environment, have shown that differences in the total hydration and state of the glass may affect the zeolitization process both in determining the type of the newly-formed zeolite and in varying the kinetics of zeolite formation. In summary, it can be concluded that the hydration state of the glass seems to be a more relevant factor than the total amount of hydration water.

ACKNOWLEDGMENTS

Thanks are due to A. Annetta for his assistance in carrying out some of the laboratory experiments.

REFERENCES

Adams, R. V. (1961) Infrared absorption due to water in glasses: *Phys. Chem. Glasses* **2**, 39-49.

Aiello, R. and Colella, C. (1973) Produzione di zeoliti a grandi dimensioni dei canali: *Ann. Chim. (Rome)* **63**, 145-153.

Drury, T., Roberts, G. J. and Roberts, J. P. (1962) Diffusion of "water" in silica glass: in *Advances in Glass Technology*, Plenum Press, New York, pp. 249-255.

Friedman, I. and Smith, R. L. (1958) The deuterium content of water in some volcanic glasses: *Geochim. Cosmochim. Acta* **15**, 218-228.

Friedman, I. and Smith, R. L. (1960) A new dating method using obsidian. Part I. The development of the method: *Amer. Antiquity* **25**, 476-522.

Friedman, I., Smith, R. L. and Long, W. D. (1966) Hydration of natural glass and formation of perlite: *Geol. Soc. Amer. Bull.* **77**, 323-328.

Nasedkin, V. V. (1964) Volatile components of volcanic glasses: *Geochem. Int.* **2**, 317-330.

Porcelli, C. (1976) Unpublished data.

Ross, C. S. and Smith, R. L. (1955) Water and other volatiles in volcanic glasses: *Amer. Mineral.* **40**, 1071-1089.

PHYSICAL AND CHEMICAL PROPERTIES

PROPERTIES OF NATURAL ZEOLITES

D. E. W. VAUGHAN

*Davison Chemical Division, W. R. Grace & Company,
7379 Route 32, Columbia, Maryland 21044, U.S.A.*

Abstract – Although there are more than thirty known natural zeolites, only seven – mordenite, clinoptilolite, chabazite, erionite, ferrierite, phillipsite, and analcime – occur in sufficient quantity and purity to be considered as exploitable natural resources. The crystal structures and properties of these available zeolites, as described in the open and patent literature, are reviewed in detail. Relevant data obtained on synthetic analogues are also included. The last 3 years have significantly changed the potential outlook for natural zeolite utilization in that there will be a developing demand for low-cost ion-exchange and sorbent materials in areas as diverse as energy development, pollution control, agriculture, and metal recovery. These industries utilize relatively low-cost products and should be receptive to the higher cost/benefit-ratio opportunities that natural zeolites have to offer. Major consideration is given to those properties that can be utilized in these fields.

INTRODUCTION

The framework aluminosilicates classified as zeolites constitute what is probably the most variable and extensive family of minerals occurring in the earth's crust. That they can be readily synthesized with great variability adds to their fascination. Their mode of occurrence and petrogenesis has been extensively documented over the last 20 years and has been reviewed by Coombs *et al.* (1959), Hay (1966), Sheppard (1970), Coombs (1970), and other authors publishing in this volume. One of the earliest recognitions of sedimentary zeolites was by Ross (1928), but it was almost 30 years later that their "rediscovery" became of significant interest to chemists and geologists (Deffeyes, 1959). Such deposits are now considered to be commercially exploitable and are the main subjects of this paper. Although there are more than thirty known natural zeolites, only seven – mordenite, clinoptilolite, ferrierite, chabazite, erionite, phillipsite, and analcime – occur in sufficient quantity and purity to be individually considered as viable mineral resources. Extensive deposits of these zeolites occur in North and Central America, Central and Southern Europe, Russia, Japan, Korea, and New Zealand.

The industrial chemists, though cognizant of the variety of zeolites, seem to have ignored all but a handful – mainly faujasite, mordenite, chabazite, and synthetic zeolite A. One of these (faujasite), in synthetic form, is an indirect but indispensable component of our daily lives in that it is used catalytically to produce gasoline. If synthetic zeolite A proves to be a suitable substitute for phosphates in detergents, it too will probably be in the same category (see Anonymous, 1976).

Early studies of natural zeolites were mainly concerned with their composition, petrogenesis, synthesis, dehydration phenomena, and ion-exchange properties and have been summarized by Doelter (1890) and Clarke (1914). In the mid-1920s, Weigel and Steinhof (1925), McBain (1926), and Sameshima (1929) investigated the novel sorption properties of several natural species, and in the 1930s, more detailed evaluations of diffusion and sorption properties were made by Tiselius (1934) and Barrer (1938). Since 1940, work on natural zeolites has progressed at a rapid pace, being given fresh impetus by the syntheses of zeolites A and X (Milton, 1959 a, b) and the subsequent rapid utilization of these synthetic zeolites in the areas of sorption and catalysis. The increasing interest in the 1950s and 1960s took place in the laboratory and in the field. Useful properties of many crystalline zeolites were known and needed by industry, but exploitable deposits had not yet materialized. These needs stimulated exploration activity throughout the western United States, and many deposits formerly designated as "clay" or "tuff" were evaluated for their natural zeolite content.

Most of the early work with natural zeolites was carried out on large crystal specimens of the type commonly found in the vugs and amygdaloids of basaltic lava in Ireland, Germany, France, Italy, and Nova Scotia. Such deposits are in sharp contrast to most of the large, economic bodies of zeolites which have been derived from altered rhyolitic and trachytic ash beds. The mineralogical differences are generally only compositional and morphological, with the more acidic rocks (i.e. higher silica) yielding zeolites of higher silica/alumina ratio.

Compositionally zeolites may be given the general formula:

$$(M^{+}{}_2, M^{2+})\,O\cdot Al_2O_3\cdot gSiO_2\cdot zH_2O$$

M^+ is usually Na or K, and M^{2+} is Mg, Ca, or Fe; more rarely Li, Sr, or Ba may substitute for M^+ or M^{2+}. Fe^{3+} is commonly assumed to substitute into the tetrahedral framework position (Sheppard and Gude, 1969). Although the M cations are exchangeable, location within the zeolite structure will determine their ease of exchange. Most iron is present as Fe_2O_3 stain and is usually readily removed by mild acid leaching. Although zeolite structures require the equivalent ratio between cations and aluminium to be unity, this is commonly obscured by the presence of salts (e.g., calcite and barite) in the sedimentary deposits, together with possible structural iron. Zeolites occurring in association with highly siliceous tuffs often have questionably high-silica analyses due to unreacted tuff or occluded silica. The significance of cation compositions is often questionable because of the multiplicity of post-crystallization conditions that could effect ion exchange and, therefore, present-day compositions.

In a series of recent reviews, Barrer summarized the thermodynamics and physical chemistry of zeolite diffusion (1971), sorption (1973), and ion exchange (1974). A series of earlier articles summarized the particular molecular sieving properties of zeolites (Barrer, 1949, 1959, 1968). Riekert (1970) reviewed sorption and diffusion in zeolites with particular regard to catalytic processes. The structural chemistry of zeolites has been greatly expanded in recent years and has been extensively reviewed by Fischer and Meier (1965), Meier (1968), and Meier and Olson (1971). Reviews of the catalytic properties of zeolites are usually strongly oriented towards synthetic faujasites (Turkevich, 1968; Venuto and Landis, 1968; Venuto, 1971; Minachev and Isakov, 1973) with little regard for the broader spectrum of potential zeolite catalysts, with the possible exception of mordenite (Burbidge *et al.*, 1971; Kouwenhoven, 1973). Three books on the chemistry of zeolites (Hersch, 1961; Sokolov *et al.*, 1965; Breck, 1974), although primarily concerned with synthetic zeolites, do cover the composition and properties of the natural zeolites. Munson and Sheppard (1974) surveyed these subjects along with the occurrences of natural zeolites. Although most of the work covered in these reviews was carried out on crystals of natural zeolites from igneous basalt vugs, it is thought to be generally applicable to the more abundant sedimentary zeolites as well.

The purpose of this paper is to review the properties of the major natural zeolites with emphasis on those properties that are exploitable in industrial utilization. Sequentially, the chemistry and structure of individual zeolites are described, followed by discussions of their sorption, ion-exchange, and catalytic properties. Specific examples are included of current and potential applications by way of illustration. The physical and chemical characteristics of analcime, chabazite, clinoptilolite, erionite, ferrierite, mordenite, and phillipsite are summarized in Tables 1 and 2.

TABLE 1. Properties of natural zeolites

Zeolite	Si/Al range	Diffusion Control		Main sorption sites	Sorption properties*				pH stability limit†
		Ring-type	Ring size, Å		H_2O	N_2	$n\text{-}C_4H_{10}$	C_6H_6	
Analcime	2	6	2.8	Channel	8	0	0	0	~5
Chabazite	1.4 – 2.8	8	4.1 × 3.7	Cavity	28	20	10	0	~4
Clinoptilolite	2.7 – 5.3	10	7.9 × 3.5	Channel	14	11	4	9	<2
		8	4.4 × 3.0						
Erionite	3 – 4	8	4.2 × 3.6	Cavity	20	15	9	0	~3
Ferrierite	3.2 – 6.2	10	5.4 × 4.2	Channel	12	11.5	–	7.5	<2
		8	4.7 × 3.4						
Mordenite	4.4 – 5.5	12	7 × 6.7	Channel	15	15.4	6.5	8.8	<2
Phillipsite	1.3 – 2.9	8	4.2 × 4.4	Channel	22	–	0	0	~4
		8	2.8 × 4.8						

*Approximate maximum reported values (wt.%).

†pH at which a structural degradation is detectable.

TABLE 2. Ion-exchange properties

	Si/Al (Exchange capacity)	Original cations	Exchange selectivities
Analcime	2	Na	Complicated by ion-sieving
Chabazite	1.4 – 2.8	Ca, K	$Tl^+ > Cs^+ > K^+ > Ag^+ > Rb^+ > NH_4^+ > Pb^{2+} > Na^+ = Ba^{2+} > Sr^{2+} > Ca^{2+} > Li^+$
Clinoptilolite	2.7 – 5.3	Ca, Na, K	$Cs^+ > K^+ > Sr^{2+} = Ba^{2+} > Ca^{2+} \gg Na^+ > Li^+$ $Pb^{2+} > Ag^+ > Cd^{2+} \sim Zn^{2+} > Cu^{2+} > Na^+$
Erionite	3 – 4	Na, K	$Cs^+ > Sr^{2+} > K^+ > Na^+$
Ferrierite	3.2 – 6.2	K, Mg	Inadequate data
Mordenite	4.4 – 5.5	Ca, Na	$Cs^+ > K^+ > NH_4^+ > Na^+ > Ba^{2+} > Li^+$ $NH_4^+ > Na^+ > Mn^{2+} > Cu^{2+} > Co^{2+} \sim Zn^{2+} > Ni^{2+}$
Phillipsite	1.3 – 2.9	K, Ca, Na	$Ba^{2+} > Rb^+ \sim Cs^+ \sim K^+ > Na^+ \gg Li^+$

ANALCIME

Chemistry

Analcime has a dense structure compared with some zeolites and is often classified as a feldspathoid. It has a wide distribution as a primary mineral in many igneous rocks and as an authigenic mineral in sedimentary rocks. In basalts, it commonly occurs as large euhedral crystals. It was the first zeolite structure to be fully evaluated (Taylor, 1930); and further structure refinements were made by Calleri and Ferraris (1964) and Ferraris *et al.* (1972). The cubic structure consists of 4-, 6-, and 8-rings linked in such a way as to give sixteen large interconnected cavities which form non-intersecting channels and a second series of twenty-four smaller individual cavities. Sixteen of the small cages are occupied by sodium cations, and the large cages are occupied by water molecules. The unit-cell formula may be written as $16(NaAlSi_2O_6 \cdot H_2O)$, although the Si/Al ratio varies (Saha, 1959). Naturally occurring analogues of analcime are leucite (K-form), pollucite (Cs-form), and wairakite (Ca-form). The first two are usually anhydrous, with the larger K and Cs cations occupying the sites filled by water in analcime. Wairakite and high-temperature leucite are hydrous with similar cation locations to analcime.

Sorption

Although several early papers described hydrocarbon sorption by analcime, the structure clearly is not accessible to molecules larger than about 3 Å in diameter. Analcime has a moderate water-sorption capacity (about 9%), and errors resulting from using incompletely dried sorbates is the most plausible explanation for apparent hydrocarbon sorption. Tiselius (1935) reported a high ammonium-sorption capacity for analcime, and Barrer (1938) confirmed that analcime will sorb ammonia, but at significantly lower levels than those reported by Tiselius. Barrer's work also showed that He, H_2, N_2, and Ar can penetrate the analcime structure.

Ion exchange

Analcime does not readily exchange at room temperature, but at elevated temperatures Na can be completely replaced by K, Ag, T1, NH_4, and Rb cations. Only negligible exchange occurs with Li, Cs, Mg, Ca, and Ba cations. Hysteresis is a common phenomenon. Early exchange studies indicated that pollucite could be formed only by a synthesis procedure (Barrer, 1950), but later work with Cs tracers (Barrer and Rees, 1960; Callahan and Kay, 1966a) showed that Cs does exchange very slowly.

Exchange in analcimes between the ion pairs Na–K, Na–T1, and K–T1 resulted in discontinuous isotherms caused by the limited mutual solid solubility of the pure cation phases (Barrer and Hinds, 1953). Further work with the K–Rb ion pair confirmed these exsolution phenomena and attributed the hysteresis in the ion-exchange isotherms to difficulties in nucleating the exsolution phase in a matrix of the end member (Barrer and Falconer, 1956). Work by Balgord and Roy (1971) showed that this limited mutual solid solubility is dependent upon the Si/Al ratio of the exchanging analcime, and hence exsolution of the exchanging phase may be controlled by the relative alumina levels. The same authors exchanged small amounts of Sr, Mg, Ni, and Co into analcime. They also observed that analcime partially exchanged with Ag would exsolve metallic silver via a photosensitized redox reaction with structural water. On the basis of his exchange work on natural and synthetic wairakite and analcime, Ames (1966) concluded that it is not possible to form stable Sr analcimes.

Barrer (1950) first noted the distinct ion-sieve effect between Rb and Cs in analcime and demonstrated that Ag-analcime can be used to purify CsCl solutions contaminated with NaCl, KCl, or RbCl (Barrer and Sammon, 1961). A more subtle ion sieving occurs with lithium isotopes (Matousova, 1968), and analcime has been shown to have a single-stage separation factor of 1.024 for ^{6}Li (Chumakov *et al.*, 1969).

CHABAZITE

Chemistry

The structure of chabazite was solved independently by Dent and Smith (1958) and Nowacki *et al.* (1958). Refinements of Si–Al ordering, cation positions as functions of temperature, and hydration effects have been made by Smith (1962), Smith *et al.* (1963), and Smith *et al.* (1964). The structure consists of stacked, double six-membered ring prisms, interconnected through four rings, in a cubic close packed array (FCC). Repetition of the ABC stacking of prisms produces cylindrical cavities, 11 × 6.6 Å, joined to adjacent cavities through six 8-membered rings having free diameters of 4.1 × 3.7 Å. The cavities may be visualized as the intersections of three 8-ring channels traversing the structure (Barrer and Kerr, 1959). It is one of the most porous natural zeolites and commonly develops surface areas of 500–600 m^2/g.

The compositional range of natural chabazites has been reviewed by Passaglia (1970). Chabazite seems to form a continuous solid solution series between calcium and sodium end members. Few high-potassium or sodium–potassium materials have been found, although in the laboratory the potassium and sodium–potassium varieties are readily synthesized. The Si/Al ratio varies between 1.4 and 2.8, the former usually being found in crystals from vesicles in basaltic lavas while the latter is characteristic of chabazite from altered rhyolitic tuff deposits.

The stability and decomposition products of natural chabazites ion exchanged with alkaline and alkaline earth cations have been extensively investigated by Barrer and Langley (1958 a, b). Large cations seem to enhance stability, and the stability sequence is Cs > Rb > K > Na > Li. Less siliceous chabazites have lower thermal stability, regardless of cationic form.

Sorption

Because of its availability in many mineral collections, chabazite was one of the first zeolites to be extensively investigated. Seeliger (1921) detailed the high sorption capacity for a range of molecules and observed that rates of sorption and sorption capacities decreased as molecular size increased, to the point where the largest molecules were completely excluded. Shortly afterwards, McBain (1926) specified the maximum dimension of the chabazite window to be about 5 Å on the basis of his own work. Further work by Weigel and Steinhof (1925), Schmidt (1928), Sameshima (1929), and Baba (1929) confirmed and extended these sorption studies. In 1938 Barrer began a series of investigations into the sorption and diffusion properties of many cationic forms of chabazite, developing along the way the modern concept of the process now called "molecular sieving" (Barrer, 1941b). Much of this earlier work has been reviewed in detail by Barrer (1941, 1959) and Hersh (1961).

The high water capacity of chabazite combined with a high silica to alumina ratio makes this zeolite interesting as a sorbent for drying gases, particularly acidic gas streams. Early work demonstrated that chabazite would desorb and resorb up to 17 wt.% water after thermal treatments of 400°C in the presence of water (Milligan and Weiser, 1937). Subsequent work by Barrer and co-workers showed that the stability of chabazite increases with the size of the alkali cation in the sequence Li to Cs, but that the retentivity for water is in the reverse order, reflecting the higher binding energy for the smaller cations. A similar trend is exhibited by bivalent cations, with Ca > Na and Ba > K (Barrer and Langley, 1958c). Further work established that the water capacity increased with decreasing silica to alumina ratio (Barrer and Langley, 1958 b, c). Detailed sorption and diffusion experiments by Barrer and Fender (1961a and 1961b, respectively) on Nova Scotia chabazite (Ca–Na form) established that the material had a water-sorption capacity of 27 wt.% (dehydrated basis) and that the diffusion rates were very fast ($\Delta E = 6.7 \pm 0.3$ kcal/mole). The water sorption as a function of cation content was shown to vary from twelve H_2O molecules/unit cell/(∿27 wt.%) for Na and Ca forms to six for the Cs form (Barrer *et al.*, 1969). The ammonium form also sorbs about twelve molecules H_2O/unit cell, one molecule of which results from dehydroxylation; but a hydrogen material made via the ammonium form only has a water-sorption capacity of 9.3 molecules/unit cell (Barrer and Davies, 1970). This decrease in capacity for hydrogen chabazite was attributed to collapse of the crystal structure during grinding. Despite its high capacity and thermal stability, chabazite has not been used in most major industrial drying applications (e.g., cracked gas, natural gas, hydrogen, LPG drying). Its only major application is for water, CO_2, and H_2S removal from sour natural gas. As suggested by Barrer and Coughlan (1967), chabazite and erionite are useful in the upgrading of low-BTU natural gas and should be used more as the wellhead price of natural gas increases.

Chabazite has a high capacity for all of the rare gases, although the larger krypton and xenon molecules require high temperatures and pressures (Barrer and Papadopoulos, 1972) to demonstrate their high adsorptivity. Barrer and Robins (1953) showed that chabazite has a high affinity for oxygen and nitrogen and that H_2, Ne, and Ar could be separated from mixtures containing O_2 and N_2. Garden and Kington (1956), Kington (1958), and Kington and MacLeod (1959) showed that preferential sorption occurs in the sequence $CO_2 > CO > N_2 > O_2 > Ar > H_2$ and that the strong quadrupoles of the preferentially sorbed gases are the major factors affecting selectivity. By suitably changing the electrostatic energy field within the zeolite by cation exchange, differences between molecular affinities can be optimized even though the size of the molecules may be similar. Such cation modification has led to the development of chabazites useful for the separation of oxygen from nitrogen. Figure 1 shows the chromatographic separation factors (ratio of retention times) for O_2 and N_2 of a wide range of cations in chabazite, and that a barium exchanged chabazite would yield almost

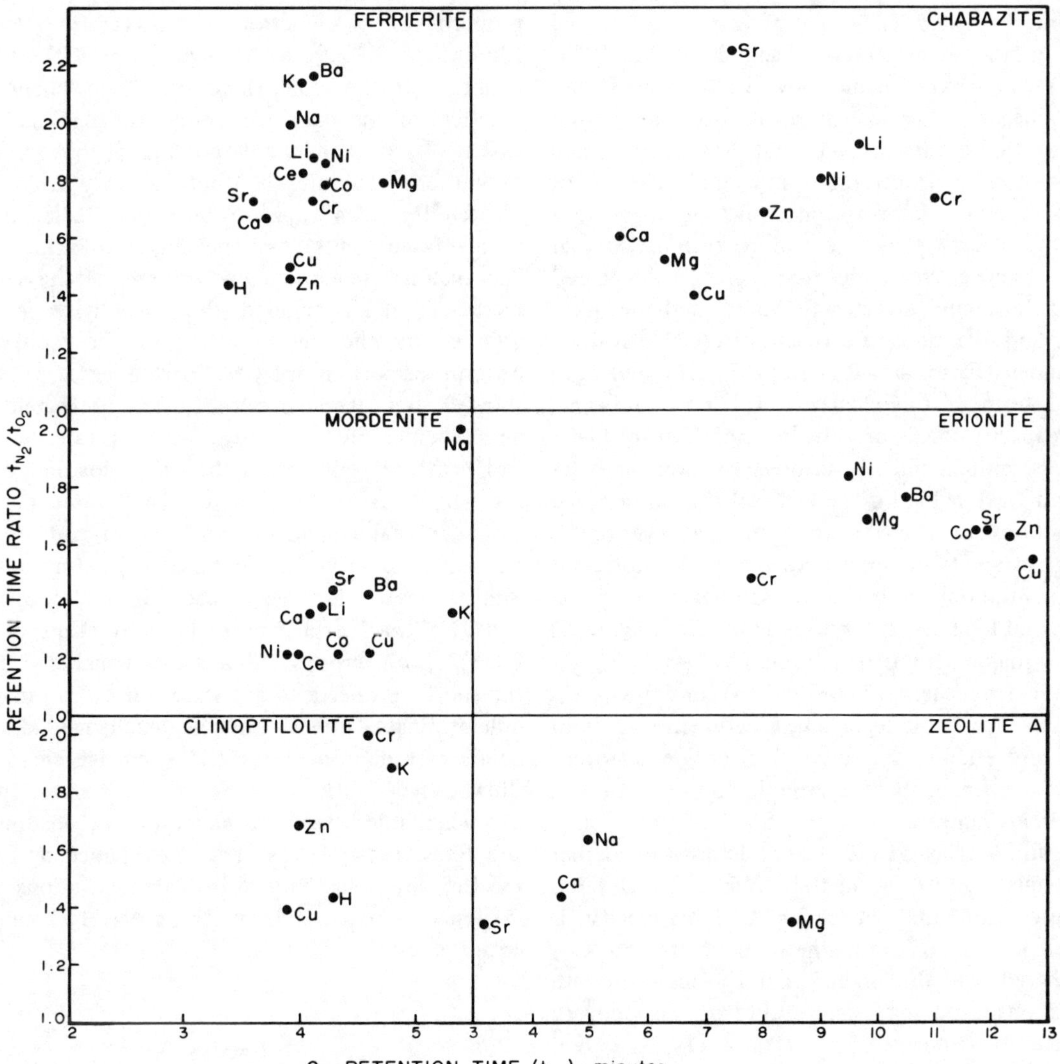

Fig. 1. Oxygen–nitrogen separation factors. Ba-chabazite (not plotted in this figure) has an oxygen-retention time of 7 min. and a retention ratio of 3.2.

pure O_2 and N_2 gas streams (Vaughan, 1974). However, the preferred method for the zeolitic separation of O_2 and N_2 is a dynamic, pressure-swing sorption system wherein zeolite sorbs N_2 selectively under one set of conditions and readily releases it when these conditions are changed. Natural chabazite, erionite, and mordenite offer viable alternatives to synthetic Ca–A zeolite in the p.s.a. separation of O_2 from air. Nitrous oxide, a molecule with physical properties similar to those of carbon dioxide, is very strongly sorbed by chabazite without decomposition. Nitric oxide, on the other hand, is strongly sorbed but disproportionates readily according to the reaction $4NO \rightarrow N_2O + N_2O_3$ (Addison and Barrer, 1955).

The earliest work on chabazite demonstrated a capability for sorbing linear paraffins, and by the early 1940s chabazite was shown to be useful for separating normal and branched chain paraffins by preferential sorption of the former (Barrer and Ibbitson, 1941, 1944 a, b). These experiments showed that Nova Scotia (Ca, Na) chabazite has a high sorption capacity, that rates of sorption are inversely proportional to the chain length of the hydrocarbons, and that degree of out-gassing and pre-sorption of hydrocarbons profoundly affect the rates of sorption. Similar molecular sieving was observed with halogenated methanes; the mono- and di-substituted methanes were occluded, and the tri- and tetra-substituted derivatives were excluded (Barrer and Brook, 1953). The hydrogen exchanged form of this chabazite has about the same sorption capacity for n-paraffins but significantly lower heats of sorption and diffusion–activation energies (Barrer and Davies, 1971). The higher diffusion rates for n-hydrocarbons in H-chabazite (and also H-erionite) improve the viability of this zeolite as a "5 Å" molecular sieve. Its main problems are low capacity and desorption rate compared to alternative synthetic zeolites (A and X) in such major industrial processes as n-C_4-C_9 purification (for chemical feedstocks) and n-C_{10}–C_{18} purification (for biodegradable detergents). Both chabazite and erionite are suitable for ethylene storage for use as a ripening agent, a growth stimulant, or as a defoliant in agriculture.

Barrer and Wasilewski (1961 a, b) demonstrated that chabazite also sorbs elemental iodine at elevated temperatures and that at >200°C its capacity is comparable to those of the synthetic zeolites A and X. A process based on the iodine adsorption of silver-exchanged chabazite may have utility in the recovery and storage of radioactive I^{129} (Pence *et al.*, 1971). Experiments with S, P and Hg (Barrer and Whiteman 1967 a,b) showed that sulfur and phosphorus are not sorbed in significant amounts; mercury is sorbed in relatively small amounts into Ag-chabazite. Zeolites A and X have high capacity for all of these elements, and it is doubtful that chabazite can compete commercially with these synthetic phases.

Ion exchange

The ion-exchange selectivities of chabazite have been evaluated by Barrer (1950), Barrer and Sammon (1955), Ames (1961a), and Barrer *et al.* (1969). The affinity sequence is $Tl^+Cs^+ > K^+ > Ag^+ > Rb^+ > NH_4{}^+ > Pb^{2+} > Na^+ = Ba^{2+} > Sr^{2+} > Ca^{2+} > Li^+$ for a natural chabazite having a Si/Al ratio of 2.5. Only Cs^+ does not fully exchange (84% replacement of Na^+). The large ions La^{3+}, $NMe_4{}^+$, and $NEt_4{}^+$ are excluded by an ion-sieve effect. Studies on a lower Si/Al ratio (1.07) synthetic chabazite indicated significant changes in exchange properties as a function of Si/Al ratio (Barrer and Klinowski, 1972). Although ion affinities were qualitatively the same, there were major changes in the kinetics of univalent–divalent exchange and in the extent of Cs^+ exchange Barrer *et al.* (1969b) showed for a synthetic faujasite that both affinity and rate were greatly dependent upon Si/Al ratios. Callahan (1966), using fused salts as exchange solvents, observed "anomalous" selectivities as the size of the salt cation increased; he concluded that selectivity is strongly affected by the sieving effect of the zeolite on the cation-transport properties. Barrer *et al.* (1963 a, b) showed that although self-diffusion coefficients follow an Arrhenius plot, exchange diffusion rates are variable and highly dependent on the cation loading. Exchange involving Ca^{2+}, Sr^{2+}, and Ba^{2+} are particularly compositional-dependent, with significant differences between the rates for forward and reverse exchange. Similar observations were made for Na^+/K^+, Na^+/Cs^+, and $2Na^+/Ba^{2+}$ exchange in chabazite (Kinshofer *et al.*, 1973). Both groups of authors concluded that the transport properties of water have a major but indeterminate effect on the exchange kinetics. A detailed evaluation of Na^+/K^+ exchange characteristics shows that both the Na- and K-tracer diffusion coefficients decrease with increasing Na^+ loading (Duffy and Rees, 1974, 1975). This is explained by proposing specific site selectivity for different cations and the existence of a particularly strong Na–H_2O complex in critical sites.

Catalysis

Ammonium chabazite has a high activity for cracking n-hexane (Miale *et al.*, 1966; Miale and Weisz, 1968) and will effectively sieve n-paraffins from mixtures with other hydrocarbons and selectively crack only the n-paraffins (Frilette and Weisz, 1966). Similar properties were reported for hydrocracking over hydrogen-chabazite (Mobil, 1964).

CLINOPTILOLITE

Chemistry

The relative merits of the names clinoptilolite and heulandite have received much attention since Mumpton (1960) proposed differentiation on the basis of Si/Al ratio and thermal stability, and Mason and Sand (1960) forwarded the alternative view that relative levels of monovalent and divalent cations should be the deciding factor. The chemistry of this zeolite type has been extended by Alietti (1972), Boles (1972), and Alberti (1975). Both Si/Al ratios and cation contents determine the properties of most zeolites, but precedents on this subject usually favor differentiation on the basis of framework composition rather than the readily variable exchangeable cations. The Si/Al ratios of these zeolites vary from a low of about 2.7 for heulandite to a high of about 5.3 for clinoptilolite. Calcium, sodium, and potassium are the major cations, with high calcium generally prevailing in the lower Si/Al ratio heulandite and high potassium in the higher Si/Al ratio clinoptilolite. A recently reported high Sr-Ba heulandite is an exception to this (Černy and Povondra, 1969). The present consensus is that heulandite and clinoptilolite are isostructural; however, clinoptilolite is the only economically important variety from the viewpoint of both stability and availability.

The structure of heulandite was determined by Merkle and Slaughter (1968), and Alberti (1975) has recently confirmed the isostructural nature of clinoptilolite. A major feature of the structure is a tight network of interlinked 4- and 5-rings in the *ac*-plane, which accounts for the characteristic cleavage of the mineral. Between these layers are open 10-ring and 8-ring channels parallel to the *c*-axis, with approximate free dimensions of 7.9 × 3.5 Å and 4.4 × 3 Å, respectively. Calcium cations in heulandite lie close to 8- and 10-ring sites in the channels, particularly in the region of high Al^{3+} substitution in the tetrahedral bridging positions between the dense sheets in the *ac*-plane.

The location of exchange sites in the open channels means that the free volume of the zeolite will be closely related to the volume of exchange cations. For this reason, the predominantly monovalent cation form has a much lower water capacity than the divalent cation form for a given Si/Al ratio heulandite (Boles, 1972). Similarly, within the series Li^+ to Cs^+, the water capacity decreases with increasing cation size (Belitskii *et al.*, 1973), and dehydration activation energies increase (Chelishchev *et al.*, 1974). As the Si/Al ratio increases for a group of zeolites, the stability to acid attack also increases. The blocking effects of cations in clinoptilolites can therefore be minimized by acid treatment, which reduces the cation-exchange capacity by leaching Al^{3+} from framework positions and substitutes H^+ into the few remaining cation positions (Barrer and Makki, 1964). Exchangeable cations are first replaced by hydrogen, followed by the hydroxylation of the Al–O bonds in the framework, and the removal of aluminum from the framework into solution. This leaves a "nest of OH groups" in the former tetrahedral Al^{3+} position, presumably hydrogen bonded to one another. A similar process in faujasite has been extensively investigated by numerous workers (e.g., Kerr, 1973), but in the latter zeolite, steam annealing of the framework is an important factor in the stabilization sequence. Acid treatment of several high-silica natural zeolites (mordenite, ferrierite, clinoptilolite) produces a range of improved or modified sorbents, both via the mechanism of decationation and dealumination, and by dissolving any silica blocking the channels in the structure.

The potassium form of zeolites of this group shows high thermal stability. The low Si/Al calcium heulandites tend to degrade between 500–550°C (Shepard and Starkey, 1964; Alietti, 1972), but potassium-exchanged heulandites and various potassium and hydrogen clinoptilolites are stable up to 800°C (Shepard and Starkey, 1964; Barrer and Makki, 1964). Although low-silica, high-calcium heulandites invert to the heulandite "B" polymorph between 250–400°C, this does not seem to be a major structural change as argon diffusion rates are only minimally effected (Barrer and Vaughan, 1969).

Sorption

Although most of the recent sorption work has been done on clinoptilolites, several investigations have been made on heulandite. Sameshima (1929) showed that heulandite would not sorb O_2, CO_2, C_2H_2, or SO_2, but that it would sorb large amounts of ammonia. The high ammonia capacity was later confirmed by Barrer (1938). Subsequent work by Barrer and Vaughan (1969) demonstrated that heulandite could sorb significant volumes of the rare gases and that argon (diameter 3.8 Å) would readily diffuse through the structure. The work of Barrer and Makki (1964) on acid-leached clinoptilolites demonstrated that the material had potential as a molecular sieve and sorbent. They showed that, depending on the concentration of acid used, the BET surface area could be increased from 30 m^2/g to 380 m^2/g and that the effective pore size could be enlarged to allow the sorption of benzene and iso-pentane. A 2 N HCl treatment readily turns a non-sorbing clinoptilolite into a material capable of separating acetone from carbon tetrachloride (Shcherbatyuk and Froer, 1973). Later work showed that increasing de-cationation and de-alumination homogenize zeolitic sorption sites towards sorption of Kr and CO_2 (Barrer and Coughlan, 1968) and H_2O (Barrer and Murphy, 1970).

Ion exchange

The ion-exchange properties of clinoptilolite have been investigated more intensively than those of any other natural zeolite, primarily because of its ability to scavenge Cs^{137} from radioactive waste solutions and ammonium ions from municipal wastewater streams. Ames demonstrated a specificity for Cs (1959, 1960, 1961a, 1962 a, b, c) and potassium (Ames and Mercer, 1962) and outlined a selectivity sequence of Cs > K > Sr = Ba > Ca > Na >> Li for material from the Hector, California deposit. Howery and Thomas (1965) reported the sequence Cs > NH_4 >> Na, and more recently Chelishchev *et al.* (1973) gave selectivities of Cs > Rb > K > Na > Sn > Li and Pb > Ag > Cd ~ Zn ~ Cu > Na (1974). In the latter case, selectivity reversals were observed for Pb at 80% exchange and for Cd, Cu, and Zn at 20% exchange.

Ames (1959, 1960) showed that both Cs^{137} and Sr^{90} can be selectively removed from radioactive wastewaters in the presence of high concentrations of competing cations, and a major amount of work was subsequently performed to optimize this system for use in nuclear facilities of the United States (e.g., Knoll, 1963; Hawkins and Short, 1965). Honstead *et al.* (1960) showed that rare-earth cations and Co^{60} can also be isolated by clinoptilolite ion-exchange processes. This and other zeolites have a major advantage over organic exchange resins in that they are especially resistant to radiation degradation. Recovered components may also be left on the exchanger and safely stored for long periods of time.

Ames (1967) and Mercer *et al.* (1969) also demonstrated the superior properties of clinoptilolite for the removal of ammonium ions from municipal sewage-treatment effluents, and several plants are currently under construction in the United States which will use this process. The high wet strength of most natural zeolite materials allow sized ore to be used directly in exchange columns at flow pressures which would cause pellets of synthetic powders to disintegrate rapidly. Clinoptilolite works best when the cation to be removed is present in low concentrations. Also, appreciable quantities of Ca^{++} and Mg^{++} have detrimental effects on the NH_4^+-exchange capacity of this zeolite. In addition to municipal wastewater treatment, recent studies on a trout-breeding system by Peters and Bose (1975) indicate that a clinoptilolite exchange process may be a superior way of controlling ammonia recirculating in fish-hatchery waters, compared with previously used bacterial oxidation methods.

Based on the results of Chelishchev *et al.* (1973) noted above and others, clinoptilolite may be a useful ion-exchanger for Pb, Cu, Cd, Zn, and Co in the final polishing step of the tertiary treatment of industrial wastewaters. Studies of exchange between Na-clinoptilolite and various alkyl-ammonium cations have revealed interesting steric and ion-sieve properties that can be readily explained on the basis of the Merkle–Slaughter structure (Barrer *et al.*, 1976). Ions small enough to enter both channels fully replace Na^+(NH_4^+, $CH_3NH_3^+$, $C_2H_5NH_3^+$, $(CH_3)_2NH_2^+$, and n-$C_3H_7NH_3^+$); those too large to enter the 8-ring channel but small enough to penetrate the 10-ring are partially exchanged ($(CH_3)_3NH^+$ and iso-$C_3H_7NH_3^+$); and the largest ions are totally excluded ($(CH_3)_4N^+$ and tert-$C_4H_9NH_3^+$). The n-$C_4H_9NH_3^+$ ion, which should penetrate both 8- and 10-ring channels, is only partially replaced. In this case (and possibly in the case of $(CH_3)_3NH^+$ and iso-$C_3H_7NH_3^+$), the free volume of the alkyl-ammonium cation exceeds the total free volume of the channels, and exchange is therefore limited by the stereochemistry of the sorbent and sorbate system. Subsequently Aitken and Keen (1971) proposed that the stereo-specific character of zeolites is useful for separating the ammonia complexes of various noble and precious metals from those of the transition metals. Barrer and Townsend (1976b) confirmed that an ammonium mordenite has stereo-specificity for the ammonia complexes of Cu^{++}, Zn^{++}, and Co^{++}.

Catalysis

The work of Barrer and Makki (1964) showing sorption of aromatic hydrocarbons by acid-treated clinoptilolite indicated the route to potential clinoptilolite catalysts. Klopp *et al.* (1967) showed that acid- or caustic-treated clinoptilolite tuff is a very active hydrogenation, dehydrogenation, and dealkylation catalyst (see also Papp *et al.*, 1971). Clinoptilolite has also been shown to catalyze many other chemical reactions, including the cracking of paraffins (Beyer, 1975), dewaxing of crudes (Foucher *et al.*, 1971), disproportionation of toluene to benzene and xylene

(Sato and Otani, 1969), and the isomerization of butenes (Kallo *et al.*, 1971; Detrekoy and Kallo, 1973; Detrekoy *et al.*, 1974). Wilson (1972) reported synergistic effects when clinoptilolite is added as a co-catalyst with other shape-selective zeolites of the A, erionite, or offretite types. This may have practical significance for "Selectoforming" natural erionite catalysts, in that several Nevada "erionites" are in reality mixtures of erionite and clinoptilolite. Natural clinoptilolite is also reported to be a useful matrix material for dispersed noble or transition metal catalysts used in isomerization, alkylation, hydrogenation, and hydrodesulphurization processes (Nagyugomasu, 1967).

ERIONITE

Chemistry

The similarities between the powder X-ray diffraction patterns and structures of erionite and offretite have led to confusion in characterizing erionite (Hey and Fejer, 1962; Bennett and Gard, 1967; Kokotailo *et al.*, 1972). The structure of erionite consists of elongated cages stacked parallel to the *c*-axis and connected to adjacent cavities in the *a–b*-plane through six elongated 8-ring windows having approximate dimensions of 5.2 × 3.6 Å (Staples and Gard, 1959). Although continuous 6-ring channels parallel the *c*-axis, the main diffusion paths are through 8-rings parallel to *c* but which are offset by every other cage (Barrer and Kerr, 1959). Sheppard and Gude (1969) reported Si/Al ratios in the range 3 to 4 and sodium and potassium as the major cations in natural erionites. From the viewpoint of high sorption capacity and diffusion rates, the hydrogen form of erionite is preferred. However, in view of the moderate acid-stability of erionite, acid treatments should be carried out at low concentrations (Piguzova *et al.*, 1968). Extended treatments caused dealumination followed by rapid lattice degradation (Ermolenko *et al.*, 1971). Ammonium exchange is probably the best approach to the H^+-form, which begins to collapse at about 800°C (Weeks *et al.*, 1975).

Sorption

Erionite sorbs up to 20 wt.% water, but its capacity is highly dependent upon cation content (Peterson *et al.*, 1965). It has a high capacity for CO_2 (Barrer and Coughlan, 1968), O_2, and N_2 and will effectively separate the last two molecules (Vaughan, 1974). The effects of cations on O_2- and N_2-separation factors are given in Fig. 1. Erionite has a high capacity for SO_2 (Gupta *et al.*, 1971) and it will selectively remove CS_2 from mixtures with hydrocarbons (Astakhov *et al.*, 1973) and NO_2 from mixtures with Cl_2 (Amirova *et al.*, 1974).

Potassium erionite has properties similar to those of a 3 Å molecular sieve (Eberly, 1964; Barrer and Galabova, 1973), and Ca-erionite readily sorbs n-paraffins and excludes iso-paraffins (Barrer and Peterson, 1964b; Eberly, 1964, 1969). Erionite has a higher Si/Al ratio than zeolite A and is therefore a more stable zeolite; this is particularly important if steam desorption is preferred for n-paraffin separation processes (Lorenz, 1968). Hydrogen or ammonium erionites are also good substitutes for these cation forms of zeolite A which are unstable (Kühl, 1973). Diffusion and rate studies have shown that lower Si/Al ratios enhance sorption rates (Barrer and Peterson, 1964) and that diffusion activation energies rapidly increase with increasing paraffin-chain length (Ruthven and Derrah, 1975).

Ion exchange

Despite the wide occurrence and ready availability of high purity erionites, there have been few systematic studies of its ion-exchange properties. Ames (1962, 1963, 1964, 1965) established the selectivity sequence Cs > Sr > K > Na and indicated that erionite was better than clinoptilolite for removing Cs from solutions containing Na and K, although Cs-exchange did not go to completion. Preparative exchanges by Eberly (1964) showed difficulty in exchanging out K, Mg, and Ca from the original zeolite.

Catalysis

Erionite has been extensively investigated by researchers at several major oil companies as a "shape selective" hydrocarbon-conversion catalyst, particularly for the hydrocracking of n-paraffins to yield C_1 to C_3 products over transition- and noble-metal erionites. Earlier work confused offretite with erionite (Chen and Garwood, 1970; Miale *et al.*, 1966), but demonstrated that hydrogen- and nickel-exchanged zeolites are very active cracking, hydrocracking, and oxidation catalysts that are highly selective for n-paraffins. The selectivity is dependent on the cage size, giving highly selective cracking of $<C_9$ and $>C_{12}$ n-paraffins but not C_{10}–C_{12} n-paraffins (Chen *et al.*, 1969; Chen and Garwood, 1973). Calcium and sodium forms of erionite are inactive catalysts (Chen and Garwood, 1970; Piguzova *et al.*, 1968). Synthetic erionite is as active as the natural zeolite (Robson *et al.*, 1971) but tends to be less stable

(Chen, 1972), probably because of its lower silica to alumina ratio. Erionite is the only natural zeolite that is presently in use as a commercial catalyst. It is used in Mobil's "Selectoforming" process in which a noble-metal-impregnated erionite selectively hydrocracks the C_5 to C_9 n-paraffins to C_1 to C_3 components, leaving unaffected the i-paraffins and aromatics in the feed. In this way, the octane rating of the feed is significantly increased, and LPG is obtained as a side product (Burd and Maziuk, 1972). Addition of erionite to faujasite-promoted cracking catalysts gives improved coke selectivity and a higher yield of light gas products (Scherzer and Albers, 1975). Kimberlin and Voorhies (1973) claim advantages for similar co-promoted hydrocracking catalysts containing Pd. H-, Ni-, and Mg-exchanged erionites are very active catalysts for the reduction of NO with ammonia (Atroshchenko *et al.*, 1974), and reduced tungsten-exchanged erionites are reported to be good hydrocracking catalysts (Neel and Joachimsmann, 1970). The effects of adding erionite to a conventional cracking catalyst are shown in Table 3.

FERRIERITE

Chemistry

The structure of ferrierite has been determined by Vaughan (1966) and Kerr (1966). The main features are elliptical 10-ring channels of dimensions 5.4 × 4.2 Å parallel to the *c*-axis and 8-ring channels (4.7 × 3.4 Å) parallel to the *c*-axis. Ferrierite became a forgotten mineral after its discovery by Graham in 1918, until new locations were described in the late 1960s in Italy (Alietti *et al.*, 1967), California (Wise *et al.*, 1969), Nevada (Sand and Regis, 1966; Regis, 1970), Japan (Yajima *et al.*, 1971), and Austria (Zirkl, 1973). Wise and Tschernich (1976) recently analyzed ferrierite from five new localities in North America and reviewed the previous published analyses. Si/Al ratios are generally in the range 3.2 and 6.2, with a tendency for the zeolite to be selective for potassium and magnesium. In the large Lovelock, Nevada, deposit, where beds of ferrierite are associated with thick beds of pure montmorillonite and amorphous ash in a vertical sequence, high magnesium contents tend to be characteristic of the montmorillonite, and high potassium characteristic of the ferrierite and ash.

Sorption

There are no published data on the sorption properties of natural ferrierite. The sorption properties of H^+-, Li^+-, Na^+-, K^+-, Ca^{2+}-, and Sr^{2+}-exchanged synthetic ferrierites have been investigated by Barrer and Lee (1969) using a range of sorbates. Acid-treated materials have the best sorption capacities and rapidly sorb pyrrole and thiophene; but xylenes are sorbed only slowly and in small quantities. At the other extreme, K^+-ferrierite does not sorb molecules larger than oxygen. More recent work on synthetic ferrierites (Kibby *et al.*, 1974) shows a sharp decrease in n-paraffin sorption at C_5, falling almost to zero at C_{10}. Iso-paraffins were excluded and ferrierite behaves essentially as a 5 Å molecular sieve. Work by the author on acid-treated Lovelock ferrierite demonstrated that the best materials develop

TABLE 3. Catalytic activity* of natural zeolites as co-promoters

	RE–Y	H^+–Ferrierite	RE–Y + H^+ Ferrierite	H-Mordenite	RE–Y + H^+–Mordenite	RE–Y + H–Erionite
Conversion, V%	69	24	71	27	68	68
H_2, W%	0.05	0.05	0.04	0.04	0.05	0.03
$C_3^=$, V%	6.5	3.3	8.3	4.0	7.3	6.2
C_3 total, V%	7.7	3.5	9.8	4.4	9.1	7.2
$C_4^=$, V%	4.4	4.2	5.1	4.7	4.3	2.7
i-C_4, V%	7.1	2.0	8.4	2.7	8.5	4.7
C_4 total, V%	12.7	6.6	14.6	7.9	14.2	8.1
C_5^+ gasoline, V%	56.8	18.2	56.7	20.5	54.7	60.2
C_4^+ gasoline, V%	69.5	24.8	71.3	28.3	69.0	68.3
C on catalyst, W%	0.4	0.1	0.3	0.1	0.3	0.4
C on feed, W %	2.4	0.6	1.8	0.8	1.9	2.3
Conversion/Coke	28.7	39.5	38.7	32.8	35.8	30.0

*Conditions: Microactivity test using West Texas Devonian Gas Oil, 900°F, 16 WHSV, 6 CO; 10% total zeolite in a semi-synthetic matrix (mixtures contain 4:1 RE–Y/natural zeolite ratio). Data after Scherzer, *et al.* (1975) and Scherzer and Albers (1975).

nitrogen surface areas up to 230 m^2/g. Isotherm and rate studies with N_2, CO_2, and Ar gave sorption values of 11.5 wt.% (−195.8°C), 15.0 wt.% (−78°C), and 6.0 wt.% (−78°C), respectively, at 500 mmHg pressure. Sorption rates were high: 7.5 wt.% benzene was sorbed at 100 mmHg pressure and 25°C. Ferrierite from Lovelock, Nevada, has good CO-sorption properties in the presence of moist air (Vaughan and Montgomery, 1974). The effects of cations on oxygen–nitrogen separation in natural ferrierite are shown in Fig. 1.

Ion exchange

There are no literature reports of systematic ion-exchange studies of ferrierite. Analyses of natural ferrierites indicate a particular selectivity for potassium and possibly magnesium. Laboratory syntheses may indicate a preference for strontium (Barrer and Marshall, 1964, 1965) and lithium (Vaughan and Edwards, 1976) on the basis of ease of formation in systems containing these cations. Experiments by Hawkins (1967) confirmed a selectivity of Sr over Ca.

Catalysis

Little catalytic work on ferrierite has been reported. As a co-promoter with faujasite, acid-treated natural ferrierite enhances the formation of olefins and light gases and improves coke selectivity in catalytic cracking processes (Scherzer *et al.*, 1975) (see Table 3). Giannetti and Perrotta (1975) reported that nickel and palladium supported on acid-leached synthetic ferrierite constitute excellent shape-selective hydrocracking catalysts. These seem to be similar to erionite "Selectoforming" catalysts. In straight cracking, in contrast to hydrocracking, ferrierite shows no particular selectivity for n-paraffin cracking unless it has first been pretreated to poison the external crystal surfaces or partially block the channels (Kirby *et al.*, 1974). Such treatments greatly increase the n-paraffin cracking selectivity over i-paraffin cracking. The present volume includes a paper by Chi *et al.* (1978) that shows acid-treated natural ferrierites to be good Claus tail-gas catalysts.

MORDENITE

Chemistry

The structure of mordenite was determined by Meier (1961) and consists of chains of 5-membered rings parallel to the *c*-axis connected through double 8-membered rings in the *a*–*c*-plane. These are joined to give 12-ring channels in the *c*-direction, with double 8-ring prisms lining the channel in the *a*-*c*-plane. As adjacent 8-ring prisms are offset about 3.7 Å in the *c*-direction (*c*/2), the path between the 12-ring channels is a very distorted 8-ring having a minimum free diameter of about 2.8 Å. Diffusion is therefore limited to the smallest gas molecules or cations. However, as about half the cations are located in these "side pockets", diffusion between the channels can occur only in the hydrogen-exchanged form of the zeolite. Other cations are located in the main channel.

Early studies of mordenite showed that access to the structure was limited to molecules having a diameter of ≤4.2 Å. Meier explained this by proposing that faulting occurs parallel to the 001 plane in the 010 direction. Such displacement replaces the 12-ring diffusion path by an 8-ring path and produces a structure consistent with much of the sorption data of natural mordenite. However, some synthetic mordenites have properties consistent with the unfaulted mordenite structure (Keough and Sand, 1961) and are now available under the trade name of Zeolon. Subsequent work by Frilette and Rubin (1965) showed that natural mordenite digested in 3 N HCl solution develops sorption and catalytic properties consistent with the 12-ring mordenite structure. The conclusion, therefore, is that the 12-ring channel of natural mordenite is blocked by acid-soluble "detritus"; as mordenites are formed in high silica environments, this is most probably SiO_2, Fe_2O_3, or other materials that may have been occluded in the mordenite channel during crystallization.

The compositions of mordenites fall within a narrow Si/Al range of 4.4 to 5.5 and generally have Ca and Na as the major cations, although K and Mg are not uncommon. The high Si/Al ratios make this one of the most thermally and acid-stable zeolites known, and like clinoptilolite (Barrer and Makki, 1964), mordenite can be readily decationated and dealuminated by digestion in acidic solutions (Barrer and Peterson, 1964). Such treatments may yield highly crystalline mordenite with Si/Al ratios greater than 50 which are of particular interest as catalysts and catalytic promoters. Further improvements in accessibility to the pores can be effected by a caustic leaching which splits apart the needle-like mordenite crystals and at the same time reduces their length to width ratio (Young, 1968). Such a treatment significantly improves the diffusion rates, and hence, the catalytic properties of the zeolite.

To differentiate between mordenites having obstructed 12-membered ring channels and those having clear channels, Sand (1968) proposed the generally accepted terms "small-port" and "large-port" mordenites. The former sorb a negligible amount of benzene or cyclohexane, while the latter sorb 6–7 wt.% of these molecules. Although all natural mordenites are small port, the synthetic varieties may be either small or large port depending on synthesis conditions and washing procedures. Many small-port varieties can be converted to the large-port form by acid treatments or ammonium exchange. Recent work on ammonium mordenites indicates that these dealuminate like acid-treated mordenites (Barrer and Klinowski, 1974).

Prior to the discovery of large-port mordenite by Sand in 1961, all the reported properties of mordenite (or ptilolite as it was often referred to) apply to the small-port variety. Subsequent to 1961 most, though not all, investigations have been concerned with the properties of the large-port mordenites since other large port zeolites have high catalytic activity (Kimberlin and Gladrow, 1961; Plank and Rozinski, 1964).

Sorption

Early sorption experiments on mordenite showed that the mineral rapidly sorbs NH_3 and CO_2 (Sameshima and Hemmi, 1934; Sameshima and Morita, 1935) and O_2, N_2, and Ar, but that sorption of CH_4 and C_2H_6 is very slow (Barrer, 1944). Barrer also showed that small polar molecules are readily occluded (NH_3, Cl_2, HCl, CS_2, CH_3, CH_3CN, CH_3NH_2, and CH_3OH), that larger molecules are excluded (I_2, Br_2, n-paraffins, iso-paraffins, and aromatic hydrocarbons), and that Ca- and Ba-exchanged mordenites have higher capacities than the original (Ca, K, Na) mineral. Later work demonstrated that a synthetic sodium mordenite can separate the components of various mixtures of permanent gases ($Ar + N_2$, $Ar + O_2$, $N_2 + H_2$, Ar + Ne, $Ar + H_2$) and that small amounts of pre-sorbed water greatly modify the gas-diffusion rates (Barrer and Robins, 1953). Subsequent work using partial sorption of polar sorbates (H_2O, NH_3, CH_3NH_2) showed that the capacity and rates of sorption of permanent gases in mordenite decrease linearly with the amount of modifier, until a critical value of the latter is attained, at which time sorption of the permanent gas rapidly decreases (Barrer and Rees, 1954 a, b). Studies of diffusion of permanent gases in various cation-exchanged forms of mordenite showed that the diffusion coefficients increased in the sequence Ca < K < Ba < Na < Li (Barrer and Brook, 1953). Control of the cationic composition and pre- and post-adsorption of polar molecules are all important methods of gaseous diffusion control in zeolites.

Studies of permanent gas sorption by large-port mordenites was first studied by Barrer and Peterson (1964), who showed that the sorption capacity is greatly increased in going from the Na to the H form; and that, although the small side pockets are partially available to N_2 in the Na-form, they are completely filled by N_2 and Ar in the H-form. The sorption capacity of the H-form for CO_2 is only about 12% greater than for the Na-form, although the heat of adsorption for the former system is much smaller than for the latter (Barrer and Coughlan, 1968 a, b). Studies of Kr, CO_2, and H_2O sorption in progressively dealuminated H-mordenites show that the zeolite sorption capacities remain constant, but the heats of sorption decrease with decreasing aluminum content (Barrer and Murphy, 1970).

The high silica to alumina ratio of mordenites make them interesting sorbents for acidic gases. Na- and H-mordenites are good sorbents for HCl and NH_3, respectively, at temperatures between 190° and 320°C (such sorption takes place by vapor-phase cation-exchange mechanism – $Na^+ + HCl \rightarrow H^+ + NaCl\downarrow$ and $H^+ + NH_3 \rightarrow NH_4{}^+$), and for NH_4Cl vapor (Barrer and Kanellopoulos, 1970 a, b). Several investigations have shown that mordenites have the acid stability, selectivity, and capacity to be practical sorbents for SO_2 and SO_3 from stack gases (Vinnikov *et al.*, 1969, 1973; Tamboli and Sand, 1970; Joubert and Zwiebel, 1971; Ma and Mancel, 1972; Roux *et al.*, 1973). Relatively little work has been done on the sorption of nitrogen oxides, although the available data indicate that mordenite may be a viable NO_x adsorbent (Krasny and Musin, 1969; Ma and Mancel, 1972). Mordenite and chabazite have been successfully used for sorption of NO_x and have the advantage of catalyzing the oxidation of NO to NO_2 (Barrer and Addison, 1958). NO_2 is recovered as dilute nitric acid and recycled to the process. The natural zeolites erionite, chabazite, and mordenite all have the capacity and acid-stability needed for NO_x recovery.

The commercial availability of synthetic large-port mordenite was a major stimulus to hydrocarbon-sorption research on mordenite. Eberly (1963) showed that 78% of the theoretical sorption capacities can be achieved for n-hexane, cyclohexane, and benzene and that all three molecules are rapidly sorbed into Na-mordenite at temperatures between 160° and 260°C. In contrast, H-mordenite sorbs these gases slowly and irreversibly, with n-hexane (or breakdown products of n-hexane) seemingly penetrating into the small side pockets lining the main channel. Even at low temperatures, the differences in catalytic activity between Na- and H-mordenites are sufficient to affect the sorption properties of the latter. At lower temperatures ($\lesssim$25°) Na-mordenite readily sorbs permanent gases up to ~4 Å diameter (Kr), but sorption of n-paraffins is slow and incomplete. In contrast, H-mordenite sorbs benzene and neo-pentane to capacity, and the side pockets are accessible to molecules smaller than n-butane (Barrer and Peterson, 1964). Satterfield and Frabetti (1967) reported rapid diffusion rates for C_1–C_4 hydrocarbons in a Na–H mordenite between 25° and 140°C, but observed significant variability in rates and capacities as functions of synthesis and preparation conditions. Na-mordenite can be used to separate isomers of xylene (Chen and Lucki, 1972) as can Na-ferrierite.

Dealumination of mordenite reduces the diffusion barriers for both paraffin and aromatic molecules (Beecher *et al.*, 1968; Eberly and Kimberlin, 1970). Nashimura and Takahashi (1971) evaluated the effects of Na, Ca, and NH_4 cations on the sorption properties of synthetic and natural mordenites using n-butane, benzene, and cumene. The natural Na- and Ca-exchanged mordenites and synthetic Ca-exchanged mordenite are small-port varieties. When the natural Na-mordenite sample was more than 40% exchanged by NH_4 ions, the mineral developed "large-port" properties: the ammonium-exchanged natural material had a hydrocarbon-sorption capacity almost identical to that of the synthetic ammonium mordenite.

Ion exchange

Ion exchange of alkali and alkaline earth elements has recently been reviewed, and the exchange isotherms re-evaluated for the Na and NH_4 forms of a synthetic mordenite having an Si/Al ratio of 5.26 (Barrer and Klinowski, 1974). The mordenite affinity sequence was found to be Cs > K > NH_4 > Na > Ba > Li, but for Ca and Sr, only 60% exchange could be effected. Similar results were reported for a natural mordenite having an identical Si/Al ratio (Rees and Rao, 1966; Rao and Rees, 1966). Barrer and Townsend (1976a) studied the exchange of transition metals in Na- and NH_4-mordenite in aqueous solutions between pH 4 and 7 and gave the selectivity sequence $NH_4^+ > Na^+ > Mn^{2+} > Cu^{2+} > Co^{2+} \sim Zn^{2+} > Ni^{2+}$. None of the transition metals exchanged beyond the 50% level, and only cations located in the main mordenite channel were exchanged. In contrast, the ammoniated complexes of Cu^{2+}, Zn^{2+}, and Co^{3+} are strongly sorbed by ammonium mordenite in ammoniacal solutions and give rectangular isotherms. Although the selectivities are increased by a factor of 100, exchange is again limited to the 50% level (Barrer and Townsend, 1976b).

It is interesting to note that molecular sieving with large transition-metal ammino complexes is a useful method for recovering valuable metal components from spent catalysts. Mordenite has a high selectivity for linear (argentous diammine complex) and square-planar (tetrammines of Pt and Pd) complexes in ammoniacal solutions containing octahedrally complexed transition metals (Fe, Ni, Co), allowing almost complete recovery of silver, platinum, and palladium from spent catalysts slurried in ammoniacal solutions (Aitken and Keen, 1971).

Catalysis

The catalytic properties of mordenite were reviewed by Burbidge *et al.*, in 1971, but since that time over eighty papers and patents have been published on the subject. These are in the areas of cracking, hydrocracking, isomerization, hydroisomerization, alkylation, hydrogenation, dehydrogenation, re-forming, disproportionation, hydrotreating, hydropolymerization, dehydration, polymerization, and transalkylation.

The main catalytic forms of mordenite are the hydrogen or dealuminated varieties, with or without reduced noble or transition metals. Since Barrer and Peterson (1964) first proposed the mordenite dealumination mechanism, numerous workers have used this as a means of "tailoring" the acidity of the zeolite to the reaction system. Details of the various methods of mordenite dealumination have been given by Belenkaya *et al.*, (1967, 1969), Chen and Smith (1970), Kranich *et al.* (1971), and Chang (1970). To achieve significant removal of lattice aluminum, several mineral-acid treatments may be necessary; weak acids either do not remove structural aluminum or do so very slowly (Belenkaya *et al.*). Gray and Cobb (1975) recently investigated several multivalent cation forms of mordenite for hydroisomerization and hydrocracking activity. They found that for both reactions with n-pentane the order of activity was: Be > La > Mg > Ca > Sr > Ba; after nickel was added and reduced, the activity order was reversed to Ba > Sr > Ca > La > Mg > Be. Product distributions changed as a function of the cation.

The main problem with mordenite catalysts results from the one-dimensional nature of the diffusion system. If reactants or products overcrack or polymerize to form materials that block the channels, diffusion rates fall because of a lack of alternate diffusion paths, and the reaction rapidly slows and becomes less selective. Eberly and Kimberlin (1970) noted such deactivation for cumene cracking over H-mordenite, and subsequent quantitative work on the same reaction showed that coke formation reduced the effective mordenite diffusivities by a factor of 2 (Butt *et al.*, 1975).

The only known commercial catalytic process using mordenite is the Shell hydro-isomerization process (Kouwenhoven and van Zijll Langhout, 1971). This process presumably uses a low sodium, noble metal-impregnated or exchanged mordenite to convert n-pentane to i-pentane. It is of some importance in view of the demand for octane numbers in the absence of tetraethyl lead in many automobile fuels (Kouwenhoven, 1973; Voorhies and Bryant, 1968). The effects of adding small amounts of mordenite to a conventional cracking catalyst as a co-promoter are noted in Table 3.

In addition, sodium-, hydrogen-mordenite (Pence and Thomas, 1974; Mays and Schwab, 1975), natural mordenite tuff (Tamura *et al.*, 1975), and Ni- and Mg-erionites (Atroshchenko *et al.*, 1974) appear to be good catalysts for the ammonia reduction of NO_x to water and hydrogen.

PHILLIPSITE

Chemistry

The structure of phillipsite was determined by Steinfink (1962) and refined by Rinaldi *et al.* (1974), who confirmed the isostructural nature of phillipsite (K-rich form) and harmotome (Ba-rich form). The structure consists of two 8-ring channels parallel to the *a* and *b* axes, having respective dimensions of 4.2 Å × 4.4 Å and 2.8 Å × 4.8 Å. There are two distinctly different cation positions – a large site occupied by K^+ or Ba^+ and a small site occupied by Na^+ or Ca^{2+} – which effectively block the channels in the structure. Recent compilations of phillipsite compositions by Sheppard *et al.* (1970) and Galli and Ghittoni (1972) show that the Si/Al ratio varies between 1.35 and 2.9, with the deep-sea phillipsites tending to be high-ratio materials; potassium is usually the major cation and is commonly associated with sodium or calcium.

Because of its ready availability in high purity, virtually all properties in the published literature were measured on a phillipsite from the Pine Valley deposit in central Nevada.

Sorption

Phillipsites contain between 8 and 22 wt.% water depending on their cation contents; the smaller the cation, the greater the water capacity (Hoss and Roy, 1960; Barrer and Munday, 1971). In nature, phillipsites contain 14–18 wt.% water, reflecting the preponderance of large potassium cations in the structure. At normal temperatures and pressures, sorption is limited to molecules having dimensions similar to the smallest window diameter in the structure – 3 Å. Regardless of earlier claims that phillipsite sorbs n-paraffins, such molecules are completely excluded from this zeolite framework (Peterson *et al.*, 1965). Sodium-rich natural phillipsite sorbs nitrogen at –78°C, but its capacity falls with increasing potassium exchange; extensive calcium exchange results in the exclusion of nitrogen (*ibid*). Barrer and Coughlan (1958 a,b) showed that CO_2 is readily sorbed by natural phillipsite with a saturation capacity of about 120 cm^3/g. Argon and krypton are sorbed in significant amounts by potassium- and rubidium-exchanged phillipsites under high gas pressures, and argon has a moderate rate of diffusion at 0°C (Barrer and Vaughan, 1971). Munson (1973) also showed a California phillipsite to sorb about 1 wt.% methane at 25°C and 1000 psig. These results confirm that the main diffusion paths are distorted 8-rings, but imply that the large cations do not cause significant blocking of the channel windows during sorption.

Ion exchange

Studies of Na–K–Cs exchange and Na–Ca–Sr exchange in phillipsite leads to the selectivity sequence Cs > Na > Sr = Ca, with the sodium zeolite having a particularly high selectivity for cesium (Ames, 1964 a, b). Barrer and Munday (1971) determined the exchange isotherms for sodium phillipsite exchanged with alkali and alkaline-earth cations. These workers confirmed the selectivities found by Ames and further showed that phillipsite has particular selectivity for all the large cations – Ba^{2+}, Cs^+, Rb^+, and K^+. All except the barium-exchange isotherm were reversible. The Ba-selectivity of phillipsite is so large that this zeolite could be used to remove that cation completely from mixed cation solutions. For cation fractions in the zeolite (Az) below 0.2, the affinity sequence was found to be: Ba > Rb ~ K ~ Cs > Na >> Li; and for Az between 0.2 and 1, Ba > Cs > K > Rb > Na >> Li. The preponderance of high potassium phillipsites in nature is not surprising in view of these reported high potassium selectivities. In ammoniacal solutions, ammonium phillipsite has a high selectivity for cuprammine and zincammine ions at low concentrations, but the ion-exchange behavior is complicated by precipitation reactions at higher concentrations of transition metals (Barrer and Townsend, 1976).

CONCLUSIONS

The recent rapid escalation in the prices of raw materials places natural zeolites in an improving price/performance position compared to synthetic materials. Those same forces also stimulate the search for alternative raw materials for manufactured products, and in this regard natural zeolites and zeolitic tuff seem to offer some potential as a silica–alumina source for the manufacture of synthetic zeolites. They may have future potential if the prices of caustic, silicates and aluminates continue to rise. Studies on the reactivity of the readily available tuff, pumice, or volcanic ash materials have been reported by several authors. Sudo and Matsuoka (1959) synthesized sodalite and probably faujasite from volcanic tuff; Tsitsishvili made sodalite (1965) and phillipsite (1966); and Aiello *et al.* (1971) reported the formation of zeolite A, sodalite, phillipsite and faujasite types from pumice. Sand (1969) synthesized mordenite, and Vaughan and Edwards (1976) ferrierite. The unique advantage of many of these materials, particularly the last two, is that the raw material already has a composition approximating that of the products. In some cases one zeolite (clinoptilolite) has been used as the raw material for the synthesis of another zeolite (faujasite) (Robson, 1973; Riley and Maness, 1975). Meta-kaolin, a very reactive raw material source for zeolite synthesis (e.g., McDaniel and Dueker, 1971; Barrer and Mainwaring, 1972), is usually more reactive than the volcanic glasses or tuff, but in the preparation of high silica zeolites it would constitute a relatively small component in the slurry. This material is very sensitive to escalating fuel prices (its price has more than doubled in the last 2 years) and cannot compete with low-cost pumice.

REFERENCES

Addison, W. E. and Barrer, R. M. (1955) Sorption and reactivity of nitrous oxide and nitric oxide in crystalline and amorphous siliceous sorbents: *J. Chem. Soc.*, pp. 757-769.

Aiello, R., Collella, C., and Sersale, R. (1971) Zeolite formation from synthetic and natural glasses: *Adv. Chem. Ser.* **101**, Amer. Chem. Soc., pp. 51-61.

Aitken, R. W. and Kenn, I. M. (1971) Improvements related to the separation of metal ions: British Patent 1,235,881; Canadian Patent 873,263.

Alberti, A. (1975) The crystal structure of two clinoptilolites: *Tschermaks. Min. Pet. Mitt.* **22**, 25-37.

Alietti, A. (1972) Polymorphism and crystal chemistry of heulandites and clinoptilolites: *Amer. Mineral.* **57**, 1448-1472.

Alietti, A., Passaglia, E., and Scaini (1967) A new occurrence of ferrierite: *Amer. Mineral.* **52**, 1562-1563.

Ames, L. L., Jr. (1959) Zeolitic extraction of cesium from aqueous solutions: U.S. At. Energy Comm. HW.62607, p. 25.

Ames, L. L., Jr. (1960) The cation sieve properties of clinoptilolite. *Amer. Mineral.* **45**, 689-700.

Ames, L. L., Jr. (1961a) Cation sieve properties of the open zeolites chabazite, mordenite, erionite and clinoptilolite: *Amer. Mineral.* **46**, 1120-1131.

Ames, L. L., Jr. (1962a) Kinetics of cesium reactions with some inorganic cation exchange materials: *Amer. Mineral.* **47**, 1067-1078.

Ames, L. L., Jr. (1962b) Removal of cesium-137 from aqueous solutions by adsorption on clinoptilolite: U.S. Patent 3,017,242.

Ames, L. L., Jr. (1962c) Effect of base cation on the cesium kinetics of clinoptilolite: *Amer. Mineral.* **47**, 1310-1316.

Ames, L. L., Jr. (1963) Mass action relationships of some zeolites in the region of high competing ion concentrations: *Amer. Mineral.* **49**, 868-882.

Ames, L. L., Jr. (1964) Some zeolite equilibria with alkali metal cations: *Amer. Mineral.* **49**, 127-145.

Ames, L. L., Jr. (1965) Self-diffusion of some cations in open zeolites: *Amer. Mineral.* **50**, 465-475.

Ames, L. L., Jr. (1967) Zeolitic removal of ammonium ions from agricultural and other wastewater: *Proc. 13th Pacific N.W. Industrial Waste Conf.*, Washington St. Univ., pp. 135-152.

Anon. (1976) P & G lowers phosphate content of Tide: *Chem. Week.* **118** (13), 27.

Anurov, S. A., Keltsev, N. V., Smola, V. I. and Torocheshnikov, N. S. (1974) Adsorption of sulphur dioxide by natural zeolites: *Zh. Fiz. Khim.* **48**, 2124-2125.

Astakhov, V. A., Lukin, V. D. and Pishchulina, T. G. (1973) Adsorptive capability of zeolite suspensions during CS_2 sorption: *Ah. Prikl. Khim. (Leningrad)* **46**, 675-678.

Atroshchenko, A. V., Tsitlin, A. N., Kirkach, L. I. and Karaputina, G. I. (1974) Catalytic activity of zeolites during the reduction of nitric oxides with ammonia: *Izv. Vyssh. Ucheb. Zaved., Khim. Khim. Technol.* **17**, 724-727.

Balgord, W. D. and Roy, R. (1971) Crystal chemical relationships in the analcime family: *Adv. Chem. Ser.* **101**, Amer. Chem. Soc., pp. 140-146.

Barrer, R. M. (1938) The sorption of polar and non-polar gases by zeolites: *Proc. Roy. Soc. Lond.* **167A**, 392-419.

Barrer, R. M. (1941b) Fractionating hydrocarbon mixtures such as butane and isopentane or heptane and toluene: U.S. Patent 2,306,610.

Barrer, R. M. (1944) Sorption by gmelenite and mordenite: *Trans. Faraday Soc.* **40**, 555-564.

Barrer, R. M. (1949b) Molecular sieve action of solids: *Quart. Rev. Chem. Soc.* **3**, 293-320.

Barrer, R. M. (1950) Ion exchange and ion sieve processes in crystalline zeolites: *J. Chem. Soc.*, pp. 2342-2350.

Barrer, R. M. (1968) Some aspects of molecular sieve science and technology: *Chem. Ind. (Lond.)*, pp. 1203-1213.

Barrer, R. M. and Addison, W. E. (1958) Disproportionation of nitric oxide to nitrous oxide using crystalline zeolites as catalysts: U.S. Patent 2,853,365.

Barrer, R. M., Bartholomew, R. F. and Rees, L. V. C. (1963) Ion exchange in porous crystals. Pt. I. Self- and exchange-diffusion of ions in chabazite: *J. Phys. Chem. Solids* **24**, 51-62.

Barrer, R. M., Bartholomew, R. F. and Rees, L. V. C. (1963) Ion exchange in porous crystals, Pt. II. The relationship between self- and exchange-diffusion coefficients: *J. Phys. Chem. Solids* **24**, 309-317.

Barrer, R. M. and Brook, D. W. (1953a) Sorption and reactivity of simple organic molecules in chabazite: *Trans. Faraday Soc.* **49**, 940-948.

Barrer, R. M. and Brook, D. W. (1953b) Molecular diffusion in chabazite, mordenite and levynite: *Trans. Faraday Soc.* **49**, 1049-1059.

Barrer, R. M. and Coughlan, B. (1968a) Influence of crystal structure of zeolitic carbon dioxide. 1. Isotherms and selectivity: *Molecular Sieves*, Soc. Chem. Ind., London, pp. 233-241.

Barrer, R. M. and Coughlan, B. (1968b) Molecular sieves derived from clinoptilolite by progressive removal of framework charge: characterization by sorption of CO_2 and Kr: *Molecular Sieves, Soc. Chem. Ind., London*, pp. 141-148.

Barrer, R. M. and Davies, J. A. (1971) Sorption in decationated zeolites. II. Simple paraffins in H forms of chabazite and zeolite L: *Proc. Roy. Soc. Lond.* **322A**, 1-19.

Barrer, R. M., Davies, J. A. and Rees, L. V. C. (1969a) Thermodynamics and thermochemistry of cation exchange in chabazite: *J. Inorg. Nucl. Chem.* **31**, 222.

Barrer, R. M., Davies, J. A. and Rees, L. V. C. (1969b) Comparison of the ion exchange properties of zeolites X and Y: *J. Inorg. Nucl. Chem.* **31**, 2599-2609.

Barrer, R. M. and Falconer, J. D. (1956) Ion exchange in felspathoids as a solid state reaction: *Proc. Roy. Soc. Lond.* **236A**, 227-240.

Barrer, R. M. and Galabova, I. M. (1973) Ion exchanged forms of zeolite L, erionite and offretite and sorption of inert gases: *Adv. Chem. Ser.* **121**, Amer. Chem. Soc., pp. 356-373.

Barrer, R. M. and Hinds, L. (1953) Ion exchange in crystals of analcite and leucite: *J. Chem. Soc.*, pp. 1879-1888.

Barrer, R. M. and Ibbitson, D. A. (1944a) Occlusion of hydrocarbons by chabazite and analcite: *Trans. Faraday Soc.* **40**, 195-206.

Barrer, R. M. and Ibbitson, D. A. (1944b) Kinetics of formation of zeolitic solids solutions: *Trans. Faraday Soc.* **40**, 206-216.

Barrer, R. M. and Kanellopoulos, A. G. (1970a) Sorption of ammonium chloride vapor on zeolites. Part 2: Mixture of hydrogen chloride and ammonia: *J. Chem. Soc.* (A), 775-782.

Barrer, R. M. and Kerr, I. S. (1959) Crystalline channels in levynite and some related zeolites: *Trans. Faraday Soc.* **55**, 1915-1923.

Barrer, R. M. and Klinowski, J. (1972) Influence of framework charge density on ion exchange properties of zeolites: *J. Chem. Soc. Faraday Trans.* 1. **68**, 1957-1963.

Barrer, R. M. and Klinowski, J. (1974) Ion exchange in mordenite: *J. Chem. Soc. Faraday Trans.* 1. **70**, 2362-2367.

Barrer, R. M. and Langley, D. A. (1958a) Reactions and stability of chabazite like phases. Pt. I: Ion exchanged forms of natural chabazite: *J. Chem. Soc.*, pp. 3804-3811.

Barrer, R. M. and Langley, D. A. (1958b) Reactions and stability of chabazite like phases. Pt. II: Ion exchanged forms of some synthetic species: *J. Chem. Soc.*, pp. 3811-3816.

Barrer, R. M. and Langley, D. A. (1958c) Reactions and stability of chabazite like phases. Pt. III: Intracrystalline water: *J. Chem. Soc.*, pp. 3817-3824.

Barrer, R. M. and Lee, J. A. (1969) Synthetic ferrierites as sorbents and molecular sieves: *J. Colloid Interfac. Sci.* **30**, 111-119.

Barrer, R. M. and Makki, M. B. (1964) Molecular sieve sorbents from clinoptilolite: *Can. J. Chem.* **42**, 1481-1487.

Barrer, R. M. and Mainwaring, D. E. (1972) Hydrothermal transformations of metakaolin in potassium hydroxide: *J. Chem. Soc.* **A**, 1254-1259.

Barrer, R. M. and Marshall, D. J. (1964) Synthetic strontium alumino silicates: *J. Chem. Soc.*, pp. 485-497.

Barrer, R. M. and Marshall, D. J. (1965) Synthetic zeolites related to ferrierite and yugawaralite: *Amer. Mineral.* **50**, 484-489.

Barrer, R. M. and Munday, B. M. (1971) Cation exchange reactions of a sedimentary phillipsite: *J. Chem. Soc.* **A**, 2904-2909.

Barrer, R. M. and Murphy, E. V. T. (1970) Influence of decationation and dealumination on sorption by mordenite and clinoptilolite: *J. Chem. Soc.* **A**, 2506-2514.

Barrer, R. M., Papadopoulos, R. and Ramsay, J. D. F. (1972) Sorption of krypton and xenon at high temperatures and pressures. II. Comparison and analysis: *Proc. Roy. Soc. Lond.* **326A**, 331-345.

Barrer, R. M. and Peterson, D. L. (1964a) Intracrystalline sorption by synthetic mordenites: *Proc. Roy. Soc. Lond.* **280A**, 466-485.

Barrer, R. M. and Peterson, D. L. (1964b) Kinetics of n-paraffin sorption in the natural zeolite erionite: *J. Phys. Chem.* **68**, 3427-3429.

Barrer, R. M. and Rees, L. V. C. (1954a) Sorption of mixtures: Part 3 – Polar sorbates as modifiers of zeolitic crystals: *Trans. Faraday Soc.* **50**, 852-863.

Barrer, R. M. and Rees, L. V. C. (1954b) Sorption of mixtures: Part 4 – Molecular diffusion in crystals modified by polar sorbates: *Trans. Faraday Soc.* **50**, 989-999.

Barrer, R. M. and Rees, L. V. C. (1960) Self diffusion of alkali metals in analcime: *Trans. Faraday Soc.* **56**, 709-721.

Barrer, R. M. and Robins, A. M. (1953) Equilibria between binary gas mixtures and some zeolites: *Trans. Faraday Soc.* **49**, 929-939.

Barrer, R. M. and Sammon, D. C. (1955) Exchange equilibria in crystals of chabazite: *J. Chem. Soc.*, pp. 2838-2849.

Barrer, R. M. and Sammon, D. C. (1961) Alkali metal ion separation using zeolites: U.S. Patent 3,010,785.

Barrer, R. M. and Townsend, R. P. (1976a) Transition metal exchange in zeolites. Part I: Thermodynamics of exchange of hydrated Mn^{2+}, Co^{2+}, Ni^{2+}, Cu^{2+} and Zn^{2+} ions in ammonium mordenite: *J. Chem. Soc. Faraday Trans. 1.* **72**, 661-673.

Barrer, R. M. and Townsend, R. P. (1976b) Transition metal exchange in zeolites. Part II: Ammines of Co^{3+}, Cu^{2+} and Zn^{2+} in clinoptilolite, mordenite and phillipsite: *J. Chem. Soc. Faraday Trans.* **1** (in press).

Barrer, R. M. and Vaughan, D. E. W. (1969) Sorption and diffusion of rare gases in heulandite and stilbite: *Surf. Sci.* **14**, 77-92.

Barrer, R. M. and Vaughan, D. E. W. (1971a) Trapping and diffusion of rare-gases in phillipsite, K–M and other silicates: *Trans. Faraday Soc.* **67**, 2129-2136.

Barrer, R. M. and Vaughan, D. E. W. (1971b) Trapping of inert gases in sodalite and cancrinite crystals: *J. Phys. Chem. Solids* **32**, 731-743.

Barrer, R. M. and Wasilewski, S. (1961a) Occlusion of iodine in some porous crystals. Part 1: *Trans. Faraday Soc.* **57**, 1140-1152.

Barrer, R. M. and Wasilewski, S. (1961b) Occlusion of iodine in some porous crystals. Part 2: *Trans. Faraday Soc.* **57**, 1153-1163.

Barrer, R. M. and Whiteman (1967a) Intracrystalline sorption of sulphur and phosphorus by some porous crystals: *J. Chem. Soc.* **A**, pp. 13-18.

Barrer, R. M. and Whiteman (1967b) Mercury uptake in various cationic forms of zeolites: *J. Chem. Soc.* **A**, pp. 19-25.

Beecher, R., Voorhies, A., Jr. and Eberly, P., Jr. (1968) Hydro-cracking and diffusion of pure compounds on mordenite catalysts: *Ind. Eng. Chem., Prod. Res. Develop.* **7**, 203-209.

Belenkaya, I. M., Dubinin, M. M. and Krishtofori, I. I. (1967) Formation and properties of the hydrogen form of mordenite. Communication 1. *Izv. Akad. Nauk. SSR, Ser. Khim.*, pp. 2082-2087.

Belenkaya, I. M., Dubinin, M. M. and Krishtofori, I. I. (1969) Formation and properties of the hydrogen form of mordenite. Communication II. *Izv. Akad. Nauk. SSR, Ser. Khim.*, pp. 2071-2076.

Belitskii, I. A., Sheherbatyuk, N. E., Krasnova, L. V., Filizova, L. D. and Tyurina, E. F. (1973) Sorption properties of cation exchanged forms of natural clinoptilolite: *Isv. Sib. Otd. Akad. Nauk SSSR, Ser. Khim. Nauk.*, pp. 84-87 (C.A. *78*, 151939).

Bennett, J. M. and Gard, J. A. (1967) Non-identity of zeolites erionite and offretite: *Nature* **214**, 1005-1006.

Beyer, H. (1975) Kinetics of the catalytic cracking of propane, butane and isobutane on H-clinoptilolite: *Acta Chim. Acad. Sci. Hung.* **84**, 25-43.

Boles, J. R. (1972) Composition, optical properties, cell dimensions, and thermal stability of some heulandite group zeolites: *Amer. Mineral.* **57**, 1463-1493.

Breck, D. W. (1974) *Zeolite Molecular Sieves*, J. Wiley & Sons, New York, 771pp.

Burbidge, B. W., Keen, I. M. and Eyles, M. K. (1971) Physical and catalytic properties of the zeolite mordenite: *Adv. Chem. Ser.* **102**, Amer. Chem. Soc., pp. 400-409.

Burd, S. D. and Maziuk, J. (1972) Selectoforming gasoline and LPG: *Hydrocarbon Process* **51** (5), 97-102.

Butt, J. B., Delgado-Diaz, S. and Muno, W. E. (1975) Effects of coking on the transport properties of H-mordenite: *J. Catal.* **37**, 158-165.

Callahan, C. M. and Kay, M. A. (1966) Ion exchange in fused salts – I. A comparison of ion exchanger characteristics of fine mineral exchangers: *J. Inorg. Nucl. Chem.* **28**, 233-244.

Calleri, M. and Ferraris, F. (1964) Crystallographic structural studies of analcime: $NaAlSi_2O_6 \cdot H_2O$: *Atti. Acad. Sci. Torino. Cl. Sci. Fis. Mat. Nat.* **98**, 424-441.

Černy, P. and Povondra, P. (1969) Polycationic heulandite; comments on crystal chemistry and classification of heulandite and clinoptilolite: *Neues Jb. Miner. Mh.*, pp. 349-361.

Chelishchev, N. F., Berenshtein, T. A., Gribanova, N. K. and Martynova, N. S. (1973) Ion Exchange properties of clinoptilolite: *Dokl. Akad. Nauk. SSSR* **210**, 110-112.

Chelishchev, N. F., Marynova, N. S., Fakina, L. K. and Berenshtein, B. G. (1974) Ion exchange of heavy metals on clinoptilolite: *Dokl. Akad. Nauk. SSSR. Chem. Phys.* **217**, 1140-1141.

Chen, N. Y. (1972) Reactivation of deactivated zeolite catalysts: U.S. Patent 3,684,738.

Chen, N. Y. and Garwood, W. E. (1970) Selective catalyst: Canadian Patent 834,931.

Chen, N. Y. and Garwood, W. E. (1973) Molecular shape selective hydrocarbon conversion over erionite: *Adv. Chem. Ser.* **121**, Amer. Chem. Soc., pp. 575-582.

Chen, N. Y. and Smith, F. A. (1970) Dealuminization of mordenite: U.S. Patent 3,551,353.

Chi, C. W., Lee, H. and Vaughan, D. E. W. (1978) Natural ferrierite as a Claus tail-gas catalyst: *Zeolite '76*, Symposium Volume (in press).

Chumakov, V. A.,. Gorshkov, V. E., Tolmacher, A. M. and Fedorov, V. A. (1969) Possible use of synthetic zeolites for the separation of lithium isotopes: *Vestn. Mosk. Univ. Khim.* **24** (5), 22-24.

Clarke, F. W. (1914) The constitution of the natural silicates: *U.S. Geol. Survey Bull.* **588**, 128 pp.

Coombs, D. S., Ellis, A. J., Fyfe, W. S. and Taylor, A. M. (1959) The zeolite facies: *Geochim. Cosmochim. Acta* **17**, 53-107.

Coombs, D. S. (1970) Present status of the zeolite facies: *Adv. Chem. Ser.* **101**, Amer. Chem. Soc., pp. 317-327.

Deffreyes, K. S. (1959) Zeolites in sedimentary rocks: *J. Sediment. Petrol.* **29**, 602-609.

Dent, L. S. and Smith, J. V. (1958) Crystal structure of chabazite, a molecular sieve: *Nature* **181**, 1794-1795.

Detrekoy, E., Jacobs, P. A., Kallo, D. and Uytterhoeven, J. B. (1974) Nature and catalytic activity of hydroxyl groups in clinoptilolite: *J. Catal.* **32**, 442-451.

Detrekoy, E. and Kallo, D. (1973) Effect of heat treatment on catalytic properties of ion exchanged clinoptilolite derivatives: *Khim. Nauk. Bulg. Akad. Nauk.* **6**, 331-336.

Doelter, C. (1890) Ueber die künstliche Darstellung und die chemische Constitution einiger Zeolither: *Neues Jahrb.* **1**, 118-139.

Duffy, S. C. and Rees, L. V. C. (1974) Ion exchange in chabazite. Pt. I. Sodium and potassium tracer diffusion studies in pure and mixed ionic forms of chabazite: *J. Chem. Soc. Faraday Trans.* 1. **70**, 777-786.

Duffy, S. C. and Rees, L. V. C. (1975) Ion exchange in chabazite. Pt. II. Comparison of theoretical and experimental rates of exchange: *J. Chem. Soc. Faraday Trans.* 1. **71**, 602-609.

Eberly, P. E., Jr. (1963) Hydrocarbon adsorption studies at low pressures on the sodium and acid forms of synthetic mordenite: *J. Phys. Chem.* **67**, 2404-2411.

Eberly, P. E., Jr. (1964) Adsorption properties of naturally occurring erionite and its cationic exchanged forms: *Amer. Mineral.* **49**, 30-40.

Eberly, P. E., Jr. (1969) Adsorption of n-paraffins in erionite and 5A molecular sieve: *Ind. Eng. Chem. Prod. Res. Develop.* **8**, 140-144.

Eberly, P. E., Jr. and Kimberlin, C. N. Jr. (1970) Mordenite, aluminum deficient mordenite and faujasite catalysts in the cracking of cumene: *Ind. Engng Chem. Prod. Res. Develop.* **9**, 335-340.

Ermolenko, N. F., Shirinskaya, L. P. and Prokopovich, A. A. (1971) Effect of aluminum removal from erionite on its structural and thermal stability: *Zh. Fiz. Khim.* **45**, 128-130.

Ferraris, G., Jones, D. W. and Yerkess, J. (1972) Neutron diffraction study of the crystal structure of analcime: *Z. Kristallogr., Kristallogeom., Kristallphys., Kristallchem.* **135**, 240-252.

Fischer, K. F. and Meier, W. M. (1965) Kristallchemie der Zeolithe: *Fortschr. Mineral.* **42**, 50-86.

Foucher, W. D., Jr., Suggitt, R. M. and Sensel, E. E. (1971) Waxlike hydrocarbon conversion using clinoptilolite: U.S. Patent 3,565,788.

Frilette, V. J. and Rubin, M. K. (1965) Sorption and catalytic properties of natural mordenite: *J. Catal.* **4**, 310-311.

Frilette, V. J. and Weitz, P. B. (1966) Selectively cracking aliphatic hydrocarbons: U.S. Patent 3,257,311.

Galli, E. and Ghittoni, A. G. L. (1972) The crystal chemistry of phillipsites: *Amer. Mineral.* **57**, 1125-1145.

Garden, L. A. and Kington, G. L. (1956) Thermodynamic properties of oxygen sorbed in chabazite: *Trans. Faraday Soc.* **52**, 1397-1408.

Giannetti, J. P. and Perotta, A. J. (1975) Selective hydrocracking with ferrierite-based catalysts: *Ind. Engng Chem., Process Des. Dev.* **14**, 86-92.

Gray, J. A. and Cobb, J. T., Jr. (1975) Hydroisomerization and hydrocracking of normal pentane over various mordenite catalysts: *J. Catal.* **36**, 125-141.

Gupta, J. C., Ma, Y. H. and Sand, L. B. (1971) Diffusion of sulphur dioxide in synthetic mordenite and natural erionite: *AIChE Symp. Ser.*, No. 117, pp. 51-57.

Hawkins, D. B. and Short, H. L. (1965) Equations for the sorption of cesium and strontium on soil and clinoptilolite: *U.S. At. Energy Comm.*, IDO-12046, p. 33.

Hay, R. L. (1966) Zeolites and zeolitic reactions in sedimentary rocks: *Geol. Soc. Amer.*, Special Paper **85**, 130 pp.

Hersh, C. K. (1961) *Molecular Sieves*: Reinhold Publishing Co., New York, Chap. 4, pp. 29-52.

Hey, M. H. and Fejer, E. E. (1962) The identity of erionite and offretite: *Mineral. Mag.* **33**, 66-67.

Honstead, J. F., Nelson, J. L., Mercer, B. W. and Haney, W. A. (1960) Waste fixation on minerals: *U.S. At. Energy Comm.*, Rept. No. TID-7613.

Hoss, H. and Roy, R. (1960) Zeolite Studies III: On natural phillipsite, gismondite, chabazite and gmelenite: *Beitr. Mineral. u. Petrogr.* **7**, 389-408.

Joubert, J. I. and Zwiebel, I. (1971) Multicomponent adsorption isotherm on hydrogen mordenite: *Adv. Chem. Ser.* **102**, Amer. Chem. Soc., pp. 209-216.

Kallo, D., Detrekoy, E. and Schay, G. (1971) Catalytic isomerization of butene: *Kem. Kozlem.* **36**, 237-245.

Keough, A. H. and Sand, L. B. (1961) A new intracrystalline catalyst: *J. Amer. Chem. Soc.* **83**, 3536-3537.

Kerr, G. T. (1973) Hydrogen Y, ultrastable Y, and aluminum deficient zeolites: *Adv. Chem. Ser.* **121**, Amer. Chem. Soc., pp. 219-229.

Kerr, I. S. (1966) Structure of ferrierite: *Nature* **210**, 294-295.

Kibby, C. L., Perrotta, A. J. and Massoth, F. E. (1974) Composition and catalytic properties of synthetic ferrierite: *J. Catal.* **35**, 256-272.

Kikotailo, G. T., Sawruk, S. and Lawton, S. L. (1972) Direct observation of staking faults in the zeolite erionite: *Amer. Mineral.* **57**, 439-444.

Kimberlin, C. N. and Gladrow, E. M. (1961) Hydrocarbon conversion catalyst: U.S. Patent 2,971,903.

Kington, G. L. (1958) Energetics of sorption: *Proc. Symp. Colston Research Soc.* **10**, 59-64.

Kington, G. L. and MacLeod, A. C. (1959) Heats of sorption of gases in chabazite, energetic heterogeneity, and role of quadrupoles in sorption: *Trans. Faraday Soc.* **55**, 1799-1814.

Kinshofer, G. S., Bunzl, K. and Sansoni, B. (1973) Differential ion exchange in a zeolite: *Zeit. Phys. Chem.* **87**, 218-232.

Klopp, G., Papp, J., Forog, J. and Mandy, T. (1967) General

purpose catalyst support: Netherlands Patent Appl. 6,610,598.

Knoll, K. C. (1963) Removal of cesium from aqueous solutions by adsorption: U.S. Patent 3,097,920.

Kouwenhoven, H. W. (1973) Isomerization of paraffins: *Adv. Chem. Ser.* **121**, Amer. Chem. Soc., pp. 529-539.

Kouwenhoven, H. W. and van Zijll Langhout, W. C. (1971) Shell's hydroisomerization process: *Chem. Engng Prog.* **67**, 65-70.

Kranich, W. L., Ma, Y. H., Sand, L. B., Weiss, A. H. and Zwiebel, I. (1971) Properties of aluminum deficient large port mordenite: *Adv. Chem. Ser.* **101**, Amer. Chem. Soc., pp. 502-512.

Krasny, E. B. and Musin, T. G. (1969) Adsorption properties of dealuminated mordenite in relation to nitrogen oxides: *Trudy Kazan. Khim.-tech. Inst.* **40**, 258-265.

Kühl, G. H. (1973) Structural stability of partially ammonium exchanged zeolite A: *J. Catal.* **29**, 270-277.

Lorenz, M. G. (1968) N-paraffin separation with erionite: U.S. Patent 3,362,904.

Ma, Y. H. and Mancel, C. (1972) Diffusion studies of carbon dioxide, nitric oxide, nitrogen dioxide and sulphur dioxide on molecular sieve zeolites: *AIChE J.* **18**, 1148-1153.

Mason, B. and Sand, L. B. (1960) Clinoptilolite from Patagonia. The relationship between clinoptilolite and heulandite: *Amer. Mineral.* **45**, 341-350.

Matousova, O. (1968) Sorption of lithium isotopes on zeolites: *Sb. Vys. Sk. Chem. – Technol. Praze, Mineral.* **10** 113-117.

Mays, E. B. and Schwab, M. R. (1975) Elimination of NO_x fumes: *Chem. Engng.* **2**, 112.

McBain, J. W. (1926) A survey of the main principles of colloid science: *Colloid Symp. Monogr.* **4**, 7-18.

McDaniel, C. V. and Dueker, H. C. (1971) Process for preparing high silica faujasite: U.S. Patent 3,574,538.

Meier, W. M. (1961) The crystal structure of mordenite (ptilolite): *Z. Kristallogr.* **115**, 439-450.

Meier, W. M. (1968) Zeolite structures: *Molecular Sieves*, Soc. Chem. Ind., pp. 10-25.

Meier, W. M. and Olson, D. H. (1971) Zeolite frameworks: *Adv. Chem. Ser.* **101**, Amer. Chem. Soc., pp. 155-169.

Merkle, A. B. and Slaughter, M. (1968) Structure of heulandite: *Amer. Mineral.* **53**, 1120-1138.

Miale, J. N., Chen, N. Y. and Weisz, P. B. (1966) Catalysis by crystalline alumino-silicates. IV. Attainable catalytic cracking rate constants, and superactivity: *J. Catal.* **6**, 278-287.

Miale, J. N. and Weisz, P. B. (1968) Production of the hydrogen ion exchange form of a crystalline zeolite: U.S. Patent 3,375,064.

Milligan, W. W. and Weiser, H. B. (1937) The mechanism of the dehydration of zeolites: *J. Phys. Chem.* **41**, 1029-1040.

Milton, R. M. (1959a) Molecular sieve adsorbents: U.S. Patent 2,882,243.

Milton, R. M. (1959b) Molecular sieve adsorbents: U.S. Patent 2,882,244.

Minachev, K. M. and Isakov, Y. I. (1973) Catalytic properties of zeolites – a general review: *Adv. Chem. Ser.* **121**, *Amer. Chem. Soc.*, pp. 451-460.

Mumpton, F. A. (1960) Clinoptilolite redefined: *Amer. Mineral.* **45**, 351-369.

Munson, R. A. (1973) Properties of natural zeolites: U.S. Nat. Tech. Info. Serv. Rept. No. 221512/7.

Munson, R. A. and Sheppard, R. A. (1974) Natural zeolites; their properties, occurrence and uses: *Mineral. Sci. Engng* **6**, 19-34.

Nagyrymasu–Kiserleti–Intezet (1967) Procedure for the production of catalysts and catalyst supports: French Patent 1,487,811.

Nashimura, Y. and Takahashi, H. (1971) Effects of exchangeable cations on the adsorption character of mordenite: *Kolloid-Z.Z. Polym.* **245**, 415-419.

Neel, E. E. A. and Joachimsmann, G. F. (1970) Tungsten erionite catalyst for selective hydrocracking: Ger. Offen. 2,145,699.

Nowacki, W., Koyama, H. and Mladeck, M. H. (1958) Über die Kristall Structur des Zeolithes Chabazit: *Experienta* **14**, 396.

Papp, J., Kallo, D. and Schay, G. (1971) Hydrodemethylation of toluene on clinoptilolite: *J. Catal.* **23**, 168-182.

Passaglia, E. (1970) Crystal chemistry of chabazite: *Amer. Mineral.* **55**, 1278-1301.

Pence, D. T., Duce, F. A. and Maeck, W. J. (1971) Metal zeolites: Iodine adsorption studies: U.S. At. Energy Comm. IN-1455, 21 pp.

Pence, D. T. and Thomas, T. R. (1974) NO_x abatement at nuclear processing plants: *Proc. 2nd A.E.C. Environmental Protection Conf.* (U.S. Govt. Printing Office).

Peters, M. D. and Bose, R. J. (1975) Clinoptilolite – a physico-chemical approach to ammonia removal in hatchery and aquaculture water re-use systems: *Can. Fisheries and Marine Service*, Tech. Rept. No. 535, 12 pp.

Peterson, D. L., Helfferich, F. and Blytas, G. C. (1965) Sorption and ion exchange in sedimentary zeolites: *J. Phys. Chem. Solids* **26**, 823-884.

Piguzova, L. I., Romanovskii, B. V., Topchieva, K. V. and Ho-Shin Tong (1968) Catalytic properties of erionite: *Dokl. Akad. Nauk. SSSR* **179**, 385-388.

Plank, C. J. and Rosinski, E. J. (1964) Catalytic cracking of hydrocarbons with a crystalline zeolite catalyst composite: U.S. Patent 3,140,249, 53 pp.

Rao, A. and Rees, L. V. C. (1966) Kinetics of ion exchange in mordenite: *Trans. Faraday Soc.* **62**, 2505-2512.

Rees, L. V. C. and Rao, A. (1966) Self diffusion of various cations in natural mordenite: *Trans. Faraday Soc.* **62**, 2103-2110.

Regis, A. J. (1970) Occurrences of ferrierite in altered pyroclastics in central Nevada: *Geol. Soc. Amer. Ann. Mtg. Program Abstr.*, p. 661.

Riekert, L. (1970) Sorption, diffusion and catalytic reaction in zeolites: *Adv. Catal.* **21**, 21, 281-322.

Riley, K. L. and Maness, D. D. (1975) Crystalline zeolite synthesis pre-treatment: U.S. Patent 3,907,711.

Rinaldi, R., Pluth, J. J. and Smith, J. V. (1974) Zeolites of the phillipsite and harmotome group: *Acta Crystallogr.* **30B**, 2426-2433.

Robson, H. E., Hamner, G. P. and Arey, W. F. (1971) Synthetic erionite and selective hydrocracking: *Adv. Chem. Ser.* No. 102, Am. Chem. Soc., pp. 417-425.

Robson, H. E. (1973) Promotion of crystalline zeolite synthesis: U.S. Patent 3,733,390.

Ross, C. S. (1928) Sedimentary analcite: *Amer. Mineral.* **13**, 195-197.

Roux, A., Huang, A. A., May, Y. H. and Zwiebel, I. (1973) Sulphur dioxide adsorption on mordenites: *AIChE Symp. Ser.*, No. 134, **69**, 46-53.

Ruthven, D. M. and Derrah, R. I. (1975) A comparative study of the diffusion of C_2H_6, C_2H_4 and nC_5H_{12} in erionite and type A zeolite: *J. Colloid Interfac. Sci.* **52**, 397-403.

Saha, P. (1959) Geochemical and X-ray investigation of natural and synthetic analcites: *Amer. Mineral.* **44**, 300-313.

Sameshima, J. (1929) Sorption of gas by mineral. I. Heulandite and chabazite: *Bull. Chem. Soc. Jap.* **4**, 96-103.

Sameshima, J. and Hemmi, H. (1934) Sorption of gases by minerals. IV. *Bull. Chem. Soc. Jap.* 9, 27-41.

Sameshima, J. and Morita, H. (1935) Sorption of gases by minerals V. *Bull. Chem. Soc. Jap.* **10**, 485-494.

Sand, L. B. (1969) Synthetic mordenite and preparation thereof: U.S. Patent 3,436,174.

Sand, L. B. and Regis, A. J. (1966) Occurrence of ferrierite, Pershing Co., Nevada (Abst.): *Geol. Soc. Amer. Annual Mtgs. Program*, p. 189.

Sato, M. and Otani, S. (1969) Disproportionation of toluene: Ger. Offen. 1,813,251.

Satterfield, C. N. and Frabetti, A. J., Jr. (1967) Sorption and diffusion of gaseous hydrocarbons in synthetic mordenite: *AIChE J.* **13**, 731-738.

Scherzer, J. and Albers, E. W. (1975) Hydrocarbon cracking process with promoter mixtures: U.S. Patent 3,925,195.

Scherzer, J., Vaughan, D. E. W. and Albers, E. W. (1975) Hydrocarbon cracking catalysts with promoter mixtures: U.S. Patent 3,894,940.

Seeliger, R. (1921) The absorption of gases by crystals: *Physik. Z.* **22**, 563-568.

Shcherbatyuk, P. E. and Froer, E. F. (1973) Adsorption of acetone and carbon tetrachloride and their liquid mixtures by natural and modified clinoptilolite: *Zh. Fiz. Khim.* **47**, 2391-2393.

Shepard, A. O. and Starkey, H. C. (1964) Effect of cation exchange on the thermal behavior of heulandite and clinoptilolite: *U.S. Geol. Survey Prof. Paper* **475-D**, D89-D92.

Sheppard, R. A. and Gude, A. J. (1969) Chemical composition and physical properties of the related zeolites offretite and erionite: *Amer. Mineral.* **54**, 875-886.

Sheppard, R. A., Gude, A. J. and Griffin, J. J. (1970) Chemical composition and physical properties of phillipsites from the Pacific and Indian Oceans: *Amer. Mineral.* **55**, 2053-2062.

Smith, J. V. (1962) Crystal structures with a chabazite framework. I. Dehydrated Ca-chabazite: *Acta Crystallogr.* **15**, 835-845.

Smith, J. V., Knowles, R. C. and Rinaldi, F. (1964) Crystal structures with a chabazite framework. III. Hydrated Ca-chabazite at +20° and −150°C: *Acta Crystallogr.* **17**, 374-384.

Smith, J. V., Rinaldi, F. and Dent-Glasser, L. S. (1963) Crystal Structures with a chabazite framework II. Hydrated Ca-chabazite at room temperature: *Acta Crystallogr.* **16**, 45-53.

Sokolov, V. A., Torosheshnikov, N. S. and Kel'tsev, N. V. (1965) *Molecular Sieves and their Use*, Transl. G.W. Hemy, Publ. J. Crosfield & Sons, Ltd., Warrington, England, Chaps. 2 and 3.

Staples, L. W. and Gard, J. A. (1959) The fibrous zeolite erionite; its occurrence, unit cell, and structure: *Mineral. Mag.* **32**, 261-281.

Steinfink, H. (1962) The crystal structure of the zeolite phillipsite: *Acta Crystallogr.* **15**, 644-651.

Sudo, T. and Matsuoka, M. (1959) Artificial crystallization of volcanic glass to sodalite and zeolite structure: *Geochim. Cosmochim. Acta* **17**, 1-5.

Tamura, T., Fujita, N. and Matsuura, C. (1975) Reduction catalyst for removing nitrogen oxides from waste gases: German Offen. 2,446,006.

Taylor, W. H. (1930) The structure of analcite: *Z. Kristallogr.* **74**, 1-19.

Tiselius, A. (1934) Diffusion of water in a zeolite crystal: *Zeit. Phys. Chem. A* **169**, 425-458.

Tiselius, A. (1935) Sorption and diffusion of ammonia in analcime: *Zeit. Phys. Chem. A* **174**, 401-423.

Tsitsishvili, G. V., Krupennikora, A. Y., Gedzhadze, T. A. and Andronikashvili, T. G. (1965) Some characteristics of crystallization of obsidian and tuffaceous rhyolite: *Izv. Akad. Nauk. SSSR Neorg. Mat.* **1**, 285-287.

Tsitsishvili, G. V., Krupennikova, A. Y., Gedzhadze, T. A. and Andronikashvili, T. G. (1966) Formation of zeolites of the phillipsite type during crystallization of natural alumino silicates: *Izv. Akad. Nauk. SSSR Neorg. Mat.* **2**, 1306-1310.

Vaughan, D. E. W. (1974) Zeolites for air separation: German Offen. 2,347,574.

Vaughan, D. E. W. and Edwards, G. C. (1976) Synthesis of lithium ferrierite: U.S. Patent 3,966,883.

Vaughan, P. A. (1966) The crystal structure of the zeolite ferrierite: *Acta Crystallogr.* **21**, 983-990.

Venuto, P. B. and Landis, P. S. (1968) Organic catalysis over crystalline alumino-silicates: *Adv. Catal.* **18**, 259-371.

Venuto, P. B. (1971) Some perspectives on zeolite catalysis: *Adv. Chem. Ser.* **102**, Amer. Chem. Soc., pp. 260-283.

Vinnikov, L. I., Mukhlenov, I. P. and Lesokhin, I. G. (1969) Kinetics of the simultaneous adsorption of sulphur dioxide and water by a bed of adsorbent: *Massobemennye Protsessy Khim. Tekhnol.* No. 4, pp. 118-119.

Vinnikov, L. I., Mukhlenov, I. P., Lesokhin, I. G. and Buzanova, G. N. (1973) Adsorption of sulphur dioxide from moist air by industrial adsorbents: *Khim. Prom.* (Moscow) **49**, 125-128.

Voorhies, A., Jr. and Bryant, P. A. (1968) Hydroisomerization of n-pentane over a zeolite catalyst: *AIChE J.* **14**, 852-856.

Weeks, T. E., Angell, C. L. and Bolton, A. P. (1975) The thermo-chemical properties of ammonium exchanged erionite: *J. Catal.* **38**, 461-468.

Weigel, O. and Steinhof, E. (1925) Adsorption of organic liquid vapors by chabazite: *Z. Kristallogr.* **61**, 125-154.

Wilson, R. C. (1972) Clinoptilolite blends with shape selective catalyst: U.S. Patent 3,640,905.

Wise, W. S., Nokleberg, W. J. and Kokinos, M. (1969) Clinoptilolite and ferrierite from Agoura, California: *Amer. Mineral.* **54**, 887-895.

Wise, W. S. and Tschernich, R. W. (1976) Chemical composition of ferrierite: *Amer. Mineral.* **61**, 60-66.

Yajima, S., Nakamura, T. and Ishii, E. (1971) New occurrences of ferrierite: *Mineral. J.* **6**, 343-364.

Young, D. A. (1968) Caustic treated zeolites: U.S. Patent 3,374,182.

Zirkl, E. J. (1973) Ferrierit im Basalt von Weitendorf, Steiermark (Austria): *Neues Jb. Miner., Mh.* **11**, 524-528.

SORPTION AND DIFFUSION PROPERTIES OF NATURAL ZEOLITES

Y. H. MA and TING YUEH LEE

Chemical Engineering Department, Worcester Polytechnic Institute, Worcester, Massachusetts 01609, U.S.A.

Abstract – The sorption and diffusion properties of natural zeolites are reviewed. The application of the idealized model of localized sorption sites (Langmuir isotherm) is discussed. Experimental methods of constant volume, constant pressure, and gas chromatography to determine diffusion coefficients are discussed. Simplified mathematical results for the evaluation of the diffusion coefficient from these methods are given. Previous results on sorption and diffusion studies of natural zeolites are presented in tabular form.

INTRODUCTION

A large variation of sorptive and diffusive properties exists in natural zeolite minerals due to their diverse origins. It is, therefore, rather difficult to provide generalized descriptions of these properties. Nevertheless, it is possible to employ some fundamental approaches to describe equilibrium sorption and diffusion in natural zeolites.

The adsorption and diffusion properties of many natural zeolites are comparable to many of the synthetic varieties. Although the utilization of natural zeolites is growing, the scope of industrial applications is somewhat limited. This is due, in part, to the lack of information concerning their equilibrium and transport properties. It is believed that information on the equilibrium adsorption and intracrystalline diffusion properties of natural zeolites will provide not only fundamental understanding of transport mechanisms of gases in a well-defined pore structure but also practical knowledge essential for the possible industrial applications of natural zeolites. The present paper gives a general review on the equilibrium adsorption and diffusion properties of natural zeolites. It is hoped that the importance of a good understanding of these properties will become evident from the present review.

Adsorption

Equilibrium adsorption isotherms of zeolites exhibit Type I isotherms which are characterized by a long flat and nearly horizontal branch. A Type I isotherm may be interpreted from the assumptions that the adsorbed layer on the walls is only one molecule thick and that the plateau region of the isotherm corresponds to the completion of this monolayer. The evaporation-condensation mechanism developed by Langmuir can generally be assumed to apply. However, it should be noted that the extension of the simple mechanism developed for adsorption on an open external surface of a microporous adsorbent represents a very long extrapolation.

The Langmuir model can be developed by considering the dynamic equilibrium existing between the gas phase and the adsorbed phase. The surface of the solid was regarded by Langmuir as an array of adsorption sites, each site being capable of adsorbing one molecule. Thus, the number of molecules adsorbed per unit area per unit time is

$$n_1 = kP\theta_0 \quad (1)$$

where

k = $N/2\pi MRT)^{1/2}$,
N = Avogadro's number,
M = molecular weight,
R = gas constant,
T = temperature,
P = pressure,
θ_0 = fraction of sites unoccupied.

A similar expression can be obtained for the number of molecules desorbed per unit area per unit time, n_1',

$$n_1' = n_m \theta_1 \nu_1 e^{-E_1/RT} \tag{2}$$

where n_m = number of molecules adsorbed per unit area in a monolayer coverage,
θ_1 = fraction of sites occupied,
ν_1 = oscillation frequency of adsorbed molecules
E_1 = energy of desorption.

At dynamic equilibrium, $n_1 = n_1'$, so that

$$kP\theta_0 = n_m \theta_1 \nu_1 e^{-E_1/RT}. \tag{3}$$

If one uses the fact that $\theta_0 = 1-\theta_1$ and rearranges equation (3) one obtains

$$\theta_1 = \frac{kP}{kP+n_m \nu_1 e^{-E_1/RT}}$$

or

$$\theta_1 = \frac{X}{X_m} = \frac{KP}{1+KP} \tag{4}$$

where $K = \dfrac{k}{n_m \nu_1 e^{-E_1/RT}}$

and X and X_m are the adsorption capacities at pressure p and at the monolayer coverage, respectively.

It should be emphasized that the following assumptions are implicit in the above derivation: (1) the gas is assumed to be ideal, (2) only monolayer adsorption is possible, (3) the solid surface is energetically homogeneous, (4) interaction between adsorbed molecules is negligible, and (5) the adsorbed molecules are localized. The first two assumptions are reasonable for adsorption on natural zeolites. However, assumptions (3) and (4) are generally not applicable due to the presence of impurities and crystal imperfections such as cracks and dislocations. The assumption of localization of the molecules is generally acceptable. However, there are evidences that in certain systems adsorbed molecules may be mobile (Peterson *et al.*, 1965).

Experimentally, one can test the Langmuir equation by rearranging equation (4) into the form

$$\frac{P}{X} = \frac{1}{KX_m} + \frac{P}{X_m}. \tag{5}$$

A plot of P/X versus P should give a straight line whose slope is $1/X_m$ and intercept is $1/KX_m$. An example of such a plot is shown in Fig. 1 for adsorption of SO_2 on natural mordenite from Union Pass, Arizona (Ma *et al.*, 1977). An excellent straight line is obtained in this case. However, in some cases the plot can be distinctly curved even though the isotherm itself undoubtedly may be of Type I. Typical equilibrium isotherms are shown in Fig. 2 (Peterson *et al.*, 1965).

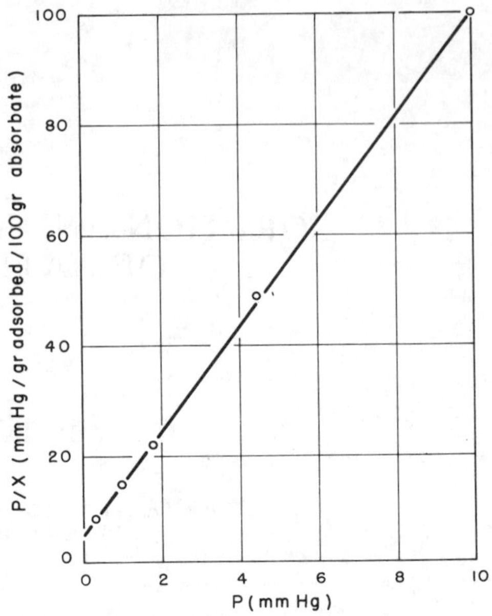

Fig. 1. Linear Langmuir plot for SO_2 adsorption on natural mordenite.

In addition to the Langmuir model described above, other models such as Volmer's equation, which considers the adsorbed molecules to be mobile, van der Waal's

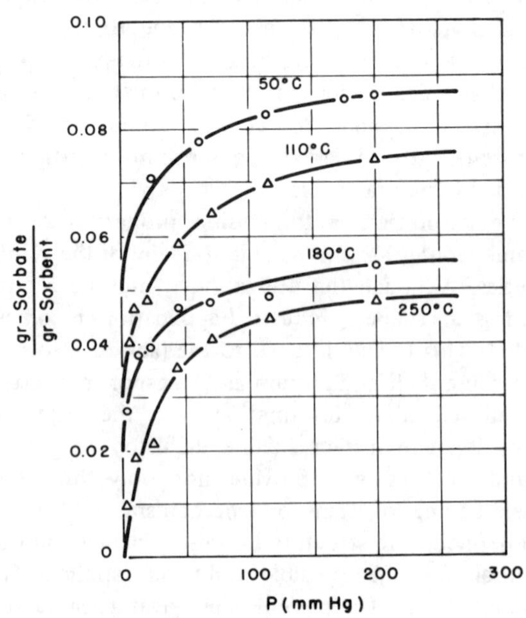

Fig. 2. n-Pentane isotherms for erionite (Peterson *et al.*, 1965).

equation, and the Virial equation also are frequently employed. The readers are referred to a standard text on physical adsorption. It should be noted that although equations such as the Langmuir isotherm can be derived theoretically, they should be regarded as empirical expressions for most cases.

Diffusion in natural zeolites

The diffusion coefficient in zeolites generally depends on (1) the specific adsorbate and zeolite system, (2) the shape and size of the diffusing molecules relative to the geometry of the zeolitic openings, (3) the polarity of the adsorbate molecules, (4) the presence of exchangeable cations in the zeolite framework, (5) defects in the structure such as stacking faults, (6) chemical and physical treatments of the zeolite, and (7) the presence of impurities. Moreover, the diffusion coefficient also depends on adsorbate concentration.

Diffusion in zeolites is generally assumed to follow the Fick's diffusion equation using chemical potential as the driving force (Barrer, 1971).

$$J_A = B_A C_A \frac{d\mu_A}{dx} \qquad (6)$$

where μ_A is the chemical potential,
$B_A C_A$ = L_A, the phenomenological coefficient of irreversible thermodynamics,
B_A is the intrinsic mobility of A.

Equation (6) may be expressed in terms of concentration gradient to give equation (7)

$$J_A = -RT(B_A)\left(\frac{d\ln a_A}{d\ln C_A}\right)\left(\frac{dC_A}{dX}\right) \qquad (7)$$

where

$$D_A = RT(B_A)\left(\frac{d\ln a_A}{d\ln C_A}\right) = RT(B_A)\left(\frac{d\ln a_A}{d\ln\theta}\right).$$

It can be seen from the above equation that D_A is a function of concentration. In the derivation of equation (7), it is assumed that the concentration dependence of the diffusion coefficient is due mainly to the non-linearity of the adsorption isotherms. This is true for certain zeolite-sorbate systems. However, in general, such a concentration dependence should be obtained experimentally.

It should be emphasized that no correlation can be found to predict satisfactorily the above-mentioned effects upon the intracrystalline diffusion coefficients. It is essential to obtain experimental data on diffusion coefficients to gain a better understanding of these factors.

EXPERIMENTAL METHODS

Most diffusion and sorption studies on zeolites used a constant-volume or a constant-pressure technique. A third powerful technique is the application of the principles of gas–solid chromatography to the measurements of effective diffusion coefficients. Discussion of these three commonly used methods follows. It should be noted that these methods can be used for the determination of both diffusion coefficients and equilibrium adsorption isotherms.

Constant-pressure system

If one assumes that diffusion in zeolites follows Fick's equation, then the following partial differential equation is obtained:

$$\frac{\partial C}{\partial t} = \mathrm{div}(D \ \mathrm{grad}\ C). \qquad (8)$$

Solution of equation (8) for a spherical particle with a uniform initial adsorbate concentration and a constant concentration at the external surface of the spherical particle of radius r_0 is given by Crank (1957):

$$\frac{M_t}{M_\infty} = 1 - \frac{6}{\pi^2} \sum_{n=1}^{\infty} e^{-Dn^2\pi^2 t/r_0^2}. \qquad (9)$$

For small time, the equation becomes

$$\frac{M_t}{M_\infty} = 6\left(\frac{Dt}{r_0^2}\right)^{1/2} \qquad (10)$$

Equation (10) shows that M_t/M_∞ is a function of $\sqrt{t}$. It is, therefore, most appropriate to plot experimental values of M_t/M_∞ versus $\sqrt{t}$. A typical plot is shown in Fig. 3 for diffusion of water in chabazite (Barrer and Fender, 1961). The diffusion coefficient can then be obtained from the slope of the initial portion of the curve. If the entire curve is used, equation (9) should be employed to fit the experimental data.

A typical experimental set-up for a constant pressure system is shown in Fig. 4 (Ma and Ho, 1974). The 8-liter total volume of the system is sufficiently large so that the system pressure remains essentially constant during

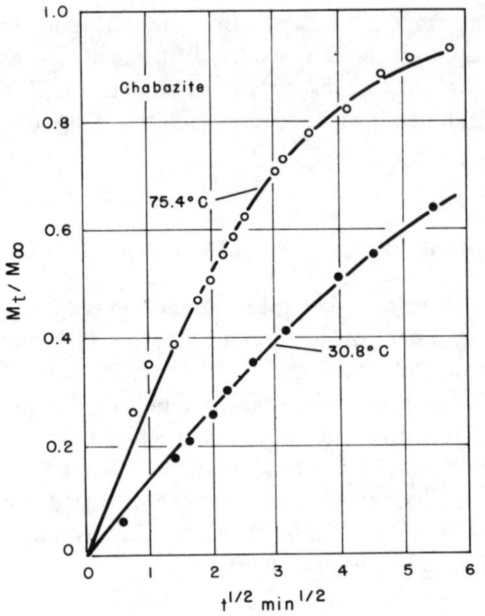

Fig. 3. M_t/M_∞ as a function of $\sqrt{t}$. Diffusion of H_2O in chabazite (Barrer and Fender, 1961).

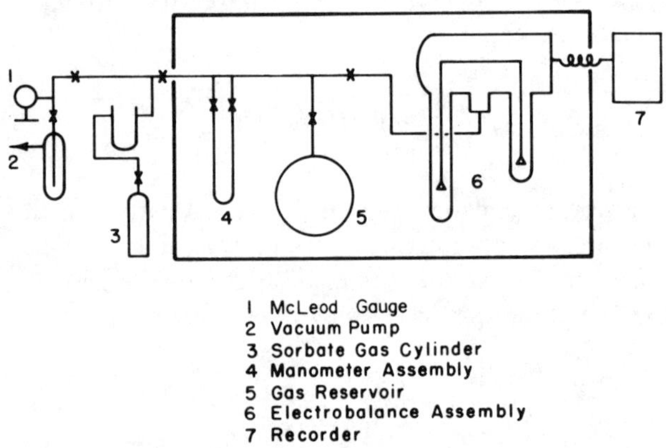

Fig. 4. Schematic diagram of a constant pressure sorption apparatus (Ma and Ho, 1974).

sorption. The sorption rate is followed by recording the weight change measured by a Cahn RG automatic recording electrobalance. Activation of sample is done under vacuum and at desired temperature (generally 300°C) for several hours. Sample injection is accomplished by opening a solenoid valve. Detailed operation of such a system may be found in Ma and Ho (1974).

Constant-volume system

Mathematics for a constant-volume system are somewhat more complicated than those for a constant-pressure system due to the time-varying boundary condition. As the system pressure varies, the adsorbate concentration at the external surface of the particle also changes with respect to time. The solution of equation (8) for a sphere of radius r_0 with a time-varying boundary condition is also given by Crank (1957).

$$\frac{M_t}{M_\infty} = 1 - \sum_{n=1}^{\infty} \frac{6X(\alpha+1)e^{-Dq_n^2 t/r_0^2}}{9+9\alpha+q_n^2\alpha^2} \tag{11}$$

where q_ns are the non-zero roots of

$$\tan q_n = \frac{3q_n}{3+\alpha q_n^2}$$

and

$$\alpha = \frac{V_g}{kV_s}.$$

For small time, equation (11) becomes

$$\frac{M_t}{M_\infty} = \frac{6}{r_0}\left(\frac{\alpha+1}{\alpha}\right)\left(\frac{Dt}{r_0^2}\right)^{1/2} \tag{12}$$

which indicates that M_t/M_∞ is proportional to $\sqrt{t}$ as in the constant pressure case. The procedure for the calculation of the diffusion coefficient is similar to that of a constant-pressure system.

A gravimetric apparatus is shown schematically in Fig. 5 (Roux *et al.*, 1973). Sorption rate is followed by recording the pressure change measured with an MKS Corporation 77H BARATRON, in conjunction with the reference side standard pressure arrangement. Detailed descriptions of the apparatus and procedure may be found in Roux *et al.* (1973).

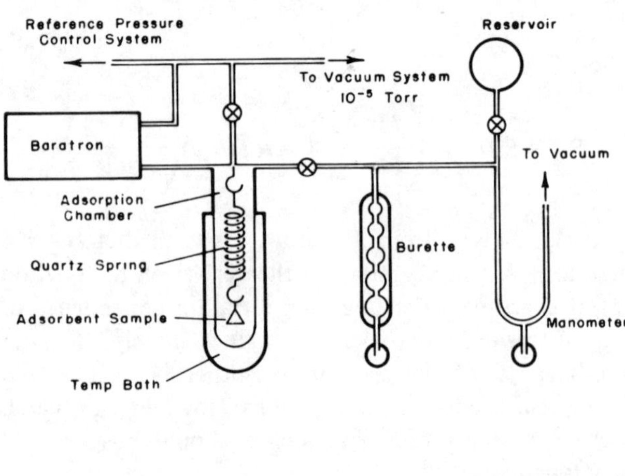

Fig. 5. Schematic of gravimetric apparatus (Roux *et al.*, 1973).

It should be noted that if only the equilibrium capacity is desired, the Cahn balance in the constant-pressure system may be replaced by a quartz spring. Measurements of the extension of the spring with a cathetometer provides the equilibrium adsorption capacity. A capillary manometer can be used to replace the BARATRON in the constant volume system for pressure measurement. Such set-ups for equilibrium adsorption measurements are much simpler and are inexpensive. Both the volumetric and gravimetric methods can be used for diffusion measurements in powder and pellets. In the case of diffusion in pellets the situation becomes more complicated due to the presence of intercrystalline pores and binder. The mathematics involved also are quite cumbersome (Ma and Ho, 1974; Ma and Lee, 1976; Ma and Mancel, 1972). However, it is sometimes possible to employ an effective diffusion coefficient for the characterization of such pellets. Thus, the methods presented above also can be applied.

Gas chromatography technique

The basic principle involved in gas chromatography is quite simple. A pulse of the gas under study is injected into a carrier gas stream at the entrance of a column packed with adsorbent particles. The broadening of the peak at the outlet of the column is then analyzed for different carrier gas-flow rates. The gas-chromatography technique is most suitable for high temperature and low sorbate-concentration measurements of diffusion coefficients and estimations of heats of adsorption. Such information is extremely difficult to obtain using constant-volume or constant-pressure techniques. Additional advantages include simple experimental set-up, simple data interpretation if one uses a simple model such as the van Deemter equation, and the use of a large number of solid pellets which provides a good representative sample of the sorbent under study.

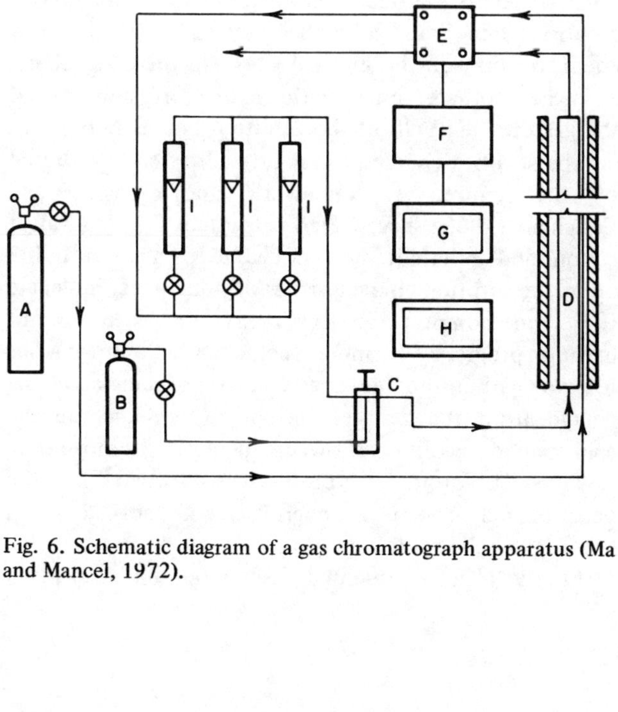

Fig. 6. Schematic diagram of a gas chromatograph apparatus (Ma and Mancel, 1972).

This technique is based on the use of van Deemter's equation (9)

$$\text{H.E.T.P.} = A + B/u + Cu \qquad (13)$$

where H.E.T.P. is the height equivalent to a theoretical plate.

Constants A, B, and C are related to mass transfer processes in the column and inside the solid. In particular, C may be related to the molecular diffusion coefficient D_{12} and the effective diffusion coefficient in the porous solids D_{eff} by

$$C = \frac{\dfrac{F_I^2 dp^2}{75(1-F_I)^2 D_{12}} + \dfrac{F_I K dp^2}{2\pi^2(1-F_I) D_{\text{eff}}}}{\left[1 + K\dfrac{F_I}{1-F_I}\right]^2}. \qquad (14)$$

C corresponds to the limiting slope of the curve of H.E.T.P. versus velocity u at high values of u. H.E.T.P. and Henry's law constant K can be calculated from the retention time and peaks at the exit of the column. Detailed determination of these values can be found in Ma and Mancel (1972).

The experimental set-up for this system is quite simple. Any existing gas chromatograph can be used for such measurements as illustrated schematically in Fig. 6 (Ma and Mancel, 1972). It should be noted that models which take into account the bipore distribution of zeolite pellets have also been developed (Ma and Mancel, 1972; Haynes and Sarma, 1973). Solution of these models by moment techniques enables one to obtain individual diffusion coefficients in micropores (intracrystalline) and macropores (intercrystalline) (Ma and Mancel, 1972; Haynes and Sarma, 1973). The shortcomings of using van Deemter's equation are presented in Haynes and Sarma (1973).

Studies on natural zeolites

A majority of the work on equilibrium adsorption and diffusion in natural zeolites was done by Barrer and his associates. Although a large variety of zeolites was investigated, most efforts were concentrated on the investigation of the equilibrium adsorption and diffusion properties of chabazite.

Sameshima and Hemmi (1934) studied the diffusion of sulfur dioxide, carbon dioxide, and ammonia into relatively long natural mordenite fibers up to 2 mm in length at 25°C. They presented their rate data, but no attempt was made to calculate the diffusion coefficient. Other materials investigated by them were laumontite, stilbite, apophyllite, inesite, analcime, and bentonite.

Barrer and Ibbitson (1944) studied the sorption of various gases on chabazite (from the Bay of Fundy and N.S.W., Australia) and analcimes (Kerguelen Island and Yavapai County, U.S.A.). They reported that in order of increasing affinity for the zeolite lattice, one may write for a series of non-polar gases and vapors: He $<$ He $<$ Ar, or $O_2 < N_2$, or $CH_4 < C_2H_6 < C_3H_8 <$ n-C_4H_{10}. They also observed that the isosteric heat of occlusion ΔH is an additive function of the molecules. They attributed the additivity in ΔH to the additive character of the dispersion forces and of polarization forces between ions in the lattice and dipoles induced in the occluded molecules.

Barrer and Brook (1953) studied the sorption kinetics of C_3H_8, n-C_4H_{10}, CH_2Cl_2, and $(CH_3)_2NH$ in chabazite, mordenite, and levynite at room temperature, –78°C, and –185°C. They reported that the diffusion coefficient is practically independent of concentration for more dilute ranges of sorbate concentration. However, over wider sorbate-concentration ranges, the mean diffusion coefficients decrease as the upper concentration limit rises. The activation energy for diffusion increases as the chain length of the sorbate molecules increases.

Iodine occlusion by chabazite was measured by Barrer and Wasilewski (1961 a, b) in the temperature range 120–300°C. The partial molal heat of occlusion $\Delta\bar{H}$ was determined from the adsorption isotherms. Very large values of $\Delta\bar{H}$ were reported when the amount of iodine sorbed was low. $\Delta\bar{H}$ decreased rapidly to a minimum as the amount of iodine sorbed increased and then rose slowly to a low maximum. Thereafter, $\Delta\bar{H}$ fell rapidly towards the heat of vaporization of iodine. The isotherm can be described satisfactorily by a Langmuir's isotherm. The preferred sites of iodine in the lattice and the paths preferred by the moving iodine molecule were discussed in terms of the iodine lattice interaction.

The diffusion and sorption of H_2O and D_2O in chabazite (from Nova Scotia, Canada), gmelinite (Antrim, Northern Ireland), and heulandite (Iceland) were investigated by Barrer and Fender (1961 a, b). The adsorption capacities, the shape of the isotherms, the isosteric heats, and the partial molar entropies are similar for all three zeolites. Self-diffusion coefficients for the zeolites have values intermediate between the self-diffusion coefficients of liquid water, at the same temperature, and ice near its melting point. Concentration dependence of the diffusion coefficients was reported to be due to non-linearity of the isotherms.

Peterson *et al.* (1965) investigated the sorption and ion-exchange characteristics of several sedimentary zeolites, chabazite, erionite, mordenite, phillipsite, and clinoptilolite. They found that erionite is unique both in its low enthalpy of water sorption and in its appreciable, selective, and rapid sorption of n-paraffins as heavy as n-tetradecane. They further found erionite to contain the most easily exchangeable cations. n-Tetradecane was also sorbed by the chabazite, and to a lesser degree by the mordenite samples, as was nitrogen in the more restrictive channels of those of phillipsite and clinoptilolite.

The origin of the energetic heterogeneity at low sorbate concentrations was discussed by Kington and Macleod (1959) in terms of heterogeneity in dispersion energy and electrostatic energy. The differential heats of sorption of nitrogen, carbon dioxide, carbon monoxide, and hydrogen in natural and calcium chabazite were determined calorimetrically. They concluded that whereas no concept of dispersion-energy heterogeneity is compatible with the experimental data, there is considerable support for the idea that the change in the heat of sorption at low sorbate concentrations is associated with the variation in the electrostatic field within the intracrystalline voids. They also demonstrated that a major component of the heterogeneity arises from the sorbate quadrupole interacting with a position-dependent electrostatic-field gradient in the solid.

Brandt and Rudloff (1964, 1968) studied the concentration dependence of sorption rates of selected gases on chabazite (Nova Scotia). They reported that argon and several other gases exhibit concentration-independent activated diffusion into the bulk of the zeolite. The rates of sorption of Freon 21 are probably controlled by Knudsen or gaseous bulk diffusion, as Freon 21 occupies only the pores or crevices of the mineral. At low surface concentrations carbon dioxide sorption rates are controlled by a relatively slow chemisorption on the internal sites. On the other hand, at higher sorbate concentrations, the diffusion rate of the gas into the bulk of the zeolite becomes important.

The sorptive characteristics of zeolites can be altered by the removal of framework charge. Barrer and Coughlan (1968) investigated the effects of removal of cations and anionic framework charge from clinoptilolite upon the sorption characteristics of polar guest molecule CO_2 and non-polar molecule krypton. The sorbent became progressively more energetically homogeneous towards carbon dioxide as a result of the removal of the charge from the zeolite. On the other hand, all the acid-treated zeolites showed energetic homogeneity towards krypton. Krypton was sorbed in the decationized sorbents although it was not sorbed within the original clinoptilolite crystals.

Eberly (1969) presented results on the adsorption

behavior of normal paraffins, n-C_5 through n-C_8 at 93°C to 207°C on erionite. He reported that rates of adsorption on erionite were low and decreased markedly with molecular weight of the paraffin. On the other hand, the rate of adsorption on 5A molecular sieve remained nearly constant. This difference was discussed in terms of differences in crystal structure of the two materials.

Mercer *et al.* (1970) employed a fixed bed packed with granular clinoptilolite to study the removal of ammonia from secondary effluent. They reported that it was possible to remove more than 99% ammonia in the laboratory with two zeolite columns in series. Regeneration can be accomplished with about 20 column volumes of recycled regenerant containing lime applied at the rate of 10 column volumes per hour. No liquid regenerant waste is produced as the ammonia is exhausted harmlessly to the atmosphere from the air stripper.

Gupta *et al.* (1971) investigated the effect of crystal size upon sorption rate of sulfur dioxide in naturally occurring fibrous erionite from Rome, Oregon. Diffusion coefficients were found to increase with increasing crystal sizes. Transient sorption rates of dilute mixtures of SO_2 and CO_2 in natural mordenite and clinoptilolite were investigated by Ma and Belmonte (1974). They experimentally observed the replacement of adsorbed CO_2 by SO_2 and reported an essentially complete replacement of CO_2 by SO_2 molecules.

Methane sorption on the naturally occurring zeolites phillipsite, clinoptilolite, erionite, and chabazite was investigated by Munson (1973). He reported that chabazite and erionite have higher methane-adsorption capacities than do the other natural zeolites. The methane adsorption capacity of natural clinoptilolite is increased by exchange with hydrogen cation but decreased by cxchange with calcium cation. Barrer and Papadopoulos (1972) obtained the equilibrium isotherms of Kr and Xe in chabazite from Bay of Fundy, Nova Scotia, in the temperature range 150–450°C and at pressures up to 100 atm. They provided an analysis of the relations between heats and entropies derived from Gibbs excess and absolute sorption isotherms.

Barrer and Vaughan (1971) investigated the sorption and diffusion characteristics of phillipsite along with zeolite K-M and other silicates for the trapping and storage of inert gases. After outgassing, the zeolite was exposed to high pressures of inert gas at high temperature and then quenched while still under pressure. Both sorption isotherm and diffusion coefficients were measured.

Adsorption and desorption curves of SO_2 gas on Futatsui zeolite (mostly clinoptilolite) from Akita Perfecture and Shiraishi zeolite (mostly mordenite) from Miyagi Perfecture were obtained by Ishikawa *et al.* (1971) to test the value of these zeolites for the removal of SO_2 from polluted air. SO_2 capacity was 4–5% at 120°C and desorption could be accomplished by heating to 350°C.

A general review of natural and synthetic zeolites was made by Saeys (1969) in terms of abundance and types of the naturally occurring zeolites and their crystal structure and uses. A summary of the previous work on natural zeolites is given in Table 1.

Recent work in the Soviet Union appears to emphasize the application of natural zeolites for the removal of pollutants from gas streams. Anurov and his co-workers (1974 a–d) investigated the sorption and diffusion characteristics of SO_2 in the natural zeolite of Azerbaidzhan deposits containing approximately 70% mordenite and natural clinoptilolite-type zeolites. The SO_2 adsorption was studied between 25–150°C. The sorbents were stable up to 700°C and were recommended for SO_2 removal from industrial waste gases. Starshov *et al.* (1975) employed a gas chromatograph column to investigate the sorption of H_2S on grained zeolites from Aidarsk deposits (Georgia). Kuliev *et al.* (1974) investigated the Tanz natural zeolite and other synthetic zeolites on the removal of H_2S and drying of natural gases. The adsorption capacities of the zeolites for sulfur compounds, the regenerability with hot gas, acid resistance and mechanical strength were studied under process conditions in a pilot plant with three adsorbers.

Belitskii *et al.* (1969) obtained the isotherms (25°C) of adsorption–desorption of water vapors by natural zeolites (chabazite, gmelinite, phillipsite, etc.). The isotherms were obtained for the range of relative water vapor pressure of 4×10^{-4} to 0.807. All isotherms were of the Langmuir type and the saturation occurred at low relative pressures 4×10^{-4} to 6×10^{-2}. Bagirov *et al.* (1974) presented the dynamic equilibrium results of H_2O and C_5 hydrocarbons on the Tanz natural zeolites and made comparisons with those of synthetic zeolites (NaA, NaX, CaX, etc.). The drying capacity of the Tanz natural zeolite was close to that of NaA zeolite. The adsorption of C_5 hydrocarbons by the natural zeolites was found to depend on the zeolite regeneration temperature. Mirasallmov *et al.* (1973) studied the sorption activity of the Uzlekistars zeolites. The H_2O sorption capacity decreased in the order: clinoptilolite > natrolite > laumonite. Sorption of the H_2O and CH_3OH vapor on natural clinoptilolite and its cation-exchanged forms with Li, Na, K, Rb, Cs, and Ti was studied by Belitskii (1973). Merabishvili (1972) obtained the heats of adsorption of benzene, toluene, and xylene on three samples of a natural rock containing 50–70% clinoptilolite.

TABLE 1. Studies on natural zeolites

Type	Location	Characterization (gas sorbed or ion exchanged)	Sorption properties	Temperature	Pressure	Reference
Chabazite (h)*	Bay of Fundy, Nova Scotia	Kr, Xe	Differential heats of adsorption and differential entropies of adsorption	150–450°C	100 atm	Barrer and Papadopoulos (1972)
Chabazite (h)	Nova Scotia	CO_2, CH_4, Ar, Freon 21	Rate curves, diffusivities change with concentrations	23–98°C	46–326 mmHg	Brandt and Rudloff (1964, 1968)
Chabazite (h)	Bay of Fundy, Nova Scotia	O_2,NH_3,HCl,CH_4, C_2H_6,C_3H_8,nC_4H_{10}, iso-C_4H_{10},nC_5H_{12}, iso-C_5H_{12},nC_7H_{16}, isoC_8H_{18}	Isotherms, thermodynamics data (heat, entropy and free energy of adsorption)	−184–204°C	40 mmHg	Barrer and Ibbitson (1944)
Chabazite, mordenite, levynite (h)		C_3H_8,nC_4H_{10},CH_2Cl_2, $(CH_2)_2NH$,Kr,O_2, Ar,N_2,H_2,Ne	Diffusivities and activation energy for diffusion	0–200°C	50 mmHg	Barrer and Brook (1953)
Chabazite (h)	Canadian origin	H_2,Ne,O_2 and N_2,Ar+O_2, Ar+N_2 mixtures	Isotherms	−183°C	50 mmHg	Barrer and Robins (1953)
Chabazite (h)	Nova Scotia	K^+,Rb^+,Cs^+,Ca^{++}, and Sr ion exchange	Self and exchange diffusivities of ions, and thermodynamic properties	0–80°C	760 mmHg	Barrer *et al.* (1963)
Chabazite and exchanged form	Canadian origin	Ar	Heat of adsorption (calorimetry) isotherms, entropy	183°C	0–225 mmHg	Garden and Kington (1956)
Chabazite		O_2	Isotherms, differential and molar heats of adsorption, differential and molar entropies of adsorption	183°C	50 mmHg	Gordon and Kington (1956)
Chabazite and its cation-exchange form		N_2,CO,CO_2,H_2	Heat of adsorption (calorimetry). Discussion of surface heterogeneity and the role of quadrupoles	0°C		Kington and Macleod (1959)
Chabazite (h)	Bay of Fundy, Nova Scotia	I_2	Isotherms, partial molal heat of adsorption	120–300°C	300 psi	Barrer and Wasilewski (1961a)
Chabazite (h)	Bay of Fundy, Nova Scotia	I_2	Partial molar heat of adsorption, partial molar thermal entropy, isotherms	120–300°C	300 psi	Barrer and Wasilewski (1961b)
Chabazite, erionite, mordenite phillipsite, clinoptilolite and their K^+, Ca^{++}, and Mg^{++} ion-exchange forms	Western U.S.A.	H_2O,nC_5H_{12}, nC_6H_{14},$nC_{14}H_{30}$	Isotherms	50–250°C	300 mmHg	Peterson *et al.* (1965)

Chabazite, clinoptilolite, erionite, phillipsite, Ca^{++} ion-exchanged erionite and mordenite	California, Arizona, Oregon, U.S.A.	CH_4,ClF_3C	Isotherms	26°C	100 psi	Manson (1973)
Clinoptilolite and its decationized derivatives	Hector, CA	CO_2,Kr	Isotherms, isosteric heat of adsorption	−100–90°C	60 mmHg	Barrer and Coughlan (1968)
Clinoptilolite	Hector, CA	removal of NH_4^+ from secondary effluents by zeolite columns	NH_4^+ breakthrough curves, elution curves for regeneration	26°C	760 mmHg	Mercer *et al.* (1970)
Laumontite, stilbite, mordenite, natrolite, analcime (h)	Takashima, Hokkaido, Japan	NH_3,CO_2,SO_2,C_2H_2, O_2,H_2	Diffusion rate curves	25°C	760 mmHg	Sameshima and Hemmi (1934)
Erionite	Rome, Oregon	SO_2	Diffusivity and its dependency on crystal sizes	25–30°C	100 mmHg	Gupta *et al.* (1971)
Erionite	Jersey Valley, Nevada, U.S.A.	nC_5H_{10} through nC_8H_{20}	Isotherms, diffusivities and effects of particle sizes	93–207°C	50 mmHg	Eberly (1969)
Phillipsite, analcime, thomsonite, mesolite, scolecite (h)	Pine Valley, Nevada, U.S.A.	He,Ne,Ar	Isotherms, diffusivities, activation energy for diffusion	100–550°C	12,000 psi	Barrer and Vaughan (1971)
Heulandite and stilbite (h)	Icelandic basaltic lava	Ne,Ar,Kr	Isotherms, diffusivities, activation energy for diffusion	21–205°C	10^4 psi	Barrer and Vaughan (1969)
Clinoptilolite, mordenite	Union Pass and Alaska Union Pass and Alaska	SO_2 on zeolites with pre-sorbed water	Diffusivities, isotherms	25°C	0.5 mmHg to 100 mmHg	Ma *et al.* (1977)
Clinoptilolite, mordenite	Oregon	SO_2 and CO_2 mixture	Sorption rate	5–45°C	40–160 mmHg	Ma and Belmonte (1974)
Chabazite, gmelinite, heulandite (h)	Nova Scotia, Canada; Antrim, Northern Ireland; Iceland	Sorption and diffusion of D_2O and H_2O	Isotherms, diffusivity	30–80°C		Barrer and Fender (1961 a, b)

*(h) = hydrothermal.

CONCLUSIONS

The present review briefly describes the fundamental theories of sorption and diffusion in natural zeolites. It also outlines methods for the determination of equilibrium adsorption isotherms and diffusion coefficients. The mathematics involved in the calculation of the diffusion coefficient are also included. Previous work on sorption and diffusion in natural zeolites is presented in a tabular form. It should be emphasized that although the potential for specific commercial applications of natural zeolites is promising, more systematic programs are needed to investigate the sorption and diffusion properties of natural zeolites. An interdisciplinary R&D program such as a joint effort among geoscientists, chemists, and chemical engineers probably would be most effective leading to utilization of natural zeolites.

REFERENCES

Anurov, S. A., Belitskii, I. A., Kravtsova, I. D., Kel'tiv, N. V. and Torocheshrikov, N. S. (1974) Dynamics of sulfur dioxide adsorption by natural zeolites: *Trudy Mosk. Khim.-tech. Inst.* **79**, 5-7.

Anurov, S. A., Kel'tev, N. V. and Torocheshrikov, N. S. (1974) Isotherms for adsorption of sulfur dioxide on natural zeolites: *Zh. Fiz. Khim.* **79**, 3-4.

Anurov, S. A., Kel'tev, N. V., Smola, V. I. and Torocheshrikov, N. S. (1974) Adsorption of sulfur dioxide by natural zeolites: *Zh. Fiz. Khim.* **48**(8), 2145-2146.

Anurov, S. A., Kel'tev, N. V., Smola, V. L. and Torocheshrikov, N. S. (1974) Kinetics of sulfur dioxide adsorption on natural zeolites: *Zh. Fiz. Khim.* **48**(10), 2124-2127.

Bagirov, R. A., Ali-Zade, E. M. and Farkhadov, T. S. (1974) Natural zeolites for an adsorption method preparing gas for transportation: *Gazov. Prom-st.* **9**, 31-33.

Barrer, R. M. (1971) Intracrystalline diffusion: *Adv. Chem. Ser.* **102**, 1-34.

Barrer, R. M., Bartholomew, R. F. and Rees, L. V. C. (1963) Ion exchange in porous crystals. I. Self- and exchange-diffusion of ions in chabazites: *J. Phys. Chem. Solids* **24**, 51-62.

Barrer, R. M. and Brook, D. W. (1953) Molecular diffusion in chabazite, mordenite, and levynite: *Trans. Farad. Soc.* **19**, 1049-1059.

Barrer, R. M. and Coughlan, B. (1968) Molecular sieves derived from clinoptilolite: *Molecular Sieves*, Soc. of Chem. Ind., London, pp. 141-148.

Barrer, R. M. and Fender, B. E. F. (1961) The diffusion and sorption of water in zeolites – I. Sorption: *J. Phys. Chem. Solids* **21**, 1-11.

Barrer, R. M. and Fender, B. E. F. (1961) The diffusion and sorption of water in zeolites – II. Intrinsic and self-diffusion: *J. Phys. Chem. Solids* **21**, 12-24.

Barrer, R. M. and Ibbitson, D. A. (1944) Occlusion of hydrocarbons by chabazite and analcite: *Trans. Farad. Soc.* **40**, 195-205.

Barrer, R. M. and Papadopoulis, R. (1972) The sorption of krypton and zenon in zeolites at high pressures and temperatures: I. Chabazite: *Proc. Roy Soc. Lond.* A, **326**, 315-330.

Barrer, R. M. and Robins, A. B. (1953) Sorption of mixtures: Part 2. Equilibrium between binary gas mixtures and some zeolites: *Trans. Farad. Soc.* **49**, 929-939.

Barrer, R. M. and Vaughan, D. E. W. (1969) Sorption and diffusion of rare gases in heulandite and stilbite: *Surface Science* **14**, 77-92.

Barrer, R. M. and Vaughan, D. E. W. (1971) Trapping and diffusion of rare gases in phillipsite, zeolite K-M and other silicates: *Trans. Farad. Soc.* **67**, 2129.

Barrer, R. M. and Wasilewski, S. (1961) Occlusion of iodine in some porous crystals, Part 1. Heat energy relationships: *Trans. Farad. Soc.* **57**, 1140-1152.

Barrer, R. M. and Wasilewski, S. (1961) Occlusion of iodine in some porous crystals, Part 2. Entropy and free energy relationships: *Trans. Farad. Soc.* **57**, 1153-1163.

Belitskii, I. A. (1969) Isotherms of water vapor sorption by natural zeolites (25°): *Eksp. Issled. Mineral.*, pp. 92-101.

Belitskii, I. A., Scherbatyuk, N. E., Krasnova, L. V., Filizova, L. D. and Tyurina, E. F. (1973) Sorption properties of cation exchanged forms of natural clinoptilolite: *Izv. Sib. Otd. Akad. Nauk SSSR, Ser. Khim. Nauk*(1), 84-87.

Brandt, W. W. and Rudloff, W. (1964) Rapid sorption processes on granular zeolite: *J. Phys. Chem. Solids* **25**, 167-176.

Brandt, W. W. and Rudloff, W. (1968) The concentration dependence of zeolitic sorption: *J. Phys. Chem.* **71**, 3948-3953.

Crank, J. (1957) *Mathematics of Diffusion*, Oxford, England.

Eberly, P. E., Jr. (1969) Adsorption of normal paraffins in erionite and 5A in erionite and 5A molecular sieve: *I&EC Prod. Res. and Dev.* **8**, 140-144.

Garden, L. A. and Kington, G. L. (1956) The thermodynamic properties of oxygen sorbed in chabazite: *Trans. Farad. Soc.* **51**, 1397-1408.

Garden, L. A., Kington, G. L. and Laing, W. (1955) The heat of sorption, isotherms and entropies of argon in natural and calcium chabazites: *Trans. Farad. Soc.* **51**, 1558-1569.

Gradev, Gradi (1973) Sorption of iridium on silver, mercury, lead, and thallium forms of natural zeolite (clinoptilolite) and vermiculite: *Radiochim. Acta* **19**, 66-69.

Gupta, J. C., Ma, Y. H. and Sand, L. B. (1971) Diffusion of sulfur dioxide in a synthetic mordenite and a natural erionite: *AIChE Symposium Series* **67**, 51-57.

Haynes, H. W. and Sarma, P. N. (1973) A model for the application of gas chromatography to measurements of diffusion in bidisperse structural catalysts: *AIChE J.* **18**, 1043-1046.

Ishikawa, H., Tanaka, H., Uchiyama, K., Kwon, S., Asayama, T. and Morimoto, T. (1971) Adsorption and desorption of sulfur dioxide gas in natural zeolites: *Waseda Daigaku Rikogaku Kenkyusho Hokoku* (51), 46-53.

Kington, G. L. and Macleod, A. C. (1959) Heats of sorption of gases in chabazite, energetic heterogeneity and the role of quadropoles in sorption: *Trans. Farad. Soc.* **55**, 1799-1814.

Kuliev, A. M., Bagirov, R. A., Kuliev, E. R., Karkhadov, T. S. and Ali-Zade, E. M. (1974) Drying and removal of hydrogen sulfide from natural gas: *Gasov. Promst.* (7), 39-41.

Ma, Y. H. and Belmonte, R. P. (1974) Experimental investigation of multicomponent rates of sorption of SO_2 and CO_2 in natural mordenite and clinoptilolite: Presented at the AIChE 76th National Meeting, Tulsa, Oklahoma, March 1974.

Ma, Y. H., Byron, R. J., Feltri, P. and Lee, T. Y. (1977) Effects of presorbed water vapor upon the sorption and diffusion of SO_2 in natural mordenites: presented at the AIChE 83rd National Meeting, March 20-23, 1977, Houston, Texas.

Ma, Y. H. and Ho, S. Y. (1974) Diffusion in synthetic faujasite powder and pellets: *AIChE J.* **20**, 279-284.

Ma, Y. H. and Lee, T. Y. (1976) Transient Diffusion in solids with a bipore distribution: *AIChE J.* **22**, 147-152.

Ma, Y. H. and Mancel, C. (1972) Diffusion studies of CO_2, NO, NO_2 and SO_2 on molecular sieve zeolites by gas chromatography: *AIChE J.* **18**, 1148-1153.

Merabishvili, M. S., Sikharylidze, N. G., Khubulava, G. K. and Bolotashvili, Z. G. (1972) Adsorption-separation properties of natural zeolites containing sorbents: *Koks. Khim.* **10**, 44-45.

Mercer, B. W., Ames, L. L., Touhill, C. J., vanSlyke, W. J. and Dean, R. B. (1970) Ammonia removal from secondary effluents by selective ion exchange: *J. Wat. Pollut. Control. Fed.* **42**, R95-R107.

Mirsalimov, A. M. and Scherbatyuk, N. E. (1973) Sorption properties of the natural zeolites of Uzbekistan: *Fiz. Khim. Suoistva Nekot. Prir. Miner. Sorbentov. Uzb.*, pp. 30-33.

Munson, R. A. (1973) Properties of natural zeolites: U.S. Nat. Tech. Inform. Service, PB Rep. No. 221512/7, 18 pp.

Peterson, D. L., Helfferich, F. and Blytas, G. C. (1965) Sorption and ion exchange in sedimentary zeolites: *J. Phys. Chem. Solids* **26**, 835-848.

Roux, A., Huang, A. A., Ma, Y. H. and Zwiebel, I. (1973) Sulfur dioxide adsorption on mordenites: *AIChE Symposium Series* **69**, 46-53.

Ruckenstein, E., Vaidjnathan, A. S. and Youngquist, G. R. (1971) Sorption by solids with bidisperse pore structure: *Chem. Engng Sci.* **261**, 1305-1318.

Saeys, H. W. (1969) Properties of natural and synthetic (zeolites) molecular sieves: *Chem. Tech. (Amsterdam)* **24**(12), 364-366.

Sameshima, J. and Hemmi, H. (1934) Sorption of gases by mineral. IV. Zeolites and bentonite: *Bull. Chem. Soc., Japan* **9**, 27-41.

Starshov, I. M., Romapova, G. Ya., Galeeva, R. G. and Sharafutdinov, R. G. (1975) Adsorption of hydrogen sulfide on natural zeolites: *Neftepererab. Neffekhim.* (Moscow) (5), 34-35.

van Deemter, J. J., Zuiderweg, F. J. and Klinkenberg, A. (1956) Longitudinal diffusion and resistance to mass transfer as causes of nonideality in chromatography: *Chem. Engng Sci.* **5**, 271-289.

CATION-EXCHANGE EQUILIBRIA IN ZEOLITES AND FELDSPATHOIDS

RICHARD M. BARRER

Physical Chemistry Laboratories, Chemistry Department, Imperial College, London SW7 2AY, England

Abstract – A summarizing account is given of equilibrium aspects of isomorphous replacement by cation exchange in zeolites and feldspathoids. Crystals studied are analcime, mordenite, chabazite, clinoptilolite, phillipsite, faujasite (as zeolites X and Y), the gismondine-type Na–P, edingtonite-type K–F, and zeolite A; and the feldspathoids ultramarine, sodalite, and cancrinite. In zeolites there is commonly more than one kind of exchange site, so that exchange can occur simultaneously on a range of site groups. Four characteristic kinds of isotherm contour have been examined and interpreted in terms of non-ideality of individual site groups and in terms of the contribution to the overall equilibrium of a range of site groups. Each of the four isotherm contours can be explained in these terms, and limited mutual solid solubility of end-members of the exchange has been given a quantitative interpretation.

Factors governing ion-exchange selectivity among the zeolites and feldspathoids studied have been shown to include: framework topology, ion size and shape, charge density, ion valency, and electrolyte concentration in the external phase. Each of these factors has been illustrated and discussed, and some consequences are described.

INTRODUCTION

An important property of zeolites is their ability to undergo ion exchange. This property is facilitated by their open tectosilicate frameworks, and is shared with feldspathoids and clay minerals. The more compact the aluminosilicate frameworks the higher the temperatures needed for ready exchange. Thus, feldspars, like zeolites, function as exchangers at elevated temperatures. The zeolites provide model systems in which cation exchanges can be studied quantitatively under readily accessible and mild conditions. They illustrate many aspects of isomorphous replacements of the types (Barrer, 1974)

$$K^+ \rightleftarrows Na^+$$

$$Ca^{2+} \rightleftarrows 2Na^+$$

The exchange capacities, illustrated in Table 1, are high, and high selectivities towards particular ions are often exhibited. In addition, ion exchange has been found to be an important way of changing molecule sieving properties and enhancing catalytic activities of zeolites.

TABLE 1. Exchange capacities of some zeolites

		Exchange capacity (meq/100 g)
Tendency to fibrous structures		
Edingtonite	$Ba[Al_2Si_3O_{10}]4H_2O$	390
Natrolite	$Na_2[Al_2Si_3O_{10}]2H_2O$	530
Scolecite	$Ca[Al_2Si_3O_{10}]3H_2O$	500
Thomsonite	$NaCa_2[Al_5Si_5O_{20}]6H_2O$	620
Tendency to lamellar structures		
Stilbite	$(Na,½Ca)[AlSi_3O_8]3H_2O$	320
Heulandite	$Ca[Al_2Si_6O_{16}]5H_2O$	330
Robust three-dimensional frameworks		
Analcime	$(Na[AlSi_2O_6]H_2O$	450
Harmotome	$(K_2,Ba)[Al_2Si_5O_{14}]5H_2O$	390
Levynite	$Ca[Al_2Si_4O_{12}]6H_2O$	400
Mordenite	$(½Ca,Na)[AlSi_5O_{12}]3.3H_2O$	230
Chabazite	$(½Ca,Na)[AlSi_2O_6]3H_2O$	400
Gmelinite	$(½Ca,Na)[AlSi_2O_6]3H_2O$	400
Faujasite	$(Ca,Na_2)[Al_2Si_5O_{14}]6.6H_2O$	390

As with other physico-chemical systems ion exchange can be studied from viewpoints of

(i) equilibrium (selectivity),
(ii) energetics, and
(iii) kinetics (exchange and tracer diffusion).

In the present paper some features will be examined of exchange equilibria studied in the author's laboratory. The studies have been made with the zeolite analcime (Barrer and Hinds, 1953), mordenite (Barrer and Klinowski, 1974b), chabazite (Barrer and Sammon, 1955; Barrer, Davies, and Rees, 1969a), clinoptilolite (Barrer, Papadopoulos, and Rees, 1967), phillipsite

(Barrer and Munday, 1971a), faujasite (Barrer, Rees, and Shamsuzzoha, 1966; Barrer, Davies, and Rees, 1968; Barrer, Davies, and Rees, 1969b) (zeolites X and Y), the gismondine-type Na-P zeolite (Barrer and Munday, 1971b), edingtonite-type K-F (Barrer and Munday, 1971c), and zeolite A (Barrer and Meier, 1959; Barrer, Rees, and Ward, 1963); and with the feldspathoids ultramarine (Barrer and Raitt, 1954), sodalite and cancrinite (Barrer and Falconer, 1956). The natural chabazite came from the Bay of Fundy and the aluminous chabazite, G1, was made in my laboratory; the clinoptilolite and phillipsite were of sedimentary origin, and their purities and compositions are described in the original papers. The analcime, mordenite, Na-P, K-F, sodalite, and cancrinite were synthetic preparations made in my laboratories. The ultramarine was made by Reckitt and Sons, and the zeolite A and faujasite (zeolites X and Y) came from Union Carbide.

TYPES OF EXCHANGE ISOTHERM

For the exchange reaction

$$A^+ + BZ \rightleftarrows B^+ + AZ \tag{1}$$

three equilibrium quotients of importance are

$$K_a = \frac{a_c^A a_s^B}{a_s^A a_c^B} = K_c \frac{f_c^A}{f_c^B} = K \frac{f_c^A f_s^B}{f_c^B f_s^A},$$

$$K_c = \frac{A_c a_s^B}{B_c a_s^A}, \tag{2}$$

$$K = \frac{A_c B_s}{A_s B_c} = \frac{A_c m_s^B}{B_c m_s^A}$$

where m_s^A, m_s^B are molalities in the aqueous phase, the a^A and a^B are thermodynamic activities, the f^A and f^B are activity coefficients, and A and B are equivalent cation fractions of ions A^+ and B^+. The subscripts c and s denote "crystals" and "solution". K_a is the thermodynamic equilibrium constant, K the mass action quotient (and for univalent ions the selectivity quotient), and K_c is the modified selectivity quotient, often termed the Kielland quotient. For ions of different valences, and of charges ν_A^+ and ν_s^+, these quotients are readily rewritten for the exchange reaction

$$\nu_s^+ A^{\nu_{A^+}} + \nu_A^+ BZ_{\nu_B^+} \rightleftarrows \nu_A^+ B^{\nu_{B^+}} + \nu_B^+ AZ_{\nu_{A^+}}. \tag{3}$$

Thus, the mass action quotient becomes

$$K = \frac{A_c^{\nu_B} B_s^{\nu_A}}{A_s^{\nu_B} B_c^{\nu_A}} \tag{4}$$

and the selectivity quotient becomes (Barrer and Klinowski, 1974a)

$$\alpha = \frac{\nu_A A_c B_s}{\nu_B A_s B_c}. \tag{5}$$

In the zeolite phases the ionic strength is fixed by the unit cell dimensions and content. In the aqueous phases the ionic strength can be varied continuously by dilution, and the overall normality as well as the equivalent cation fractions are required for complete specification of the equilibria.

Consideration of the mass action quotient indicates that the exchange isotherm is fully described if A_s is plotted against A_c because $A_s = (1 - B_s)$ and $A_c = (1 - B_c)$. At least four kinds of exchange isotherm contour have been observed with zeolite exchangers which are illustrated in Fig. 1 a, b, c, and d respectively, for sodalite hydrate, cancrinite hydrate, the gismondine-type Na-P, and the edingtonite type K-F.

To understand these isotherm contours, the structures (Meier and Olson, 1971) of zeolite crystals must be

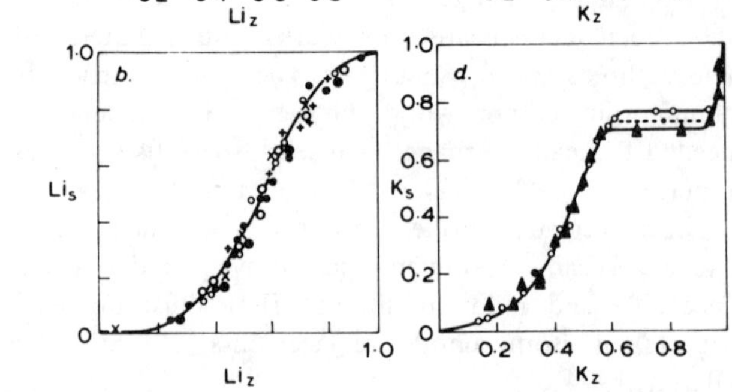

Fig. 1. Characteristic contours of exchange isotherms in some zeolites. Ordinate: cation fraction in solution; abscissa: cation fraction in crystals. (a) Li⇄Na in sodalite hydrate (Barrer and Falconer, 1956). (b) Na⇄Li in cancrinite hydrate (Barrer and Falconer, 1956): ○, 25°C; o, 85°C; ×, 110°C. Li→Na: ●, 25°C; •, 85°C; +, 110°C. (c) Na⇄K at 25°C in zeolite Na-P (Barrer and Munday, 1971b). ○, K entering crystals; ▲, Na entering crystals. (d) Na⇄K at 25°C in zeolite K-F (Barrer and Munday, 1971c). ○, K entering crystals; ▲, Na entering crystals.

considered. In many zeolites there is more than one kind of crystallographic site which can be occupied by cations (Smith, 1971). All sites of the same crystallographic type constitute one site group, and there may be n kinds of the sites are occupied. Thus, the zeolites may present a cation sites are mixed with one another throughout the crystal. Moreover, the anionic charge of the alumino-silicate frameworks may be neutralized when only some of the sites are occupied. Thus the zeolites may present a high degree of cationic disorder, both in terms of unoccupied sites and in terms of different distributions of cations of different kinds among the site groups.

One site group

In a few aluminosilicate exchangers there is only one site group to consider. An example is found in sodalite hydrate where the cation sites are close to the 6-ring windows of the 14-hedral (8 × 6-ring, 6 × 4-ring) cages of which the structure is composed. Only three-quarters of these window sites are occupied for the unit-cell composition $6(NaAlSiO_4)\cdot 8H_2O$. The Li $\rightleftarrows$ Na exchange isotherm in Fig. 1a, corresponding with this situation, shows no inflection. The mass action quotients for this exchange, for K $\rightleftarrows$ Na, and for K $\rightleftarrows$ Li were nearly constant, independent of the cationic composition of the exchanger, as was shown from plots of log (A_s/B_s) against log (A_c/B_c), which were straight lines with slopes close to unity (Barrer and Falconer, 1956). Thus, the quotients $f_c^A f_s^B / f_c^B f_s^A$ are constant and so presumably near to one, and $K \sim K_c \sim K_a$. This behavior is further illustrated in Table 2. Any one of the three equilibrium quotients can be calculated from the other two by means of the triangle rule.

TABLE 2. Equilibria in sodalite hydrate

Ion A	pair B	$\dfrac{A_c B_s}{B_c A_s} = K \sim K_a$	$\Delta G^\theta = -RT \ln K_a$ (cal/g ion)
Li^+	Na^+	0.46	540
K^+	Na^+	0.063	1970
K^+	Li^+	0.135	1430

It is not, however, always true that, when there is only one site group, the exchange behavior is as ideal as that reported for sodalite hydrate. In the exchange Na $\rightleftarrows$ K in the much tighter feldspar framework the K^+ ions do not fit well in the unaltered framework typical of sodium feldspar. Thus, the sodium feldspar framework tolerates only a limited potassium content before a new phase, still with the feldspar topology but rich in potassium, nucleates on or in a matrix of the parent sodium-rich phase (Orville, 1963). Over the resultant immiscibility gap both these phases co-exist. Finally, as the average K-content of the mixture increases, only the potassium-rich phase remains, and a continuous exchange isotherm once more results.

In 1956 (Barrer and Falconer, 1956) this behavior was first given a quantitative basis for uni-univalent exchanges in terms of a model based upon the statistical thermodynamics of localized sorption with interaction for an exchanger which provided one kind of cation site. When ions of A occupy adjacent sites, an extra energy term was assumed to arise, which was taken to be pair-wise additive. This term, which did not occur with B–B or A–B neighbors, was wN_A^2/N where N is the total number of cation sites, and N_A is the number of these occupied by ions A^+. The partition functions of the anionic framework and of the cations A^+ and B^+ were taken to be the same in the A- and B-forms of the exchanger, except for the term arising from the extra energy. The treatment gave

$$\ln K_c = \ln \frac{j_A(T)}{j_B(T)} + \frac{(\mu_A^\theta - \mu_B^\theta)}{kT} + \frac{(E_A - E_B)}{kT} - \frac{2N_A w}{NkT} \quad (6)$$

for the Kielland quotient $K_c = A_c a_s^B / B_c a_s^A$. In this relation $j_A(T)$ and $j_B(T)$ are the partition functions of ions A^+ and B^+ on the exchange sites, excluding the terms $\exp -E_A/RT$ and $\exp -E_B/RT$, the corresponding binding energies being E_A and E_B. The μ^θ are the standard chemical potentials of ions A^+ and B^+ in the external aqueous phase.

Since for uni-univalent exchanges N_A is related to A_c and B_c by

$$A_c = (1 - B_c) = N_A/(N_A + N_B) \quad (7)$$

it follows that when log K_c is plotted against A_c or B_c a straight line should be obtained. Nearly linear plots of log K_c against B_c were first obtained by Kielland (Kielland, 1935) for gel exchangers. Figure 2 shows a Kielland plot for the Na $\rightleftarrows$ Li exchange in cancrinite hydrate (Barrer and Falconer, 1956). Equations (6) and (7) can be written as

$$\log K_c = C_0' + C_1(1 - 2B_c) = C_0 - 2C_1 A_c \quad (8)$$

where C_0', C_0, and C_1 are coefficients and

$$\left.\begin{aligned} 2.303\, C_1 &= -\frac{w(N_A + N_B)}{NkT}\,;\; C_0 = C_0' - C_1\,, \\ 2.303\, C_0 &= \ln \frac{j_A(T)}{j_B(T)} + \frac{\mu_A^\theta - \mu_B^\theta}{kT} + \frac{(E_A - E_B)}{kT} \end{aligned}\right\} \quad (9)$$

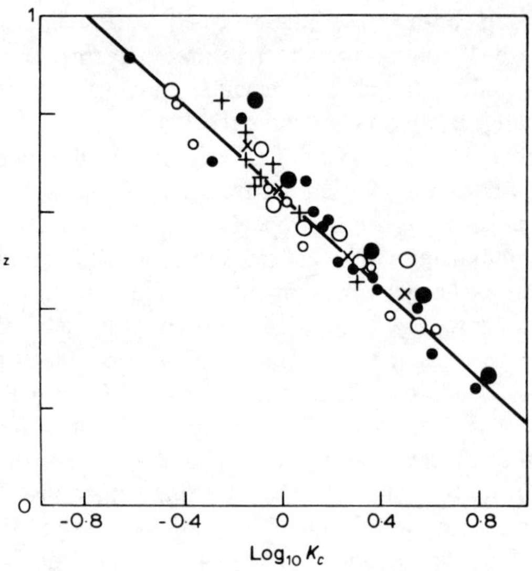

Fig. 2. Plot of log K_c against the cation fraction Li_c, for Na⇄Li in cancrinite hydrate (Barrer and Falconer, 1956). Points are as designated in Fig. 1b.

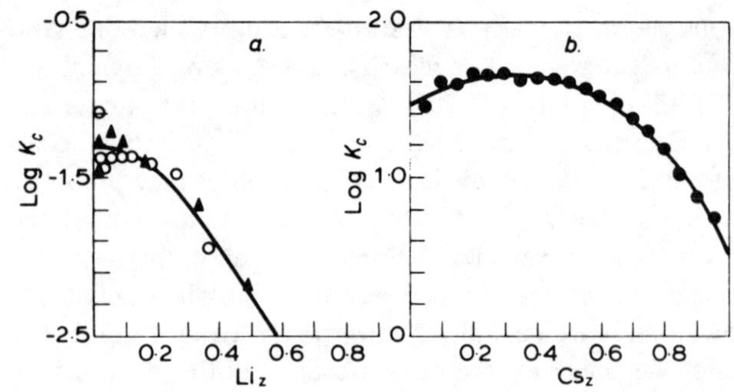

Fig. 3. Plots of log K_c against the cation fraction of (a) Li and (b) Cs for the exchanges Na⇄Li and Na⇄Cs in phillipsite (Barrer and Munday, 1971a).

Some examples of isotherms interpreted in this way gave the values of C_1 and $w(N_A + N_B)/N$ shown in Table 3. However, the Kielland plots are often not straight lines over an extended range in composition, as shown in Fig. 3 for the exchanges Na ⇄ Li and Na ⇄ Cs in phillipsite (Barrer and Munday, 1971a). Then eqn. (8) can formally be replaced by a polynomial expression.

Isotherms for K_a = 1 and corresponding with linear Kielland plots are illustrated in Fig. 4. Their shapes are determined by the values of C_1. Negative values of C_1 give sigmoid isotherms like that of Fig. 1b. For uni-univalent exchanges with $C_1 > +0.87$ the isotherms develop a maximum followed by a minimum. This is a metastable condition and involves a two-phase region. For ions of different valence the positive values which C_1 must exceed for extrema to appear in the isotherms are given in Table 4 (Barrer and Klinowski, 1972a). Thus,

TABLE 3. Coefficients C_1 and values of $w(N_A + N_B)/N$

Exchanger	Exchange	T°C	C_1	$w(N_A + N_B)/N$ (J/g ion)
Chabazite	Na → Rb	84	−1.0	6810
	Na → Tl	45	−0.30	1840
	Na → Ag	45	0.0	0
	K → Ag	25	−0.44	2510
Cancrinite	Li → Na	~68	−1.08	7070
Faujasite	Na → Ag	25	−0.48	2760
	K → Ag	25	−0.49	2810
	Na → K	25	−0.52	2980
	Tl → Ag	25	−0.68	3890

the treatment of Barrer and Falconer can produce isotherm contours typical of Figs. 1 a, b, and c and can also account for immiscibility gaps (Fig. 1d), when only one kind of site-group is involved.

This treatment assumed that the disposition of A^+ cations on the sites was random, despite the extra energy for an A–A pair. However, the limitation of randomness has been removed, using the rigorous method of Lacher (Lacher, 1937; Fowler and Guggenheim, 1939) for localized adsorption with pair-wise additive interaction. The refined treatment (Barrer and Klinowski, 1977) gave Kielland plots which were not strictly linear, but which otherwise gave the same general behavior of the isotherm contours and of immiscibility gaps as the approximate treatment.

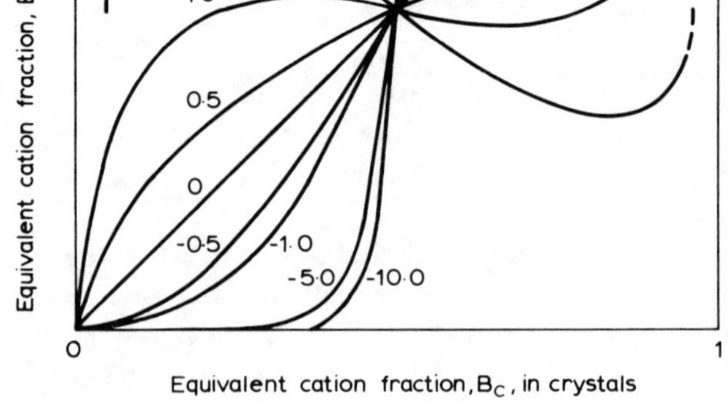

Fig. 4. Calculated exchange isotherms for univalent ion pairs and for different values of C in the Kielland equation

$$\log K_c = \log K_a + C(1-2B_c)$$

where K_a was taken as unity (Barrer and Falconer, 1956).

TABLE 4. Values which C_1 must exceed for extrema in isotherms

Z_A	Z_B	C_1
1	1	> 0.87
1 2	2 1	> 1.27
1 3	3 1	> 1.62
2	2	> 1.74
3 2	2 3	> 2.16
3	3	> 2.61

Immiscibility gaps are usually associated with hysteresis loops, as illustrated in Fig. 1d. Sometimes, as in the Rb–K exchange in the analcime framework, shown in Fig. 5, this loop can be remarkably wide. The figure also shows scanning curves within the loop. The loops occur in the two-phase region only, the isotherm outside this region being reversible. The hysteresis was attributed to the two extra positive-free-energy contributions due to interfacial tension and to strain when a somewhat misfitting nucleus of the A-rich form develops on or in a matrix of the parent B-rich zeolite. This behavior can be encountered in other solid-state reactions involving nucleation of new phases. In the direction B-zeolite → A-zeolite, the extra positive free energy terms delay appearance of A-rich nuclei beyond the true thermodynamic two-phase equilibrium between bulk B-rich and separate bulk A-rich forms. Conversely, in the direction A-zeolite → B-zeolite the same two free

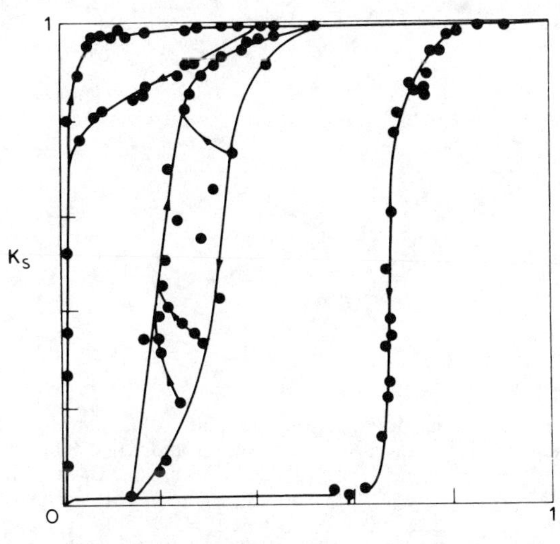

Fig. 5. The exchange K⇄Rb at 95°C in analcime, showing an extreme hysteresis loop, and scanning curves within the loop (Barrer and Falconer, 1956).

energy terms again delay nucleation of the B-rich form in the matrix of the parent form. The hysteresis introduced some notable history-dependent properties into the exchanges in analcime (Barrer and Hinds, 1953).

Two or more site groups

As noted earlier, in many zeolites there are site groups of more than one kind. If each of n-site groups is in equilibrium with the external electrolyte, then (Barrer and Klinowski, 1972a)

$$\frac{a_s^B}{a_s^A} = K_a \cdot \frac{a_c^B}{a_c^A} = \ldots = K_i \frac{a_i^B}{a_i^A} = \ldots = K_n \frac{a_n^B}{a_n^A} \qquad (10)$$

where K_i and the a_i are the thermodynamic equilibrium constant and activities on the ith site group. Likewise, for the Kielland quotients,

$$K_c \frac{B_c}{A_c} = \ldots = K_{ci} \frac{B_i}{A_i} = \ldots = K_{cn} \frac{B_n}{A_n} \qquad (11)$$

where B_i and A_i are the cation fractions of ions B^+ and A^+ on the ith site group. Also, if the standard states are chosen as pure AZ and pure BZ (in equilibrium with infinitely dilute solutions of salts of ions A^+ and B^+, respectively), then for exchanges going from pure AZ to pure BZ (Barrer, Klinowski, and Sherry, 1973),

$$\begin{aligned} \Delta G^\theta &= \sum_{i=1}^{n} X_i \, \Delta G_i^\theta \\ \Delta H^\theta &= \sum_{i=1}^{n} X_i \, \Delta H_i^\theta \\ \Delta S^\theta &= \sum_{i=1}^{n} X_i \, \Delta S_i^\theta \end{aligned} \qquad (12)$$

and thus

$$K_a = K_1^{X_1} K_2^{X_2} \ldots K_n^{X_n} \qquad (13)$$

where ΔG^θ, ΔH^θ, and ΔS^θ are the overall standard free energy, enthalpy, and entropy of exchange; and ΔG_i^θ, ΔH_i^θ, and ΔS_i^θ are the corresponding quantities for the ith site group. X_i is the fraction of the total exchange ions associated with the ith site group. These relations show how the component site groups contribute to the overall thermodynamics of exchange and to the overall equilibrium constant.

One may next inquire what can happen to the overall isotherm contours and to the form of the Kielland plots when exchange on each of the separate site groups is ideal, so that $f_i^A = f_i^B = 1$, and $K_{ci} = K_i$ = constant. Model calculations were therefore made (Barrer and Klinowski, 1972a) which showed that:

(i) For all $K_i < 1$ the isotherm has the form of Fig. 1a. If all $K_i > 1$ the isotherm also has no inflections but is convex in the opposite sense, the entering ion showing preference for the exchanger.

(ii) When for simplicity two site groups only are considered and one has $K_1 > 1$ and the other $K_2 < 1$, the resultant isotherm is sigmoid in form, like Fig. 1b. The Kielland plots are also sigmoid and the overall isotherm resulting from exchange on the ideal site groups is itself non-ideal since K_c is a function of composition. The behavior is illustrated in Fig. 6 for different values of the K_i, and different proportions X_i.

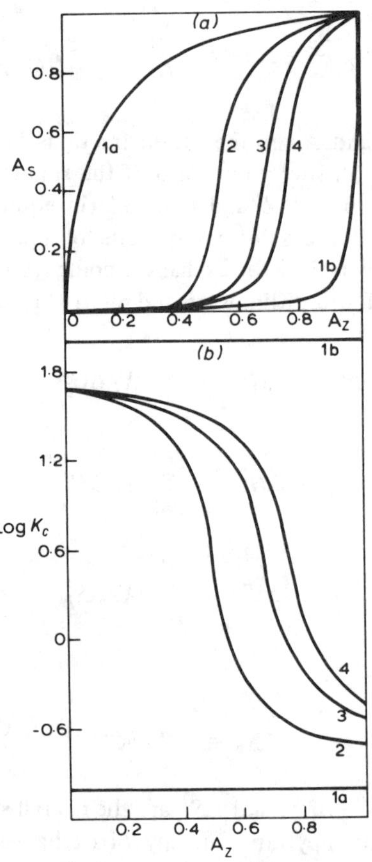

Fig. 6. (a) Ion exchange isotherms and (b) plots of log K_C against $A_C(=A_Z)$ for exchangers (Barrer and Klinowski, 1972a) containing, respectively, 1, 2, 3, and 4 different types of ideal site in a uni-univalent exchange. Curve 1a, K_a=0.1; 1b, K_a=100, 2, K_1=0.1, K_2=100, $X_1=X_2$=0.5; 3, K_1=0.1, K_2=50, K_3=100, $X_1=X_2=X_3$=0.33; 4, K_1=0.1, K_2=33.33, K_3=66.66, K_4=100, $X_1=X_2=X_3=X_4$=0.25.

(iii) Under no circumstances can exchange in mixtures of n ideal site groups give isotherm contours such as those in Fig. 1d. Kielland plots are never linear and never show the maxima or minima sometimes observed. Thus, isotherms on ideal site groups cannot reproduce all experimental features of exchange isotherms.

(iv) It is impossible, by analyzing a sigmoid isotherm, to say how many ideal site groups (e.g., 2, 3, or 4) contribute. A given sigmoid isotherm can be represented equally well by different numbers of site groups with different K_i and in different proportions X_i [cf. eqn. (13)]. Fortunately, at least in principle, X-ray crystallography can give the needed information as to the number of site groups and the fractions X_i for each such group.

So far only ideal component site groups have been considered. If these site groups are individually non-ideal the overall exchange isotherms can be more complex. For example, one may assume that each site group gives a linear Kielland plot, with coefficients C_{0i} and C_{1i} for the ith site group given by

$$\log K_{ci} = C_{0i} + 2C_{1i}\,A_{ci}. \tag{14}$$

Then for the resultant isotherms due to two or more such site groups the Kielland plots may show maxima, minima, be nearly linear, or have maxima followed by minima, as illustrated in Fig. 7 (Barrer and Klinowski,

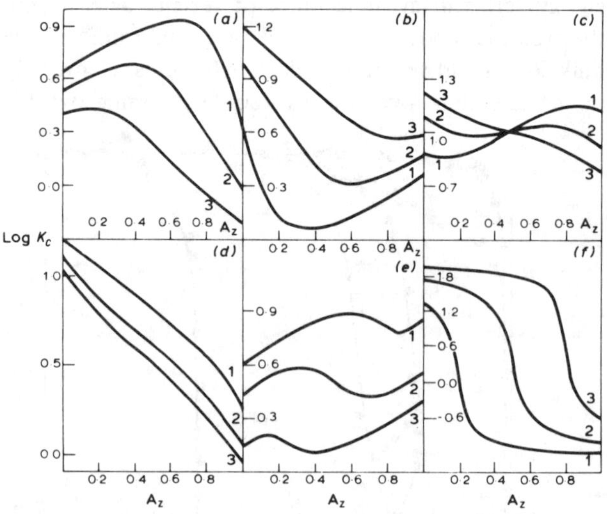

Fig. 7. Kielland plots for uni-univalent exchanges in a zeolite having two kinds of cation sites (Barrer and Klinowski, 1972a). For (a)–(e) Kielland-type non-ideality is assumed where C_{11} and $C_{12} \neq 0$ (log $K_{ci}=C_{oi} + 2C_{1i}$ A_C with i=1 and 2). In each case, for curves 1 X_1=0.2, X_2=0.8; for curves 2, $X_1=X_2$=0.5; for curves 3, X_1=0.8, X_2=0.2. (a) K_1=1, K_2=10, C_{11}=–0.3, C_{12}=0.3. (b) K_1=10, K_2=1, C_{11}=0.3, C_{12}=0.3. (c) K_1=10, K_2=10, C_{11}=–0.3, C_{12}=0.3. (d) K_1=1, K_2=10, $C_{11}=C_{12}$=–0.3. (e) K_1=1, K_2=10, $C_{11}=C_{12}$=0.3. (f) K_1=100, K_2=0.05, $C_{11}=C_{12}$=0.

1972a). Moreover, if the values of C_{1i} are sufficiently large and positive, one or more of the site groups may give miscibility gaps, and these will be reflected in the overall isotherm. By using eqn. (14), Barrer and Munday (1971c) analyzed the isotherm for Na ⇄ K in the edingtonite-type zeolite K–F shown in Fig. 1d, in terms of two kinds of site group, the first with $X_1 = 0.64$ and $C_1 < 0.87$ and the second with $X_2 = 0.36$ and $C_1 > 0.87$. Thus, the second site group gave a miscibility gap [cf. Table 4]. It was assumed that the true thermodynamic two-phase line was the dashed line in Fig. 1d which bisected the hysteresis loop. A narrow hysteresis loop was also observed in the same zeolite for Na ⇄ Li, and, with the same assumptions, this was also analyzed. Figure 8a shows the isotherm; Fig. 8b, curves 1 and 2, show the component isotherms on the two site groups, and curve 3 shows the resultant calculated isotherm for $X_1 = 0.64$ and $X_2 = 0.36$. Figure 8c shows the experimental Keilland plot. The relevant numerical data for this analysis of Na ⇄ Li and Na ⇄ K exchanges in zeolite K–F are collected in Table 5. In connection with the assumed values of X_1 and X_2 it is of interest that a subsequent X-ray diffraction study of K–F and of edingtonite-type zeolites derived from K–F by exchange confirmed that two kinds of cation site occurred in the proportions 2:1 (Baerlocher and Barrer, 1974).

ION SELECTIVITY

Earlier the interpretation of exchange isotherm contours was considered, and it was shown that very reasonable explanations can be provided. In the present section aspects of ion-exchange selectivity will be described, since this is one of the areas in which ion-exchangers have potential applicability.

The selectivity is conveniently defined as the quotient of eqn. (5). For ideal exchange involving uni-univalent ion pairs the value of α will be independent of the composition of the zeolite, and it can sometimes be very large for the entering ion, A. Thus, for Na ⇄ Ag, Li ⇄ Ag, and K ⇄ Ag in sodalite hydrate, silver was strongly preferred in the zeolite phase, with α for Na ⇄ Ag equal to about 335. Sodalite hydrate was used for quantitative estimation of silver (Barrer and Falconer, 1956).

In many other zeolite exchangers the selectivity is strongly dependent upon cationic composition, as illustrated in Fig. 9 for Na ⇄ alkylammonium in

TABLE 5. Analysis of isotherms Na ⇄ Li and Na ⇄ K in zeolite K–F (edingtonite-type)

Exchange	C_{11}	C_{12}	K_{a1}	K_{a2}	ΔG_1^θ (kJ/equiv.)	ΔG_2^θ (kJ/equiv.)
Na ⇄ K	−0.33	1.55	4.57	0.374	−3770	+3280
Na ⇄ Li	0	1.27	1.11	0.89	− 250	+ 290

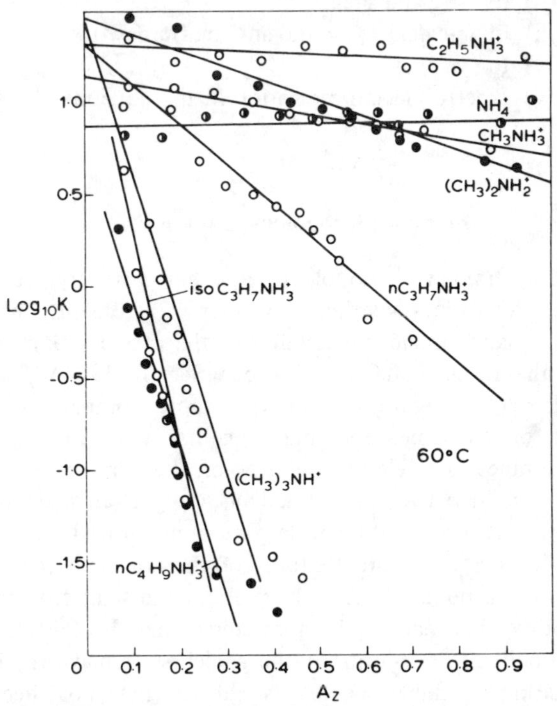

Fig. 9. Plots of log mass-action-quotient against cation fraction of the organic cation in clinoptilolite at 60°C. The exchanges are all Na⁺ ⇄ organic cation (Barrer, Papadopoulos, and Rees, 1967).

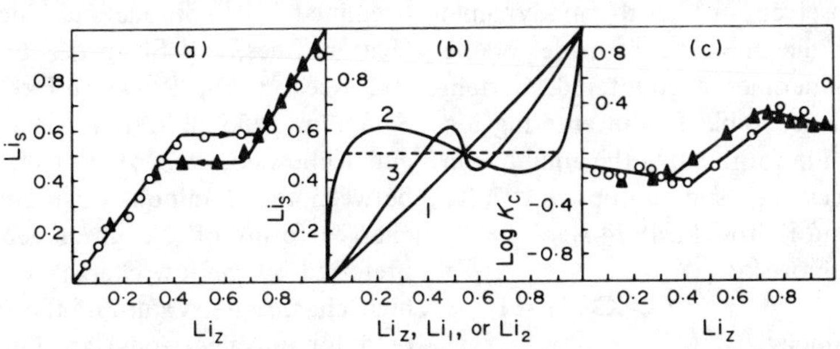

Fig. 8. (a) Na⇄Li at 25°C in zeolite K–F, showing the same kind of hysteresis loop as in Fig. 1d (Barrer and Munday, 1971c). (b) Curve 1, isotherm for sites of the first kind; curve 2, isotherm for sites of the second kind; curve 3, the resultant isotherm. (c) The Kielland plot for the experimental isotherm.

clinoptilolite (Barrer, Papadopoulos, and Rees, 1967). This must be the case for all sigmoid isotherms like that in Fig. 1b, where initially the exchange medium is selective for the entering ion A^+ but later becomes selective for ion B^+. One may distinguish between true thermodynamic affinity sequences, according to which the ions are arranged in order of descending values of K_a, and selectivity sequences, in which descending values of α are considered. These latter values may differ among the ions for different degrees of exchange. From the practical viewpoint of ion separation the selectivity sequences are the relevant ones.

Among the factors which influence selectivity may be included:

(i) Framework topology.
(ii) Ion size and shape
(iii) Charge density on the anionic framework.
(iv) Ion valency.
(v) Electrolyte concentration in the aqueous phase.

Framework topology and ion size

The framework topology, together with ion size, often determines whether a given ion can fit readily into the framework and the stability of the product (Barrer, Baynham, and McCallum, 1953; Barrer, 1966). The feldspar framework is somewhat more compact than that of analcime, and this, together with ion size, determines the relative stability of the same cationic form of these two structures. Thus, on ignition analcime (Na-form) loses zeolitic water and transforms to albite (Na-feldspar). Leucite (K-form of the analcime framework) and potash feldspar have about the same thermal stability; Rb-analcime is more stable than Rb-feldspar; and pollucite (Cs-form of the analcime framework) is remarkably stable while Cs-feldspar has not been synthesized. The Cs ion is too large to fit in the feldspar framework, but fits well with the analcime topology.

Among the more open zeolite structures considerable regularity has been observed in the thermodynamic affinity sequences, when due allowance is made for the incomplete exchange sometimes encountered. Certain site groups may prove inaccessible to the entering ion and are then not included in formulating the equilibrium on the remaining accessible site groups. Affinity sequences were (Barrer and Klinowski, 1974b):

Mordenite (Barrer and Klinowski, 1974b)	Cs>K>Na>Li
Phillipsite (Barrer and Munday, 1971a)	Cs>K>Rb>Na>Li
Gismondine-type Na–P (Barrer and Munday, 1971b)	Cs, Rb, K>Na>Li
Faujasite-type Na–X (Barrer, Rees, and Shamsuzzoha, 1966)	Rb>Cs>K>Na>Li
Faujasite-type Na–X (Sherry, 1966)	Cs>Rb>Na>K>Li
Faujasite-type Na–Y (Barrer, Davies, and Rees, 1968)	Cs>Rb≥K>Na>Li
Zeolite A (Barrer and Meier, 1959; Barrer, Rees, and Ward, 1963)	Cs>Na>Rb>K>Li
Chabazite (Barrer and Sammon, 1955; Barrer, Davies, and Rees, 1969)	Cs>K>Rb>Na>Li
Chabazite-type zeolite K-Gl (Barrer and Klinowski, 1972b)	Cs>K>Na>Li

Although there are irregularities, the trend is very striking; the heavy, polarizable, alkali metal ions of large ionic radius tend to be preferred by the zeolite and the smaller alkali metal ions by the aqueous phase, so far as the values of K_a and of ΔG^θ are concerned.

For the selectivities α [eqn. (5)] the frequently encountered dependence of α upon the cationic composition of the zeolite makes it more difficult to present generalizations. However, for low concentrations of the entering ion within the zeolite, the ion of greater size tends to be concentrated in the exchanger crystals. This means that zeolites such as mordenite are effective in concentrating Cs over Na or K. In chabazite K is concentrated well in presence of Na ($\alpha \sim 13$), indeed, about as well as the heavier, more polarizable, but smaller silver ion (Barrer and Sammon, 1955).

Charge density

In zeolite exchangers it is often possible to vary the $SiO_2:Al_2O_3$ ratio of a given framework either by synthesis or by finding natural occurrences of different compositions. Chabazites have been synthesized or found in nature in which the above ratio varies between 2.14 (Barrer and Baynham, 1956) and 7.8 (Gude and Sheppard, 1966). This represents a large change in the charge density, which could affect the selectivity shown towards a given ion pair. This behavior has been examined in aluminous and siliceous forms of faujasite (Barrer, Rees, and Shamsuzzoha, 1966; Barrer, Davies, and Rees, 1968, 1969) and of chabazite (Barrer and Sammon, 1955; Barrer, Davies, and Rees, 1969; Barrer and Klinowski, 1972b). For exchange of a pair of ions between the aluminous (high charge) and siliceous (low charge) forms of the given zeolite the larger ion was preferred by the low charge form and the smaller by the high charge one. Values of the mass action quotients K (= α for univalent ions) are illustrated in Table 6. The aluminous chabazite, G1, had $SiO_2:Al_2O_3$ = 2.14, and the siliceous natural crystals had $SiO_2:Al_2O_3$ = 5.13. The observed dielectric constant, ϵ, is higher for a zeolite

TABLE 6. Mass action quotients, K, for exchange between aluminous chabazite ($A_{G,1}$) and siliceous chabazite (Barrer and Klinowski, 1972b)

A = Li	B = Na	Quotient K	A = K	B = Na	Quotient K
$A_{chab.}$	$A_{G,1}$		$A_{chab.}$	$A_{G,1}$	
0.054	0.100	0.514	0.982	0.937	3.67
0.100	0.231	0.370	0.938	0.832	3.06
0.190	0.360	0.417	0.905	0.750	3.18
0.240	0.420	0.436	0.813	0.520	4.01
0.310	0.500	0.449	0.780	0.430	4.70
0.460	0.680	0.401	0.745	0.350	5.42

of fixed Si:Al ratio the larger the cation; and for a given cationic form is higher the larger the charge density. These assumptions were supported by measurements of the refractive indices, n, if $n^2 = \epsilon$ (Barrer and Klinowski, 1972b).

Ion sieving

As the size of the cation increases, in many zeolites ion sieving can become as clearly defined as molecule sieving. Ions must traverse the same channels as molecules, and may not be able to penetrate some, or indeed any, parts of the framework. Because many inorganic cations are considerably smaller than molecules, ion-sieve effects for such ions were observed in compact frameworks. For organic ions, however, ion sieving is found among open frameworks. Some examples are included in Table 7. As a rule high temperatures will offset the effect of larger ion radius. Thus, at ~200°C Cs^+ begins to migrate in analcime or sodalite. Double ion-sieve effects are also observable in zeolite structures where small cages with restricted windows co-exist with wider channels and windows in other parts of the same framework. Thus, in erionite 11-hedral cancrinite cages and hexagonal prisms co-exist with 23-hedra having some 8-ring windows. Large ions such as cesium can exchange in the 23-hedra but near room temperature cannot penetrate the 11-hedra or hexagonal prisms. To obtain more complete exchange such zeolites may be part exchanged at normal temperatures and then outgassed at high temperatures to promote ion mixing on all sites. The zeolite is then cooled to room temperature and again exchanged with the large ion; re-heated to mix the ions once again; re-exchanged at room temperature, and so on.

Ion sieving can result in some notable selectivities (Barrer and Sammon, 1956). Thus, Ag-analcime can readily and quantitatively remove Na salts from admixture with Cs salts. If Z denotes the analcime framework one has the reaction

$$\mathrm{Ag} - Z + \mathrm{NaCl} \rightarrow x\mathrm{AgCl}\downarrow + \mathrm{Ag}_{(1-x)}\mathrm{Na}_x{-}Z.$$

The solids may be filtered off and the AgCl removed with dilute ammonia from the (Ag,Na)-analcime. When there is no other exchangeable cation in the solution except one which will not penetrate the zeolite, one may hydrolyze salts of such cations even when these salts arise from strong acids and strong bases, for example, CsCl, by using the silver-zeolite. This excludes Cs^+ but precipitates insoluble AgCl and takes up hydronium ions from the water:

$$2x\mathrm{H_2O} \rightarrow x\mathrm{H_3O^+} + x\mathrm{OH^-}$$

$$x\mathrm{Cl^-} + x\mathrm{H_3O} + \mathrm{Ag} - \mathrm{Z} \rightarrow \mathrm{Ag}_{(1-x)}(\mathrm{H_3O}) - \mathrm{Z}$$

$$2x\mathrm{H_2O} + x\mathrm{Cl^-} + \mathrm{Ag} - \mathrm{Z} \rightarrow \mathrm{Ag}_{1-x}(\mathrm{H_3O})_x - \mathrm{Z}$$

$$+ x\mathrm{AgCl}\downarrow$$

$$+ x\mathrm{OH^-} + x\mathrm{AgCl}\downarrow$$

For more open zeolites, such as chabazite, the same behavior may be observed if for CsCl one substitutes

TABLE 7. Some ion-sieve processes near room temperature (Barrer, 1962)

Exchanger	Windows (dimensions in Å)	Exchangeable ions	Non-exchangeable ions
Analcime	6- and 8-rings (~2.2)	K^+ and smaller	Cs^+ and larger
Ultramarine	6-rings (~2.3)	K^+ and smaller	Cs^+ and larger
Sodalite hydrate	6-rings (~2.3)	K^+ and smaller	Cs^+ and larger
Chabazite	8-rings (3.7 × 4.2)	n-alkyl-NH_3^+	iso-alkyl-NH_3^+, NMe_4^+, etc.
Zeolite A	8-rings (4.3)	n-alkyl-NH_3^+	iso-alkyl-NH_3^+, NMe_4^+, etc.
Clinoptilolite	10-rings (3.5 × 7.9 Å) 8-rings (3.0 × 4.4 Å)	n-alkyl-NH_3^+	NMe_4^+, iso-butyl-NH_3^+

NMe_4Cl, the tetramethyl ammonium ion, NMe_4^+, being excluded. This behavior has been suggested as a field test for zeolites (Helfferich, private communication).

Steric limitations

A further profound effect upon exchange of large ions in open tectosilicates is imposed by the rigid nature of the zeolite frameworks and hence the nearly fixed pore volumes in zeolites. This effect is wholly distinct from ion sieving. Faujasite is one of the most porous zeolites with approximately 53% of the volume of each crystal accessible to water molecules. However, faujasite is saturated when it has sorbed ~30 neo-pentane molecules per unit cell. Tetramethylammonium (N^+Me_4) and neo-pentane (CMe_4) have about the same volume, so that about 30 N^+Me_4 ions would also fill the accessible pore volume in faujasite. However, in aluminous synthetic faujasite (zeolite X) there may be ~90 framework charges per unit cell. It follows that exchange, for steric reasons not associated with molecule sieving, must be limited when Na^+ is replaced by NMe_4^+ or other large organic cations. The upper limit referred to above will not be reached for several additional reasons. Firstly, all water in the 26-hedra would have to be displaced, but water has itself a strong affinity for the zeolite. Also clusters of the positive organic ions would be crowded together in the 26-hedra without being solvated. This could produce an energetically unfavorable situation. Finally, many of the Na ions in the faujasite are in sites inaccessible to large organic ions [e.g., in 14-hedra (sodalite cages) and in hexagonal prisms] so that ion-sieve effects also operate for such sites. Some examples of observed limitations in the exchange of large ions by faujasite (zeolite X) are illustrated in Table 8. The NH_4^+ ion exchanged nearly completely, but the other ions exchanged to decreasing extents as their sizes increased.

TABLE 8. Exchange limits in faujasite (zeolite X) near room temperature (Barrer, Buser, and Grutter, 1956)

Ion	Exchange solution	% Exchange
NH_4^+	2N chloride	92
$CH_3NH_3^+$	2N "	58
$(CH_3)_2NH_2^+$	1N "	37
$(CH_3)_3NH^+$	2N "	28
$(CH_3)_4N^+$	2N "	23

Electrolyte concentration

For exchanges involving only univalent ions the concentration of the external electrolyte need have no large effect upon the selectivity. However, when the ions have different charges, the situation becomes more complex. The behavior to be expected has been investigated quantitatively (Barrer and Klinowski, 1974). It has been shown

(i) that selectivity becomes progressively greater for the ion of higher valence the more dilute the exchanging solution; and

(ii) that isotherms without inflections and convex upwards (Fig. 1a) may change with increasing dilution to sigmoid curves (Fig. 1b) and finally back once more to curves without inflection but with the opposite curvature, i.e. convex downwards.

The second of these observations means that, when the ions exchanging do not have the same valence, sigmoid isotherms cannot be taken unambiguously as evidence that site groups of at least two kinds are involved. On the other hand, when the exchanging cations have the same charge this ambiguity does not arise, although of course sigmoid isotherms can still occur on a single kind of site group if the sites behave non-ideally, as considered above.

CONCLUSIONS

Zeolites provide excellent model systems in which to study ion exchange and isomorphous replacement under mild, easily accessible experimental conditions. Their selectivity can be extreme from total to partial ion sieving, from steric limitations due to the restricted numbers of larger organic ions which fill the intracrystalline pore volumes, to specific selectivities dependent upon neither of the above effects. A number of types of exchange isotherm have been observed and can be interpreted in terms of ideality and/or non-ideality of site groups, of which there may be several kinds in any one zeolite exchanger. The equilibria on each kind of site group can be related quantitatively to the equilibrium for the crystal as a whole. Thus, encouraging progress has been made in a field of common interest to earth scientists, chemists, and engineers.

REFERENCES

Baerlocher, Ch. and Barrer, R. M. (1974) The crystal structure of synthetic zeolite F: *Z. Kristallogr.* **140**, 10.

Barrer, R. M. (1962) Some features of ion exchange in crystals: *Chem. and Ind.*, p. 1258.

Barrer, R. M. (1966) Mineral synthesis by the hydrothermal technique: *Chemistry in Britain*, p. 380.

Barrer, R. M. (1974) Isomorphous replacement by ion exchange: some equilibrium aspects: *Bull. Soc. Fr. Mineral. Cristallogr.* 97, 89.

Barrer, R. M. and Baynham, J. W. (1956) Hydrothermal chemistry of silicates. Part VII. Synthetic potassium aluminosilicates: *J. Chem. Soc.*, p. 2882.

Barrer, R. M., Baynham, J. W. and McCallum, N. (1953) Hydrothermal chemistry of silicates. Part V. Compounds structurally related to analcite: *J. Chem. Soc.*, p. 4035.

Barrer, R. M., Buser, W. and Grutter, W. F. (1956) Synthetischer faujasite. I. Eigenschaften und ionenaustauschcharakter: *Helv. Chim. Acta* **39**, 518.

Barrer, R. M., Davies, J. A. and Rees, L. V. C. (1968) Thermodynamics and thermochemistry of cation exchange in zeolite Y: *J. Inorg. Nucl. Chem.* **30**, 3333.

Barrer, R. M., Davies, J. A. and Rees, L. V. C. (1969a) Thermodynamics and thermochemistry of cation exchange in chabazite: *J. Inorg. Nucl. Chem.* **31**, 219.

Barrer, R. M., Davies, J. A. and Rees, L. V. C. (1969b) Comparison of the ion exchange properties of zeolites X and Y: *J. Inorg. Nucl. Chem.* **31**, 2599.

Barrer, R. M. and Falconer, J. D. (1956) Ion exchange in feldspathoids as a solid state reaction: *Proc. Roy. Soc.* **236A**, 227.

Barrer, R. M. and Hinds, L. (1953) Ion exchange in crystals of analcite and leucite: *J. Chem. Soc.*, p. 1879.

Barrer, R. M. and Klinowski, J. (1972a) Ion exchange involving several groups of sites: *J. Chem. Soc., Faraday. I* **68**, 73.

Barrer, R. M. and Klinowski, J. (1972b) Influence of framework charge density on ion exchange properties of zeolites: *J. Chem. Soc. Faraday. I* **68**, 1956.

Barrer, R. M. and Klinowski, J. (1974a) Ion exchange selectivity and electrolyte concentration: *J. Chem. Soc. Faraday. I* **70**, 2080.

Barrer, R. M. and Klinowski, J. (1974b) Ion exchange in mordenite: *J. Chem. Soc. Faraday. I* **70**, 2362.

Barrer, R. M. and Klinowski, J. (1977) Theory of isomorphous replacement in aluminosilicates: Phil. Trans. of Roy. Soc. London, **285A**, 637.

Barrer, R. M., Klinowski, J. and Sherry, H. S. (1973) Zeolite exchangers. Thermodynamic treatment when not all ions are exchangeable: *J. Chem. Soc. Faraday. II* **69**, 1669.

Barrer, R. M. and Meier, W. M. (1959) Exchange equilibria in a synthetic crystalline exchanger: *Trans. Faraday Soc.* **55**, 130.

Barrer, R. M. and Munday, B. M. (1971a) Cation exchange reactions of a sedimentary phillipsite: *J. Chem. Soc.* **A**, 2904.

Barrer, R. M. and Munday, B. M. (1971b) Cation exchange reactions of zeolite Na–P: *J. Chem. Soc.* **A**, 2909.

Barrer, R. M. and Munday, B. M. (1971c) Cation exchange in the synthetic zeolite K–F: *J. Chem. Soc.* **A**, 2914.

Barrer, R. M., Papadopoulos, R. and Rees, L. V. C. (1967) Exchange of sodium in clinoptilolite by organic cations: *J. Inorg. Nucl. Chem.* **29**, 2047.

Barrer, R. M. and Raitt, J. S. (1954) Ion exchange in ultramarine: *J. Chem. Soc.*, p. 4641.

Barrer, R. M., Rees, L. V. C. and Shamsuzzoha, M. (1966) Thermochemistry and thermodynamics of ion exchange in a near-faujasite: *J. Inorg. Nucl. Chem.* **28**, 629.

Barrer, R. M., Rees, L. V. C. and Ward, D. J. (1963) Thermochemistry and thermodynamics of ion exchange in a crystalline exchange medium: *Proc. Roy. Soc.* **273A**, 180.

Barrer, R. M. and Sammon, D. C. (1955) Exchange equilibria in crystals of chabazite: *J. Chem. Soc.*, p. 2838.

Barrer, R. M. and Sammon, D. C. (1956) An ion sieve reagent for caesium-alkali metal separations: *J. Chem. Soc.*, p. 675.

Fowler, R. H. and Guggenheim, E. A. (1939) *Statistical Thermodynamics*, Cambridge Univ. Press, p. 437.

Gude, A. J. III. and Sheppard, R. A. (1966) Silica-rich chabazite from the Barstow Formation, San Bernardino County, Southern California: *Amer. Mineral.* **51**, 909.

Kielland, J. (1935) Thermodynamics of base-exchange equilibria of some different kinds of clays: *J. Soc. Chem. Ind.* **54**, 232T.

Lacher, J. R. (1937) The statistics of the hydrogen–palladium system: *Proc. Camb. Phil. Soc.* **33**, 518.

Meier, W. M. and Olson, D. H. (1971) Zeolite Frameworks, in *Molecular Sieve Zeolites – I*: Amer. Chem. Soc., Advances in Chem. Series **101**, 155.

Orville, P. M. (1963) Alkali ion exchange between vapor and feldspar phases: *Amer. J. Sci.* **261**, 201.

Sherry, H. S. (1966) Ion exchange properties of zeolites. I. Univalent ion exchange in synthetic faujasite: *J. Phys. Chem.* **70**, 1158.

Smith, J. V. (1971) Faujasite-type structures: the aluminosilicate framework: positions of cations and molecules: nomenclature: *Molecular Sieve Zeolites – I* Amer. Chem. Soc., Advances in Chem. Series **101**, 171.

ADSORPTION AND CATALYTIC PROPERTIES OF SOME SOVIET NATURAL ZEOLITES

G. V. TSITSISHVILI

Institute of Physical and Organic Chemistry
Academy of Sciences of the Georgian SSR, Tbilisi, U.S.S.R.

Abstract – This paper presents the main results of studies of the physiochemical properties (adsorption, chromatographic, thermal) of natural zeolites from the Transcaucasia, the Ukraine, the Far East, Siberia, Middle Asia, and other regions of the Soviet Union. Studies of gas adsorption over a wide range of relative pressures, of the chromatographic separation of mixtures of gases, and of the thermal properties of natural and modified clinoptilolites, mordenites, and other zeolites have shown that their physical properties are strongly dependent upon the nature and extent of cation substitution.

Excellent chromatographic separations of many gas streams are achieved with natural zeolites; however, both the nature and degree of chemical modification and the preliminary thermal treatment influence their separation properties. Natural, high-silica zeolites can be used successfully in purification processes in acid media. The adsorption, catalytic, drying, and separation properties of Soviet natural zeolites (e.g., clinoptilolite and mordenite) are equal to those of the highest grade natural zeolites of other countries and promise to find wide practical utilization.

INTRODUCTION

The Soviet Union is rich in natural adsorbents including members of the zeolite group of minerals (Fersman, 1962; Gvakharia, 1952; Senderov and Khitarov, 1970; Kossovskaya, 1973). In addition to the geological and mineralogical studies made of natural zeolites, many of their adsorption, chromatographic, and thermal properties have now been determined. These latter studies and the results of spectroscopic and catalytic activity studies are reviewed.

ADSORPTION AND CHROMATOGRAPHIC PROPERTIES

The adsorption of many gases and vapors by Soviet natural zeolites has been determined over a wide range of pressures. The adsorption of water vapor has been studied in detail. Because the diameter of the entrance windows into the pore structure of many zeolites varies from 3 Å to 5 Å, their intracrystalline pore volume is easily accessible to water molecules which have a critical diameter of 3.15 Å.

Natural zeolites, as well as synthetic molecular sieves, are microporous adsorbents with bidispersional pore structures. The adsorption isotherms generally have a rectangular shape; the principal adsorption occurs in the primary pore structure, while adsorption in the secondary pore structure is considerably less important.

The adsorption of water vapor on heulandite, thomsonite, desmine, laumontite, and mordenite has been studied by the isopiestic method (Bikov *et al.*, 1965; Bikov and Scherbatyuk, 1967). It was shown that all isotherms are convex with respect to the pressure axis and rise steeply. It was established that dehydrated analcime does not rehydrate. Heating to 270°C with the loss of 8% of water leads to irreversible changes in the structure.

The adsorption properties of laumontite, mordenite, and rocks containing mordenite and other zeolites were investigated by Mirsalimov (1970), Tarasevich and Suprichev (1975), and Tsitsishvili (1977). The effect of chemical modification on the adsorption properties has been reported by Korneichuk *et al.* (1973), Tsitsishvili *et al.* (1973a), Chelishchev (1973), and Barnabishvili *et al.* (1977). The acid treatment of natural clinoptilolites leads to the removal of cations and dealuminization with unscreened cations being the first to be removed.

Investigations of the adsorption of water vapor on Dzegvi, GSSR, clinoptilolite with a microbalance have shown that acid treatment leads to a decrease of adsorption ability proportional to an increasing concentration of acid. Such a mechanism is clear if one takes into account the fact that unscreened cations are the main energetic centers for adsorption.

Water molecules are adsorbed in the primary pores of the zeolite, and at $P/P_s \approx 0.1\text{-}0.2$, the adsorption space is nearly filled. Vapors of benzol and n-hexane are adsorbed into secondary pores which have a surface area of nearly 20 m^2/g. Acid modification promotes the development of secondary porosity. Modifying clinoptilolite by potassium ion-exchange produces a decrease in its ability to adsorb water vapor.

Data obtained on the adsorption of nitrogen by clinoptilolite at low temperature show that the process occurs on the external crystal surface. Acid modification produces a considerable increase in the adsorption capacity, i.e. opening of the pore structure for nitrogen molecules occurs.

The clearly pronounced microporosity of natural zeolites results in a specific character of the adsorption process as illustrated by the results obtained from studies of carbon dioxide and methyl alcohol adsorption on clinoptilolite (Dubinin *et al.*, 1977). It was shown that at low relative pressures ($P/P_s = 5 \times 10^{-4}$), the adsorption of carbon dioxide is much higher on clinoptilolite than on a NaX zeolite. The same result was observed for the adsorption of methyl alcohol on clinoptilolite and on a carbon molecular sieve from Japan. For clinoptilolite, the value of the characteristic energy of CO_2 adsorption is almost 7000 cal/mole, while for NaX it does not exceed 4000 cal/mole, and for carbon sorbents it is 2500–2700 cal/mole.

Relatively high values of the characteristic energies of vapor adsorption on clinoptilolite reflect relatively lower micropore volumes. This fact is of great importance, because it determines the ability of natural zeolites to compete with synthetic molecular sieves. Data on adsorption properties of natural zeolites are given in Table 1.

Recently, chromatographic properties of natural zeolites have been studied using a mixture of argon, oxygen, and nitrogen as a model system (Tsitsishvili *et al.*, 1973b). It was shown that separation does not occur

TABLE 1. Water-vapor adsorption of some natural zeolites at $P/P_x = 0.40$ and 20–25°C

Zeolite, locality*	Adsorption a $\frac{\text{mmole}}{\text{g}}$	Adsorption V $\frac{\text{cm}^3}{\text{g}}$	Reference
Desmine, Primorskiy Kray	7.25	0.131	Bikov and Scherbatyuk (1967)
Natrolite, Khibini	6.0	0.108	,, ,,
Heulandite, Akhaltsikhe	8.10	0.146	Bikov *et al.* (1965)
Heulandite, Siberian platform	7.10	0.128	,, ,,
Thomsonite, Siberian platform	5.17	0.093	,, ,,
Stellerite, Zabaykalie	8.32	0.150	,, ,,
Mordenite, Karadag	5.56	0.10	Tarasevich and Suprichev (1975)
Mordenite, Siberian platform	5.89	0.106	,, ,,
Mordenite-containing tuff, Bolnisi	4.19	0.075	Tsitsishvili (1977)
Mordenite, Transcarpathians	6.10	0.110	,, ,,
Mordenite, Siberian platform	8.42	0.152	,, ,,
Chabazite, USA No. W7-6A	10.82	0.195	,, ,,
Phillipsite, USA No. W7-25A	7.69	0.139	,, ,,
Laumontite, Pritashkentskiy district	3.15	0.056	Mirsalimov (1970)
Clinoptilolite, Kermine	7.20	0.130	,, ,,
Clinoptilolite, Hector, California	7.20	0.130	Barrer and Makki (1964)
Clinoptilolite, USA No. M4-36	8.42	0.152	Barnabishvili *et al.* (1977)
Clinoptilolite, Khekordzula	7.93	0.143	,, ,, ,,
Clinoptilolite, Ai-Dag	6.28	0.113	,, ,, ,,
Clinoptilolite, Valley of Gheizers	7.75	0.140	,, ,, ,,
Clinoptilolite, Dzegvi	6.62	0.119	,, ,, ,,
Clinoptilolite, Badkhiz	5.50	0.099	,, ,, ,,
Clinoptilolite, Sakirnitsa	3.30	0.059	,, ,, ,,
"Klinosorb", Tokaj, Hungary	4.90	0.088	,, ,, ,,

*All localities are in the USSR unless otherwise indicated.

on the sample of calcium clinoptilolite studied. It is known that the separation selectivity of clinoptilolite for this mixture essentially depends on the potassium and calcium content. Introduction of potassium ions into the zeolite leads to a separation and to the following sequence of elution: argon, nitrogen, oxygen. With an increasing amount of potassium, oxygen is eluted after argon. With further substitution, entrance windows of the zeolite are apparently broadened, and not only oxygen molecules, but also nitrogen molecules can penetrate the inner zeolite structure. Enrichment of clinoptilolite by potassium ions also leads to amplification of the dispersive interaction. The presence of a quadrupole moment in the nitrogen molecule causes its intensive interaction with clinoptilolite, and the elution of this substance is last.

Studies of the separation properties of mordenite-containing tuffs from the Bolnisi region with the mixture oxygen–nitrogen–methane (Tsitsishvili *et al.*, 1975) showed that the separation becomes better when the activation temperature is higher (from 300°C to 450°C). It is appropriate to note that mordenite is potentially one of the most attractive adsorbents for the oxygen-enrichment of air.

CATALYTIC PROPERTIES

It was shown by studies of the catalytic activity of clinoptilolites in the reaction of isopropyl alcohol dehydration that the hydrogen forms of these zeolites are characterized by high selectivity and dehydrating ability (Tsintskaladze and Sidamonidze, 1977). The Georgian clinoptilolite was found to be an active catalyst in the reaction between methyl alcohol and hydrogen sulphide (Areshidze and Gadziev, 1977). As a result of interaction of these substances, methyl mercaptan and methyl sulphate are formed, which are the initial substances for the synthesis of many valuable compounds.

SPECTRAL STUDIES

The infrared spectra of heat-treated zeolites provide new data concerning their dehydration and rehydration, the state of water, and the existence of hydroxyl groups and hydronium ion. The infrared spectra of clinoptilolites and of their ion-exchange forms showed that, depending on the radius, atomic number, and valence of the exchange cations, changes in the positions of maxima and of the intensities of high-frequency bands take place in the range 3100–3700 cm^{-1}. The water content is also affected. Cations have little effect on the clinoptilolite spectrum in the range of the natural oscillations of the framework (400–800 cm^{-1}) and of the Al–O and Si–O bands (900–1200 cm^{-1}). Natrolite and analcime, which contain smaller intracrystalline pores, show a stronger interaction between cations and the framework (Belitskiy and Golubova, 1968).

For the quantitative determination of molecular water and hydroxyl groups from the infrared absorption bands, a band corresponding to the deformation oscillation was chosen. The relative content of water in the specimens was estimated by the ratio of the optical densities of the deformation oscillation at the studied temperature to its value for an initial air-dry specimen. Studies of clinoptilolite from Khekordzula and of heulandite from the Akhaltsikhe region (Charkviani *et al.*, 1977) show that most of the water is lost during pumping out at room temperature. The water is completely removed at 300–400°C. Unlike clinoptilolite, a change of the spectrum of heulandite was observed at low temperatures due to the phase transition. After vacuum treatment at 300°C, there remains very little molecular water; the weak band at 1690 cm^{-1} apparently belongs to the hydronium ion.

The presence of hydroxyl groups in clinoptilolite and heulandite was not established by previous investigators because they used only very small amounts of the specimen embedded in KBr pellets for obtaining the spectrum. Calculations of the number of hydroxyl groups based on the data of frequency intensities of valent oscillations give a value of 0.5 mole OH per gram of clinoptilolite, or one hydroxyl group per 12–16 atoms of silicon. A distinct separation of the bands of hydroxyl groups (maximum at 3633 cm^{-1}) and of the molecular water (maximum at 3390 cm^{-1}) was obtained by means of a lithium fluoride prism.

Thus, in addition to the hydroxyls (maximum at 3749 cm^{-1} for the clinoptilolite) on the external surface of the microcrystals, there are also hydroxyls in the micropores (maximum at about 3702 and 3633 cm^{-1} in the clinoptilolite spectrum and at 3556 and 3620 cm^{-1} in the heulandite spectrum). An interesting peculiarity of the spectra of clinoptilolite and of heulandite is the presence of a band at 1693 cm^{-1}, which is caused, apparently, by deformation oscillations of the hydronium ion $(H_3O)^+$.

The proton magnetic resonance in zeolites gives important clues concerning the mobility of water molecules. The molecular mobility leads to a sharp narrowing of the spectrum, which is determined by the total complex of all real positions occupied by water molecules in the zeolite structure.

From the temperature dependence of the components of the local magnetic field tensor, it can be concluded that in clinoptilolite the positions occupied by water molecules change with temperature. The absence of such changes in heulandite shows a profound difference between the structures of these two zeolites (Belitskiy *et al.*, 1973). Studies of laumontite show that there are two types of water molecules with different mobilities in this zeolite. The molecules of one type diffuse, while the others only rotate, keeping the center of gravity at rest (Belitskiy *et al.*, 1964).

ACID RESISTANCE

Based on their solubility in 1N hydrochloric acid, zeolites can be subdivided into four groups: highly resistant, resistant, poorly resistant, and unresistant. In a study by Bogdanova and Belitskiy (1968), 0.1 g of zeolite and 50 ml of solution were heated on a steam-bath for 1 hour. Such an acid solution failed to remove channel cations or to dissolve aluminum and silicon from zeolites of the first group. The presence of dissolved aluminum ions, exchangeable cations, and negligible silicon was characteristic of the reaction of zeolites of the second group. In the case of the third group, the channel cations and aluminum dissolved readily. Zeolites of the fourth group dissolved without residue. Ferrierite and mordenite are highly resistant; clinoptilolite, desmine, and heulandite are resistant; phillipsite, erionite, and chabazite are poorly resistant; and natrolite is unresistant.

Detailed studies of the acid resistance of clinoptilolite have been reported by Chelishchev and Berenshtein (1973), Tsitsishvili *et al.* (1973a), and Koridze *et al.* (1977). It was shown, in agreement with the previous work of Barrer and Makki (1964), that in relatively dilute solutions, decationization takes place, i.e. it is possible to prepare a hydrogen form of the zeolite, and dealuminization takes place with increasing concentration. It is noteworthy that in the latter case the amount of aluminum removed is equal to the sum of the exchange cations removed from the zeolite by the acids.

THERMAL STUDIES

Thermal studies provide important information concerning the dehydration, rehydration, and thermal stability of zeolites. Studies in this field have shown that thermal stability depends on the type of the zeolite, on the Si/Al ratio, and on the nature of the cation that compensates the negative charge of the aluminum-silicon-oxygen crystalline framework.

Zeolites can be classified on the basis of their endothermal effects into several groups: zeolites with sharply pronounced effects (natrolite, mesolite, scolecite, thomsonite), those with less resolved effects (edingtonite, laumontite, heulandite, and others), and those with a single broad effect (chabazite, erionite, ferrierite, mordenite, and others). Such a peculiarity of the dehydration process is explained by the degree of "tightness" of the crystalline structure of zeolites. Exothermal effects in a number of cases are manifested above 700°C. At a lower temperature they are rarely observed, because, if present, they are obscured by the endothermal dehydration effects.

Natural and K-exchanged clinoptilolite from Dzegvi and the Hungarian "Klinosorb" are characterized by similar thermograms. The greatest part of water is lost between 80°C to 350°C. Introduction of potassium ions into the crystalline lattice promotes a decrease in the number of different single peaks, homogenization, and an increase of the thermal stability of the zeolite (Tsitsishvili *et al.*, 1973b).

The DTA curves of ammonium forms of clinoptilolites show a bifurcated endothermal effect caused by the separation of water and ammonia (the reaction range is 50-500°C). An endoeffect leads into an exoeffect which is caused by the completion of deammonization, by the chemical transformation of the separated ammonia, and by the dehydration of the zeolite (Kvantaliani and Tsitsishvili, 1974). The ability of the zeolite to rehydrate is closely connected to its thermal stability (Belitskiy, 1971 a, b).

Soviet natural zeolites (in particular, clinoptilolite) have been examined for their utility in the following processes: drying of air, drying of natural gas, drying of the returned products of butyl rubber and xylene fraction, purification of waste gases by removal of sulphur dioxide, fine purification of chlorine and different acid media, modification and structurization of rubbers, as well as in other applications.

REFERENCES

Areshidze, Kh. I. and Gadziev, M. K. (1977) Studies of catalytic properties of aluminosilicates: in *Clinoptilolite*, Proc. Symp. Problems of Clinoptilolite Studies and Application, Nov. 1974, Tbilisi, G.S.S.R. 213-219.

Barnabishvili, D. N. *et al.* (1977) Sorption properties of natural and modified clinoptilolites from Georgian deposits: in *Clinoptilolite*, Proc. Symp. Problems of Clinoptilolite Studies and Application, Nov. 1974, Tbilisi, G.S.S.R. 148-154.

Barrer, R. M. and Makki, M. B. (1964) Molecular sieve sorbents from clinoptilolite: *Can. Chem. J.* **42**, 1481-1487.

Belitskiy, I. A. (1971a) *Materials on Genetics and Experimental Mineralogy* **6**: Nauka, Moscow.

Belitskiy, I. A. (1971b) Rehydration of natural zeolites: *Geology and Geophysics* **11**, 12-21.

Belitskiy, I. A. and Golubova, G. A. (1968) Infra-red spectra of cation exchanged forms of natural zeolites: *Geology and Geophysics* **12**, 40-47.

Belitskiy, I. A. *et al.* (1964) Investigation of laumontite by the nuclear magnetic method: *Dokl. Akad. Nauk SSSR* **159**,

1038-1041.

Belitskiy, I. A. *et al.* (1973) Proton magnetic resonance in plate zeolites: *Zh. Strukt. Khim.* **14**, 1024-1027.

Bikov, V. T. *et al.* (1965) Investigation of adsorption of water vapor of natural zeolites: *Zeolites, their Synthesis, Properties and Utilization*, Nauka, Moscow, p. 360.

Bikov, V. T. and Scherbatyuk, N. E. (1967) Sorption of water vapor of natural zeolites and influence of the mobility of water molecules in zeolites upon the desorption process: *Natural Sorbents*, Nauka, Moscow, p. 156.

Bogdanova, V. I. and Belitskiy, I. A. (1968) The resistance of natural zeolites in hydrochloric acid: *Geology and Geophysics* **4**, 44-53.

Charkviani, M. K., Tsitsishvili, G. V. and Tsintskaladze, G. P. (1977) Infra-red spectra of clinoptilolite from the Khekordzula deposit: in *Clinoptilolite*, Proc. Symp. Problems of Clinoptilolite Studies and Application, Nov. 1974, Tbilisi, G.S.S.R. 90-95.

Chelishchev, N. F. (1973) *Ion Exchange Properties of Minerals*, Nauka, Moscow, p. 124.

Chelischev, N. F. and Berenshtein, B. G. (1973) Ion exchange properties of clinoptilolite: *Dokl. Akad. Nauk SSSR* **210**, 1110-1112.

Dubinin, M. M., Lozhkina, N. S. and Onusaitis, B. A. (1977) Peculiarities of clinoptilolite adsorption properties: in *Clinoptilolite*, Proc. Symp. Problems of Clinoptilolite Studies and Application, Nov. 1974, Tbilisi, G.S.S.R. 5-11.

Fersman, A. E. (1962) *Zeolites of Russia and their Mineralogy*, Vol. 1, Akad. Nauk SSSR, Moscow.

Gvakharia, G. V. (1952) *Zeolites of Georgia*, Akad. Nauk GSSR, Tbilisi.

Koridze, Z. I., Krupennikova, A. Yu. and Andronikashvili, T. G. (1977) Effect of aggressive media on clinoptilolite composition: in *Clinoptilolite*, Proc. Symp. Problems of Clinoptilolite Studies and Application, Nov. 1974, Tbilisi, G.S.S.R. 96-101.

Korneichuk, G. K., Scherbatyuk, N. E. and Froer, E. F. (1973) The adsorption of acetone and carbon tetrachloride, and their liquid mixtures on natural and modified clinoptilolites: *Zh. Fiz. Khim.* **47**, 2391-2393.

Kossovskaya, A. C. (1973) Genetic associations of sedimentary zeolites in the Soviet Union: *Molecular Sieves*, Advances in Chemistry Series **121**, 200-208.

Kvantaliani, E. K. and Tsitsishvili, G. V. (1974) Thermographic study of ammonium forms of clinoptilolite: *Bull. Acad. Sci. GSSR* **73**, 89-92.

Mirsalimov, A. M. (1970) The sorption of water vapor of natural zeolites from Uzbekistan: *Uzbekskiy Chimicheskiy Zhurnal* **5**, 44-46.

Senderov, E. E. and Khitarov, N. I. (1970) *Zeolites, their Synthesis and Conditions of Formation in Nature*, Nauka, Moscow.

Tarasevich, Yu. N. and Suprichev, V. A. (1975) Sorption properties of natural mordenite: *Colloid. Zh.* **37**, 812-815.

Tsintskaladze, Z. P. *et al.* (1976) Investigation of dehydration activity properties of some high-siliceous native and synthetic zeolites: *Bull. Acad. Sci. GSSR* **83**, 93-96.

Tsintskaladze, S. P. and Sidamonidze, Sh. I. (1977) *Bull. Acad. Sci., GSSR* (in press).

Tsitsishvili, G. V. (1977) Clinoptilolite and other natural zeolites: in *Clinoptilolite*, Proc. Symp. Problems of Clinoptilolite Studies and Application, Nov. 1974, Tbilisi, G.S.S.R. 72-21.

Tsitsishvili, G. V. *et al.* (1973a) The effect of chemical modification on adsorption properties of clinoptilolite: *Bull. Acad. Sci. GSSR* **69**, 345-349.

Tsitsishvili, G. V. *et al.* (1973b) Physical-chemical properties of modified clinoptilolite: *Izv. Akad. Nauk SSSR* **9**, 1394-1398.

Tsitsishvili, G. V. *et al.* (1975) Gas chromatographic properties of tuffs containing mordenite: *Dokl. Akad. Nauk* **225**, 587-589.

SORPTION AND DIFFUSION IN NATURAL ERIONITE

DOUGLAS M. RUTHVEN and RUSSELL I. DERRAH*

Department of Chemical Engineering, University of New Brunswick, Fredericton, N.B., Canada

Abstract – The structure of natural erionite is similar to that of the A zeolites in that both frameworks contain discrete cavities interconnected through 8-membered oxygen windows (~ 4 Å free diameter). A comparative study of the sorption of three representative hydrocarbons (C_2H_6, C_2H_4, and nC_5H_{12}) in natural erionite from Rome, Oregon, and in types 4A and 5A zeolites was undertaken to determine the extent to which the structural similarities and differences are reflected in the sorptive properties.

The cavities of erionite (406 Å^3) are much smaller than those of the A zeolites (776 Å^3), and as a result of this difference the erionite isotherms are more rectangular and the heats of sorption are larger. However, the equilibrium data for both zeolites can be correlated by the same statistical thermodynamic model.

Diffusivities in both zeolites are strongly dependent on concentration, but the concentration dependence can be accounted for entirely by the non-linear relationship between sorbate activity and concentration as defined by the equilibrium isotherm. Corrected diffusivities (D_0) calculated according to Darken's equation ($D = D_0 \, d \ln p / d \ln c$) are essentially independent of concentration. The orders of magnitude of the diffusivities for both erionite and 5A zeolite are correctly predicted by transition state theory. Diffusional activation energies for C_2H_6 and nC_5H_{12} in erionite are somewhat greater than in 5A while for the planar C_2H_4 molecule the diffusivity in erionite is slightly smaller. This is consistent with the difference in the shape of the windows (3.6 × 5.2 Å for erionite, 4.2 × 4.2 Å for 5A). The diffusivities for nC_6H_{12} are consistent with values obtained independently from measurements of the Thiele modulus for the pentane cracking reaction.

NOTATION

c	sorbate concentration (molecules/cavity)
D	Fickian diffusion coefficient (cm^2 sec^{-1})
D_0	Corrected diffusivity calculated according to equation (4)
D_*	Pre-exponential factor [equation (5)]
E	Diffusional activation energy
f^*_z	Partition function for diffusing molecule in transition state
f'_g	Partition function per unit volume for free gas phase molecule
f'_{trans}	Translational partition function per unit volume for free gas phase molecule
f_{rot}	Rotational partition function
h	Planck's constant
K	Henry's law constant defined by $c = Kp$
K_0	Pre-exponential factor in equation (1)
k	Boltzmann's constant
m	Integer given by $m \leqslant \nu/\beta$
n	Number of cavities per cm^3 of zeolite crystal
p	Sorbate pressure (torr)
q_0	Limiting heat of sorption
R	Gas constant
ΔS°	Standard entropy change on sorption defined by equation (2)
β	Effective molecular volume of sorbate
δ	Lattice parameter

*Canadian Industrial Ltd., McMasterville, Quebec.

INTRODUCTION

The pore structure of erionite is in some respects similar to that of the type A zeolites, as both structures consist of discrete cavities interconnected through 8-membered oxygen windows. There are, however, important differences. The type A lattice is cubic, and the cavities, which are nearly spherical (free diameter ~ 11.4 Å, volume 776 Å^3), are interconnected through perfectly regular octagonal windows (free diameter ~4.2 Å in the 5A form). By contrast, the lattice of erionite is hexagonal with smaller ellipsoidal cavities (~ 13 × 6.3 Å, volume 406 Å^3) interconnected through distorted octagonal windows (free dimensions 3.6 × 5.2 Å). More complete structural details have been given by Barrer and Kerr (1959) and Gorring (1973). Sorption and diffusion in the type A zeolites have been widely studied, but comparable information concerning the sorptive properties of erionite is less extensive. A comparative study of the diffusion of C_5–C_8 hydrocarbons in 5A zeolite and in natural erionite has been reported by Eberly (1969), but the data obtained were not sufficiently extensive to allow the relationship between structure and sorptive properties to be examined in detail. The present investigation of the sorption of three representative hydrocarbons in a sample of natural erionite was undertaken in order to provide the experimental data necessary for a detailed comparison of the equilibrium isotherms and diffusivities with those previously obtained for the 4A and 5A zeolites.

EXPERIMENTAL

The sample of erionite was from Rome, Oregon. X-ray powder diffraction measurements gave for the unit cell dimensions $a_0 = 13.17$ Å, $c_0 = 15.03$ Å. These values are close to the values given by Breck (1974) (13.26 Å, 15.12 Å). The mean diameter of the needle-like erionite crystals, determined from electron micrographs, was 3.8 μm with a standard deviation of 1.1 μm.

Equilibrium isotherms and diffusivities were determined gravimetrically using a Cahn vacuum-microbalance system to follow the change in weight of a small sample of the zeolite in response to small step changes of sorbate pressure. Diffusivities were calculated by matching the experimental uptake curves to the transient solution of the diffusion equation for a set of infinite cylinders, duly corrected to take account of the distribution of crystal diameters. The method was essentially the same as has been previously used to study diffusion in the 5A sieve, and a more complete description of the experimental method has been given elsewhere (Loughlin *et al.*, 1971).

Prior to a run the sample was degassed at 400°C and 10^{-5} torr for several hours.

RESULTS

Equilibrium isotherms

Examination of the sample by optical and electron microscopy revealed the presence of a significant quantity of non-zeolitic impurity, tentatively identified as quartz. Detailed characterization of the sample was not attempted, but the proportion of erionite present was estimated from the equilibrium isotherms on the assumption that the impurity does not contribute significantly to the sorptive capacity.

For all three hydrocarbons the saturation capacity, on a sample weight basis, was found to be half that of 5A zeolite crystals at the same temperature. The specific void volumes of the two zeolites, calculated from the crystal structures on the assumption that there is no adsorption within the sodalite or cancrinite cages, are 0.2 cm^3/g (erionite) and 0.265 cm^3/g (5A). The purity of the sample was thus estimated as 0.5 × 0.265/0.20 = 66%wt. erionite. From the density (1.76 g/cm^3) given by Eberly (1964) for a similar sample of erionite and the unit-cell volume (2300 Å^3 or 1150 Å^3 per cavity) it follows that a sorbate concentration of 1 mmole/g sample is equivalent to 1.87 molecules per cavity. This factor was therefore used in order to express the concentrations on a comparable basis.

The experimental isotherms for C_2H_4, C_2H_6, and nC_5H_{12} are shown in Fig. 1. Henry's law constants (defined by the relation $c = Kp$) were extracted from the isotherms by extrapolation of virial plots of $\log(c/p)$ vs. c to zero concentration after the method of Barrer and

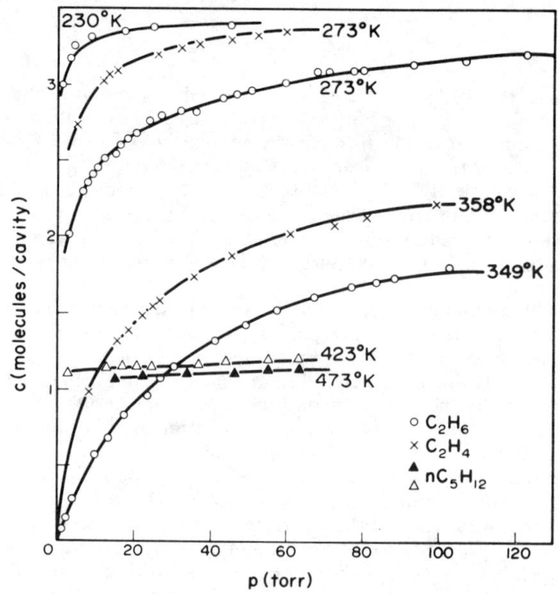

Fig. 1. Experimental equilibrium isotherms for sorption of C_2H_4, C_2H_6, and nC_5H_{12} in natural erionite.

TABLE 1. Parameters K_0 and q_0 giving temperature dependence of Henry constants according to equation (1) with derived values of ΔS^o

Sorbate	Zeolite	$K_0 \times 10^8$ (molecule/cavity torr)	q_0 (kcal)	ΔS^o (e.u.)
C_2H_6	5A*	77	6.6	−7.02
	Erionite	44	8.7	−7.20
	4A†	2.9	10	−13.4
C_2H_4	5A‡	18	9.7	−9.95
	Erionite	7	11.0	−10.9

*Data of Ruthven and Loughlin (1972).

†Data of Derrah obtained in these laboratories.

‡Data of Holborrow (1974).

Davies (1971). The parameters K_0 and q_0 giving the temperature dependence of the Henry constants according to the equation:

$$K = K_0 \exp(q_0/RT) \tag{1}$$

are listed in Table 1 together with the corresponding parameters for sorption in the 4A and 5A zeolites. The range of concentrations covered by the pentane isotherms was not sufficiently wide to allow the Henry constants to be determined with any accuracy. For both C_2H_4 and C_2H_6 heats of sorption in erionite are somewhat larger than in the A zeolites, reflecting the stronger interaction arising from the greater density of the erionite framework.

Standard entropies of sorption (referred to standard states of unit concentration in both phases) are also given in Table 1, calculated according to the equation:

$$\Delta S^\circ = R \ln\ [nK_0kTe]. \tag{2}$$

The values of ΔS° for both C_2H_4 and C_2H_6 are essentially the same for both erionite and type A zeolite suggesting similarity in the states of the adsorbed molecules. The value of ΔS° is numerically larger for C_2H_4 than for C_2H_6 suggesting some degree of localization or loss of rotational freedom for the adsorbed C_2H_4 molecules.

To account for the form of the isotherms in the higher concentration range a theoretical model is required. The simple Langmuir model has been shown to be unsatisfactory for most zeolitic systems because the basic assumptions of the model (ideal adsorption on a set of identical sites, one molecule per site and no interaction between molecules adsorbed on adjacent sites) are generally not fulfilled. A more realistic model has been developed on the basis of the following assumptions (Ruthven, 1971):

(1) Adsorbed molecules are confined within particular cavities but are not adsorbed at specific localized sites within a cavity.
(2) The interaction between a molecule and the zeolite framework is characterized by the Henry's law constant and is not affected by the presence of more than one molecule within a cavity.
(3) When two or more molecules occupy the same cavity the sorbate–sorbate molecular interaction is represented simply by a reduction in the free volume available to each molecule. Attractive forces between sorbate molecules are neglected.
(4) Each cavity is treated as an independent identical sub-system which is not affected by the presence of sorbate molecules in the neighboring cavities.

Subject to these approximations the equation for the isotherm is obtained in the form:

$$c = \frac{Kp + (Kp)^2(1-2\beta/v)^2 + \ldots \dfrac{(Kp)^m}{(m-1)!}(1-m\beta/v)^m}{1 + Kp + ½!\,(Kp)^2(1-2\beta/v)^2 + \ldots \dfrac{(Kp)^m}{m!}(1-m\beta/v)^m} \tag{3}$$

where m (integer) is given by $m \leqslant v/\beta$. This expression, which contains two parameters – the Henry constant (K) and the ratio of the sorbate molecular volume to the cavity volume (β/v) – provides a very satisfactory correlation of the experimental isotherms for sorption of hydrocarbons and other simple sorbates in zeolite A and in H-chabazite over a wide range of conditions (Ruthven and Loughlin, 1972; Derrah *et al.*, 1972; Coughlan *et al.*, 1975). In the original formulation of the model, exponential factors were included to take account of intermolecular attraction between adsorbed molecules. However, for all systems investigated the effect of these terms was found to be negligible and the important features of the sorption equilibria (such as the variation

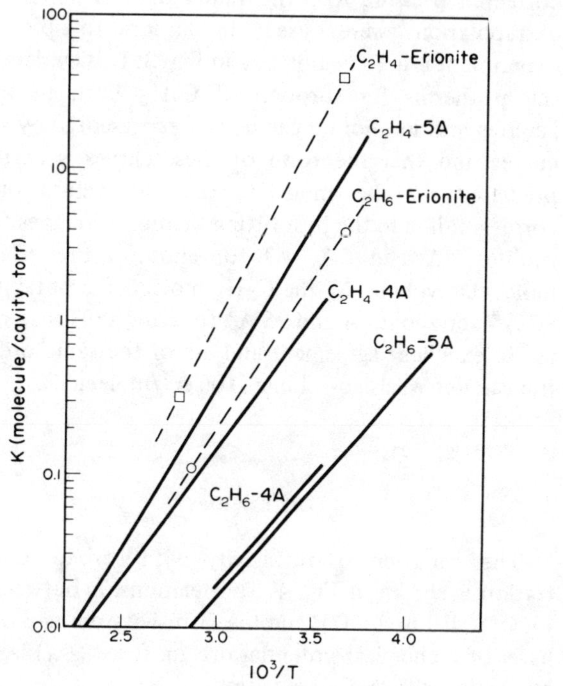

Fig. 2. Vant Hoff plot showing temperature dependence of Henry's law constants.

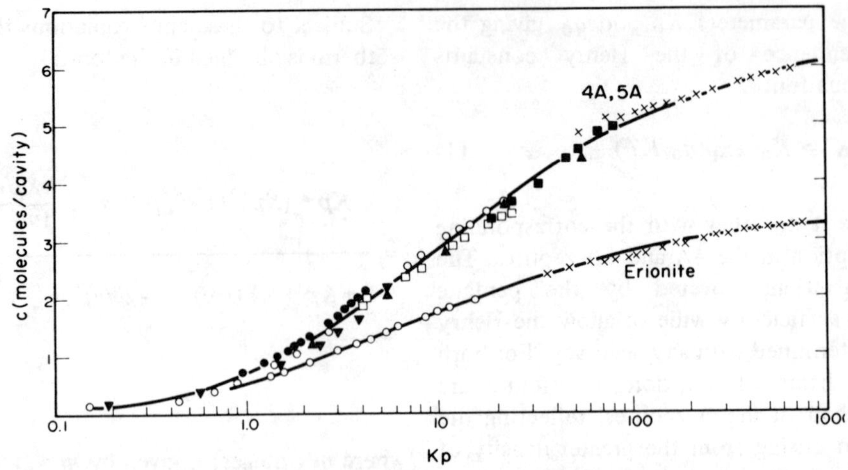

Fig. 3. Equilibrium isotherm for C_2H_6 in type A zeolite and in erionite plotted against corrected pressure Kp. (4A zeolite: ■, 273°K; ▲, 298°K; ▼, 348°K. 5A zeolite: ×, 230°K; □, 273°K; ○, 308°K; ●, 324°K. Erionite: ×, 273°K; ○, 349°K. Theoretical lines are for ν/β=8 and ν/β=4.3.)

of heat of sorption with sorbate concentration) could be accounted for by considering only the repulsion effect due to the finite size of the adsorbed molecules. These terms have therefore been omitted.

The isotherms for C_2H_6 cover a wide range of concentrations and can therefore be used to test the theoretical model. It follows from equation (3) that if the temperature dependence of sorbate molecular volume can be neglected, the isotherm data should follow a single curve when plotted in terms of the corrected pressure Kp. The values of K from the virial extrapolation were used in order to plot the experimental data in this way in Fig. 3. It is evident that the isotherms for sorption of C_2H_6 both in the A zeolites and in erionite can be well represented by single curves and that the form of these curves is correctly predicted by the model. From the values of ν/β corresponding to the best fitting theoretical curves (8 for zeolites 4A and 5A, 4.3 for erionite) the effective molecular volume of the C_2H_6 molecule is obtained as 97 Å^3 for zeolite A and 95 Å^3 for erionite These values are in excellent agreement and are of the same order as the van der Waals co-volume (106 Å^3/molecule).

Diffusivities

The variation of diffusivity with sorbate concentration is shown in Fig. 4. The relationship between the Fickian diffusivity (D) and the diffusivity defined on the basis of a chemical potential driving force (D_0) is given by the familiar Darken equation:

$$D = D_0\ (d \ln p/d \ln c) \qquad (4)$$

in which ($d \ln p/d \ln c$) is the gradient of the equilibrium isotherm in logarithmic coordinates. Values of D_0 calculated according to equation (4) showed no significant trend with concentration. The theoretical lines shown in Fig. 4, which are calculated according to

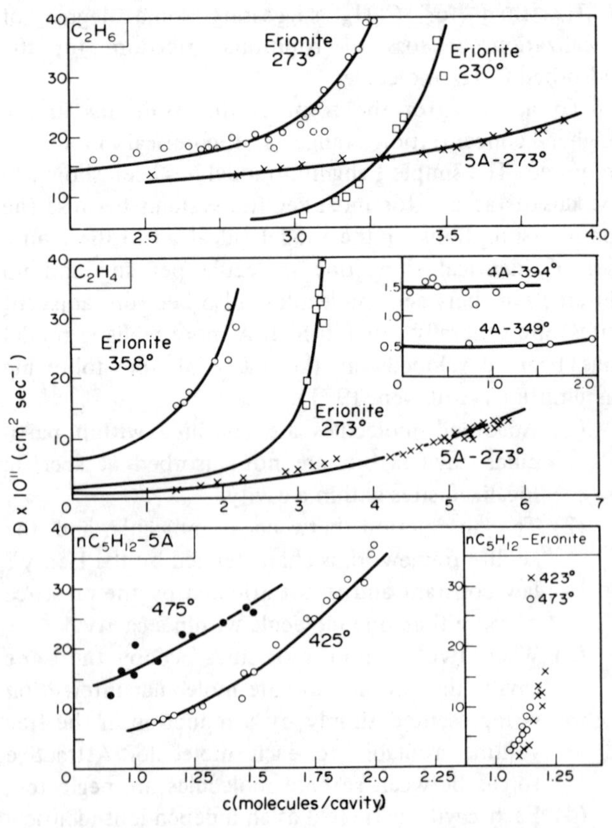

Fig. 4. Variation of diffusivity with sorbate concentration. Comparison of data for erionite, 4A and 5A zeolites.

equation (4) using the average values of D_0, and the values of $d \ln p/d \ln c$ from the experimental isotherms give a satisfactory representation of the data, suggesting that the concentration dependence of the Fickian diffusivity arises entirely from the non-linearity of the relationship between activity and concentration as defined by the equilibrium isotherm. The corrected diffusivity (D_0) appears to be independent of concentration. This is the same form of behavior as has been observed for the diffusion of light hydrocarbons in 5A zeolite (Ruthven *et al.*, 1973) as well as for monatomic and diatomic gases in 4A (Ruthven and Derrah, 1975a).

For measurements made over small differential concentration steps, the diffusivities obtained from adsorption and desorption runs were essentially equal. However, for large step changes the rate of adsorption becomes much faster than desorption as observed by Eberly (1969). This effect can be quantitatively accounted for by considering the concentration dependence of the diffusivity (Garg and Ruthven, 1972).

The temperature dependence of the corrected diffusivity is shown in Fig. 5, and the Arrhenius parameters, calculated from the equation

$$D_0 = D_* \exp(-E/RT) \quad (5)$$

are compared in Table 2 with the corresponding values for diffusion in 4A and 5A zeolites. Activation energies for diffusion in erionite are similar to the values for 5A and smaller than the values for 4A. This suggests that the erionite windows, like the 5A windows, are not obstructed by cations. Detailed comparison of the diffusional activation energies for C_2H_4 and C_2H_6 in 5A zeolite and erionite show precisely the trends which are to be expected from considerations of the shapes of the diffusing molecules and the zeolite windows. The planar C_2H_4 molecule can pass more easily through the distorted octagonal window of the 5A zeolite, leading to a lower diffusional activation energy for erionite. For the

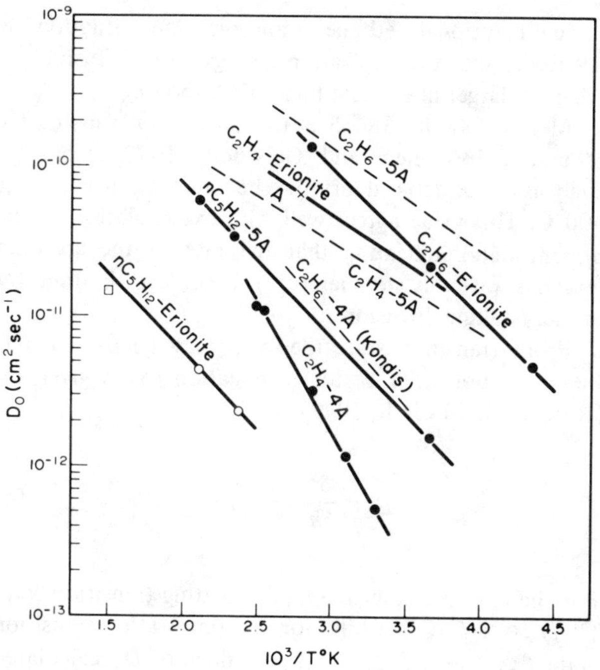

Fig. 5. Temperature dependence of corrected diffusivity D_0. Comparison of data for erionite, 4A and 5A zeolites.

TABLE 2. Parameters D_* and E giving temperature dependence of the limiting diffusivities according to equation (5)

Sorbate	Zeolite	E (kcal)	Experimental $D_* \times 10^8$ (cm² sec⁻¹)	Theoretical values $D_* \times 10^8$ (cm² sec⁻¹) No rotation	Free rotation
C_2H_6	5A*	3.02	1.28	1.01	332
	4A†	5.66	6.24	–	–
	Erionite	4.3	6.6	3.1	1100
C_2H_4	4A	8.5	61	5	3550
	5A	2.75	0.20	0.79	565
	Erionite	2.84	0.36	1.26	830
nC_5H_{12}	5A	4.6	0.63	–	–
	Erionite	(5.0)	(0.09)	–	–

Theoretical values are calculated at the mean of the experimental temperature range.

*Data of Ruthven *et al.* (1973).

†Data of Kondis and Dranoff (1971).

three-dimensional ethane molecule the situation is reversed, and the activation energy for diffusion of ethane is larger in erionite than in 5A zeolite.

Also shown in Fig. 5 is the diffusivity for nC_5H_{12} estimated by Chen and Garwood (1973) from an analysis of kinetic data for the cracking reaction at 400°C. This value agrees well with extrapolation of the present data, indicating that the rate of the cracking reaction (at this temperature) is indeed limited by intracrystalline diffusion.

From transition state theory it may be shown that the pre-exponential factor in equation (5) is given by (Ruthven and Derrah, 1972)

$$D_* = \frac{\delta^2}{6h} \cdot \frac{1}{K_0} \frac{f^*_z}{f'_g} . \qquad (6)$$

For the extreme case of a freely rotating transition state $f^*_z/f'_g \approx 1/f'_{\text{trans}}$ while for a non-rotating transition state $f^*_z/f'_g \approx 1/f'_{\text{trans}} \cdot f_{\text{rot}}$. Values of D_* calculated according to equation (6) are included in Table 2. Equation (6) is strictly valid only for a cubic lattice as in the type A zeolites ($\delta \approx 12.3$ Å). The diffusivities in the *a* and *c* directions in erionite may be expected to be different, but as the length to diameter ratio of the crystals is large, one is measuring diffusion essentially in the *a* direction. The value of δ for erionite was therefore taken as 7.0 Å, this being the distance between successive windows in the *a* direction.

For the diffusion of both C_2H_4 and C_2H_6 in erionite and in the A zeolites, the experimental values of D_* lie close to the theoretical values for a non-rotating transition state suggesting that a molecule at the center of the window has little freedom of movement. In view of the small free diameter of the windows (~4.2 Å) such a conclusion seems entirely reasonable. A more detailed discussion of these diffusivity data has been given previously by Ruthven and Derrah (1975b).

CONCLUSIONS

The similiarities and differences between the structures of the A zeolites and natural erionite are clearly reflected in their adsorptive properties. The equilibrium isotherms for both zeolites are consistent with the same general theoretical model in which the parameters are the Henry's law constant and the ratio of zeolitic cavity volume to sorbate molecular volume. The ability of this model to predict the form of the isotherms for these two zeolites with quite different cavity volumes (776 Å^3 vs. 409 Å^3) provides strong evidence that the basic assumptions of the model are essentially correct.

The diffusivity data for both erionite and zeolite A show the same form of concentration dependence and are consistent with a simple activated diffusion model. The general similarity in the size of the 8-membered oxygen windows of erionite and 5A zeolite is reflected in similar diffusivities for each species in the two zeolites. Furthermore, the observed differences in diffusional activation energies are consistent with the difference in the shapes of the windows.

Both the kinetic and equilibrium data for sorption of simple hydrocarbons in erionite are thus entirely consistent with the conclusions reached from earlier studies of sorption and diffusion in the A zeolites. Many of the important features of the adsorption behavior can be successfully explained by quite simple theoretical models.

ACKNOWLEDGMENT

The authors are grateful to Professor L. B. Sand of the Worcester Polytechnic Institute for providing the erionite sample.

REFERENCES

Barrer, R. M. and Kerr, I. S. (1959) Intracrystalline channels in levynite and some related zeolites: *Trans. Faraday Soc.* **55**, 1915-1923.

Barrer, R. M. and Davies, J. A. (1971) Sorption in decationated zeolites: *Proc. Roy. Soc.* A, **320**, 289-308.

Breck, D. W. (1974) *Zeolite Molecular Sieves*: Wiley, New York, p. 43.

Chen, N. Y. and Garwood, W. E. (1973) Molecular shape selective hydrocarbon conversion over erionite: *Adv. Chem.* **121**, 575-582.

Coughlan, B., Kilmartin, S., McEntee, J. and Shaw, R. G. (1975) Interpretation of zeolitic CO_2 equilibria using a theoretical isotherm equation: *J. Colloid Interface Sci.* **52**, 386-390.

Derrah, R. I., Loughlin, K. F. and Ruthven, D. M. (1972) Sorption of ethylene, propylene and cyclopropane in 5A zeolite: *J. Chem. Soc. Faraday Trans. I.* **68**, 1947-1955.

Eberly, P. E. (1964) Adsorption properties of natural erionite and its cation exchanged forms: *Amer. Mineral.* **49**, 30-40.

Eberly, P. E. (1969) Adsorption of normal paraffins in erionite and 5A molecular sieve: *Ind. Eng. Chem. Prod. Res. Develop.* **8**, 140-144.

Garg, D. R. and Ruthven, D. M. (1972) The effect of concentration dependence of the diffusivity on zeolitic sorption curves: *Chem. Engng Sci.* **27**, 417-423.

Gorring, R. L. (1973) Diffusion of normal paraffins in zeolite T: *J. Catalysis* **31**, 13-26.

Holborow, K. A. (1974) Multicomponent sorption equilibria of gases in 5A zeolite: Ph.D. thesis, University of New Brunswick, Fredericton, N.B., Canada.

Kondis, E. F. and Dranoff, J. S. (1971) Kinetics of ethane sorption on 4A molecular sieve: *Adv. Chem.* **102**, 171-179.

Loughlin, K. F., Derrah, R. I. and Ruthven, D. M. (1971) Measurement of zeolitic diffusion coefficients: *Can. J. Chem. Engng* **49**, 66-70.

Ruthven, D. M. (1971) A simple theoretical isotherm for zeolites: *Nature, Phys. Sci.* **232** (29) 70-71.

Ruthven, D. M. and Loughlin, K. F. (1972) Sorption of light paraffins in type A zeolite: *J. Chem. Soc. Faraday Trans. I* **68**, 696-708.

Ruthven, D. M. and Derrah, R. I. (1972) Transition state theory of zeolitic diffusion: *J. Chem. Soc. Faraday Trans. I* **68**, 2332-2343.

Ruthven, D. M., Derrah, R. I. and Loughlin, K. F. (1973) Diffusion of light hydrocarbons in 5A zeolite: *Can. J. Chem.* **51**, 3514-3519.

Ruthven, D. M. and Derrah, R. I. (1975a) Diffusion of monatomic and diatomic gases in 4A and 5A zeolites: *J. Chem. Soc. Faraday Trans. I* 71, 2031-2044.

Ruthven, D. M. and Derrah, R. I. (1975b) Comparative study of diffusion in erionite and type A zeolite: *J. Colloid Interface Sci.* **52**, 397-403.

A SURVEY OF CATALYTIC PROPERTIES OF NORTH AMERICAN CLINOPTILOLITES

N. Y. CHEN, W. J. REAGAN, G. T. KOKOTAILO, and L. P. CHILDS

Mobil Research and Development Corporation
Princeton and Paulsboro, New Jersey, U.S.A.

Abstract – Clinoptilolite, a member of the heulandite family of zeolites, has both eight-membered and ten-membered oxygen-ring openings in its framework structure. It is among the most abundant of the naturally occurring zeolites. We have examined the catalytic cracking properties of seven clinoptilolite samples from different locations in California and Nevada. The test, which involves the conversion of a mixture of hexane isomers, provides structural information since the eight-membered oxygen-ring openings will admit only n-hexane and exclude the branched isomers. Our results showed that these zeolites, in the hydrogen form, differ significantly in activity, shape selectivity, and rate of deactivation. A relationship between these catalytic data and certain features in the detailed X-ray diffraction patterns of these samples was found. This observation led to the proposal that the heulandite–clinoptilolite family of zeolites consists of two distinct structures and their intergrowth. The samples from Nevada have sharp odd *k*-lines in their X-ray diffraction patterns similar to those observed for heulandite. These materials preferentially convert n-hexane. This suggests that the active catalytic sites are accessible through the eight-membered oxygen-ring pores. Two samples from California have weak and/or broad odd *k*-lines. These samples convert both the normal and branched hexanes. This suggests fewer eight-membered oxygen-ring pores and the availability of catalytic sites through the ten-membered oxygen-ring pore openings. Other samples from California have intermediate properties in odd *k*-line intensity and in catalytic shape selectivity.

We propose that the established heulandite structure, with an equal number of eight- and ten-membered oxygen-ring pores, represents one end member of a family of zeolites. The other end member, a hypothetical zeolite, has no odd *k*-reflections in its X-ray diffraction pattern. Its pore openings parallel to the 001 plane consist of only ten-membered oxygen-rings. Other members of this series, with varying proportions of eight- and ten-membered ring pore openings, result from stacking faults in the structure. This would explain the wide variation in catalytic properties among the natural clinoptilolite samples.

INTRODUCTION

Since their discovery, the natural zeolites clinoptilolite and heulandite have received considerable attention from geologists and mineralogists. An important question concerning these zeolites is whether clinoptilolite is isostructural with heulandite. Although the X-ray diffraction patterns of these zeolites are similar, a wide variation in structural stability of natural samples has been observed (Hey and Banister, 1934; Mumpton, 1960).

Mumpton (1960) suggested that the major difference between these two zeolites is their silicon to aluminum ratio. He also introduced a simple experimental test to distinguish these two minerals – their structural stability at 450°C. Mason and Sand (1960) suggested that the thermal stability of the natural clinoptilolite and heulandite ores was related to the type of cations present in the zeolite. The important role of the cations in the zeolites was confirmed by the ion-exchange studies of Shepard and Starky (1964).

Recently, Alietti (1972) and Boles (1972) proposed additional refinements in the classification of minerals in the heulandite zeolite group. This new definition includes both silicon to aluminum ratio and cation population as parameters. Efforts to classify these zeolites according to differences in framework structure or cell dimensions have been unsuccessful (Alietti, 1972).

The crystal structure of heulandite was published in 1968 by Merkle and Slaughter (1968). The heulandite structure contains four-, five-, and six-membered oxygen-ring building units. Open, elliptical channels, in three dimensions, are formed by one ten- and three types of eight-membered oxygen rings. These elliptical openings have estimated sizes ranging from 5.2 × 3.9 Å to 7.05 × 4.25 Å. The effective pore size of the largest opening is greater than those of most shape-selective zeolites, such as zeolite A or erionite, but smaller than the pore-size dimensions of faujasite or mordenite.

During a survey of the shape-selective properties of some clinoptilolite samples in the hydrogen form, we found a variation in their catalytic behavior. To assess the connection between the observed catalytic properties and structural variations in these clinoptilolite samples, a more thorough catalytic study of additional materials was made and coupled with crystallographic

Table 1. Composition of natural clinoptilolites

As received ores

	Hector Calif.	Baroid Mine, Hector, Calif.	Opal Mountain, Calif.	Barstow, Calif.	Shoshone, Calif.	Jersey Valley, Nevada	Eastgate, Nevada
Si/Al	5.15	5.20	4.40	3.90	3.90	4.05	4.10
K/Al	0.14	0.10	0.23	0.08	0.45	0.38	0.29
Na/Al	0.79	0.94	0.21	0.62	0.52	0.05	0.35
Ca/Al	0.09	0.19	0.35	0.29	0.04	0.35	0.10
Mg/Al	0.06	0.10	–	–	0.14	0.27	0.08
	1.08	1.33	0.79	0.99	1.15	1.05	0.82

NH_4-exchanged samples

	Hector Calif.	Baroid Mine, Hector, Calif.	Opal Mountain, Calif.	Barstow, Calif.	Shoshone, Calif.	Jersey Valley, Nevada	Eastgate, Nevada
Si/Al	5.15	5.60	4.45	4.35	4.05	4.35	4.10
K/Al	0.04	0.04	0.07	0.08	0.07	0.07	0.05
Na/Al	0.03	0.03	0.06	0.02	0.01	0	0.01
Ca/Al	0.02	0.02	0.02	0.01	0	0.01	0.10
Mg/Al	0.05	0.05	0.05	–	–	–	0.08
NH_4/Al	0.86	0.86	0.80	0.89	0.92	0.92	0.76

TABLE 2. X-ray data for Jersey Valley and Baroid Mine clinoptilolite

°2θ	d(Å)	hkl	Jersey Valley, I/I_o	Baroid Mine, I/I_o	°2θ	d(Å)	hkl	Jersey Valley, I/I_o	Baroid Mine, I/I_o
6.6	13.4	–	–	2	27.70	3.21	151	4	11*
7.6	11.6	110	2	2*	28.17	3.17	$42\bar{2}$	34	24
8.7	10.2	–	–	4	28.56	3.12	510	18	9*
9.9	8.93	020	100	100	29.00	3.08	$13\bar{2}$	18	12*
11.14	7.94	200	42	38	30.00	2.98	151	52	28*
13.00	6.81	$11\bar{1}$	22	20*	30.87	2.90	401	2	3
13.28	6.67	001	10	9	31.80	2.81	530	26	15*
14.87	5.96	220	10	4	32.72	2.74	532	20	10*
15.93	5.56	130	4	1*	33.38	2.68	132	6	4*
16.37	5.42	$\bar{2}21$	4	2	34.90	2.57	222	6	4
16.65	5.32	021	6	4	35.28	2.54	$35\bar{2}$	6	4
16.90	5.25	$\bar{3}11$	14	10*	36.60	2.46	403	8	8
17.25	5.14	111,310	24	19*	36.90	2.44	$20\bar{3}$	8	8
19.08	4.65	$\bar{1}31$	18	9*	37.65	2.39	$32\bar{3}$	4	2
19.70	4.51	–	2	9	39.47	2.28	640	2	3
20.35	4.36	$\bar{4}01$	4	4	40.40	2.23	080	2	1
20.93	4.24	201	4	9	40.80	2.21	332	–	1
22.37	3.97	400	98	97	41.30	2.19	$\bar{1}33$	2	1
22.70	3.92	330	57	41*	42.50	2.13	402	4	3
23.07	3.86	221	2	10	43.15	2.10	621	6	2
23.70	3.75	$24\bar{1}$	6	5	43.80	2.07	461	4	2
24.00	3.71	041	10	8	44.23	2.05	371	2	2
25.10	3.55	$31\bar{2}$	14	8*	44.85	2.02	$\bar{6}43$	6	3
25.88	3.44	$22\bar{2}$	6	24	45.60	1.989	800	2	2
26.10	3.41	311	44	26*	46.18	1.966	133	6	2
26.25	3.39	$40\bar{2}$	10	9	46.58	1.950	480	2	2
26.75	3.33	002	18	59					

*Most prominent broad odd *k*-lines.

analyses.

This paper presents the results of such a study of seven clinoptilolite samples from California and Nevada.

EXPERIMENTAL

Materials

The zeolite samples were obtained from various deposits in California and Nevada. When necessary, the ores were crushed or milled to a powder form. Elemental analyses of the "as received" samples are presented in Table 1.

The zeolite ores were ammonium exchanged by two successive reflux treatments (16 and 4 hours) with 20 cm^3 of 5 M ammonium chloride solution per gram of zeolite. Analyses of the products are also listed in Table 1. The hydrogen forms of the zeolite samples were obtained by air calcination of the ammonium-exchanged materials.

TABLE 3. Relative crystallinity* and thermal stability

Source	NH_4^+	H^+	% Loss†
Hector, California	57	45	21
Baroid Mine, Hector, California	47	38	19
Opal Mountain, California	74	54	27
Barstow, California	83	45	46
Shoshone, California‡	54	30	44
Jersey Valley, Nevada	100	40	60
Eastgate, Nevada‡	90	–	–

*Jersey Valley, Nevada, NH_4^+ form, taken as 100%. Peak heights and/or peak areas were used for comparison with this standard.

†After 2 hours in air at 480°C.

‡Contains phillipsite and/or erionite.

X-ray diffraction techniques

X-ray powder diffraction studies were carried out on both ammonium-exchanged and air-calcined (2 hours at 480°C) zeolite samples (Tables 2 and 3) with a Siemens X-ray diffractometer and CuKα radiation.

Catalytic test equipment

Catalytic activity was measured with a pulse-alpha test, a modification of the alpha test published in 1966 (Miale *et al.*, 1966). A schematic diagram of the reactor configuration is shown in Fig. 1. Hydrocarbon feed, an equimolar mixture of n-hexane, 3-methylpentane, and 2,3-dimethylbutane, is delivered at 0.5 cm^3/hr by a syringe pump into a flowing stream of helium (8 cm^3/min). The molar ratio of helium to hydrocarbon is about 5. This feed stream flows continuously through the heated Perkin–Elmer gas-sampling valve. A separate carrier stream of helium, connected to the other side of the gas-sampling valve, flows through the reactor which is connected with heat-traced lines to a gas chromatograph (G.C.) containing a 30-foot benzoquinoline column. The flow rate of the second carrier gas is maintained at 55 cm^3/min with a flow controller.

In practice the zeolite catalyst (1–2 cm^3, 20X30 mesh particles) is packed in the glass reactor which has a thermowell for temperature measurement. The catalyst is pretreated at 538°C for 30 minutes with air at a flow rate of 60 cm^3/min. After air is vented through valve 1 (Fig. 1) the valving arrangement is switched, and the reactor is connected to the G.C. The reactor is cooled to the reaction temperature by flowing helium. The sequence of events for the analysis is controlled by a 1-hour programmer. A block diagram illustrating the elements of the analysis system is shown in Fig. 2 and the timing sequence of the programmer is presented in Fig. 3. Normally, the gas-sampling valve is in the "fill" position, with the He/hydrocarbon feed flowing through

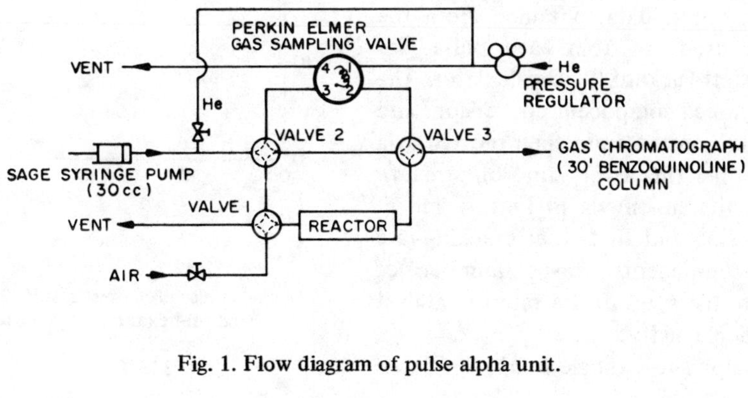

Fig. 1. Flow diagram of pulse alpha unit.

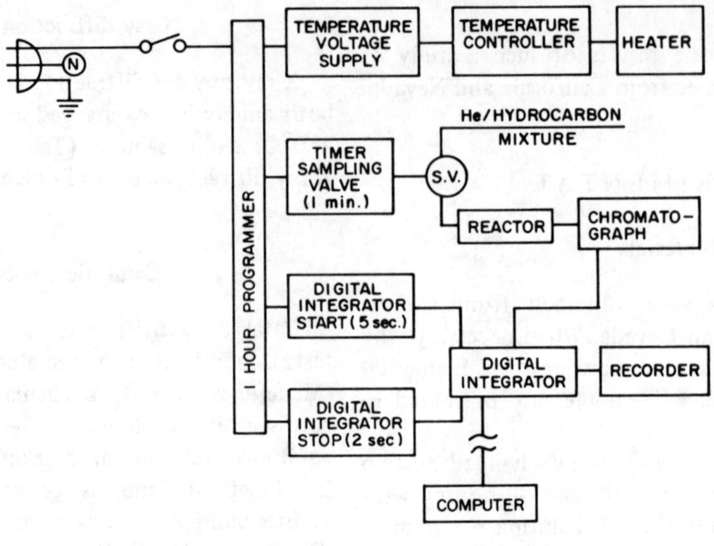

Fig. 2. Block diagram of programmer.

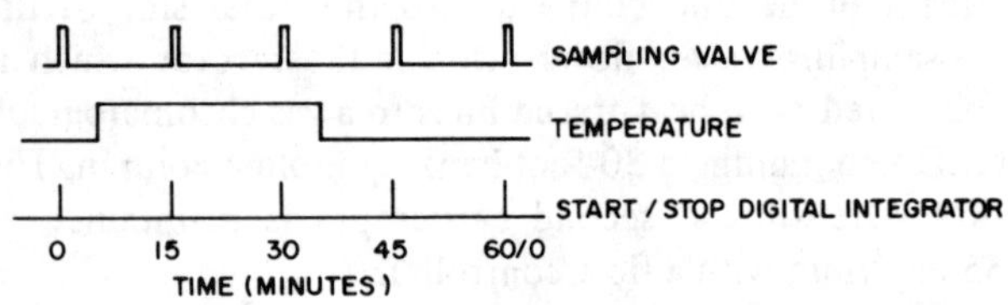

Fig. 3. Timing sequence.

one side of the valve. A command signal from the programmer pneumatically actuates the sampling valve, and a pulse of hydrocarbon is swept into the reactor. A digital integrator is also started at this time.

The reactor temperature can be changed by inserting a millivolt "bucking voltage" in series with the controlling thermocouple. As shown in Fig. 3, the programmer applies or bypasses the "bucking voltage"; thus, pairs of hydrocarbon pulses alternate between the two temperatures (normally 28°C apart). This feature allows the calculation of an apparent activation energy for the cracking of each of the hydrocarbon-feed components. The conversion data, obtained from the G.C. analysis of the products from each pulse, are processed with a Hewlett-Packard 9830 calculator. The individual hydrocarbon-feed component conversions and first-order cracking-rate constants are then plotted as a function of pulse number (in effect, time on stream). Examples of these results are shown in Figs. 4 and 5. These plots of conversion and first-order cracking-rate constants of feed components versus number of hydrocarbon pulses are measures of the rate of catalyst deactivation and shape selectivity.

As in the original alpha test (Miale *et al.*, 1966) we

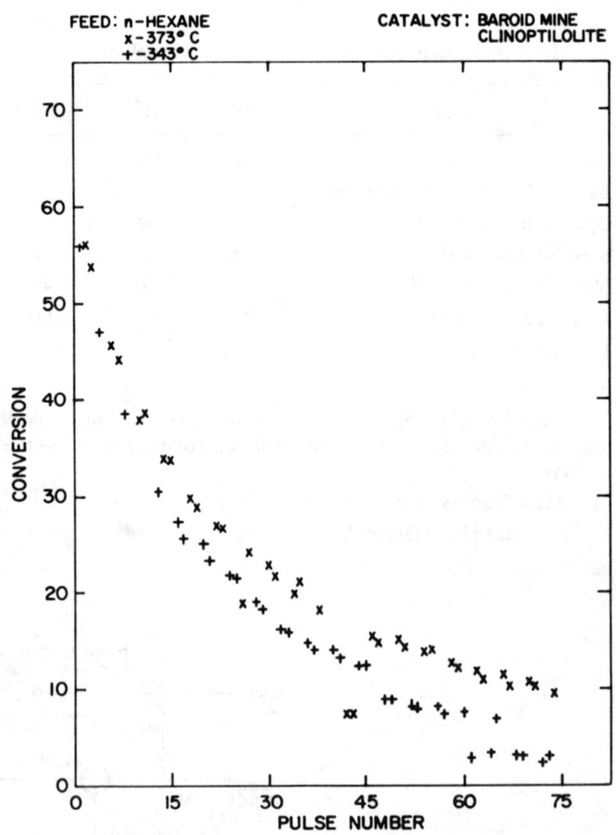

Fig. 4. Pulse-alpha test result.
Feed: n-hexane X=373°C +=343°C
Catalyst: Baroid Mine Clinoptilolite

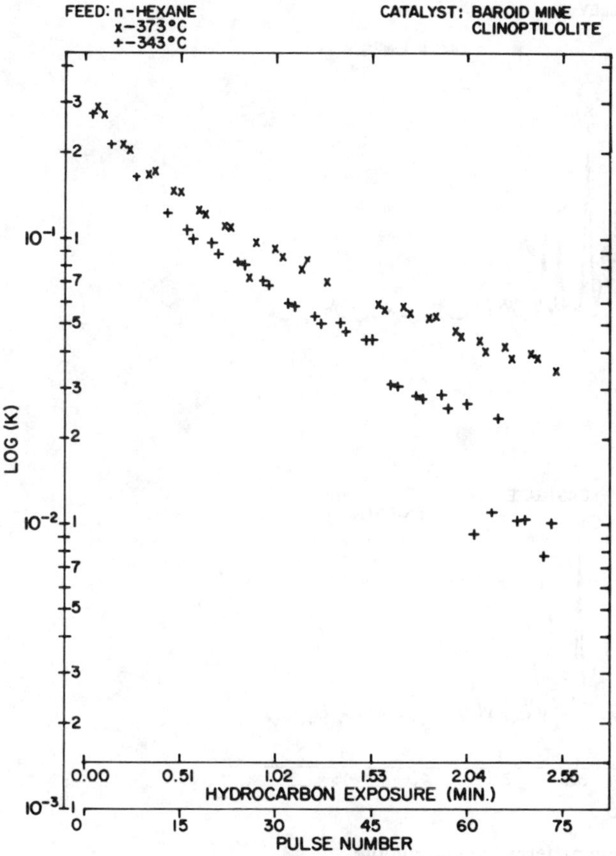

Fig. 5. Pulse-alpha test result.
Feed: n=hexane Catalyst: Baroid Mine Clinoptilolite
X=373°C
+=343°C

have used "alpha" values to represent the activity of the various catalysts compared to a standard, amorphous silica-alumina gel (alpha = 1, by definition).

Since the test is made at different temperatures depending on the activity of the sample, extrapolation of the rate constant measured at a kinetically acceptable temperature to that of the standard catalyst (538°C) is based on an Arrhenius plot with a 30 kcal/mol temperature-dependence.

RESULTS AND DISCUSSION

Physical characterization of zeolite samples

The elemental analyses of the "as received" and ammonium-exchanged samples of clinoptilolite are presented in Table 1. The high Si/Al ratios and completeness of exchange of the "as received" and ammonium-exchanged samples can be noted in Table 1. The samples cannot be readily distinguished from these results.

The X-ray diffraction patterns of six of the samples are shown in Figs. 6, 7, and 8. A listing of the major *d*-spacings and relative intensities are presented in Table 2. The major component of each sample is clinoptilolite, although there is some erionite and phillipsite in the Shoshone sample, some erionite in the Eastgate sample, some mordenite and quartz in the Baroid Mine sample, and some unidentified crystalline material in the Barstow sample.

These X-ray diffraction patterns are similar yet show significant differences, particularly in the width of the odd *k*-lines. The odd *k*-lines are strong and sharp for the Nevada samples; while they are weak and broad for the California samples from the Baroid Mine and Opal Mountain. The other two California samples, Barstow and Shoshone, fall between these two groups. The significance of these observations will be discussed in more detail following the presentation of the catalytic data.

The relative crystallinites of the clinoptilolite samples in the ammonium form and after air-calcination for 2 hours at 480°C are presented in Table 3. The zeolite materials from Nevada are more crystalline before thermal treatment than those from California. However, the Nevada samples are less thermally stable.

In view of the similarity in cation population and silicon to aluminum ratios of these samples, these two variables are not sufficient to explain the marked difference in thermal stability, as suggested by other workers (Mason and Sand, 1960; Alietti, 1972; Boles, 1972), between the California and the Nevada samples. The observed differences in thermal stability may in fact be related to subtle differences in the framework structure of these minerals.

Catalytic characterization of zeolite samples

The results of the shape-selectivity pulse-alpha test for the clinoptilolite samples are presented in Table 4. These include the reaction temperatures, first-order cracking-rate constants for each hydrocarbon feed component, and the ratios of these rate constants, a measure of a catalyst's shape-selectivity.

An Arrhenius plot, shown in Fig. 9, presents the relationship between the first-order cracking-rate constants, alpha values, and reaction temperature for n-hexane conversion. As expected, the alpha values for most of the samples are independent of temperature. The two samples from Hector, California, can be distinguished by their markedly lower apparent activation energies and higher catalytic activity (alpha values).

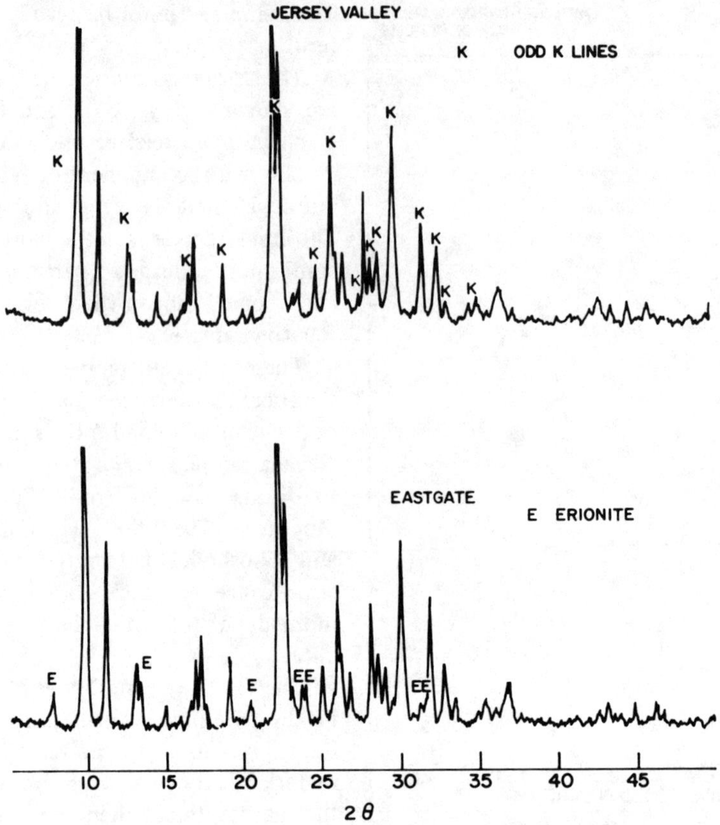

Fig. 6. Powder X-ray diffraction patterns of clinoptilolite.

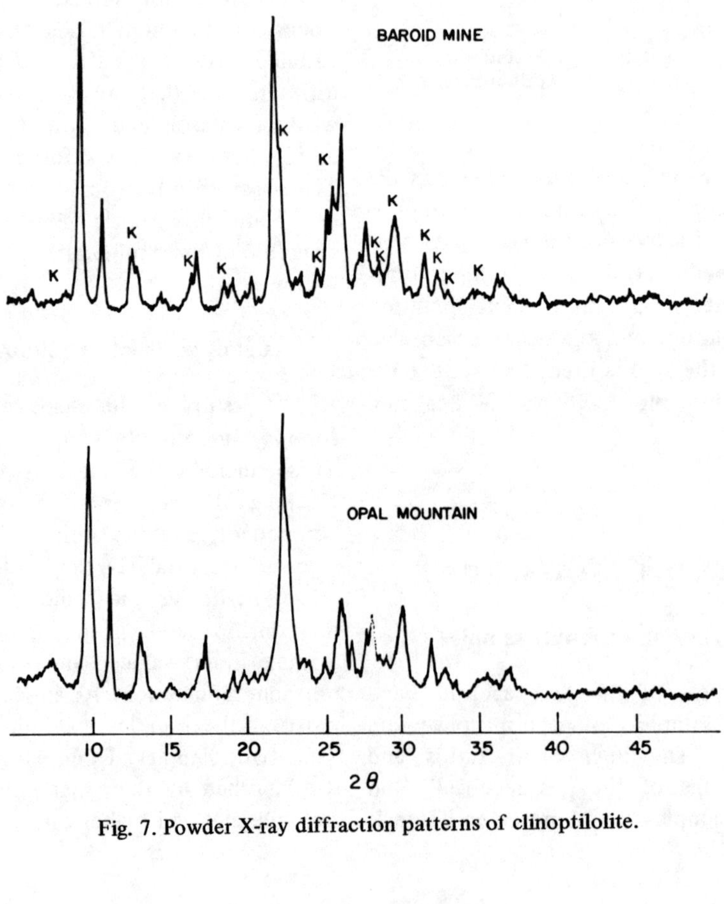

Fig. 7. Powder X-ray diffraction patterns of clinoptilolite.

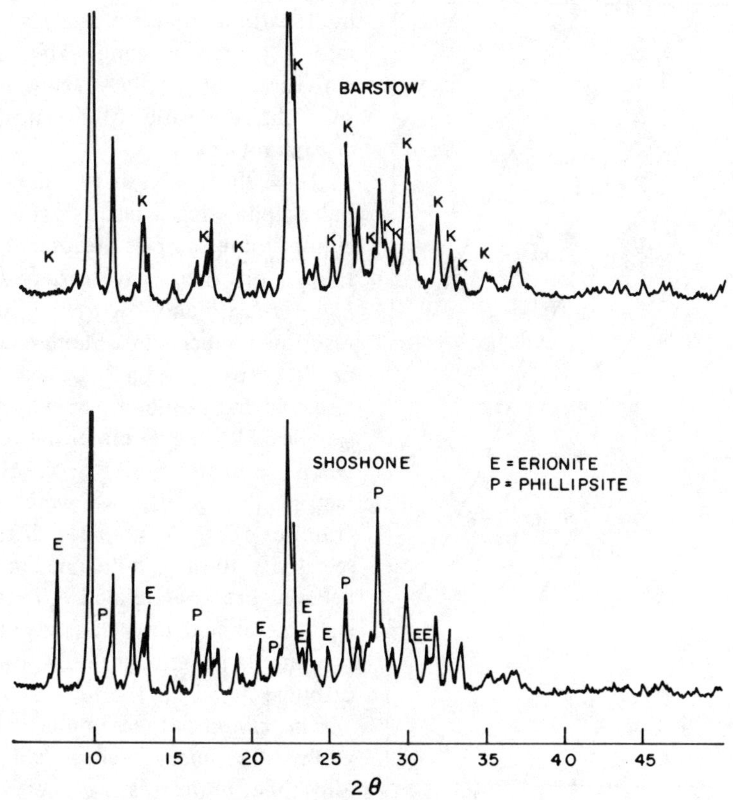

Fig. 8. Powder X-ray diffraction patterns of clinoptilolite.

TABLE 4. Shape selective alpha test: clinoptilolite

Catalyst	T°C		nC_6 $k \times 10$		2, 3DMB $k \times 10$		3MP $k \times 10$		knC_6 / k3MP		k3MP / k2, 3DMB	
		PLS no.	1	75	1	75	1	75	1	75	1	75
		Min*	.03	2.5	.03	2.5	.03	2.5	.03	2.5	.03	2.5
Baroid Mine,	343		1.24	.104	1.56	.242	.95	.057	1.31	1.82	.79	.24
Hector, Calif.	319		.95	.078	1.34	.188	.95	.051	1.00	1.53	.71	.27
Hector, Calif.	397		1.20	1.10	1.11	1.06	.71	.57	1.70	1.90	.64	.54
	371		.92	.92	.85	.85	.64	.48	1.40	1.90	.75	.58
Shoshone,	537		3.3	3.2	.85	.80	.50	.50	6.6	6.4	.59	.62
California	511		2.1	2.1	.60	.60	.34	.31	6.2	6.7	.57	.52
Opal Mountain,	485		.63	.36	.98	.62	.65	.37	.97	.99	.66	.60
California	456		.33	.20	.60	.34	.30	.19	1.1	1.0	.50	.56
Eastgate,	577		3.1	2.0	.50	.31	.40	.29	7.8	6.9	.8	.93
Nevada	549		2.0	1.1	.41	.18	.34	.16	5.9	6.5	.83	.89
Barstow,	493		.30	–	–	–	.16	–	1.88	–	–	–
California	469		.19	–	–	–	.08	–	2.4	–	–	–
Jersey Valley,	548		.91	.37	.58	.38	.31	.13	2.9	2.8	.53	.34
Nevada	526		.67	.27	.42	.20	.18	.07	3.7	3.9	.43	.35

*Equivalent hydrocarbon exposure time.

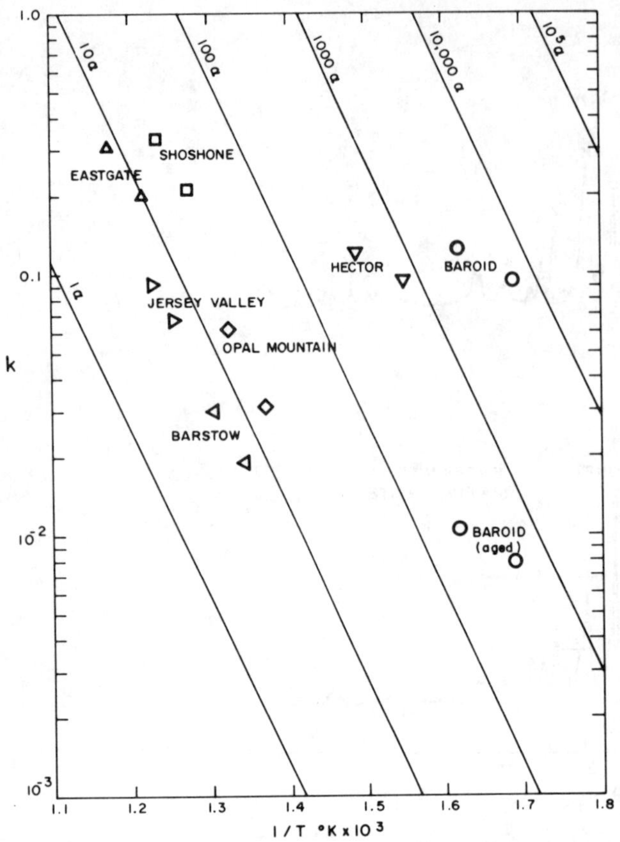

Fig. 9. Arrhenius plot clinoptilolites, n=hexane.

The relative activity (alpha values) and aging characteristics of the clinoptilolites samples are summarized in Table 5. The change in the first-order cracking-rate constant for n-hexane conversion between the first hydrocarbon pulse and the seventy-fifth pulse (corresponding to about 2.5 minutes of total-hydrocarbon exposure time) is taken as a measure of the rate of catalyst aging. The wide range of catalyst activities and catalyst aging rates was not expected from the similarities of chemical composition and X-ray crystallinity data.

The shape-selectivity data obtained from the pulse-alpha test results (Pulse No. 1) of these clinoptilolite samples are presented in Figs. 10 and 11 as ratios of the first order cracking-rate constants k_{n-C_6}/k_{3MP} and $k_{3MP}/k_{2,3DMB}$ as a function of the reaction temperature. Similar ratios for a non-selective zeolite, RE-Y, and a shape-selective zeolite, erionite, are also plotted for comparison. All of the clinoptilolite samples showed preferential conversion of n-hexane when compared to RE-Y. However, the selectivity factors (k_{n-C_6}/k_{3MP}) were much lower than for erionite (Fig. 10). The differences in the shape selectivity towards n-hexane among the samples tested is believed to be related to the availability of the eight-membered oxygen-ring pore system which has an opening large enough only for straight chain hydrocarbons.

The clinoptilolite samples can be divided into three groups: (1) Jersey Valley and Eastgate, Nevada, and Shoshone, California, samples showed higher selectivity

TABLE 5. Activity and aging characteristics of $H^{\pm}$ clinoptilolites

Source	Activity α at 427°C	Aging* k_1/k_{75}
Baroid Mine, Hector, California	700	18.8
Hector, California	200	1.1
Shoshone, California	24	1.4
Opal Mountain, California	15	2.0
Eastgate, Nevada	10	1.6
Barstow, California	6	–
Jersey Valley, Nevada	6	3.2

*k = first-order cracking rate constant for n-hexane, subscript denotes pulse number. k_{75} corresponds to about 2.5 minutes total hydrocarbon exposure time.

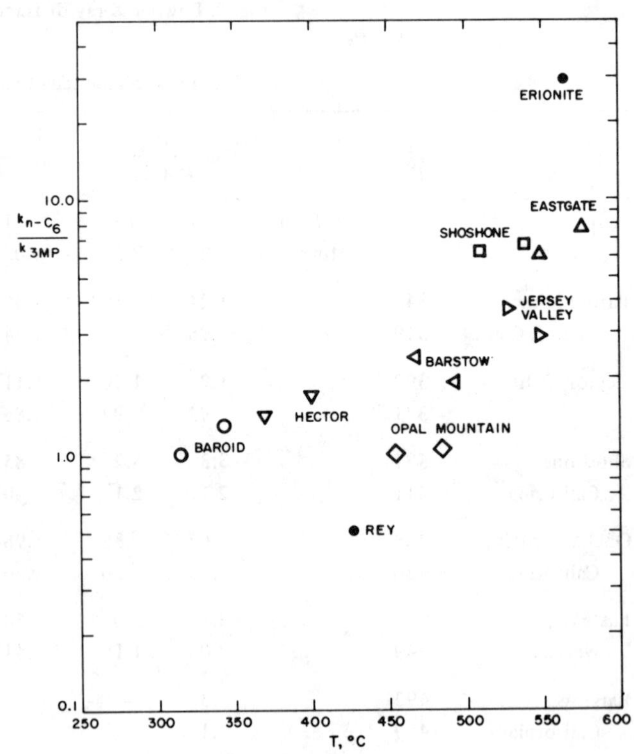

Fig. 10. Shape-selectivity vs. temperature.

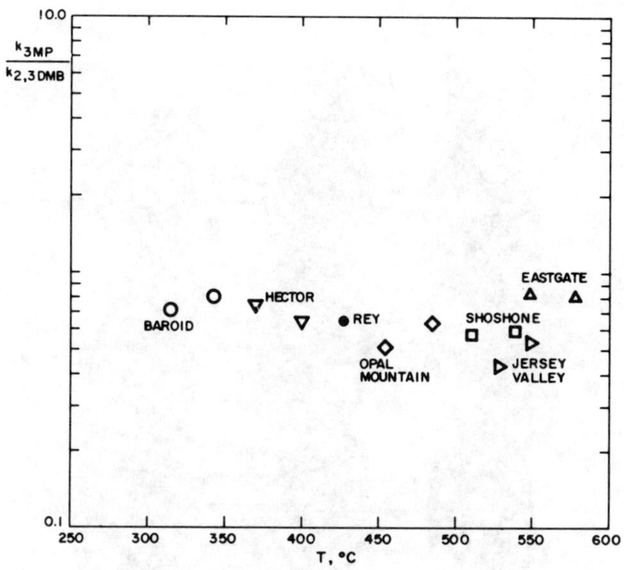

Fig. 11. Shape-selectivity vs. temperature.

values, i.e. $k_{n-C_6}/k_{3MP} > 3$; (2) Opal Mountain and Hector, California, samples had the lowest selectivity, i.e. $k_{n-C_3}/k_{3MP} \sim 1$; and (3) the Barstow, California, sample was intermediate between the above two groups. This catalytic shape-selectivity ranking coincides with the X-ray diffraction grouping, except for the Shoshone, California sample. The discrepancy is believed to be due to the presence of erionite in the Shoshone sample. Erionite, which contains eight-membered oxygen-ring pores, could shift the selectivity to higher values. With this exception, then, the high shape-selectivity of the Nevada samples appears to be related to the presence of sharp odd k-lines as in the published structure of heulandite (Merkle and Slaughter, 1968), which has both eight-membered and ten-membered oxygen ring pores. A hypothetical structure can be constructed from the basic structure of heulandite by displacing every second plane of chains along the 010-plane by $a/2$. This results in a framework structure with half the a-parameter of heulandite, the other parameters remaining the same. This model should yield a diffraction pattern with all the odd k reflections missing. If the displacement or stacking of these planes is random, the odd k reflections will be broad. A similar situation exists for the odd l reflections in the erionite–offretite family of zeolites. Zeolite T, which has a random rotation of layers perpendicular to the c-axis, has broad, odd l reflections (Bennett and Gard, 1967; Gard and Tait, 1971; Sherry, 1970).

The significance of these fine details in the X-ray diffraction patterns of the clinoptilolite samples can be explained by reference to the framework structure of these zeolites. Heulandite has a *Cm* space group, and its pore system consists of channels of alternate eight- and ten-membered oxygen-ring openings parallel to the 001-plane. If alternate 010-planes are displaced by $a/2$, a framework with a space group P_2 results, and its channel system consists of only ten-membered oxygen-ring channels parallel to the 001-plane. Thus, the presence of faults in heulandite will increase the number of ten-membered ring channels in its structure. This is illustrated in Fig. 12, with views along the 001-plane of the framework structures of heulandite and the hypothetical model.

Thus, we propose that the Nevada samples which have sharp odd k-lines are nearly isostructural with heulandite, while the California samples which have varying degrees of broad and weak odd k-lines are closer to the hypothetical model. The added availability of more larger ten-membered oxygen-ring pores helps to explain the markedly lower shape-selectivity of the California samples toward straight-chain hydrocarbons.

From Fig. 11 it can be seen that the clinoptilolite samples showed $k_{3MP}/k_{2,3DMB}$ ratios similar to those of the RE–Y sample, a large-pore zeolite with twelve-membered oxygen rings. This suggests that the ten-membered oxygen-ring pores in these samples are accessible to singly and doubly branched hexane isomers. No additional information was obtained from this specific pair of hydrocarbons to distinguish possible size differences of the ten-membered oxygen-ring pores among these samples.

CONCLUSIONS

The seven clinoptilolite samples from Nevada and California show a wide range of catalyst activities and aging rates in the shape-selective cracking test. In particular, these samples exhibit varying degrees of preferential conversion of n-hexane. The relationship between this selective conversion of n-hexane and the odd k reflections in the X-ray diffraction pattern of these materials suggests that the established heulandite structure could be only one end member of a family of zeolites of similar structure. The other end member of this family has no odd k reflections in its X-ray diffraction pattern, and its pore openings parallel to the 001 plane consist of channels of ten-membered oxygen rings only. Such a zeolite has neither been found in nature nor synthesized to date. Natural clinoptilolites from Hector, California, are closer to this hypothetical structure than those found in Nevada. Most of the clinoptilolite samples appear to have intermediate structures. An analogous situation exists for the odd l reflections in the erionite–offretite family of zeolites.

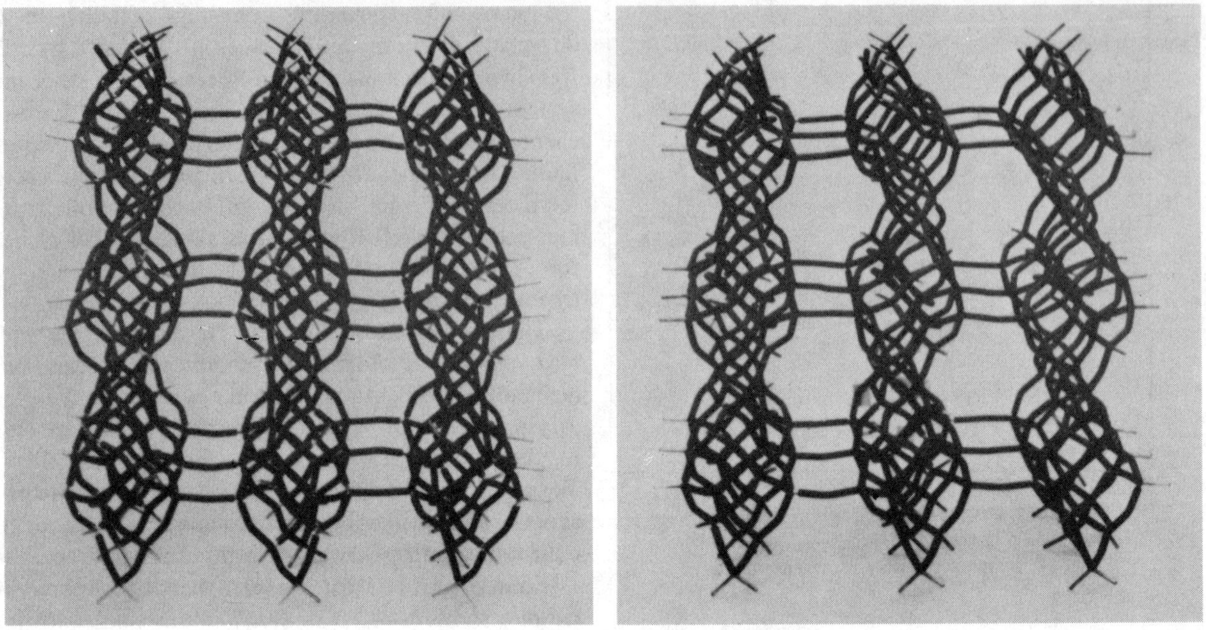

Fig. 12. Framework of heulandite and a hypothetical structure viewed along (001).

ACKNOWLEDGMENTS

D. H. Olson of this laboratory kindly supplied some of the data listed in Table 3. W. P. Burgess assisted with the pulse a test apparatus.

REFERENCES

Alietti, A. (1972) Polymorphism and crystal chemistry of heulandite and clinoptilolites: *Amer. Mineral.* **57**, 1448.

Bennet, J. M. and Gard, J. A. (1967) The solid state non-identity of the zeolites erionite and offretite: *Nature* **214**, 1005.

Boles, J. R. (1972) Composition, optical properties, cell dimensions, and thermal stability of some heulandite group zeolites: *Amer. Mineral.* **57**, 1463.

Gard, J. A. and Tait, J. M. (1971) Structural studies on erionite and offretite: *Adv. Chem. Ser.* **101**, 230.

Hey, M. H. and Banister, F. A. (1934) Clinoptilolite, a silica-rich variety of heulandite – studies on the zeolites. VII: *Mineral. Mag.* **23**, 556-559.

Mason, B. and Sand, L. B. (1960) Clinoptilolite from Patagonia, the relationship between clinoptilolite and heulandite: *Amer. Mineral.* **45**, 341-350.

Merkle, A. B. and Slaughter, M. (1968) Determination and refinement of the structure of heulandite: *Amer. Mineral.* **53**, 1120.

Miale, J. N., Chen, N. Y. and Weisz, P. B. (1966) Catalysis by crystalline aluminosilicates: *J. Catal.* **6**, 278.

Mumpton, F. A. (1960) Clinoptilolite redefined: *Amer. Mineral.* **45**, 351-369.

Shepard, A. O. and Starkey, H. C. (1964) Effect of cation exchange on the thermal behavior of heulandite and clinoptilolite: *U.S. Geol. Surv. Prof. Paper* **475D**, 89.

Sherry, H. S. (1970) Ion exchange properties of the synthetic zeolite, Linde T: *Proc. Int. Conf. Ion Exch. Proc. Ind. July 1969*; Soc. Chem. Ind., p. 332.

THERMAL STABILITY OF NATURAL GMELINITE AND SOME OF ITS ION-EXCHANGED FORMS

G. H. KÜHL and J. N. MIALE

Mobil Research and Development Corporation, Paulsboro, New Jersey 08066, U.S.A.

Abstract – Gmelinite from Prospect Park, New Jersey, has a SiO_2/Al_2O_3 molar ratio of 4.4. The hydrogen form, obtained by calcining the ammonium form at 20°C/min to 550°C, sorbed initially 12 g of cyclohexane/100 g of calcined zeolite. Recalcination caused complete loss of sorption capacity, and the product was amorphous. Hydrogen gmelinite obtained by fast calcination of ammonium gmelinite has a high n-hexane cracking activity of α = 1800 after 5 minutes on stream. The products indicate shape selectivity. After regeneration, the α-value dropped to 4.

A mixed calcium–ammonium form was investigated by thermal analysis. Dehydroxylation started before decomposition of the ammonium ions was complete. The product sorbed 9 g of cyclohexane/100 g of zeolite and after recalcination, 13.9 g of H_2O/100 g of zeolite. Dehydration of the sample containing sorbed water caused complete collapse of the crystal structure.

A completely calcium-exchanged form showed the highest water-sorption capacity after calcination at 220°C. The dehydration of this sample was followed by DTA, TGA, DTG, and X-ray diffraction.

The potassium form of gmelinite is considerably more stable to thermal treatment. Water sorption was 16 g/100 g of zeolite with a sorption curve of the usual S-shape. However, water continued to be sorbed at a slow rate. The highly hydrated form was considerably expanded in the direction of the *c*-axis.

INTRODUCTION

The crystal structure of gmelinite was established by Fischer (1966) on a gmelinite sample from Nova Scotia. The ideal gmelinite structure contains straight channels formed by 12-member rings, but these channels are usually blocked at irregular intervals by double-six rings, so that hydrocarbons are excluded (Barrer, 1944). A gmelinite without these stacking faults should be a good catalyst for hydrocarbon conversions provided it has the required thermal stability.

The secondary pore structure of gmelinite is characterized by non-planar 8-member rings. This pore system should behave similarly to chabazite; the pores are expected to become more circular upon heating, and adsorption of straight-chain hydrocarbons would increase initially upon heating.

Pécsi-Donáth (1966) studied the dehydration of natural gmelinite and other zeolites by thermal analysis, X-ray diffraction, and infrared spectrometry and found good thermal stability of the gmelinite up to 800°C. However, the chemical composition of the gmelinite sample appears questionable. The author explained the high SiO_2/Al_2O_3 ratio (5.87) of the sample with the presence of quartz. Subtraction of the quartz content yielded a SiO_2/Al_2O_3 of 3.63, too low for gmelinite, especially if the high thermal stability is taken into account. The ratio of the cation equivalents to aluminum is only 0.735, making it likely that the Al_2O_3 content reported is too high. Thus, it appears that the gmelinite studied by Pécsi-Donáth apparently had both an unusually high SiO_2/Al_2O_3 ratio and an unusually low CaO content. Pécsi-Donáth found that the water removed from natural gmelinite at temperatures up to about 200°C was completely readsorbed at ambient temperature upon exposure to water vapor. After calcination at higher temperatures, the water sorption did not exceed 10 wt%.

METHODS

The sorption capacities were measured with a DuPont 950 TGA unit modified according to Kerr (1969). In a dry helium stream, about 40 mg of the zeolite sample was calcined by heating at 5°C/min to the selected activation temperature and subsequently allowed to cool to ambient temperature. The helium stream was then loaded with the sorbate, so that p/p_0 was near 1, and passed over the sample at 190 ml/min. From the weight increase the sorption capacity in grams of sorbate per

100 g of zeolite was calculated.

The degree of crystallinity of the calcined sample was estimated from the sorption capacity and checked by Debye-Scherrer X-ray diffraction powder photographs.

The change in crystal structure of certain samples was followed by X-ray powder diffraction using a Siemens Kristalloflex IV X-ray diffractometer equipped with scintillation counter, pulse height analyzer, and strip chart recorder. Lattice parameters (a_o, c_o) were calculated from the 410 and 213 lines, as measured from a goniometer scan at $^1/_8{}^\circ$ 2θ/min and a chart speed of 1 cm/min. The position of the lines at positive and negative angles was averaged.

The n-hexane cracking activity of some samples was determined according to Weisz and Miale (1965). The alpha values reported represent the relative rate constants of the catalytic n-hexane cracking reaction at 538°C after 5 minutes on stream, based on $\alpha = 1$ for a silica–alumina activity index of 46.

TABLE 1. X-ray diffraction data of natural gmelinites

Prospect Park			Iceland (Strunz)	
d,Å	I/I_1	*hkl*	*d*,Å	I/I_1
11.93	42	100	12.0	90
7.70	27	101	7.69	60
6.905	18	110	6.81	40
5.985	8	200	5.99	10
5.135	27	201 }	5.07	60
5.045	43	002 }		
4.65	4	102 }	4.53	20
4.50	15	210 }		
4.11	100	211	4.10	100
3.97	5	300	–	–
3.44	28	220	3.44	20
3.30	9	310	–	–
3.23	48	103	3.22	60
–	–	302	3.09	5
2.981	93	400	2.96	80
2.927	31	203	–	–
2.859	49	401	2.85	60
2.691	59	213	2.67	60
2.601	20	410 }	2.57	10
2.565	3	402,303 }		

RESULTS

Gmelinite mineral

The gmelinite sample used in this investigation was a section of a large geode from Prospect Park, New Jersey. It contained about 50 g of gmelinite, much of it in large crystals (>0.5 inch), and some calcite. The material was identified by X-ray diffraction of a powdered sample with CuK_α radiation, and the observed lines (Table 1) were compared with those for an Iceland specimen published by Strunz (1956). In addition to the lines reported in Table 1, the X-ray diffraction pattern of our gmelinite sample contained a few weak lines caused by an unidentified impurity:

d,Å	I/I_1
9.36	3
8.91	7
4.33	3
3.58	9

The lattice parameters of gmelinite were found to be $a_o = 13.76_6$; $c_o = 10.06_7$; $c_o/a_o = 0.731$, in good agreement with Fischer (1966).

The chemical composition, as determined by X-ray fluorescence spectrometry, is given in Table 2. The sorption capacities of the mineral at 25°C after calcination at 350°C were:

	g/100 g of zeolite
Cyclohexane at 20 torr	0.97
n-Hexane at 20 torr	0.95
Water at 12 torr	13.8

TABLE 2. Composition of gmelinite samples

Sample	Weight %							$SiO_2/$ Al_2O_3	% $CaCO_3$
	SiO_2	Al_2O_3	Na_2O	K_2O	CaO	NH_3	H_2O		
Mineral*	61.9	23.8	10.9	n.d.	3.2	–	–	4.42	0
Ca-NH_4-gmelinite	50.0	19.3	0	n.d.	1.3	4.6	16.35	4.40	6.25
Ca-gmelinite	46.5	18.2	0	n.d.	10.0	–	19.0	4.34	5.0
K-gmelinite	49.1	19.5	0.56	15.85	0.78	–	13.85	4.28	0

*Based on solids.

Gmelinite is structurally related to chabazite and, therefore, is expected to sorb 16–20 g of water per 100 g of dry zeolite. The sorption found is considerably lower, but higher than that found by Pécsi-Donáth (1966) after calcination at 250°C. The low hydrocarbon sorption indicates that the large channels are blocked (Barrer, 1944). The n-hexane cracking activity of this gmelinite in its natural form was $\alpha = 0.4$.

Ammonium gmelinite

When the preparation of the hydrogen form was attempted by calcination of ammonium gmelinite, a complete collapse of the crystal structure was observed. A sample of ammonium gmelinite was calcined in the TGA unit at 20°C/min to 500°C, cooled initially to 200°C, and then continued to cool while in contact with cyclohexane vapor in a stream of dry helium. The sample sorbed slowly 12.2 g of cyclohexane per 100 g of zeolite. After recalcination at 20°C/min to 350°C and cooling to ambient temperature, the material sorbed only 0.7% n-hexane. X-ray examination revealed that the solid was amorphous. Various amounts of cyclohexane and n-hexane were sorbed in subsequent experiments, but the high sorption was never reproduced. It is hypothesized that a very labile, partially collapsed structure was formed that was capable of sorbing the large amount of cyclohexane; recalcination probably caused complete breakdown. The water-sorption capacity decreased from 6.3 g/100 g of zeolite after calcination at 5°C/min to 550°C to 1.7 g/100 g of zeolite after recalcination. The X-ray diffraction pattern taken after the sorption showed that the sample was amorphous. A fast heating rate in the initial calcination causes a somewhat greater, but temporary retention of crystallinity, as observed previously for zeolites A (Kühl, 1973) and X (Kühl and Schweizer, 1975). A sample calcined at 20°C/min to 500°C had a sorption capacity for n-hexane of 6.1 g/100 g of zeolite. After recalcination at 20°C/min to 350°C, the water-sorption capacity was 14.3 g/100 g. The material was then examined by X-ray diffraction and found to be only partially crystalline. The same ammonium gmelinite calcined initially at a heating rate of 5°C/min to 550°C sorbed only 4.8 g H_2O/100 g of zeolite; this value decreased after recalcination to 0.5 g H_2O/100 g of zeolite.

In the n-hexane cracking test (Weisz and Miale, 1965), a sample of ammonium gmelinite calcined at a fast heating rate to 550°C in the reactor gave a relatively high α-value of 1800 after 5 minutes on stream. The products of the α-test indicate that the zeolite is shape selective (no branched-chain hydrocarbons). However, after regeneration, the α-value dropped to 4 as a result of further collapse of the crystal structure.

Calcium–ammonium gmelinite

In an attempt to prepare an acid-gmelinite form that is more stable, ammonium gmelinite was competitively exchanged with a mixture of ammonium chloride and calcium chloride. The composition of this material, as determined by X-ray fluorescence spectrometry, is summarized in Table 2.

The X-ray diffraction pattern of the Ca-NH_4-gmelinite is listed in Table 3. The lattice parameters were $a_o = 13.78_9$ and $c_o = 10.04_4$, as determined from the (410) and (213) lines, respectively. After calcination at 15°C/min to 500°C, the material sorbed 9.25 g of cyclohexane/100 g of zeolite. After recalcination (15°C/min) to 350°C, it sorbed 13.9 g H_2O/100 g. An attempt was made to recalcine again and repeat the water sorption, but no water was sorbed. This result indicates complete collapse of the zeolitic structure upon heating the hydrogen zeolite in the presence of water. X-ray diffraction proved that the product was amorphous. The unexpectedly high cyclohexane

TABLE 3. X-ray diffraction data of Ca-NH_4-gmelinite

d,Å	I/I_1	hkl
11.945	38	100
7.694	56	101
6.915	40	110
5.985	15	200
5.140	22	201
5.025	100	002
4.634	10	102
4.518	33	210
4.123	93	211
3.986	3	300
3.850	3	202
3.446	25	220
3.357	3	212
3.311	4	310
3.226	85	103
3.142	3	311
3.110	3	302
2.986	49	400
2.921	19	203
2.862	60	401
2.690	47	213
2.644	4	321
2.606	9	410

sorption capacity of a transient form corroborates the finding obtained with ammonium gmelinite. In a similar experiment, water was sorbed after the initial calcination; a capacity of 17.6 g H_2O/100 g of zeolite was found. Recalcination caused again almost complete collapse of the structure as evidenced by X-ray diffraction.

Both calcium-exchanged samples contained some calcite from the geode specimen; however, since calcite does not decompose in the temperature range studied, it did not interfere with the investigation.

Figure 1 shows the thermogravimetric analysis of the calcium-ammonium-gmelinite at a heating rate of 10°C/min. The evolved ammonia was periodically titrated as described by Kerr (1969). In the lower plot the total weight loss is separated into weight of ammonia and weight of other volatiles lost. The decomposition of the ammonium ion began at 200-250°C and was complete at about 550°C. All zeolitic water was given off by 375°C. Dehydroxylation started at about 450°C, long before the decomposition of the ammonium ions was complete. When dehydroxylation occurs at so low a temperature, it is usually accompanied by a collapse of the crystal structure. Indeed a sample calcined to 500°C at 15°C/min in the TGA sorbed 17.6 g H_2O/100 g of zeolite, whereas a sample calcined for 1 hour at 550°C sorbed only 4 g H_2O/100 g of zeolite, a clear indication of loss of structure The calculated loss of chemical water, equivalent to the loss of ammonia, is 0.66 mg (Fig. 1) and should be complete at a total water loss of 4.26 mg. However, the ammonium ion exchange was accompanied by hydronium ion exchange, as was previously observed with zeolites A (Kühl, 1973) and X (Kühl and Schweizer, 1975). These hydronium ions may have caused removal of framework-aluminum at elevated temperature. Dehydroxylation of framework Si-OH groups generated in this manner, or of hydroxylated aluminum cations, occurs at this temperature in addition to the dehydroxylation of acidic Si-OH groups formed by decomposition of NH_4^+ ions. These dehydroxylation reactions overlap, thus causing a larger weight loss than calculated from the ammonium content.

When the sample that sorbed 17.6 g H_2O/100 g of zeolite was recalcined to 500°C at 15°C/min, cooled to ambient temperature and contacted with water vapor, no weight increase was observed. X-ray diffraction

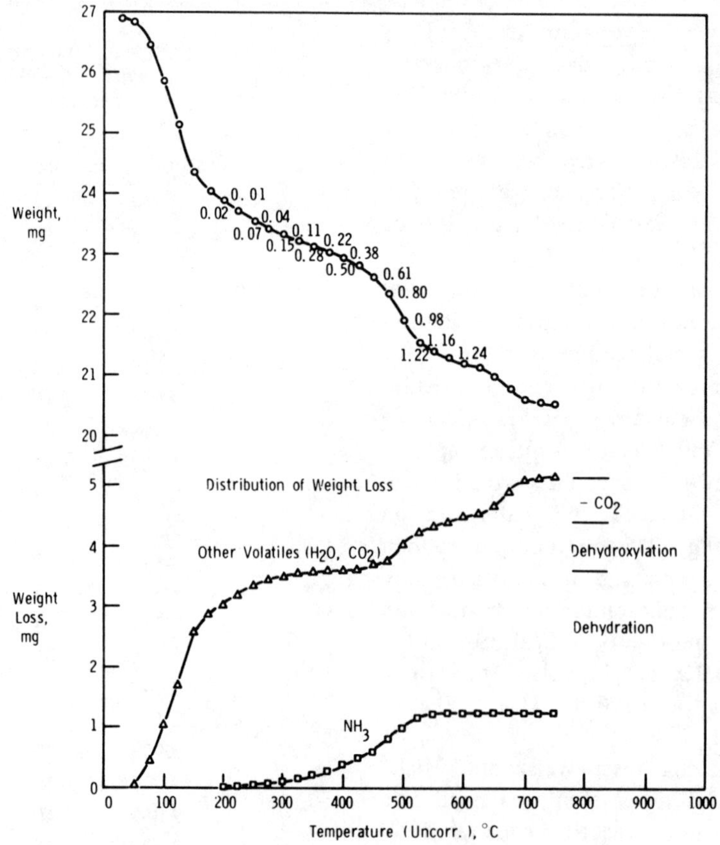

Fig. 1. Thermogravimetric analysis of Ca-NH_4-gmelinite; heating rate: 10°C/min; flowing He; numbers are mg NH_3 lost (cumulative).

revealed that the sample was amorphous. It is believed that the collapse of the crystal structure was caused by hydrolytic attack of the sorbed water on the acid sites resulting in removal of Al from the framework at elevated temperature. Such a reaction was previously observed with other zeolites.

The n-hexane cracking activity of this sample cannot be high because of the low thermal stability. Indeed, an α-value of 1.1 was found.

Calcium gmelinite

A completely calcium-exchanged sample was obtained by exhaustive exchange with 1 N $CaCl_2$ solution at 80°C. The lattice parameters of the hydrated material were found to be $a_o = 13.83_5$ Å, $c_o = 9.79_0$ Å. The chemical composition is listed in Table 2. The unit-cell composition of this Ca-gmelinite is $Ca_{3.785}(AlO_2)_{7.57}(SiO_2)_{16.43}(H_2O)_{22.38}$. The sample contained 5.0% calcite as determined from the CO_2 liberated in the TGA, from X-ray examination, and from the Ca/Al balance in the sample. The water-sorption capacity after calcination at various temperatures was:

After calcination at °C	Water-sorption capacity g/100 g of zeolite
175	15.1
220	17.2
350	16.1
550	15.2

Figure 2 shows the TGA and the differential thermogram of calcium gmelinite at a heating rate of 30°C/min in a helium stream. The corresponding DTA curve, taken at the same heating rate, is shown in Fig. 3. The major portion of water ($16H_2O$/unit cell) is lost in two overlapping steps of $8H_2O$ each with maximum rates at 117° and 197°C (TGA) and the corresponding endotherms at 130° and 210°C (DTA). About $3H_2O$ are then liberated by a rather slow reaction with maximum at 314° and a broad endotherm at 320°. Another $2H_2O$ are lost in a faster reaction at 419° with endotherm at 430°. The remaining water is finally given off at 588°. The shallow endotherm at 530° is apparently not associated with a weight loss and may indicate a change in crystal structure. Samples of calcium gmelinite were calcined for ½ hour at 260°, 460°, 620°, and 760°C; the calcined materials were investigated by X-ray diffraction. The results are shown in Fig. 4. The changes in the line intensities obtained after calcination at 260°C were caused by the dehydration of the zeolite. Calcination for ½ hour at 460°C caused not only further decrease in the line intensities, with the exception of the (100) line, which was enhanced further, but also the appearance of lines (marked X) foreign to gmelinite. This trend continued for the 620°C calcination. No further changes were observed in the X-ray diffraction pattern of the sample calcined at 760°C (calcite lines were subtracted from all patterns). A sample calcined for ½ hour at 350°C showed no significant difference from that calcined at 260°C. However, when the calcination was extended to 3 hours, the new lines observed in the

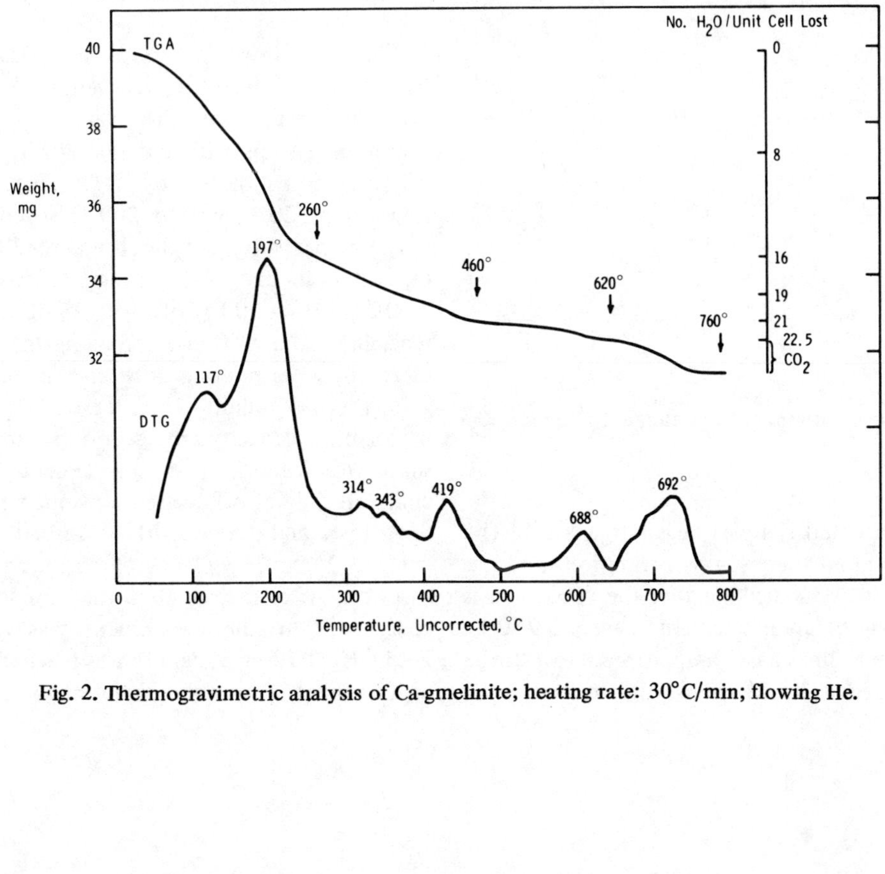

Fig. 2. Thermogravimetric analysis of Ca-gmelinite; heating rate: 30°C/min; flowing He.

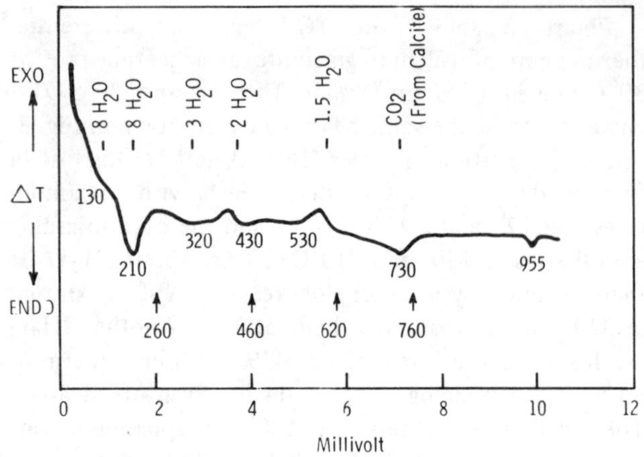

Fig. 3. Differential thermal analysis of Ca-gmelinite; heating rate: 30°C/min; He atmosphere. Thermocouple: Pt/Pt–13% Rh. Numbers are temperatures (°C).

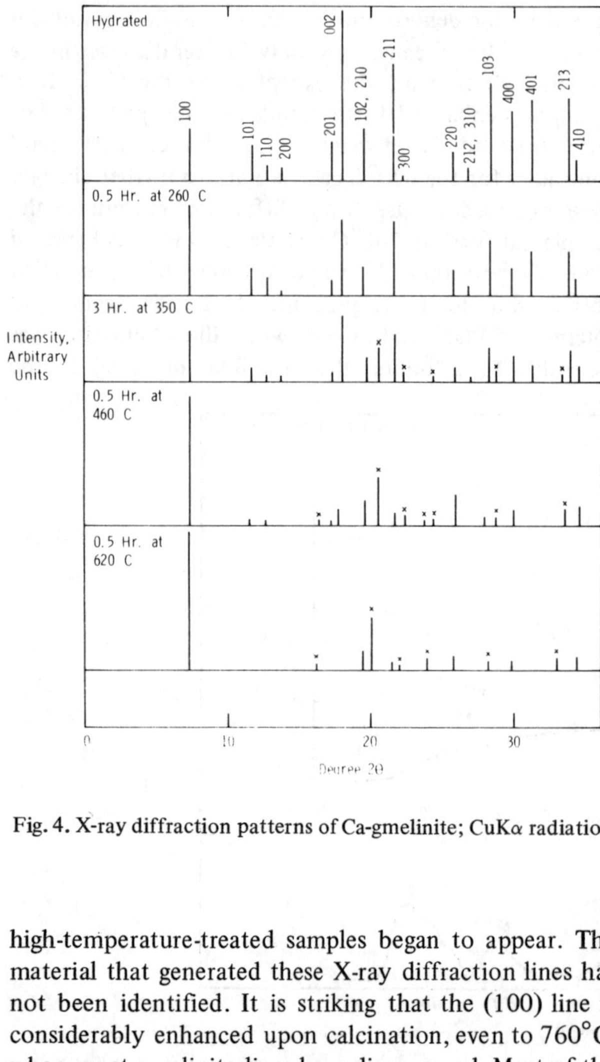

Fig. 4. X-ray diffraction patterns of Ca-gmelinite; CuKα radiation.

high-temperature-treated samples began to appear. The material that generated these X-ray diffraction lines has not been identified. It is striking that the (100) line is considerably enhanced upon calcination, even to 760°C, when most gmelinite lines have disappeared. Most of the retained, but weakened, lines have hk0 indices.

The apparent activation energies of the dehydration reactions were estimated from the TGA according to Dave and Chopra (1966):

Peak temperature, °C	No. H_2O lost	E_A Kcal/mol
17	8	3.8
197	8	9.2
314	3	10.7
419	2	34.2
588	1.5	49.6
692	CO_2	56.3

The activation energy found for the decomposition of calcite agrees with the value obtained by Dave and Chopra (1966).

Potassium gmelinite

Chabazite is thermally most stable in its potassium form. As the gmelinite structure is closely related to that of chabazite, the potassium form of our gmelinite sample was prepared by ion exchange at ambient temperature and examined by thermal analysis. The composition of the potassium gmelinite is summarized in Table 2. The material has the approximate unit-cell formula

$$Ca_{0.27}Na_{0.36}K_{6.74}(AlO_2)_{7.64}(SiO_2)_{16.36}\cdot 26H_2O.$$

Figure 5 shows the TGA and DTG of K-gmelinite. Essentially all the water is given off below 300°C. The DTG indicates that the water is released in two overlapping steps with maxima at 80° and 150°C. This result is corroborated by DTA (Fig. 6). The sample calcined at 20°C/min to 500°C sorbed only 1.2 g of n-hexane per 100 g of solid, but sorbed the same amount of water it had lost during calcination, 15.9 g (vs. 16.05 g) H_2O/100 g of dry solid. The sample was recalcined to 550°C and sorbed again 16.0 g H_2O/100 g of zeolite after cooling to room temperature, indicating good thermal stability to at least 500°C. Up to this level of sorption capacity, the sorption curve had the usual shape for zeolitic sorption, except that it did not completely level off. Water continued to be sorbed at a slow rate, and the weight of sorbed water eventually reached 24.0 g H_2O/100 g of zeolite in the initial water sorption. After recalcination the slow sorption was again observed, but the experiment was terminated when 20.5 g H_2O/100 g of zeolite was reached.

To investigate this phenomenon further, an un-

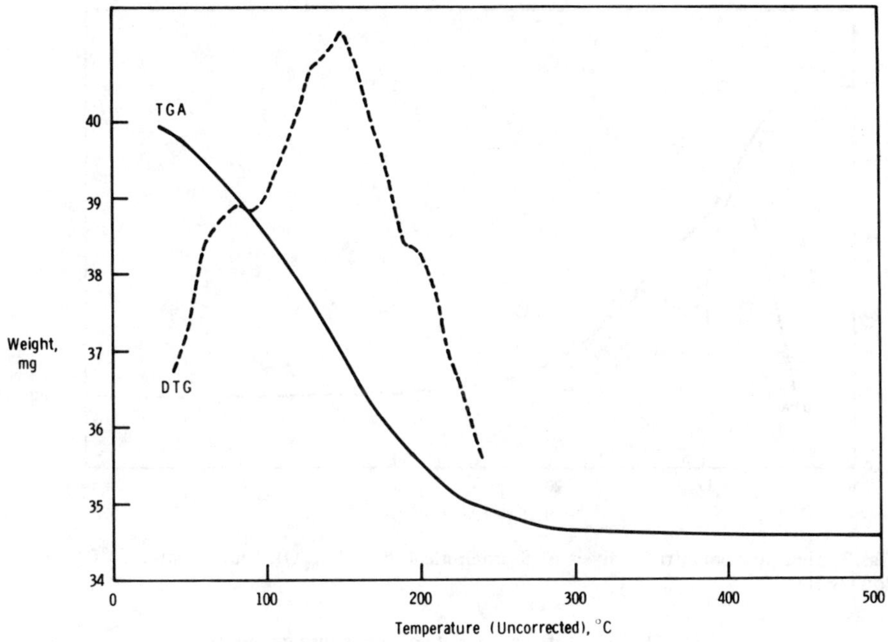

Fig. 5. Thermogravimetric analysis of K-gmelinite (13.84% H_2O); heating rate: 20°C/min; flowing He.

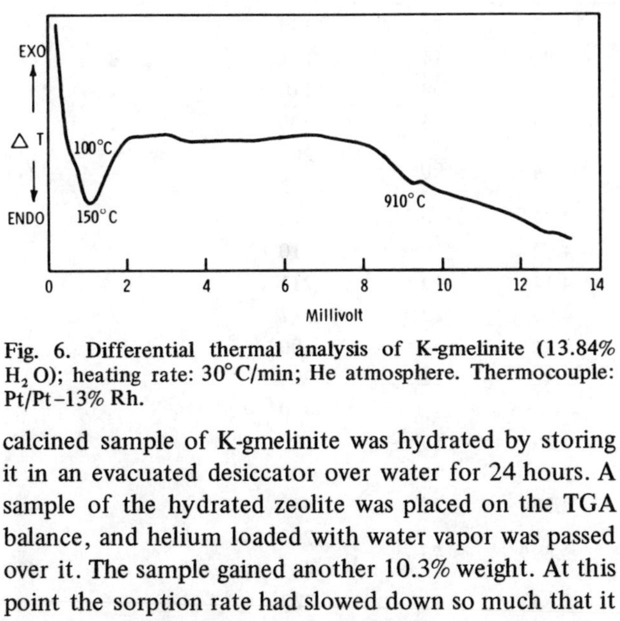

Fig. 6. Differential thermal analysis of K-gmelinite (13.84% H_2O); heating rate: 30°C/min; He atmosphere. Thermocouple: Pt/Pt–13% Rh.

calcined sample of K-gmelinite was hydrated by storing it in an evacuated desiccator over water for 24 hours. A sample of the hydrated zeolite was placed on the TGA balance, and helium loaded with water vapor was passed over it. The sample gained another 10.3% weight. At this point the sorption rate had slowed down so much that it took 50 minutes to sorb 1.0 g H_2O/100 g of zeolite. In a subsequent thermogravimetric analysis of the product, all of the sorbed water was released again (Fig. 7), and it became apparent that the zeolite had contained 38.25 g H_2O/100 g of dry zeolite. The enhanced DTG peak occurs in a temperature range in which free water might be expected to evaporate in a stream of dry helium. To distinguish between free water and water in the zeolite structure, the sample hydrated in the desiccator and containing 25.35 g H_2O/100 g of zeolite was examined by X-ray diffraction, and its pattern was compared with that of the original sample containing 16.05 g H_2O/100 g of zeolite (Table 4). The surprising result was that not only was a change in the relative intensities of the X-ray diffraction lines found, but also a considerable expansion (1.8%) in the direction of the *c*-axis had occurred in the highly hydrated sample. In summary, the additional water has the following properties:

1. It is sorbed very slowly at ambient temperature.
2. It is loosely bound and is released readily at rather low temperature. The desorption is reversible.
3. It causes an expansion of the gmelinite lattice in the direction of the *c*-axis.

These observations are consistent with the following sorption scheme: Water is initially sorbed fast into the large channels formed by 12-member rings. Migration of this water into gmelinite cages is blocked by cations located in non-planar 8-member rings which border the large channels. These cations are rather tightly bound to the framework and therefore not very mobile. The rate at which water molecules temporarily displace them and slip into the gmelinite cages is slow. At the same rate, space in the large channels becomes available for more water to be sorbed. Hydration of cations in the gmelinite cages causes the observed expansion in the direction of the *c*-axis. Upon heating, the water is readily desorbed because the hydrated cations are much more mobile and may not be in blocking positions at all. Calcination at 550°C of the highly hydrated zeolite (25.35 g H_2O/100 g of zeolite) causes a decrease of the *c* lattice parameter of 0.34 Å, or 3.35%.

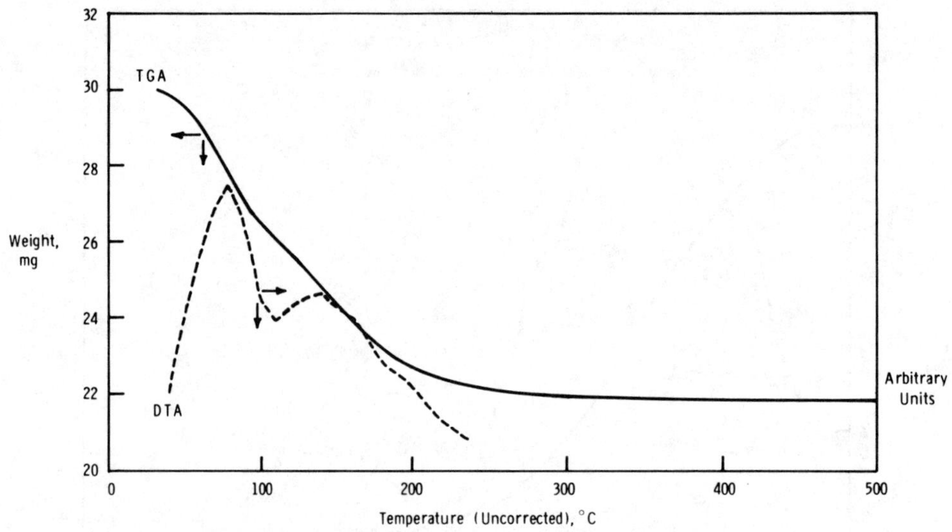

Fig. 7. Thermogravimetric analysis of K-gmelinite (38.25% H_2O); heating rate: 30°C/min; flowing He.

TABLE 4. X-ray diffraction data of potassium gmelinite

16.05 g H_2O/100 g of solid		25.35 g H_2O/100 g of solid		
d,Å	I/I_1	*d*,Å	I/I_1	*hkl*
12.08	79	12.08	75	100
7.74	13	7.80	14	101
6.94	21	6.88	8	110
6.01	6	5.97	5	200
5.15	12	5.14	100	201
5.04	93			002
4.64	11	4.72	8	102
4.54	29	4.53	18	210
4.14	65	4.11	49	211
4.06	10	4.07	22	112
4.00	5	3.98	4	300
3.47	21	3.46	22	220
3.32	4	–	–	310
3.23	100	3.29	59	103
3.13	19	3.13	25	302
2.998	56	2.984	22	400
2.926	26	2.699	43	203
2.870	39			401
2.853	8	2.853	23	222
2.694	64	2.721	76	213
2.647	7	2.649	7	321
2.619	21	2.612	18	410
$a_o \sim 13.85$		$a_o \sim 13.82$		
$c_o \sim 10.04$		$c_o \sim 10.22$		

ACKNOWLEDGMENT

The authors are indebted to Miss Carolann Craig for her able assistance with thermal analysis and X-ray diffraction.

REFERENCES

Barrer, R. M. (1944) Sorption by gmelinite and mordenite: *Trans. Farad. Soc.* **40**, 555-564.

Dave, N. G. and Chopra, S. K. (1966) A method for the evaluation of kinetics of solid state reactions from thermogravimetric curves: *Z. Phys. Chem.*, Neue Folge, **48**, 257-266.

Fischer, K. (1966) Untersuchung der Kristallstruktur von Gmelinit: *Neues Jahrbuch für Mineralogie, Monatsh.* **I**, 1-13.

Kerr, G. T. (1969) Chemistry of crystalline aluminosilicates. VII. Thermal decomposition products of ammonium zeolite Y: *J. Catal.* **15**, 200-204.

Kühl, G. H. (1973) Structural stability of partially ammonium-exchanged zeolite A: *J. Catal.* **29**, 270-277.

Kühl, G. H. and Schweizer, A. E. (1975) Structural stability of sodium ammonium zeolite X: *J. Catal.* **38**, 469-476.

Pécsi-Donáth, É. (1966) On the relationships between lattice structure and "zeolite water" in gmelinite, heulandite, and scolecite: *Acta Universitatis Szegediensis, Acta Mineralogica-Petrographica* **17**, 143-158.

Strunz, H. (1956) Die Zeolithe Gmelinit, Chabasit, Levyn (Phakolith, Herschelit, Seebachit, Offretit): *Neues Jahrbuch f. Mineralogie A, Monatsh.* **II**, 250-259.

Weisz, P. B. and Miale, J. N. (1965) Superactive crystalline aluminosilicate hydrocarbon catalyst: *J. Catal.* **4**, 527-529.

OXYGEN ENRICHMENT OF AIR USING BULGARIAN CLINOPTILOLITE

I. M. GALABOVA, G. A. HARALAMPIEV, and B. ALEXIEV

Department of Physical Chemistry, Higher Chemical Technological Institute, Sofia, Bulgaria

Abstract – The possibility of air enrichment using Bulgarian natural clinoptilolite was reported in 1973. Improvement was achieved by acid treatment and by ion exchange of the starting material to give sorbents of much higher nitrogen selectivity. Samples from Oligocene 1st horizon (Ol_1) and Oligocene 3rd horizon (Ol_3) in Northeastern Rhodope were examined. Both samples contain between 85 and 95% clinoptilolite, along with small amounts of celadonite in sample Ol_1 and montmorillonite and cristobalite in sample Ol_3.

Air enrichment was carried out at room temperature in an adsorption column of 400 mm length and 14 mm diameter. Initial gas pressures were 3 to 5 atmospheres, and the contact time was 30 to 120 seconds. Under fixed experimental conditions, different sorbents were compared by their coefficients of oxygen production (a), expressed in "over-atmospheric oxygen" produced by a unit weight of the zeolite. Each sample was checked in its natural form and in potassium-enriched forms. The natural forms were found to be fairly poor in air enrichment, yielding values of 0.20 cm^3/g and 0.35 cm^3/g for Ol_1 and Ol_3, respectively. For potassium-enriched forms, the corresponding values were found to be 0.50 cm^3/g and 0.54 cm^3/g, respectively, indicating a greater enrichment and a minimized difference between the two samples.

INTRODUCTION

The discovery of large sedimentary zeolite deposits in Bulgaria by Alexiev (1968) became a basis and a start for intensive investigations of their properties directly connected with possible industrial application. Some modest achievement in Bulgarian clinoptilolite application were made by Gradev and Daier (1972) in cleaning radioactive waste waters, by Haralampiev *et al.* (1975) in applying pressure swing process in gas drying, and by Galabova *et al.* (1975) in drying organic solvents. The problem of oxygen enrichment of air takes a special place among all projects on Bulgarian clinoptilolite both from industrial and scientific points of view. The use of zeolites as convenient sorbents for air enrichment is well known (Breck, 1974), but the use of clinoptilolite for this application is not defined in this respect. Torii *et al.* (1973) and Galabova *et al.* (1973) give some information on the possibility of using clinoptilolites for oxygen enrichment of air.

The present paper describes some Bulgarian clinoptilolite deposits and characterizes them with respect to air-enrichment processes.

DEPOSITS AND MATERIALS

Sedimentary and volcano-sedimentary rocks of Oligocene age, formed under marine conditions, are spread over hundreds of square kilometers in south Bulgaria (Northeast Rhodopes). They are divided (Goranov, 1960) into eight horizons. Much of the vitroclastic rocks of the first and third Oligocene horizons (Ol_1 and Ol_3) are highly altered and transformed into zeolitites (rocks containing zeolite mineral above 50%). Analcime, clinoptilolite, and mordenite zeolites are found in the Ol_1 profile. In the Ol_3 profile, only clinoptilolite zeolites are present.

First Oligocene horizon (Ol_1)

Analcime zeolites occur in the lowest part of the Ol_1 profile according to Djuorova (1976). They consist of analcime, quartz, adularia, and chlorite that was generated by the alteration of volcanic glass. Sanidine, andesine, biotite, zircon, titanite, garnet, epidote, cristobalite, and muscovite are present in small quantities. The structure of the zeolites is fine-grained and relicto-vitroclastic. The thickness of the analcime zeolites reaches 190 meters. Analcime zeolites are replaced by clinoptilolite zeolites vertically. Analcime zeolites, however, are not found elsewhere in the Ol_1 profile. Commonly the Ol_1 profile starts with clinoptilolite zeolites instead of analcime.

Clinoptilolite zeolites of Ol_1 overlie sedimentary rocks of late Eocene age and analcime zeolites of Ol_1. These clinoptilolite zeolites are similar in appearance to analcime zeolites, as well as in the mineral composition of the crystaloclasts and in texture. The two types of rocks differ only in the zeolite mineral present.

The vertical profile of clinoptilolite zeolites, overlying late Eocene sedimentary rocks, can be divided into four zones. The thickness of the first zone reaches 25 to 30 m and is grey or white. The second zone is 90–100 m thick and is mottled in appearance. Light-pink rocks comprise the third zone and consist of clinoptilolite zeolites and partially altered tuffs with unaltered volcanic glass. This zone is about 25 m thick. Fresh vitreous rocks (typical vitroclastic tuffs) form the fourth zone, and in general they are light grey or white. Changeable thickness, sometimes reaching 30–40 m, is typical for this zone.

Clinoptilolite zeolites of Ol_1 consist of zeolitized volcanic glass and single grains of muscovite, biotite, hornblende, sanidine, andesine, zircon, epidote, titanite, garnet, apatite, quartz, and tourmaline. The amount of crystaloclasts is up to 15%. Zeolitized volcanic glass grains range up to 1.5–2 cm. Rarely zeolitized pumice pieces of 8 to 10 cm in size can be found. Despite zeolitization, volcanic glass fragments have preserved their different textures. Volcanic glass alteration has a zonal character. A thin film (about 0.002 mm) of celadonite or a clay mineral (montmorillonite) is formed around each piece. A compact zone of zeolite crystals radially oriented and with positive elongation projects inwards. The internal part of the fragments is made up of fairly large clinoptilolite crystals (up to 0.05 mm).

Division of clinoptilolite zeolites of Ol_1 in four zones is based not only on their color but also on their mineralogical composition. Clinoptilolite in all zones is colorless. The mineral outlining the zeolitized pieces of volcanic glass determines the zone color. In the first zone this is a colorless phyllosilicate, in the second – celadonite, and in the third – pink montmorillonite. "Fresh-vitreous" (unaltered) rocks from the fourth zone of Ol_1 contain pieces of unaltered volcanic glass. The shape, size, and structures of the unaltered pieces of volcanic glass are analogous to zeolitized ones, except that their periphery is outlined by a narrow band of clay mineral (montmorillonite).

An examination of the full profile of zeolites in Ol_1 gives the following vertical zonal distribution with respect to zeolite minerals: analcime – calcium clinoptilolite – potassium clinoptilolite – unaltered (unzeolitized) volcanic glass. Mordenite zeolites are included in the extreme north of the clinoptilolite zone. They formed by the hydrothermal alteration of clinoptilolite zeolites (Alexiev and Djuorova, 1976).

Third oligocene horizon (Ol_3)

Two zones are observed in the vertical profile of Ol_3: white clinoptilolite zeolites underlying "fresh-vitreous" rocks (tuffs). White zeolites of the lower part of profile are coarse with many crystaloclasts. Going upwards the grain size gradually decreases and crystaloclasts virtually disappear. Seven cycles are found in some profiles of zeolite horizon. Such a cyclicity is also observed in "fresh-vitreous" rock horizons.

Zeolites of Ol_3 are analogous to those in Ol_1. The amount of zeolite present varies but it is always above 50%, and in some places reaches 90%.

Analcime and clinoptilolite zeolites of Ol_1 and clinoptilolite zeolites of Ol_3 are formed by alteration of vitroclastic sedimentary rocks (Alexiev and Djuorova, 1975). Strong volcanic activity with ejection of enormous amounts of vitroclastic materials took place in the Northeast Rhodopes at the time of Ol_1 and Ol_3 origin and was followed by rapid sedimentation of vitroclastic pieces in a water basin. The high temperature and thermal-isolation properties of these materials created conditions considered similar to those that might exist in autoclaves of enormous size, with more or less equal physical-chemical parameters over vast areas; these were important in the zeolitization process.

Zeolites in the Northeast Rhodopes resemble ignimbrites, as far as their material composition and the manner of transfer and sedimentation are concerned. Differences in these two types of rocks are controlled by the site of their formation – ignimorites are formed on the land and zeolites in a water basin.

Experiments were carried out on a sample from the first zone of Ol_1 and from a clinoptilolite zeolite of the middle part of Ol_3. They are referred to as (Ol_1) and (Ol_3), respectively.

EXPERIMENTAL

Sample preparation and characterization

Samples Ol_1 and Ol_3 were obtained as fairly large, irregularly shaped pieces weighing about 1 to 3 kg. They were ground and classified by sieving. The 1 to 3 mm was separated as a convenient one for the adsorption experiments.

Screened material from each sample was separated into two portions – one for direct characterization by chemical and X-ray diffraction analyses and for enrichment experiments and the other for modification. Gas chromatographic investigations (Galabova *et al.*, 1976) showed that oxygen enrichment of air on Bulgarian clinoptilolite can be considerably improved by acid treatment of the native material. Oxygen concentrations of 96% and separation coefficients of 0.98% were reached for samples of moderate decationation (about 40%) and of low dealumination (below 7%). Ion-exchange of various metal cations (Galabova and Haralampiev, 1975) resulted in further improvement. Among the investigated cationic forms, potassium forms gave satisfactory oxygen enrichment characteristics. In view of this, potassium ion-exchange was chosen as a method of modifying the Ol_1 and Ol_3 samples. Aqueous solutions of 0.5 N KCl were used in double stoichiometric excess. A hundred grams of each ground sample was put in a glass ion-exchange column of 30 mm diameter. The solution was percolated through the sorbent for 6 hours. Ion-exchanged material was washed with distilled water and dried first at room temperature and then at 110°C for 24 hours. Samples for chemical and X-ray diffraction analyses were rehydrated.

TABLE 1. Chemical composition of native and K-exchanged clinoptilolites (wt. %)*

Component	Ol_1	K-Ol_1	Ol_3	K-Ol_3
SiO_2	66.18	–	67.30	–
Al_2O_3	11.69	11.15	12.17	11.86
Fe_2O_3	0.90	0.84	1.02	1.09
CaO	3.90	2.34	2.22	1.43
MgO	0.19	0.30	0.69	0.24
Na_2O	1.05	0.30	1.86	0.40
K_2O	3.21	6.20	4.16	7.83
H_2O^+ H_2O^-	12.85	–	10.65	–
sum	99.97	–	100.07	–

*Ol_1 = clinoptilolite, first Oligocene horizon.
Ol_3 = clinoptilolite, third Oligocene horizon.
K-Ol_1 and K-Ol_3 = potassium ion-exchanged forms.

Chemical and X-ray analyses

Chemical analyses of native samples (Ol_1 and Ol_3) and their potassium enriched forms (K-Ol_1 and K-Ol_3) are given in Table 1. A higher content of CaO is found in Ol_1, compared with Ol_3, while the alkali cation content in Ol_3 exceeds that in Ol_1. According to the classification of Gottardi *et al.* (1973), sample Ol_3 is closer to typical clinoptilolites. A thermal stability check (Galabova *et al.*, 1976) showed Ol_3 to be stable and to exhibit thermal behavior of Gottardi's type 3.

Potassium content of K-Ol_3 is still higher than that of K-Ol_1. It is interesting to note the equal degree of sodium removal from Ol_1 and Ol_3. Calcium, considerably decreased in the two ion-exchanged samples, is higher in K-Ol_1 than in K-Ol_3.

X-ray diffraction patterns of Ol_1, K-Ol_1, Ol_3, and K-Ol_3 were taken on a TUR-M61 unit with a HCG-3 goniometer, using CoK_α radiation; X-ray diffraction patterns are shown in Fig. 1. Comparison of the two native samples reveals one difference: Ol_3 possesses an additional line (θ = 12.80°), which we have assigned to α-cristobalite.

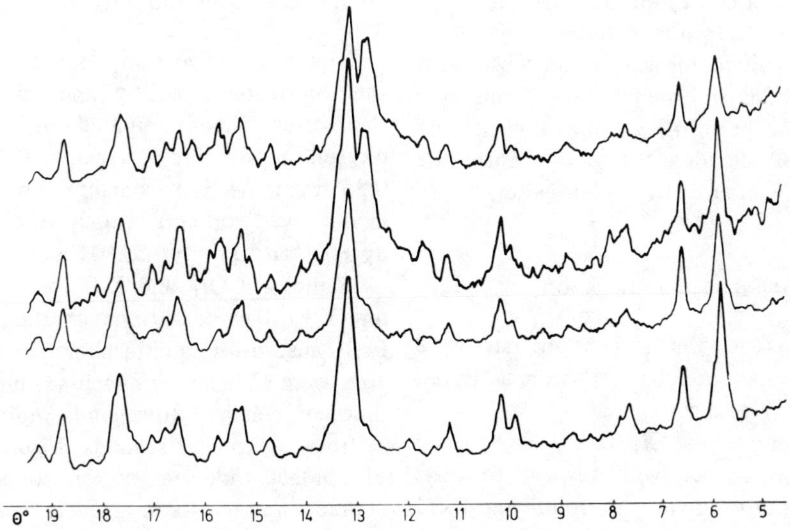

Fig. 1. X-ray diffraction patterns of native and K-exchanged clinoptilolite. (a) native Ol_1; (b) K-exchanged Ol_1; (c) native Ol_3; (d) K-exchanged Ol_3.

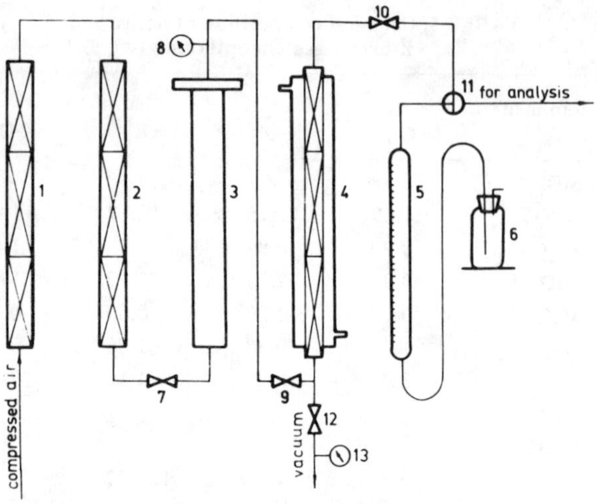

Fig. 2. Equipment for oxygen enrichment of air. Drying columns (1, 2); dozing column (3); adsorption column (4); graduated gas burette (5); vessel for keeping atmospheric pressure (6); manometer (8); stopcock (11); valves (7, 9, 10, 12); vacuummanometer (13).

The intensity of the reflections at $\theta = 5.80°$ and $\theta = 13.15°$ are slightly decreased in K-Ol_1 although for K-Ol_3 these lines are weaker. Noticeable replacement of diffraction peaks was not found.

Adsorption equipment

Adsorption equipment used for the experiments is shown schematically in Fig. 2. Air, compressed up to desired pressure, enters the system through two drying columns loaded by native clinoptilolite and goes into a previously calibrated dozing column. From column 3 air is allowed to pass into the adsorption column whose length (L) = 400 mm and ϕ= 12 mm. Air enriched in O_2 coming out of the adsorption column goes to a graduated burette for volume measuring and subsequent analysis. The adsorbent is activated between runs by a rotary vacuum pump. The initial gas pressure and the pressure reached after allowing the gas to enter the adsorption column are measured by a manometer.

Experimental data and discussion

Experiments in oxygen enrichment of air were carried out at room temperature. Samples were activated beforehand at 400°C for 4 hours.

All investigated samples – Ol_1, Ol_3, K-Ol_1, and K-Ol_3 – were first compared with respect to their oxygen-enrichment ability under the following fixed experimental conditions: initial gas pressure, P, equal to 4 atm, and contact time for contact sorbent/sorbate, τ = 60 sec; time for allowing gas to the sorbent, τ^i, and desorption time, τ^d equal to 30 seconds. The oxygen-enrichment process was characterized by "final" pressure, P^f (atm), reached in time τ^i and by volume V(cm^3) and concentration C_{O_2} (%) of the outgoing gas obtained for τ^d. The volume of the overatmospheric oxygen (gas of concentration above 21%) was calculated on the basis of the above direct data and used to calculate coefficients of oxygen production, a(cm^3 g^{-1}). The latter expresses the volume of overatmospheric oxygen produced per cycle by unit weight of dehydrated zeolite. Table 2 presents data on oxygen enrichment of air on native Ol_1 and Ol_3 samples as well as on their K-enriched forms.

TABLE 2. Oxygen enrichment parameters for native and K-exchanged clinoptilolites at initial pressure 4 atm*

Sample	P^f atm	C_{O_2} %	α cm^3/g
Ol_1	1.60	29.2	0.20
K-Ol_1	1.05	37.2	0.52
Ol_3	1.75	30.5	0.35
K-Ol_3	1.30	35.5	0.58

*Parameters: final pressure (P^f); oxygen concentration (C_{O_2}); coefficient of oxygen production (a). Samples are identified in footnote to Table 1.

$$a = \frac{V(C_{O_2} - 2L)}{G_V}, \text{ where}$$

G = dehydrated weight of the sorbent, V and C_{O_2} are volume and concentration.

The two native samples are on the whole poor in their enrichment ability and fairly different between themselves when compared with their coefficient of oxygen production (a_{Ol_1} = 0.20 cm^3 g^{-1}; a_{Ol_3} = 0.35 cm^3 g^{-1}). Ion exchange towards potassium results in improved sorbents, which behave in a similar way – $a_{K\text{-}Ol_1} \approx a_{K\text{-}Ol_3} \approx 0.52\text{–}0.58$ cm^3 g^{-1}.

Samples K-Ol_1 and K-Ol_3 were further investigated under various conditions of the enrichment process. Four adsorption quasi-isobars, in the region 3.5 to 5 atm, were obtained for each sample, keeping desorption time, τ^d, equal to 30 seconds and varying contact time, τ, from 30 to 120 seconds. Figure 3 shows the effect of contact time on oxygen concentration. Increasing contact time results in an increase in oxygen concentration, but that does not necessarily mean that increased contact time favors the enrichment process. Figure 4 shows a drop in the coefficient of oxygen production,

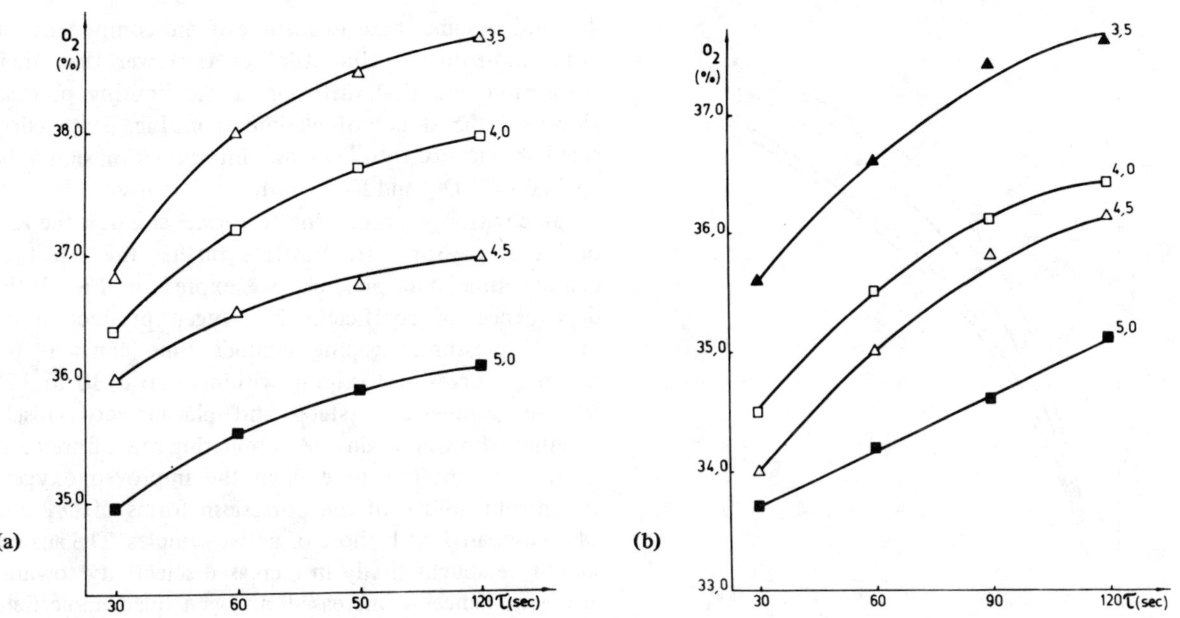

Fig. 3. Oxygen concentration vs. contact time. Oxygen concentration in enriched air (O_2%); contact time (τ sec); K-exchanged $O1_1$ (3a); K-exchanged $O1_3$ (3b). Initial gas pressure is constant for each curve and varies from 3.5 to 5 atm.

α, while contact time, τ, is raised. In fact, contact time should be increased when high oxygen concentrations are desired and total oxygen production, defined by α, is of secondary importance.

The monotonic increase in Fig. 3 of oxygen concentration with contact time implies that the nearer to equilibrium the system the higher will be the oxygen concentrations. The opposite statement is true for the coefficients of oxygen production.

The increase in oxygen concentration with contact time describes samples K-Ol_1 and K-Ol_3 as air-component sorbents much more similar to 4A than to 5A and 13X, according to Dominé and Montgareul (1960). Therefore we may accept their interpretation for

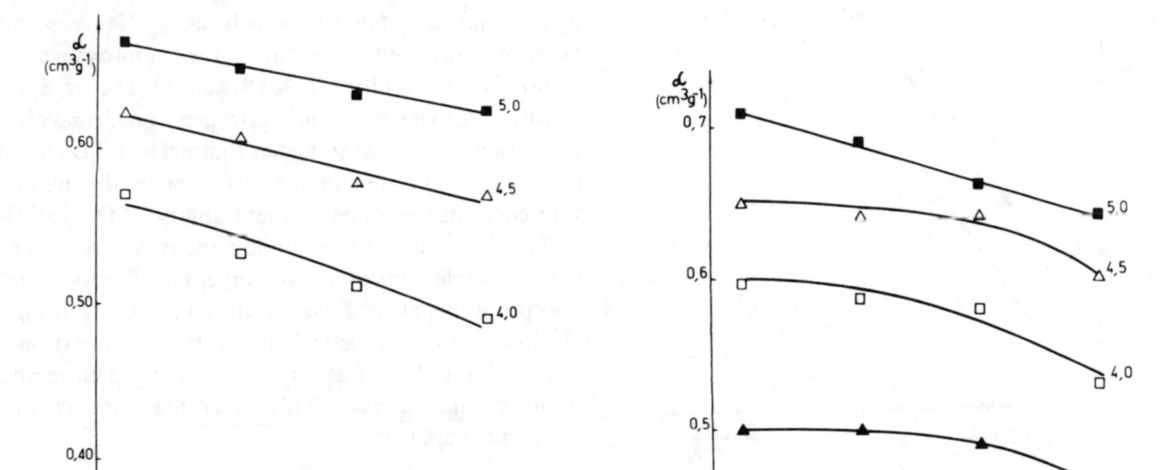

Fig. 4. Coefficient of oxygen production vs. contact time. Coefficient of oxygen production (α cm^3 g^{-1}); contact time (τ sec); K-exchanged $O1_1$ (4a); K-exchanged $O1_3$ (4b). Initial gas pressure is constant for each curve and varies from 3.5 to 5 atm.

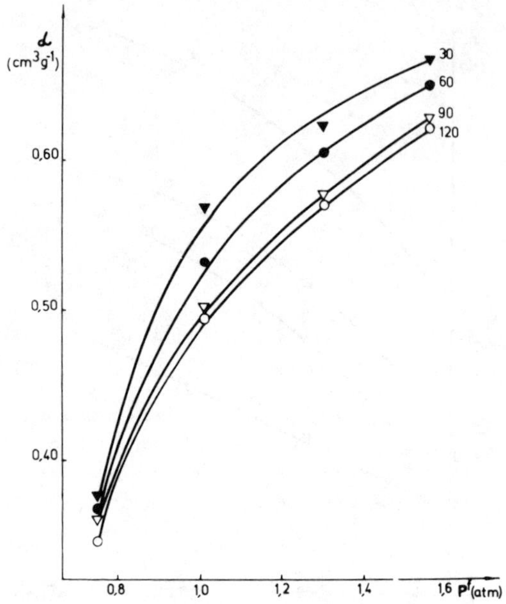

(a)

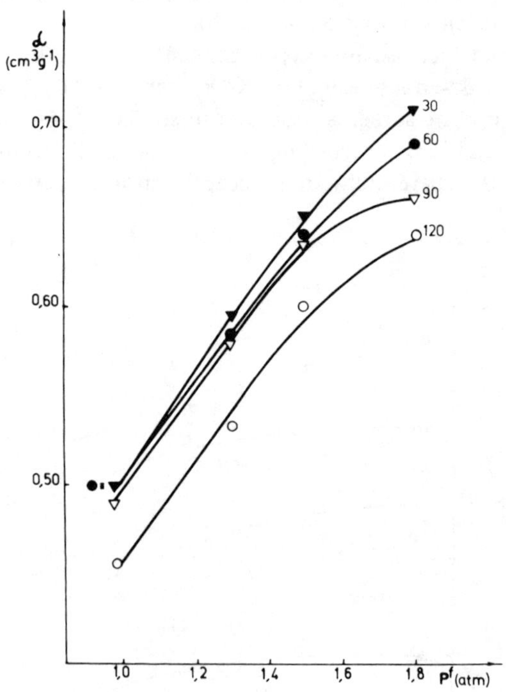

(b)

Fig. 5. Coefficient of oxygen production vs. final gas pressure. Coefficient of oxygen production (α cm^3 g^{-1}); final gas pressure (*Pf* atm). Contact time is constant for each curve and varies from 3.5 to 5 atm.

4A and assume that diffusion of air components in potassium-enriched clinoptilolites is slower than their adsorption and that diffusion is the limiting process. However, the slopes of all curves in Fig. 3 are fairly small. In view of this, Dominé's interpretation should be applied to K-Ol_1 and K-Ol_3 rather cautiously.

Much stronger, according to Figs. 3 and 4, is the role of initial pressure. To illustrate further the effect of contact time and pressure, we express in Fig. 5 the dependence of coefficient of oxygen production on "final" pressure, keeping contact time constant for different curves and staying within interval 30 to 120 seconds. Curves are steep and placed very closely together, showing again the dominating role of pressure.

It is of interest to explain the improved oxygen-enrichment ability of the potassium forms of Ol_1 and Ol_3, compared with those of native samples. The answer should be sought firstly in increased selectivity towards nitrogen, hence – increased nitrogen quadrupole/field interaction (Barrer and Stuart, 1959). Indeed, potassium exchange affects the chemical composition – therefore, field is also modified. Some interpretations (Galabova and Haralampiev, 1975) suggest a relation between the coefficient of oxygen production, a, and some electrostatic and polarization characteristics of the main cations contained in clinoptilolite structure. The potassium forms agree well with such a relation, but this hardly covers all complicated aspects of the phenomenon. For instance, the presence of potassium in clinoptilolite structure is usually connected (Mason and Sand, 1960; Mumpton, 1960; Gottardi *et al.*, 1973) with higher thermal stability which also affects sorbent behavior. It could be that the improvement in air-enrichment ability of K-Ol_1 and K-Ol_3 is due to positive changes in the nitrogen quadrupole/field interaction, as well as to higher thermal stability. Similar arguments could be applied to explain the observed differences in the air-enrichment ability of Ol_1 and Ol_3.

The fact that potassium ion-exchange of different native samples resulted in sorbents of very similar adsorption behavior towards air components brings a conclusion of a practical interest – comparison of natural clinoptilolites as sorbents for oxygen enrichment of air should be made only after the same chemical treatment is applied.

REFERENCES

Alexiev, B. (1968) Clinoptilolite des Rhodopes du nord-est: *Compt. Rend. Acad. Bulg.* **21**, 1093-1095.

Alexiev, B. and Djuorova, E. (1975) On the origin of zeolite rocks: *Compt. Rend. Acad. Bulg.* **28**, 517-520.

Alexiev, B. and Djuorova, E. (1976) Mordenite zeolites in the NE Rhodopes: *Compt. Rend. Acad. Bulg.* **29** (in press).

Barrer, R. M. and Stuart, W. I. (1959) Energetics of occlusion of Ar and N_2 by faujasite-type crystals: *Proc. Roy. Soc.* A, **249**, 464-483.

Breck, D. W. (1974) *Zeolite Molecular Sieves*, Wiley, New York, pp. 710-717.

Djuorova, E. (1976) Analcime zeolite in the NE Rhodopes: *Compt. Rend. Acad. Bulg.* **29** (in press).

Dominé, D. and Montgareul, P. G. (1960) Procede de separation d'une melange gazeux binaire par adsorption: Brevet D'invention, Fr. P. No. 1223261.

Galabova, I. M. *et al.* (1973) Method for air separation and oxygen enrichment of air: Bulgaria National Bureau for Inventions No. 18223.

Galabova, I. M. and Haralampiev, G. A. (1975) Electrostatic characterization of zeolite sorbent and selectivity towards nonpolar sorbates: *Proceedings of Scientific Session on Kinetics & Mechanism*, Higher Chem. Technol. Inst., Sofia, pp. 287-291.

Galabova, I. M., Haralampiev, G. A. and Petkova, L. (1976) Frontal gas-chromatography analyses on acid modified Bulgarian clinoptilolite: *Compt. Rend. Acad. Bulg.* (in press).

Galabova, I. M., Vassileva, B. and Kantcheva, V. (1975) Drying of organic solvents by Bulgarian clinoptilolite: Proceedings of annual scientific session of Higher Chemical Technological Institute, Sofia (in press).

Goranov, A. (1960) Lithology of paleographic deposits in areas of East Rhodopes: Works on Bulgarian geology: *Geochemistry* **1**, 259-310.

Gottardi, G., Alietti, A. and Poppi, P. (1973) The behavior on heating of the zeolites with heulandite structure: *Proc. Third Intn. Conf. Molec. Sieves, Zurich*, pp. 178-181.

Gradev, G. and Daiev, H. (1972) Bulgarian natural clinoptilolites used as Cs^{137} scavengers: Scientific meeting of Council for Economic Help; Kolobzeck; Poland.

Haralampiev, G. A. *et al.* (1975) Possibilities of oxygen enrichment of air for the metallurgy using Bulgarian clinoptilolite, Annual report, Scientific research center, Higher Chem. Technol. Inst., Sofia, pp. 60-75.

Mason, B. and Sand, L. B. (1960) Clinoptilolite from Patagonia. The relationship between clinoptilolite and heulandite: *Amer. Mineral.* **45**, 341-350.

Mumpton, F. (1960) Clinoptilolite redefined: *Amer. Mineral.* **45**, 351-369.

Torii, K. *et al.* (1973) Adsorption of nitrogen–oxygen on zeolite tuffs: *J. Chem. Soc. Japan, Chem. and Ind. Chem.*, No. 2, 225-232.

APPLICATIONS

UTILIZATION OF NATURAL ZEOLITES IN JAPAN

KAZUO TORII

National Industrial Research Institute, Tohoku, Nigatake, Haranomachi, Sendai, Japan

Abstract – In recent years, applications and uses of natural zeolites have attracted much attention, and considerable progress has been made in the utilization of sedimentary zeolites in many phases of industrial and agricultural technology in Japan. Dehydrated zeolite-tuffs are used as adsorbents for water removal from refrigerants, in air separation, and in SO_2-removal processes. Clinoptilolite-tuff is used for ammonia-nitrogen removal from waste water from factories and barns, from fish-breeding tanks, and from secondary effluent of sewage. Clinoptilolite-tuff is also used for the removal of ferrous ion from well water and as a catalyst in the deodorization of fish-manure factories.

At present, depending on physical properties, about 4000 tons of clinoptilolite-tuff per month is used industrially in Japan as fillers in paper, carriers of agricultural chemicals, caking-proof powders for fertilizer, in plywood manufacturing, in detergents, as filters in resins, and as starting material for the synthesis of zeolite A. Lightweight bricks from mordenite-tuff are also used for the lining of high chimneys. Because they increase the growth rate and decrease the feed-conversion value, zeolite-tuffs have been used as dietary supplements for livestock. Zeolite-tuffs are also used as soil conditioners.

Estimated amounts and prices of natural zeolites used each month in Japan in several uses are listed below:

Use	Amount (tons/month)	Price ($/ton)
Fillers in paper		
SGW for groundwood	2500 to 3000	100
HiZ for wood-free paper	800 to 1000	117
Other industrial uses	800 to 1000	67 to 200
Fish breeding	200 to 400	100 to 200
Dietary supplements in animal feed	400 to 500	67 to 100
Soil improver	400 to 500	50 to 67

INTRODUCTION

Since the original discovery of zeolite minerals in a volcanogenic sedimentary rock (Ota and Sudo, 1949), zeolitic tuffs have been found in numerous places in Japan, and their utilization has attracted much attention. Although mordenite-tuff and clinoptilolite-tuff are both abundant, the latter is more commonly used in industrial applications. Five to six thousand tons of zeolitic tuff is currently produced each month from more than ten open-pit mining operations. In Japan this material is used as adsorbents in drying, in air separation, as ion exchangers in pollution-abatement processes, as catalysts in deodorization facilities, as fillers in paper, as lightweight bricks, as dietary supplements for livestock, and as soil conditioners. The monthly usage of natural zeolites in Japan and the prices of several products are listed in Table 1.

INDUSTRIAL USES

Adsorption processes

It is well known that dehydrated zeolites have excellent adsorption properties and that synthetic zeolites are widely used in the fields of drying and gas separation (Breck, 1974). Both mordenite-tuff and clinoptilolite-tuff found in Japan show excellent adsorptive properties compared with the commercial adsorbents, such as synthetic zeolites, silica gel, and activated alumina (Torii *et al.*, 1971). Figure 1 indicates that zeolitic tuffs have significantly larger water-adsorption capacities in comparison with synthetic zeolite 5A or activated alumina. Because of this high affinity for water, dehydrated clinoptilolite-tuff is primarily used as a desiccant, e.g., in the packaging of caking-proof powders.

Adsorbents from clinoptilolite-tuff are also utilized in the drying of Freon gases used in refrigerators (Tombo, 1973). These adsorbents are made from the tuff, activated alumina, and glue and have been sold under the trade name "Ref-clean". As shown in Table 2, the equilibrium water-loading values for "Ref-clean" are about twice those of an imported synthetic adsorbent.

TABLE 1. Uses of natural zeolites in Japan

Uses	Estimated amount of zeolites used (tons/month)	Price (US$/ton)
Fillers in paper		
SGW for groundwood paper	2500–3000	100
HiZ for wood free paper	800–1000	117
Other industrial uses	800–1000	67–200
Fish breeding	200–400	100–200
Dietary supplements in feed	400–500	67–100
Soil conditioner	400–500	50–67

Conversion factor: 300 yen = US $1.00.

TABLE 2. Equilibrium water loadings for several adsorbents*†

Refrigerant	Adsorbent	Water-adsorptive capacity ($g\ H_2O/160$ g adsorbent)	
		75°F (24°C)	125°F (52°C)
Freon-12	Ref-clean(a)	10.7	9.7
EPD: 15 ppm	Imported desiccant	5.1	4.6
Freon-22	Ref-clean(a)	8.5	7.7
EPD: 60 ppm	Ref-clean(b)	7.6	7.0
	Imported desiccant	4.0	3.6

*Data from (Tombo, 1973).

†Testing method: American Society of Heating, Refrigerating and Air-conditioning Engineers Standard 35 B-56, Air-conditioning and Refrigeration Institute Standard.

EPD: Equilibrium point dryness.

Activation temperature: 300°C.

Imported desiccant: made from activated alumina.

Ratio of clinoptilolite-tuff to activated alumina in Ref-clean: (a) = 2:1, (b) = 1:1.

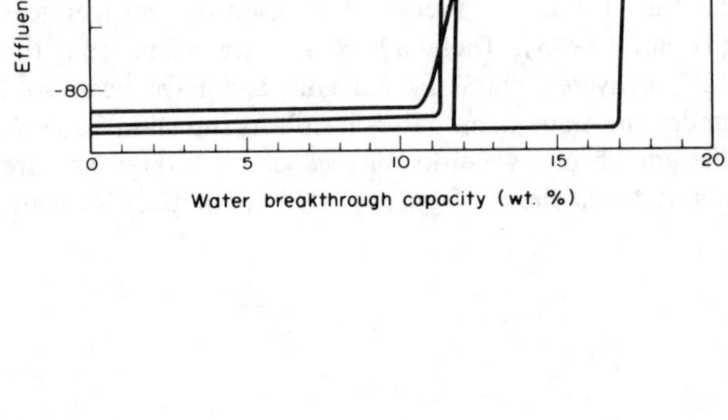

Fig. 1. Drying air by several adsorbents at room temperature. Activation temperature: zeolites = 400°C. Activated alumina = 200°C. 1 : Mordenite-tuff from Shirasawa, Miyagi Prefecture, 2 : Clinoptilolite-tuff from Futatsui, Akita Prefecture, 3 : Activated alumina, 4 : LMS 5A. Data from Torii *et al.* (1971).

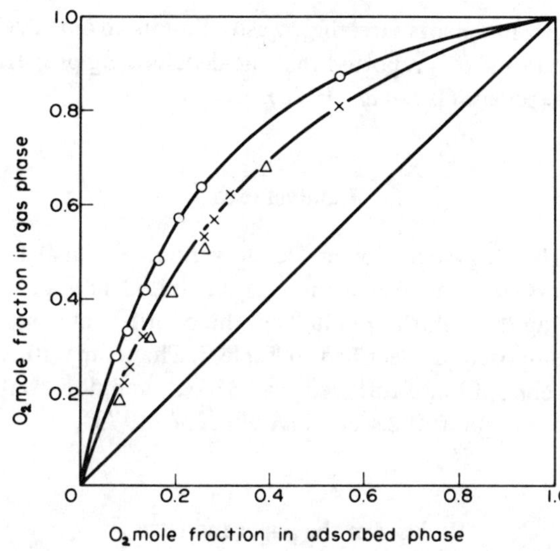

Fig. 2. Equilibrium composition diagrams for N_2–O_2 binary mixtures on several zeolites at 25°C and 3 kg/cm². Activation temperature : 400°C. ○ : Mordenite-tuff from Shirasawa, Miyagi Prefecture, △ : Clinoptilolite-tuff from Futatsui, Akita Prefecture, × : LMS 5A. Data from Torii *et al.* (1973).

The binary adsorption equilibria in Fig. 2 show that the separation factor for O_2–N_2 adsorption on mordenite-tuff is greater than that determined on zeolite 5A or on clinoptilolite-tuff (Torii *et al.*, 1973). Since 1968, a plant capable of producing 300 cubic meters of oxygen-enriched air per hour has operated using preferential adsorption of nitrogen from air on dehydrated mordenite-tuff. Smaller plants producing 15 to 300 liters per hour have been sold for fish breeding and to laboratories (Minato, 1975).

Clinoptilolite-tuff is also used to remove SO_2 from vent gas in a sulfuric acid plant (Minato, 1975).

Water-treatment processes

Zeolitic tuffs have superior ion-exchange properties, with clinoptilolite-tuffs showing high selectivity for ammonium ions (Ames, 1960). As ammonia can be toxic to fish life, fish culturists who breed eels, carp, and sweet fish must remove ammonia from the water of breeding tanks or ponds. In Japan 200–400 tons of clinoptilolite-tuff per month are used for this purpose. Clinoptilolite-tuff is also used for ammonia-nitrogen removal in wastewater-treatment systems, such as effluents from factories and barns and from secondary sewage streams. Since December 1971, a plant with a complete ion-exchange and regeneration system for the removal of ammonium ions from the wastewater of a soap and detergent factory has been in operation. It handles 500 cubic meters of water per day. Operational data are shown in Table 3. Although ammonia-nitrogen cannot be removed by sand filtration, it can be almost completely removed by an ion-exchange process using clinoptilolite-tuff.

TABLE 3. Properties of wastewater and treated effluents from a soap and detergent factory*

Analysis	Wastewater	Sand-filtration effluent	Ion-exchange effluent
pH	6.8	7.2	7.8
COD (ppm)	132	84	70
BOD (ppm)	209	29	24
SS (ppm)	98	5	3
NH_4-N (ppm)	23	22	0.68

*Unpublished data. Date: September 9, 1973.

In Toba, Mie Prefecture, there is an observatory where tourists are able to enjoy a fine view. To protect the fish resources of the area, about 80 cubic meters of wastewater per day from the observatory must be treated. An ammonium-ion removal facility with two ion-exchange columns has been in operation since April, 1973. The water quality demanded by the Toba fishermen's union and the results of water analyses after ammonia removal are shown in Table 4 (Toba, 1973). At present there are more than thirty such facilities in Japan and others are under construction. These facilities are small-scale plants which treat 80 to 2000 cubic meters of water per day and are used by factories, hotels, observatories, or small community sewage systems where residents must deal with wastewater by themselves.

Clinoptilolite-tuff from Futatsui, Akita Prefecture, is used as a carrier of manganese and utilized in the removal of iron from water of boilers in order to prevent scale problems. Recently, untreated clinoptilolite-tuff has been used directly for iron removal. In this case,

TABLE 4. Required and actual properties of wastewater from Toba Observatory*

Analysis	Water quality demanded	Ion-exchange effluent
pH	6.0–8.5	7.25
SS (ppm)	<3	2.4
BOD (ppm)	<5	2.3
COD (ppm)	<3	2.65
NH_4-N (ppm)	<1	0.06
Free-Cl (ppm)	<0.02	0.01

*Data from (Toba, 1973). Date: May 29, 1973.

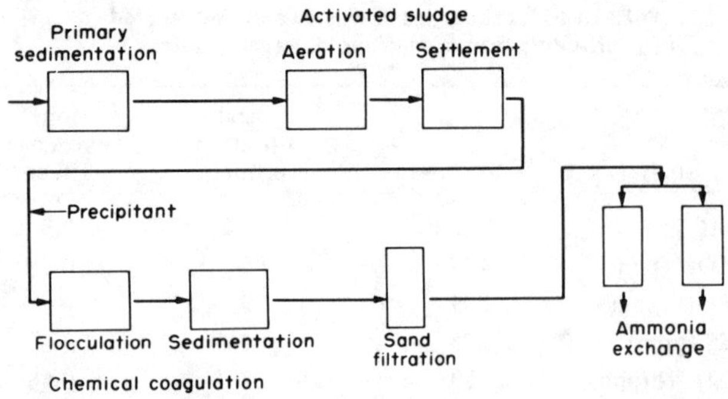

Fig. 3. Flowsheet of the water-treatment system at Toba. Data from Toba (1973).

clinoptilolite-tuff is used in contact filtration after aeration. It is assumed that the zeolite acts as an oxidation catalyst and as a filtration medium. As such, it is used for iron removal from drinking water, from water used in the food industry, and from car-washing water for buses.

Calcined clinoptilolite-tuff is utilized with aluminum sludge in plants treating waste liquors from dyeing operations. It is reported that the decolorizing properties are excellent (Iso *et al.*, 1976).

Catalyst uses

The flowsheet of a deodorization facility for a factory treating fish manure is illustrated in Fig. 4. In passing through the zeolite bed, the odoriferous gases are decomposed, as decribed in Table 5. The temperature of the clinoptilolite-tuff bed was 315°C, and the contact time was about 0.2 second (Araki *et al.*, 1973).

Uses in the paper industry

At the present time, about 3300–4000 tons per month of clinoptilolite-tuff is sold under the trade names "SGW" and "HiZ" as filler for groundwood and wood-free paper, respectively, by the Zieklite Chemical Company (Zieklite, 1974). Since February, 1964, SGW

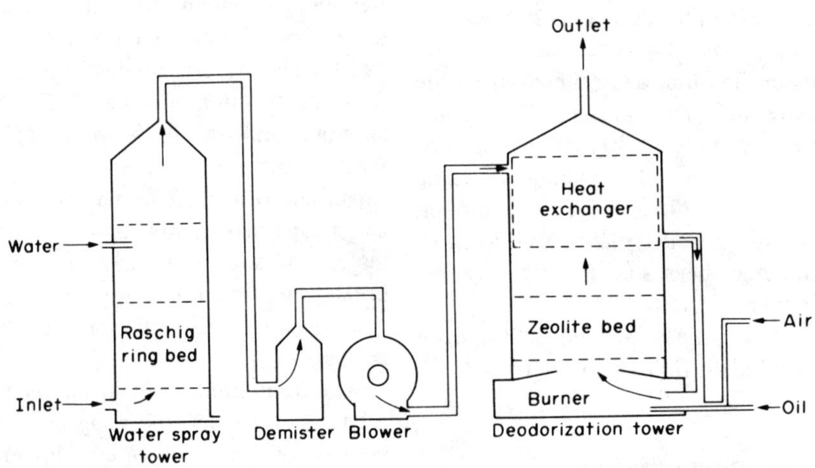

Fig. 4. Deodorization flowsheet. Data from Araki *et al.* (1973).

TABLE 5. Composition of waste gas from a fish-manure factory*

	NH_3 (ppm)	Amine-N (ppm)	Mercaptane (ppm)	H_2S (ppm)
Waste gas from a fish-manure factory	154.17	95.61	1.94	0.93
After water-spray tower	12.67	9.60	1.66	0.84
After deodorization tower	0.01	0.00	0.01	0.03

*Data from Araki *et al.*, 1973. Date: July 14, 1972.

has been produced by simple dry grinding and air classification at the Itaya plant in Yamagata Prefecture. The flowsheet of the SGW production process is shown in Fig. 5. Although this dry production process can handle large amounts of material, the classification efficiency is not good, and it is too difficult to treat the zeolite products further by chemical means. Consequently, the quality of the SGW products are closely dependent on the properties of the raw clinoptilolite-tuff.

Since August, 1972, however, the HiZ product has been manufactured by a wet process. The flowsheet of the HiZ production process is shown in Fig. 6. After wet grinding and classification of the ore, three products are obtained. The coarsest product contains most of the feldspar, quartz, and opaque minerals present in the ore, while the finest product contains most of the cristobalite, ferrous oxide, and organic matter contained in the ore. Clinoptilolite is concentrated in the 0.7-15-micrometer fraction. Impurities, such as small amounts of ferrous oxide and organic materials, which decrease the brightness of the zeolite product, are dissociated by oxidation and reduction-bleaching. High-purity clinoptilolite powder in the form of filter cake is extruded into pellets and dried. The yield of the HiZ process is about 70%.

About 1000 tons of SGW powder per month is transported to the paper manufacturing companies by pneumatic systems. The rest of the SGW product and the HiZ product is packaged and sent to users by means of freight cars or trucks. The properties of the SGW and HiZ products are listed in Table 6. As shown in Table 7, the use of SGW as a filler makes the paper bulkier than clay-filled paper. The smoothness of SGW-filled paper is also not as good under the same calender conditions, but the printing opacity is better than that of clay-filled products. SGW is as soft as kaolin or talc, and thus the life of movable type of a rotary press is prolonged by more than 60%.

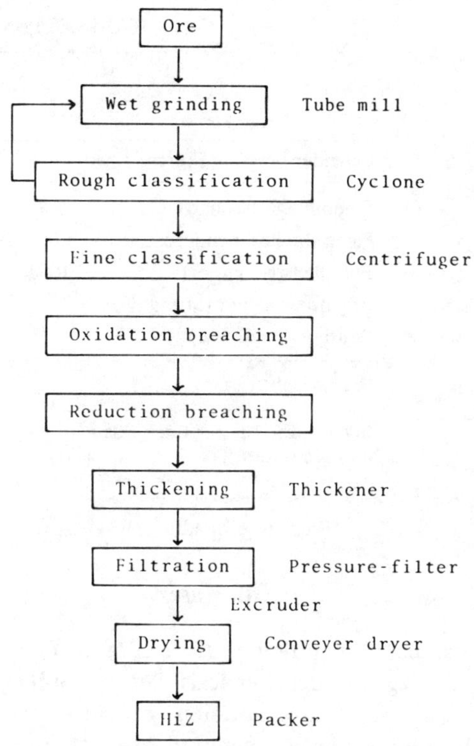

Fig. 6. Flowsheet of the HiZ production process. Data from Zieklite (1974).

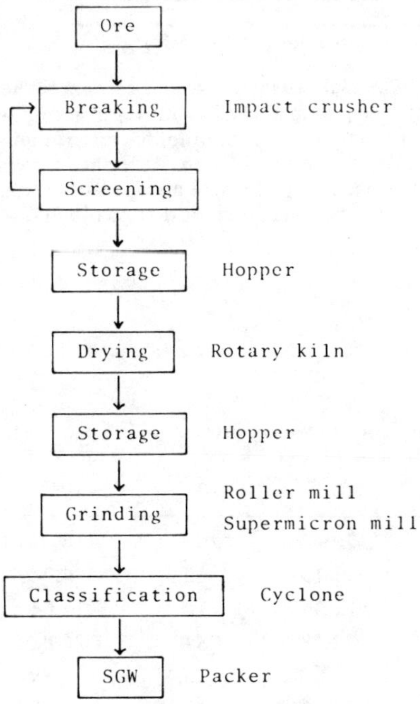

Fig. 5. Flowsheet of the SGW production process. Data from Zieklite (1974).

TABLE 6. Properties of SGW and HiZ

Parameter	SGW*	HiZ†
Brightness (%)	78.5	84.2
Water content (%)	5.0	12.9
pH	9.6	4.5
Abrasion resistance (mg)	70.5	85.2
% >10 micrometer	16.45	14.7
% 10–5 micrometer	20.05	25.1
% 5–2 micrometer	24.54	27.4
% <2 micrometer	38.96	32.6

*Data from Zieklite (1971).
†Data from Zieklite (1974).

TABLE 7. Properties of clay- and SGW-filled papers*

Parameter	Filler							
	Clay (Japan)				SGW			
Calender pressure (kg/cm²)	0	10	20	30	0	10	20	30
Smoothness (second)	7.3	33	62	84	7.6	30	49	64
Paper thickness (micrometer)	82	77	73	69	98	88	83	75
Bulk density (g/cm³)	0.81	0.86	0.91	0.96	0.70	0.78	0.82	0.91
Brightness after printing (%)		39.2	28.0	24.7		37.1	32.1	29.5
Printing opacity (%)		81.5	76.3	77.2		82.6	83.7	84.9
Basis weight (g/m²)	66.2				68.2			
Ink absorbed paper thickness (micrometer)	30				20			

*Data from Zieklite (1974). Amount of ink used: 0.8 ml.

Other uses

Granular clinoptilolite-tuff can also function as a carrier of agricultural chemicals. Because of its resistance to powdering, its low abrasion resistance, and its high oil-absorption capacity, clinoptilolite-tuff from Futatsui is more suitable for carriers of agricultural chemicals than pumice or limestone. For this use, about 1000 tons of clinoptilolite-tuff was produced by Sun Zeolite Industrial Corporation in 1975. Effects of the particle size of clinoptilolite-tuff carrier and the application rate of the carrier on the herbicidal effect in a paddy field are shown in Fig. 7 (Mori *et al.*, 1974).

Two or three hundred tons per month of clinoptilolite-tuff are used as caking-proof powders for fertilizers. By covering granular fertilizers with a small amount of powdered zeolite, it is possible to prevent grains from adhering to one another.

To adhere plywood initially before hot pressing, wheaten paste is used. To reduce the cost while maintaining the quality of the plywood, about 400 tons of clinoptilolite-tuff were added to such paste in 1975.

One or two hundred tons of clinoptilolite-tuff per

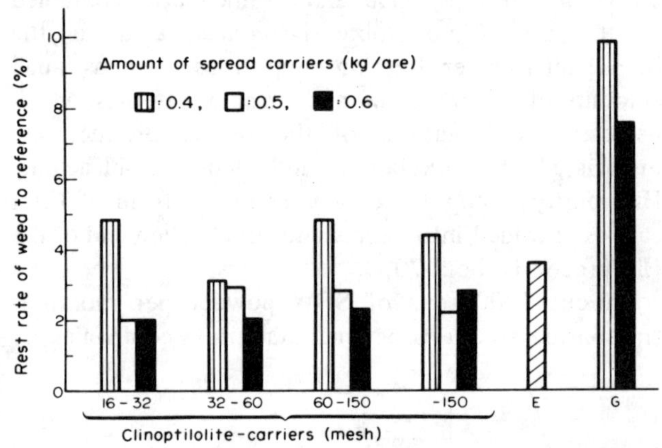

Fig. 7. Herbicidal effects of several carriers of Benthiocarb. E : emulsion; G : granular carriers on the market, amount of used benthiocarb emulsion : clinoptilolite-carriers and emulsion = 40 gare, granular carriers = 80 gare. Reference : average amount of weed in the paddy field, where no weed-killer was used, were 49 per square meters. Data from Mori *et al.* (1974).

TABLE 8. Properties of mordenite-filled melamine-formaldehyde resin*

Amount of mordenite-tuff (wt.%)	0	5	10	20	40
Rockwell hardness (HRM)	122	122	120	120	118
Tensile strength (kg/mm²)	5.66	3.85	3.19	4.09	4.78
Cross-breaking strength (kg/mm²)	11.74	12.11	11.36	11.17	11.64
Compression strength (kg/mm²)	30.19	29.00	29.18	28.43	26.82
Izod impact strength (kg/cm²)	1.87	1.76	1.89	1.84	1.89
Specific gravity	1.541	1.567	1.581	1.619	1.680

*Data from Shimada *et al.* (1973). Testing pieces were made by compression molding at 200°C and pressure of 70 kg/cm² for 6 minutes.

TABLE 9. Properties of clinoptilolite- and water-ground lime-filled polypropylene resin*

Sample	Tensile strength	Cross-breaking strength	Impact strength
100% Polypropylene	100	100	100
50% Polypropylene + 50% clinoptilolite-tuff	74	121	88
50% Polypropylene + 50% water-ground lime	63	69	37

*Data from Matsuda (1975).

month are utilized as mixtures with detergents. Clinoptilolite-tuff is also used as carriers of manganese zeolite, fireworks, etc.

Zeolitic tuffs have been investigated as possible fillers in resins. The effects of adding mordenite-tuff in a thermosetting resin are shown in Table 8 (Shimada *et al.*, 1973). The effects of using clinoptilolite-tuff and water-ground lime as fillers in thermoplastic resin are compared in Table 9 (Matsuda, 1975). Although these uses are still in their infancy, it is anticipated that the demand for zeolites as fillers in resins will increase in the near future.

From August 1973 to May 1975 about 3000 tons of lightweight bricks were produced from mordenite-tuff by the Iwao Giki Corporation. These bricks were used mainly for internal linings of high chimneys of thermal power plants and chemical factories. Table 10 shows that lightweight bricks from mordenite-tuff have superior physical strength and greater resistance to strong acid and alkali in comparison with other lightweight bricks (Iwao, 1973).

Using clinoptilolite as a source of silica, alumina, and other oxides, about 20 tons per month of synthetic zeolite A is produced and sold under the trade name "Nikka Pellet KZ" by Nippon Kasseihakudo Corporation.

AGRICULTURAL USES

Dietary supplements

Zeolitic tuffs are used in livestock farming as dietary supplements in the feed. When zeolitic tuffs are introduced into the feed for chickens, swine, and cattle, the following effects are noted: (1) there is a prevention and curing of diseases of the digestive organs, such as diarrhoea; (2) the growth rate is increased; (3) feed-conversion values decrease; (4) offensive odors from excrement are controlled; and (5) the feed is prevented from molding during storage. The statistical data of disease prevention at the Keai farm, near Morioka, Iwate Prefecture, where about 4000 swine are bred, are shown in Table 11 (Keai, 1974). Once clinoptilolite-tuff was introduced into the feed, it is evident that both the number of cases of disease and the death rate were

TABLE 10. Properties of lightweight bricks from mordenite-tuff*

Parameter	IF-100	IF-70	Other product
Apparent porosity (%)	7.04	5.41	5.24
Bulk specific gravity (kg/l)	1.0	0.71	0.51
Bending strength (kg/cm^2)	63.7	44.6	34.1
Compressive strength (kg/cm^2)	235	150	126
Abrasion resistance (mg)	74	90	230
Thermal conductivity (kcal/m,hr,°C)	0.320	0.262	0.150
Acid resistance (%)†	0.061	0.122	0.214
Alkali resistance (%)‡	13.6		46.5

*Data from Iwao (1973).

†Acid resistance: decreased gravity % in acid equivalent mixture solution (10% HCl + 10% HNO_3 + 10% H_2SO_4) at 90°C for 5 hours.

‡Alkali resistance: decreased gravity % in 10% NaOH solution at boiling point for 48 hours.

TABLE 11. Health effects of clinoptilolite-tuff in swine rations*

Amount of zeolite in the feed (%)	Period	Gastric ulcer cases	Pneumonia cases	Heart dilation cases	Death rate (%)	Medicine cost (yen/head)
0	Feb. 72–Jan. 73	77	128	6	4.0	700
6	Feb. 73–Jan. 74	22	51	4	2.6	500

*Keai-farm, unpublished data.

reduced, and that the medicine cost per head also decreased.

The effect on newborn pigs when clinoptilolite-tuff was given to female swine which had conceived is shown in Table 12 (Buto, 1967). All pigs in control groups were attached by scours during 21 to 35 days after birth and were therefore undergrown. Experimental pigs, on the other hand, were essentially free of diarrhoea and were all well grown. Thus, a large difference in the weight of 35-day-old pigs was noted between two groups.

When 5% clinoptilolite-tuff was introduced into swine rations in powdered form, the growth rate of pigs was clearly increased (Kondo and Wagai, 1968). The values of weight gain, intake feed, and feed conversion in these experiments are described in Table 13. The average weight gain of pigs in the experimental groups was larger

TABLE 12. Effect on newborn pigs when clinoptilolite-tuff is given to pregnant sows*†

Species	Number of animals	Average weight of pig (kg)		
		Newborn	21-day-old	35-day-old
Experimental group (a) Yorkshire	6	1.25	4.30	7.83
Experimental group (b) Laundry	6	1.20	4.70	8.68
Comparison group (c) Yorkshire	10	1.10	4.20	4.81
Comparison group (d) Laundry	10	1.10	4.00	4.67

*Data from Buto (1967).

†35-day-old: weaning period. Since a month ago before mating, 400 g of clinoptilolite-tuff per day had been given to the sows in the experimental groups.

TABLE 13. Effect of clinoptilolite-tuff on growth rate and feed-conversion value of swine*

Group	Number of animals	Average weight-gain (kg)	Weight-gain index	Average feed intake (kg)	Feed-conversion value (feed intake / weight-gain)
First experiment†					
Experimental group §	4	29.00	113	85.00	2.93
Comparison group	4	22.93	100	90.60	3.95
Second experiment‡					
Experimental group §	10	54.57	130	167.62	3.07
Comparison group	10	42.30	100	137.20	3.24

*Data from Kondo *et al.*, (1968).

†First experiment period: August 15, 1966–October 15, 1966.

‡Second experiment period: January 7, 1967–March 27, 1967.

§ Five percent of clinoptilolite-tuff was added to the feed in the experimental groups.

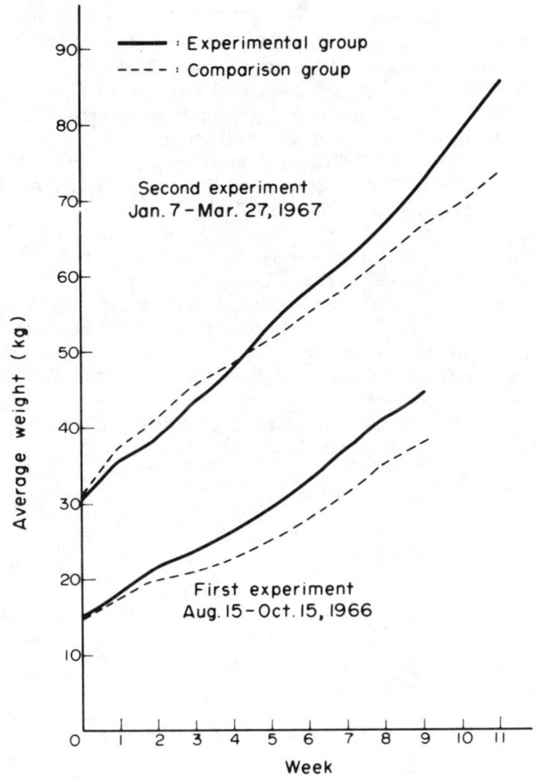

Fig. 8. Growth-rates of swine fed a zeolite-supplemented diet. In the first experiment eight 60-day-old pigs and in the second experiment twenty 98-day-old pigs were used. Data from Kondo and Wagai (1968).

TABLE 14. Effects of clinoptilolite-tuff as soil conditioners on yields of several farm products*

Crops	Year	Amount of zeolite used (tons/10 ares)†	Ratio of yield index‡ (%)
Wheat	1962	1	113
		2	115
Paddy	1964	0.5	106
		1	103
Eggplant	1964	1	155
		2	119
Apple	1964	1	113
		2	128
Paddy	1965	0.5	102
		1	117
Carrot	1965	1	163
Apple	1965	1	114
		2	110

*Data from Yamagata (1966).
†10 ares = 0.25 acre.
‡Compared with control plots.

than those of pigs in the comparison groups (see Fig. 8). It is therefore apparent that zeolitic tuff in swine rations is effective in increasing the growth rate and decreasing the feed-conversion values.

Excrement treatment

Two or three hundred tons per year of clinoptilolite-tuff from Itaya are mixed with the loose droppings of chickens to remove offensive odors and to dry the droppings rapidly. About 40 tons per month of activated zeolitic tuff, heated at 450°C and 600°C, have also been used as cleaning materials for pig houses in Sapporo without the use of water. Waste mixtures of urine-soaked zeolitic tuff and excrement have been sold as fertilizer. This same process has been adopted at Ikawa, Akita Prefecture.

Soil conditioners

The effects of clinoptilolite-tuff from Itaya as soil conditioners on yields of several farm crops are shown in Table 14 (Yamagata, 1966). The additions of clinoptilolite powder to paddy fields and plowed fields improved strikingly the growth and yield of the crops. It is assumed that these effects are due to the adsorption and retention capacities for ammonia-nitrogen and potassium ion, resulting in a good maintenance of water, the prevention of root decay, and the supplying of necessary nourishment subcompositions.

ACKNOWLEDGMENT

Mr. Charles Narupaves read the manuscript. Special thanks are due Prof. F. A. Mumpton for his valuable comments and encouragement.

REFERENCES

Ames, L. L. (1960) The cation sieve properties of clinoptilolite: *Amer. Mineral.* **45**, 189-200.

Araki, K., Honda, S. and Tanaka, Y. (1973) Deodorization facility by contact oxidation catalytic method: *Report of Hokkaido Industrial Research Institute* **223**, 2-10.

Breck, D. W. (1974) *Zeolite Molecular Sieves*: Wiley, New York.

Buto, K. (1967) Supplement test of clinoptilolite-tuff for gravid swine: Internal Report of Ichikawa-City Livestock Experiment Station.

Iso, F., Shibata, T. and Okonogi, T. (1976) Treatment of dyeing waste liquor with waste sludge from aluminium production factory: *Pollution Control* **11**, 8-13.

Iwao (1973) Light brick, I-Foam: Company Catalogue of Iwao Giki Corporation.

Keai-farm: Unpublished data.

Kondo, T. and Wagai, B. (1968) Experimental use of clinoptilolite-tuff as dietary supplements for pigs: Yotonkai, May 1968.

Matsuda, S. (1975) Unpublished data of Composite Research Group (private).

Minato, H. (1975) Zeolite, its natural resources and utilization: *Ceramics Japan* **10**, 941-957.

Mori, Y., Edo, Y., Toryo, H. and Ito, T. (1974) Effects of the particle size and the application rate of carrier on the herbicidal effect and the growth of dry-seeded rice: *Zasso Kenkyu* **18**, 21-26.

Ota, S. and Sudo, T. (1949) Studies on Oya-ishi, Part 2, Mineralogical composition: *J. Geol. Soc. Japan* **55**, 242-246.

Shimada, T. and Imai, E. (1973) Effect of natural zeolite as fillers in melamine-formaldehyde resin: *Report of Tochigi-ken Kennan Industrial Laboratory*, pp. 47-54.

Toba (1973) Toba Observatory: travel brochure of Toba-city.

Tombo (1973) Filtering and drying unit, Ref-clean: Company Catalogue of Tombo Pencil Corporation.

Torii, K., Hotta, M., Onodera, Y. and Asaka, M. (1971) Effects of dehydration on adsorption properties of zeolite tuff: *Kogyo Kagaku Zasshi* **74**, 2012-2017.

Torii, K., Hotta, M., Onodera, Y. and Asaka, M. (1973) Adsorption of nitrogen-oxygen on zeolite tuff: *Nippon Kagaku Kaishi* **1973**, 225-232.

Yamagata (1967) Effect of zeolite as soil conditioners: Internal Report of Agricultural Improvement Section, Yamagata Prefectural Government.

Zieklite (1971) Guidance of Zieklite Chemical Company: Company Catalogue.

Zieklite (1974) Study on natural zeolite as fillers in paper: *Technology on Paper and Pulp* **24**, 2-6.

ZEOLITE ION EXCHANGE IN RADIOACTIVE AND MUNICIPAL WASTEWATER TREATMENT

B. W. MERCER and L. L. AMES

Battelle-Northwest, P.O. Box 999, Richland, Washington 99352, U.S.A.

Abstract – The desirable characteristics of high ion-exchange selectivity and resistance to radiolytic degradation have made certain zeolites quite useful for the separation and purification of cesium-137 from radioactive wastewater. Natural zeolites which have these characteristics include mordenite, clinoptilolite, erionite, and chabazite. Millions of gallons of highly radioactive wastes stored on the Hanford Reservation have been processed through an 8.4-m^3 zeolite bed to remove cesium-137. The recovered cesium is currently being purified by ion exchange using a synthetic mordenite. Natural and synthetic zeolites have also been effectively used for over a decade in the nuclear industry for the treatment of low- and intermediate-level radioactive wastes.

The same zeolites which exhibit high cesium-ion selectivity have also proved to be effective for selectively removing ammonium ions from wastewater. The natural zeolite, clinoptilolite, is currently the leading candidate for use in municipal sewage-treatment plants for ammonia removal. An advanced municipal wastewater-treatment plant in Rosemount, Minnesota, is currently in operation using 40,000 kg of clinoptilolite to treat a design flow of 2300 m^3 of wastewater per day. The Upper Occoquan Water Reclamation Plant near Washington, D.C. is expected to be in operation late in 1977 and will use 340,000 kg of clinoptilolite to treat 85,000 m^3 of wastewater per day at the maximum design flow rate. Two additional treatment plants in the United States are also planning to use clinoptilolite for ammonia removal.

Desirable properties of zeolites for use in ion exchange columns include high ion-exchange capacity and selectivity, rapid exchange rates and high resistance to attrition during backwash operations. The ammonium ion-exchange capacity for clinoptilolite should be a minimum of 1.7 milliequivalents per gram. Ion-exchange grade 20 × 50-mesh clinoptilolite should also generate less than 3% fines passing a 100-mesh screen after vigorous mixing with water for 5 minutes on a paint shaker. The ion-exchange rate should be sufficient to permit utilization of at least half of the equilibrium ammonium capacity of the zeolite bed at a flow of six bed volumes per hour and 95% removal of ammonia from the feed stream.

INTRODUCTION

The use of zeolites for the treatment and disposal of radioactive wastes has been extensively studied in the United States over the past 17 years. The studies have encompassed three general areas: (1) removal of cesium-137 from high-level radioactive wastes; (2) decontamination of low- and intermediate-level wastes; and (3) fixation of radioactive wastes for long-term storage. Radioactive waste-treatment processes utilizing zeolites have been developed and successfully used at several atomic energy installations in the United States.

The principal reasons for using zeolites in radioactive waste treatment are their high ion-exchange selectivity and resistance to degradation from radiation. The selectivity for cesium ion is of particular importance, as the removal of radioactive cesium from waste solution is greatly facilitated by ion-exchange processing. Zeolites which exhibit high cesium selectivity include clinoptilolite, chabazite, erionite, and mordenite (see Ames, 1960; 1961; 1967a). Gamma radiation has no effect on the basic structure and ion-exchange characteristics of clinoptilolite to an exposure level of 10^{10} roentgens (Fullerton, 1961).

Research on zeolite ion exchange of cesium from radioactive wastes, conducted under the auspices of the United States Atomic Energy Commission, has led to an important development in the municipal waste treatment field. The zeolites possessing high cesium-ion selectivity are also potentially useful for removing ammonium ion from wastewater (Ames, 1967b). A natural zeolite, clinoptilolite, is very effective as an ion exchanger for removing ammonium ion from municipal wastewater (Mercer *et al.*, 1970a).

A number of minerals, including zeolites, possess desirable ion-exchange characteristics but their physical form is not readily adapted for use in an ion-exchange column. Sufficient physical stability of the granular form of the ion-exchange media is required to withstand the rigor of fluidization during the backwashing cycle of an ion-exchange process. In addition to a high capacity or selectivity, the ion-exchange media should allow rapid intragranular diffusion of ions. Otherwise, the rate of ion exchange will be too slow for most dynamic ion-exchange applications. Because natural zeolites vary widely in quality with respect to ion-exchange characteristics and physical stability, it is important to have testing procedures available to determine the adequacy of a particular zeolite for a specific ion-exchange application. Some of these procedures are outlined in the present paper.

ZEOLITE CHARACTERIZATION FOR USE AS A CATION EXCHANGER

Physical properties

The amount of zeolites found in large crystals in basaltic rock cavities is relatively insignificant compared to the extensive, easily mined, bedded tuff deposits. Several simple tests have therefore been devised to evaluate critical physical and ion-exchanger properties of the latter type of zeolites. Most ion-exchange applications require the use of fixed beds or columns of sized zeolite. Moving-bed contactors containing zeolites are generally not practical due to relatively high zeolite attrition losses. To achieve particle sizes which allow optimum flow rates and contact times, some processing of the zeolite is required. The crushing and screening characteristics of bedded zeolites are therefore of interest. A crushing study of mine-run Hector clinoptilolite from Hector, California, will be used as an example.

Beginning with 3- to 6-inch lumps of ore, crushing and screening were carried out so as to obtain the maximum recovery of 10 X 50-mesh zeolite. The mine-run ore was passed through a jaw crusher to prepare it as feed for a roll mill. After roll milling, the clinoptilolite was dry screened on a triple-deck screen consisting of 10-, 20-, and 50-mesh screens. Three products were collected and the oversize material (+10-mesh) was returned to the roll mill (see Table 1). Size-distribution data were obtained by dry sieving of the three products (Skarpelos, 1963) and are listed in Table 2.

A series of tests was made grinding the clinoptilolite in a laboratory rod mill in a simulated closed circuit with a 10-mesh screen to determine the optimum size distribution that could be obtained during crushing. During milling, coarse particles protect finer particles by wedging the rods apart, resulting in products with a minimum of fines. In the rod mill experiments, 1 kg of 4 X 8-mesh Hector clinoptilolite was milled for 2 minutes and screened on a 10-mesh sieve. The oversize (+10-mesh material) was returned to the mill with enough new feed to make a 1-kg charge. The process was repeated until a steady state was reached, resulting in a circulating load of 110%. A similar series of tests with 4-minute cycles gave a circulating load of 23%. The products from both tests were identical. The products

TABLE 1. Weight distribution of the three products

Size fraction	Weight (lb)	Weight distribution (%)
10 X 20 mesh	1677	47.1
20 X 50 mesh	823	23.1
−50 mesh	1061	29.8
	3561	

TABLE 2. Size distribution of crushed and screened clinoptilolite product

Size fraction, Tyler mesh	Weight %		
	10 X 20 mesh fraction	20 X 50 mesh fraction	−50 mesh fraction
10 X 14	60.1	0.1	
14 X 20	35.7	0.4	
20 X 28	3.4	42.1	
28 X 35	0.2	38.4	0.0
35 X 48	0.1	18.0	8.4
48 X 65	0.1	0.6	21.5
65 X 100	0.1	0.2	21.2
100 X 150	0.1	0.1	13.5
150 X 200	0.0	0.1	15.6
200 X D	0.2	0.1	19.8
	100.0	100.0	100.0

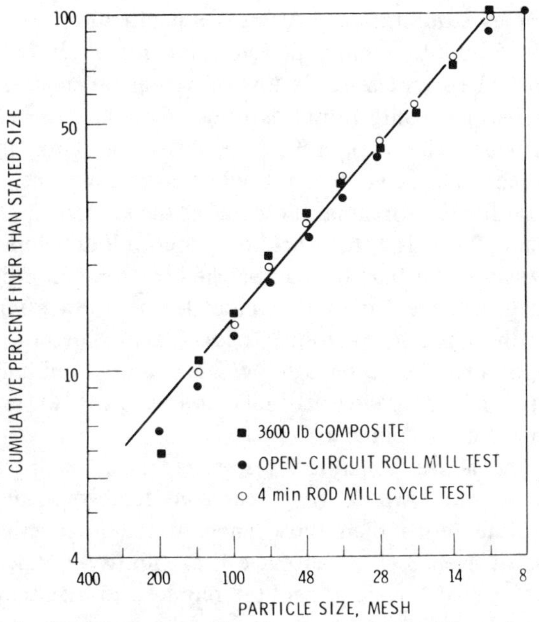

Fig. 1. Cumulative size distribution of Hector clinoptilolite from crushing tests (Skarpelos, 1963).

from these tests should yield a size distribution that is a straight line when the cumulative percent passing a certain screen size is plotted on log-log paper vs. the size because the sieve series is logarithmic as shown in Fig. 1.

In order that the size distribution follow such a straight line, the sieving should be done wet to wash the fine particles adhering to the coarse particles through a 400-mesh screen. After the +400-mesh material has been dried, it should then be sieved in the usual manner. The same cumulative size distributions are listed in Table 3 for the 4-minute, closed-circuit rod milling test, the open-circuit, roll mill crushing test, and the composite of the crushed 3600-lb sample.

The size distributions graphed in Fig. 1 are all quite similar. The equation of the line is $y = 100\,(x/1650)^{0.78}$, where y is the cumulative percent of crushed clinoptilolite finer than size x, and x is the size in micrometers. A slope as large as 0.78 is surprising. Usually materials such as clinoptilolite yield an excessive quantity of fines, and an exponent closer to 0.4. This may be due to the tendency of clinoptilolite to break into two pieces rather than into fragments having a complete size distribution. Higher circulating loads would take advantage of this unusual fracture mechanism and would result in the production of slightly less fine material.

A second important physical property of a sized zeolite is its resistance to wet attrition. Commonly hydraulic lifting or backwashing of the zeolite column or bed is required during the regeneration cycle to remove solids caught in the bed during downflow loading operations. Bed backwashing causes the zeolite particles to move violently and abrade during the process. Thus, fines are produced that are either lost with the backwash solution or contribute to an increased pressure drop across the column when it is returned to a downflow loading mode. The cost of makeup zeolite can become a major operating expense.

A wet attrition test has been designed to establish a means of comparison of the amount of −100-mesh fines produced when sized materials are abraded in an aqueous environment under standard conditions. The procedure of this wet attrition test is given in Appendix I. In this test, 25 g of the sized zeolite is placed in a 4-oz

TABLE 3. Cumulative size distribution of hector clinoptilolite from crushing tests

Size, Tyler mesh	4 min Rod mill cycle test (%)	Open-circuit roll mill test (%)	3600 lb composite (%)
8	100.0	100.0	100.0
10	99.5	88.2	99.7
14	74.6	72.6	71.4
20	55.7	54.4	54.6
28	43.6	40.0	43.3
35	35.1	30.4	34.3
48	26.2	23.6	27.6
65	20.0	17.8	21.0
100	13.8	12.5	14.6
150	10.0	8.9	10.6
200	7.0	6.8	5.9
270	0.5	3.2	1.1
400	0.2	1.5	0.3

($118.3\ cm^3$) jar with 75 ml of distilled water and agitated on a paint shaker for 5 minutes. The −100-mesh fines are removed by wet screening, dried, weighed, and expressed as a percent of the original 25 g zeolite sample.

As an example of the results of the wet attrition test outlined in Appendix I, a batch of mine-run, Hector clinoptilolite lost an average of 10.1 wt.% −100-mesh fines. Hand-picking the cleanest and hardest chunks of clinoptilolite in the batch upgraded the material and reduced fine losses to 7.5 wt.%. Much of the weight loss, even on the hand-picked material, represented losses of admixed clay, calcite, and materials other than clinoptilolite. A thorough and vigorous acidic and water wash can be used to reduce the fine losses on mine-run clinoptilolite to less than 3.0 wt.% for the standard wet attrition test. For comparison, wet attrition tests under the same conditions of 20 X 50-mesh Linde AW-500, a pelletized chabazite, and Zeolon 900, a synthetic mordenite, yielded an average 7.6 wt.% and less than 3 wt.%, respectively.

A series of wet attrition tests were made on 20 X 50-mesh samples of Zeolon, hand-picked Hector clinoptilolite, and clinoptilolite from Buckhorn, New Mexico, to determine the decrease in attrition with mixing time. After the first wet attrition test, the zeolite samples were resieved to remove −50-mesh material, dried, and weighed for the second test. Zeolite lost in the first test was replaced from a separate sample which had received the same treatment. The tests were repeated several times. The decrease in attrition with mixing time is shown in Fig. 2. Both clinoptilolite samples exhibited large decreases in attrition after the first 5-minute mixing period. This initial decrease is believed to represent the loss of extraneous matter and the loss of zeolite from fracturing of friable edges of the particles. The Zeolon 900 exhibited no large initial decrease although it stabilized at attrition values near those for Hector clinoptilolite after the second and third mixing periods. The Buckhorn clinoptilolite appeared initially softer than the hand-picked Hector clinoptilolite but gave a wet attrition value of less than 3% after the fourth mixing period. These data illustrate the importance of thorough washing and tumbling of crushed and graded natural zeolites prior to use in ion-exchange columns.

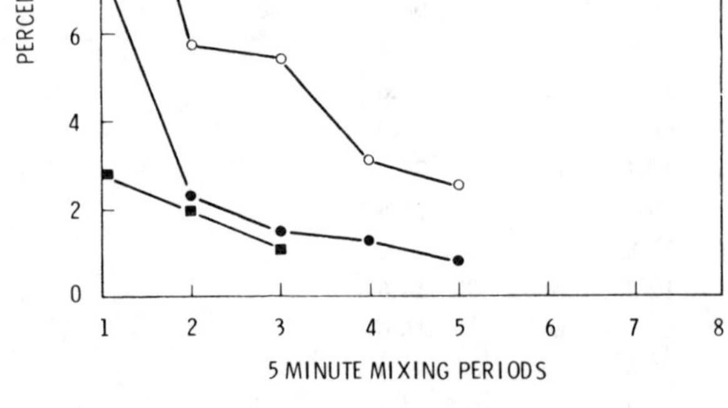

Fig. 2. Effect of mixing time on attrition.

The above weight-loss results represent comparative values only which may be considerably higher in absolute terms than those encountered in practice. A loss of even 3 wt.% per cycle is prohibitively expensive in terms of "makeup zeolite" required to maintain the desired exchange capacity of the column. Actual makeup requirements must be determined in use because wet attrition losses are a function of a given physical system.

Ion-exchange properties

Assuming that the zeolite is physically or mechanically stable in the fluid with which it is contacted, three main properties are important in zeolite ion-exchange applications: ion-exchange kinetics, ion-exchange capacity, and cation selectivity. Ion-exchange kinetics concern the time required for the counter ion to travel to the exchange site and displace a cation in the structure. The capacity refers to the number of milliequivalents of a given cation per gram of zeolite that the zeolite can retain on its fully loaded exchange sites. For some large or multivalent cations, the zeolite capacity may be less than for sodium, for example. Cation selectivity refers to the preference order of a zeolite for cations, based on the energetics of distance between anionic sites, cationic radii, and cationic hydration energies.

A standard cesium-adsorption test was devised that allows measurement of all effects of the above three chemical properties for a sized zeolite in a single test. The equipment required and procedures are given in Appendix II. Mixing with a propeller stirrer was found to be superior to magnetic stirring or a wrist-action shaker because zeolite particle disintegration was less. As shown in Fig. 3, no significant effect of stirring rate on cesium adsorption by AW-500 over the range used was observed. Thirty minutes was selected as the optimum mixing time and 450 rpm as the stirring rate for the test

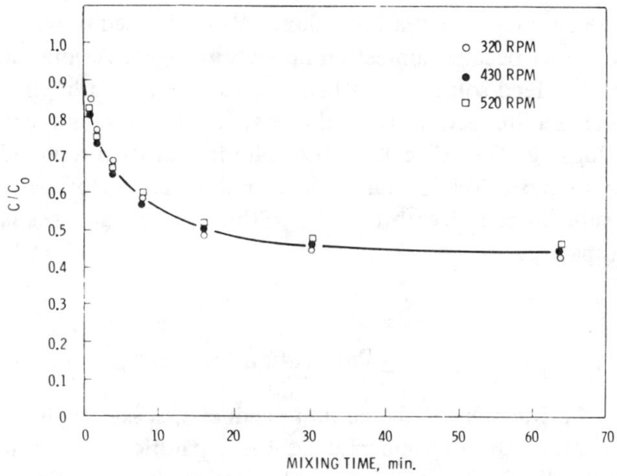

Fig. 3. Uptake of cesium by 20 to 25 mesh AW-500 as a function of time. C/C_0 equals the fraction of the original cesium removed from solution.

procedure.

Sodium is included in the test solutions to measure the selectivity of the zeolite for cesium in sodium-cesium solutions; however, other cation combinations can be substituted, such as NH_4Cl–NaCl. The zeolite must first be based in the cation not being measured, sodium in this case. The solution also must contain a sufficient concentration of the cation being measured to ensure that diffusion of cations within the zeolite is the controlling factor for exchange rate. Concentration is important because solution–zeolite mixing rates do not affect cation-exchange rates as long as ion diffusion in the particle is the rate-controlling step. If less than expected adsorption capacities are obtained, the problem may be one of poor kinetics. This can be tested by rerunning the sample for several hours. Progressively larger extractions of the measured cation should occur with longer contact time. If the zeolite capacity is inherently low, there will be little change with increased contact times. Zeolite cation-selectivity is measured by the ratio of the two cation concentrations left in solution.

The test is also useful for selecting a zeolite lot with the best ion-exchange characteristics from among several lots. An example of this is shown in Table 4. The removal of cesium was greatest by AW-500 lot number 9154-60A for both the 30-minute and 16-hour contact times.

Wet attrition resistance should be measured on such samples as well. Enhanced kinetics or capacity commonly are found for the softer zeolite lots, most likely due to increased surface areas produced by the mechanical breakdown of particles. Because the mineralogical characteristics such as cementation, porosity, and

TABLE 4. Cesium adsorption (C/C_0) by different lots of AW-500

AW-500 Lot no.	Cs C/Co,* 30 min	Cs C/Co,* 16 hrs
23	0.46	0.34
565001	0.46	0.37
565002	0.47	0.35
9154-60A	0.35	0.25

*Contact time.

zeolite content of bedded zeolite deposits may vary from one part of the deposit to another, and between different deposits of the same mineral, it is essential that such tests be made on all materials of potential interest. Wet attrition and cation-adsorption tests are vital to maintain physical and ion-exchange specifications.

In some instances where zeolite cation selectivity is of little concern, a simple ion-exchange capacity determination would suffice to set specifications. For example, if the influent consists of essentially a single cation, and the columns are to be loaded in series, zeolite capacity alone becomes the most important chemical characteristic. A standard test of ammonium ion-exchange capacity is given in Appendix III. The method is relatively rapid and allows the capacity determinations on several samples at the same time.

SEPARATION AND PURIFICATION OF RADIOACTIVE CESIUM

A unique zeolite ion-exchange process has been developed and used on the Hanford Atomic Energy Project for the separation of radioactive cesium from highly radioactive processing wastes (Nelson and Mercer, 1963; Brandt, 1970). These wastes contain large concentrations of sodium salts, principally sodium nitrate, nitrite, and carbonate; the total sodium-ion concentration in the waste solution is approximately 4.5 molar, compared to a cesium concentration of about 10^{-4} molar. The separation of the radioactive cesium from the high-level radioactive wastes takes place prior to evaporation and solidification of the wastes in large underground storage tanks ranging in size from 2 million liters to 3.8 million liters. Excessive heat generation in the resultant salt cakes would occur if the radioactive cesium were not removed from the wastes.

Separation of radioactive cesium has been accomplished using AW-500 extrudates containing chabazite. Selection of AW-500 for process use was made on the basis of its high ion-exchange capacity and cesium selectivity and of its commercial availability at the time the separation plant went into operation.

The cesium separation process utilized a 1.83-m diameter by 3.16-m deep bed of 20 X 50-mesh AW-500 (see flow schematic in Fig. 4). The waste solutions were pumped downflow through the zeolite bed until breakthrough of the cesium occurred. The bed of zeolite was then washed with water to remove the salt waste solution. The water wash was followed by a dilute ammonium carbonate-ammonium hydroxide solution which served to elute most of the sodium ion from the zeolite while leaving most of the cesium on the zeolite bed. A warm concentrated solution of ammonium carbonate was then pumped upwards through the bed to elute the cesium. Upflow regeneration was employed to keep the effluent side of the bed relatively free of cesium and thus avoid high cesium losses during the loading cycle. The regenerant was followed by a water wash to remove the regenerant prior to the next loading cycle.

The spent regenerant was routed to an evaporator where the ammonium carbonate and ammonium hydroxide were distilled and recovered for reuse. The evaporator bottoms consisted of a solution of sodium, cesium, potassium, and rubidium carbonates which were subsequently treated with dilute nitric acid to form nitrate salts. The sodium:cesium ratio in the cesium product solution was normally less than 10:1, which represents a decontamination factor ($Na^+:Cs^+$ in feed/$Na^+:Cs^+$ in product) of more than 500. The product solution from the separation process was retained for further processing to purify the cesium.

Approximately 3 million gallons of supernatant waste were processed through the initial charge of AW-500. At this point the breakthrough capacity of the zeolite bed had decreased to less than half the initial value of about 250,000 curies of cesium-137 (Buckingham, 1970). The pressure drop across the column also increased to where the bed became almost completely plugged. Aluminum in the feed solution was believed to have precipitated in the zeolite bed causing the loss in capacity and bed plugging. The AW-500 in the column was then replaced with fresh AW-500 and several million gallons of waste were processed with this bed without significant loss in capacity.

Purification

Purification of the cesium product was accomplished in much the same manner as the separation of cesium from the high-salt supernatant waste as shown in Fig. 4. In addition to removal of sodium impurity, the purification process must also remove potassium and rubidium impurities. A synthetic mordenite, Zeolon 900 (product of Norton Company), was selected for this process in preference to chabazite, erionite, and clinoptilolite because of its superior cesium capacity and rubidium-separation capability over the other zeolites (Mercer *et al.*, 1970b). The purification process required careful attention to the loading and scrub cycles. The feed solution to the Zeolon 900 column was terminated just prior to breakthrough to avoid high cesium losses during the scrub cycle. Sufficient dilute ammonium carbonate scrub solution was then pumped through the zeolite bed to remove much of the impurities while avoiding a high cesium loss. Pilot plant runs conducted in a hot cell with actual product solution from the separation process and a 1.0-liter Zeolon 900 column produced a purified cesium product which contained 98.6 wt.% Cs, 0.07 wt.% Rb, 1.13 wt.% K, and 0.19 wt.% Na. The decontamination factors for rubidium, potassium, and sodium were 9, 87, and 600,

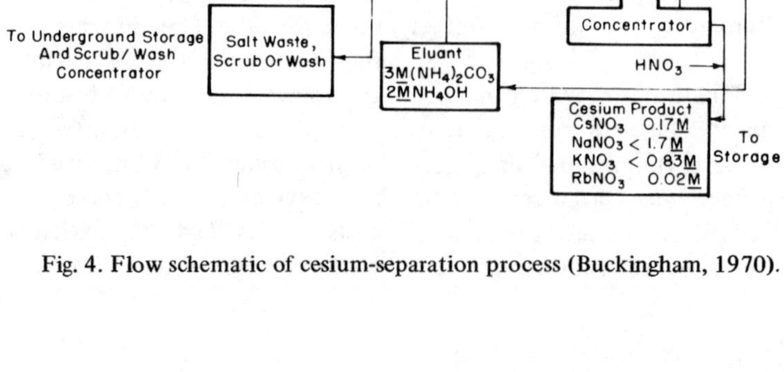

Fig. 4. Flow schematic of cesium-separation process (Buckingham, 1970).

respectively (Bray and Fullam, 1971). The cesium purification facility is now in operation at Hanford producing cesium-137 for encapsulation as the chloride salt in high integrity containers.

LOW-LEVEL RADIOACTIVE-WASTE TREATMENT

Idaho National Engineering Laboratory

A local clinoptilolite was used for several years at the Idaho National Engineering Laboratory near Idaho Falls, Idaho, to decontaminate low-level radioactive wastewater from an irradiated fuel-storage basin. Several locally available earth materials were evaluated for their evaluate strontium-90 and cesium-137 adsorption characteristics, and clinoptilolite was selected on the basis of its superior equilibrium-distribution coefficients for both strontium and cesium (Wilding and Rhodes, 1963).

The process utilized four clinoptilolite columns in parallel, each column consisting of two 200-liter steel drums connected in series (IAEA, 1972). A flow schematic of the two drums in series is presented in Fig. 5. Each drum contained 0.15 m^3 of clinoptilolite in a bed 60 cm deep. When radioactivity broke through the bottom drum, the drum was drained and used to replace the top drum. The top drum was removed from the system, sealed, and buried as solid waste. A new drum of fresh clinoptilolite replaced the bottom drum. The volume capacity of each drum was 120 m^3 of wastewater and the average percent removal of radioactivity was 99.5%. The average composition of waste is given in Table 5.

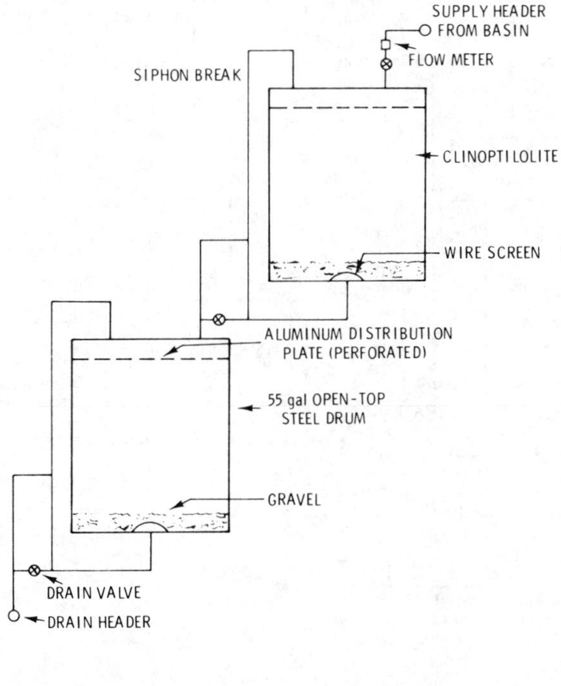

Fig. 5. Clinoptilolite column consisting of two drums connected in series (IAEA, 1972).

TABLE 5. Composition of waste treated by clinoptilolite at the Idaho chemical processing plant

Chloride	0.5 mg/l
Nitrate	200 mg/l
Calcium	40 mg/l
Magnesium	12 mg/l
Sodium	95 mg/l
Cesium-137	(0.8–1.3) × 10^{-3} μCi/ml
Strontium-90	(3.5–10) × 10^{-4} μCi/ml
pH	8.2

Savannah River Plant

Evaporator overheads and miscellaneous wastewaters were treated at the Savannah River Plant near Aiken, South Carolina, with a column of AW-500 to remove cesium-137 (IAEA, 1967). Total dissolved solids in the overheads were usually less than 100 mg/l but reached 10,000 mg/l in the miscellaneous wastes. The cesium-137 concentration in these wastewaters was approximately 0.1 μCi/ml. The wastewater was pumped through a 0.26 m^3 bed of AW-500 at a flow rate of about 38 l/min. When the cesium breakthrough reached 1%, the flow was stopped, and the AW-500 was discharged to a storage tank. A fresh batch of AW-500 was put into the ion-exchange vessel for further processing. Cesium-137 removal exceeded 99%, and a bed of AW-500 was found to handle from 800 to 2650 m^3 of overheads or 265–400 m^3 of miscellaneous wastewater.

Hanford

An ion-exchange column with a 1.8 m deep layer of Zeolon 900 topped by a 0.6 m layer of organic ion exchange resin was used to treat a process condensate wastewater (Hanson, 1973). The column removed 98% of the strontium-90 and over 99.5% of the cesium-137. The Zeolon 900 was used for cesium removal and the organic resin was added for strontium removal. When radioactivity breakthrough occurred, the dual purpose ion-exchange bed was regenerated with sodium nitrate solution.

AMMONIA REMOVAL FROM WASTEWATER

The development of an ammonia-removal process for wastewater using zeolites is an outgrowth of research conducted for the Atomic Energy Commission on cesium removal from radioactive wastes. Zeolites which exhibit high cesium selectivity are also selective for the ammonium ion in the presence of other common cations in natural waters such as Mg^{+2}, Ca^{+2}, and Na^{+}. The degree of selectivity is less than that for cesium but it is sufficient to render a zeolite such as clinoptilolite useful for removing ammonia from wastewater.

Initial studies on ammonia removal from municipal wastewater with an organic ion-exchange resin (Schlecta and Culp, 1967) disclosed that a considerable cost would be incurred for salt consumed during resin regeneration. Further, disposal of the spent regenerant posed a particularly difficult problem. Therefore, ion-exchange removal of ammonia with conventional organic ion-exchange resins did not appear to be an attractive process for municipal wastewater.

A conventional organic cation-exchange resin has considerable selectivity towards divalent ions such as Ca^{+2} and Mg^{+2} in a dilute electrolyte solution such as municipal wastewater. Ammonium ion, therefore, breaks through ahead of magnesium and calcium in a column of organic ion-exchange resin receiving clarified municipal wastewater. In the case of a zeolite selective for ammonium ion, the magnesium and calcium ions will break through before the ammonium ion, thus reducing the amount of regenerant required for eluting the sorbed calcium and magnesium ions from the ion-exchange column. This, however, does not eliminate the problem of disposing of spent regenerant.

The problem of spent regenerant disposal was solved by utilizing chemical characteristics specific to ammonium ion. The first of these is the conversion of ammonium ion to dissolved ammonia gas by increasing the pH of the spent regenerant. Lime or sodium hydroxide is added to the regenerant to elevate the pH to a range of 10–12 whereupon the ammonia is stripped from the regenerant by countercurrently contacting the regenerant in a stripping column with a large flow of air. This approach allows the recycling and reuse of the regenerant, thereby eliminating the disposal problem and greatly reducing the amount of regenerant salt used.

The second desirable characteristic of ammonia with respect to waste treatment is its reaction with chlorine to form nitrogen gas. This is called breakpoint chlorination which occurs when 7.5 parts by weight of chlorine is added to a solution containing one part by weight of ammonia–nitrogen. A process was developed whereby chlorine is produced electrolytically by passing the spent regenerant from a clinoptilolite bed through an electrolysis cell which generates sufficient chlorine to convert the ammonia–nitrogen to elemental nitrogen. This approach also renovates the spent regenerant which is then recycled for reuse (Battelle-Northwest/South Tahoe Public Utility District, 1971).

The zeolite–ammonia-removal process with air stripping has been further refined to permit recovery of the ammonia, thus avoiding atmospheric discharge of this substance. The exhaust gas from the air stripper is routed to an absorption column where it is contacted with a dilute sulfuric acid stream (Kepple, 1974). This concept has been incorporated into an advanced wastewater-treatment plant for the Upper Occoquan Sewage Authority near Washington, D.C. (Suhr and Kepple, 1975), and another advanced wastewater-treatment plant at the north shore of Lake Tahoe, California. Figure 6 depicts a simplified flow schematic of the regeneration process used at the Occoquan plant. A split regeneration scheme is employed to minimize the volume of regenerant sent to the ammonia stripper. The first 11 bed volumes of regenerant pumped through an exhausted clinoptilolite bed has been used in the previous two regeneration cycles – first as recovered regenerant and second as twice-used regenerant. The regenerant solution under equilibrium conditions con-

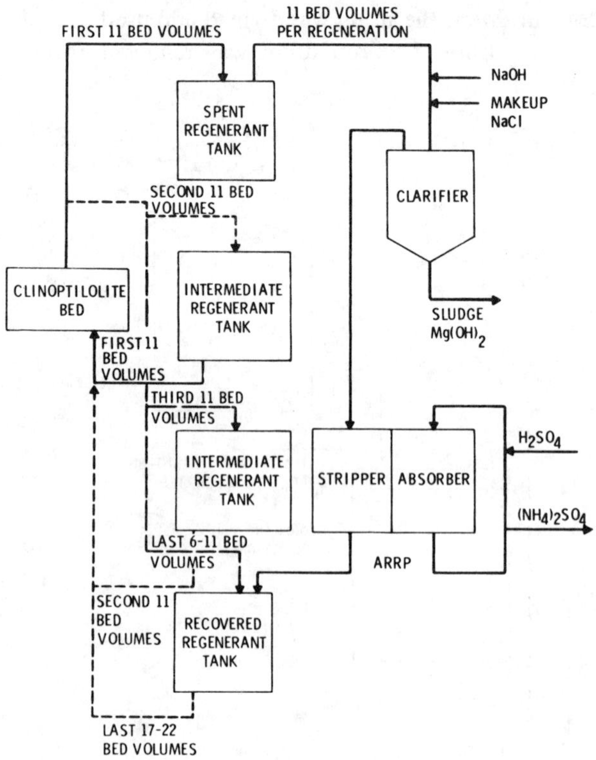

Fig. 6. Flow schematic of regenerant recovery system (Suhr and Kepple, 1975).

tains approximately 2.0% sodium chloride, 1.0% calcium chloride, and about 600 mg/l potassium chloride. The regenerant is initially 2% sodium chloride – however, calcium chloride and potassium chloride are added through repeated use of the regenerant. The spent regenerant contains about 250 mg/l ammonia, while recovered regenerant contains about 50 mg/l. The pH varies from about 7.0 in the spent regenerant to about 9.5 in the recovered regenerant. The residual ammonia (50 mg/l) in the recovered regenerant does not interfere with regeneration of the clinoptilolite. Sodium hydroxide is added to the spent regenerant to give a pH of 10.7–11.0 prior to stripping. The pH is reduced to 9.5 by recycling prior to transfer to the regenerant recovery tank. Some magnesium hydroxide precipitates following the sodium hydroxide addition and is removed as sludge from a clarifier. A 40% ammonium sulfate solution is produced by recycling a sulfuric acid solution at pH 1.5 through the absorber.

The Upper Occoquan Wastewater Treatment Plant is designed to process initially 57,000 m^3 of wastewater per day with an ultimate capacity of 85,000 m^3 of wastewater per day. The ammonia-removal process employs eight beds of clinoptilolite, four of which are in service and two in regeneration at any given time. Two of the clinoptilolite beds are held in spare at the 57,000 m^3-per-day level. Each clinoptilolite bed measures 3 m in width, 15 m in length, and 1.2 m in depth. The total weight of clinoptilolite used in the plant is 34,000 kg (375 short tons).

A 2300 m^3-per-day wastewater treatment plant is currently in operation at Rosemount, Minnesota (Blain and Vanderboort, 1973). This plant uses 40,000 kg (44 tons) of clinoptilolite in three beds each having a diameter of 3.4 m and a depth of 1.8 m. The beds are operated in series, two or three at a time. The regenerant is treated with sodium hydroxide to elevate the pH, and the ammonia is removed by steam stripping.

A zeolite ammonia removal plant capable of treating 204,000 m^3 per day of waterwater is currently in the planning stage for Alexandria, Virginia (Wilson *et al.*, 1972).

REFERENCES

Ames, L. L., Jr., Sand, L. B. and Goldich, S. S. (1968) A contribution on the Hector, California, bentonite deposit: *Econ. Geol.* **53**, 22-37.

Ames, L. L., Jr. (1960) Cation sieve properties of clinoptilolite: *Amer. Mineral.* **45**, 689-700.

Ames, L. L., Jr. (1961) Cation sieve properties of the open zeolites chabazite, mordenite, erionite and clinoptilolite: *Amer. Mineral.* **42**, 1120-1131.

Ames, L. L., Jr. (1967a) Alkali metal cation equilibrium with chabazite: *Can. Mineral.* 8, 572-581.

Ames, L. L., Jr. (1967b) Zeolite removal of ammonium ions from agricultural and other wastewaters: *Proc. of the 13th Pacific Northwest Industrial Waste Conference, April 6-7, Washington State University*, Pullman, Washington, pp. 135-152.

Battelle-Northwest/South Tahoe Public Utility District (1971) Wastewater ammonia removal by ion exchange: U.S. Environmental Prot. Agency, Water Poll. Cont. Res. Ser., Report No. 17010 EC2 02/71.

Blain, W. A. and Vandervoort, T. J. (1973) The Rosemount advanced wastewater treatment plant: Brochure published by the Metropolitan Sewer Board, St. Paul, Minnesota.

Brandt, H. L. (1970) B-Plant recovery of cesium from purex supernatant: U.S. Atomic Energy Commission Document No. ARH-1639.

Bray, L. A. and Fullman, H. T. (1971) Recovery and purification of cesium-137 from purex waste using synthetic zeolites: Advances in Chemistry Series No. 101, *Molecular Sieve Zeolites*–1: Amer. Chem. Soc., pp. 450-455.

Buckingham, J. S. (1970) Laboratory evaluation of zeolitic material for removing radioactive cesium from alkaline waste solutions: U.S. Atomic Energy Comm. Doc. No. ARH-SA-49.

Fullerton, R. (1961) The effect of gamma radiation on clinoptilolite: U.S. Atomic Energy Comm. Doc. No. HW-69256.

Hanson, G. L. (1973) Reuse of aqueous wastes in the Hanford chemical processing facilities: *Proc. National Conference on Complete Water Reuse, April 23-27, Washington, D.C.*, AIChE-EPA, pp. 360-366.

IAEA (1967) Operation and control of ion exchange processes for treatment of radioactive wastes: *Tech. Rep. Ser.* No. **78**, International Atomic Energy Agency, Vienna, p. 73.

IAEA (1972) Use of local minerals in the treatment of radioactive waste: *Tech. Rep. Ser.* No. **136**, International Atomic Energy Agency, Vienna, pp. 61 and 64.

Kepple, L. G. (1974) Ammonia removal and recovery becomes feasible: *Water and Sewage Works* **121**, No. 4, 42-43.

Mercer, B. W., Ames, L. L., Touhill, C. J., VanSlyke, W. J. and Dean, R. B. (1970a) Ammonia removal from secondary effluents by selective ion exchange: *J. Wat. Pollut. Control Fed.* **42**, Part 2, R95-R107.

Mercer, B. W., Ames, L. L. and Smith, P. W. (1970b) Cesium purification by zeolite ion exchange: *Nucl. Appl. & Technol.* **8**, 62-69.

Nelson, J. L. and Mercer, B. W. (1963) Ion exchange separation of cesium from alkaline waste supernatant solutions: U.S. Atomic Energy Comm. Doc. No. HW-76449.

Schlecta, A. F. and Culp, G. L. (1967) Water reclamation studies at the South Tahoe Public Utility District: *J. Wat. Pollut. Control Fed.* **39**, 787-813.

Skarpelos, J. M. (1963) Progress in the treatment of a radioactive condensate waste: U.S. Atomic Energy Comm. Doc. No. HW-79174.

Suhr, L. G. and Kepple, L. G. (1975) AWT plant gets tough with ammonia: *Water and Wastes Engng*, March 1975, pp. 29-30.

Wilding, M. W. and Rhodes, D. W. (1963) Removal of radioisotopes from solution by earth materials from eastern Idaho: U.S. Atomic Energy Comm. Doc. No. IDO-14624.

Wilson, T. E., Riddell, M. D. R. and Newton, D. (1972) Standards dictate design: *Water and Wastes Engng*, Nov. 1972, pp. 47-49.

APPENDIX I
WET ATTRITION TEST

Materials

1. Granular zeolite.
2. Distilled water.

Apparatus

1. Paint shaker (Red-Devil Model No. 30 Paint Conditioner by Red-Devil Tools, Union, New Jersey) with frequency of approximately 700 cpm.
2. Glass jar, wide mouth, 4-oz. capacity, No. 6734 with 50-mm metal lids, and sealed-in rubber gasket. Dimensions are approximately 80-mm high, 44-mm i.d., 50-mm o.d.
3. A jar holder consisting of an 18-cm diameter by 2.5-cm thick wooden disc with four 5-cm diameter holes drilled 1-cm deep. The center points of the holes form the corners of rectangle 11.8-cm long by 5.8-cm wide.
4. Metal funnel with a 2-inch diameter 100-mesh screen.
5. Tall form beakers, 500 ml.
6. Balance, accuracy of ±0.95 g.
7. Electric oven, gravity convection, maximum operating temperature of 300°C.

Procedure

1. Weigh out 25 g of the granular zeolite (as received) ±0.05 g.
2. Place sample into a 4-oz jar and add 75 ml of distilled water.
3. Seal jar tightly with cap and test for leakage by inverting briefly.
4. Place the sealed jars in the holder on the Red-Devil Paint Shaker. Each jar should be spaced 5-cm from the center line of the jar to the center line of the shaft of the shaker. Be sure jars are secured.
5. Shake for 5 minutes ± 3 seconds.
6. Remove jars from Paint Shaker and filter the sample through a 100-mesh screen collecting the water in a *tared* beaker. Rinse the jar and cup with distilled water and pour over the sample on the 100-mesh screen.
7. Place the beaker containing the fines plus water in a 220°C (approximately) oven for at least 4½ hours.
8. Weigh the beaker and fines immediately after removal from the oven to ±0.05 g and record the weight.

Calculation

$$\text{Wet attrition loss} = \frac{(\text{Weight of beaker} + -100\text{-mesh fines}) - (\text{Empty weight of beaker})}{\text{Original weight of sample}} \times 100.$$

APPENDIX II
CESIUM-ADSORPTION TEST

Apparatus

1. Flame emission spectrophotometer.
2. Propeller stirrer consisting of a 45-mm long by 9-mm wide stainless-steel propeller, 0.5-mm thick, with a 30° pitch mounted on a 7-mm diameter shaft.
3. Beakers, 150 ml.
4. Pipettes, 40 ml and 1 ml.
5. Filters flask 25 ml.
6. Funnel, Buchner type with medium porosity fritted disc, 15 ml.
7. Volumetric flask, 50 ml.
8. Sieves, U.S. Bureau of Standards size 20 and 25 with pan and lid.
9. Sieve shaker, Ro-Tap.
10. Balance.

Reagents

1. Test solution 2.00 M NaCl–0.200 M CsCl.
2. Standard CsCl solutions in 0.04 M NaCl.
   0.004 M CsCl,
   0.003 M CsCl,
   0.002 M CsCl,
   0.0014 M CsCl,
   0.001 M CsCl.

Procedure

1. Place a sufficient quantity of granular zeolite (as received) on a 20-mesh sieve stacked with a 25-mesh sieve to yield at least 10 g of 20 X 25-mesh zeolite.
2. Shake on a sieve shaker for 5 minutes and transfer the 20 X 25-mesh fraction to a clean dry container.
3. Weight 5 g of activated 20 X 25-mesh zeolite on a balance to ±0.01 g.
4. Pipette 50 ml of test solution 2 M NaCl–0.2 M CsCl into a 150 ml beaker.
5. Place the beaker with the 50 ml of test solution under the stirrer and position the propeller about

3-mm above the bottom of the beaker.

6. Start the stirrer and run at a speed of about 450 rpm. (This stirring should be sufficient to cause the zeolite particles to move around freely without stacking up.)
7. Place the 5-g sample of zeolite in the stirred solution and mix for 30 minutes at 25°±2°C.
8. At the end of the stirring period stop the stirrer and immediately (within 1 minute) filter a portion of the solution through a 15-ml medium porosity, fritted glass filter on a 25-ml filter flask.
9. Dilute 1-ml of the filtrate to 50-ml with distilled water with a volumetric flask and determine Cs^+ concentration with a flame emission spectrophotometer at 852.1 mμ.

Calculation

$$\frac{C}{Co} = \frac{\text{Cesium concentration in test solution after contact with zeolite}}{\text{Initial cesium concentration in test solution}}$$

APPENDIX III
DETERMINATION OF AMMONIA ION-EXCHANGE CAPACITY OF ZEOLITES

Apparatus

1. Spectrophotometer.
2. Polyethylene test tubes, 20 ml.
3. Fraction collector.
4. Sieves, U.S. Bureau of Standards, size 20 and 50 with pan and lid.
5. Sieve shaker, Ro-Tap.
6. Polyethylene bottles, 500 ml.
7. Peristaltic pump fitted with plastic tubing.
8. Glass vials, 18 mm i.d., 25-mm high.
9. Volumetric flask, 500 ml.
10. Volumetric flasks, 50 ml.
11. Glass wool.

Reagents

1. 1 M NH_4Cl.
2. 1 M NaCl.
3. 0.01 M $AgNO_3$.
4. Nessler reagent.
5. Ammonia-free distilled water.

Procedure

1. Crush and sieve the zeolite to obtain 20 X 50-mesh material.
2. Wash the granular zeolite with tap water followed by distilled water to remove dust and fines.
3. Dry the zeolite sample for 16 hours at 110°C.
4. Cool the zeolite and weigh 3 g into a 20 ml polyethylene test tube which contains a glass wool plug (packed to ∼¼ inch depth) at the bottom. The test tube has been previously perforated with a sharp knife to allow liquid drainage.
5. The test tube containing the zeolite sample is put in a fraction collector with the bottom of the test tube placed in a vial of sufficient depth such that the top of the vial is above the top of the zeolite bed.
6. Distilled water is added to the test tube until drainage from the test tube fills the vial. The water is jetted into it from the zeolite bed to agitate the bed and remove air bubbles from the bed. The vial beneath the test tube serves to keep the liquid level above the top of the zeolite bed at all times.
7. A solution of 1 M NH_4Cl is percolated through the zeolite bed with the aid of a peristaltic pump over a 2-day period. A large number of samples can be processed with the aid of the fraction collector. Each sample is positioned under the 1 M NH_4Cl outlet for a short period of time (e.g., 0.3 minute). A fraction collector with a reel that holds thirty samples in a row will require about 10 minutes to complete one full rotation of the reel. The flow rate is set to allow about 500 ml 1 M NH_4Cl to flow through each sample over the 2-day period.
8. Upon completion of the ammonium ion loading, the samples are washed with distilled water until a test with silver nitrate solution shows no Cl^- in the wash water. The wash may be accomplished with the fraction collector if desired, using the same flow rate as that used for the 1 M NH_4Cl. The fraction collector and tubes should be thoroughly rinsed with water to remove the bulk of the NH_4Cl both inside and outside the tubes prior to starting the wash cycle.
9. The sample test tubes are then placed in a 500 ml polyethylene bottle, and 1 M NaCl is added over a

period of ½ to 1 hour until the level of the 1 M NaCl eluate is just above the zeolite bed (trapped air is removed from the bed). The remaining 1 M NaCl to make the volume just under 500 ml is added slowly over a 16 hour period.

Calculation

$$\text{Ammonium ion-exchange capacity, meq/g} = \frac{(\text{Ammonia molarity in eluate})\,(500)}{3}$$

ION-EXCHANGE REACTIONS OF PROCESSED ZEOLITE AND ITS APPLICATION TO THE REMOVAL OF AMMONIA—NITROGEN IN WASTES

ZENZI HAGIWARA and MASASHI UCHIDA

Central Research Laboratory, Osaka Oxygen Industries Ltd., 1095 Katsube-Cho, Moriyama-city, Shiga-Pref., Japan

Abstract – Using processed zeolite obtained by chemical and physical treatment of a natural mordenite, cation-exchange equilibria were investigated for the ion pairs Na^+–NH_4^+, Na^+–K^+, and K^+–NH_4^+ at ionic strengths of 0.05–0.10 M over a temperature range of 15° to 60°C. The deduced thermodynamic quantities, $K_{M_1}^{M_2}$, $\Delta G°$, $\Delta H°$, and $\Delta S°$, were calculated. Separation factors for the systems Na^+–NH_4^+, Na^+–K^+, Na^+–Mg^{2+} and Na^+–Ca^{2+} were measured, and the ion-exchange reactions of heavy metal ions in wastewater treatment were studied.

A mobile pilot plant for the removal of NH_3–N in waters has been operated as a function of flow rate, employing various simulated waters, and the adsorption and desorption curves of NH_3–N and other constituents were determined. The experimental results show that the removal of NH_3–N below 3 ppm in the effluent from the pilot column is easily achieved under a high flow rate, and the retention of NH_4^+ in the exchange site of the processed mordenite is fairly high even in the presence of other metal ions.

INTRODUCTION

Mordenite is one of the most interesting of the naturally occurring zeolites, and it has been synthesized by several workers (Barrer, 1948; Sand, 1969). Mordenite has been extensively investigated as an adsorbent for gas adsorption, and the ion-exchange properties have been studied from the viewpoint of an inorganic ion exchanger. The exchange isotherms were measured for a number of ion pairs. The present paper reports fundamental measurements of the ion-exchange equilibria of several ion pairs on a processed mordenite (zeoharb) in relation to wastewater treatment. Thermodynamic analysis of the equilibria also are reported. Another purpose of this investigation has been to develop a process of NH_3–N removal from water employing the processed zeolite and to compare the observed results with those obtained by other workers (Koon and Kaufman, 1975; McLaren and Farquhar, 1973; Slechta and Culp, 1967; Shindola, 1972).

EXPERIMENTAL

Ion-exchange materials used

Commercial zeoharbs developed by Osaka Oxygen Industries Ltd. were used for the present experiments. Zeoharbs are produced in various available shapes such as powder, particle, and pellet through the fabrication and calcination processes, after chemical and physical treatments have been made for natural mordenite obtained from Shiroishi, Miyagi-Prefecture, Japan. Zeoharb ZE-500 and ZE-505 described in this paper have the same properties, and the difference between them is only in particle size. The former has a coarse particle distribution, and the latter a fine-particle distribution of 200 X 250-mesh. Both are prepared by calcination after an ion-exchange treatment of the washed natural mordenite by sodium chloride solution. The treated mordenite products have effective pore diameters of 5.1 Å. The X-ray diffraction of powdered zeoharb ZE-505 shows a typical mordenite structure (Fig. 1). In general, zeoharbs, like other molecular sieves, show an ion-exchange property.

Conditioning of inorganic exchanger

Prior to the experiments 30 X 60-m aggregates of zeoharb ZE-500 were placed in a Pyrex column containing a sintered glass filter and treated with excess 2 M NaCl in order to produce the sodium-exchange form. After washing with water, the exchanger was suctioned with aspirator to remove water and dried for at least 24 hours at 110°C under a reduced pressure. The exchange capacity of the treated zeoharb for ammonium ion was

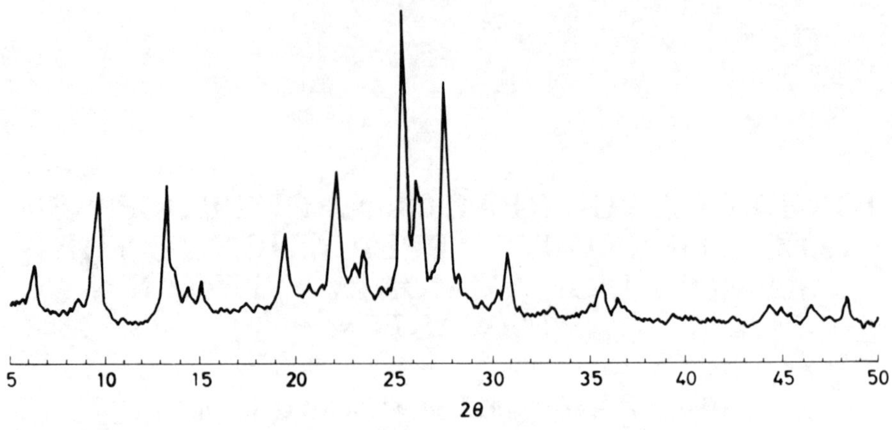

Fig. 1. X-ray diffraction pattern of powdered zeoharb ZE-505 using CuK_{α} radiation.

1.8 meq/g of the zeoharb exchanger. KZ and NH_4Z also were prepared according to the same procedure as above, using load solutions of 2 M KCl and NH_4Cl, respectively.

Equilibrium experiments

Equilibrium ion-exchange data for NH_4^+ and K^+ on the Na-form of zeoharb ZE-500 (30 X 60-mesh) were performed at ionic strengths of 0.05 and 0.10 M and at 15°, 25°, 40°, and 60°C using the method of Hagiwara and Shinada (1970). Additional data for NH_4^+ on the K-form of zeoharb ZE-500 were collected at an ionic strength of 0.1 M at 25°C. The experiments were carried out in the directions: $NaZ \leftarrow NH_4^+$, $NaZ \leftarrow K^+$, $KZ \leftarrow NH_4^+$. Some experiments were made from the reverse direction in order to confirm the ion-exchange isotherms. It was found that the reversibility of the present exchange system was excellent within the experimental range studied.

In each series of experiments, 100 ml of standard chloride solution was placed in a 200-ml glass-stoppered flask with 0.3–20 g zeoharb and allowed to stand for at least 20 hours. The mixture was stirred occasionally and kept at 15° to 60°C in a constant-temperature bath. The total concentrations of the aqueous phase at equilibrium were held at either 0.5 M or 0.10 M. The establishment of equilibrium was virtually complete within several hours, except in the case of extremly low concentration in the aqueous phase. After equilibration was obtained, the aqueous and solid phases were separated for analysis. Prior to analysis, the adsorbed constituents in the solid phase were eluted with salt solutions.

The determination of Na and K in both phases was performed by atomic absorption, using wavelengths of 589 and 766.5 nm, respectively. Spectrophotometric determination of ammonium in the two phases was made at 400 nm incorporating Nessler's method.

Small pilot-plant system

A small, mobile pilot plant was constructed consisting of three exchange columns filled with zeoharb, a pretreatment column, a stripping tower, a regenerant storage tank, an air blower (800 l/min), an air heater, a regenerant heater, a water-conditioning cell, and a water-storage cell. A schematic flowsheet of the pilot plant is shown in Fig. 11 along with the adsorption and desorption systems of NH_3–N. The rubber-lined exchange columns are 200 mm in diameter and 2000 mm in height. Each column was packed with about 25 kg of process zeolite (zeoharb ZE-500, 10 X 30-mesh, NaZ-type) and operated as a function of flow rate. The pretreatment column with a dimension of 300 mm (i.d.) X 1200 mm (height) was employed as a filter to remove suspended solids from test solutions. The rubber-lined stripping tower is approximately 200 mm in diameter and was packed with 1780 mm height of 1-inch netrings.

The influent stream of a simulated wastewater was controlled to a proper value of pH and pumped by gravity flow through the filter bed and then through the two zeoharb exchange columns in series at a controlled flow rate. During the NH_3–N adsorption step, down flow was employed for the two exchange columns while the third column was regenerated. After the first column in the adsorption step was saturated with NH_4^+, it was removed from the system, and the second zeoharb column was connected to a freshly regenerated column. In regeneration, using upflow, the saturated zeoharb column was eluted by NaCl or a mixed solution of NaCl and NaOH. The regenerant was recycled through an exchange bed and a stripping tower until the NH_3–N concentration had been reduced to a desired value.

Most of the fundamental data related to adsorption or desorption were obtained by the use of individual zeoharb columns equipped in the small pilot plant.

Analytical method used

Metal-ion concentrations were mainly determined by atomic absorption, and the following wavelengths were employed: Na, 589 nm; K, 766.5 nm; Ca, 422.7 nm; Mg, 285.2 nm; Cu, 324.7 nm; Zn, 213.9 nm; Cd, 228.8 nm; Ag, 328.1 nm; Pb, 217.0 nm. Further, sometimes, chelometry was employed for the determination of Ca and Mg using EDTA standard (ethylenediamine-tetraacetic acid–Na salt) and EBT indicator.

RESULTS AND DISCUSSION

Ion-exchange equilibria

The ion-exchange reaction for two univalent counterions A^+ and B^+ on an inorganic exchanger such as zeolite is expressed by

$$\bar{A}^+ + B^+ \rightleftharpoons \bar{B}^+ + A^+ \quad (1)$$

where the superscript bar indicates the solid phase, and ion hydration is omitted for simplicity. Thus the thermodynamic equilibrium constant may be defined by

$$K_A^B = \frac{\bar{a}_B a_A}{\bar{a}_A a_B} = \frac{[\bar{B}^+][A^+]\bar{f}_B f_A}{[\bar{A}^+][B^+]\bar{f}_A f_B} \quad (2)$$

where the brackets denote the concentration of the indicated species, and the a and f symbols represent activity and activity coefficient, respectively. For equation (1), the exchange quotient may be written in the form

$$K_A^B = \frac{\bar{X}_B \bar{f}_B m_A f_A}{\bar{X}_A \bar{f}_A m_B f_B} \quad (3)$$

Here, m represents the molarity, and X the equivalent fraction. The selectivity coefficient can be written in the form

$$K_{sA}^{\ B} = \frac{\bar{X}_B m_A}{\bar{X}_A m_B}, \quad (4)$$

Further, corrected selectivity coefficient may be written

$$K_{cA}^{\ B} = \bar{X}_B \cdot a_A / \bar{X}_A \cdot a_B . \quad (5)$$

With the purpose of finding the thermodynamic equilibrium constants for the exchange reactions, theoretical studies have been made by Gregor (1951), Bonner and Argersinger (1952), and Gaines and Thomas (1953). Application of the Gibbs–Duhem equation leads to the following relations for the ion-exchange reaction represented by equation (1):

$$\ln\bar{f}_A = (1 - \bar{X}_A)\ln K_{cA}^{\ B} - \int_{X_A}^{1} \ln K_{cA}^{\ B} d\bar{X}_A, \quad (6)$$

$$\ln\bar{f}_B = -\bar{X}_A \ln K_{cA}^{\ B} + \int_{0}^{X_A} \ln K_{cA}^{\ B} d\bar{X}_A. \quad (7)$$

Introducing the above relationships into $K_A^B = K_{cA}^{\ B}(\bar{f}_B/\bar{f}_A)$, we have

$$\ln K_A^B = \int_0^1 \ln K_{cA}^{\ B} d\bar{X}_A. \quad (8)$$

Considering account of the contributions of sorbed electrolyte and solvent to ion exchange, Gaines and Thomas (1953) give us

$$\ln K_A^B = \int_0^1 \ln K_{cA}^{\ B} d\bar{X}_A + \ln\frac{\bar{f}_B(a)}{\bar{f}_A(b)} - \int_a^b n_s d\ln a_s. \quad (9)$$

The first terms on the left-hand side of equation (9) can be found graphically from a plot of $\ln K_{cA}^{\ B}$ with respect to $\bar{X}_A$; the plot of $K_{cA}^{\ B}$ against $\bar{X}_A$ should be associated with the corresponding plot of a_S with respect to $\bar{X}_A$. Further, the last integral on the right-hand side of equation (9) should be performed from the extrapolated point at which $\bar{X}_A = 0$ (point a) to the extrapolated point at which $\bar{X}_A = 1$ (point b). The symbols $\bar{f}_A$ and $\bar{f}_B$ are the activity coefficients of the pure forms of AZ to BZ in solutions in the concentrations used for the measurement of $K_{cA}^{\ B}$, while n_S represents the amount of water entering into the exchanger, the unit being the number of moles per equivalent of the exchanger; a_S represents the activity of water. Actually, in a monovalent exchange, the contribution of the second and third terms in equation (9) may be insignificant. Therefore, in the present investigation, the values of equilibrium constant were calculated by the use of equation (8).

Typical ion-exchange isotherms of NH_4^+ on processed zeolite (zeoharb ZE-500, Na-form) at ionic strengths of 0.05 and 0.10 M and at 15°C are presented graphically in Fig. 2, where the aqueous and zeolite phases are expressed by X_{NH_4} and $\bar{X}_{NH_4}$, respectively, in equivalent fraction. Further, the ion-exchange isotherms

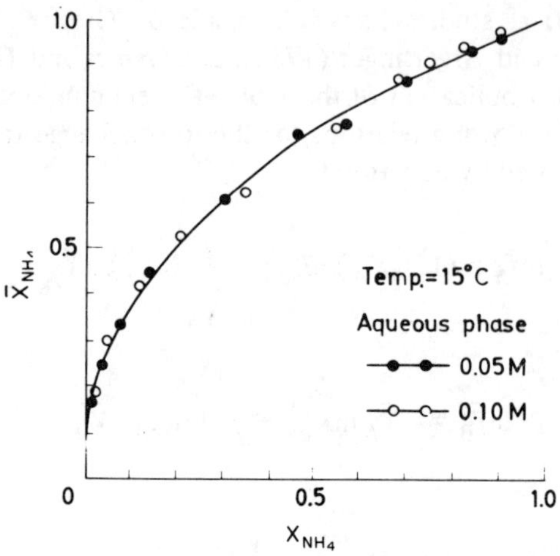

Fig. 2. Ion-exchange isotherm of ammonium ion on zeoharb ZE-500 at 15°C.

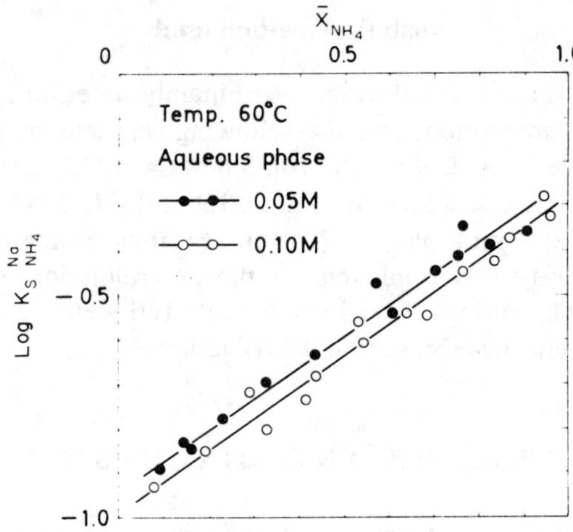

Fig. 4. Selectivity coefficients for the NH_4^+–Na^+ exchange as a function of $\bar{X}_{NH_4}$ (60°C).

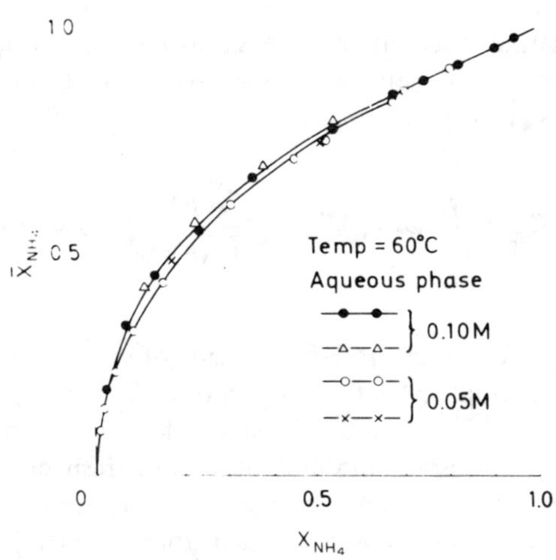

Fig. 3. Ion-exchange isotherms of ammonium ion on zeoharb ZE-500 at 60°C.

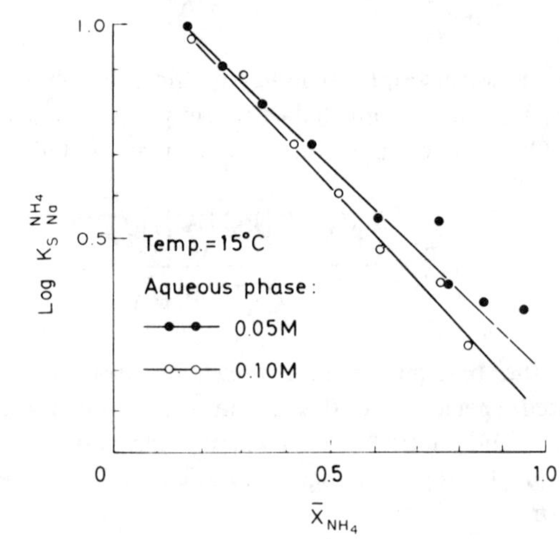

Fig. 5. Selectivity coefficients for the Na^+–NH_4^+ exchange as a function of $\bar{X}_{NH_4}$ (15°C).

obtained under the same condition except for employing a temperature of 60°C are given in Fig. 3, in which the exchanges $NaZ \leftarrow NH_4^+$ (symbol: ● and ○) and $NH_4Z \leftarrow Na^+$ (symbol: Δ and X) are clearly reversible and a single isotherm fits the forward and reverse exchanges.

The selectivity coefficient of the Na–NH_4 system can be calculated from $K_{S\,NH_4}^{Na} = \bar{X}_{Na} X_{NH_4} / \bar{X}_{NH_4} X_{Na}$, and the plots of $K_{S\,NH_4}^{Na}$ with respect to $\bar{X}_{NH_4}$ are shown in Fig. 4, where the $K_{S\,NH_4}^{Na}$ value increases with $\bar{X}_{NH_4}$. Analogous plots of $K_{s\,Na}^{NH_4}$ appear in Fig. 5. The thermodynamic equilibrium constants $K_{Na}^{NH_4}$ and K_{Na}^{K} can be found from equation (8), assuming $K_{c\,A}^{B} = K_{s\,A}^{B}$ for the uni-univalent exchange (Hagiwara and Shinada, 1970), and the resulting values are tabulated in Table 1.

TABLE 1. Log $K_{Na}^{NH_4}$ and log K_{Na}^{K} values obtained using zeoharb ZE-500, 30 × 60-mesh

Temp. (°C)	Log $K_{Na}^{NH_4}$			log K_{Na}^{K}
	0.05 M	0.10 M	Mean	
15	0.656	0.640	0.648	–
25	0.661	–	0.661	0.733
40	–	0.606	0.606	0.720
60	0.594	0.627	0.611	0.655

Na–NH_4 system: M = total concentration of external solution.
Na–K system: total concentration of external solution = 0.10 M.

The heat of exchange ΔH° is evaluated from the slope of d log $K_A^B/d(1/T) = -\Delta H^\circ/2.303$ R, and the values of −485 and −1880 cal/mole are found for the Na-NH_4 and Na-K systems, respectively (Fig. 6). Further, the standard free energy ΔG° and entropy ΔS° have been calculated with the reasonable results listed in Tables 2 and 3. Also, the value of $K_{NH_4}^K$ was derived to be 1.83 from a series of the experimental data obtained in the NH_4-K system at 25°C; total concentration of external solution at equilibrium was held to 0.10 M, and K-type zeoharb and NH_4Cl solution were employed through the experiments.

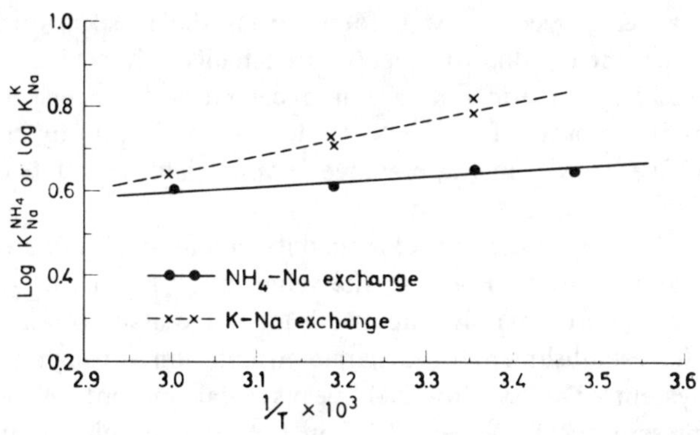

Fig. 6. Temperature dependence of $K_{Na}^{NH_4}$ for the NH_4^+–Na^+ exchange and K_{Na}^K for the K^+–Na^+ exchange.

TABLE 2. Thermodynamic data obtained in the Na–NH_4 system with zeoharb

Temp. (°C)	$K_{Na}^{NH_4}$	ΔG° (cal/mole)	ΔH° (cal/mole)	ΔS° (e.u.)
15	4.45	−854	−485	1.3
25	4.58	−901	−485	1.4
40	4.04	−868	−485	1.2
60	4.08	−931	−485	1.3

TABLE 3. Thermodynamic data obtained in the Na–K system with zeoharb

Temp. (°C)	K_{Na}^K	ΔG° (cal/mole)	ΔH° (cal/mole)	ΔS° (e.u.)
25	5.41	−999	−1880	−3.0
40	5.25	−1031	−1880	−2.7
60	4.52	−998	−1880	−2.7

Water treatment

A potential application of the processed zeolite is in the field of water treatment such as water softening, and tertiary treatment of industrial and municipal wastewaters. In the present research, the fundamental data related to water treatment are mainly described.

The sodium form of zeoharb displays a selective ion exchange for ammonium ion, as well as for certain heavy metal ions. Some experimental results on the adsorbability of univalent and divalent metal ions on zeoharb ZE-500, 10 X 25-mesh, are shown graphically in Fig. 7. It can be seen that the zeoharb product is an efficient material for the removal of trace amounts of metal ions.

The separation factors (α) for binary mixtures concerning water treatment were measured at 20°C by the aid of a column method using zeoharb ZE-500, 20 X 30-mesh, NaZ form; the following values were obtained:

$\alpha_{Na}^{NH_4} = 4.97$ (0.101 M NaCl–0.105 M NH_4Cl);
$\alpha_{Na}^{NH_4} = 4.75$ (0.0514 M NaCl–0.0525 M NH_4Cl);
$\alpha_{Na}^{K} = 5.79$ (0.0514 M NaCl–0.0501 M KCl);
$\alpha_{Na}^{Mg} = 0.340$ (0.101 M NaCl–0.049 M $MgCl_2$);
$\alpha_{Na}^{Mg} = 0.410$ (0.514 M NaCl–0.0244 M $MgCl_2$);
$\alpha_{Na}^{Ca} < 1.0$ (0.0514 M NaCl–0.0246 M $CaCl_2$).

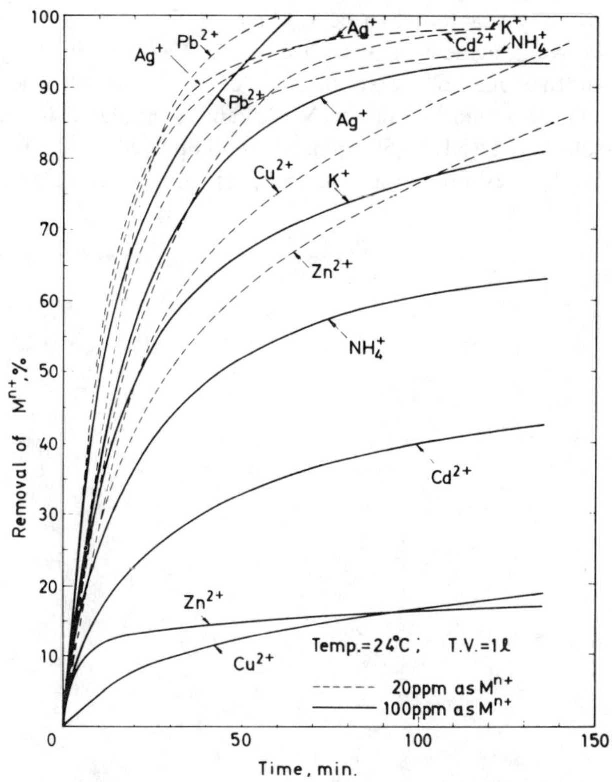

Fig. 7. Removal of M^{n+} by processed zeolite. Temperature = 24°C; initial total volume (T.V.) = 1 *l*; adsorbent = zeoharb 500, 10–25-mesh 3.00 g; initial concn. of M^{n+} used, 20 ppm (broken line), 100 ppm (solid line). Initial pH of M^{n+} (20 ppm): 4.95 ($AgNO_3$ soln.); 5.05 ($Pb(NO_3)_2$ soln.); 5.22 ($Cu(SO_4)_2$ soln.); 5.64 ($CdCl_2$ soln.); 4.32 ($ZnCl_2$ soln.). Initial pH of M^{n+} (100 ppm): 4.95 ($AgNO_3$ soln.); 5.05 ($Pb(NO_3)_2$ soln.); 5.08 ($Cu(SO_4)_2$ soln.); 5.60 ($CdCl_2$ soln.); 3.74 ($ZnCl_2$ soln.). Each experiment was carried out under stirring (150 rpm) at 24°C in a constant-temperature bath, and 10 ml of sample solution was taken at regular intervals for analysis.

The compositions of load solutions used for the column experiments are given in parentheses. Considering the thermodynamic equilibrium constants and α-values, the selectivity sequence should exist:

$$K^+ \geqslant NH_4^+ > Na^+ > Ca^{2+} > Mg^{2+}.$$

The NH_3–N removal in simulated water solutions was performed employing individual zeoharb columns (i.d. 200 mm) equipped in the mobile pilot plant (Fig. 11). Fairly sharp break-through curves of ammonium ion are presented graphically in Fig. 8, where curves 1 and 2 were obtained by the use of flow rates of 10.4 and 25.8 cm/min, respectively. In the above column experiments, test solution A, with a composition of 50 ppm as NH_4^+, was employed. The exchange capacity of NH_4^+ for the Na-type of zeoharb ZE-500, 10–30-mesh, was calculated by integrating curve 1 in Fig. 8, in which the zeoharb column containing 23.66 kg NaZ, had bed dimensions of 200 X 1350 mm. The capacity was found to be 1.78 meq/g NaZ. This value is in good agreement with 1.8 meq/g NaZ obtained by another method. Breakthrough curve 3 was obtained using a zeoharb column (No. 2) and a simulated water solution A (NH_4^+ = 50 ppm; Na^+ = 60 ppm; K^+ = 10 ppm; Mg^{2+} = 10 ppm; Ca^{2+} = 15 ppm). A similar column experiment was performed as shown in Fig. 9, where the indicated species (shown as solid lines) were obtained by passing a simulated water solution B (NH_4^+ = 26.5 ppm; Na^+ = 30 ppm; K^+ = 12.6 ppm; Mg^{2+} = 10 ppm; Ca^{2+} = 15 ppm) through the pilot column (No. 3: bed volume (B.V.) of Na-form of zeoharb ZE-500, 10–30-mesh = 43.65 l) at a flow rate of 25.04 cm/min. On the other hand, another curve in Fig. 9 (shown as a broken line) was obtained by passing simulated water solution C (NH_4^+ = 20.8 ppm; Na^+ = 10 ppm; K^+ = 1 ppm; Mg^{2+} = 5 ppm; Ca^{2+} = 25 ppm) through one of the pilot columns (No. 2: B.V. = 42.39 l) at a flow rate of 22.18 cm/min. In the above experiments, gravity flow was employed. It was found from the breakthrough experiments that the zeoharb exchanger showed good exchange kinetic and gave an excellent performance for the removal of NH_3–N to less than 3 ppm in the effluent, even in the presence of appreciable amounts of other metal ions.

In the adsorption experiments employing simulated water solutions B and C, the values of $\eta_{\overline{NH}_4}$ and $\alpha^{NH_4}_{\Sigma M}$ were approximately calculated under the assumption of the establishment of dynamic equilibrium through the system. The symbol ΣM means total amount of the present species except NH_4^+ in the aqueous phase, and $\Sigma\overline{M}$ total amount of the present species except NH_4^+ in

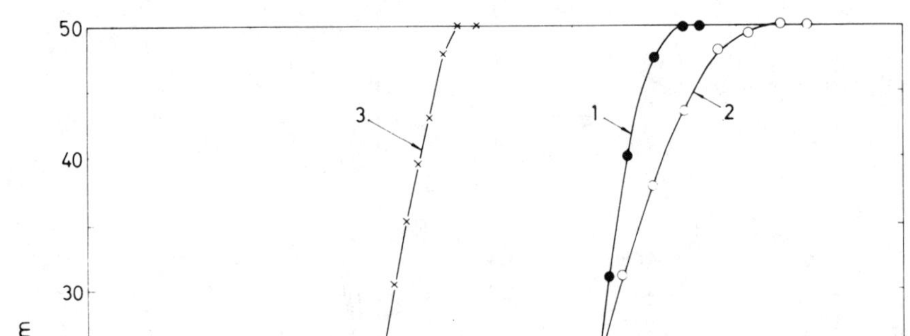

Fig. 8. Break-through curves of NH_3–N obtained by using simulated solutions. Curve-1: bed volume=42.39 1; flow rate=10.4 cm/min (3.25 1/min). Curve-2: bed volume = 43.65 1; flow rate=25.8 cm/min (8.10 1/min). Curve-3: bed volume=42.39 1; flow rate= 20.7 cm/min (6.49 1/min). The flow rate was measured at rhe bottom of the i.d.–200 mm column.

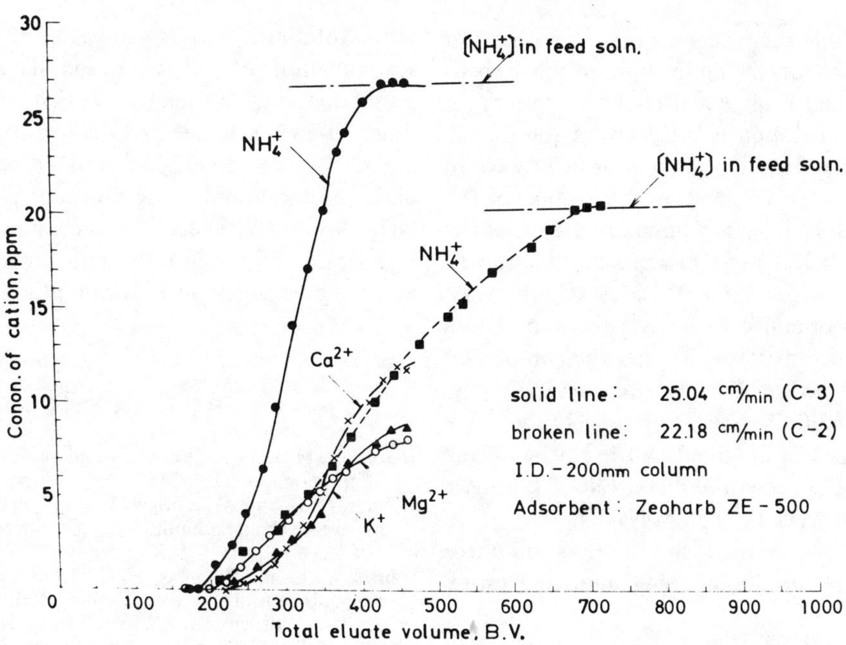

Fig. 9. Break-through curves of metal ions obtained by using simulated water solutions.

the zeoharb phase. Further, $\eta_{\overline{NH}_4}$ expresses the availability of NH_4^+ in the exchange site of zeoharb, in equivalent fraction, and $a_{\Sigma M}^{NH_4}$ the separation factor. The following values were obtained at 15°C:

Adsorption experiment using simulated water solution C:

$$\eta_{\overline{NH}_4} = 0.494; \eta_{\Sigma \bar{M}} = 0.506; a_{\Sigma M}^{NH_4} = 1.8.$$

Adsorption experiment using simulated water solution B:

$$\eta_{\overline{NH}_4} = 0.433; \eta_{\Sigma \bar{M}} = 0.567; a_{\Sigma M}^{NH_4} = 1.5.$$

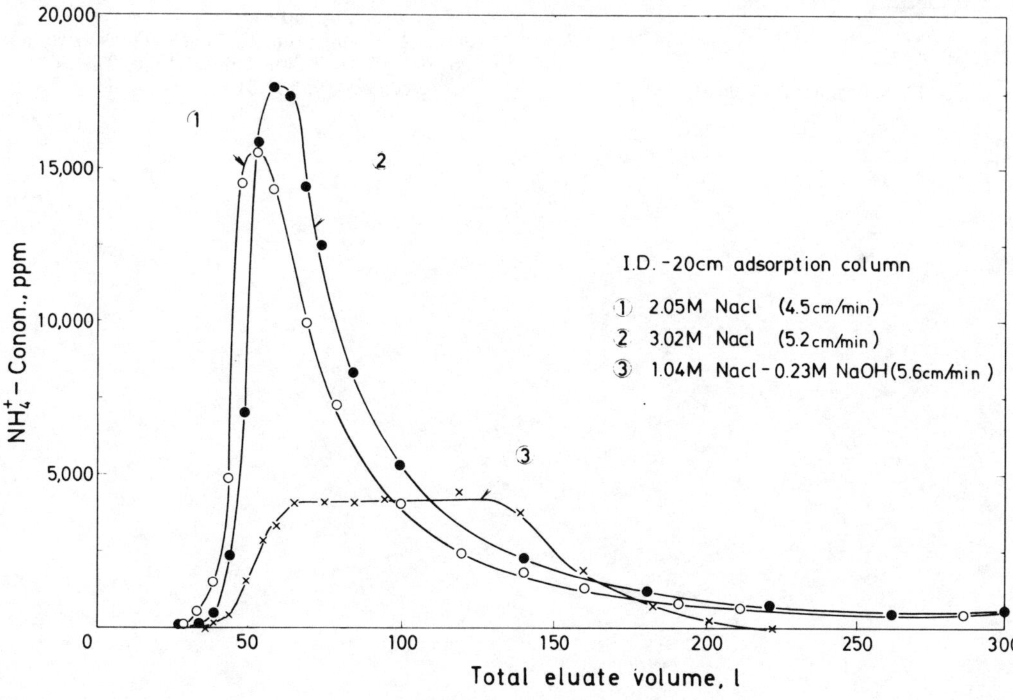

Fig. 10. Elution curves of ammonium.

In order to confirm the accuracy of the observed α-values, another set of dynamic equilibrium experiments were performed using small columns, the $\alpha_{\Sigma M}^{NH_4}$ values were 1.47 for solution B and 1.65 for solution C.

Elution curves of ammonium ion appear in Fig. 10. In this figure, curve 1 was obtained by the elution of the ammonium ion in the zeoharb column (No. 2) after the completion of the adsorption experiment relating to curve 1 in Fig. 5, using 2.05 M NaCl eluent, while elution curve 2 was obtained by eluting the ammonium ion in the zeoharb column (No. 3) after the completion of the adsorption experiment relating to curve 2 in Fig. 8, using an eluent of 3.02 M NaCl. Further, elution curve 3, with a flat region, was obtained by the elution of the ammonium ion in the zeoharb column (No. 2), using a mixed eluent of 1.04 M NaCl-0.23 M NaOH.

The constructed pilot plant is in full operation under varying experimental conditions, using simulated waste-water solution. In order to get a final decision, further accumulation of the operation data as well as cost estimation may be necessary. However, at the present time, an ion-exchange process with zeoharb (NaZ-type) can remove successfully at least 97% of influent NH_3-N under optimum operating procedures. In the removal of NH_3-N in waste water, the processed zeoharbs having a high degree of structural stability seem to be as suitable as other exchangers such as clinoptilolite.

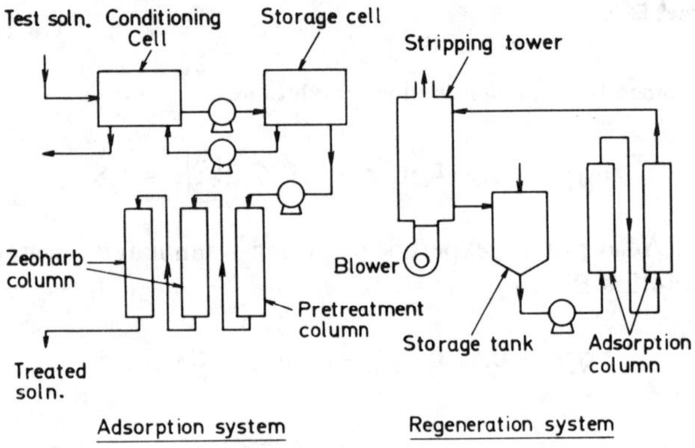

Fig. 11. Schematic flow sheet.

REFERENCES

Barrer, R. M. (1948) Syntheses and reactions of mordenite: *J. Chem. Soc.* pp. 2158-2163.

Bonner, O. D. and Argersinger, W. J., Jr. (1952) Factors involved in ion-exchange equilibria: *J. Amer. Chem. Soc.* **74**, 1044-1047.

Gaines, G. L. and Thomas, H. C. (1953) Adsorption studies on clay minerals. II. A formulation of the thermodynamics of exchange adsorption: *J. Chem. Phys.* **21**, 714-718.

Gregor, H. P. (1951) Gibbs–Donnan equilibria in ion exchange resin: *J. Amer. Chem. Soc.* **73**, 642-650.

Hagiwara, Z. and Shinada, Y. (1970) Studies on ion-exchange equilibria for Cs–H system: *J. Nucl. Sci. Technol. Japan* **7**, 219-226.

Koon, J. H. and Kaufman, W. J. (1975) Ammonia removal from municipal waste waters by ion exchange: *J. Wat. Pollut. Control Fed.* **47**, 448-464.

McLaren, J. R. and Farquhar, G. J. (1973) Factors affecting ammonia removal by clinoptilolite: *J. Environmental Engineering Division*, pp. 429-444.

Sand, L. B. (1969) Synthetic mordenite and preparation: U.S. Pat. 3,436,174.

Shindola, A. (1972) Nitrogen and phosphorus removal from waste waters. Part 2 – Chemical–biological: *Water & Sewage Works*, July, pp. 60-67.

Slechta, A. F. and Culp, G. L. (1967) Water reclamation studies at the South Tahoe Public Utility District: *J. Wat. Pollut. Control Fed.* **39**, 787-814.

NATURAL ZEOLITES: NOVEL USES AND REGENERATION IN WASTEWATER TREATMENT

C. B. MURPHY, O. HRYCYK, and W. T. GLEASON

O'Brien & Gere Engineers, Inc.,
1304 Buckley Road, Syracuse, New York 13201, U.S.A.

Abstract – The use of natural zeolites, and in particular clinoptilolite, for the removal of ammonia from municipal wastewaters is drawing increased interest. The effects of grain size, hardness of the treated wastewater, and contact time have been established; however, one factor that has not been thoroughly investigated is the effect of the source of the clinoptilolite ore on the ammonia-removal capacity.

Effective or break-through capacities for a number of western clinoptilolite samples show a considerable range. Samples from Owyhee County, Idaho, have the highest capacities, 0.14 meq/g; whereas samples from Washoe County, Nevada, exhibit the lowest capacities, 0.06 meq/g. These data indicate an urgent need to define those clinoptilolite deposits having optimum exchange capacities in order to reduce operating costs and make clinoptilolite systems even more competitive with biological processes for ammonia removal.

Demonstration-scale evaluations, conducted for the U.S. Environmental Protection Agency on the removal of nutrients from combined sewer overflows, demonstrated that a natural zeolite ion-exchange system can have a dual function in wastewater treatment. Using 20 × 50-mesh clinoptilolite, an ammonia-exchange system was utilized successfully to destabilize and filter an in-line, alum-treated wastewater. Significant reductions in nutrient, oxygen-demanding constituents, bacteria, viruses, and other pollutants were realized.

Ammonia-saturated clinoptilolite can be regenerated by thermal activation to 500°C. The principal gaseous degradation products are N_2 and NH_3, with only traces of NO_2. The offgas composition appears to be most dependent on the temperature of the reactivation and the oxygen content of the applied regeneration-gas streams. X-ray diffraction studies indicate little or no structural degradation of the thermally regenerated clinoptilolite. The ion-exchange capacity of thermally regenerated clinoptilolite is considerably higher than that obtained by effective wet processes.

Biological regeneration of spent clinoptilolite was achieved in an aerated-column mode of application with and without the innoculation of acclimated nitrogen. Pilot-scale evaluations indicate an approximate 75% regeneration after 35 days of aeration without an initial acclimated mass of nitrifiers. If the system is innoculated with nitrifiers, bioregeneration is ten times as fast.

INTRODUCTION

The use of natural zeolites, in particular clinoptilolite, for the removal of ammonia from municipal wastewaters is drawing increased interest. One novel application of clinoptilolite is its use in the treatment of combined sewer overflows. Because the intermittent nature of combined sewer overflows and the space limitations of typical overflow sites preclude the use of conventional biological treatment systems, an ion-exchange system using a naturally occurring zeolite ideally lends itself to the point-source treatment of such overflows.

In a demonstration project conducted for the United States Environmental Protection Agency in Syracuse, New York, clinoptilolite served as both an exchange and secondary filtration medium. Ammonia reductions to levels below 0.02 mg/l, as well as significant reductions in the oxygen-demanding constituents, bacteria, and other pollutants, were realized in the exchange-filtration system.

In order for such an ion-exchange system to be economically practical, the exchange capacity of the natural zeolite must be such that its use can be justified over the use of synthetic zeolites which have considerably higher exchange capacities. The second factor determining the system's practicability is the ease and effectiveness involved in the regeneration of the exhausted exchange material. With these factors in mind, preliminary investigations of the "effective" ammonia-exchange capacities of clinoptilolite from various deposits, as well as investigations comparing several regeneration techniques, were conducted.

MATERIALS AND METHODS

Sample preparation

Eight samples of clinoptilolite were obtained from the following areas of the western United States: Grant County, New Mexico; Owyhee County, Idaho; Fremont

County, Wyoming; Lander County, Nevada; Washoe County, Nevada; Malheur County, Oregon; Maricopa County, Arizona; and San Bernardino County, California. The latter sample is from the well-known locality at Hector, California. The samples were supplied by F. A. Mumpton.

All samples were sieved to a 30 X 40-mesh (standard screen scale) to minimize the effect of varying particle sizes. The samples were washed with deionized water (D.I. water) to remove any fine material, dried at 105°C for 18 hours, and then resieved to 30 X 40-mesh.

Breakthrough ion-exchange capacity

Breakthrough exchange capacity is defined as the milliequivalents of ammonia nitrogen removed per gram of clinoptilolite at that point in time when the influent ammonia concentration is equal to the effluent ammonia concentration. To determine the exchange capacity for each of the eight samples, static tests were performed utilizing an equipoise shaker. Two grams of each sample were placed in a 1-quart container to which 500 ml of primary wastewater (previously passed through a 200-mesh screen) was added. The ammonia-nitrogen NH_3(N) concentration of the wastewater was approximately 20 mg/l in all tests. The applied wastewater was analyzed for NH_3(N), total Kjeldahl nitrogen (TKN), K, Ba, Ca, Na, Fe, and Mg. The containers were shaken for 2 hours after which a sample of the wastewater in each container was quantitatively analyzed for each of the above species. The clinoptilolite was then filtered and contacted with an additional 500 ml of wastewater. The above procedure was repeated to a maximum of three times. The resulting reduction in NH_3(N) yielded the effective exchange capacity for each clinoptilolite sample.

A second set of tests was conducted to determine if pretreatment of the clinoptilolite would increase its exchange capacity. Preliminary tests were conducted utilizing Hector clinoptilolite to determine the most effective pretreatment solution. Ten-gram samples of Hector clinoptilolite were contacted with 500 ml of each of the following pretreatment solutions; 72 g/l HCl, 80 g/l NaOH, 0.5 g/l NaOH-30 g/l NaCl, a combination of 72 g/l HCl and 0.5 g/l NaOH-30 g/l NaCl, and a control solution of D.I. water. The containers were then placed on the equipoise shaker for 2 hours. In the case of combination pretreatment, the sample was shaken for 2 hours in the 72 g/l HCl, filtered, washed with D.I. water, and then shaken for 2 hours in the 0.5 g/l NaOH-30 g/l NaCl solution. Samples were taken of the pretreatment solutions and analyzed for K, Ba, Na, Ca, Fe, Mg, Al, and Si after contact with the clinoptilolite samples. After pretreatment, the clinoptilolite samples were washed thoroughly with D.I. water and then dried at 105°C for 18 hours.

A 2 g sample of Hector clinoptilolite from each pretreatment solution was contacted with 500 ml of municipal wastewater. The effective exchange capacity of each sample was determined as previously described. Subsequently, 2 g samples of all eight clinoptilolite were pretreated using 72 g/l HCl followed by 0.5 g/l NaOH-30 g/l NaCl, in the method previously described. The effective exchange capacity of each sample after pretreatment was then determined.

Chemical regeneration

Approximately 30 g of previously exhausted Hector clinoptilolite (with 24 mg NH_3(N) exchanged/g clinoptilolite) was contacted with 100 ml of D.I. water for 1 minute and then drained. The clinoptilolite was rinsed two more times with 50-ml aliquots of D.I. water. After draining on a fine screen for 15 minutes, the clinoptilolite was dried at 105°C for 2 hours and finally cooled in a desiccator.

Seven 2 g samples of the dried, spent clinoptilolite were placed in separate 500 ml plastic bottles. Each sample was contacted with 100 ml of one of the following regenerant solutions: (1) 49 g/l H_2SO_4, (2) 36 g/l HCl, (3) 63 g/l HNO_3, (4) 40 g/l NaOH, (5) 58 g/l NaCl, (6) 5.25% NaOCL, and (7) D.I. water (control).

After a contact time of 10 minutes, the regenerant solution was decanted and analyzed for NH_3N and cations such as Na, Al, and Si. The clinoptilolite was then rinsed with three 100-ml aliquots of D.I. water and dried for 2 hours at 105°C. The effective ammonia exchange capacity was subsequently determined on each of the regenerated clinoptilolite samples in the manner above.

Biological regeneration

A 300 g sample of Hector clinoptilolite which had been previously exhausted (to an effluent NH_3(N) concentration of 13 mg/l) in the course of operating the demonstration scale facilities was placed in a 2-foot-long column 2 inches in diameter. One liter of a pH buffer solution (Table 1) was added to the exhausted clinoptilolite in the column; nitrifying bacteria naturally present in the stormwater overflow were utilized without concentration in the initial studies.

The column was aerated for 40 days by means of a diffuser stone at the bottom at a sufficiently slow rate to

TABLE 1. Buffer solution for biological regeneration of clinoptilolite

Buffer solution	Composition (Norris and Ribbons, 1969)
Na_2CO_3	1000 mg/l
$NaHCO_3$	500 mg/l
$MgSO_4 \cdot 7H_2O$	140 mg/l
KH_2PO_4	50 mg/l
$CaCl_2 \cdot 2H_2O$	10 mg/l
$FeSO_4$	7.5 mg/l

mix the contents of the column. Throughout aeration, samples were periodically taken, filtered, and analyzed for NH_3(N), NO_3(N), and NO_2(N).

A second bio-regeneration test involved the addition of a mass of acclimated sludge which had been batch-fed in an NH_3(N) buffer solution. The column was aerated at a flow rate of 2 l/min through a diffuser in the bottom of the column for 24 days. This aeration rate resulted in fairly good mixing of the column contents. Evaporation was reduced by wetting the air prior to column aeration. Samples were taken at random intervals, filtered, and analyzed for NH_3(N), NO_3(N), and NO_2(N).

Thermal regeneration

Thermal regeneration was conducted on Hector clinoptilolite samples at temperatures of 300°C, 400°C, 500°C, and 600°C in a Lindberg rotary tube furnace as shown schematically in Fig. 1. The spent clinoptilolite samples were placed in a stainless-steel tube which could be inserted into the furnace. The furnace was designed to allow a carrier gas to pass through the stainless-steel tube, thereby conveying the off-gases through a trap containing 500 ml of 120 g/l H_2SO_4.

The exchange capacity of the clinoptilolite was determined by passing a solution containing 3000 mg/l NH_3(N) through a column containing 400 g of clinoptilolite until the effluent NH_3(N) concentration was equal to the influent NH_3(N) concentration. Following exhaustion, 2 l of D.I. water were passed through the column to remove any residual NH_3(N) not actually adsorbed on the clinoptilolite. The clinoptilolite was then air dried for 36 hours.

A 15 g sample of the dried, spent clinoptilolite was placed in the stainless-steel tube and inserted into the furnace, previously heated to the desired temperature. Air was blown through the tube at a constant rate of 1 l/min as the tube was rotating at 10 rpm. Samples of

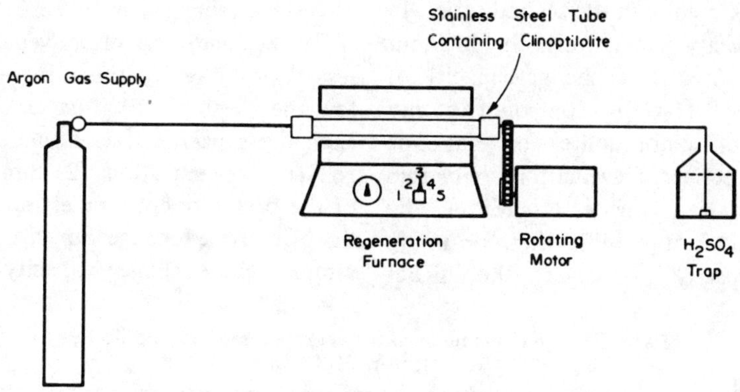

Fig. 1. Schematic diagram of zeolite regeneration system.

TABLE 2. Effective exchange capacities of clinoptilolite samples

Sample	Series 1–18 mg/l NH_3(N) applied		Series 2–21 mg/l NH_3(N)	
	mg NH_3(N) adsorbed	meq NH_3(N)/g clinoptilolite	mg NH_3(N) adsorbed	meq NH_3(N)/g clinoptilolite
Grant Co., NM	8.05	0.288	9.28	0.331
Owyhee Co., ID	6.92	0.247	9.61	0.343
Fremont Co., WY	5.24	0.187	5.25	0.187
Lander Co., NV	6.31	0.225	7.00	0.250
Washoe Co., NV	3.80	0.140	4.10	0.146
Malheur Co., OR	10.57	0.377	10.92	0.390
Maricopa Co., AZ	3.90	0.139	4.55	0.162
Hector clinoptilolite	8.01	0.286	8.72	0.311

the ammonia trap were taken initially and every 15 minutes for 2 hours. From the data, the amount of NH_3(N) desorbed could be determined.

RESULTS AND DISCUSSION

Effective exchange capacity of clinoptilolite samples from varying deposits

The data indicate a substantial difference in the effective exchange capacities of clinoptilolite samples from different deposits. Results of the first and second series of tests are presented in Table 2. Although influent NH_3(N) concentrations varied from 18 mg/l to 21 mg/l, the clinoptilolite exchange capacities did not vary. The samples which showed the highest effective exchange capacities were obtained from Malheur County, Oregon; Owyhee County, Idaho; and Grant County, New Mexico. Those samples exhibiting the lowest effective exchange capacity were obtained from Maricopa County, Arizona; and Washoe County, Nevada. The Malheur County sample's effective exchange capacity was more than 170% greater than the sample obtained from Maricopa County in the first series of tests and more than 160% greater in the second series. In the initial exchange-capacity evaluations, the Malheur County sample had an effective exchange capacity of approximately 0.377 meq NH_3(N)/g, compared to only 0.140 meq NH_3(N)/g of clinoptilolite for Maricopa County. In the second series of evaluations only two runs (2 hours each) were made, therefore the clinoptilolite samples did not exhibit their optimum effective exchange capacity. However, the higher influent NH_3(N) concentration, 21 mg/l, is probably responsible for the increased exchange capacities through system dynamics. After only two runs the partial exchange capacity of the Malheur County clinoptilolite was determined to be 0.390 meq/g while that of the Maricopa County sample was established as 0.162 meq/g. The increased effective exchange capacity of the clinoptilolite exhibited as a result of increased influent NH_3(N) concentrations is consistent with the results of others (Koon and Kaufman, 1975).

The results indicate that, of the eight clinoptilolite samples tested, four (Malheur, Grant, Owyhee, and Hector) had the higher effective exchange capacities while the Maricopa, Washoe, Fremont, and Lander samples had substantially lower effective exchange capacities.

Data from the preliminary pretreatment tests using Hector clinoptilolite showed that the samples treated with 72 g/l HCl for 2 hours and then a caustic brine solution (0.5 g NaOH/l and 30 g NaCl/l) for 2 hours resulted in the greatest increase in effective exchange capacity. This pretreatment increased the effective exchange capacity by over 20% when compared with samples lacking any pretreatment. All other pretreatment solutions resulted in an increase in the effective exchange capacity of 5% or less.

To evaluate the effectiveness of pretreatment, the results of the pretreatment investigations were compared to the results of the second set of effective exchange-capacity tests because of the similar influent NH_3(N) concentrations (21.8 mg/l vs. 21 mg/l). Analysis of the pretreatment studies indicated a 116% increase in the effective exchange capacity of the Maricopa County sample. The exchange capacity of the Malheur County

TABLE 3. Ammonia removal by wet-regenerated clinoptilolite from Hector, California*

Regenerating solutions	mg/l NH_3(N) removed after First contact	Second contact	Third contact	Total
(Virgin ore)	11.4	5.0	0.1	16.5
NaOH	11.9	4.4	1.6	17.9
NaOCl	11.3	3.0	0.6	14.9
NaCl	2.8	−2.3	−3.4	−2.9
H_2SO_4	0.4	−5.7	−3.9	−9.2
HCL	2.2	−4.5	−2.8	−5.1
HNO_3	2.8	−4.5	−0.8	−2.5
H_2O	−19.8	−13.3	−9.7	−42.8
Initial NH_3(N) conc. of contacting solution	16.8	12.3	12.9	42

*Clinoptilolite weight: 2.0 g. Contact time: 1 hour per contact.

sample also increased, however, only by 17%. Both the Owyhee County and Washoe County samples had negligible increases in exchange capacity.

Wet regeneration

Cation analyses of the regenerant solutions revealed a significant increase in the concentration of a number of the structural elements of clinoptilolite. The acidic regenerants from exhausted Hector clinoptilolite contained more than 200 mg/l of aluminum, while the caustic regenerants contained an elevated level of silica. Aluminum and silica were not detected in the NaCl regenerant. These data strongly suggest that some degradation of the crystalline structure of clinoptilolite occurs during wet regeneration with alkaline or acidic solutions. This is in agreement with Breck (1974) who reported high silica concentrations in zeolite alkaline lakes which have a pH greater than 9.0.

Each of the regenerant solutions contained more than 250 mg/l of NH_3(N). This is compared to a potential 480 mg/l of NH_3(N) if complete desorption of the Hector clinoptilolite had occurred. It is possible that more ammonia could have been desorbed from the clinoptilolite had the regeneration contact time been increased. Although the regenerant solutions were essentially equivalent in their ability to deamminate the zeolite, their effect on the exchange capacity of clinoptilolite varied. Alkaline-regenerated clinoptilolite samples were more efficient in removing ammonia than were the acid- and NaCl-regenerated samples. The acid- and NaCl-regenerated clinoptilolite samples lost their ability to remove ammonia from a wastewater containing approximately 13 mg/l NH_3(N). In fact, these regenerated samples of Hector clinoptilolite desorbed ammonia, releasing it into contacting wastewater solution.

The Hector clinoptilolite regenerated with NaOH had an effective exchange capacity slightly greater than that exhibited by virgin ore, whereas the sodium hypochlorite (NaOCl)-regenerated clinoptilolite exhibited an effective exchange capacity equivalent to 90% of that determined for the virgin material. A summary of the results of these tests is listed in Table 4.

TABLE 4. Effective ion-exchange capacity of thermally regenerated clinoptilolite from Hector, California

Sample	mg NH_3(N) adsorbed	meq NH_3(N)/g clinoptilolite
Virgin ore	6.38	0.228
300°C regenerated clinoptilolite	−24.63	–
400°C regenerated clinoptilolite	−3.37	–
500°C regenerated clinoptilolite	6.86	0.245
600°C regenerated clinoptilolite	5.71	0.204

Bioregeneration

An increase in the NO_3(N) concentration of the liquid in contact with the Hector clinoptilolite during the bioregeneration investigation was indicative of conversion of ammonia to nitrate. After 4 days of aerating clinoptilolite without any seeding of nitrifying bacteria at temperatures of 18°C, the NO_3(N) concentration in surrounding medium increased by 1.57 mg/l. On the 32nd day of aeration the NO_3(N) concentration had increased from 13.3 mg/l on the starting day to 142 mg/l. By the 35th day of continuous aeration the NO_3(N) concentration was 168 mg/l. These data suggest that the conversion of ammonia to nitrate can proceed with the existing microbial population encountered in combined sewer overflows. In monitoring several other parameters during the bioregeneration evaluations, it was discovered that the ammonia concentration had increased from 33 mg/l on the 32nd day of aeration to 102 mg/l on the 35th day. During this same period the nitrite (NO_2) concentration had increased from 0.25 to 4.7 mg/l.

It is postulated that during the bioregeneration process, nitrifying bacteria utilize only the ammonia in solution; the ammonia exchanged in the pore spaces in the zeolite is unattainable to the microbes. As the ammonia is depleted from solution by conversion to nitrite and finally nitrate, the ammonia in the clinoptilolite is leached into solution in order to maintain equilibrium in the system. The sudden increase in the ammonia concentration noted above may be a result of a toxic agent inhibiting the complex microbial mechanism of oxidizing ammonia to nitrate.

Static ammonia-exchange tests indicated that bioregenerated clinoptilolite regains approximately 80% of its ability to remove ammonia, compared to that of virgin clinoptilolite. After contact with an influent solution containing 9.45 mg/l NH_3(N), the bioregenerated clinoptilolite removed 7.07 mg/l NH_3(N) from solution compared with 8.61 mg/l NH_3(N) removed by virgin Hector clinoptilolite.

In order to accelerate the bioregeneration process, the testing column filled with spent Hector clinoptilolite was seeded with a culture of nitrifying bacteria. Initially, the production of nitrates was accelerated such that 106 mg/l of nitrate was produced after only 15 days, compared with 28 days required to attain an equivalent level of nitrates in the unseeded test. However, after the 26th day of aeration the level of nitrate appeared to taper off, again suggesting the presence of an inhibition mechanism or the presence of denitrifying organisms. Unlike the results of previous tests, the ammonia concentration in the surrounding medium remained below 5.0 mg/l. Also, the nitrite level in the liquid medium was considerably higher (4.4–7.4 mg/l) throughout this testing period. Information regarding the ammonia exchange-capacity associated with the accelerated bioregenerated clinoptilolite will be available in the near future.

Thermal regeneration

The results of the thermal regeneration using compressed air as the carrier gas were non-reproducible in duplicate runs. The variations in the results were most likely caused by the oxidation of ammonia between the temperature of 500–600°C in an atmosphere of air. Therefore, it was probable that an undeterminable quantity of ammonia which was desorbed from the

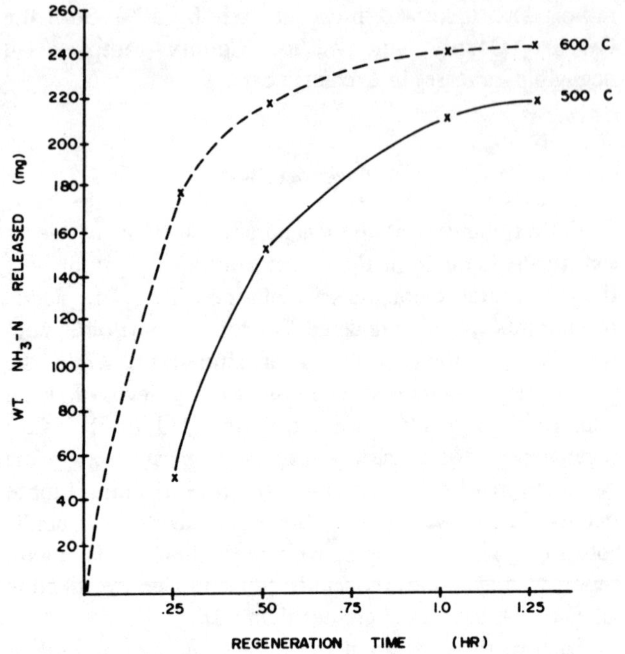

Fig. 2. Ammonia recovery from thermally regenerated clinoptilolite from Hector, California.

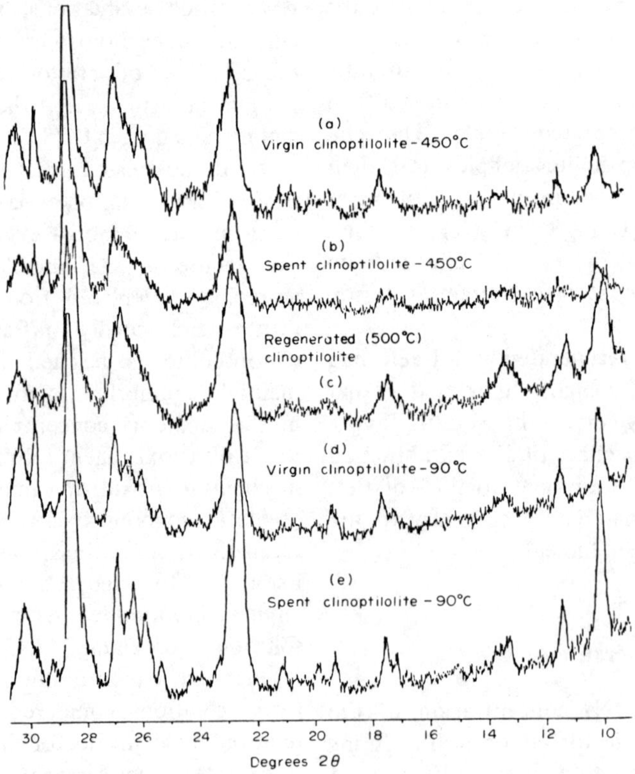

Fig. 3. X-ray diffraction patterns of treated Hector clinoptilolite (CuKd radiation).

clinoptilolite was oxidized to N_2 before being trapped in the 98 g/l H_2SO_4 solution. Unfortunately, the analytical control was not designed to overcome this problem and it was therefore impossible to determine the exact quantity of ammonia oxidized at the higher temperatures.

The highest percentage of adsorbed ammonia recovered was only 53% at 500°C. The results also show that only 35% of the adsorbed ammonia was recovered at 600°C, indicating that the difference may be due to the oxidization of a larger quantity of ammonia.

Since there was so much variation in the results of the recovered ammonia in the thermal regeneration tests using compressed air, an inert gas (argon) was used as the carrier gas. The results obtained using argon are presented in Fig. 2. It was apparent that the thermal regeneration was much more effective at 500°C and 600°C than 300°C or 400°C. At 300°C and 400°C only 25% and 65%, respectively, of the adsorbed NH_3(N) was desorbed and recovered, while 80% and 90% of the NH_3(N) was desorbed and recovered at 500°C and 600°C, respectively. The results also indicate that at 500°C and 600°C there was little additional ammonia desorbed after 90 minutes of regeneration.

Results from tests on thermally regenerated clinoptilolite from Hector indicate that the samples regenerated at 500°C had an effective exchange capacity similar to that of the virgin clinoptilolite. The data are presented in Table 4. With an influent ammonia concentration of approximately 13 mg/l, the partial effective exchange capacity of the virgin clinoptilolite was 0.228 meq NH_3(N)/g clinoptilolite, while the clinoptilolite regenerated at 500°C and 600°C had partial effective exchange capacities of 0.244 and 0.203 meq NH_3(N)/g clinoptilolite, respectively. The possibility exists that the reduction in effective exchange capacity of the 600°C sample could have been caused by an alteration in the structure of the clinoptilolite. Additional tests are being conducted to determine if thermal regeneration at 600°C causes large reductions in the effective exchange capacity after several regeneration cycles and subsequent wastewater contact. The samples regenerated at 300°C and 400°C showed no ammonia uptake and, in fact, desorbed ammonia to the contacted wastewater.

X-ray diffraction analyses of Hector clinoptilolite samples were performed following the methods of Mumpton (1960) and Rooney and Kerr (1964). Analyses of virgin, exhausted, and thermally regenerated (500°C) Hector clinoptilolite samples indicate that there was essentially no structural degradation of the thermally regenerated clinoptilolite. Several X-ray diffraction spectra are shown in Fig. 3. From Fig. 3 it is evident that thermal regeneration results in the broadening and elimination of a number of diffraction spectral lines indicating the loss of adsorbed constituents, occluded water, and possibly some minimal structural alteration. A summary of "d" spacings and relative sharpness is listed in Table 5, along with the peak positions and relative intensities (I/I_0) of Hector clinoptilolite (Mumpton, 1960).

CONCLUSIONS

Samples of clinoptilolite obtained from various areas of the western United States were found to have different effective ammonia-exchange capacities. The sample with the greatest effective exchange capacity (meq/g), obtained from Malheur County, was 170% greater than the sample, obtained from Washoe County, which exhibited the lowest exchange capacity (meq/g).

Pretreatment of clinoptilolite samples, in an attempt to increase their effective ammonia exchange capacities, proved to be effective only for the Maricopa County sample, which exhibited a 100% increase in its ammonia-exchange capacity.

Results of bioregeneration of ammonia-exhausted clinoptilolite indicate that long aeration times are required to realize approximately 80% of the original exchange capacity. Therefore, until the bioregeneration process can be accelerated considerably, its application would be limited to regenerating clinoptilolite used to treat waste streams with intermittent flows.

Clinoptilolite thermally regenerated at 500°C retained almost 100% of its original effective exchange capacity.

ACKNOWLEDGMENTS

The data presented herein were obtained from the project "Nutrient Removal Using Existing Combined Sewer Overflow Treatment Facilities", which was jointly funded by the U.S. Environmental Protection Agency and Onondaga County, New York. Special thanks are extended to Mr. Richard I. Field, EPA Chief of the Storm and Combined Sewer Section, Edison, New Jersey, and Messrs. John J. Hennigan, Director, and John M. Karanik, Special Projects Manager, of the Onondaga County Department of Drainage and Sanitation.

REFERENCES

Breck, D. W. (1974) *Zeolite Molecular Sieves: Structure, Chemistry, and Uses*: John Wiley & Sons, New York, p. 204.

Koon, J. H. and Kaufman, W. J. (1975) Ammonia removal from municipal wastewaters by ion exchange: *J. Wat. Poll. Control*

Fed. **47**, 448-464.

Mumpton, F. A. (1960) Clinoptilolite redefined: *Amer. Mineral.* **45**, 351-369.

Norris, J. R. and Ribbons, D. W. (1969) *Methods in Microbiology*: Academic Press, New York, **3B**, 30.

Rooney, T. P. and Kerr, P. F. (1964) Clinoptilolite: a new occurrence in North Carolina phosphorite: *Science* **148**, 1453.

TABLE 5. X-ray diffraction data for virgin, spent, and regenerated clinoptilolite from Hector, California

Hector clinoptilolite (Mumpton, 1960) d(Å)	°2θ	I/I_o	Virgin clinoptilolite °2θ	Spent clinoptilolite °2θ	Regenerated clinoptilolite (Δ500°C) °2θ
9.00	9.82	10	10.0 (sharp)	10.0 (sharp)	10.15 (sharp)
7.94	11.15	4	11.5 (sharp)	11.45 (sharp)	11.25 (sharp)
5.91	14.99	1	13.2–13.8 (broad)	13.2–13.8 (broad)	13.0–13.7 (broad)
5.24	16.91	3	17.25 (broad)	17.2 (broad)	17–18 (broad)
			17.65 (doublet)	17.55 (doublet)	
4.48	19.82	2	19.35 (broad)	19.3	19.2–19.8 (broad)
				19.9	
				21.15	
3.96	22.45	10	22.7 (sharp)	22.6 (sharp)	22.6–23.0 (broad)
3.90	22.80	8	23.05	22.95 (sharp)	
3.73	23.85	1		23.8–24.3 (broad)	
3.46	25.75	2	25.35	25.35	
			26.0	25.9	26–27 (broad)
			26.3–25.6 (broad)	26.35 (broad)	
			26.9 (sharp)	26.9 (sharp)	
3.12	28.61	3	28.6 (sharp)	28.55 (sharp)	28.6 (sharp)
			29.3 (sharp)	29.2 (broad)	
2.97	30.09	5	29.8 (sharp)		
			30.3 (sharp)	30.2 (sharp)	30.1–30.6 (broad)
				31.1 (broad)	
2.82	31.73	3	32.25 (sharp)	32.1 (sharp)	32.0–32.5 (broad)
2.72	32.90	1	33.1 (sharp)	32.95 (sharp)	32.8–33.4 (broad)
2.68	33.43	1		33.6 (broad)	

SOLAR ENERGY APPLICATION OF NATURAL ZEOLITES*

D. I. TCHERNEV

Lincoln Laboratory, Massachusetts Institute of Technology, Lexington, Massachusetts 02173, U.S.A.

Abstract – The utilization of solar energy for cooling is usually achieved by means of sorption-refrigeration cycles. However, the conventional cycles using ammonia–water or lithium bromide–water solutions are inefficient because of the low solution temperatures obtainable with solar radiation and the high condenser temperatures required by air-cooled condensers. Zeolites provide a unique opportunity for a solid–gas adsorption cooling system because of their extremely nonlinear adsorption isotherms. We have demonstrated the feasibility of using a zeolite system to provide domestic hot water and space heating with overall efficiencies above 75% and space cooling with an overall efficiency above 50%. The system uses natural chabazite or clinoptilolite as the solid adsorber and water vapor as the working fluid. The operation of the system is described and the experimental results are discussed. Preliminary estimates indicate that such a system will be economically competitive.

INTRODUCTION

There are many different ways to obtain cooling from solar energy. For example, one can utilize photovoltaic solar cells to convert the solar energy to electricity and then use conventional refrigerators. Such a system, in addition to being expensive, will have a small overall efficiency of less than 15%. Another method is to convert the energy from the sun into mechanical form by means of one of the numerous cycles available and then to use the mechanical energy to drive the compressor in refrigeration equipment. However, the Carnot efficiency of all such cycles is low because of the low temperatures obtainable from the sun without extensive concentration and complicated sun-following techniques.

The simplest and potentially most efficient method therefore utilizes the well-known sorption-refrigeration cycle. Faraday was the first to discover that some gases, like ammonia, are sorbed by solids or liquids at low temperature and desorbed at high temperature. Numerous solid–gas sorption-refrigeration systems utilizing alkali-metal chlorides and ammonia gas have been constructed, mostly in the 1920s and 1930s in Europe. In the United States the majority of systems have utilized instead the variation with temperature of the solubility of a gas in a liquid, initially ammonia in water and later water vapor in lithium bromide. Many of the solid–gas and liquid–gas systems have recently been revived for the purpose of using solar energy for cooling.

In sorption-refrigeration cycles the liquid or solid sorbent is heated and desorbs the working gas. The partial pressure of the gas thereby increases, and it can be liquefied in a condenser, rejecting heat to the outside. The liquified gas is evaporated at lower temperature and partial pressure, while absorbing heat from the volume to be cooled. After the depleted sorbent is cooled down, it resorbs the evaporated gas at lower partial pressures. Thus, in each complete cycle heat is withdrawn by the evaporator from the refrigerated volume and rejected by the condenser to the outside world. Solid sorbents have to be cycled in time while liquid sorbents can be pumped through a heater and cooler, thereby providing a continuous cooling process.

Such sorption systems are used extensively for commercial air conditioning and refrigeration whenever an inexpensive supply of heat (usually steam) is available. They can achieve an overall efficiency (also called coefficient of performance) of 70–80% if the source of heat is at about 250–300°F and the condenser is water-cooled to below 85°F. Solar collectors, however, usually provide hot water at temperatures of less than 190°F. Furthermore, an air-cooled condenser is necessary for residential applications, and such a condenser can reach 120°F on days when the air temperature is 100°F. Under such conditions the efficiency of conventional sorption systems drops to below 30%. Since solar collectors are only 50% efficient at such operating temperatures, the overall efficiencies would be on the order of 10 to 15%.

* This work was supported by NSF-RANN.

OPERATING PRINCIPLE OF THE ZEOLITE SYSTEM

Zeolites provide a unique solution to the above problems of solar cooling because of their unusual sorption properties. Zeolites are capable of adsorbing large quantities of a variety of refrigerant gases, ranging from water vapor and ammonia to carbon oxides and freons. Interestingly enough, for most of these gases the amount sorbed is about the same – 30 wt.%. Since the heat of vaporization of water is the largest of any common refrigerant and about 10 times larger than that of the freons, the zeolite-water vapor combination can provide the most efficient system and requires the smallest quantity of zeolite for its operation. Zeolites also have the unique property, which is of critical importance for solar applications, that their adsorption isotherms have an extremely *nonlinear* pressure dependence.

In contrast with zeolites, other sorbents have almost *linear* pressure isotherms. For example, Fig. 1 shows a family of adsorption isotherms for water vapor on silica gel; these are similar to those for most solid-gas and liquid-gas systems. It is seen that the quantity of adsorbed water vapor is a monotonic, almost linear function of partial pressure. For this reason in the refrigeration cycle, from sorption at low temperature and low partial pressure to desorption at high temperature and high partial pressure, there is only a small change in the amount of adsorbed gas. This accounts for the low efficiency of solar cooling with conventional sorbents. Actually, under the extremely unfavorable conditions of low desorption temperature available from the sun and high condenser pressure for air-cooled condensers, the efficiency can even go to zero. This is also the case for the lithium bromide-water and ammonia-water systems.

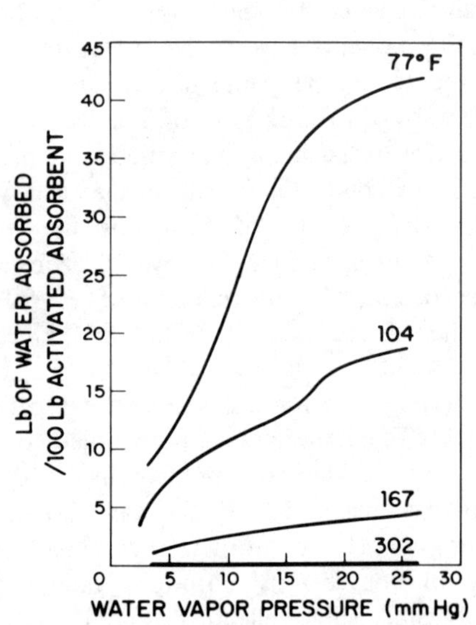

Fig. 1. Adsorption isotherms for water vapor on silica gel, showing quantity of water adsorbed as a function of the partial pressure of water vapor.

The contrasting behavior of zeolites is shown in Fig. 2, which gives typical isotherms for the sorption of water vapor on a zeolite. The extreme nonlinearity of the amount adsorbed as a function of partial pressure is obvious. The isotherms saturate at low partial pressure, after which the amount adsorbed is almost independent of pressure. Therefore, when the zeolite is at ambient temperature it can adsorb large quantities of water vapor even at low partial pressures. On the other hand, when the zeolite is heated it desorbs most of the water vapor even at high partial pressures, corresponding to high condenser temperature. Therefore, the difference in adsorbed gas between the high- and low-temperature states is large and depends only slightly on condenser pressure. This makes it possible to achieve high efficiencies under conditions typical of solar applications. Thus, the shape of the adsorption isotherms is critical for the superiority of the zeolite system over conventional sorption-refrigeration cycles.

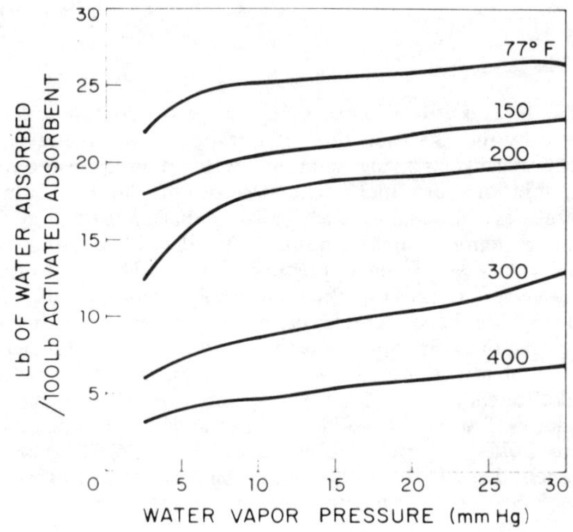

Fig. 2. Adsorption isotherms for water vapor on a zeolite.

To describe the situation in another way, one can compare the thermal activation of the different processes involved. The solubility of ammonia in water, the solubility of water vapor in lithium bromide, and the adsorption of water vapor on silica gel or activated alumina all depend exponentially on $\Delta H/RT$, where ΔH is the heat of solution or adsorption and T is the absolute temperature (i.e. they obey the Arrhenius equation). Adsorption on zeolites, on the other hand, has been shown by Dubinin (1971) to depend exponentially on at least the second, and as high as the fifth, power of $\Delta H/RT$. It is this extreme nonlinearity of thermal activation that makes zeolites so well suited for cooling.

Figure 3 illustrates the principle of operation of a zeolite solar-cooling system. The zeolite is sealed in an airtight container that is irradiated by the sun. During the day cycle, shown on the left side of Fig. 3, the zeolite and its container are heated to a maximum temperature of 250°F. At about 100°F water vapor starts desorbing from the zeolite, and its partial pressure begins to rise. When the pressure reaches the value determined by the condenser temperature, for example 1 psia for 100°F, the vapor begins to liquefy, heat is rejected to the outside, and the liquid water is stored in a storage tank. During the night cycle, on the right side of Fig. 3, the zeolite is cooled by convection cooling to ambient temperature and is ready to adsorb water vapor even at low partial pressures. Liquid water from the storage tank is introduced into the evaporator, where it absorbs heat from the space to be cooled and is converted into water vapor. If the partial pressure can be maintained at 0.1 psia, the water in the evaporator will boil at 35°F. The function of the zeolite is to adsorb the water vapor produced by the evaporator, maintaining the partial pressure below 0.1 psia and rejecting the heat of adsorption to the atmosphere. At the end of the night cycle the zeolite is loaded with all the water it will adsorb at 0.1 psia and is ready for the beginning of a new day cycle.

A zeolite system, like any refrigeration system, has to be free of air. Therefore the zeolite is placed in hermetically sealed metal panels. These panels, which are painted black for maximum solar absorption, are connected through ordinary plumbing to the condenser, water-storage tank, and evaporator. After the zeolite is charged with water vapor, the whole system is outgassed, evacuated, and sealed off.

In the operation of the system in Fig. 3, cooling is produced during the night. Although this is quite satisfactory for many applications, such as the production of ice and the refrigeration of food, in the air conditioning of buildings the demand for cooling occurs mostly during the day. For this reason it is necessary to provide some type of storage, a feature typical of all solar systems.

One storage method is shown in Fig. 4. In this system cooling is produced during the day by operating the evaporator, and the water vapor is then held in a buffer storage until evening, when the zeolite is cool enough to be recharged. The buffer storage in principle could be just an empty tank with sufficient volume to keep the water partial pressure below 0.1 psia until the zeolite can begin adsorption. For a system of large cooling capacity,

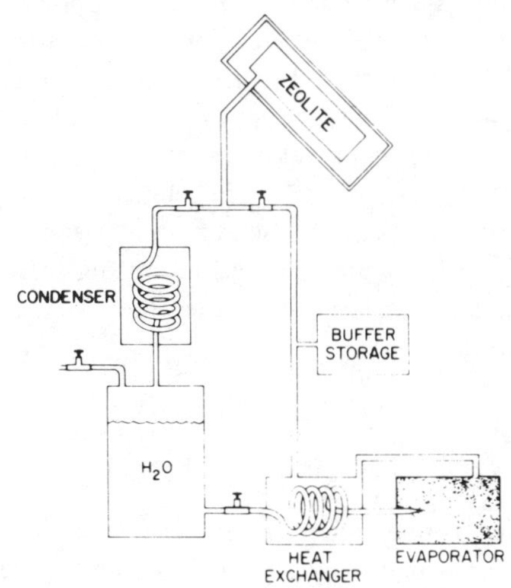

Fig. 4. Schematic diagram of zeolite system with buffer storage for water vapor.

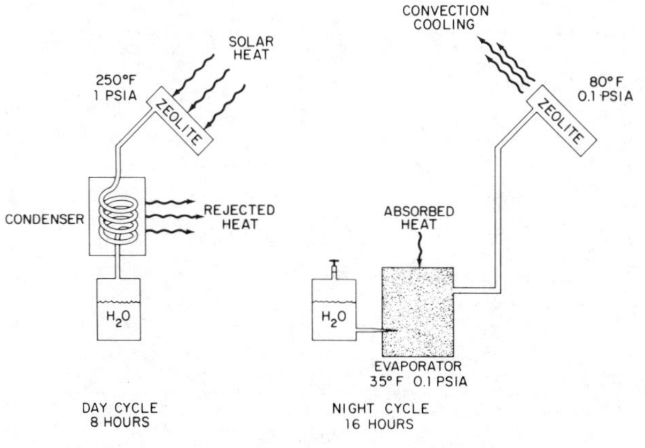

Fig. 3. Schematic diagram illustrating day and night cycles for zeolite–water system.

however, the size required for such an empty tank would be impractical. In order to reduce its size, the tank can be filled with a sorbent, such as silica gel, whose affinity for water vapor is less than the zeolite's. During the day, because of the temperature difference between the two sorbents, the water vapor is adsorbed by the silica gel, but at night, when the zeolite cools down, the vapor is transferred from the storage tank to the zeolite.

Another method for providing buffer storage is especially suitable for complete systems utilizing solar energy for both heating and cooling. One such system is shown in Fig. 5. Here the condenser and evaporator are combined into a single unit that is cooled by an external water loop. During the day water vapor desorbed from the solar-heated zeolite is condensed in this unit, and the liquid water is stored below the condenser in the usual condensate storage tank until evening. The heat of condensation is rejected to the external water loop. This heat can be used for providing domestic hot water and during the winter season for space heating as well. During the heating season hot water is stored during the day to provide heat during the night, as in any conventional solar system. Whenever there is a demand for heat, hot water from the storage tank is circulated through a coil located in the air ducts of the forced air-system, and the heated air is distributed throughout the building. If too much heat is liberated in the condenser, as for example in the spring and summer, the excess can be rejected to the outside by an air-cooled coil.

During the night, water from the condensate storage tank is circulated back into the condenser–evaporator unit, where it evaporates from the same surfaces on which it originally condensed. The vapor is then adsorbed on the cool zeolite. Since evaporation takes place at night, the system does not require buffer storage for water vapor. The external water loop provides the necessary heat of vaporization, producing chilled water for use in air conditioning. The chilled water can be stored during the summer season in the same storage tank used for hot water during the heating season. The changeover from one season to the other is achieved by simple valving.

EXPERIMENTAL RESULTS

In order to demonstrate the efficiency of zeolite systems for solar heating and cooling, we have constructed a number of 1- and 2-ft² test panels from various metals, including brass, copper, aluminium, and tin-plated steel. (We expect panels up to 16 ft² in area to be used in full-scale systems.) For condensation and evaporation pressures of 1 and 0.1 psia, respectively, for most zeolites the differential water loading is about 10% by weight between ambient temperature and 250°F, which is about the maximum temperature attainable with present flat-plate solar collectors. Since the average daily amount of incident solar energy is sufficient to cause the desorption of about 1 lb of water per square foot, the panels were designed to have a zeolite capacity of 10 lb/ft², requiring a depth of 2½ inches. The upper surface of each panel is painted black, and a single pipe is provided at the side wall to permit the water vapor to enter and leave. Each panel was leak-tested prior to use.

Some initial tests were conducted with various synthetic zeolites, but the efficiencies obtained were quite low. The maximum theoretical efficiency for cooling with a zeolite is roughly equal to the heat of vaporization of water – about 1000 BTU/lb – divided by the heat of adsorption of water on the zeolite. Since synthetic zeolites used for drying gases have heats of adsorption of about 1800 BTU/lb of water, their maximum theoretical efficiency is only about 55%. On the other hand, natural zeolites such as mordenite, with heats of adsorption of only 1200 BTU/lb of water, have maximum theoretical efficiencies of about 80%.

Most of our tests were performed with two natural zeolites obtained from Letcher and Associates of Lancaster, California: Zeosorb 3.5 A, a clinoptilolite from Horseshoe Dam, Arizona, and Zeosorb 5.0 A, a chabazite/erionite from Bowie, Arizona. Figure 6 shows

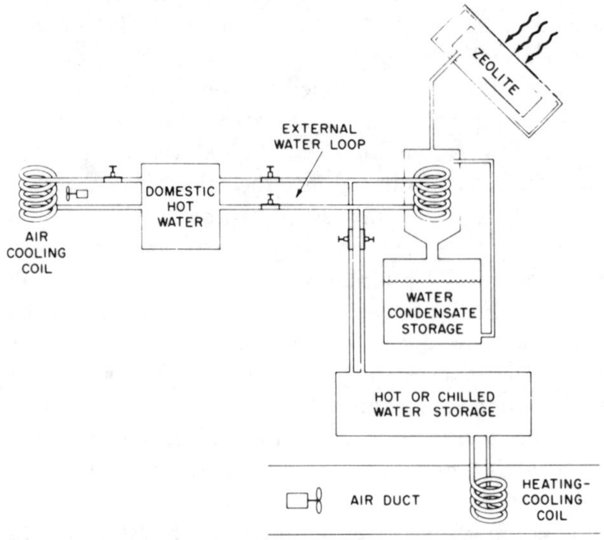

Fig. 5. Schematic diagram of zeolite combined heating/cooling system.

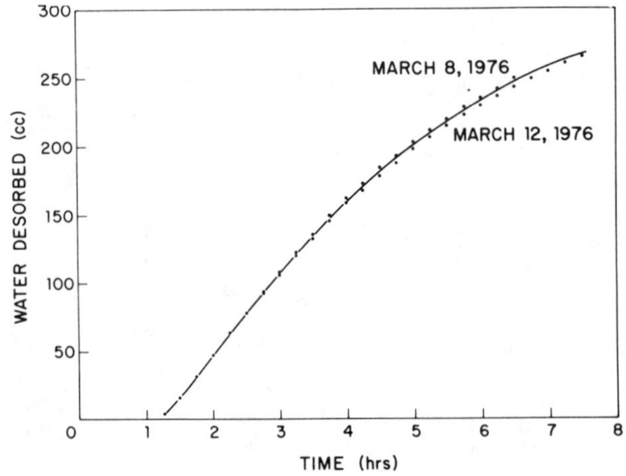

Fig. 6. Amount of water desorbed from panel containing 8 lb of zeolite for input power density of 1 kW/m².

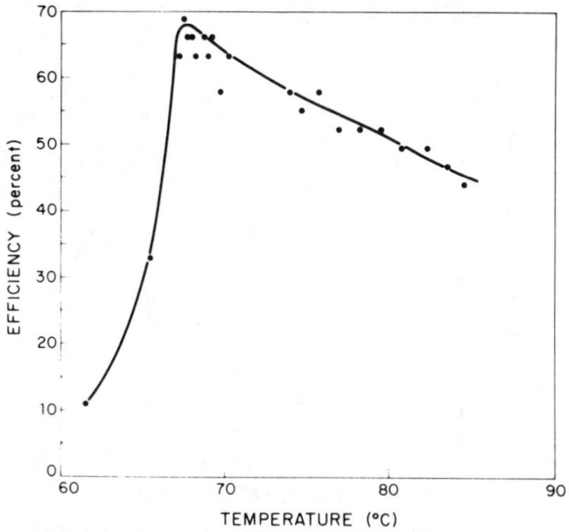

Fig. 7. Heating efficiency as a function of panel temperature for Zeosorb 5.0 A with input power density of 1 kW/m².

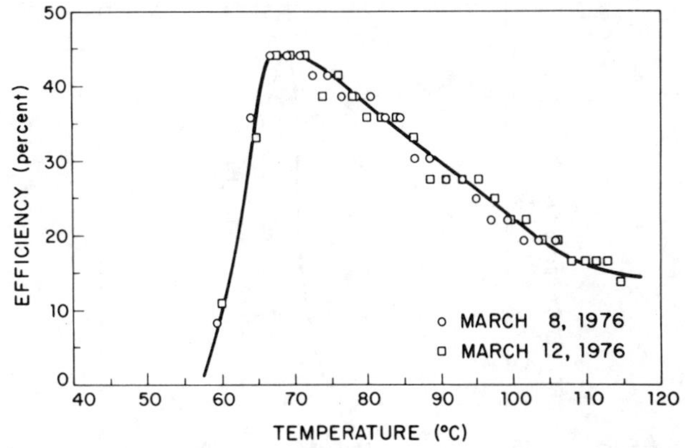

Fig. 8. Cooling efficiency for Zeosorb 5.0 A with input power density of 1 kW/m².

the weight of water desorbed (and condensed) from a 1-ft² aluminum panel containing about 8 lb of Zeosorb 5.0 A as a function of time for an initial zeolite temperature of 70°F(20°C), a condenser temperature of 120°F(50°C), and a simulated solar input of 1 kW/m² (or 320 BTU/ft²-hr). The experimental points are from two runs made 4 days apart. It is seen that in 7½ hours over 8% by weight of water is desorbed and condensed.

Figure 7 shows the heating efficiency for Zeosorb 5.0 A as a function of panel temperature for a simulated solar input of 1 kW/m² and a condenser temperature of 140°F(60°C). The measured engineering efficiency, defined as the ratio of the heat of condensation tó the radiant energy incident on the panel, is 68% for a panel temperature of 160°F. In these experiments the zeolite is loaded with water at a vapor pressure of 15 psia, since the evaporator can be bypassed during the heating season by pumping liquid water from the condensate storage tank directly into the zeolite panels.

In the cooling experiments the zeolite is loaded at room temperature with water vapor at a partial pressure of 0.1 psia, corresponding to an evaporator temperature of 35°F. Figure 8 shows the efficiency of Zeosorb 5.0 A as a function of panel temperature for a simulated solar input of 1 kW/m² and a condenser temperature of 120°F(50°C). The value of 44% obtained when the panel is heated to 160°F is two to three times larger than the efficiencies possible with present lithium bromide-water systems under the same conditions.

Outdoor tests of cooling and heating with solar energy were conducted in November and February on the roof of our laboratory in Boston. Figures 9 and 10 show the heating efficiencies obtained with Zeosorb 3.5 A and 5.0 A, respectively, as a function of panel temperature for a condenser temperature of 140°F. The overall heating efficiency of 70% obtained with Zeosorb 5.0 A in February, without selective absorbing coatings for the panels, is better than the values so far reported for conventional flat-plate collectors.

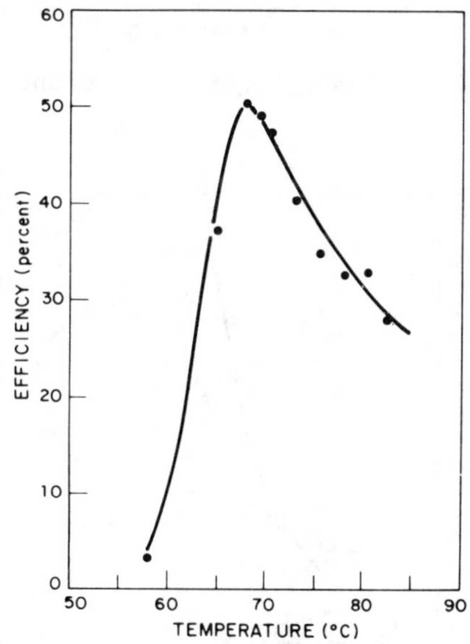

Fig. 9. Heating efficiency for Zeosorb 3.5 A heated by the sun.

Finally, in Figs. 11 and 12 the cooling efficiencies obtained with solar energy are plotted as a function of panel temperature for Zeosorb 3.5 A and 5.0 A, respectively, for a condenser temperature of 120°F. Again the chabazite-rich 5.0 A shows superior performance, achieving an overall engineering efficiency of 42%, including all losses. Even higher efficiencies would be obtained for lower condenser temperatures.

For tests on a larger scale, a 1-square-yard system has been constructed from nine panels of 1 ft² each. The system is insulated with 2-inch polyurethane foam and

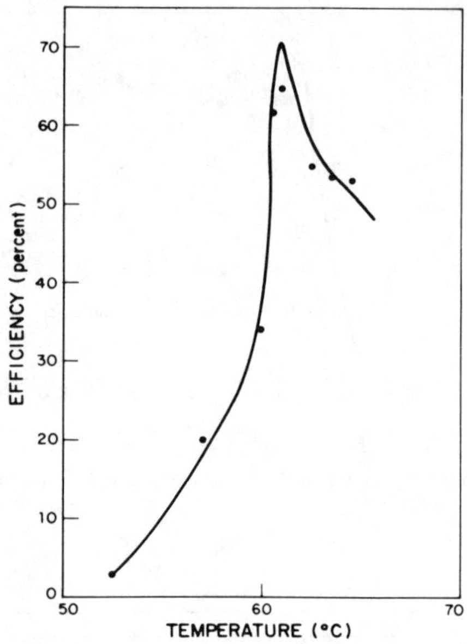

Fig. 10. Heating efficiency for Zeosorb 5.0 A heated by the sun.

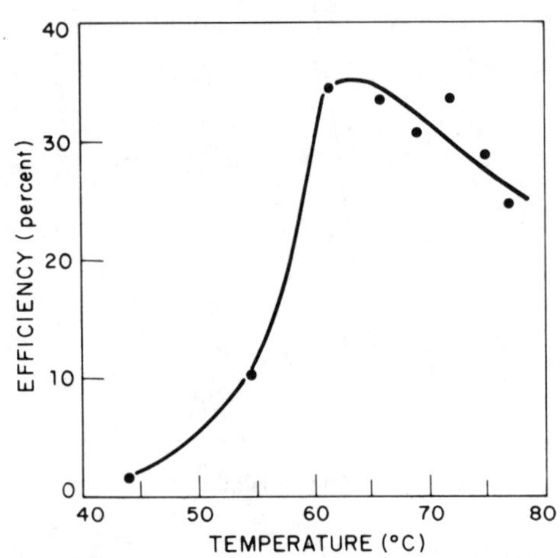

Fig. 11. Cooling efficiency for Zeosorb 3.5 A heated by the sun.

has double-layer glazing. A view of the partially assembled system is shown on Fig. 13. With a solar input of 20,000 BTU/day it produces 9000 BTU/day cooling. The only moving part is the fan used to blow the cold air out of the evaporator. The system has been operated for one summer season without any difficulties, and we expect operating lifetimes of over 20 years for full-scale systems.

CONCLUSIONS

We have demonstrated that natural zeolites can be used successfully for solar heating and cooling. The

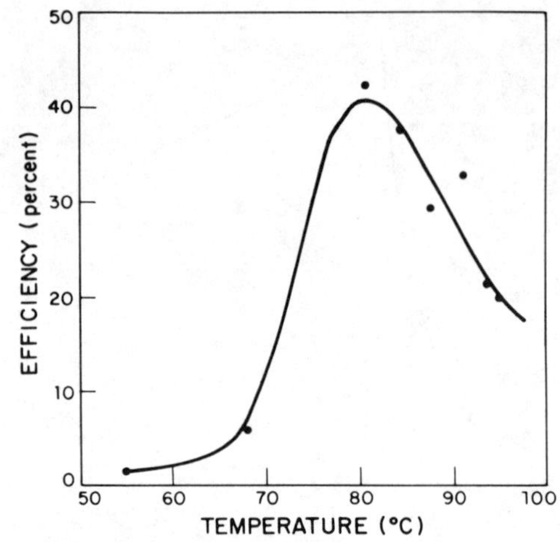

Fig. 12. Cooling efficiency for Zeosorb 5.0 A heated by the sun.

efficiencies already obtained are at least as good as those of any solar systems now available. They can be increased further by using zeolites that have smaller heats of adsorption, such as mordenite, particularly if their adsorption isotherms more closely approach a step-function.

It will require about 1 ton of zeolite, distributed over 200 ft^2, to produce 1 ton of air conditioning. For a typical house with a 3.5-ton air conditioner, less than half of the roof area will have to be covered with zeolite panels. We estimate that in large-scale production these panels will cost \$1 to \$2 more per square foot than the conventional flat-plate collectors used for solar heating. However, with the addition of a condenser–evaporator unit costing only a few hundred dollars a zeolite system will provide cooling as well as heat and hot water. Because of this additional capability, the year-round reduction in energy costs will permit the repayment of the initial investment on the zeolite system in only 10 to 12 years, rather than the 25 to 30 years typically projected for solar-heated homes. We believe that with a small additional development effort, economically feasible zeolite systems could be available in only a few years.

ACKNOWLEDGMENTS

The author is grateful to W. J. LaFleur, D. F. Kolesar, and R. E. Fahey for their help in the construction and testing of the zeolite system.

REFERENCE

Dubinin, M. M. and Astakhov, V. A. (1971) Description of adsorption equilibria of vapors on zeolites over wide ranges of temperature and pressure: In *Molecular Sieve Zeolites–II*, American Chemical Society, Washington, D.C., pp. 69-85.

Fig. 13. Square-yard zeolite system, consisting of nine panels each 1 foot square.

CLEANSING ACTION OF NATURAL ZEOLITES IN DETERGENTS

H. G. SMOLKA and M. J. SCHWUGER

Henkel & Cie, GmbH, Düsseldorf, Germany

Abstract – Recently there have been intensive discussions concerning the application of water-insoluble ion-exchangers in the washing process to soften water, to adsorb dye-stuffs and pigment dirt, and to act as a substrate for the deposition of sparingly soluble salts. The use of such substances is advantageous, not only from technical, hygienic, toxicological, and ecological viewpoints, but also from the economic point of view. Previous work was mainly concerned with synthetic, amorphous, and crystalline Na–Al-silicates, especially those of the NaA- and NaX-types. The possibilities of using natural zeolites in comparison to the synthetic types is discussed in the present publication.

The following factors are decisive for the application of natural zeolites: (1) particle-size distribution, (2) ion-exchange capacity and kinetics, and (3) purity. A suitable particle size ($<50\ \mu m$) which is not prone to deposit on textiles and in washing machines can be attained by grinding.

As many naturally occurring zeolites contain significant amounts of bivalent ions, such as calcium and barium, it is necessary that they be first converted into the Na-form before use. Only those natural zeolites which possess a pore radius, after conversion to the Na-form, which is larger than the effective hydrated radii of the Ca^{+2} and Mg^{+2} ions to be removed during washing, are suitable. The following natural zeolites have been examined: faujasite, phillipsite, mordenite, analcime, harmotome, desmine (stilbite), natrolite, chabazite, and thomsonite. The best results have been obtained with faujasite.

It is necessary for the stability of the washing powder that the concentration of heavy metals not exceed a certain limit (e.g., 100 ppm Fe). This is because they catalyze the decomposition of certain substances present in the detergent (e.g. perborate in European detergents). For American washing conditions (detergents contain no perborate) the amount of heavy metals is not so critical. As most of the heavy metal ions are firmly imbedded in the crystal lattice and occur exclusively in the interior of the crystals, it can be expected that only a small fraction of them will be catalytically active. Because natural zeolites usually contain heavy metal ions in excess of 100 ppm, masking or purification procedures are necessary. The following masking procedures have been investigated: (1) coating with pure inorganic silicates, for example, $MgSiO_3$ and amorphous Al-silicates, and (2) complexing the heavy metal ions with agents such as EDTA.

INTRODUCTION

In the last 2–3 years the use of synthetic, crystalline Na–Al-silicates for washing and cleaning processes has been discussed (Schwuger *et al.*, 1974; Smolka and Schwuger, 1974). Results up to the present indicate satisfactory technical, hygienic, toxicological, and ecological properties (Berth *et al.*, 1975). These Na–Al-silicates contributed to the washing process by virtue of their ion-exchange capacity for multivalent cations (mainly calcium), by their ability to adsorb various soiling constituents (e.g., special dye-stuffs and pigments), and their effectiveness as substrates for the deposition of sparingly soluble salts (Schwuger and Smolka, 1976). The result is that such insoluble silicates not only show favorable effects during the primary process of soil-dissolution during washing, but also reduce undesirable greying and incrustation of textiles during repeated washings.

This work describes the special problems associated with the use of natural zeolites, as opposed to the use of synthetic zeolites.

RESULTS AND DISCUSSION

Resulting from work carried out with synthetic Na–Al-silicates, it is known that ion-exchange capacity, ion-exchange kinetics, and particle-size are the most important factors for their use in washing. The well-controlled product quality for synthetic zeolites as far as impurities are concerned is not present in the case of natural zeolites, which can show large variations, depending upon their source of origin. Therefore, in the case of natural zeolites, this problem must be given special attention.

Particle-size

In order to avoid deposition of water-insoluble particles on the textiles, an unhindered flow of the particles through the textile-material must be obtained.

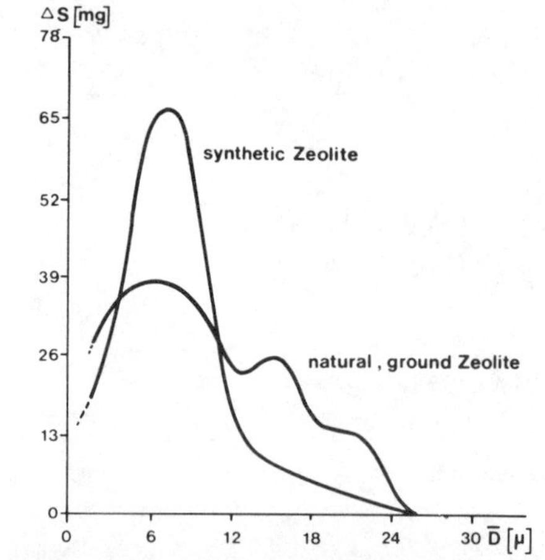

This means basically that the particle-size of the water-insoluble products must be smaller than the mesh-size of the textile. In Fig. 1 the particle size distribution of a synthetically prepared zeolite and a naturally occurring ground zeolite (desmine) is compared.

In comparison to the synthetic zeolites, whose particle size and form can be somewhat controlled by

Fig. 1. Particle size distribution of zeolites.

Examination method:	Sedimentation analysis.
Apparatus:	Sedimentation balance (Firm Cahn).
Temperature:	25°C.

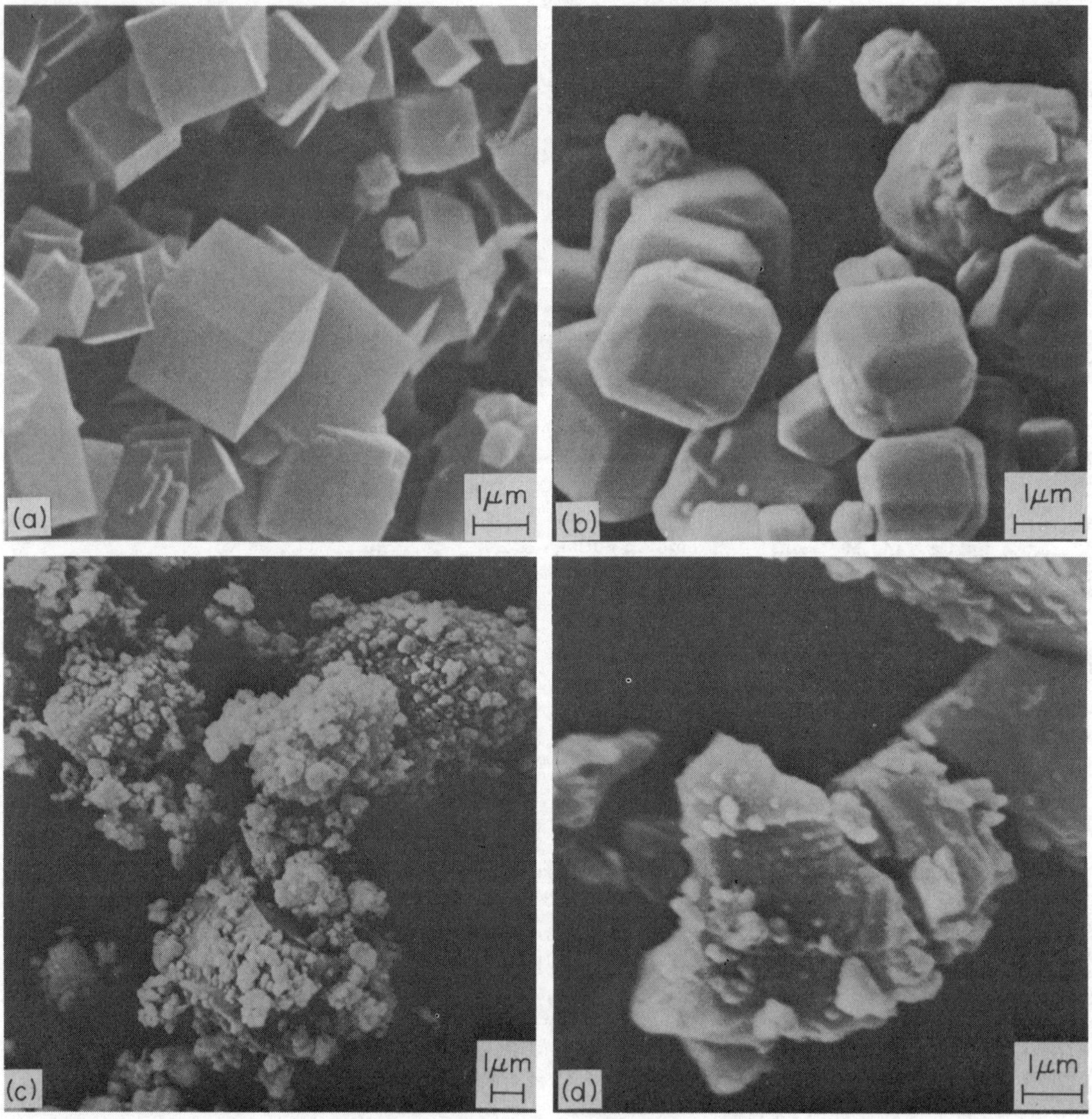

Fig. 2. Scanning electron micrographs of: (a) Synthetic zeolite (Type A) sharp-edged. (b) Synthetic zeolite (Type A) rounded-off. (c) Natural faujasite-ground. (d) Natural analcime-ground.

the conditions of synthesis, natural zeolites possess much larger particles. In addition, the crystallites can be admixed with foreign material, so that suitable products must be obtained by means of grinding and separation processes. As a result of these processes, natural products of a comparable particle size differ from synthetic products in terms of their morphology. Finely ground natural zeolites mainly possess an irregular particle shape (shell-like fracture-surfaces) as shown in Figs. 2c and 2d. Specifically prepared synthetic zeolites of type A with rounded-off corners and edges and with sharp edges are shown in Figs. 2a and 2b, respectively. Independent of their chemical composition or crystal structure, all ground natural zeolites show hardly any differences in their appearance after the grinding process.

In the case of a favorable ratio of particle size to mesh size of the textile, no deposition on the textiles has been observed when tested under European washing conditions, irrespective of all particle shapes. Under critical conditions, i.e. when the particle size approaches the mesh size of the textiles, the tendency to deposition decreases in the following order:

sharp-edged synthesized zeolites >
ground natural zeolites >
round-off synthetic zeolites

In the formulation of a detergent, the morphology of the particles plays a more important role, especially when considered in relation to the abrasive attack on various machine parts. With reference to this problem, ground natural products are less abrasive than rounded-off, synthetic zeolites and especially less abrasive than the sharp-edged type (Table 1).

Ion exchange and the washing process

The two important components of a modern detergent are surfactants and complexing-agents (mainly sodium triphosphate).* Earlier studies by the present authors carried out with synthetic, crystalline Na–Al-silicates have shown that a part of the complexing agent can be replaced. The remaining part of the water-soluble sequestering agent has the effect of dissolving the polyvalent metal ions by means of specific adsorption and desorption processes on the fiber by the complexing agent and dirt. This results in a loosening of the soiling material and an increase in its electrical charge, thus resulting in an easing of the process of soil-loosening (Jakobi and Schwuger, 1975). The remaining complexing agent in the detergent therefore acts as a carrier for the transport of multivalent ions from the soiling material and fibers into the aqueous phase. The Na–Al-silicate used as replacement for sodium triphosphate aids in the elimination of multivalent ions, especially Ca^{++}, from the washing liquid, mainly by ion exchange.

The ion exchange depends on both the chemical

TABLE 1. Influence of morphology upon abrasiveness

Product appearance	Average particle-radii (μm)	Abrasiveness (mg)			
		Steel (V 2A)	Zinc	Polypropylene	Plexiglas
Sharp-edged zeolite A	3.7	2.0	31.9	1.5	29.9
Rounded-off zeolite A	3.7	1.3	19.3	0.8	17.9
Shell-like (ground) faujasite†	5.4	0.5	4.7	0.3	5.2

†Natural Faujasite: Na-enriched faujasite from the Kaiserstuhl ground to powder with particle-size distribution similar to that for desmine in Fig. 1.

Test-conditions: Test apparatus: Taber Model 503 Abraser (Firm Table Instr. Corp.).

Bearing weight:	1000 g/roll
Test-time:	41 min, 40 sec
Speed:	60 rpm
Concentration:	31 (%)
Temperature:	295 ± 2 (°K)

†A comprehensive review concerning the mechanism of action of water-soluble complexing agents in the washing-process is to be found in Jakobi and Schwuger (1975).

composition and the cavities present in the crystal structure. Naturally occurring zeolites show marked variations in their ratio $Me_{2/n}O \cdot Al_2O_3 \cdot SiO_2$. As shown in Fig. 3, in which synthetic zeolite A and hydrosodalite are compared with different natural zeolites; this ratio affects the ion-exchange capacity. The maximum theoretical values for Ca-exchange (continuous line) are compared to the measured values obtained for several naturally occurring zeolites. The large differences between the calculated and the measured values can be traced back to calculated and the measured values can be traced back to differences in the cavities of the crystallites. For maximum exchange, the porous structure must be so constructed that unhindered diffusion of the Ca-ions into the cavity structures can occur. An especially favorable pore structure is found in faujasite, which has a ratio of the diameter of its entry pores to the diameter of hydrated Ca-ions significantly >1 (Wolf and Fürtig, 1975). In Table 2 a survey of the data important for the ion-exchange for some Na–Al-silicates is given. Products with relatively small pores, e.g., sodalite, show the largest deviations from the theoretical values for ion-exchange capacity in spite of their favorable chemical composition. All the Na-forms of naturally occurring zeolites examined lie between these two extremes. Also of importance for ion-exchange kinetics and ion-exchange capacity are the symmetry of the pore-system and its uniformity. This is exemplified by means of a comparison between faujasite and zeolite A (Fig. 4). For these types of zeolites, the

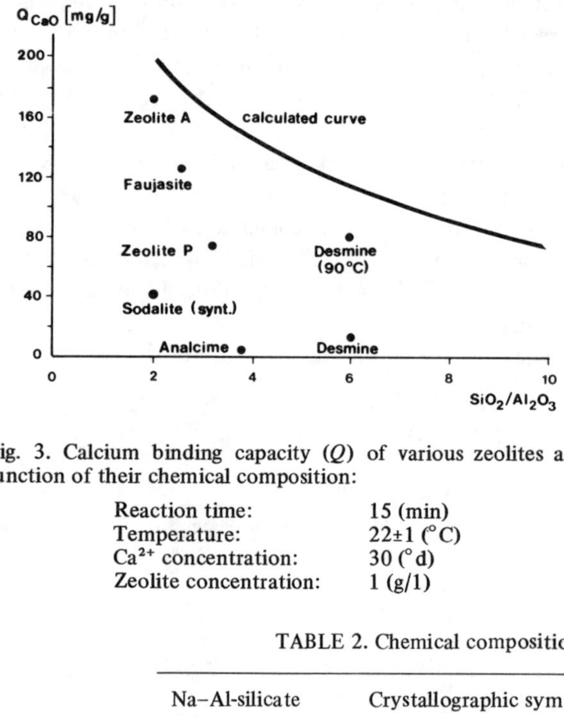

Fig. 3. Calcium binding capacity (Q) of various zeolites as a function of their chemical composition:

Reaction time:	15 (min)
Temperature:	22±1 (°C)
Ca^{2+} concentration:	30 (°d)
Zeolite concentration:	1 (g/l)

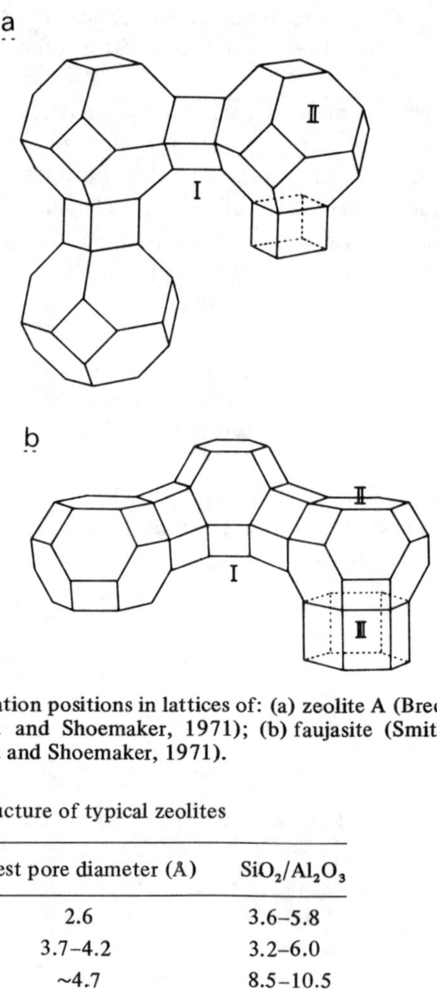

Fig. 4. Cation positions in lattices of: (a) zeolite A (Breck, 1964; Broussard and Shoemaker, 1971); (b) faujasite (Smith, 1971; Broussard and Shoemaker, 1971).

TABLE 2. Chemical composition and pore structure of typical zeolites

Na–Al-silicate	Crystallographic symmetry	Largest pore diameter (Å)	SiO_2/Al_2O_3
Analcime	cubic	2.6	3.6–5.8
Chabazite	rhombic	3.7–4.2	3.2–6.0
Clinoptilolite	monoclinic	~4.7	8.5–10.5
Desmine (stilbite)	monoclinic	4.1–6.2	3.0–7.7
Erionite	hexahedral	3.6–4.8	5.8–7.4
Faujasite	cubic	7.4	4.2–4.6
Harmotome	monoclinic	4.2–4.4	4.6–5.0
Mordenite	orthorhombic	6.7–7.0	8.2–10.0
Natrolite	orthorhombic	2.6–3.9	2.8–3.2
Phillipsite	orthorhombic	4.2–4.4	2.8–4.8
Sodalite	cubic	2.2	2.0
Zeolite A	cubic	4.1	2.0–2.5

exact positions of the cations in the lattice have been intensively studied and described in detail (Breck, 1964; Lechert, 1973; Smith, 1971). Whereas all the positions for exchange in zeolite A are accessible (uniform pore system of 4.2 Å), position III in the faujasite structure is not so readily accessible (pore system with 7.4 Å channels, as well as a pore system with 2.2 Å channels). For the majority of zeolites in the literature similar to faujasite several values for their pore-openings can be found (Barrer, 1973). Consequently, the majority of natural zeolites possess pore systems which are only partially accessible, thus resulting in unfavorable consequences for the ion-exchange capacity and, partially, for the kinetics.

These theoretical predictions are confirmed by practical washing tests. In Fig. 5 the washing effectiveness of various zeolites in water with a hardness of 16°d is given as a function of the zeolite concentration. It can be seen that the best results are obtained with zeolite A and the least favorable with analcime. Although only a minimal ion-exchange occurs for calcium in the case of sodalite and desmine, a small but measurable increase in the cleansing action of these substances can be observed at higher concentrations.

The positive contribution is to be seen in the context of the properties mentioned above concerning heterogeneous substances as substrates for sparingly soluble salts in competition with the textile. This is especially noticeable in the case of the stronger alkaline sodalite when compared to analcime. The result indicates that various solids can achieve a positive, but low washing effect, without possessing a high degree of ion-exchange. Independent of ion-exchange processes, these heterogeneous substances function as substrates for the deposition of sparingly soluble salts (e.g., Ca-phosphate) in competition with the textile fiber. Thus, in the case of over-stoichiometric ratios of Ca^{+2}/phosphate, the phosphate salts become sparingly soluble leading to depositions. This can occur especially during the dilution that occurs during rinsing with hard water. Under such conditions the presence of water-insoluble zeolites is favorable.

However, the washing-effectiveness of zeolites in combination with other constituents of a detergent is of importance. The additional increase of the cleansing effects by means of combination with water-soluble complexing-agents and surfactants should also be mentioned. Specifically, it can be shown that relatively small amounts of water-soluble complexing agents accelerate the loosening of soiling by means of adsorption and desorption processes on the textile fibers (Schwuger and Smolka, 1976). Furthermore, these substances functioned as quasi-carriers for Ca and other polyvalent ions from the soiled article to the aqueous phase. The limited increase of the washing-effect during a 15–45-minute washing process can only be reached in conjunction with a high ion exchange of the zeolite. This is clearly shown in Fig. 6 using analcime and zeolite A. The STP–zeolite A mixture is to be contrasted with the significantly smaller increase in the washing effectiveness

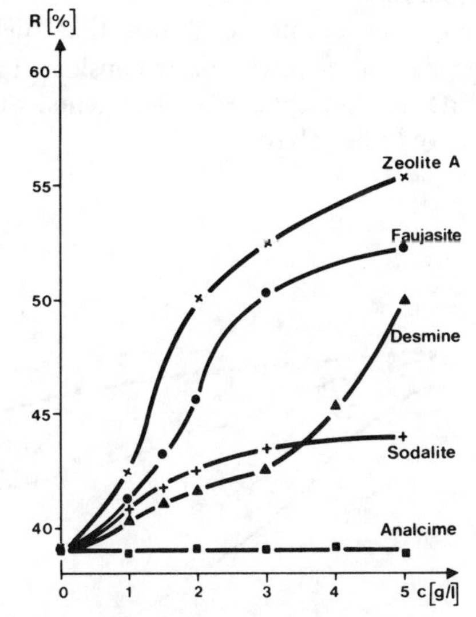

Fig. 5. Comparison of the washing effectiveness of various zeolites.

Apparatus:	Launder-o-meter.
Water hardness:	16 (°d).
Temperature:	95 (°C).
Washing-time:	30 (min) with heating.
Fabric:	non-finished cotton.

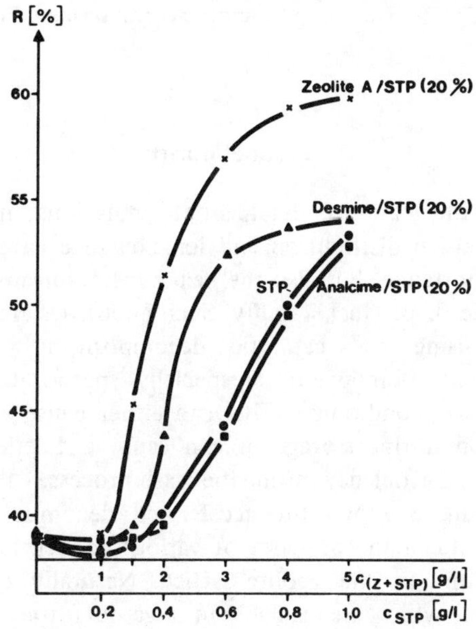

Fig. 6. Performance of zeolite/STP-mixtures. Experimental conditions: see Fig. 5.

of the analcime–STP mixture. The washing effectiveness of the analcime–STP mixture is identical with that of the separately investigated pure STP. In the case of good ion-exchange properties of all components together is greater than the sum of the individual contribution of the ion exchanger and the STP.

This increase of the washing effect by addition of zeolite to STP is shown in Fig. 7 as a function of concentration. The larger the proportion of STP in the mixture, the larger the contribution of STP becomes to the washing efficiency, so that at high STP contents, ΔR approaches the value 0. For these reasons, the curves pass through a maximum, which, with increasing concentration of STP in the mixture, largely takes over the washing efficiency.

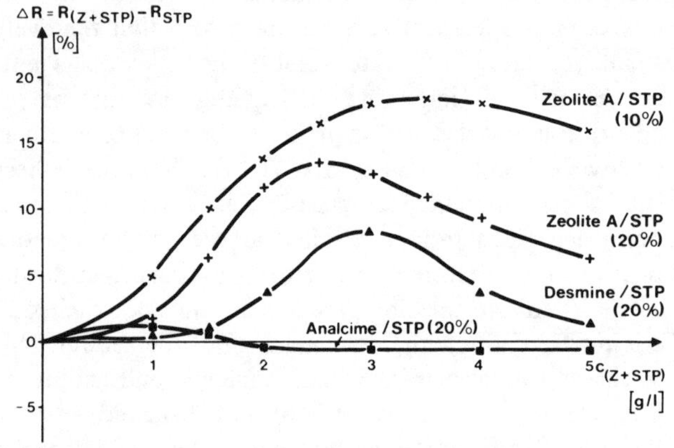

Fig. 7. Increase in the performance in zeolite/STP mixtures: $\Delta R = R(Z/\mathrm{STP}) - R(\mathrm{STP})$. Experimental conditions: see Fig. 5.

Product purity

In the case of detergent formulations, not only particle-size distribution and ion-exchange capacity are of importance, but also the heavy metal-ion impurities, because of product stability. Such impurities are capable of causing the catalytic decomposition of other detergent components (especially perborate under European conditions). This can either cause a decomposition during storage or can cause a drastic loss of bleaching efficiency during the wash process.

Figure 8 shows the accelerated decomposition of perborate in the presence of various concentrations of iron ions in the zeolite lattice. Naturally occurring zeolites, which are available in large quantities, possess, in general, a much higher concentration of heavy metal ions than do the synthetic substances which are prepared from sodium silicate and sodium aluminate solutions. The destabilization of the perborate is so strong (Fig. 8)

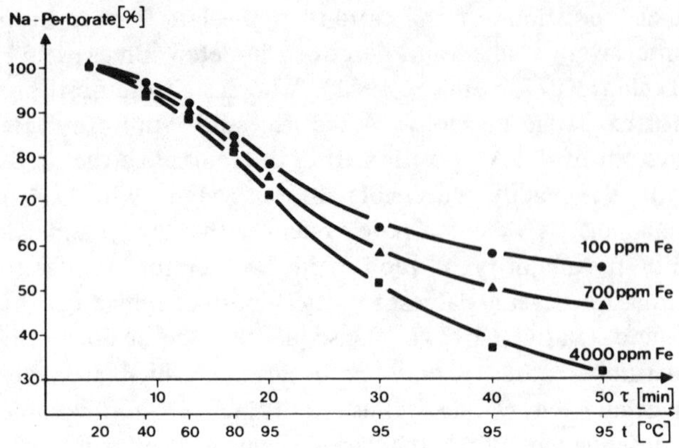

Fig. 8. Stability of Na-perborate in the presence of zeolite containing various iron concentrations.

that the use of such Fe-containing zeolites is not possible without the use of either special washing procedures for heavy metals, or the use of stabilizers for perborate. In the case of synthetic Na–Al-silicates, small additions of magnesium silicate and EDTA, as already present in modern detergents, are sufficient.

Comparable, high perborate stabilities (Fig. 9) can be obtained with high Fe-content zeolites by mixing them with 5–10% pure, amorphous silicates (e.g., magnesium silicate) in combination with specific complexing agents for heavy metals, such as EDTA. The importance of the degree of purity of the products under European conditions is not so critical for American ones, where the bleaching process is carried out separately to the washing process.

Summing up, it can be shown that high Ca-ion exchanging natural zeolites can be considered as partial replacements for phosphate in detergents. Still larger problems have to be solved.

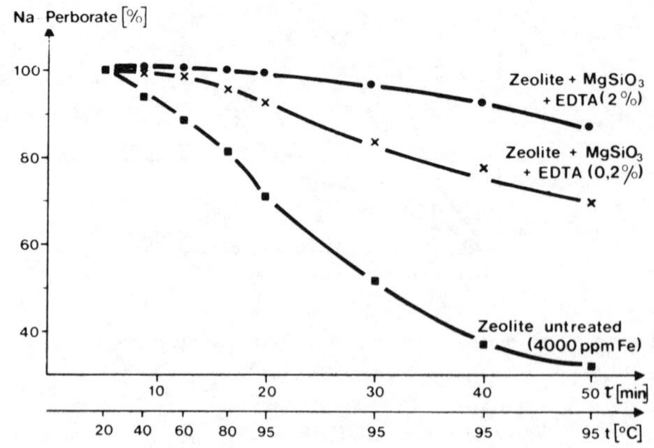

Fig. 9. Stability of Na-perborate in the presence of variously treated zeolites containing large amounts of iron.

REFERENCES

Barrer, R. M. (1973) Clathration, trapping, and zeolitic sorption: *Adv. Chem. Ser.* **121**, 1-28.

Berth, P., Jakobi, G., Schmadel, E., Schwuger, M. J. and Krauch, C. H. (1975) Möglichkeiten und grenzen des Ersatzes von Phophat in Waschmittein: *Angewandte Chemie* **87**, 115-123.

Breck, D. W. (1964) Crystalline molecular sieves: *J. Chem. Educ.* **41**, 678-689.

Broussard, L. and Shoemaker, H. S. (1971) The structure of synthetic molecular sieves: *J. Amer. Chem. Soc.* **82**, 1041-1051.

Jakobi, G. and Schwuger, M. J. (1975) Die Bedeutung von Komplexbildnern und Ionenaustauschern für den Waschprozeb: *Chemiker Zeitung* **99**, 182-193.

Lechert, H. (1973) NMR investigations of the framework cations of various faujasite-type zeolites and their interpretation by model calculations: *Adv. Chem. Ser.* **121**, 74-86.

Schwuger, M. J., Smolka, H. G. and Rostek, H. M. (1974) Henkel & Cie GmbH, Verfahren zum Waschen und Reinigen der Oberflächen von festen Werkstoffen, insbesondere von Textilien, sowie Mittel zur Durchführung des Verfahrens: *DOS* **2**, 412, 837, S. 1-61.

Schwuger, M. J. and Smolka, H. G. (1976) Na–Al-silicates in the washing process. Part I – Physiochemical aspects of phosphates substitution by heterogeneous ion exchanges in the washing process: *Colloid Polymer Sci.* (in press).

Smith, J. V. (1971) Faujasite-type structures. Aluminosilicate framework. Positions of cations and molecules. Nomenclature: *Adv. Chem. Ser.* **101**, 171-220.

Smolka, H. G. and Schwuger, M. J. (1974) Henkel & Cie GmbH, Verfahren zur Herstellung von festen, schüttfähigen Wasch – oder Reinigungsmitteln mit einen Gehalt an Calcium bindenden Substanzen: *DOS* **2**, 412, 838, S. 1-41.

Wolf, F. and Fürtig, H. (1965) Untersuchung von Ionenaustauschgleichgewichten an synthetischen Zeolithen des Types A: *Kolloid Z.Z. Poylmere* **206**, 48-55.

EVALUATION OF NATURAL FERRIERITE AS A CLAUS TAIL-GAS CATALYST

CHANG W. CHI, HANJU LEE,* and DAVID E. W. VAUGHAN

Davison Chemical Division, W. R. Grace & Co.
7379 Route 32, Columbia, Maryland 21044, U.S.A.

Abstract – Natural ferrierite from Lovelock, Nevada, has been evaluated as an adsorbent/catalyst for the Claus reaction:

$$2H_2S + SO_2 \xrightarrow{catalyst} 3\,S + 2H_2O.$$

This process requires an acid-stable catalyst with moderate thermal stability; activated alumina is the present catalyst of choice. Activity for converting sulfur compounds was tested using a "Claus Tail-gas Stream" [2.6 V% H_2S, 1.3 V% SO_2, 5 V% CO_2, 91.1 V% N_2 (dry basis), and 10% H_2O] and a "Concentrated Acid-gas stream (16 V% H_2S, 8 V% SO_2, 20 V% CO_2, 56 V% N_2) in a 1-inch diameter and 13-inch long reactor, using a reaction temperature of 600°F and a regeneration temperature of 900°F. Sulfur-breakthrough times ($H_2S + SO_2$) at 250, 500, and 1000 ppm were used to measure performance.

Dynamic test data for acid-washed ferrierite were obtained through ten regenerations for break points of 250, 500, and 1000 ppm S; and H-ferrierite was evaluated against Pechiney CR alumina for removing S from both Claus and acid-gas streams. The data indicate that these materials are essentially equivalent in the Claus tail-gas clean-up process, and that natural ferrierite could be a competitive catalyst. Comparative test data show that the activities of H- and mixed H- and Na-ferrierites are equivalent, but that both exchanged forms are significantly more active than the untreated mine-run material, containing mostly K, Mg, and Ca cations. NaX tested under identical conditions had good activity, but rapidly degraded in the acidic, hydrothermal environment.

INTRODUCTION

The Claus process, and Farben's modified Claus processes, are well established and widely used in industry to recover elemental sulfur from sulfur-containing gas streams (Valdes, 1964; Grekel *et al.*, 1965; Goar, 1968; Maddox, 1974). The original Claus process comprised oxidation of hydrogen sulfide with air over a bauxite or iron ore catalyst in a single reactor. The true mechanisms of the Claus reaction are not yet fully understood, but are written as:

$$2H_2S + SO_2 \rightleftharpoons \frac{3}{x}S_x + 2H_2O$$

where x could be 2, 4, 6, or 8.

Numerous process and mechanical improvements have been made. Perhaps the first significant advance, made by Farben, was that instead of burning the H_2S directly over the catalyst, one-third of the hydrogen sulfide was burned to sulfur dioxide in a waste-heat boiler; then the sulfur dioxide was reacted with the remaining hydrogen sulfide in a catalytic converter over bauxite to recover elemental sulfur. The principal reactions are:

$$H_2S + 3/2\,O_2 \rightarrow SO_2 + H_2O$$

$$2H_2S + SO_2 \rightleftharpoons \frac{3}{x}S_x + 2H_2O.$$

Generally, the degree of sulfur recovery of Claus plants is in the range of about 80 to 95%. A properly designed plant, in good operating condition, will yield a sulfur recovery of 97%. The United States Environmental Protection Agency requires Claus sulfur plants to reduce the in-stack sulfur emission to about 1300 ppmv. It is anticipated that this pollution restriction will continue, and certain state and local governments have set the limit even lower, to 250 or 500 ppmv. This corresponds to the overall sulfur-recovery efficiency of a Claus unit in the range of 99.5 to 99.9%. Conventional Claus units, without tail-gas clean-up facilities, cannot conform to the stringent air-quality standards already enacted. Furthermore, all Claus units are potential offenders of ever-tightening EPA regulations, making the installation of "add-on" tail-gas processes mandatory.

There are a number of commercial processes available for Claus tail-gas clean-up (Chalmers *et al.*, 1971; Hyne, 1972; Barry, 1972; Davis, 1972; Maddox, 1974; Goar, 1975 a, b); both wet-scrubbing processes (Beavon, 1971; Coady, 1971) and dry-bed processes (SNPA, 1976). The

*Now with Agency for Defense Development, Seoul, Korea.

SNPA/Lurgi's Sulfreen process (Martin and Guynot, 1971) and Amoco's CBA process (Goddin *et al.*, 1974) are well-known dry-bed processes. They mainly employ alumina as an adsorbent-catalyst to extend the Claus reaction at a lower reaction temperature and remove the residual sulfur produced from the bed. The operating conditions, especially modes of regeneration, are key differences between the processes.

This paper is concerned with evaluating natural ferrierites as a Claus tail-gas treatment catalyst for the dry-bed Claus reaction. Pechiney CR activated alumina was tested in parallel to provide a base case, and also a synthetic NaX sieve was given preliminary tests before it was clearly shown to be unstable, although this zeolite is a very active catalyst through the first two to three cycles.

EXPERIMENTAL METHODS

Materials

Ferrierite was considered to be a relatively rare zeolite until Sand and Regis (1966) and Regis (1970) reported major occurrences in northern Nevada. One of these (located about 10 miles north of Lovelock, Nevada) is presently leased by W. R. Grace & Co. and was the source of the samples used in this study. The ferrierite crystal structure has been described by Kerr (1966) and Vaughan (1966) and consists of parallel 6-ring and 10-ring channels in the *a–b*-plane, interconnected by 8-ring windows in the *a–c*-plane. Previous reports (Graham, 1918; Vaughan, 1966; Wise *et al.*, 1969; Sand and Regis, 1966; Regis, 1970) show the mineral to have a silica:alumina ratio in the range of 8.4 to 15. Analyses of three pure ferrierite samples from different parts of the Lovelock deposit yielded silica:alumina ratios of 11.42 ± 0.2.

Preparation of samples

Ferrierite. This was a pure ferrierite sample taken from surface outcrops in the middle of the deposit. X-ray diffraction analyses and scanning electron microscopy showed the sample to be monomineralic with virtually no visible glassy material. The rock was crushed, sieved, and washed. Granules of 8 X 12-mesh were used in these experiments. The chemical analysis of this sample is given in Table 1.

H^+-Ferrierite. A 100-g sample of 8 X 12-mesh granules of the ferrierite described above was added to 5 l of 1 N hydrochloric acid; the slurry was heated at 80°C for 3 hours with mild agitation. The product was filtered, washed free of chloride ion with copious volumes of deionized water, and dried for 4 hours at 150°C. Through 12-mesh fines were removed by sieving. The chemical analysis of this sample is given in Table 1. Similar materials were made using 0.1 N, 0.3 N, and 2 N hydrochloric acid solutions.

Na^+-Ferrierite. A 1000-g sample of H^+-ferrierite was made in the manner described above using 1 N acid and placed in a glass column. The sample was then transformed into the sodium form by countercurrent exchange using 30 l of 10% NaCl solution at 90°C (90 cm^3/min, 3 hours) and washed for 4 hours at the same rate using deionized water adjusted to a pH of 7 with NaOH. The sample was then dried at 150°C for 4 hours.

CR Alumina. This is a premium alumina made by Pechiney (France) and was obtained in the form of 8 X 12-mesh beads. Either this alumina, or a lower grade (DR alumina), is used as the Claus catalyst in many sulfur-recovery systems. A typical analysis is given in Table 1.

TABLE 1. Chemical analyses of materials

Material	SiO_2 (wt.%)	Al_2O_3 (wt.%)	Na_2O (wt.%)	K_2O (wt.%)	MgO (wt.%)	CaO (wt.%)	Fe_2O_3 (wt.%)	N_2 Surface Area (m^2/g)	H_2O pore volume (wt.%) (50% RH)	Density (g/cm^3)
Pechiney CR alumina (8–12-mesh spheres)	0.98	98.57	0.03	0.08	–	<.01	–	314	–	0.71
Davison NaX 544 (8–12-mesh spheres)	49.20	27.89	16.90	<.01	2.01	0.68	0.55	650	20.8	0.69
Ferrierite (8–12-mesh granules)	75.99	11.16	0.79	5.88	1.27	3.87	1.48	10	–	0.84
H^+ ferrierite	88.9	10.9	–	–	–	–	–	210	–	–

All analyses are on a dry basis. Samples pre-calcined at 1700°F for 30 minutes.

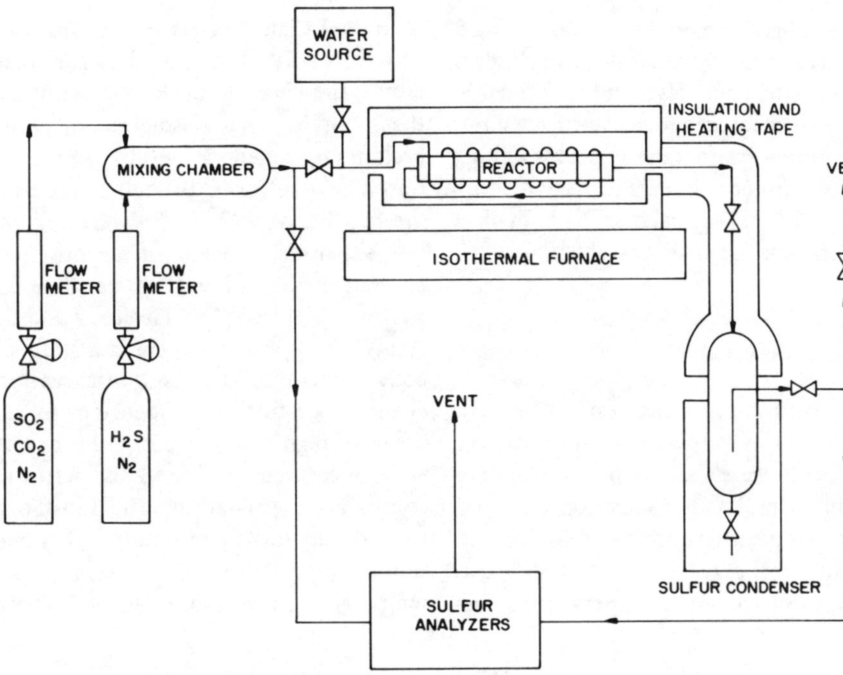

Fig. 1. Schematic of experimental Claus tail-gas catalyst screening apparatus.

NaX *Zeolite.* This is a conventional commercial NaX product obtained in the form of 8 X 12-mesh beads (Davison C-542). A typical analysis is given in Table 1.

Equipment

The apparatus consists of a metal Claus tail-gas reactor charged with the adsorbent-catalyst, an isothermal furnace, a sulfur condenser, a water-injection pump, feed-flow meters, a gas-mixing chamber, and sulfur analyzers as shown in Fig. 1. Two sizes of reactors (1 inch in diameter by 13 inches long; 2 inches in diameter by 26.5 inches long) were used; one for screening various adsorbents of potential interest and the other for further testing of promising adsorbents. The sulfur analyzers were a Barton Electrolytic Titrator (Model No. 286) and a Traco Gas Chromatograph (Model No. 550) with a Flame Photometric Detector (Model No. 12003).

Feedstock compositions

Simulated gas-feed compositions were obtained by mixing equal volumes of two stock compositions – one cylinder contained a mixture of SO_2 and CO_2 with the balance nitrogen and the second H_2S with the balance nitrogen. Both gas sources were dry. To minimize condensation and corrosion problems, water was injected by a calibrated syringe pump immediately in front of the reactor where instantaneous vaporization

TABLE 2. Standard test conditions and simulated feed gas compositions (dry basis)

Test condition		
Pressure	20 psig	
Temperature (reaction)	600°F	
(regeneration)	900°F	
Hourly space velocity	150–200 vol./vol.	
Unit No. 1	1 inch diameter 13 inches long	
Unit No. 2	2 inches diameter 26.5 inches long	
Gas composition		
	Claus tail gas (C.T.G.) (vol.%)	High acid gas (H.A.G.) (vol.%)
H_2S	2.6 (D.B.)	16 (D.B.)
SO_2	1.3 "	8 "
CO_2	5.0 "	20 "
N_2	Balance "	Balance "
H_2O	10 "	–

occurred. The "Claus tail-gas" composition was attained by mixing equal volumes from two separate gas cylinders (2.6% SO_2, 10% CO_2, balance N_2; and 5.2% H_2S, balance N_2). Water was injected with a syringe pump. The "high acid gas" composition was made by mixing equal volumes from two other cylinders (16% SO_2, 40% CO_2, balance N_2; and 16% H_2S, balance N_2). Typical analyses of these gas streams are given in Table 2.

Procedure

An experimental procedure was established to determine the catalytic activity and stability of the catalyst. The experiments were carried out at standard conditions employing simulated Claus tail-gas and concentrated acid-gas compositions presented in Table 2.

The adsorbent-catalyst charged in the test unit was regenerated *in situ* prior to each experiment. The simulated Claus tail-gas or acid-gas was fed to a test unit at 160 VHSV and 20 psig. The reactor temperature during the clean-up cycle was maintained at 600°F, and the spent bed was regenerated at 900°F with nitrogen. Preliminary regeneration at lower temperatures demonstrated that sulfur is difficult to remove completely from the bed below 900°F. The inlet and initial conditions of the experiments were kept constant, and the sulfur content in the effluent was continuously measured by a Barton Electrolytic Titrator and a Traco Sulfur Analyzer. A process cycle consisted of adsorption and conversion at 600°F and regeneration at 900°F. This cyclic operation was repeated to obtain the sulfur-breakthrough times at 250, 500, and 1000 ppmv, which were measured and used as performance indices for comparative evaluation. The most promising materials were tested further in a unit scaled approximately eight times larger than the screening unit at the same temperature, space-velocity, and feedstock conditions.

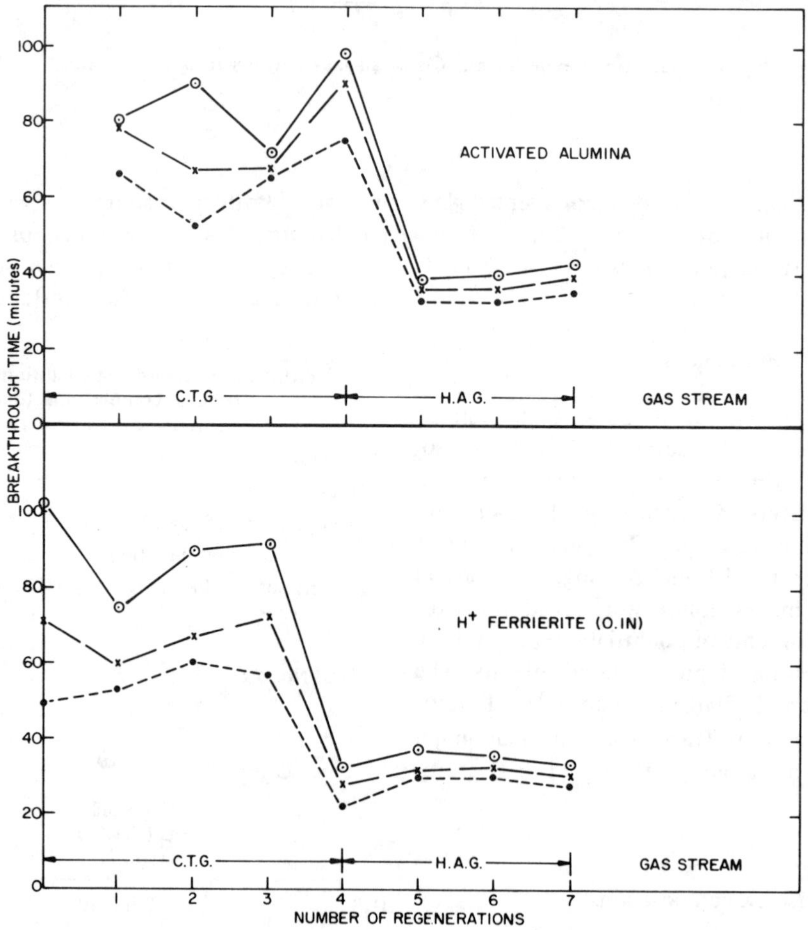

Fig. 2. Breakthrough times at 250 (•), 500 (X), and 1000 (O) ppm sulfur, measured as a function of number of regenerations for CR activated alumina and natural ferrierite washed with 0.1 N HCl. "Claus tail-gas" and "high acid gas" simulated feedstocks were used with a 12-inch-long screening unit.

DISCUSSION OF RESULTS

Screening tests (Unit No. 1)

To establish a base case, the activated alumina was tested first, using the Claus tail-gas (C.T.G.) through four regenerations, followed by the more concentrated high-acid sulfur gas stream (H.A.G.) through three further regenerations. The data in Fig. 2 show the 250 and 1000 ppm average breakthrough times to be, respectively, 65 and 80 minutes for the first feed gas, and 33 and 40 minutes for the second feed gas which had a fourfold increase in sulfur concentration. However, one notes that the high acid gas contains no water; as water is a product of the Claus reaction, a dry gas would react more completely than one containing a high water loading. Figure 2 also shows data for acid-treated ferrierite (0.1 N HCl exchanged). In this case the 250 and 1000 ppm breakthrough times were 55 and 90 minutes for the Claus tail-gas and 28 and 35 minutes for the high-acid gas. The major decrease in breakthrough times as the sulfur concentration increases reflects the importance of sulfur adsorption. Experiments with 600°F regenerations on both alumina and ferrierite clearly demonstrated the sulfur is not completely removed at 600°F. Consequently, breakthrough times were reduced by about one-half, but after one subsequent 900°F regeneration the breakthrough times returned to normal.

A second preparation of hydrogen-exchanged ferrierite was made using 2 N HCl and tested using both conventional and dry Claus tail-gas. The results (Fig. 3) are essentially identical to those for activated alumina with the Claus tail-gas. There seem to be negligible effects from water in these runs, possibly indicating that water does not inhibit reactivity in zeolite Claus catalysts. Further experimentation is needed to clarify this observation.

To determine the effects of cations on ferrierite performance, an original untreated rock sample, a

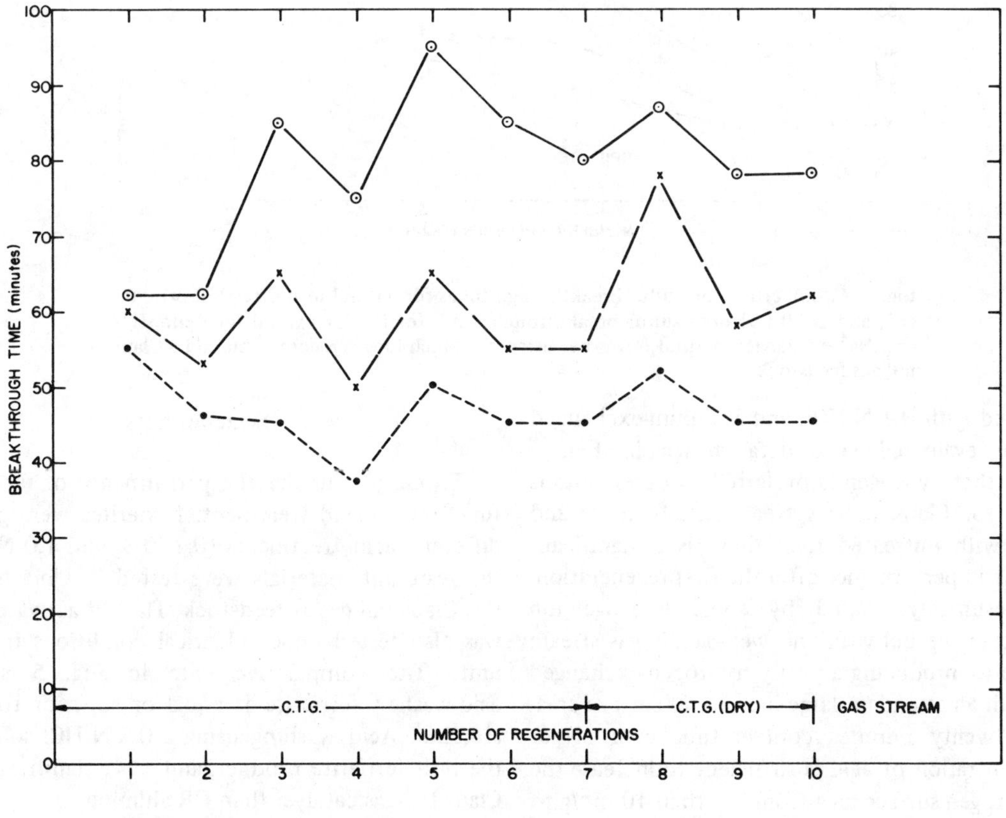

Fig. 3. Ten-cycle evaluation of natural ferrierite pretreated with 2 N HCl, using a wet (CTG) and dry (CTG-dry) Claus tail-gas, at 250 (•), 500 (X), and 1000 (O) ppm sulfur breakthrough times, using a 12-inch-long screening unit.

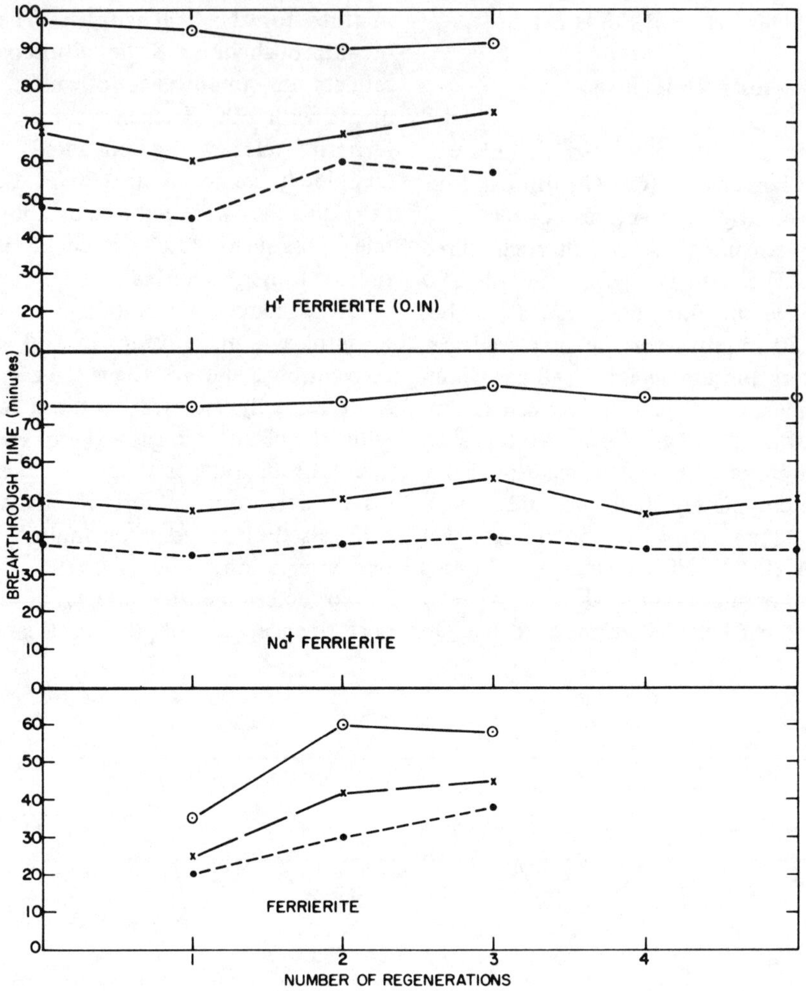

Fig. 4. Cation effects on sulfur breakthrough times for natural ferrierite, 250 (●), 500 (X), and 1000 (O) ppm sulfur breakthrough times for (K,Ca) natural ferrierite, H^+- and Na^+-exchanged natural ferrierite, using a 12-inch-long screening unit and a Claus tail-gas feedstock.

sample treated with 0.1 N HCl, and a sodium-exchanged sample were evaluated. The data shown in Fig. 4 demonstrate that hydrogen is preferred to other cations in ferrierite for Claus tail-gas treatment. In these and other tests with untreated rock there is a significant improvement in performance after the first regeneration. This is presumably caused by some hydrogen-ion exchange occurring between the wet, acidic gas stream and the zeolite, producing a partly hydrogen-exchanged ferrierite with an attendant large increase in surface area. Fifteen to twenty minutes contact time in a 2N or higher concentration of acid is sufficient to increase the ferrierite nitrogen surface area from less than 10 m^2/g to 200 m^2/g. A comparison of Figs. 2, 3, and 4 indicates that a 1 N acid treatment is preferable to either 2 N or 0.1 N. The performance of 1 N acid-treated ferrierite is superior to activated alumina at the 1000 ppm sulfur breakthrough but inferior at 250 ppm.

Scale-up tests

To clarify further the performance of ferrierite as a function of acid treatment, ferrierites were given three different acid treatments (0.1, 0.3, and 1.0 N HCl) and the resultant materials were tested in Unit No. 2 with the Claus tail-gas as feedstock. The CR activated alumina was also tested under identical conditions in the same unit. The comparative data in Fig. 5 show that acid-washed ferrierite is equal or superior to activated alumina. Acid washing using a 0.3 N HCl solution gave the best ferrierite product, and it is a significantly better Claus tail-gas catalyst than CR alumina.

The performance of materials in Unit No. 2 was significantly different from Unit No. 1. Breakthrough times were shorter in all cases, and the interval between the 1000 and 250 ppm breakthrough times was very narrow in all cases. Whether this reflects a heat-transfer

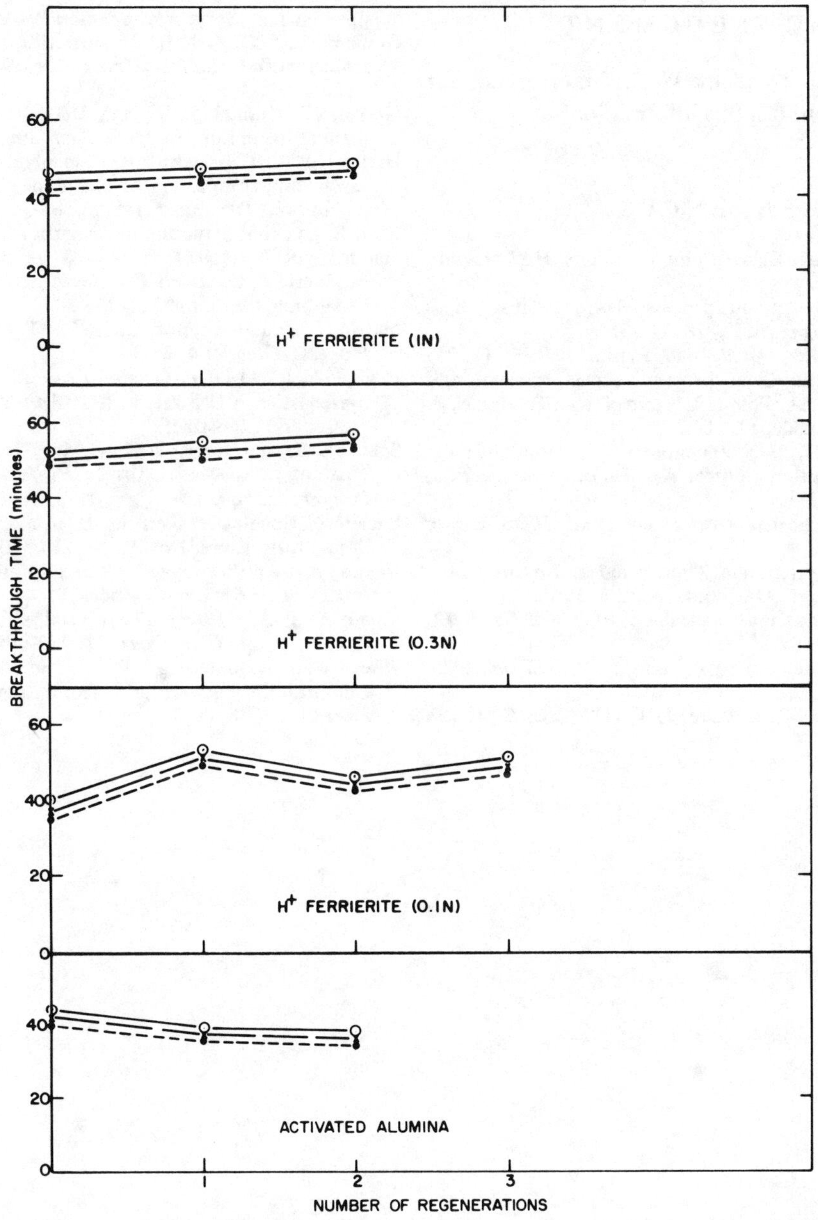

Fig. 5. Scale-up Claus tail-gas tests for acid-treated natural ferrierites and CR-activated alumina. Evaluation of ferrierite 250 (●), 500 (X), and 1000 (O) ppm sulfur breakthrough times as a function of acid pre-treatment of the natural ferrierite.

problem causing more rapid sulfur deposition, or channeling in the bed, was not established.

CONCLUSIONS

The data given in this paper show that acid-washed natural ferrierite may be considered as a material competitive with activated alumina for the catalyst-adsorbent in the dry-bed Claus tail-gas clean-up process. The extent to which untreated raw ferrierite rock would be a suitable catalyst is questionable and needs further clarification. Some experiments show that ferrierite may be less susceptible to water poisoning in wet gas streams. Experiments with NaX show that it has very high activity but low stability.

Preliminary experiments with synthetic mordenite indicate that this material is superior to both ferrierite and alumina for Claus tail-gas clean-up.

ACKNOWLEDGMENT

The authors wish to thank W. R. Grace & Co. for their permission to publish this information.

REFERENCES

Barry, C. B. (1972) Reduce Claus sulphur emissions: *Hydrocarbon Process.* **51**, 102-106.

Beavon, D. K. (1971) Add-on process slashes Claus tailgas pollution: *Chem. Engng (N.Y.)* **78**, 71-73.

Chalmers, W. W., Becker, H. W. and Rankine, R. P. (1971) Improvements to the Claus process – past, present and future: Ann. Mtg. Air Pollution Control Ass. Pacific N.W. Section, Calgary, Canada, 11/71.

Coady, A. B. (1971) SO_2 stack emissions – 1000 ppm a reality: Ann. Mtg. Air Pollution Control Ass. Pacific N.W. Section, Calgary, Canada, 11/71.

Davis, C. J. (1972) Add-on processes stem H_2S: *Chem. Engng (N.Y.)* **79**, 66-68.

Goar, B. G. (1968) Contemporary recovered sulfur processes: Gas Conditioning Conf., Oklahoma, April 3, 1968.

Goar, G. G. (1975a) Claus tail-gas clean-up, Pt. I: *Oil Gas J.* **73**, 109-112.

Goar, G. G. (1975b) Claus tail-gas cleanup, Pt. II: *Oil Gas J.* **73**, 96-103.

Goddin, C. S., Hunt, E. B. and Palm, J. W. (1974) CBA process ups Claus recovery: *Hydrocarbon Process.* **53**, 122-124.

Graham, R. P. D. (1918) On ferrierite, a new zeolitic mineral from British Columbia: *Trans. Roy. Soc. Canada*, Ser. 3, **12**, 185.

Grekel, H., Kunkel, L. V., and McGalliard, R. (1965) Package plants for sulphur recovery: *Chem. Engng Prog.* **61**, 70-73.

Hyne, J. B. (1972) Desulfurization of effluent gas streams – a review and comparison of techniques: *Proc. 51st Ann. Conv. Natural Gas Processors Ass.*, pp. 85-94.

Kerr, I. S. (1966) Structure of ferrierite: *Nature* **210**, 294.

Maddox, R. N. (1974) Gas and liquid sweetening; sulfur production: *Campbell Petroleum Series*, 2nd edn., Norman, Oklahoma, Ch. 9, pp. 221-256.

Martin, J. E. and Guynot, G. (1971) The Sulfreen process: *Gas Process. (Canada)* **4**, 26-30.

Regis, A. J. (1970) Occurrences of ferrierite in altered pyroclastics in Central Nevada: *Ann. Mtg. Geol. Soc. Amer.*, Pt. 7, p. 661 (abstract).

Sand, L. B. and Regis, A. J. (1966) Occurrence of ferrierite, Pershing Co., Nevada (Abst.): *Geol. Soc. Amer. Annual Mtgs. Program*, p. 189.

Société Nationale Des Petroles D'Aquitaine (1976) Process for desulphurising gas: Brit. Pat. 1,420,270.

Valdes, A. R. (1964) New look at sulphur plants: *Hydrocarbon Process. Pet. Refiner* **43**, 104-108.

Vaughan, P. A. (1966) The crystal structure of the zeolite ferrierite: *Acta Crystallogr.* **21**, 983-990.

Wise, W. S., Nokleberg, W. J. and Kokinos, M. (1969) Clinoptilolite and ferrierite from Agoura, California: *Amer. Mineral.* **54**, 887-895.

THE POTENTIAL USE OF NATURAL ZEOLITES FOR AMMONIA REMOVAL DURING COAL-GASIFICATION

D. T. HAYHURST*

Department of Chemical Engineering, Worcester Polytechnic Institute, Worcester, Massachusetts 01609, U.S.A.

Abstract – A large number of coal-gasification plants are planned for the conversion of coal to synthetic natural gas. Each plant would react about 12,000 tons of high-quality eastern coal with steam and oxygen to produce 250 million cubic feet of raw gas each day. After removal of NH_3 and SO_2, the purified stream would be methanated to produce synthetic pipeline gas. It is estimated that in one plant, 100 tons of ammonia must be removed per day.

The adsorption selectivity for ammonia over methane was determined for five natural zeolites. At 100 mm Hg pressure and 25°C, adsorption capacities ranged from 4.03 to 9.65 wt.% for ammonia and 0.35 to 0.79 wt.% for methane. The zeolites tested were mordenite, chabazite, clinoptilolite, ferrierite, and phillipsite. Phillipsite demonstrated particular potential applicability in this process, as it will not adsorb significant amounts of most of the reactants in a gasification stream, including CO, CO_2, CH_4, and O_2. The most significant result observed was the very slow rate of water adsorption by phillipsite as compared to its adsorption rate for ammonia. The adsorption-rate curve for saturated water vapor at 25°C on phillipsite is linear with a slope of 0.033 wt.% per minute; while the rate curve for ammonia is "knee shaped" with an initial slope of 4.80 wt.% per minute and a capacity of 8.78 wt.% at 100 mm Hg pressure and 25°C.

INTRODUCTION

The objective of this study was to demonstrate the application of natural zeolites in the selective removal of a polluting species, ammonia, produced in the gasification of coal to methane, a synthetic natural gas (SNG). The process of coal gasification is of current interest because of its unique problems with respect to environmental safeguards. During coal gasification significant amounts of nitrogen and sulfur compounds and toxic trace elements are produced which must be removed from a final SNG product stream. The nitrogen and sulfur in the coal are hydrogenated during gasification to ammonia and hydrogen sulfide. The 2.6 Å kinetic diameter (Breck, 1974) of ammonia is smaller than that of any other molecule produced during gasification, making selective adsorption by natural molecular-sieve zeolites of potential significance.

Coal gasification

The current U.S. energy goal is to replace 20% of the country's oil consumption with clean fuel from coal or oil shale. A number of large plants are planned for converting coal to synthetic natural gas. It is reported (Jahnig and Bertrand, 1975) that a typical plant would produce 250 million cubic feet per day of a high BTU SNG and consume in the process about 12,000 tons per day of high-quality eastern coal. With western coal or lignite, consumption would be increased. It would take eighty of these large plants to provide clean fuel equivalent to 20% of the U.S. current oil consumption.

For the purposes of this discussion, a generalized coal gasification operation will be described. Most SNG processes gasify coal with steam and oxygen, as shown in the process flow diagram (Fig. 1). The coal first is gasified using steam and oxygen. In the next stage, this raw gas is shifted by adding steam (the water–gas shift reaction) to provide the required hydrogen to carbon monoxide ratio of 3 to 1. From this synthesis gas, ammonia is scrubbed out and any remaining water is removed.

The next step is acid-gas removal, in which H_2S is separated out and sent to sulfur recovery. Carbon dioxide is removed as well and is rejected to the atmosphere.

The final step is methanation to make the product, a high-BTU pipeline gas. This step produces a large amount of heat which is used to generate steam. The reaction also produces water which is recovered for

*Present address: Department of Chemical Engineering, Cleveland State University, Cleveland, Ohio 44115, U.S.A.

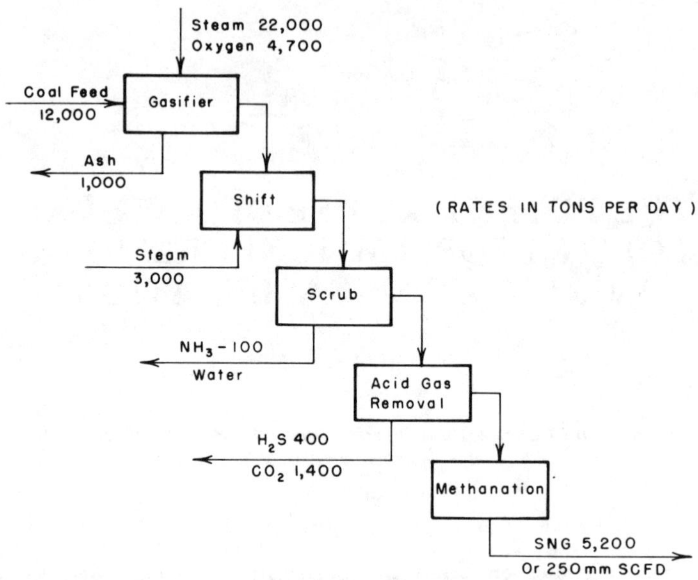

Fig. 1. Flow diagram of a typical coal-gasification operation.

subsequent process use.

As is indicated by the flow rates in Fig. 1, a large amount of ammonia (100 tons per day) is produced, which must be removed. To date, much more attention has been focused on the sulfur-recovery operation (Wall, 1973) as compared to the ammonia recovery. However, this report hopes to demonstrate that the use of natural zeolites may provide an attractive alternative to the scrubbing process for ammonia recovery.

EXPERIMENTAL

Materials

Samples of clinoptilolite, mordenite, phillipsite, chabazite, and ferrierite from various sedimentary deposits were used in this study and are listed in Table 1. No attempt was made to upgrade the zeolite mineral content of the samples.

Analytic methods

Quantitative phase identification. Powder X-ray diffraction was used to identify crystalline phases and estimate the percent crystallinity. X-ray diffractograms were taken on a Norelco XRG-3000 X-ray diffractometer using nickel-filtered CuK_{a1}, (λ = 1.54040 Å) radiation with settings of 40 kV–20 mA, scanning speed of 0.01° per 0.1 second over a range of 40° to 5°. For the estimation of the concentration of a single phase, diffraction peak intensities were summed and compared to a well-characterized standard. A minimum of six peaks were summed to minimize statistical error.

Adsorption. Adsorption experiments were carried out in a Cahn Model RG Electrobalance and in a mechanical lever balance designed and built in our department. The experimental procedures and operation of both balances essentially were identical. A sample of zeolite was placed into the balance sample holder and

TABLE 1. Estimate of purity and equilibrium adsorption capacities (at 100 mmHg pressure, 25°C)

Type	Locality	Estimate of purity	gNH_3 sorbed per 100 g sorbent	gCH_4 sorbed per 100 g sorbent	Samples obtained from
Mordenite	Talakeetna Mts., Alaska	95%	9.65	0.799	D. B. Hawkins
Chabazite	Bowie, Arizona	97%	9.28	0.785	L. B. Sand
Phillipsite	Rome, Oregon	84%	8.78	0.082	L. B. Sand
Chabazite	Durkee, Oregon	62%	6.94	0.432	L. B. Sand
Ferrierite	South Korea	71%	6.02	0.460	L. B. Sand
Mordenite	Union Pass, Arizona	69%	5.87	0.717	L. B. Sand
Clinoptilolite	Sheaville, Oregon	94%	5.23	0.480	L. B. Sand
Ferrierite	Lovelock, Nevada	60%	4.03	0.349	L. B. Sand

material activated by heating to 300°C *in vacuo* (under 0.003 mmHg) maintained for at least 2 hours. After activation, the sample was allowed to cool to the desired adsorption temperature, usually room temperature, and then was exposed to the sorbate. The rate of weight change of the sample as it adsorbed was recorded. The details of the experimental procedures are given elsewhere (Hayhurst, 1976).

RESULTS

Adsorption

Selected samples of natural zeolites were analyzed with respect to their characteristics, especially effective port diameter, adsorption rate, and adsorption capacity, to evaluate their potential for ammonia removal in coal gasification. Each variety tested rapidly adsorbed ammonia, and adsorption capacities varied from 4 to 10% by weight at a gas pressure of 100 mmHg. A typical adsorption-rate curve for ammonia rises steeply and then becomes horizontal as the equilibrium capacity is reached, as illustrated in Fig. 2. This contrasts with the adsorption-rate curve for methane, also shown in Fig. 2. Methane adsorbs slowly, and the corresponding capacity is much lower (0.10–0.80 wt.%) than that found for the ionic ammonia molecule. Equilibrium capacities for both ammonia and methane sorption into several natural zeolites are listed in Table 1.

The two smallest molecules found in the gasification process stream are ammonia, with a minimum cross-sectional diameter of 2.60 Å, and water, with a diameter of 2.65 Å (Breck, 1974). Comparative rate data were determined for the adsorption of these two molecules into several natural zeolites. Phillipsites were found to be of particular interest because of their very small port diameter. Rate data, therefore, were obtained for ammonia and water at two adsorption temperatures and are plotted in Figs. 3 and 4 for phillipsite from Rome, Oregon.

In Fig. 3 the rate of ammonia adsorption by phillipsite at 25°C and 65°C is shown. For both temperatures the sorption contours were found to be "knee-shaped" with a very rapid initial rate and most of the adsorption taking place within the first 8 minutes. The capacity for ammonia was reduced at the higher temperature.

Water was found to be the only molecule, other than ammonia, that will be adsorbed by phillipsite, so the rate of adsorption for water vapor into phillipsite was measured at the same two temperatures, 25 °C and 65 °C, and is illustrated in Fig. 4. At the higher temperature, the rate of water adsorption is very rapid with a capacity of 9.6 wt.%; however, at room temperature, the rate is much slower, and it appears that phillipsite almost excludes water vapor at room temperature.

Further adsorption studies were conducted on a system where both water and ammonia were present and

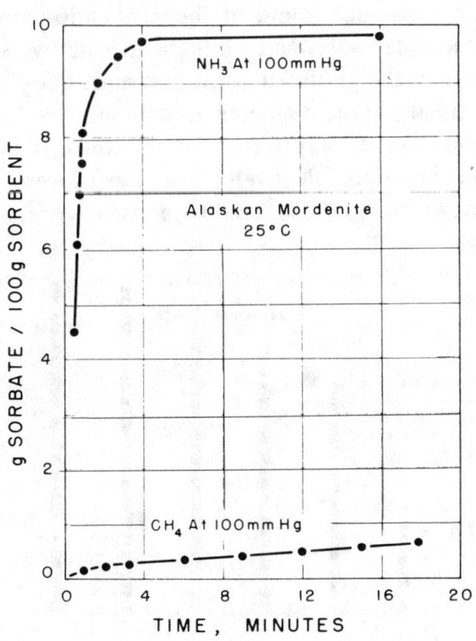

Fig. 2. Rate of adsorption for Alaskan mordenite.

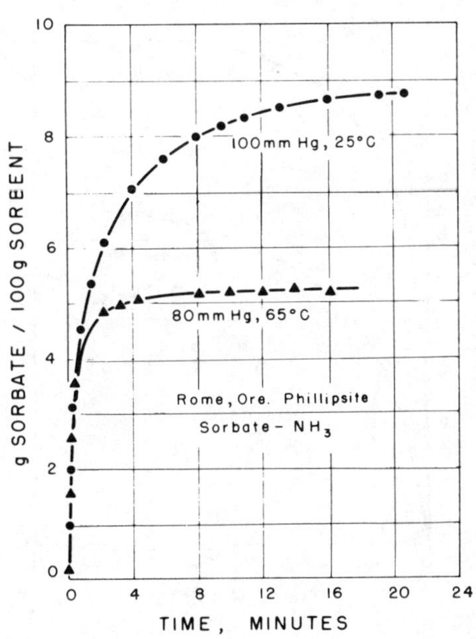

Fig. 3. Adsorption-rate determination for ammonia into phillipsite.

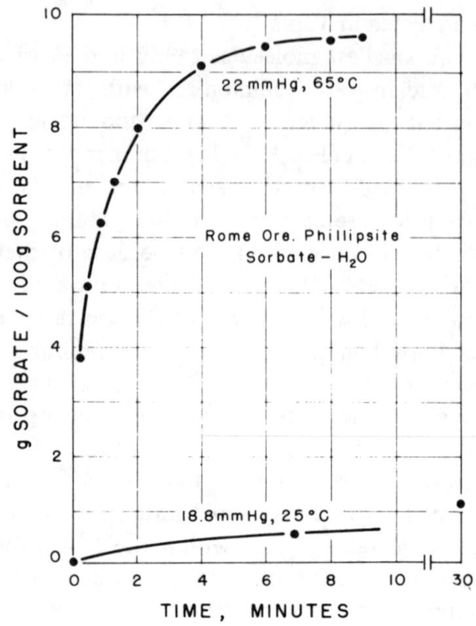

Fig. 4. Adsorption-rate determination for water vapor into phillipsite.

the results are plotted in Fig. 5. In this case, a mixture of water vapor and ammonia was exposed to the phillipsite sample, and the rate of adsorption was measured. Also plotted in the same figure is the adsorption rate for pure ammonia at a pressure identical to the partial pressure of ammonia in the mixture. The rate and capacity for adsorption of the mixture is somewhat lower than that for pure ammonia adsorption, but significantly higher than what was observed for adsorption of pure water.

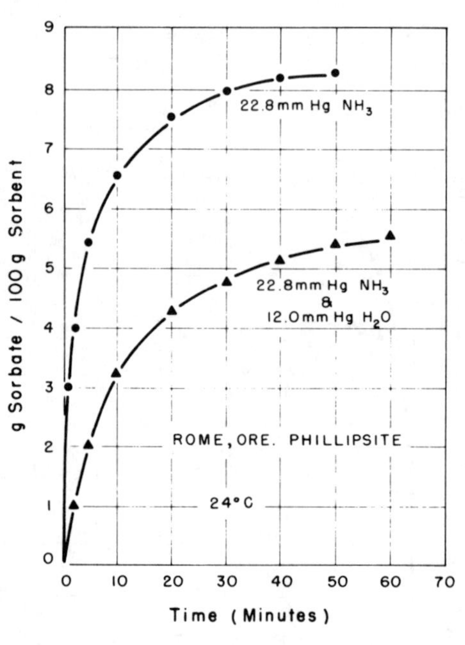

Fig. 5. Adsorption-rate determination for both ammonia and water into phillipsite.

DISCUSSION AND CONCLUSIONS

A basic problem in coal gasification is that the carbon to hydrogen ratio in the coal is approximately 1:1, while the ratio in the desired methane product is 1:4. Therefore, it is necessary either to add hydrogen or to remove carbon, or both. Hydrogen produced in the gas-shift reaction reacts with the other elements found in coal, nitrogen, and sulfur. They produce ammonia and hydrogen sulfide, and these pollutants must be removed from the final product. What is desired is to purge the product gas of NH_3 and H_2S without removing methane. Every variety of natural zeolite tested exhibited a strong adsorption preference for ammonia over methane as illustrated in Fig. 2 and Table 1. A coal-gasification process could be envisioned where coal is gasified, shifted, methanated, and purified by zeolite adsorption to produce the desired pipeline SNG. Any of the natural zeolite tested is of potential usefulness since each exhibited a high sorptive preference for ammonia. However, methanation catalysts are poisoned readily by nitrogen compounds, so it therefore is desirable to remove all ammonia prior to the methanation stage.

In Fig. 6, the kinetic port diameters of the most abundant natural molecular sieve zeolites are plotted along with the minimum cross-sectional diameters of the common molecules found in the gasification stream. It can be seen that water and ammonia are the two smallest molecules in the gasification process and that phillipsite has the smallest pore diameter. Experiments verified that phillipsite adsorbs only water and ammonia. If ammonia could be removed selectively from the process stream, then an adsorption bed could be used to replace the

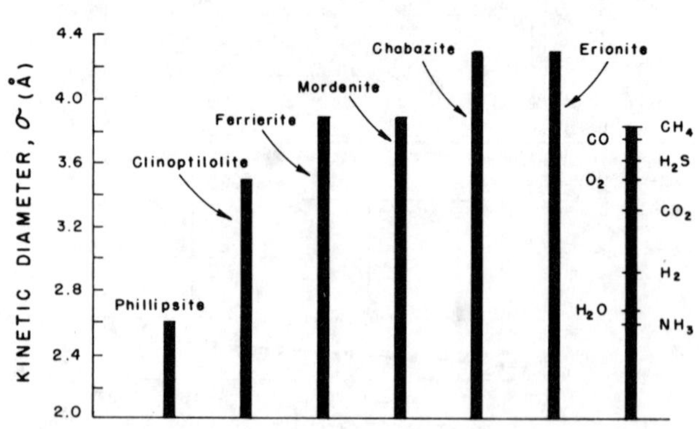

Fig. 6. Kinetic pore diameters for various molecules and natural zeolites (partly after Breck, 1974).

scrubbing operation in coal gasification. The results plotted in Figs. 3 and 4 show that at 25°C phillipsite adsorbs NH_3 rapidly but almost excludes water. In practice, synthesis gas is cooled to 25°C prior to scrubbing; so an adsorption process, utilizing natural phillipsite, might be used instead of the present scrubbing process. Although only coal gasification was cited as an example, phillipsite potentially is useful in any application where it is desired to remove ammonia from a water-containing system.

At 65°C phillipsite rapidly adsorbs water, while at 25°C the rate is much less (see Fig. 4). These data suggest a molecular sieving effect by the phillipsite. At room temperature, the kinetic diameter of the water molecule is slightly larger than the effective port diameter for phillipsite, but at the higher temperature, thermal energy may open the framework structure of the phillipsite enough to admit water molecules.

As shown in Fig. 5, when both ammonia and water are present in the gas phase, the sorption rate and capacity are both reduced from that found for pure ammonia adsorption. It is possible that water molecules adsorb at the entrances to the phillipsite channels, essentially blocking these channels from further ammonia adsorption. Further investigation will be required to clarify this.

ACKNOWLEDGMENTS

I would like to thank Professor L. B. Sand for the use of his laboratory facilities for this investigation and for his invaluable advice. I would also like to thank Mr. Mario P. DiGiovanni and Miss Christin E. Walker for their assistance in the laboratory.

REFERENCES

Breck, D. W. (1974) *Zeolite Molecular Sieves*: John Wiley & Sons, New York.

Hayhurst, D. T. (1976) Ph.D. Thesis: *Crystallization Kinetics and Properties of Na,K-Phillipsites*: Worcester Polytechnic Institute, Worcester, Massachusetts.

Jahnig, C. E. and Bertrand, R. R. (1975) *Environmental Aspects of Coal Gasification*: AIChE 80th National Meeting, Boston, Massachusetts.

Wall, J. (1973) Review of gas processing: *Hydrocarbon Processing*, April, pp. 108-110.

PRODUCTION OF OXYGEN AND NITROGEN WITH NATURAL ZEOLITES

HIDEO MINATO

Institute of Earth Science and Astronomy, College of General Education, University of Tokyo, Japan

and

TAKAAKI TAMURA

Nuclear Engineering Research Laboratory, University of Tokyo, Japan

Abstract – Investigations on the utilization of natural zeolites have been carried out for about 20 years in Japan. There are many zeolite occurrences in the country, and the quality of the material is suitable for numerous uses, such as paper filler, gas adsorbent, catalyst, cation exchanger, and soil conditioner. This paper reports on the use of natural mordenite as a selective gas adsorbent in the production of oxygen from air. Tuffaceous rocks rich in mordenite occur at Itado, Akita Prefecture, and at Maji, Shiname Prefecture. They have high adsorption capacities for several gases and are excellent adsorbers of nitrogen compared with mordenites from several other localities in Japan. After activation at 500°C, the Itado mordenite adsorbs about 30 cm^3 N_2/g at room temperature and atmospheric pressure.

An oxygen generator has been constructed using the Itado mordenite and consists of a pretreatment column for the removal of moisture, CO_2, and SO_2, and a zeolite-adsorption column filled with granular mordenite. The pretreated air enters the mordenite column, and oxygen with a small amount of argon and nitrogen exits from the generator. Nitrogen can also be produced by this apparatus by evacuating the nitrogen-saturated zeolite column during a pressure-swing cycle. In practice, three sets of parallel columns are used in on-stream, regeneration, and stand-by modes. Generators have been designed in sizes ranging in oxygen outputs from 15 l/hr to 300 m^3/hr. Products containing up to 93% O_2 or 99.99% N_2 can be obtained.

INTRODUCTION

Sedimentary zeolites are formed mainly by diagenetic processes in thick marine and freshwater deposits or in bedded, alkaline lake deposits. In Japan, the genesis of sedimentary zeolites is restricted to the former case, and occurrences of zeolites have been reported from about 100 localities (Minato and Utada, 1969). The zeolites are clinoptilolite, mordenite, analcime, laumontite, and wairakite; clinoptilolite and mordenite are predominant. The principal deposits of clinoptilolite and mordenite are in the "Green Tuff" region of Japan, a broad belt of Neogene rocks in the southwestern part of Hokkaido and in the western part of Honshu, facing the Japan Sea. These rocks contain large amounts of green, altered pyroclastics and are therefore referred to as "Green Tuffs".

Currently, about 4000 to 6000 tons of zeolite are mined each month in Japan; the main zeolite mines are listed in Table 1. Mordenite tuffs from Itado, Akita Prefecture, and Maji, Shimane Prefecture, have high adsorption capacities for nitrogen and are useful in oxygen and nitrogen generators. Zeolite tuffs from Ten'ei, Fukushima Prefecture, have smaller adsorption capacities, while those from other localities in Japan have even lower capacities and are not suitable adsorbents for this purpose. Mordenite tuff from the Itado mine has been found to be the best for oxygen/nitrogen generation.

MINERALOGICAL PROPERTIES OF THE ITADO MORDENITE

The best mordenite ore from the Itado area occurs in a white, tuffaceous bed named the Sanzugawa bed. This bed is of Neogene age and is distributed widely in Minase-mura, Okachi-gun, Akita Prefecture, on northern Honshu and is mined at Itado. Here, the bed is pale

TABLE 1. Zeolite mines in Japan

Mine	Prefecture	Amount of ore (t/month)	Zeolite
Osiyamambe	Hokkaido	100–200	Clinoptilolite
Futatsui	Akita	150–300	Clinoptilolite
Yasawagi	Akita	100–200	Clinoptilolite and mordenite
Itado	Akita	Spot (about 300 t)	Mordenite
Itaya	Yamagata	4000–5000	Clinoptilolite
Shiroishi	Miyagi	100–200	Mordenite
Ten'ei	Fukushima	100–200	Mordenite
Maji	Shimane	Spot (about 300 t)	Mordenite
Iwami mine	Shimane	50–100	Clinoptilolite

TABLE 2. X-ray powder diffraction data for mordenite

Itado		ASTM card (6-0239)		
d(A)	I/I_0	d(A)	I/I_0	*hkl*
13.69	46	13.70	50	110
9.09	68	9.10	90	200
6.62	52	6.61	90	111
4.53	80	4.53	80	330
4.00	98	4.00	90	150
3.84	100	3.84	60	241
3.48	80	3.48	100	202
3.39	74	3.39	90	060
3.22	86	3.22	100	530
2.894	46	2.896	60	402
2.520	50	2.522	50	442

TABLE 3. Chemical analysis of mordenite rock from Itado

	Wt.%
SiO_2	67.90
Al_2O_3	13.09
Fe_2O_3	1.48
MgO	0.38
CaO	1.05
Na_2O	3.12
K_2O	2.83
$H_2O(+)$	6.10
$H_2O(-)$	4.10
Total	100.05

brown with fractures. When activated, it has a high adsorption capacity for gases such as nitrogen, carbon dioxide, and sulfur dioxide. X-ray powder diffraction results using CuKα_1 radiation are given in Table 2. The ore consists mainly of mordenite with small amounts of quartz, plagioclase, and α-cristobalite; the X-ray tracing shows that the material has a high degree of crystallinity compared with that of mordenites from other localities. Chemical analyses in Table 3 show that the mordenite is rich in Na_2O. The weight-loss curve of the powdered material is shown in Fig. 1. About 90% of the water was lost below 400°C.

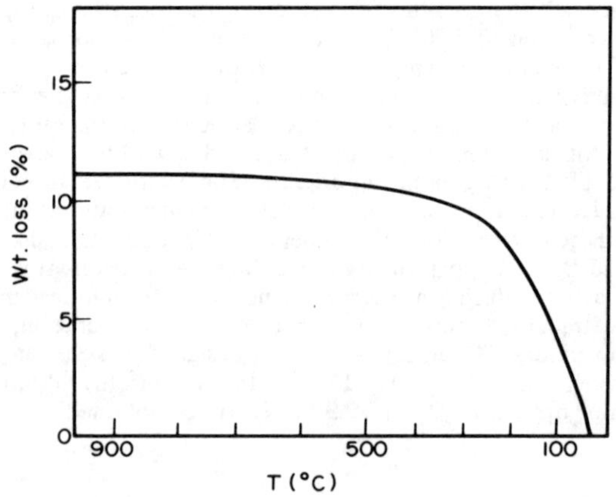

Fig. 1. Weight-loss curve of mordenite from Itado.

The adsorption capacity of the mordenite was measured with nitrogen and carbon dioxide. Pulverized zeolite was loaded into a "Pyrex" glass tube and heated under vacuum to 400°C. After the zeolite cooled, gas was adsorbed at different pressures. The amount of gas adsorbed was measured by increases in weight at a given pressure. Adsorption capacities were obtained at room temperatures (10°C) and at 0°C for each gas; the results are shown in Fig. 2 and compared with mordenite from Union Pass, Arizona, U.S.A., and with clinoptilolite from Shizume, Shiname Prefecture, Japan (see Minato and Utade, 1971). The adsorption capacity of the Itado material is two to three times greater than that of any of the other zeolites tested.

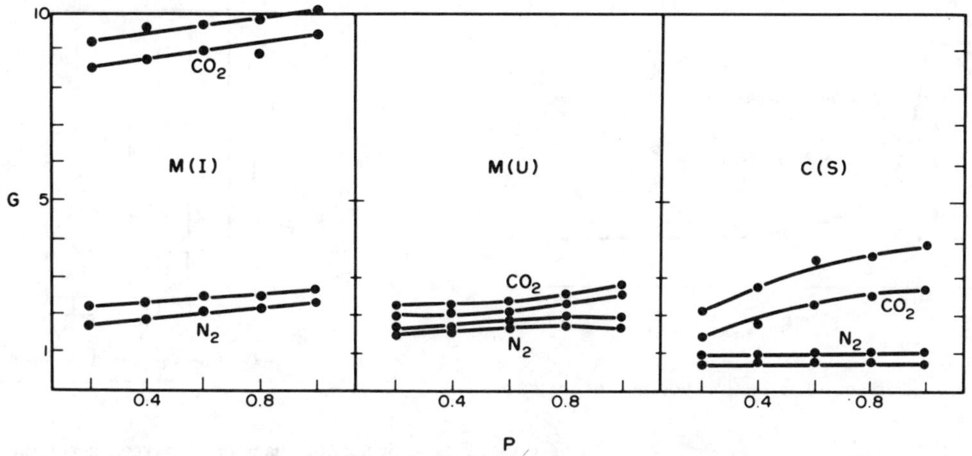

Fig. 2. Adsorption curves for CO_2 and N_2 by mordenites and clinoptilolites. G = wt.% adsorbed; P = gauge pressure (kg/cm^2); M(I) = mordenite from Itado; M(U) = mordenite from Union Pass, Arizona; C(S) = clinoptilolite from Shizuma. The upper curve of each pair was obtained at 0°C. The lower curve of each pair was obtained at 10°C.

TABLE 4. Oxygen generating capacity of zeolite

Zeolite (dried at 500°C)	Volume of oxygen (cm^3/g)
Mordenite (Itado)	2.44
Clinoptilolite (Itaya)	1.21
Mordenite (Ten'ei)	1.40
Chabazite (U.S.A.)	1.49
Mordenite (U.S.A.)	1.10
Mordenite (U.S.A.)	0.69
Clinoptilolite (U.S.A.)	1.53
Clinoptilolite (U.S.A.)	0.62
Synthetic molecular sieve 3A	0.00
,, ,, 4A	0.92
,, ,, 5A	1.39
Synthetic mordenite (Zeolon, Norton Co.)	2.30

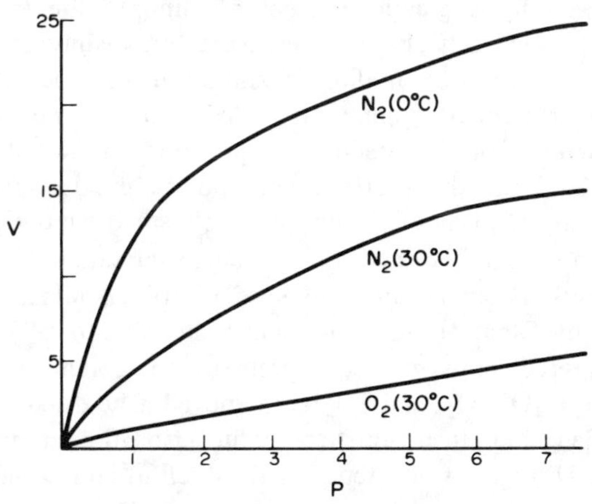

Fig. 3. Adsorption curves of N_2 and O_2 by mordenite from Itado. V = volume adsorbed (cm^3/g); P = pressure (×100 mm Hg).

FUNDAMENTAL STUDY OF OXYGEN GENERATION

The volume of oxygen generated per unit weight of zeolite was measured by the following methods:

1. Adsorption capacities for nitrogen and oxygen were obtained at 30° and 0°C under pressures between 0–1 atmosphere and are shown in Fig. 3. The production of oxygen in a generator is related to the difference between the adsorption capacity of the zeolite for these two gases.

2. The effective volume of oxygen adsorbed by the zeolite was measured on eight natural zeolites and four synthetic zeolites. Dried zeolite grains were packed in a column and the pressure brought to 0.1 mm Hg. Dried air was then drawn through the column at 1 atmosphere pressure, and the production of oxygen measured by the volume of the exhaust oxygen. The production of oxygen is shown schematically in Fig. 4. The capacities of several samples are listed in Table 3. In a typical generator, air is supplied to the first zeolite column between times t_0 and t_1. As demanded by the concentration of oxygen in the off-gas, the air supply is switched to the second zeolite column. The actual production of oxygen is represented as the area *a–c–d–b* in Fig. 4.

OXYGEN AND NITROGEN GENERATORS

Both oxygen and nitrogen generators operate by pressure-swing adsorption methods, with pressure changes from 1 atmosphere to a partial vacuum in the zeolite column at ambient temperatures. As shown schematically in Fig. 5, the generator consists of a

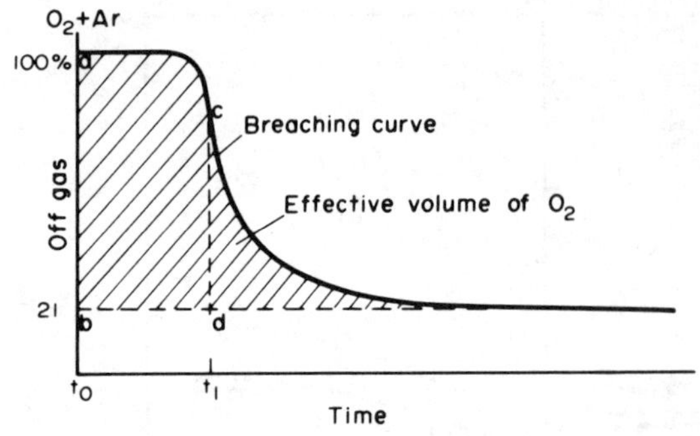

Fig. 4. Schematic diagram of effective volume of oxygen produced.

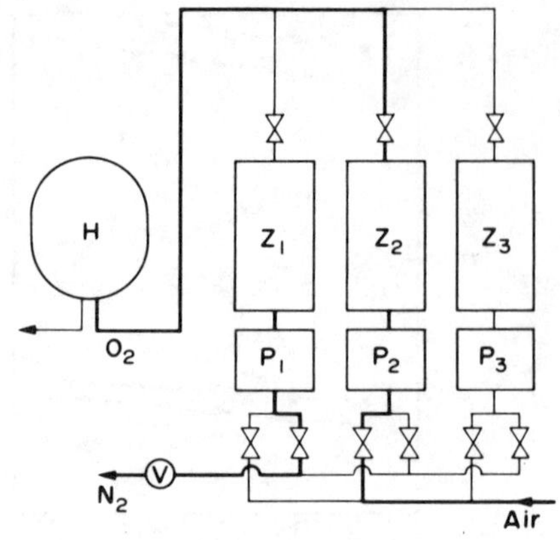

Fig. 5. Schematic diagram of oxygen generator. Z_1, Z_2, Z_3 = zeolite adsorption columns; P_1, P_2, P_3 = pretreatment columns; H = gas holder; V = vacuum pump.

pretreatment column, a zeolite-adsorption column, a vacuum pump, and a blower with a gas reservoir. The pretreatment column is filled with silica gel desiccant followed by a granulated zeolite (clinoptilolite from Itaya) which acts also as a desiccant but mainly as an adsorbent for carbon dioxide and sulfur dioxide. The main adsorbent column is filled with granulated mordenite from the Itado mine, previously activated at 500°C. Both the pretreatment and main adsorption columns are made of steel or plastic (in smaller models), and in general, series of three parallel pretreatment and adsorption columns are employed. Additional details are given by Tamura (1969) and Tamura and Minato (1973).

Filtered air is sent to a pretreatment column (P_1) where H_2O, CO_2, and SO_2 are removed. The dried and purified air is then sent to a zeolite-adsorption column (Z_1). Nitrogen is adsorbed on the zeolite, and oxygen with a small amount of argon and nitrogen is collected in the gas-holding tank (H). Before the zeolite-adsorption column becomes saturated with nitrogen, the purified air is switched to a second pretreatment (P_2) and zeolite adsorption column (Z_2), and the columns P_1 and Z_1 are connected automatically to the vacuum pump. While the second pair of columns generates oxygen, the first pair of columns is regenerated by the pressure swing, exhausting the nitrogen. Oxygen is generated continuously by the alternate working of three parallel sets of columns.

Oxygen enrichment varies from 50% to 93%, with 70% usually being suitable for most commercial uses. Electric-power requirements of the generator vary from 0.4 kWh/m^3 to 6 kWh/m^3, depending on the oxygen purity demanded and the capacity of the generator. The concentration of oxygen in the product gas depends mainly on the amount of nitrogen remaining in the zeolite column after regeneration; the degree of vacuum imposed on the column directly affects the oxygen content of the product.

The mechanism of nitrogen generation is similar to that of oxygen generation. Nitrogen is collected from the adsorbed nitrogen in the zeolite adsorption column through exhausting with the vacuum pump. The resultant product varies from 99.5 to 99.9% N_2. Nitrogen is obtained from the middle stage of the vacuum regeneration; in the early stage, water vapor and carbon dioxide are included with the nitrogen, whereas in the later stage, oxygen gas is included with the nitrogen.

Oxygen generators utilizing natural zeolites are presently being used in Japan for the melting of pig iron, in the treatment of municipal wastewaters, and for the oxygenation of fish tanks during the transportation of live fish. Nitrogen generators are used to provide a high purity, inert gas used for the sealing of various materials during canning operations. The capacities of several oxygen and nitrogen generators are listed in Table 5. Examples of various installations using natural zeolites are illustrated in Figs. 6, 7, and 8.

ACKNOWLEDGMENTS

The writers would like to express their sincere thanks to Dr. Minoru Utada of the Department of Geology, University of Tokyo, for his data on the natural zeolites used in this investigation. The generous support of the Kyatarakogyo Company, the Koyokaihatsu Company, and the Zeeklite Chemical and Mining Company is also gratefully acknowledged.

REFERENCES

Minato, H. (1975) Zeolite, its natural resources and utilizations: *Ceramics in Japan* **10**, 941-957.

Minato, H. and Utada, M. (1969) Zeolite: in the *Clays of Japan, Geological Survey of Japan*, pp. 121-134.

TABLE 5. Capacities of oxygen and nitrogen generators

Oxygen generators

Type	Volume per hour	O_2 content of product	Weight of zeolite per column
LP015	15 l	50%	200 g
SP30	300 l	90%	3.5 kg
SP60	600 l	90%	8 kg
SA10	1000 l	90%	8 kg
SA30	3 m^3	90%	30 kg
SA50	5 m^3	90%	64 kg
LC30	300 m^3	70%	13 T

Nitrogen generators

Type	Volume per hour	N_2 content of product	Weight of zeolite per column
NT15	150 l	99.5%	1.2 kg
NA30	3 m^3	99.5%	30 kg
SB10	10 m^3	99.9%	64 kg
SA15	15 m^3	99.9%	84 kg
SB30	30 m^3	99.9%	180 kg
NC10	100 m^3	99.9%	1.8 T

Minato, H. and Utada, M. (1971) Clinoptilolite from Japan: *Molecular Sieve Zeolites – I*: in *Advances in Chemistry Series* **101**, 311-316.

Tamura, T. (1969) Adsorption process for gas separation: U.S. Patent 3,4,73,296, Oct. 21, 5 pp.

Tamura, T. and Minato, H. (1973) Process for the separation or concentration of gas mixture: U.S. Patent 3,772,852, Nov. 20, 4 pp.

Fig. 6. Oxygen plant LC 30 of Takaokakogyo Company at Toyohashi City, Japan. Abbreviations are the same as in Fig. 5.

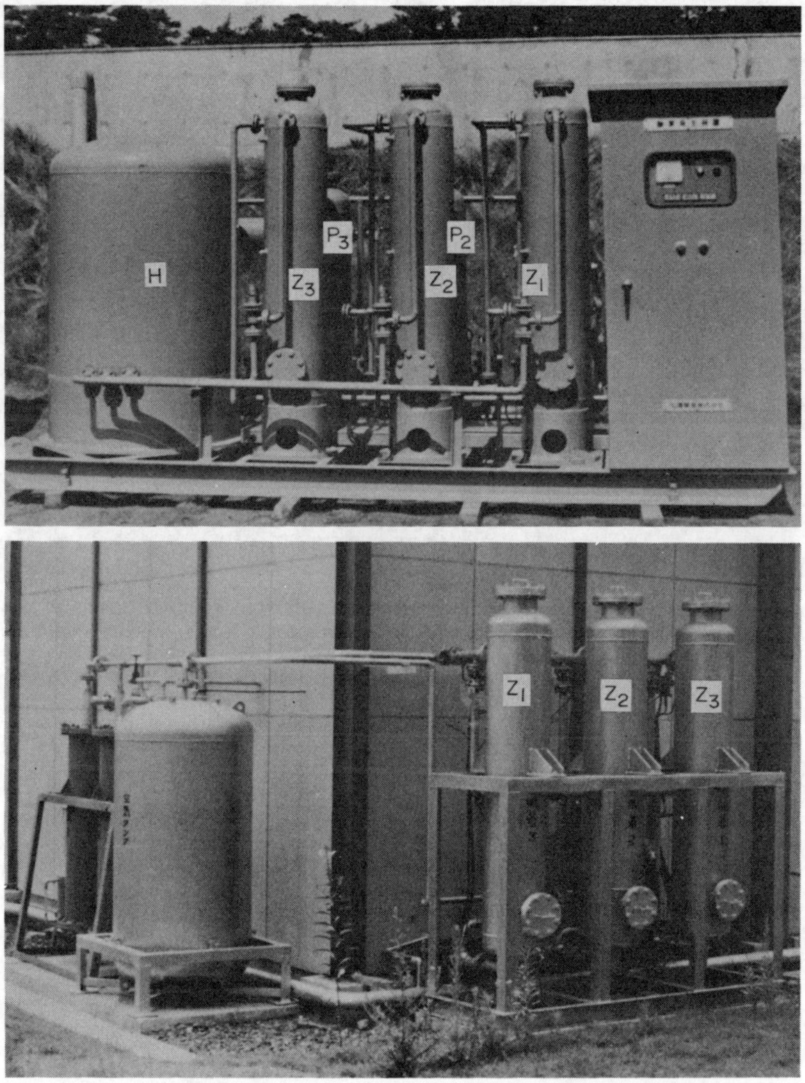

Fig. 7. Upper: oxygen generator SA 50. Lower: nitrogen generator SB 30.

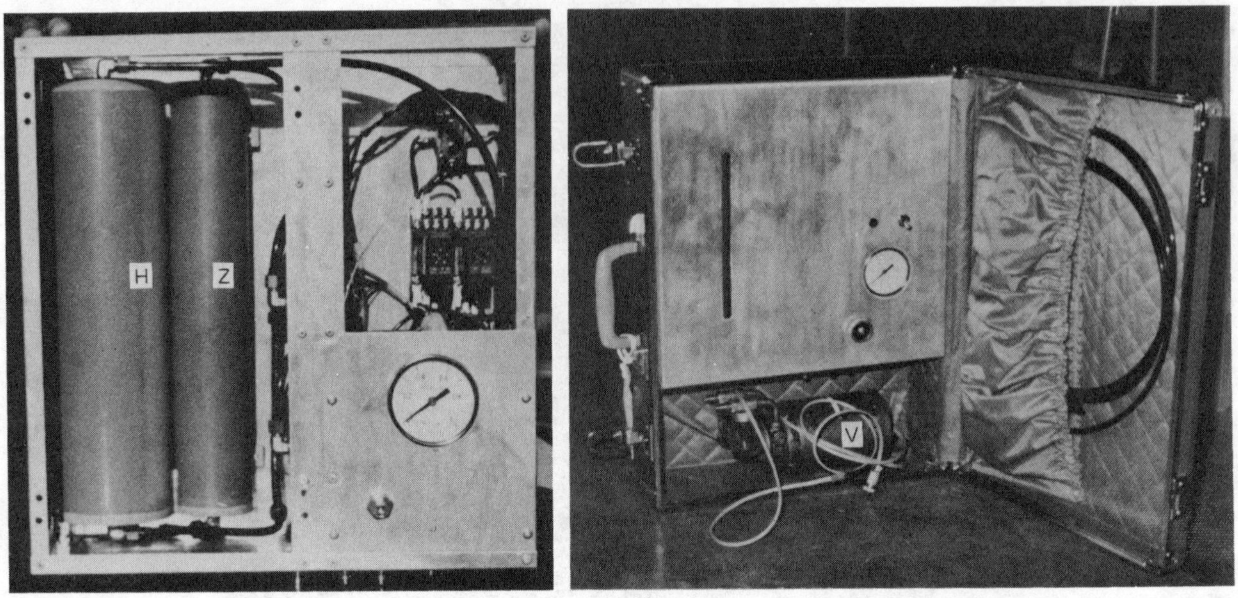

Fig. 8. Portable oxygen generator LP 015. Single column system; weight = 11 kg. Left: inside of generator. A pretreatment column is set behind the zeolite adsorption column (Z). Right: generator set in suitcase, showing vacuum pump (V).

THE SELECTIVITY OF CLINOPTILOLITE FOR CERTAIN HEAVY METALS

M. J. SEMMENS

Department of Civil Engineering, University of Illinois, Urbana, Illinois 61801, U.S.A.

and

M. SEYFARTH

Environmental Protection Agency, Rockford, Illinois 61100, U.S.A.

Abstract – The ammonium capacities of variously prepared samples of clinoptilolite varied between 1.70 and 2.25 meq/g depending upon the pretreatment technique.
pretreatment technique.
Samples of conditioned and acid-washed clinoptilolite in the sodium form were exposed to solutions of approximately 2×10^{-3} M Ba^{+2}, Pb^{++}, Cu^{+2}, Zn^{+2}, and Cd^{+2}. Both samples of clinoptilolite exhibited very high selectivities for barium and lead and somewhat lower selectivities for copper, cadmium, and zinc.

The clinoptilolite, which was not acid-treated, leached significant concentrations of sodium bicarbonate into solution, causing the solution pH to rise during equilibration. Nevertheless, analysis of the initial and final metal concentrations in solution suggested that only exchange took place. Measured selectivities with this zeolite sample were higher than those measured with the acid-washed clinoptilolite.

INTRODUCTION

Studies were initiated in 1974 to determine the feasibility of employing nitrifying bacteria to regenerate clinoptilolite that had been used to remove ammonium from wastewaters. Nitrifying bacteria oxidize ammonium to nitrate, and in the presence of sodium salts, the clinoptilolite is regenerated to the sodium form (Semmens and Goodrich, 1977; Semmens, Warg and Booth, 1977). In addition to ammonium, many wastewaters contain low concentrations of heavy metals. These metals can also be concentrated by the clinoptilolite during ion-exchange processes and perhaps released during regeneration. As in high concentrations these metals are toxic to nitrifying bacteria, they may seriously affect the efficiency of the biological regeneration process. For this reason the selectivity of clinoptilolite for a number of heavy metals was investigated.

Ames (1960) identified the selectivity series for the alkali metal cations, the alkaline earth metal cations, and trivalent aluminum and iron. The preference of clinoptilolite for various ions decreases in the order Rb^+, K^+, NH_4^+, Ba^+, Sr^{++}, Na^+, Ca^{+2}, Fe^{+2}, Al^{+3}, Mg^{+2}, and Li^+ in exchange for cesium. However, there is no information available concerning the selectivity of the sodium form of clinoptilolite for heavy metal ions.

Equilibrium studies were therefore conducted to determine the selectivities of clinoptilolite for Ba^{+2}, Cd^{+2}, Cu^{+2}, Pb^{+2}, and Zn^{+2} in exchange with sodium.

EXPERIMENTAL

Sample preparation

Clinoptilolite from Buckhorn, New Mexico, was provided by the Double Eagle Petroleum and Mining Company, Casper, Wyoming. The crushed ore was graded using U.S. standard sieve sizes, the 20 X 25-mesh (0.0701–0.840 mm) fraction being retained for use in this study. Most of the past studies of clinoptilolite selectivities for metal ions have employed acid-treated samples, as the acid treatment is intended to rid the zeolite of basic impurities such as calcite. Because the zeolite is not acid-treated prior to use in wastewater treatment, it is questionable whether selectivities measured with acid-treated zeolite accurately reflect the selectivities that would be observed in practice with untreated zeolite. For this reason samples of clinoptilolite were prepared in several different ways for this study.

A 180 g sample of *conditioned* zeolite was prepared by exposing washed zeolite to two sequential treatments of 0.5 M NH_4Cl and 1 M NaCl. Two liters of each solution was contacted with the zeolite at a flowrate of 20 ml/min (4.5 bed volumes/hr).

A 50 g sample of the conditioned zeolite was given an additional treatment with 2000 ml of 0.5 M NH_4Cl followed by elution with 7 liters of 1 M NaCl and another 7 liters of 1 M NaCl adjusted to pH 10 at the same flowrate. The zeolite sample is referred to as *reconditioned*.

A sample of *acid-washed zeolite* was prepared by paddle stirring 130 g of conditioned clinoptilolite in 1.5 liters of 10% nitric acid for an hour, just vigorously enough to keep the clinoptilolite in suspension (Ames, 1963). The acid-washed clinoptilolite sample was then converted to the sodium form by treatment with 6 liters of NaCl at a flow rate of 2 ml/min; this process took 2 days.

The sodium zeolite samples were rinsed with deionized water, oven-dried overnight at 103°C, and then stored in a desiccator in capped glass bottles.

Two methods were employed for measuring the ion-exchange of the zeolite: a known weight of clinoptilolite was treated with an excess of 0.5 M NH_4Cl and rinsed well with deionized water. The ammonium was then eluted from the clinoptilolite with 1 M NaCl and its concentration in the eluate was measured by means of an ammonium-selective electrode; additional ion-exchange capacity measurements were made by eluting the sodium zeolite with 1 M KCl and subsequent measurement of Na by flame photometry.

Analytical equipment

Ammonium concentrations were determined using an ammonium-selective electrode (ORION 95-10) and an ORION Digital pH meter (Model 701) in conjunction with an ORION electrode switch (Model 605). The meter was sensitive to 0.01 millivolts or pH units.

Samples were analyzed for metals at the Illinois Environmental Protection Agency Laboratory in Champaign, Illinois. Sodium concentrations were determined on a Perkin Elmer atomic absorption spectrophotometer (Model 403) connected to a Perkin Elmer recorder (Model 65). The divalent metals (Ba, Cd, Cu, Pb, Zn, Ca, and Mg) were analyzed on a Jarrel Ash atomic absorption spectrophotometer (Model 810) coupled with a Fisher Recordall Series 500.

Experimental techniques

Three equilibrium studies were performed using five divalent metal salts: $Ba(NO_3)_2$, $CdCl_2{\cdot}2.5H_2O$, $CuCl_2{\cdot}2H_2O$, $Pb(NO_3)_2$, and $Zn(NO_3)_2{\cdot}6H_2O$.

In the ***first equilibrium study*** 0.1–1.0 g amounts of reconditioned zeolite were shaken with 125 ml of 10^{-3} M solutions of each metal salt at 160 c/min on a rotary shaker for 2 days. The solution concentrations of the heavy metals were then measured. The equilibrium pH and temperature were also measured.

In the *second equilibrium study* the amount of zeolite was fixed, and the initial solution concentration of the heavy metals was varied. One hundred milliliters of 1.6×10^{-9} M to 1×10^{-2} M solutions adjusted to pH of 4.5 with nitric acid and caustic were added to 0.2 g of *acid-washed* zeolite in each bottle. Nine tests were conducted with each metal to cover a wide range of initial ionic fractions. Nine controls consisting only of 100-ml samples of each solution were also carried.

Equilibration of the samples and controls was achieved by shaking at 160 c/min for 48 hours. After 1 day of shaking, the pH of the solutions was measured and adjusted to pH 4.5 with NaOH and after the second day of shaking equilibrium pH and temperature were recorded and the metal concentrations of the solutions were measured.

The *third equilibrium study* was the reverse of the second study. Here 100 ml of 10^{-2} M $NaNO_3$ (pH 4.51) was added to the 0.2 g samples of zeolite from the second study. Only those samples exposed to the five most concentrated solutions of each heavy metal were tested. These twenty-five samples, plus a control consisting of 100 ml of $NaNO_3$ solution without any zeolite, were brought to equilibrium by shaking at 160 c/min for 2 days and evaluated in the manner described above.

All solutions were decanted into capped, polyethylene bottles containing 0.8 ml of nitric acid for every 100 ml of liquid. Five milliliters of 60 g/l KCl was added to all samples to be analyzed for barium and/or sodium to control ionization interferences (Kahn, 1968). For this reason the samples were analyzed for sodium last.

Finally, a *leaching study* was conducted. One hundred milliliters of deionized water was added to two 1-g samples of reconditioned zeolites and to two 1-g samples of acid-washed zeolite. After measuring the initial pH of the samples and of two controls containing no zeolite, the samples were shaken at 160 c/min for 1 week. Equilibrium pH and temperatures were measured and recorded. Following the procedures in Standard Methods (1971), the solutions were analyzed for alkalinity by the potentiometric method, for chloride by argentometric titration using potassium chromate indicator, and for suspended solids by drying at 103°C using glass fiber filters. Sodium, calcium, and magnesium were determined by atomic absorption analysis of 25-ml subsamples to which 1 ml of 60 g/l KCl was added to control ionization interferences.

RESULTS

Ion-exchange capacity

Measured ion-exchange capacities are presented in Table 1. The results proved more variable than anticipated, and a number of tests were made to identify the cause of this variation. The volume of ammonium solution employed to convert the clinoptilolite to the ammonium form was varied from 100 to 1000 ml but had no effect on either the magnitude or the variance of the measured capacities. However, the reconditioning process and repeated capacity determinations upon a single sample of clinoptilolite caused a significant increase in measured capacity. For example, the average capacity of conditioned clinoptilolite increased from 1.69 meq/g during the first capacity measurement to 2.23 meq/g when the capacities of individual samples were measured for the third time.

The acid-washed clinoptilolite had an ion-exchange capacity approximately equal to that of the reconditioned clinoptilolite when the first capacity tests of the two samples were compared. The measured ammonium and sodium capacities also agreed very well for the samples of acid-washed zeolite. Since samples of reconditioned and acid-washed clinoptilolite were employed in the equilibrium studies, the ion-exchange capacity was taken as 1.85 meq/g.

Equilibrium studies

The results of the first equilibrium study are plotted in Figs. 1 to 5. The ordinate axis in each diagram, $\bar{X}_M^{++}$, represents the computed fractional ion-exchange capacity of the clinoptilolite occupied by the individual heavy metal ions, and the abscissa, X_M^{++}, depicts the equivalent ionic fraction of heavy metal in solution at equilibrium. These diagrams indicate that heavy metals are concentrated well by clinoptilolite at low solution fractions of the heavy metals.

The pH of the solutions ranged between 5.5 and 6.5 initially but increased during equilibration. The extent of the pH increase correlates with the amount of reconditioned clinoptilolite in each bottle. Figure 6

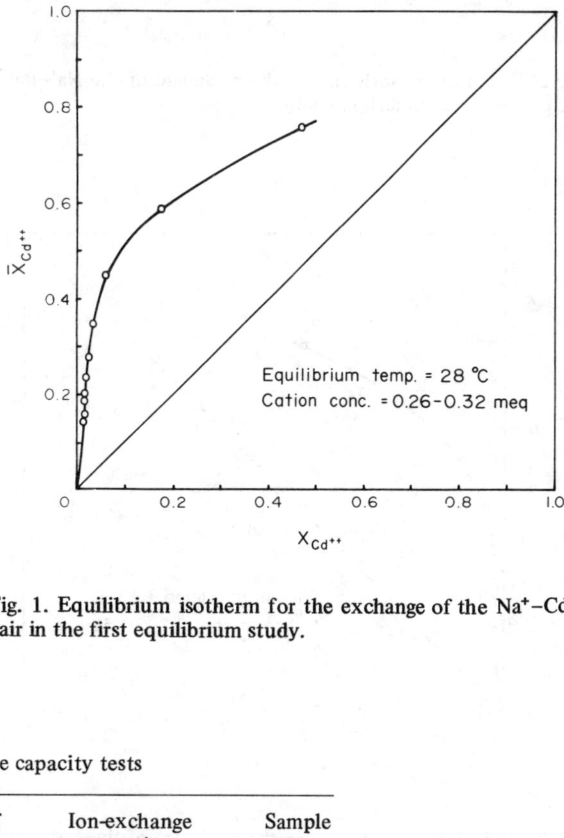

Fig. 1. Equilibrium isotherm for the exchange of the Na^+–Cd^{+2} pair in the first equilibrium study.

TABLE 1. Results of the capacity tests

Type of clinoptilolite sample*	Number of samples tested	Ion-exchange capacity (meq/g)	Sample test number
Conditioned	5	1.69 ± 0.08	1
Conditioned	3	2.11 ± 0.10	2
Conditioned	2	2.23 ± 0.01	3
Reconditioned	6	1.82 ± 0.23	1
Reconditioned	2	2.25 ± 0.07	2
Acid washed	4†	1.85 ± 0.05	1

*Clinoptilolite from Buckhorn, New Mexico; supplied by Double Eagle Petroleum and Mining Company.

†Capacities measured by both ammonium- and sodium-elution studies.

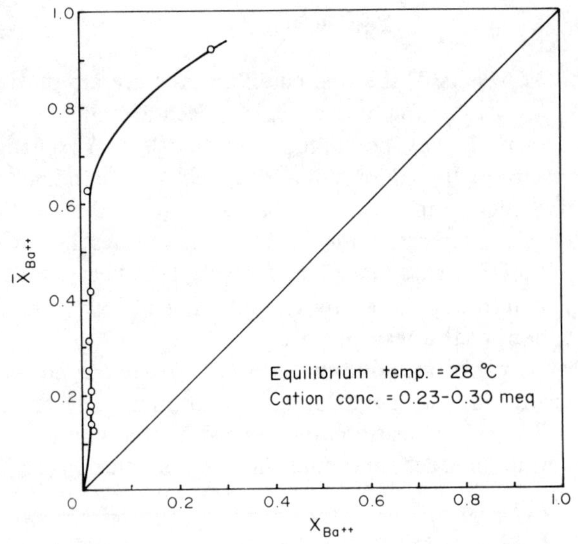

Fig. 2. Equilibrium isotherm for the exchange of the Na^+–Ba^{+2} pair in the first equilibrium study.

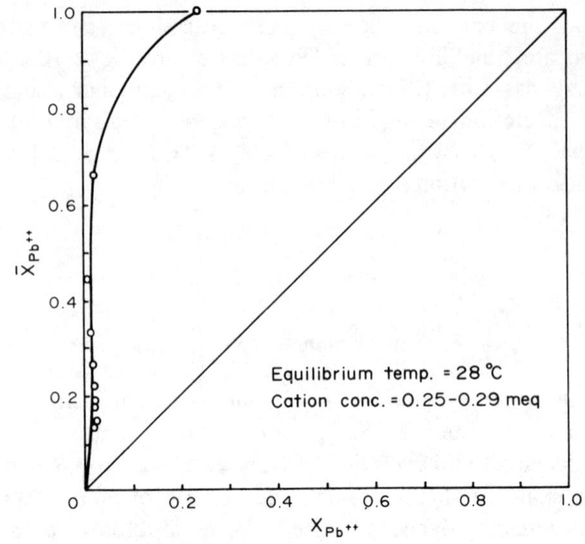

Fig. 4. Equilibrium isotherm for the exchange of the Na^+–Pb^{+2} in the first equilibrium study

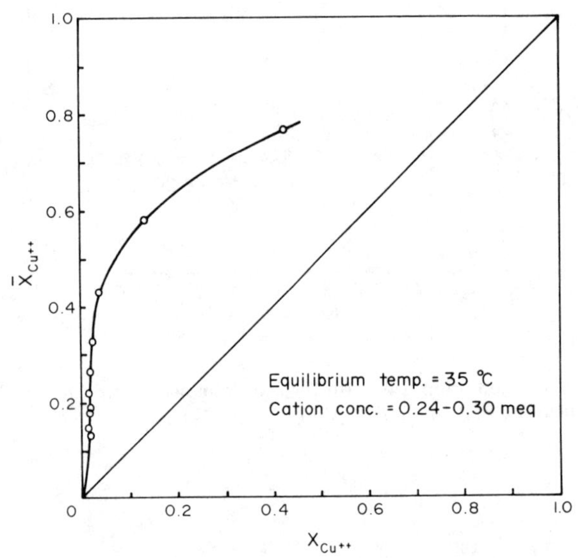

Fig. 3. Equilibrium isotherm for the exchange of the Na^+–Cu^{+2} pair in the first equilibrium study.

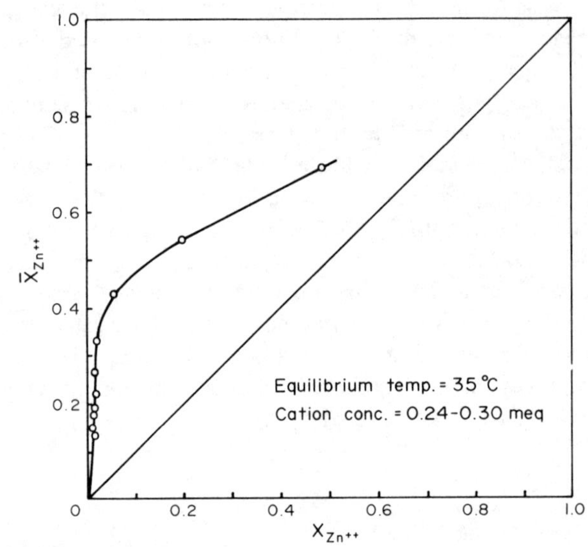

Fig. 5. Equilibrium isotherm for the exchange of the Na^+–Zn^{+2} pair in the first equilibrium study.

depicts the final measured pH values for the studies conducted with barium, cadmium, and lead.

The results of the second and third equilibrium studies are plotted in Figs. 7 to 11. These studies were conducted at a lower pH (4.5) with acid-washed clinoptilolite and show a pattern similar to the results of the first equilibrium test discussed above. Lead and barium are much more preferred by clinoptilolite than are copper, cadmium, and zinc; the last three ions exhibit almost identical equilibrium curves. When acid-washed clinoptilolite was employed the maximum values of $X_M{}^{+2}$ were significantly lower.

The results of the third equilibrium study demonstrate that, with the exception of the zinc and lead isotherms, the curves depicted in Figs. 7 to 11 show good reversibility. In the case of lead and zinc, however,

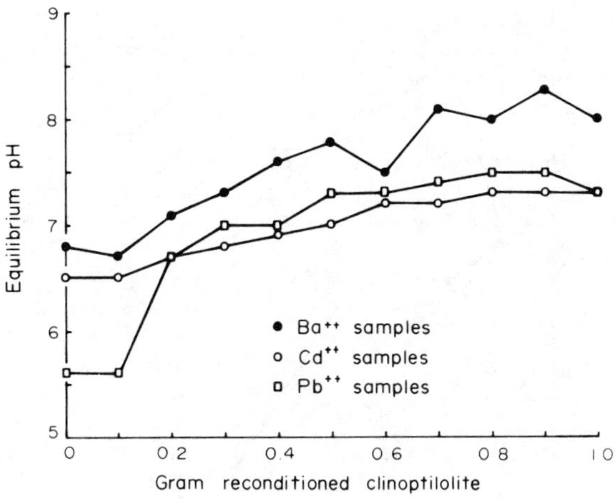

Fig. 6. Equilibrium pH of the barium, cadmium, and lead samples from the first equilibrium study.

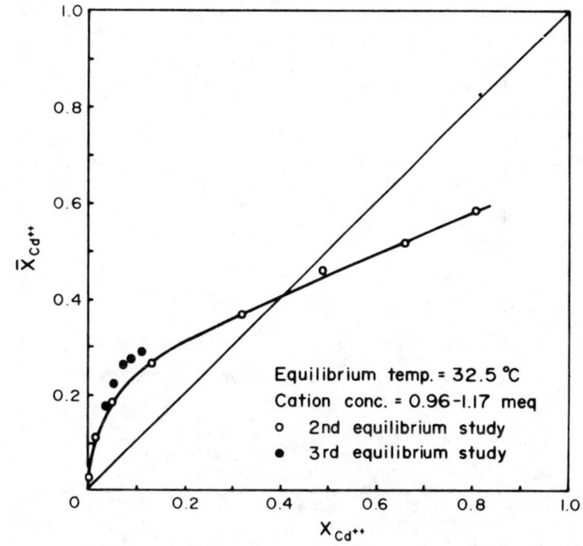

Fig. 8. Equilibrium isotherm for the exchange of the Na^+–Cd^{+2} pair in the reverse equilibrium studies.

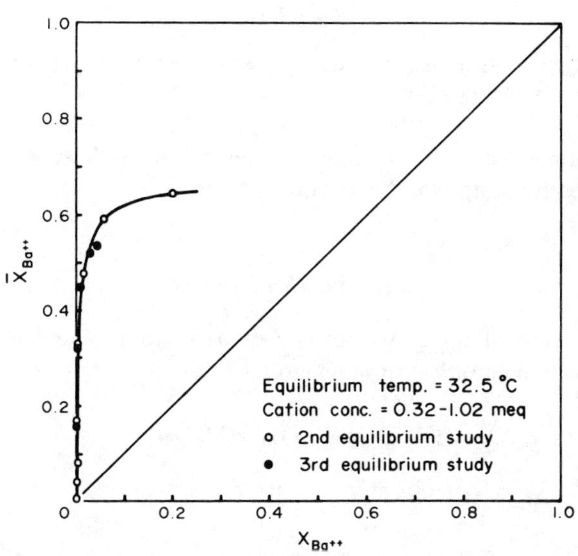

Fig. 7. Equilibrium isotherm for the exchange of the Na^+–Ba^{+2} pair in the reverse equilibrium studies.

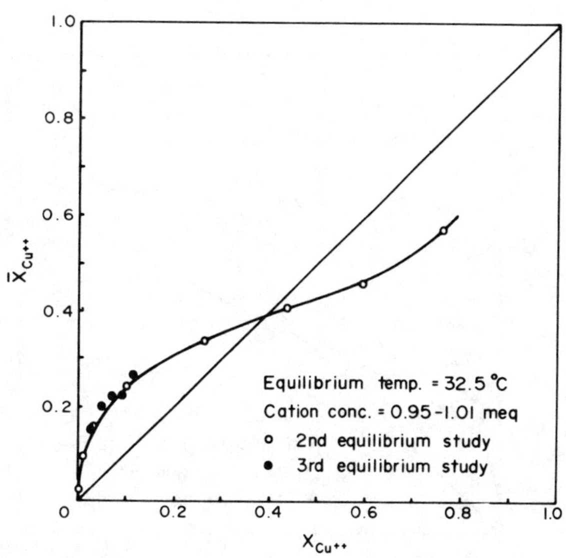

Fig. 9. Equilibrium isotherm for the exchange of the Na^+–Cu^{+2} pair in the reverse equilibrium studies.

the metal ions were apparently not released from the clinoptilolite to the extent indicated in the curves in Figs. 10 and 11.

In the second equilibrium study the measured pH declined during the 2-day equilibration period. Intermediate pH values were recorded after 1 day, and the pH was adjusted to 4.5. The final equilibrium pH values were also recorded. The results of lead–sodium exchange are shown in Fig. 12. The magnitude of the pH drop increased as the initial concentration of the heavy metal increased.

All equilibrated samples were slightly milky in appearance due to attrition of clinoptilolite during shaking.

Table 2 presents the results of the leaching study in which samples of clinoptilolite were shaken with deionized water for a week. The final pH and alkalinity values of the controls and of the samples of acid-washed zeolite are identical; no measurable change was detected. The reconditioned clinoptilolite, however, caused the pH and alkalinity to increase significantly. In addition, calcium, sodium, and magnesium ions leached into solution in varying concentrations. Chloride was not detected in any of the solutions. Sodium was displaced

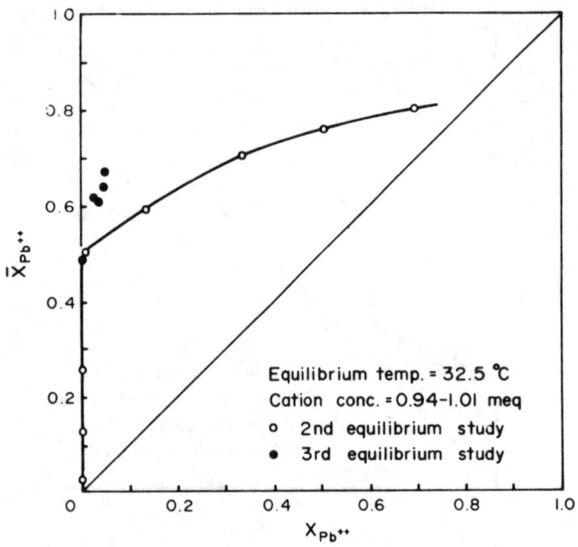

Fig. 10. Equilibrium isotherm for the exchange of the Na^+–Pb^{+2} pair in the reverse equilibrium studies.

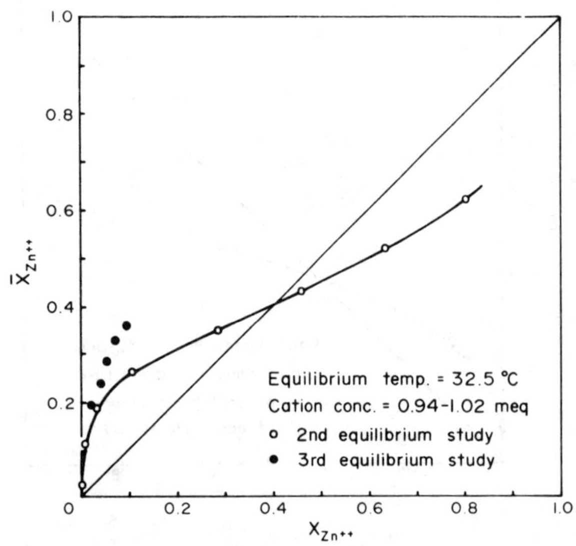

Fig. 11. Equilibrium isotherm for the exchange of the Na^+–Zn^{+2} pair in the reverse equilibrium studies.

in the largest amount, with a maximum concentration of 35 mg/l being found following equilibration with the reconditioned zeolite. Calcium and magnesium were never found in concentrations exceeding 0.3 and 2.2 mg/l, respectively. In all cases much lower concentrations of the ions were found in solutions exposed to acid-washed zeolite than in solutions contacted with the reconditioned zeolite.

The cation concentrations reported in Table 2 are larger than the reported alkalinities, and other anions, such as sulfate for which analyses were not made, must also have been present in solution. The large amount of suspended solids recorded in tests with both types of zeolite samples is due to zeolite fines.

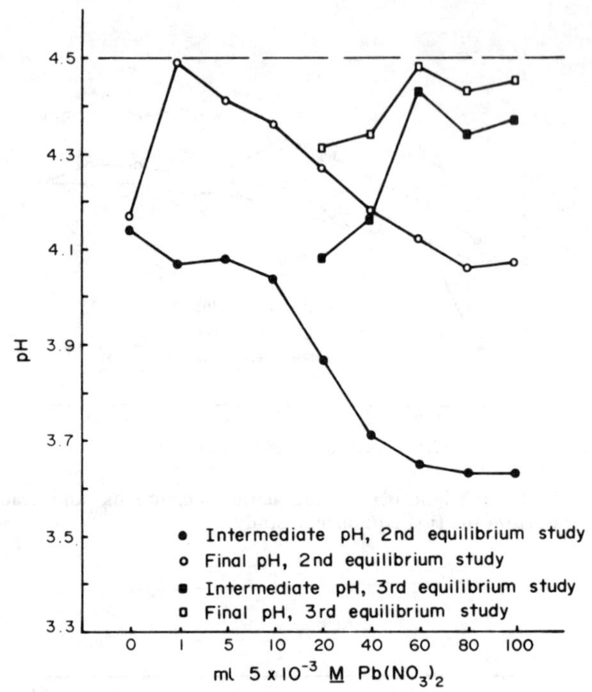

Fig. 12. pH during the exchange of the Na^+–Pb^{+2} pair in the first equilibrium study.

DISCUSSION

According to Mumpton (1960) clinoptilolite has a general empirical formula close to

$$(Na_2O)_{0.70}(CaO)_{0.10}(K_2O)_{0.15}(MgO)_{0.05}$$

$$\cdot Al_2O_3 \cdot 8.5\text{–}10.5SiO_2 \cdot 6\text{–}7H_2O$$

Both SiO_4 and AlO_4 tetrahedra are arranged to form a repeating series of cavities or cages with interconnecting channels. Recent studies by Alberti (1975) suggest that the structure of clinoptilolite is similar to that of heulandite to which it is closely related. The heulandite structure has elliptical windows of two different sizes, with free dimensions of approximately 3.5 X 7.9 and 3.0 X 4.4 Å (Merkle and Slaughter, 1967).

In the preparation of the zeolite, acid washing may have dissolved some of the clinoptilolite. However, clinoptilolite is more acid resistant than other zeolites (Ames, 1960), and the extent of dissolution was probably minor. The use of sodium chloride at pH 10 was the harshest alkaline condition, and chemical attrition of the zeolite at this pH should also be negligible (Barrer *et al.*, 1967; Koon and Kaufman, 1971).

TABLE 2. Results of the leaching study in which clinoptilolite samples were shaken with deionized water for 1 week*

Sample†	Initial pH	Final pH	Alkalinity (mg/l as $CaCO_3$)	Cl (mg/l)	Suspended solids (mg/l)	Ca (mg/l)	Mg (mg/l)	Na (mg/l)
1	5.77	5.20	2.6	0	16	0.1	0.0	0
2	5.79	5.95	2.6	0	32	0.2	0.1	0
3	6.73	8.10	15.5	0	404	0.3	2.2	35
4	6.52	8.24	10.4	0	252	0.3	1.9	32
5	5.77	5.68	2.6	0	216	0.1	0.8	14
6	5.74	5.50	2.6	0	172	0.1	0.7	13

*Equilibrium temperature = 33°C.

†Samples 1 and 2 were controls without zeolite; 3 and 4 were samples of reconditioned clinoptilolite; and 5 and 6 were samples of acid-washed clinoptilolite.

Ion-exchange measurements

The average ion-exchange capacities of the several different preparations of clinoptilolite are not the same. The capacity of the conditioned zeolite (1.69 ± 0.08 meq/g) is smaller than those of the reconditioned and acid-washed zeolite (1.82 ± 0.23 and 1.86 ± 0.05 meq/g, respectively). Although the latter two preparations have approximately the same average capacity, the variability for the reconditioned zeolite is much greater than for the acid-washed zeolite. Reconditioning with a salt solution adjusted to pH 1.0 should have removed acidic impurites from the zeolite. Similarly, acid-washing should have removed any basic impurities.

The ion-exchange capacity of a single zeolite sample increased with each repetition of the determination and appeared to level off at about 2.24 meq/g. Jorgensen *et al.* (1976) also noted that the capacity of a sample of clinoptilolite increases with the number of regenerations. Their regeneration process consisted of treatment with 4% NaOH for 30 minutes followed by a rinse with one of deionized water for 40 minutes. After repeating this procedure four times, the measured capacity was found to be constant. They concluded that sodium hydroxide "activates" the clinoptilolite. The results of the present study suggest that it is not necessarily the sodium hydroxide that is responsible for the increased capacity, since in this study neutral sodium chloride was used. The increase may be related to the number of exchange cycles that the zeolite is exposed to after being oven dried at 103°C. This may be seen by comparing the conditioned and reconditioned zeolite capacities. Since the reconditioned zeolite was exposed to two additional conditioning cycles, one might expect the first measured capacity of this sample to be approximately equal to the third measured capacity of the conditioned zeolite. These values, however, are significantly different; and the first measured capacity values of the conditioned and reconditioned samples are close to one another.

Equilibrium studies

The increase in solution pH during the first equilibrium study was clearly caused by the leaching of basic impurities from the clinoptilolite into solution. This was also demonstrated in the leaching study on reconditioned clinoptilolite. Analysis of deionized water following exposure to reconditioned clinoptilolite samples for 1 week revealed that high concentrations of sodium and bicarbonate were present. The elevated pH caused some concern as to whether or not the metals were precipitating in the system. Using the solubility products of the metal hydroxides and the metal carbonates, it is possible to calculate the pH at which precipitation occurs in a 10^{-3} M solution of the individual metals. For calculating carbonate solubilities, the total carbonate concentration was estimated from the results of the leaching study to be approximately $10^{-3.6}$ M. The results listed in Table 3 indicate that in some of the tests precipitation may have occurred. For example, lead and cadmium carbonates are the least soluble, and a 10^{-3} M solution of these metals should have precipitated at pH of approximately 3.7 and 4.8, respectively.

No metal precipitates were identified visually; with the exception of copper hydroxide, they are white or colorless and zeolite fines would have masked their presence. The suspended solids in the equilibrated solution consist of clinoptilolite fines only, and have the same beige color as the clinoptilolite particles.

TABLE 3. Carbonate and hydroxide solubility products and calculated pH and pCO_3 values required to form a precipitate in a 10^{-3} M solution of the specified metal

Precipitate	pK_{so}*	pCO_3	pH†
$Ba(OH)_2 \cdot 8H_2O$	2.98		14.01
$Cd(OH)_2$	14.19		8.40
$Cu(OH)_2$	18.79		6.10
$Pb(OH)_2$	14.36		8.32
$Zn(OH)_2$ or ZnO	15.74–17.04		6.98–7.63
$BaCO_3$	8.67	5.67	8.23
$CdCO_3$	12.07	9.07	4.83
$CuCO_3$	9.63	6.63	7.27
$PbCO_3$	13.17	10.17	3.73
$ZnCO_3$	10.27	7.27	6.63

*Values of pK_{so} for hydroxide precipitates calculated from equations given by Pourbaix (1966); and those for carbonate precipitates are averages of data from *Stability Constants* (1964).

†Calculation of the pH for metal carbonate precipitation assumed an alkalinity equal to 12 mg/l as $CaCO_3$ (see Table 2).

Although the solubility data listed in Table 3 indicates the possibility of carbonate precipitation in some of the systems, one must remember that bicarbonate is released to solution slowly over the length of the equilibrium study. The metal ions, therefore, have an opportunity to exchange with sodium ions of the zeolite before precipitation occurs. The metal ions may, however, precipitate at the zeolite surface or within the zeolite itself as others have observed (Vaughan, 1976; Barrer and Townsend, 1976). When the metal is concentrated on the zeolite, the solubility product(s) may be exceeded, and precipitation may follow.

Analysis of the initial and final metal- and sodium-ion concentrations reveal that the final total cation concentration was slightly higher than the initial concentration. Final cationic concentrations can be calculated assuming stoichiometric exchange and using data for sodium leaching taken from Table 2. This suggests that no precipitation occurred within the zeolite. However, since the final metal concentrations were measured in decanted and acidified solutions, the above mass balance does not preclude the possibility of metal precipitation in solution. If precipitation did occur in solution, the measured solution concentrations would represent the sum of metal as precipitate and metal in true solution. If this were so, the true values of X_M^{++} in Figs. 1 to 5 should be lower. This would, of course, mean that the clinoptilolite is more selective for the metals than Figs. 1 to 5 suggest.

In the second equilibrium study acid-washed clinoptilolite was employed at an ambient pH of approximately 4.5. Under these conditions all of the heavy metals were soluble. The analyses indicated that equivalent amounts of sodium were released by the exchange of heavy metals. Unlike the first equilibrium study, there was no net increase in cation concentration during equilibration.

The equilibrium curves obtained in the second study were similar to those of the first study. In both studies lead and barium are much more preferred by the clinoptilolite than are copper, cadmium, and zinc. With respect to barium and lead, much lower metal concentrations remain in solution at low values of X_M^{++} in the second study than in the first study. For example, the concentration of lead in equilibrium with $X_{Pb}^{++} \simeq 0.1$ was approximately 2 mg/l in the first study and 0.1 mg/l in the second. Precipitation in solution during the first study may be responsible for the difference. However, both barium and lead were taken up to a greater extent by the reconditioned clinoptilolite at higher solution concentrations of these metals.

Reconditioned clinoptilolite shows better uptake of Cd, Zn, and Cu over the whole range of concentrations investigated. Why the reconditioned clinoptilolite should be more selective for copper, zinc, and cadmium than the acid-washed clinoptilolite is not clear. If the clinoptilolite ore is treated with mineral acid, basic impurities such as calcite are removed if they are present. In addition, hydronium ions may exchange for cations on the zeolite and hydrolysis of Al–O–Si bonds may take place. Barrer (1964) showed that refluxing clinoptilolite with 1 N acid solutions at 100°C for 4 hours tended to increase the pore volume of the zeolite making the structure capable of sorbing polar molecules of larger molecular dimensions. Although the acid-treated clinoptilolite in the present study was not prepared under the same conditions, the action of the acid in removing impurities would improve the accessibility of the exchange sites to large, hydrated heavy metals. Precipitation of the metals within the zeolite pores would explain these differences.

The decrease in pH during the second equilibrium study may be connected with the lower selectivity of acid-washed clinoptilolite for the metals. A lower pH can be explained if metal hydroxides were precipitating; however, this does not agree with the observed stoichiometric exchange. Since the acid-washing clinoptilolite caused no similar pH decrease in the leaching study when it was contacted with deionized water for 1 week, lower pH cannot be ascribed to the displacement of excess acid. Furthermore, such an explanation would not account for the observed correlation of pH with metal-ion concentration. The data, therefore, suggest that some hydronium ions occupy exchange sites even after extended exposure to sodium chloride and that these ions were displaced by the heavy metals during the second equilibrium study. The reverse equilibrium study shows a marginal pH decrease as well, weakening such an argument.

The reverse equilibrium study in which the metal ions were exchanged off the clinoptilolite indicates that equilibrium may not have been achieved in the case of lead, zinc, and cadmium. The data plotted in Fig. 7 correspond closely with the curve obtained in the first equilibrium study. The zinc and cadmium data fall between the curves for the first and second studies. It is therefore probable that more than a few days of contact are required before true equilibrium is achieved.

Because of their size, certain heavy metals may not be able to exchange with all of the sodium ions within the crystal structure of the zeolite. In Figs. 7 and 8 the equilibrium curves for barium and cadmium level off rapidly at $X_M{}^{++} \simeq 0.6$–0.8. Sherry (1969) showed that Ba^{++} could not replace 16 of the 85 Na^+ ions in the unit cell of Linde X zeolite. Only 81% barium loading could be achieved. With lead and barium exchange on acid-washed clinoptilolite, the maximum capacity for these metals may also be less than 100%.

Inspection of Figs. 2 and 4 indicates that the metal uptakes of the reconditioned zeolite approach the maximum measured capacity. The value chosen for the maximum capacity is clearly very important, given the variability in measured capacities, and it is not really possible to draw conclusions from this observation.

Selectivity series

Table 4 presents ionic radii, hydrated ionic radii, and free energies of hydration for the ions employed in this study. If the ionic radii are compared with the free dimensions of the clinoptilolite channels, it is apparent that all of the unhydrated ions can pass readily through the channels, but since the hydrated ions are approximately the same size as the channel dimensions, they may exchange only with difficulty. It follows that for exchange to take place, at least some of the waters of hydration must be stripped from the solvated ions.

Eisenman (1962) proposed a very simple ion-exchange model for predicting selectivity sequences. In this model, selectivity of exchange cations is accounted for solely in terms of their hydration free energies and of their energies of electrostatic interaction with the zeolite-fixed anions. Eisenman's original model was developed for the exchange of monovalent cations, but Sherry (1969) extended the model to the exchange of a divalent cation for two monovalent cations. Clinoptilolite has a silicon to aluminum atom ratio of approximately 4.3 to 5.3 and a correspondingly low ion-exchange capacity. The low capacity provides a weak anionic field within the zeolite. Consideration of Eisenman's model in the case of exchange on a zeolite in the presence of a weekly anionic field indicates that selectivity is predominantly determined by the free energies of hydration of the competing ions. For a more detailed discussion the reader is referred to Sherry (1969).

TABLE 4. Properties of cations used in this study

Cation	Ionic radius (Å)	Hydrated ionic radius† (Å)	Free energy of hydration (kcal/g-ion)
Ba^{+2}	1.43	4.04	−315.1
Cd^{+2}	1.03	4.26	−430.5
Cu^{+2}	0.82*	4.19	−498.7
Pb^{+2}	1.32	4.01	−357.8
Zn^{+2}	0.83	4.30	−484.6
Na^+	0.98	3.58	−98.2

*Ionic radius from thermochemical data. All other ionic radii from Goldschmidt (1929).

†Hydrated ionic radii from Nightingale (1959).

The free energies of hydration listed in Table 4 indicate the following selectivity sequence: $Ba^{+2} > Pb^{+2} > Cd^{+2} > Zn^{+2} > Cu^{+2}$. Thus copper, with the largest hydration energy, prefers the solution phase where it may satisfy its hydration requirements, and barium, with the least hydration energy, prefers the zeolite phase. The observed selectivity sequences show $Pb \simeq Ba >> Cu, Zn, Cd$.

CONCLUSIONS

The ion-exchange capacity of clinoptilolite depends significantly on the way in which it is pretreated. The capacity tends to increase with repeated capacity determinations on the same zeolite sample. It is recommended, therefore, that zeolite samples which have not been acid-washed be conditioned with at least two exhaustion–regeneration cycles before equilibrium capacities are measured.

Significant differences were recorded in the selectivities of Ba^{+2}, Pb^{+2}, Cu^{+2}, and Zn^{+2} for the reconditioned and acid-washed clinoptilolite; the reconditioned clinoptilolite was apparently more selective for the metals studied than the acid-washed sample.

Clinoptilolite is highly selective for barium and lead and should remove these metals from water and wastewater. It is considerably less selective for copper, cadmium, and zinc, although these metals are preferred over sodium.

ACKNOWLEDGMENT

The support of Illinois Water Resources Center is gratefully acknowledged. This research was conducted by M. Seyfarth in partial fulfilment of her Masters Degree in Environmental Engineering at the University of Illinois, 1975.

REFERENCES

Alberti, A. (1975) The crystal structure of two clinoptilolites: *Tscher. Min. Petrol. Mitt.* **22**, 25-37.

Ames, L. L., Jr. (1960) The cation sieve properties of clinoptilolite: *Amer. Mineral.* **45**, 689-700.

Ames, L. L., Jr. (1963) Mass action relationships of some zeolites in the region of high competing cation concentration: *Amer. Mineral.* **48**, 868-882.

Barrer, R. M. (1964) Molecular sieves: *Endeavour* **23**, 122-130.

Barrer, R. M., Papadopoulos, R. and Rees, L. V. C. (1967) Exchange of sodium in clinoptilolite by organic cations: *J. In. Nucl. Chem.* **29**, 2047-2063.

Barrer, R. M. and Townsend, R. P. (1976) Transition metal ion exchange in zeolite. Part 1. Thermodynamics of exchange of hydrated manganese (II), cobalt (II), copper (II), nickel (II) and zinc (II) ions in ammonium mardertiles: *J. Chem. Soc. Fara. Trans.* I. **72**(3), 661-673.

Eisenman, G. (1962) Cation selective glass electrodes and their mode of operation: *Biophys. J. Supp.* **2**(8), 258-323.

Goldschmidt, V. M. (1929) Crystal structure and chemical composition: *Trans. Farad. Soc.* **25**, 253-283.

Jorgenson, S. E., Libor, O., LeaGraber, K. and Barkacs, K. (1976) Ammonia removal by the use of clinoptilolite: *Wat. Res.* **10**(3), 213-224.

Koon, J. H. and Kaufman, W. J. (1971) *Optimization of Ammonia Removal by Ion Exchange Using Clinoptilolite*: S.E.R.L., Report No. 71-5, University of California at Berkeley.

Merkle, A. B. and Slaughter, M. (1967) The crystal structure of heulandite: *Amer. Mineral.* **52**, 273-276.

Mumpton, F. A. (1960) Clinoptilolite redefined: *Amer. Mineral.* **45**, 351-369.

Nightingale, E. R., Jr. (1959) Phenomenological theory of ion solvation. Effective radii of hydrated ions: *J. Phys. Chem.* **63**, 1381-1387.

Pourbaix, M. (1966) *Atlas of Electrochemical Equilibria in Aqueous Solutions*: Pergamon Press, New York.

No author (1964) *Stability Constants of Metal-Ion Complexes,* Special Publications No. 17, The Chemical Society, London, 1964.

No author (1971) *Standard Methods for the Examination of Water and Wastewater*, American Public Health Ass., 13th ed.

Semmens, M. J. and Goodrich, R. R., Jr. (1977) The biological regeneration of ammonium saturated clinoptilolite. Part. I. Initial observations; *Env. Sci. Tech.* **11**, p. 255.

Semmens, M. J., J. T. Warg and A. C. Booth (1977) Biological Regeneration of Ammonium-Saturated Clinoptilolite. II. Mechanism of Regeneration and Influence of Salt Concentration., *Env. Sci. Tech.* **11**, p. 260.

Sherry, H. S. (1969) The ion-exchange properties of zeolite: Chapter 3 in *Ion Exchange* **2**, ed. by Jacob A. Marinsky, Marcel Dekker, New York.

Vaughan, D. E. W. (1976) W. R. Grace & Co., Columbia, Maryland (personal communication).

BENEFICIATION OF NATURAL ZEOLITES FROM BOWIE, ARIZONA: A PRELIMINARY REPORT

K. D. MONDALE,*[1] F. A. MUMPTON,[2] and F. F. APLAN[1]

[1] *Mineral Processing Section, Pennsylvania State University, University Park, Pennsylvania 16802, U.S.A.*

[2] *Department of the Earth Sciences, State University College, Brockport, New York 14420, U.S.A.*

Abstract – Zeolite ore from a 4- to 5-foot thick series of volcanic tuffs is mined in several localities about 11 miles north of Bowie, Arizona. The zeolitic tuff consists of two distinct units: an upper, lower-grade, 4-foot thick series of thinly bedded layers, and a lower, higher-grade, 6-inch thick, massive "ore bed". Where the upper unit has been altered to zeolites, it consists of chabazite, erionite, clinoptilolite, and traces to minor amounts of quartz, feldspar, calcite, volcanic glass, smectite clay, and other gangue minerals. The lower "ore bed", however, is decidedly richer in chabazite, with subordinate amounts of erionite and clinoptilolite. The two samples tested here were found to contain approximately 60–70% and 95% total zeolites, respectively.

Beneficiation tests were carried out on both materials in an attempt to upgrade the chabazite and/or the total zeolite content. By size classification using wet cycloning and the shaking table, products containing up to 85% total zeolite were produced from the upper, lower-grade unit in approximately 50% of the total weight. Only a slight increase in the total zeolite content of the higher-grade "ore bed" was achieved using the same techniques. The effectiveness of the "ore bed" and beneficiated products from the upper, lower-grade unit for removing heavy metal ions from aqueous solution was demonstrated. "Ore-bed" samples took up more than 95% of the Cu^{+2} from 0.001 M solutions. Beneficiated samples of both ores removed more than 90% of the Pb^{+2} from 0.01 M solutions.

INTRODUCTION

The Bowie zeolite deposit is one of the largest known deposits of chabazite in the United States and currently is the only one being mined on a commercial scale. It is located largely in Graham County, Arizona, about 11 miles north of the town of Bowie. As described by Sheppard *et al.* (1978), zeolitic material is contained in a 4- to 5-foot thick, flat-lying sedimentary unit known as the "marker tuff". The "marker tuff" underlies an area of approximately 12 km^2 and generally consists of two distinct units: an upper, thin-bedded, lower-grade unit approximately 4 feet thick, and a lower, massive, higher-grade unit about 6 inches thick. The latter unit is often referred to as the "ore bed" and consists of over 90% total zeolites. The total zeolite content of the upper, thin-bedded unit is closer to 60%, the remainder consisting chiefly of volcanic glass, quartz, feldspar, calcite, illitic mica, and smectite. The principal zeolites in both units are chabazite and erionite, with smaller amounts of clinoptilolite and rarely analcime in different parts of the deposit. The erionite and clinoptilolite content of the upper, thin-bedded unit is noticeably higher than that of the "ore bed", and although it is currently stockpiled in favor of the higher-grade "ore bed", the upper unit represents a potential source of several million tons of chabazite-rich ore.

To determine the applicability of ore-dressing techniques for upgrading the zeolite content of both the lower-grade, thin-bedded unit and the higher-grade "ore bed" underlying it in the Bowie zeolite deposit, the following study was carried out, utilizing standard classification and concentration procedures along with shaking table and hindered settling techniques. This report is part of a much broader study of the beneficiation and ion-exchange characteristics of ten sedimentary zeolite ores from the western United States.

MINERALOGY OF THE ORE

All mineral separation processes are based on inherent differences in the physical or chemical properties of the constituent minerals of an ore, such as hardness, density, particle size, particle shape, etc. The Bowie zeolite ore is

Present address: Agrico Chemical Company Mulberry, Florida 33867, U.S.A.

no exception, and mineralogical examinations were therefore made of two, 25 lb samples of "ore bed" and the overlying, thin-bedded unit to be used as starting samples for the beneficiation studies. The samples were collected in 1974 from separate locations in the southeastern part of the deposit; and although they are not intended to duplicate ores from any one property, they are thought to be representative of the kind of ore that might be encountered anywhere in the deposit.

Both samples of ore consist of finely crystalline aggregates of zeolite minerals admixed with varying amounts of coarser gangue phases. Except for crude estimates of the amount of some of the coarse gangue materials present (especially volcanic glass), standard petrographic techniques were of limited use; and most identifications and attempts at quantification were made by powder X-ray diffractometry, using Ni-filtered, CuK_{α} radiation. Because of the lesser degree of crystallinity and sub-micrometer particle size of the Bowie zeolites, quantitative estimates of individual zeolites could not be made by a straightforward comparison of the peak heights on Bowie X-ray tracings with those of pure zeolites from other deposits. It was necessary to purify individual zeolites from the Bowie ore itself for use as X-ray standards. Accordingly, Bowie samples rich in chabazite or erionite were purified by repeated water classification and concentration. The amount of gangue remaining was estimated by careful X-ray diffraction and microscopic examination, and the zeolite percent determined by difference. By diluting these "standards" with amorphous volcanic glass, calibration curves were prepared and used to estimate the zeolite content of unknown samples and products. Additional details of this procedure are given by Mondale (1978).

Using the above techniques, supplemented with microscopic estimates of certain gangue constituents, the mineralogical composition of the two starting ores was determined, as listed in Table 1. The total zeolite content of the "ore bed" is between 90 and 95%, with chabazite being the principal zeolite present. Only traces of erionite, clinoptilolite, and gangue phases were noted. In contrast, the overlying, thin-bedded unit contains about 60–70% total zeolite and a significant amount of gangue, mainly unaltered volcanic glass. Chabazite is the predominant zeolite in this sample also, but slightly larger amounts of erionite and clinoptilolite were detected. Trace to minor quantities of quartz, feldspar, smectite, and mica are also present. Although it was not detected in X-ray patterns of the starting ore, magnetite was identified in tailings products of this ore sample.

The texture and morphological relationships of the zeolites in the Bowie ore are illustrated by the scanning electron micrograph of Fig. 1. Chabazite occurs as aggregates of extremely fine-grained rhombs, a few tenths of a micrometer in diameter. The rhombs form platelets and strings of crystals up to several micrometers in length. Erionite occurs as bundles of rod-shaped crystals, 30–40 μm long, with each bundle containing several score of individual rods approximately 1 μm in diameter. Commonly, erionite bundles are surrounded by clusters of chabazite. Clinoptilolite is not common in the samples examined in this study, but where present it occurs as isolated laths, or tablets, 4–5 μm thick and 30–40 μm in length. Smectite is intimately associated with the zeolites in the Bowie ore; and, according to Sheppard *et al.* (1978), it was generally the first authigenic mineral to form from the parent volcanic glass. It occurs both as substrata for later-formed crystals of zeolites and as coatings on earlier-formed zeolites.

Although crystal densities of zeolites are readily available from the literature, such data are commonly of little use in beneficiation processes. Here, "bulk" or "aggregate" densities are much more important, especially if separations are carried out on mineral aggregates rather than on completely liberated particles. Accordingly, the densities of several chabazite-rich grains (a few millimeters in size) from both starting ores and of volcanic glass-rich grains from the upper, thin-bedded unit were determined by use of a Berman balance. As shown in Table 2, the measured "bulk" densities are significantly less than the textbook values for the chabazite aggregates.

TABLE 1. Mineralogical composition of natural zeolite ores from Bowie, Arizona*

	High-grade† "ore bed"	Low-grade‡ thin-bedded unit
Chabazite	85–90	55–60
Erionite	5–10	5–10
Clinoptilolite	trace	~5
Volcanic glass	~5	10–15
Quartz	trace	~10
Feldspar	–	~10
Mica (illite)	trace	trace
Smectite (poorly crystalline)	trace	~2
Heavy minerals (magnetite, etc.)	trace	~1

*Zeolites determined by X-ray diffraction estimates; volcanic glass and other gangue phases estimated by optical microscopy.

†The "ore bed" is about 6 inches thick, homogeneous, and underlies the lower-grade, thin-bedded unit.

‡The thin-bedded unit is about 4 feet thick, consists of individual tuff units 1–4 inches thick, and overlies the "ore bed".

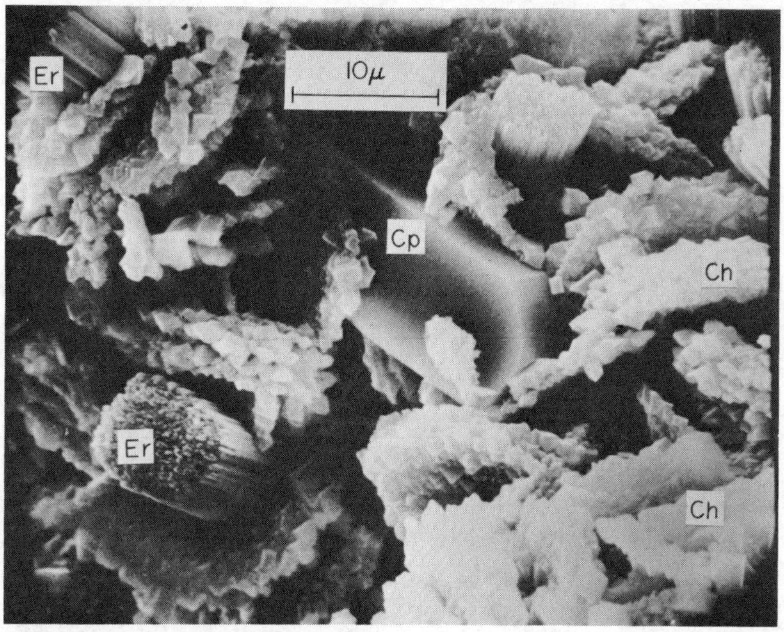

Fig. 1. Scanning electron micrograph of typical zeolite ore from Bowie, Arizona. Note the tabular nature of clinoptilolite, the bundles of erionite rods, and the sub-micrometer size of rhombic chabazite particles (micrograph courtesy of W. C. Ormsby).

TABLE 2. Density of chabazite-rich ore and volcanic glass from Bowie, Arizona

	Average measured* density (g/cm³)	Literature†,‡ density (g/cm³)
Chabazite (from "ore bed")	1.73	2.05–2.10
Chabazite (from lower-grade, thin-bedded unit)	1.62	2.05–2.10
Volcanic glass (from lower-grade, thin-bedded unit).	2.21–2.40	2.2–2.7

*Measured on millimeter-size aggregates by means of a Berman balance.
†Breck (1974, p. 212).
‡Rogers (1937, p. 516).

BENEFICIATION TESTS

Samples of both the higher-grade "ore bed" and the lower-grade, thin-bedded unit were crushed to a nominal −20 mesh, a size believed from preliminary studies to be sufficient to accomplish a substantial separation of gangue materials from zeolite. Screen analyses of crushed products are listed in Table 3. Note that both ores are relatively soft and that 20–25% of each reported to the −200-mesh-size fractions. From these results and the mineralogical data discussed above, it appeared that gangue materials could be separated from zeolites on the basis of their inherently greater hardness, higher densities, and coarser particle sizes. The flat nature of micaceous and clay minerals was also thought to offer a means of separation, as such materials characteristically settle at about one-half the rate of an equivalent spherical particle. The separation scheme shown in Fig. 2 was therefore employed.

In this procedure, initial comminution in a jaw crusher and crushing rolls was designed to achieve a significant liberation of gangue and zeolite. Hydrocycloning was carried out in a dilute pulp (about 10% solids) using a 30 mm Liquid-Solid Separation, Ltd. laboratory cyclone. The cyclone-overflow was flocculated, settled, and filtered. The underflow was transferred to a hindered-settling classifier where the sample was split into nominal +65- and −65-mesh fractions. Both fractions were fed separately to a one-sixteenth size, DECO Wilfley shaking table equipped with a standard sand deck, and concentrate, tailings, and middlings fractions were collected for each. Because the desired

Table 3. Screen analysis of natural zeolite ores from Bowie, Arizona*

	"Ore bed"			Low-grade, thin-bedded unit		
Tyler screen size	wt.% direct	Cum.% passing	Cum.% retained	Wt.% direct	Cum.% passing	Cum.% retained
20 X 28	13.4	86.6	13.4	6.1	93.9	6.1
28 X 35	15.0	71.6	28.4	9.7	84.2	15.8
35 X 48	12.0	59.6	40.4	10.1	74.1	25.9
48 X 65	10.0	49.6	50.4	11.8	62.3	37.7
65 X 100	8.3	41.3	58.7	12.8	49.5	50.5
100 X 150	7.9	33.4	66.6	13.7	35.8	64.2
150 X 200	8.5	24.9	75.1	13.6	22.2	77.8
–200	24.9			22.2		
Total	100.0			100.0		

*Crushed to –20 mesh and dry screened.

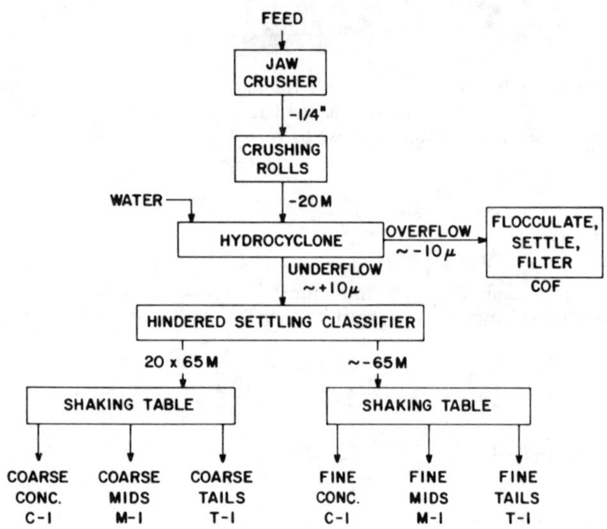

Fig. 2. Flowsheet for the beneficiation of Bowie, Arizona, zeolite ore.

zeolite has a lower specific gravity than the gangue minerals, the products are removed from the table in a reverse sequence to that commonly found in metalliferous ore dressing. Thus, the tailings discharge from the upper part of the deck, where normally the "heavy" minerals concentrate when treating metalliferous ores.

Beneficiation of the "ore bed"

The results of beneficiation studies of the "ore bed" are listed in Table 4; scanning electron micrographs of various products are shown in Fig. 3. For all practical purposes, no separation was accomplished by gravity concentration of either the coarse or the fine table feeds. This is probably because the gangue is extremely fine-grained in this sample and intimately associated with zeolites. The percentage of gangue is essentialy the same in all fractions, and montmorillonite, which normally has an inherent particle size of less than 1 μm, is present at about 2% in all fractions. The close association of zeolites and volcanic glass is illustrated in Fig. 3b. To liberate particles in this size range, the ore would have to be ground to a very fine particle size, thus lowering substantially the efficiency of most concentrating devices, such as a shaking table. In addition, the high-grade nature of this ore as received also makes concentration procedures not only difficult, but impractical.

The –10 μm cyclone-overflow fraction was somewhat enriched in zeolite. This was achieved by the near total elimination of volcanic glass. As might be expected, smectite clay was concentrated in cyclone-overflow fractions; however, in the finest portion of the cyclone overflow (nominal –2 μm material which remained in suspension for 2 days and had to be flocculated), there was no additional evidence of a clay–zeolite separation. This may be attributed to the intimate association of these two minerals or to a situation where both minerals have approximately the same particle size.

There was some evidence of a slight chabazite-erionite separation in the –10 μm-size fraction. This was probably due to liberation of individual chabazite rhombs and the preservation of larger bundles of erionite rods (see Fig. 1).

As seen from Table 4, nearly 50% of the starting feed disaggregated into –10 μm-size particles, compared with the 25% of –200 (74 μm) mesh material determined by dry screen analysis (Table 3). This is striking evidence of the ineffectiveness of dry-screening techniques as a means of estimating the amount of fines that may be encountered in a practical processing operation.

TABLE 4. Metallurgical balance for the beneficiation of Bowie "ore bed"

	X-ray diffraction estimates (%)							Distribution (%)	
Beneficiation fraction	Wt.%	Chab.	Erion.	Tot. zeol.	Glass	Smect.	Tot. gangue	Tot. zeol.	Tot. gangue
Coarse table fractions									
Concentrate	25.9	85	8	93	5	2	7	26	30
Middlings	5.4	85	8	93	5	2	7	5	7
Tailings	7.7	85	8	93	5	2	7	8	9
Feed†	39.0	85	8	93	5	2	7	39	46
Fine table fractions									
Concentrate	7.6	85	8	93	5	2	7	7	9
Middlings	4.0	88	7	95	3	2	5	4	3
Tailings	2.0	85	8	93	5	2	7	2	2
Feed†	13.6	86	8	94	4	2	6	13	14
Cyclone overflow fractions									
~ −10 μm	29.9	90	5	95	–	5	5	30	25
~ −2 μm (floc.)	17.5	90	5	95	–	5	5	18	15
Combined†	47.4	90	5	95	–	5	5	48	40
Calculated heads†	100.0	87	7	94	3	3	6	100	100

*Chab. = chabazite; Erion. = erionite; Tot. zeol. = total zeolite; Smect. = smectite; Tot. gangue = total gangue.
†Calculated values.

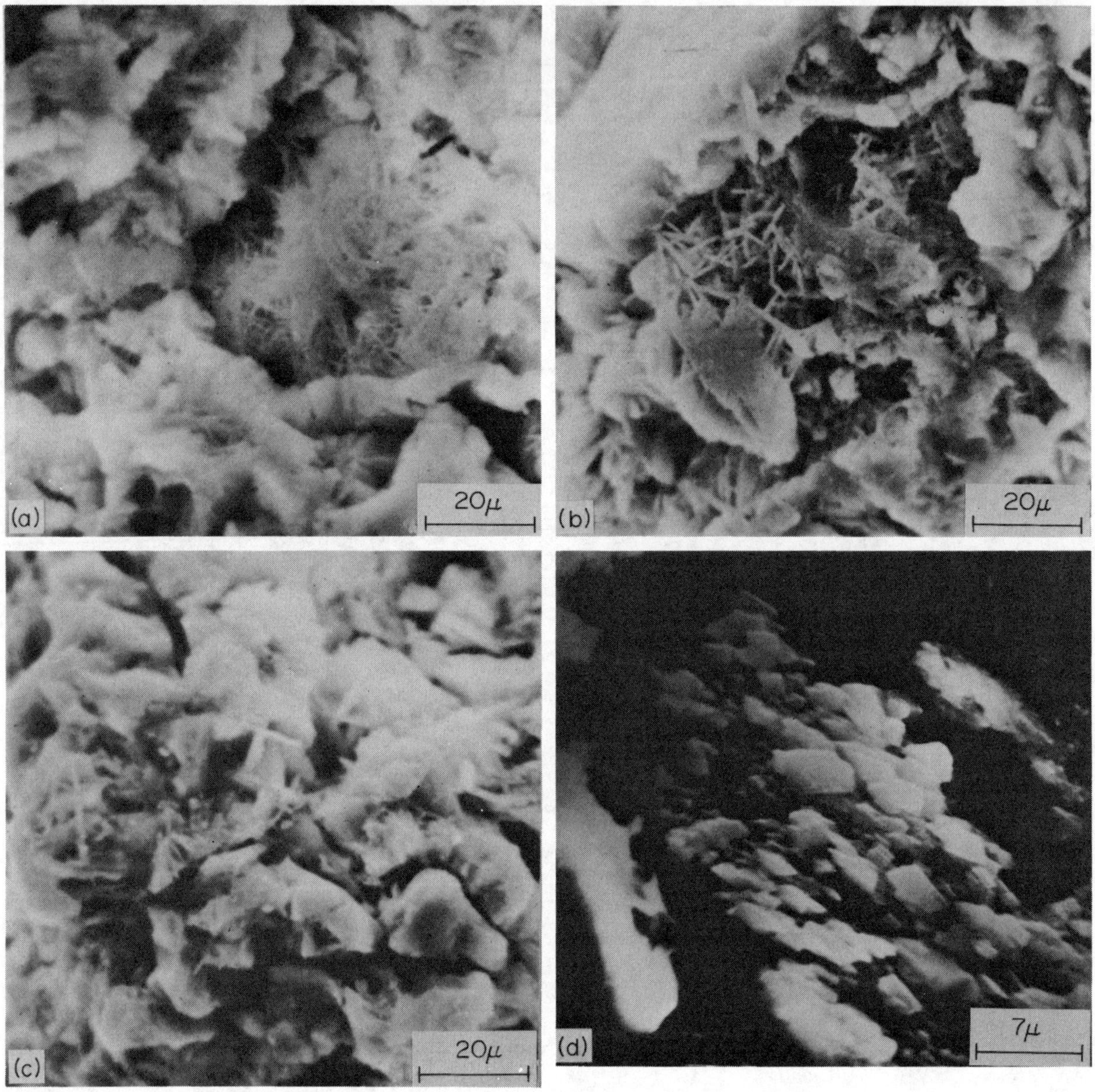

Fig. 3. Scanning electron micrographs of beneficiated products from the "ore bed" of the Bowie, Arizona, zeolite ore. (a) Coarse table concentrate showing aggregates of sub-micrometer-size chabazite crystals. (b) Coarse table middlings showing sub-micrometer-size chabazite crystals and unaltered volcanic glass about 20 μm in size. (c) Coarse table tailings similar to concentrate. (d) Cyclone overflow product showing chabazite crystals 2–3 μm in size.

Beneficiation of the lower-grade, thin-bedded unit

The results of beneficiation studies of the lower-grade, thin-bedded unit are listed in Table 5. Here, the concentration process was much more effective. The separation scheme shown in Fig. 2 was used, except that the coarse table middlings and tailings were screened to remove large aggregates of zeolite. These aggregates were quite porous and, as a result, their effective density was much lower than expected. Because of their large size and low effective densities, such particles therefore were misplaced in both the hindered-settling classifier and on the shaking table and reported to the middlings and tailings fractions with tabling.

As seen from Table 5, tabling appears to be an effective means of removing coarse gangue materials, such as volcanic glass, quartz, and feldspar. Note that the coarse table tailings amounted to 25% of the total weight and contained 54% of the total gangue, whereas

TABLE 5. Metallurgical balance for the beneficiation of Bowie lower-grade, thin-bedded unit

	X-ray diffraction estimates (%)										Distribution (%)	
Beneficiation fractions	Wt.%	Chab.	Erion.	Tot. zeol.	Qtz.	Feld.	Glass	Smect.	Heav.	Tot. gangue	Tot. zeol.	Tot. gangue
Coarse table fractions												
Concentrate	2.3	75	7	82	1	2	15	–	–	18	3	1
Middlings (+35 mesh)	3.6	73	7	80	5	10	5	–	–	20	5	2
Middlings (–35 mesh)	11.1	53	7	60	12	10	18	–	–	40	10	12
Tailings (+35 mesh)	2.0	64	7	71	11	8	10	–	–	29	2	2
Tailings (–35 mesh)	25.1	12	7	19	29	25	24	–	3	81	8	54
Feed†	44.1	33	7	40	21	18	19	–	2	60	28	71
Fine table fractions												
Concentrate	9.1	58	7	65	5	10	20	–	–	35	9	9
Middlings (+35 mesh)	0.3	75	7	82	3	2	13	–	–	18	1	–
Middlings (–35 mesh)	3.3	53	7	60	13	10	17	–	–	40	3	4
Tailings (+35 mesh)	0.1	73	7	80	5	5	10	–	–	20	–	–
Tailings (–35 mesh)	0.9	46	7	53	10	30	7	–	–	47	1	1
Feed†	13.7	57	7	64	7	11	18	–	–	36	14	14
Cyclone overflow fractions												
~ –10 μm	41.3	82	5	87	1	–	7	5	–	13	57	14
~ –2 μm (floc.)	0.9	92	5	97	–	–	–	3	–	3	1	1
Combined†	42.2	82	5	87	1	–	7	5	–	13	58	15
Calculated heads†	100.0	57	6	63	11	10	13	2	1	37	100	100
Combined concentrate (all fractions containing more than 70% zeolite)	50.5			86						14	69	20

*Chab. = chabazite; Erion. = erionite; Tot. zeol. = total zeolite; Qtz. = quartz; Feld. = feldspar; Smect. = smectite; Heav. = heavy minerals; Tot. gangue = total gangue.

†Calculated values.

Fig. 4. Scanning electron micrographs of beneficiated products from the lower-grade, thin-bedded unit of the Bowie, Arizona, zeolite deposit. (a) Coarse table concentrate, similar to assemblage shown in Fig. 3a. (b) Coarse table tailings showing coarse-grained gangue material.

only 8% of the total zeolite was lost in this step. Tabling of the finer feed yielded products of lower overall zeolite content. Because liberation would be expected to be greater in the finer feed, these results must be due to an intimate association of zeolite and fine-grained gangue in this size range. Fortunately, the fine table fraction amounted to less than 15% of the total weight.

Glass was identified in both shaking table concentrates and in the -10 μm-size cyclone-overflow product. Only the -2 μm-size cyclone product was free of this contaminant, containing an estimated 97% zeolite. Unfortunately, this fraction represented only about 0.1% of the total weight. Glass and zeolite are almost always associated with one another suggesting an intimate intergrowth of the two materials in the starting ore. Smectite was identified only in the two cyclone-overflow products. From the abundance of -10 μm-size products listed in Table 5, it is obvious that dry-screen analyses give a poor indication of the particle-size distribution in the Bowie zeolite ore (*vide supra*).

The highest grade concentrate of any significant sample weight was achieved in the -10 μm portion of the cyclone overflow and contained an estimated 87% zeolite. This fraction represented 42% of the total weight and 58% of the total zeolite in the starting feed. It is obvious that selective grinding results in a concentration of the softer zeolite particles into the finer size fractions. If all of the concentrates containing more than 70% zeolite were combined, it would be possible to obtain a combined concentrate containing about 86% zeolite in 50% of the weight. This would represent a 69% recovery of the zeolites from the starting feed and a significant upgrading of the lower-grade, thin-bedded unit which is currently removed and for the most part wasted in exposing the higher-grade "ore bed".

Although little success was achieved in attempting to upgrade the chabazite-rich "ore bed", substantial beneficiation of the lower-grade, thin-bedded unit is indeed possible. The procedure indicated by the present study would be to make use of selective grinding to reduce the ore (and especially the zeolite portion of the ore) to some fine-particle size. This would be followed by a size classification and gravity concentration to produce a high-grade zeolite fraction. The gravity concentration would provide an outlet for the coarser, harder gangue particles. Dry separation is also a distinct possibility and is currently under investigation.

Chemical analyses of beneficiated products

Chemical analyses of selected products from the beneficiation study are listed in Table 6. It should be noted that these fractions are not monomineralic or even monozeolitic, but contain between 82 and 93% total zeolites (mostly chabazite). As has been noted in earlier works, the Si/Al ratios of these materials are notably larger than those of their "igneous" counterparts (see Breck, 1974; Sheppard, 1971; Deer *et al.*, 1963). The zeolites contain predominantly calcium and sodium with lesser amounts of potassium, comparing favorably with the data reported by Sheppard *et al.*

TABLE 6. Chemical analyses of beneficiated natural zeolites, Bowie, Arizona

Zeolite	Beneficiated fraction	SiO_2	Al_2O_3	Fe_2O_3	Na_2O	CaO	K_2O	Atom ratio Si/Al
Chabazite ("ore bed")	Coarse table concentrate (~93% zeolite)	53.5	12.2	3.36	2.96	3.05	0.82	3.72
Chabazite (lower-grade, thin-bedded unit)	Coarse table concentrate (~82% zeolite)	56.5	11.4	3.43	2.15	3.97	1.20	4.21
Chabazite (lower-grade, thin-bedded unit)	Cyclone overflow (~87% zeolite)	56.0	13.0	4.07	2.47	3.40	1.27	3.65

(1976), especially sample "A" listed in their Table 2. It should also be noted that on a weight basis, the chabazite-rich product from the "ore bed" contains about an equal amount of sodium and calcium, whereas the chabazite-rich products from the lower-grade, thin-bedded unit are decidedly richer in calcium than sodium.

METAL-ION SORPTION TESTS

The effect of beneficiation on the uptake of heavy metal ions can be seen from the data listed in Table 7. In all cases, the beneficiated products were superior in their sorption of cupric ions to untreated materials. The slight apparent superiority of the upgraded product of the lower grade, thin-bedded unit over the ore bed material for the extraction of Cu^{+2} from a 0.001 M/L solution (99% vs. 97% removal) is within the estimated reproducibility of these tests (± 2%).

Using 1 g samples and 50 ml of 0.001 M/L solution, all samples had the ability to remove almost all of the Cu^{+2} presented to them during a 10-minute contact period. However, the superiority of the ore-bed materials to sorb Cu^{+2} was clearly evident with the stronger 0.01 M/L solutions where the copper content exceeds the sorption capacity of the samples. Beneficiation also appears to increase the ability of the ores to sorb Cu^{+2} for both materials, and it was not in direct proportion to the amount of gangue minerals removed. A particle-size/zeolite content bias and the washing out of clay-size materials by attritioning during the cycloning step are probably responsible in part also. At least for the lower-grade, thin-bedded ore sample, the data of Table 5 indicate that the selection of the 20 X 35-mesh fraction of itself leads to some up-grading of the zeolite content.

Expanding on these results, the uptake of other metal ions by beneficiated products of both ores was investigated. The following results are part of a much broader study (Mondale, 1978, Mondale *et al.*, 1977 a, b), and experimental details and additional sorption data are being presented elsewhere. As seen from the data of Table 8, the beneficiated product from the "ore bed" appears to be more effective for heavy metal-ion removal than does the product from the lower-grade, thin-bedded unit, even after correcting for the lesser amount of zeolite in the latter product. Again, caution must be used in interpreting these results because the available Cu^{+2} in the solutions exceeded the sorption capacity of the samples. When the concentration of the solution was lowered to 0.001 M/L (Table 9), the zeolite removed nearly all (97%) of the copper ions present. Only in some circumstances, such as dump-leaching of copper ores where solutions may

TABLE 7. Copper sorption by beneficiated and unbeneficiated Bowie zeolite ores*

Zeolite sample	Initial solution concentration (M/L)	Percent Cu^{+2} removed	
		Unbeneficiated	Beneficiated
"Ore bed"	0.001	87 ± 2	97 ± 2
	0.01	52	59
Lower grade, thin-bedded unit	0.001	84	99
	0.01	41	44

*1 g of 20 × 35-mesh sample, 50 ml of solution, 10 minutes contact time.

TABLE 8. Sorption of heavy metal ions by beneficiated Bowie zeolite ores*

		Percent sorbed from 0.01 M solution‡			
Zeolite sample	Zeolite content†	Cu^{+2}	Zn^{+2}	Hg^{+2}	Pb^{+2}
"Ore bed"	93	59	58	3.4	96
Lower grade, thin-bedded unit	80	44	44	6.1	92

*1 g of 20 × 35-mesh sample, 50 ml of solution, 10 minutes contact time.
†Largely chabazite, with about 8% erionite. See Tables 4 and 5.
‡Hg as the chloride.

contain as much as 1 g copper/liter, would solutions as strong as 0.01 M/L be encountered in practice. In spite of these shortcomings, the data of Table 8 suggest that the sorption of Pb^{+2} is nearly complete under these conditions, and that the sorption of Cu^{+2} and Zn^{+2} is substantial. Of the metal ions tested, only Hg^{+2} was rejected by these zeolites, probably due to formation of a mercuric chloride complex whose ionic dimensions are too large to enter the zeolite channels.

Table 9 also compared the uptake of Cu^{+2} by the beneficiated Bowie "ore bed" material to those of several synthetic zeolites. From the less concentrated solution (0.001 M/L), the sorption capacity of the Bowie product is superior to the synthetic zeolites 3A, 4A, and 5A, and only slightly inferior to zeolite 13X. Differences between the Bowie product and zeolite 13X for copper uptake, especially in the more concentrated solution, are probably due to partial channel blockage in the natural product.

SUMMARY

This study revealed the following important features:

1. The Bowie ores used in the study are predominantly chabazite with much smaller amounts of erionite.

2. The bulk density of the Bowie ores is lower than the mineral density reported in standard texts.

3. Beneficiation can substantially extend the reserves of the Bowie deposit by providing a method of increasing the zeolite content of the lower-grade, 4-foot thick, thin-bedded unit from about 60% to about 85%, with a 70% recovery.

4. Gravity and classification concentration is not effective in increasing the zeolite content of the high-grade "ore bed" above its natural 95% zeolite content.

5. There is some evidence of chabazite-enrichment in the finest size-fractions during beneficiation.

6. The heavy metal-ion uptake capabilities of chabazite-rich products of Bowie ores are superior to those of synthetic zeolites 3A, 4A, and 5A, and nearly as good as those of zeolite 13X for the removal of Cu^{+2}. The Bowie zeolite products are also shown to remove Pb^{+2} and Zn^{+2} from solution.

REFERENCES

Breck, D. W. (1974) *Zeolite Molecular Sieves*: Wiley, New York.

Deer, W. A., Howie, R. A. and Zussman, J. (1962-3)

TABLE 9. Sorption of copper ions by beneficiated Bowie zeolite ore and several synthetic zeolites*

Zeolite sample	% Cu^{+2} removed 0.001 M/L solution	% Cu^{+2} removed 0.01 M/L solution
Beneficiated product "ore bed", about 93% chabazite	97	59
Type 13X	99	78
Type 3A	61	
Type 4A	64	
Type 5A	56	

*1 g sample, 50 ml $CuSO_4$ solution, 10 minutes contact time.

Rock-forming Minerals, Vol. 4, London, Longmans.

Mondale, K. D. (1978) *The Beneficiation, Characterization and Ion Exchange Properties of Natural, Sedimentary Zeolites*, M.S. Thesis in Mineral Processing, The Pennsylvania State University.

Mondale, K. D., Mumpton, F. A. and Aplan, F. F. (1977a) The beneficiation and characterization of natural, sedimentary zeolites, submitted to *Trans. AIME.*

Mondale, K. D., Mumpton, F. A. and Aplan, F. F. (1977b) Metal ion-exchange properties of natural, sedimentary zeolites, submitted to *Trans. AIME.*

Sheppard, R. A. (1971) Zeolites in sedimentary deposits of the United States – A review: *Molecular Sieve Zeolites*, Vol. 1, Advances in Chemistry Series 101, Washington, D.C., American Chemical Society, pp. 279-310.

Sheppard, R. A., Gude, A. J. and Edson, G. M. (1978) Bowie zeolite deposits, Cochise and Graham Counties, Arizona: this volume.

AUTHOR INDEX

SUBJECT INDEX